T0180430

Mario Bunge: A Centenary Festschrift

Michael R. Matthews

Editor

Mario Bunge:
A Centenary Festschrift

 Springer

Editor
Michael R. Matthews
School of Education
University of New South Wales
Sydney, NSW, Australia

ISBN 978-3-030-16675-5   ISBN 978-3-030-16673-1   (eBook)
https://doi.org/10.1007/978-3-030-16673-1

This Springer imprint is published by the registered company Springer Nature Switzerland AG.
The registered company address is: Gewerbestrasse 11, 6330 Cham, Switzerland

Photo 1 Mario Bunge, Montreal, August 2018

Foreword

I feel lucky, as well as honoured, for this homage organized by my good friend, the learned and tireless Michael Matthews, who rounded up so many friends both old and new, and invited them to write something from the bridge that joins science to philosophy.

If you wonder how I succeeded in reaching a century, my answer is: through luck, hard but exhilarating work, tenacity, optimism, and comparatively clean living. Far from ignoring the many political and ideological horrors of my time, I take pride in having been an *engagé* intellectual, but I have not allowed those horrors to smother me. I kept thinking of scientific, philosophical and social problems even under duress and while escaping my creditors. This symbiosis did not take courage: it is the way my brain works.

I have kept my curiosity by reading scientific journals and consulting and arguing with experts in many fields, and by learning from some authorities while questioning others. For instance, I started out by admiring Marxism-Leninism, but ended up criticising its ontology and politics. Likewise, at the beginning of my research work in quantum microphysics I accepted the dogmas of the standard interpretation of quantum mechanics, only to subject them to harsh criticisms later on.

In 1946, when I wrote the last salvo of my philosophical journal *Minerva*, I stated that, while the military conflict had ceased, the ideological war was still on. But I did not foresee that Nietzsche and Heidegger would become even more popular, or that a whole new troop of intellectuals would endorse the most anti-popular ideologies in the name of freedom. Nor did I anticipate then the postmodern attacks on rationality, objectivity, and science, typical of Nazism, which became so popular in the humanities faculties ever since Thomas Kuhn proclaimed that truth is a matter of opinion. Let us not lament the fact that those faculties have little in common with their science sisters, for the anti-science they spout does not spatter on white coats. The distance between the two cultures, which Charles Snow regretted, protects those who practice intellectual rigor.

At this time the enlightened crew has to fight on two fronts: the internal one of mostly honest but naive torchbearers, and the external one bent on hoarding us back into the dark cave. The cave is now inhabited by the scholars who hold that there is opinion but not truth, that anything goes, that the world is our own creation, and so on – in short, by the constructivist-relativist party, popular among the people who praise "weak thinking" and consequently reject all intellectual discipline. Surely, this postwar rebellion against rationality is not new: it was inherent in the Romantic Counter Enlightenment. What is new about it is the claim that it is progressive.

To conclude, the defence of rationality is today just as necessary as it was in past epochs. Remember that Athena wore a helmet.

Philosophy Department, McGill University,   Mario Augusto Bunge
Montreal, QC, Canada

Preface and Acknowledgements

Although personal, it is worth telling a little of the story of how an Australian educator has had the good fortune to edit this festschrift for Mario Bunge, the renowned Argentine/Canadian physicist and philosopher. Hopefully, the personal story has some wider, non-personal lessons for the better preparation of science teachers and the need for science-informed teaching of philosophy.

I first became aware of Mario Bunge's work in the early 1990s, twenty years after my appointment at the University of New South Wales and thirty years after beginning my own science, philosophy and education studies at the University of Sydney. For me, this is a clear case of being better late than never. I had been editing the Springer journal *Science & Education: Contributions of History, Philosophy and Sociology of Science* since its foundation in 1992.[1] In 1995, two researchers who I did not know, Martin Mahner and Mario Bunge, submitted a manuscript on 'Religion and Science Teaching' (Mahner and Bunge 1996a).

On account of the manuscript being so comprehensive, informed, clearly argued, and on a much-discussed and debated educational topic, I invited a group of six philosophers, theologians and educators to comment on it and for Martin and Mario to respond (Mahner & Bunge 1996b). We had a good deal of correspondence back and forth about the original submission, reviews, and the responses. The papers were aggregated into a journal special issue (volume 5 number 2, 1996) that was separately sold, widely read and much cited. It introduced Mario to a wide international science teaching community, but especially to that segment concerned with the utilisation of history and philosophy of science in dealing with theoretical, curricular and pedagogical issues in science education. I subsequently published other papers of Mario's in the journal (Bunge 2000, 2003a,b, 2012a).

My own background prepared me to appreciate the initial Mahner and Bunge manuscript and to see its connection to a core theoretical issue in the teaching of science. After a Catholic schooling, I had completed a science degree at the University of Sydney (1965–1967), during which time I was involved in Catholic

---

[1] An account of my 25 years of editorship is given in Matthews (2015b).

student affairs, to the extent of being President of the university Newman Society. The staff and fellow-student members of the society had a significant intellectual and personal impact on my undergraduate years; they shaped that important and life-directing experience.

This was followed by a teacher education degree at Sydney Teachers' College (1968). There I did a semester course in philosophy of education, and this is where this autobiographical story connects with a larger, more objective lesson about philosophy in teacher education.

The course was taught by Anna Hogg who had completed her PhD in philosophy of education with Richard Peters at the, then famed, London Institute of Education. The entire course was based upon a detailed reading of the substantial, just-published book of Peters – *Ethics and Education* (Peters 1966). This 'theoretical' course was the most practical part of my teacher education; it was a course that shaped my entire educational trajectory. There is some parallel with Mario Bunge's oft-made remark that there is nothing so practical for physicists as good philosophy, and nothing so impractical as bad philosophy. More is the pity that such courses have now all but disappeared from Anglo-American teacher education programmes, being replaced by 'how to teach' and supposed psychology courses that mostly struggle to rise above passing fads (Hirst 2008). Teacher education has been dumbed down; in many places it is reduced to on-the-job training with barely a nod towards any disciplinary competence.

Peters' arguments, in the Analytic Philosophy tradition, were very simple; they were concerned with identifying the *process* of education and its *outcome*, namely the characteristics of an educated person (Peters 1967, 1973). For an educational process to be taking place – as distinct from just a teaching, training, indoctrinating, or coaching process – students should be learning something valuable, and the aim should be for them to understand, in the long term, what they are being taught; not merely to repeat what they are taught. Students being educated should end up with a cognitive grasp of the subject, how it all 'hangs together', how its truth claims are settled, and how it connects with other topics and subjects. Consequently, teachers aspiring to educate students need a good or deeper understanding of the subject matter they are teaching. And for this to happen, as was comprehensively argued by Israel Scheffler (Scheffler 1970, Matthews 1997), they need some appreciation of the history and philosophy of their subject, no matter what that might be – economics, history, mathematics, theology, literature, or anything else.

Apart from a *cognitive* requirement, for Peters, educational processes need to meet *moral* requirements; education needs to be conducted in an ethical manner. Students need to be respected; methods cannot be demeaning, discrimination cannot occur.

The hoped-for outcome of educational processes is the formation of an educated person. Such a person is characterised by *cognitive* qualities, namely a certain breadth and depth of knowledge; and by *moral* qualities as manifest in their life and decision making.

I was an immediate convert to these ideals of Liberal Education and absorbed the view that if I were to be a good teacher in that tradition, I needed to improve my own subject-matter education and be more conscious of the ethical dimension of education, and act accordingly both in the classroom, the staffroom, and in the profession. These were simple and obvious implications of Peters' argument.

Consequently, while I was a young high-school science teacher and energetically supporting educational ventures in the school (debating, General Studies classes, visiting speakers, etc.), I returned to Sydney University and completed part-time degrees in philosophy, psychology and education. To teach a discipline meant you had to know something about it, and about how students learn.

The Sydney Philosophy Department was arguably the best in Australia; some staff thought it was the best for a considerable distance beyond Australia. Among its more science-orientated faculty were Alan Chalmers, David Armstrong, David Stove, Michael Devitt and Wallis Suchting. All valued science and clear argument; they decried obfuscation, weasel words and slogans. Suchting was a philosophy teacher who became a close friend. I was privileged to be able to publish a number of his papers in the early years of my editorship of *Science & Education* (Suchting 1992, 1994, 1995). They are among the most philosophically informed and sophisticated papers to appear in a science education journal, or indeed in any education journal.

The Sydney philosophy faculty were writing on many of the questions that Mario Bunge was working on, but the work was done in parallel worlds, with little if any cross-fertilisation. The exception was Mario's early awareness of David Armstrong's materialism and psychoneural monism that he faulted for its 'inexactness and radical reductionism' (Bunge 2016, p. 213). Characteristically, Mario knew more about Sydney philosophers than they knew about him. The disjunct can retrospectively be seen by the fact that a course on Causality that I completed in the early 1970s was taught without reference to Mario's groundbreaking book on the subject (Bunge 1959). My own philosophical education proceeded without awareness or benefit of Mario's herculean contributions to so many fields in which I was taking an amateur interest: he did not figure in philosophy of science or philosophy of mind courses.

I was appointed as a lecturer in philosophy of education first at Sydney Teachers' College (1972), then at University of New South Wales (1975), having special responsibility for teaching philosophy to trainee science teachers. There are general issues about philosophy of education with which all teachers need to engage; but there are also disciplinary-specific philosophical matters with which they should engage. This was the beginning of my teaching and writing on how the history and philosophy of science could inform theoretical, curricular and pedagogical issues in science teaching. The constant refrain in my courses was that history and philosophy did not have to be brought into the science classroom, it was already there. HPS was there whenever inquiry was conducted, experiments were done, observations made, instruments used, causes listed, explanations given, laws related, models elaborated, when names such as Galileo, Newton, Boyle, Dalton, Darwin, Mendel, Einstein, Bohr, Heisenberg, Schrödinger mentioned, or scientific-technological controversies were discussed. Teachers just had to know something about philosophy and history in order to bring it to student awareness and engage them.

In contrast, the Sydney University Psychology Department was immersed in the Behaviourism of the time. The Head of the department used to only half-jokingly declare that 'I would like to study human beings, but what do they tell us about rats?' So, the empirical thesis of my honour's year was a study of 'The Safety-Signal Account of Bar-Press Avoidance Learning'. This, decades later, gave me grounds for appreciating Mario's dismissal of behaviourism as philosophically ill-informed 'brainless psychology'. The theoretical thesis was an aspect of intentions as causes of behaviour, a routine topic in philosophical psychology of the time.

Unbeknown to me, Mario had for decades been writing in these fields. He simply embodied all the liberal education ideals; he was at once a philosophical physicist and a scientific philosopher.

Our paths almost crossed at Boston University in 1978. I had gone to the Boston University Center for History and Philosophy of Science for my first sabbatical leave. The Philosophy Department had stellar quality staff, including Michael Martin, Robert Cohen, Marx Wartofsky, Joseph Agassi and Abner Shimony. All were committed to science, and to illuminating the connections of science with its history, and its philosophy. This commitment is manifest in volume after volume of the *Boston Studies in Philosophy of Science,* first published in 1963 by Reidel, then Kluwer, then Springer.

While there I had the good fortune to take a graduate course on Marxism with Cohen and Wartofsky, and a Galileo-based course on philosophy of science with Shimony. Like Bunge, Cohen and Shimony were jointly professors of physics and philosophy; and both shared Bunge's commitment to defending the Enlightenment project against its detractors. It was a profound lesson to see how Shimony, a significant contributor to contemporary quantum theory (Myrvold and Christian 2009), and to philosophy (Shimony 1993a,b) devoted an entire graduate course on philosophy of science to just careful reading and elaboration of Galileo's *Dialogue Concerning Two Chief World Systems* (Galileo 1633/1953).

Having a philosophy course begin with the texts and achievements of the great scientists was a lesson taken into my subsequent teaching of the subject to trainee teachers; it is an approach that well resonates with science teachers (Matthews 1990). Pre-service and in-service teachers can be bewildered and impatient with reading the standard big names in philosophy of science, but not with reading the scientists whose methodologies and achievements were being disputed. They are reduced to spectators watching the prominent philosophers – Nagel, Hempel, Kuhn, Popper, Lakatos, Toulmin, Laudan, Feyerabend – debate about the scientists, but not knowing what the scientists ever wrote or achieved. This is akin to geology students studying rocks but never picking one up; or students reading about experiments but never conducting one. Mach and Dewey were right about the importance of experiential learning. Again, this is something that from the outset Mario has affirmed: to philosophise about science, it helps to have practiced science; to interpret the history of science, it is helpful to have read the texts.

One outcome of this Boston University leave was my source book on *The Scientific Background to Modern Philosophy* (Matthews 1989). This was compiled so that philosophy students might see that the history of philosophy is not a long disciplinary soliloquy, but rather a long dialogue with the science of the time. A second outcome was the commencement of my pendulum motion studies; and detailing how history and philosophy can enrich its teaching (Matthews 2015a, Chap. 6). Both outcomes were of a Bungean kind, but did not benefit from his work, as I had not at that point read any of it. Unfortunately, Bunge's work was rarely read in even the best Anglo-American philosophy programmes. More is the pity.

Bunge also went to Boston University in 1978 but arrived after I had left. While there, among other things, he clashed with Stephen Jay Gould over the latter's account of species as individuals, and of species rather than populations evolving (Bunge 2016, pp. 219, 285–286). An excellent photo of Marx Wartofsky, Joseph Agassi and Bunge by Charlie Sawyer is reproduced below. A trio of exceptional philosophers, with Wartofsky (1928–1997) sadly dying early.[2]

For the two years 1992–1993, I was Foundation Professor of Science Education at The University of Auckland. This was a tumultuous couple of years because the country's science education, including National Curriculum writing and examinations had been taken over by Waikato University constructivists with support from University of Auckland postmodernists. I was involved in debates up and down the country, on radio, TV and newspapers, defending realism, rationality, reason, and liberal education (Matthews 1995). It was clear that all

Photo 2 Marx Wartofsky, Joseph Agassi, Mario Bunge. Boston University, 1969. (Charlie Sawyer, photographer)

---

[2]For some of Wartofsky's illuminating studies of the interplay of science and philosophy see Wartofsky (1968a,b, 1976). For Agassi's contributions to the subject see Agassi (1964, 1968, 1975, 1981, 2013).

the big national issues in science education were at base philosophical issues; it was a case, again, of bad philosophy having unfortunate educational and national consequences. The constructivist establishment embraced and promoted relativism, idealism, irrationalism (called localism), and the multi-science thesis.

As just one of scores of examples of the practical ill-effects of this philosophy, at the University of Auckland completion of an anthropology course on Maori science was deemed to satisfy the science requirement for trainee elementary teachers, and the science requirement of the university's General Education programme. But this policy decision raises the question: Is Maori science, science? This was a philosophical matter of great consequence. Unfortunately, faculty, administrators and students were poorly prepared to grapple with it. There could have been social, political, or cultural reasons for the anthropology allowance, but these were not argued; instead the epistemological claim was made that Maori science was science. A consequence was that Maori education students avoided science, and this fed on down through the school system. Philosophical decisions had consequences.

Again, unbeknown to me Mario was publishing on these very subjects at the very time of the New Zealand debates (Bunge 1991a,b, 1992, 1993, 1994). His arguments could have greatly enlightened discussion, but they simply were not read; neither inside nor outside of philosophy. New Zealand was fortunate in having good local philosophical defenders of realism, rationality and universalism – in particular, Robert Nola (Nola 1988, 2003) and Alan Musgrave (Musgrave 1993) – but these also were not read by educators nor were they invited on to national curriculum committees. The whole experience underwrites Mario's lifelong conviction that philosophy is important, and needs to be seriously attended to. There are unfortunate personal, cultural and social consequences of embracing faulty or discredited philosophy. In the New Zealand case, the effects reached all the way up to the constructivist-inspired National Science Curriculum.

My 'at a distance' relationship with Mario moved to a personal one in 2001 when Marta Bunge accepted a Visiting Fellowship in the UNSW School of Mathematics. Mario, who travelled with her, asked me if he might be attached as an Honorary Visitor to the School of Education. The six months that he and Marta spent in Sydney were very happy and productive for both of them and for my own family, who shared many times and occasions with them. Mario describes this period in his *Memoirs* (Bunge 2016, pp. 371–75). Incidentally, given Mario's reputation for combative argument, the administrative staff said after he left: 'Mario was the most polite visitor that the School of Education ever had'. He was unfailing in his courtesy and consideration of the office staff.

Mario's UNSW sojourn laid the foundation for two decades of personal friendship and collaborative work, including the publication of two thematic issues of *Science & Education*. The first issue was devoted to appraisals of his accounts of Quantum Theory (vol.12 nos.5–6, 2003); the second issue appraised his Systemic Philosophy (vol.21 no.10, 2012).

My own contribution to science education debates had independently, but in a much scaled-down form, mirrored Mario's eight-decade long defence of realism, rationality and science; and his criticisms of constructivist epistemology, idealist ontology, and pseudoscientific pretence (Matthews 1998, 2009, 2015a, 2018, 2019a).

In 2015 having again met Mario and Marta, this time at their Montreal home, when he was 95 years, and seeing him in such good health, spirits and with his ever-lively mind functioning so well – the prospect of him celebrating his 100th birthday was very real. Thinking ahead, this was something that I thought should be suitably celebrated by the academic community – it is a rare enough event for anyone to reach five-score years, much less for an academic to be still writing and publishing as the occasion draws near (Bunge 2012 a,b, 2013, 2016, 2017, 2018).

In late 2016 I contacted Lucy Fleet, a Springer philosophy editor who had overseen some of Mario's earlier Springer publications, including his *Matter and Mind* (Bunge 2010) and his absorbing and informative *Memoirs of a Philosophical-Scientist* (Bunge 2016), and proposed the idea of a Centenary Festschrift to suitably mark the occasion. Lucy was enthusiastic and after internal discussion and external review, Springer accepted the idea and issued a contract for the volume to appear in the *Boston Studies in the Philosophy of Science* series. There followed almost two years of advertising, soliciting, letter writing, reviewing, settling on accepted contributions, and structuring the volume. It was Lucy's immediate enthusiasm that set the festschrift project in train, and that has subsequently guided it through to fruition. The number, variety and quality of papers finally published well testify to the international esteem in which Mario is held. Hopefully, this festschrift will contribute a little to enhancing that esteem and appreciation.

Returning now to the beginning of these remarks. If this volume does have merit and does make a positive contribution to philosophy, then it needs be recognised that the volume was only possible because fifty years ago philosophy was included in the science teacher education programme at Sydney Teachers College (Matthews 2019b). Would that more teacher education programmes valued and included philosophy (Colgan and Maxwell 2019); and that more teacher educators might thus have the good fortune of meeting, learning from, and working with first-rate historians and philosophers of science such as Mario Bunge.

Understandably, the festschrift project has taken a great deal of – happily given – editorial work and time. In the final stage I have received the wonderful assistance of a team of friends and colleagues who have copyedited all final submissions, with most having been copyedited twice by different readers. Copyediting used to be a routine part of all major publishers' operations. It no longer is. Doubtless this collection will have typos, missing references, poor punctuation, and unclear sentences, but without the heroic labours of the following there would have been so many more of them giving irritation to readers: Don Allen (Texas A.&M. University), Robert Carson (Montana State University), John Forge (University of Sydney), Ron Good (Louisiana State University), Walter Jarvis (University of Technology, Sydney), Jim Mackenzie (University of Sydney), Mitch O'Toole (University of Newcastle), Stuart Rowlands (University of Plymouth), Roland Schulz (Simon Fraser University), Wendy Sherman-Heckler (Otterbein College), Roger Wescombe, Kay Wilson, and Robyn Yucel (La Trobe University).

Paul McColl – a retired Australian science teacher, a graduate of the University of Melbourne, and part-time lecturer at La Trobe University – warrants special thanks for the large number of chapters he meticulously read and corrected whilst meeting tight deadlines. Would that all authors and editors could have such assistance as I have had.

Sydney, Australia   Michael R. Matthews
February 2019

References

Agassi, J. (1964). The nature of scientific problems and their roots in metaphysics. In M. Bunge (Ed.), *The critical approach*. Glencoe: Free Press. Reprinted in Agassi, J. (1975). *Science in Flux* (pp. 208–239). Boston: Reidel.
Agassi, J. (1968). *The continuing revolution: A history of physics from the Greeks to Einstein*. New York: McGraw Hill.
Agassi, J. (1975). *Science in flux*. Dordrecht: Reidel Publishing Company.
Agassi, J. (1981). *Science and society: Studies in the sociology of science*. Dordrecht: Reidel.
Agassi, J. (2013). *The very idea of modern science: Francis Bacon and Robert Boyle*. Dordrecht: Springer.
Bunge, M. (1959). *Causality and modern science*. Cambridge: Harvard University Press. Third Revised Edition, Dover Publications, New York, 1979.
Bunge, M. (1991a). What is science? Does it matter to distinguish it from pseudoscience? *New Ideas in Psychology, 9*(2), 245–283. [special issue on pseudoscience in psychology].
Bunge, M. (1991b). A critical examination of the new sociology of science: Part 1. *Philosophy of the Social Sciences, 21*(4), 524–560.
Bunge, M. (1992). A critical examination of the new sociology of science: Part 2. *Philosophy of the Social Sciences, 22*(1), 46–76.
Bunge, M. (1993). Realism and antirealism in social science. *Theory and Decision, 35*, 207–235. Reprinted in Mahner, M. (Ed.) *Scientific realism: Selected essays of Mario Bunge* (pp. 320–342). Amherst: Prometheus Press.
Bunge, M. (1994). Counter-enlightenment in contemporary social studies. In P. Kurtz & T. J. Madigan (Eds.), *Challenges to the enlightenment: In defense of reason and science* (pp. 25–42). Buffalo: Prometheus Books.
Bunge, M. (2000). Energy: Between physics and metaphysics. *Science & Education, 9*(5), 457–461.
Bunge, M. (2003a). Twenty-five centuries of quantum physics: From Pythagoras to us, and from subjectivism to realism. *Science & Education, 12*(5–6), 445–466.
Bunge, M. (2003b). Quantons are quaint but basic and real. *Science & Education, 12*, 587–597.
Bunge, M. (2012). Does quantum physics refute realism, materialism and determinism? *Science & Education, 21*(10), 1601–1610.
Bunge, M. (2016). *Between two worlds: Memoirs of a philosopher-scientist*. Dordrecht: Springer.
Colgan, A. D., & Maxwell, B. (Eds.). (2019). *Philosophical thinking in teacher education*. New York: Routledge.
Galileo, G. (1633/1953). *Dialogue concerning the two chief world systems* (S. Drake, Trans.). Berkeley: University of California Press. (2nd Rev. ed., 1967).

Hirst, P. H. (2008). Philosophy of education in the UK. The institutional context. In L. J. Waks (Ed.), *Leaders in philosophy of education. Intellectual self portraits* (pp. 305–310). Rotterdam: Sense Publishers.

Mahner, M., & Bunge, M. (1996a). Is religious education compatible with science education? *Science & Education, 5*(2), 101–123.

Mahner, M., & Bunge, M. (1996b). The incompatibility of science and religion sustained: A reply to our critics. *Science & Education, 5*(2), 189–199.

Matthews, M. R. (Ed.). (1989). *The scientific background to modern philosophy*. Indianapolis: Hackett Publishing Company.

Matthews, M. R. (1990). History, philosophy, and science teaching: What can be done in an undergraduate course? *Studies in Philosophy and Education, 10*(1), 93–98.

Matthews, M. R. (Ed.). (1998). *Constructivism and science education: A philosophical examination*. Dordrecht: Kluwer Academic Publishers.

Matthews, M. R. (Ed.). (2018). *History, philosophy and science teaching: New perspectives*. Dordrecht: Springer.

Matthews, M. R. (1994). *Science teaching: The role of history and philosophy of science*. New York: Routledge.

Matthews, M. R. (1995). *Challenging New Zealand science education*. Palmerston North: Dunmore Press.

Matthews, M. R. (1997). Scheffler revisited on the role of history and philosophy of science in science teacher education. In H. Siegel (Ed.), *Reason and education: Essays in honor of Israel Scheffler* (pp. 159–173). Dordrecht: Kluwer Academic Publishers.

Matthews, M. R. (2015a). Reflections on 25-years of journal editorship. *Science & Education, 24*(5–6), 749–805.

Matthews, M. R. (2015b). *Science teaching: The contribution of history and philosophy of science: 20th anniversary revised and enlarged edition*. New York: Routledge.

Matthews, M. R. (2019a). *Feng Shui: Teaching about science and pseudoscience*. Dordrecht: Springer.

Matthews, M. R. (2019b). The contribution of philosophy to science teacher education. In A. D. Colgan & B. Maxwell (Eds.), *Philosophical thinking in teacher education*. New York: Routledge.

Musgrave, A. (1993). *Common sense, science and scepticism*. New York: Cambridge University Press.

Myrvold, W. C., & Christian, J. (Eds.). (2009). *Quantum reality, relativistic causality, and closing the epistemic circle: Essays in honour of Abner Shimony*. Dordrecht: Springer.

Nola, R. (Ed.). (1988). *Relativism and realism in science*. Dordrecht: Reidel Academic Publishers.

Nola, R. (2003). *Rescuing reason: A critique of anti-rationalist views of science and knowledge*. Dordrecht: Kluwer Academic Publishers.

Peters, R. S. (Ed.). (1973). *The philosophy of education*. Oxford: Oxford University Press.

Peters, R. S. (1966). *Ethics and education*. London: George Allen and Unwin.

Peters, R. S. (1967). What is an educational process? In R. S. Peters (Ed.), *The concept of education* (pp. 1–23). London: Routledge & Kegan Paul.

Scheffler, I. (1970). Philosophy and the curriculum. In *Reason and teaching*. London: Routledge. 1973, pp. 31–44. Reprinted in *Science & Education, 1*(4), 385–394, 1992.

Shimony, A. (1993a). *Search for a naturalistic world view Vol. I Scientific method and epistemology*. Cambridge: Cambridge University Press.

Shimony, A. (1993b). *Search for a naturalistic world view Vol. II Natural sciences and metaphysics*. Cambridge: Cambridge University Press.

Suchting, W. A. (1992). Constructivism deconstructed. *Science & Education, 1*(3), 223–254. Reprinted in Matthews, M. R. (Ed.) (1998). *Constructivism in science education: A philosophical examination* (pp. 61–92). Dordrecht: Kluwer Academic Publishers.

Suchting, W. A. (1994). Notes on the cultural significance of the sciences. *Science & Education, 3*(1), 1–56.

Suchting, W. A. (1995). The nature of scientific thought. *Science & Education, 4*(1), 1–22.

Wartofsky, M. W. (1968a). Metaphysics as a heuristic for science. In R. S. Cohen, & M. W. Wartofsky (Eds.), *Boston studies in the philosophy of science* (Vol. 3, pp. 123–172). Republished in his *Models*, Reidel, Dordrecht, 1979, pp. 40–89.

Wartofsky, M. W. (1968b). *Conceptual foundations of scientific thought: An introduction to the philosophy of science*. New York: Macmillan.

Wartofsky, M. W. (1976). The relation between philosophy of science and history of science. In R. S. Cohen, P. K. Feyerabend, & M. W. Wartofsky (Eds.), *Essays in memory of Imre Lakatos* (pp. 717–738). Dordrecht: Reidel. (*Boston Studies in the Philosophy of Science* 39.) Republished in his *Models*, Reidel, Dordrecht, 1979.

Contents

Contributors

Joseph Agassi is Professor of Philosophy Emeritus at Tel Aviv University and York University, Toronto. He is Fellow of the American Association for the Advancement of Science, the Royal Society of Canada, and the World Academy of Art and Science; he is a former Senior Fellow of the Alexander von Humboldt Stiftung. He has published more than 600 papers in diverse fields. He has authored 20 books and edited 10.

Evandro Agazzi is Director of the Interdisciplinary Center for Bioethics of the Panamerican University of Mexico City, and Emeritus Professor of the Universities of Genoa (Italy) and Fribourg (Switzerland). He is Honorary President of the International Academy of Philosophy of Science (Brussel), the International Federation of the Philosophical Societies, the International Institute of Philosophy (Paris). He has published in several languages, as author or editor, more than 80 books and about 1000 papers and book chapters. His main fields of research are logic, philosophy of mathematics, philosophy of physics, general philosophy of science, ethics of science, metaphysics, and bioethics.

Omar Ahmad is an Internal Medicine physician and Clinical Informatics Lead at Stanton Hospital in Yellowknife, Canada. His main research interests are genetic epidemiology, clinical informatics, and biophysics.

Richard T. W. Arthur is Professor of Philosophy at McMaster University, Canada. He researches in early modern natural philosophy and mathematics, the philosophy of physics, and scientific epistemology, specializing in the theory of time, the infinite, and thought experiments. He is author of *G. W. Leibniz: The Labyrinth of the Continuum* (Yale UP 2001), *Leibniz* (Polity 2014), *Natural Deduction* (Broadview 2011), *Introduction to Logic* (Broadview 2017), and over 60 articles and book chapters.

Russell Blackford holds a conjoint research appointment at the University of Newcastle, Australia. He is the author or editor of numerous books, including

Freedom of Religion and the Secular State (Wiley-Blackwell 2012), *Humanity Enhanced: Genetic Choice and the Challenge for Liberal Democracies* (MIT Press 2014), *The Mystery of Moral Authority* (Palgrave 2016), and *Philosophy's Future: The Problem of Philosophical Progress* (co-edited with Damien Broderick; Wiley-Blackwell 2017).

Mauro A. E. Chaparro is Professor at the Facultad de Ciencias Exactas and Naturales of Universidad Nacional de Mar del Plata and Assistant Researcher of the National Research Council (CONICET) of Argentina. He has published several papers on mathematicals models of gastrointestinal nematodes and magnetic monitoring. His main current interest is on geostatistics and mathematicals models of atmospheric pollution.

Alberto Cordero is Professor of Philosophy and History, City University of New York. Numerary Member of the *Academie Internationale de Philosophie des Sciences* and of the *Institute de Hautes Sciences Theoriques, Brussels*. He has published extensively on philosophy of science, the foundations of physics, naturalism, scientific realism, and the philosophical history of science. He has edited *Philosophy and the Origin and Evolution of the Universe* (with Evandro Agazzi; Synthese Library), *Reflections on Naturalism* (with J.I. Galparsoro; Sense Publishers), and the forthcoming *Philosophers Look at Quantum Mechanics*.

Marta Crivos is a researcher at National Scientific and Technical Research Council (CONICET), Argentina, and Professor of Anthropological Theoretical Orientations in the Department of Anthropology at the Faculty of Natural Sciences and Museum (UNLP). She has published more than 100 scholarly works, most of them with focus on the foundations of the ethnographic methodology and the integration of this perspective in research on different issues, especially those involving the relationship between human communities and their natural environment.

Alberto Cupani is a retired professor of the Philosophy Department at the Federal University at Santa Catarina, Brazil, and former researcher of the National Scientific and Technological Research Council (CNPq) of Brazil. He is former president of the Association of History and Philosophy of Science of the Southern Cone (AFHIC). He has published numerous articles and five books on the philosophy of science and technology. His main interests are the rationality and objectivity of scientific research, the value of science for society and the conditioning of human life by technology.

Pierre Deleporte is retired French CNRS researcher in biology at University of Rennes (UMR 6552 Ethologie Animale et Humaine). While his scientific work concerned mainly animal social behavior and evolution, he is also former President of the French Systematics Society. His main current interest is materialist philosophy. He translated to French several of Mario Bunge's works (including *Matérialisme Scientifique, Entre Deux Mondes* and *Philosophie de la Médecine*).

Guillermo M. Denegri is Associate Professor in the Department of Biology at the Universidad Nacional de Mar del Plata (UNMdP) and Principal Researcher of the National Research Council (CONICET) of Argentina. His field of study is parasitic diseases of public health importance. He gives the annual Permanent Seminar of the Biophilosophy of the FCEyN, addressed to PhD students in science. He was President of Federación Latinoamericana de Parasitología (FLAP).

Heinz W. Droste is a psychological consultant and does industry training. He was trained as a sociologist, philosopher and psychologist at German universities. He has published a series of surveys on opinion formation strategies used in various social fields to assert political and economic interests. In addition, he has authored seven books (including *Turn of the Tide – Gezeitenwechsel* a semi-popular introduction to Mario Bunge's philosophy). He is co-founder of the Public Relations – Akademie, Wiesbaden.

Olival Freire Junior is Professor of History of Science at the Federal University of Bahia and Researcher at the National Council for Scientific and Technological Development (CNPq) in Brazil. He was elected to the council of the History of Science Society (2018–2020). He is the author of *The Quantum Dissidents – Rebuilding the Foundations of Quantum Mechanics 1950–1990*, Springer, 2015. His main current interest is on the history of quantum physics, history of physics in Brazil, and the uses of history and philosophy in science teaching.

Carolina I. García Curilaf is a graduate assistant in Philosophy of Science at the University of Mar del Plata and scholar of the National Research Council (CONICET) of Argentina. She is a member of the research group in production, health and environment (IIPROSAM), Facultad de Ciencias Exactas y Naturales, Universidad Nacional de Mar del Plata. She has published about 10 papers on philosophy of ecology. Her main interests at present are the epistemological problems of ecology.

José María Gil is Independent Researcher of the National Research Council (CONICET) of Argentina and Professor of Logic at the University of Mar del Plata. He has published articles, books and chapters on linguistics, philosophy, and education. His main interests are curriculum design in language teaching and sport education.

Rafael González del Solar is an independent philosopher and freelance translator based in Barcelona. He has conducted research and published in both ecology and the philosophy of science. As a translator, he has rendered into Spanish nine of Bunge's works, including *Chasing Reality, Emergence and Convergence*, and the four first volumes of the *Treatise on Basic Philosophy*. He is currently in the process of translating the remaining four volumes of the *Treatise* and expanding his research interests to the problem of human nature.

Ibrahim A. Halloun is Professor of Physics and Education at Lebanese University. In addition to physics and education, his research interests include history and philosophy of science, and cognitive sciences and neuroscience. Throughout his career, Prof. Halloun contributed to curriculum reform in many countries around the world. Through classroom-based research, he has developed, among others,

Modeling Theory in science education that evolved into Systemic Cognition and Education (SCE), a generic pedagogical framework for student and teacher education.

Art Hobson is Professor Emeritus of Physics at the University of Arkansas, USA. He has published *Tales of the Quantum: Understanding Physics' Most Fundamental Theory* (Oxford University Press, 2017) as well as many physics papers on topics in quantum physics. He is also interested in physics literacy for the general public and has published *Physics: Concepts & Connections* (Pearson, 5th edition 2010), a physics-literacy textbook for non-science college students.

Rögnvaldur D. Ingthorsson is a researcher at Lund University, Sweden. He is the author of *McTaggart's Paradox*, Routledge, 2016, and co-editor of *Mental Causation and Ontology*, with S. C. Gibb and E. J. Lowe, OUP, 2013. He has published numerous articles addressing central issues in the philosophy of time, persistence, causation and truth. He is the primary investigator of the project *Scientific Essentialism: Modernising the Aristotelian View*, funded by Riksbankens Jubileumsfond: Swedish Foundation for Humanities and Social Sciences.

Leonardo Ivarola is Assistant Professor at the University of Buenos Aires and a researcher of the National Scientific and Technical Research Council of Argentina (CONICET) and at the Interdisciplinary Institute of Political Economy (IIEP). He has published papers in both national and international journals; has been UBACyT doctoral fellow and postdoctoral fellow at CONICE; a member of the Research Center in Epistemology of Economic Sciences (CIECE); and member of the Executive Committee of the Conference on Epistemology of the Economic Sciences.

Ingvar Johansson is Professor Emeritus in Theoretical Philosophy at Umeå University, Sweden. He is author of *A Critique of Karl Popper's Methodology* (Esselte Studium 1975) and *Ontological Investigations. An Inquiry into the Categories of Nature, Man and Society* (Routledge, 1989; enlarged edition, Ontos Verlag, 2004). Apart from book introductions to the philosophy of science in Swedish, he has in English (together with Prof. em. N. Lynøe) written *Medicine & Philosophy. A Twenty-First Century Introduction* (Ontos Verlag 2008). He has published papers on many philosophical topics in both Swedish and international philosophical journals.

Reinhard Kahle is Professor of Mathematical Logic at the Universidade Nova de Lisboa in Portugal and member of the *Académie Internationale de Philosophie des Sciences*. His main research interests are proof theory and the history and philosophy of modern mathematical logic, in particular, of the Hilbert School. His publications include also areas like Proof-Theoretic Semantics, Intensionality, and Computational Complexity. He has (co-)edited ten books and special issues as, for instance, *Gentzen's Centenary: The quest for consistency* (Springer, 2015, with Michael Rathjen).

Byron Kaldis is Professor of Philosophy and Head of the Humanities Department at the National Technical University of Athens, Greece, and Distinguished Xiaoxiang

Professor of Ethics at the Moral Culture Institute of the Hunan Normal University, China. He works in the philosophy of science and social science, metaphysics, political philosophy, applied ethics, and the philosophy of technology. He has published articles, books and edited volumes in these areas and was the editor of the Sage *Encyclopedia of Philosophy and the Social Sciences* (2013) and co-editor of *Wealth, Commerce and Philosophy* (Chicago, 2017).

Michael Kary studied undergraduate philosophy under Mario Bunge. He received his PhD in mathematics from Boston University. He has worked in industry, taught mathematics at Dawson College in Montreal, and authored or co-authored publications appearing in diverse journals including *Spine*, *The Journal of Theoretical Biology*, and *Injury Prevention*.

Javier Lopez de Casenave is Associate Professor of Biological Sciences at the University of Buenos Aires and Principal Researcher of the National Research Council (CONICET) of Argentina. His main current interest is the community ecology of ants and birds, especially in desert environments. He is former President of the Argentine Ecological Society.

María Manzano is Professor of Logic and Philosophy of Science at the University of Salamanca, Spain. She has held visiting positions at California State University (Berkeley) and Stanford University. Her publications include *Model Theory* (Oxford University Press), *Extensions of First-Order Logic* (Cambridge University Press), *Lógica para principiantes*, (Alianza Editorial) and *The Life and Work of Leon Henkin: Essays on His Contributions* (Springer). Her pedagogical initiatives include *Tools for Teaching Logic* (ALFA project of UE).

Luis Marone is Professor of Evolution and Epistemology at the University of Cuyo and Principal Researcher of the National Research Council (CONICET), both of Argentina. He was a fellow of the International Council for Canadian Studies, and the John Simon Guggenheim Memorial Foundation. He is interested in community ecology, and, especially, the epistemology and methodology of ecology, wildlife science and environmental sciences.

Michael R. Matthews is an Honorary Associate Professor at the University of New South Wales, Australia. He has published books, anthologies and articles in the areas of philosophy of education, history and philosophy of science, and science education. His recent book *Science Teaching: The Contribution of History and Philosophy of Science* (Routledge 2015) has been published in Greek, Spanish, Korean, Chinese and Turkish. He was Foundation Editor of the Springer journal *Science & Education: Contributions from the History and Philosophy of Science*.

Manuel Crescencio Moreno completed his doctoral dissertation in Logic and Philosophy of Science under the supervision of María Manzano at the University of Salamanca. He graduated in Philosophy (specialization in 'Science and Philosophy') and in Theology at the Pontifical Gregorian University in Rome. His PhD thesis, entitled 'Intensions, Types and Existence', is focused on intensional logic, type theory and ontological arguments. He teaches at Cervantes School in

Cordoba (Spain) and is actively involved in some research projects at the University of Salamanca.

Ignacio Morgado-Bernal is Professor of Psychobiology and Director of the Institute of Neurosciences at the Autonomous University of Barcelona in Spain. He has been a member of the Executive Committee of the European Brain and Behaviour Society. He has published more than 100 scientific papers. In the last five years, he has also developed an important activity publishing newspapers articles and books about popular neuroscience.

A. Z. Obiedat is Assistant Professor of Arabic at Wake Forest University. He received his PhD in the area of Arab-Islamic Philosophy and Secularism from McGill University. His research specialization reflects interest in contemporary science-oriented philosophies and classical Arab-Islamic scholasticism. The common thread between the two domains is the 'intellectual and political strife over modernity in Western and Middle Eastern contexts'. Studying the proposals of Bunge's secular modernism versus Taha 'Abd al-Rahman's Islamic modernism, is his current research focus.

Íñigo Ongay de Felipe teaches philosophy at the University Deusto in Spain and is an associate researcher in philosophy with the Fundación Gustavo Bueno. He has had teaching positions in Mexico, China and Costa Rica. His research covers a broad variety of issues ranging from the general philosophy of science with particular attention to the philosophy of biology and life sciences to the history of modern and contemporary philosophy.

Martín Orensanz is a Licentiate in Philosophy by the National University of Mar del Plata. He has a Doctoral Scholarship awarded by the National Scientific and Technical Research Council (CONICET) of Argentina, and is currently finishing his Doctorate in Philosophy. He has published several papers on different topics as well as a book on Argentine philosophy. His main interests are philosophy of science, contemporary philosophy, and Argentine philosophy.

Eduardo L. Ortiz is Emeritus Professor, Imperial College London. He is a fellow of the Institute of Mathematics, Great Britain; foreign fellow of the Royal Academy of Sciences, Spain; National Academy of Sciences, Argentina, National Academy of Exact and Physical Sciences, Argentina. He has been a visiting professor at Massachusetts Institute of Technology, Université de Rouen, Université d'Orleans. He has been a Guggenheim Fellow, Harvard University. He is a recipient of the José Babini History of Science Prize (Ministry of Science and Technology/CONICET, Argentina).

Andreas Pickel is Professor of Global Politics at Trent University, Peterborough, Canada. He has published in the areas of post-communist transformation, nationalism, and the philosophy of social science. He is the editor of two special issues of the journal *Philosophy of the Social Sciences* (2004, 2007) devoted to Mario Bunge's philosophy, in particular systems and mechanisms.

Dominique Raynaud is Associate Professor at University Grenoble Alpes, France. Among his recent books are: *Sociologie des controverses scientifiques* (Éditions matériologiques, 2018), *A Critical Edition of Ibn al-Haytham's Epistle on the Shape of the Eclipse: The First Experimental Study of the Camera Obscura* (Springer, 2016), *Studies on Binocular Vision: Optics, Vision and Perspective from the Thirteenth to the Seventeenth Centuries* (Springer, 2016), *Qu'est-ce que la technologie?* (Éditions matériologiques, 2016), *Scientific Controversies. A Sociohistorical Perspective on the Advancement of Science* (Transaction, 2015; Routledge 2017), *Géométrie pratique. Géomètres, Ingénieurs, architectes, XVIe-XVIIIe siècle* (PUFC, 2015), *Optics and the Rise of Perspective* (Bardwell Press, 2014).

Nicholas Rescher Distinguished University Professor of Philosophy at the University of Pittsburgh, is a much-published philosopher. He has been awarded the Aquinas Medal of the American Catholic Philosophical Association, the Helmholtz Medal of the Berlin/Brandenburg Academy of Sciences, and the Order of Merit of the Federal Republic of Germany. Honorary degrees have been awarded to him by eight universities on three continents.

Andrés Rivadulla is an Emeritus Professor of Philosophy of Science at the Faculty of Philosophy of the Complutense University of Madrid. He has published more than a hundred articles in general philosophy of science, epistemology, methodology of science, philosophy and history of physics, philosophy and history of probability, and theoretical statistics. He has published or edited eight books. His current research is in the field of scientific discovery, particularly theoretical prediction and sophisticated abduction, as well as in the theoretical models of physics and scientific explanation.

Gustavo E. Romero is Professor of Relativistic Astrophysics at the University of La Plata and Superior Researcher of the National Research Council (CONICET) of Argentina. He is former President of the Argentine Astronomical Society and Helmholtz International Fellow. He has published more than 400 papers on astrophysics, gravitation, and the foundations of physics. He has authored or edited 10 books (including *Introduction to Black Hole Astrophysics*, with G.S. Vila, Springer, 2014 and the forthcoming *Scientific Philosophy*). His main current interest is on high-energy astrophysics, black hole physics and ontological problems of space–time theories.

Peter Slezak is Associate Professor of philosophy at the University of New South Wales and formerly director of the university's Cognitive Science Centre. He graduated in sociology, and completed his PhD in philosophy at Columbia University. He has published in philosophy, philosophy of science, sociology of science, cognitive science, theoretical psychology and science education.

Marc Silberstein is an editor for scientific and philosophical publishing (Éditions Matériologiques, Paris). For many years, he has helped to make known in France the idea of Mario Bunge being his main publisher. He also co-directed with T. Heams, P. Huneman, G. Lecointre, *Handbook of Evolutionary Thinking in the*

Sciences (Springer, 2013); with G. Lambert & P. Huneman, *Disease, Classification and Evidence: New Essays in the Philosophy of Medicine* (Springer, 2013). In particular, Marc Silberstein (eds.), *Matériaux philosophiques et scientifiques pour un matérialisme contemporain* (Editions Matériologiques, 2013).

José Geiser Villavicencio-Pulido is Professor-Researcher of Department of Environmental Sciences of Universidad Autónoma Metropolitana (UAM), México. He is Member of Mexican Research System (SNI-CONACYT). He was President of Sociedad Latinoamericana de Biología Matemática (SOLABIMA). He has published several papers, mainly on Biomathematics. His current interests are in issues of mathematical epidemiology.

Francisco Yannarella (1939–2017) was Professor of Parasitology and Parasitic Diseases of the Faculty of Veterinary Medicine of the University of La Plata, Argentina. His doctoral thesis director was Prof. J.J. Boero. Unfortunately, Dr. Yannarella died on May 1, 2017 in his hometown, Lobos, province of Buenos Aires, Argentina. He was one of the first veterinary parasitologists who thought and approached the phenomenon of parasitism in an original and creative way.

Chapter 1
Mario Bunge: An Introduction to His Life, Work and Achievements

Michael R. Matthews

Abstract This chapter outlines something of Mario Bunge's long life and career as a physicist-philosopher originally living and working in Argentina for 40 years, then in Canada for nearly 60 years. It indicates the extraordinary breadth, depth and quantity of his research publications. It deals briefly with some key components of his work, such as: systemism, causation, theory analysis, axiomatization, ontology, epistemology, physics, psychology and philosophy of mind, social science, probability and Bayesianism, defence of the Enlightenment project, and education. Finally, the chapter gives an account of the structure of the festschrift, and an indication of each of the 41 contributions.

Mario Bunge is a physics-trained philosopher who has made significant contributions to an extraordinarily wide range of disciplines. He was born in Buenos Aires, Argentina on 21st September 1919. This Festschrift celebrates his 100-year life, and his contributions to so many scholarly disciplines: physics, philosophy, sociology, psychology, cognitive science, and more. In terms of longevity, productivity, and liveliness of mind, he is in the same small and exclusive league as his own philosophical hero, Bertrand Russell. Bunge held chairs in physics and in philosophy at universities in Argentina (University of Buenos Aires, *Universidad Nacional de La Plata*), and visiting professorships in the USA (University of Texas, University of Delaware, University of Pennsylvania and Temple University) before his appointment as professor of philosophy at McGill University in Montreal in 1966. He held this chair, and later the Frothingham Chair in Logic and Metaphysics, until his retirement in 2009, when he became McGill's Frothingham Professor Emeritus. He has had visiting professorships at major universities in Europe, Australasia, as well as North and South America. He has published 70 books (many with revised editions and translations) and 540 articles (including translations). Age has not wearied him. After celebrating his 95th birthday in 2014, he published three

M. R. Matthews (✉)
School of Education, University of New South Wales, Sydney, NSW, Australia
e-mail: m.matthews@unsw.edu.au

© Springer Nature Switzerland AG 2019
M. R. Matthews (ed.), *Mario Bunge: A Centenary Festschrift*,
https://doi.org/10.1007/978-3-030-16673-1_1

1

books (Bunge 2016, 2017a, 2018a) and a good many articles (Bunge 2014a, b, 2015, 2017b, c, 2019). All titles and details are in this Festschrift's 'Bunge Bibliography'.

1.1 Recognition

Bunge has been awarded many prestigious fellowships and prizes. In 1965 he received the German government's Alexander von Humboldt fellowship for work on the axiomatic foundation of physics at the Institute of Theoretical Physics in Freiburg. In 1969 he received a Canada Council for the Arts Killam Fellowship, awarded to 'outstanding scholars to carry out their ground-breaking projects', the bequest aiming 'to promote sympathetic understanding between Canadians and the peoples of other countries'. In 1971 he received a Guggenheim Fellowship, awarded for 'exceptionally productive scholarship'. In 1982 he became a Prince of Asturias Laureate for Communication and Humanities. In 2014 the Bertalanffy Center for the Study of Systems Science (BCSSS) in Vienna awarded him the Ludwig von Bertalanffy Award in Complexity Thinking. Bunge is one of just two philosophers in the Science Hall of Fame of the American Association for the Advancement of Science: the other is Bertrand Russell.

Bunge's work has been celebrated in festschrifts of 40 years ago (Agassi and Cohen 1982) and 30 years ago (Weingartner and Dorn 1990); more recently in Spanish anthologies (Denegri and Martinez 2000; Denegri 2014); and appraised in at least three journal thematic issues (Matthews 2003, 2012; Pickel 2004). Bunge briefly surveyed his own life and work in a chapter in an anthology on Latin American philosophy (Bunge 2003c), and later in a wonderful and engaging 500-page autobiography *Between Two Worlds: Memoirs of a Philosopher-Scientist* (Bunge 2016).

1.2 Family and Education

The Bunge family had its origins on the island of Gotland, off the Swedish coast where the village of Bunge remains. Ancestors moved to Unna in Westphalia, and then to Argentina in the early nineteenth century, soon after independence (Bunge 2016, chaps.1 and 2). Bunge's father, Augusto Bunge (1877–1943), and three of his father's eight siblings, distinguished themselves in various fields: economics, sociology, medicine, philosophy, law and literature.

Mario's father, Augusto, attended a Jesuit school, where he won all the prizes, but at 14 he lost the faith and became an atheist. He studied medicine, and in 1900 he graduated as a medical doctor with the gold medal. His doctoral thesis dealt with tuberculosis as a social disease, for it affected far more the poor than the rich. The Argentine government sent him to Germany and France to study public health policies. On his return, he published two thick tomes expounding the state of public health in those countries. During his student days he joined the young Socialist Party, and in 1916 he became a Socialist congressman, an office that he held for 20 years. In his parliamentary career he promoted several worker welfare bills, and in 1936 he introduced a national medical insurance bill whose provisions were advanced even by contemporary standards.

Augusto and his wife Mariechen (1882–1977) created a home, *El Ombú,* outside the village of Florida on the outskirts of Buenos Aires. They were avid gardeners with a 6000 m^2 plot of grape vines, fruit trees, vegetables, and 130 rose varieties. Their home was the centre for a liberal, intellectual salon including scholars and professionals from many fields. In 1943 Augusto was briefly jailed for raising funds for the Allied war effort at a time when the government supported the Nazis; shortly after his release he suffered a stroke and died at age 66 years. Mariechen was jailed for a month for criticising the newly installed military dictatorship (Bunge 2016, pp. 69–71). When released from jail, she had just one tooth remaining in her mouth.

Mario's parents wanted their son to be 'a citizen of the world'. From an early age he was set a demanding schedule of reading literature in six languages: Spanish, English, French, Italian, German and Latin, with Chinese read in translation. This early multi-lingualism was of inestimable benefit to his education, allowing him to read the classics and the best moderns in their own words. It also freed him from dependence on commercial, political and ideological judgements about what books would be translated and published in Spanish. His reading of Heisenberg did not have to wait upon Spanish translations; nor his reading of the major European and Anglo philosophers, and important Enlightenment texts whose translations were prohibited in Argentina.

One consequence of the demanding multi-lingual reading regime his parents fostered is Bunge's critical judgement of the mono-lingual limitations of the bulk of Anglo-American scholarship. In a critical review of a major book (1120 pages) by Randall Collins on the sociology of philosophies—which has the less than modest subtitle of *A Global Theory of Intellectual Change* (Collins 1998)—Bunge laments that Collins ignores Descartes' central scientific works because 'they were not available in English translation until recently' (Bunge 1999e, p. 281); that his secondary sources are all English (ibid p. 280); and that he exclusively uses English translations of European philosophers even when the translations are notoriously unreliable.

At age 12, he gained entry to the prestigious *Colegio Nacional de Buenos Aires.* The *Colegio* was a disappointment. He relates that teachers 'instilled more fear than respect', and 'Most of our professors were not interested in teaching, and some of them were frankly incompetent' (Bunge 2016, p. 27). He completed his undergraduate physics degree at the *Universidad Nacional de La Plata,* where subsequently he became a professor of physics.

1.3 Breadth and Coherence

Bunge has been enormously productive as a researcher in physics, philosophy, social science, and other fields. Many of his books have appeared variously in Spanish, Portuguese, German, Italian, French, Polish, Russian, Chinese, Arabic, Japanese, Farsi, Romanian, and Hungarian editions. Additionally, he has published books in Spanish and French that have not appeared in English.

Bunge has made substantial contributions to a remarkably wide range of fields: physics, philosophy of physics, metaphysics, methodology and philosophy of science, philosophy of mathematics, philosophy of psychology, philosophy of social science, philosophy of biology, philosophy of technology, moral philosophy, social and political philosophy, medical philosophy, criminology, legal philosophy, and education.

Beyond breadth, Bunge's work is noteworthy for its coherence. In the past half-century, the pursuit of systemic philosophy, 'big pictures', 'grand narratives' or even cross-disciplinary understanding has waned, with fewer and fewer scholars having serious competence beyond their own narrow field of research. As Susan Haack wrote:

> Our discipline becomes every day more specialized, more fragmented into cliques, niches, cartels, and fiefdoms, and more determinedly forgetful of its own history. (Haack 2016, p. 39)

The disciplinary norm has shrunk from scientifically-informed philosophers with wide systemic concerns, to those with narrow-focus pursuits.

Philosophers of science are usually, and understandably, just philosophers of science; it is uncommon for them to also be scientists, much less to make contributions to other areas of philosophy, and other disciplines. The pattern of graduate studies, and the pressures of finding a position and securing tenure, fuel this move to specialization and discipline-specific research programmes; to a narrowing of the disciplinary mind. Bunge defied this trend maintaining that:

> A philosophy without ontology is invertebrate; it is acephalous without epistemology, confused without semantics, and limbless without axiology, praxeology, and ethics. Because it is systemic, my philosophy can help cultivate all the fields of knowledge and action, as well as propose constructive and plausible alternatives in all scientific controversies. (Bunge 2016, p. 406)

1.4 Vocation of an Academic

As an academic, Bunge has had a life-long commitment not just to research, but also to the social and cultural responsibility of academics; he has never been seduced by the 'Ivory Tower' option, comfortable though it would have been at many stages of his life. In other contexts, and in former ages, his version of academic commitment might be called 'a vocation'.

While in high school Bunge became interested in physics, philosophy, and psychoanalysis, and wrote a book-length criticism of the latter. In 1938 he was admitted to the *Universidad Nacional de La Plata*, where he studied physics and mathematics. Shortly thereafter he founded a workers school (the *Universidad Obrera Argentina*). In doing this he was inspired by the Mexican socialist and educator, Vicente Lombardo Toledano (1894–1968), who had established in 1936 the Workers University of Mexico (still in existence today as part of Mexico's national university system). This was quintessential Enlightenment thinking and practice about education. The school's effectiveness prompted its closure by the government 5 years later in 1943. At the time it had 1000 students enrolled.

In 1944, along with involvement in the UOA, Bunge founded the journal *Minerva: Revista Continental de Filosofía*, in order to facilitate the development of contemporary, science-informed, modern philosophy in Latin America. It did not have just a scholarly purpose. The first issue announced that the journal was 'armed and in combat: armed of reason and in combat for reason and against irrationalism'. In the subsequent 80 years, Bunge has never wavered in this commitment. As he said in his *Memoirs*:

> I had the idea of organizing a sort of rationalist common front to fight irrationalism, in particular existentialism. This pseudo-philosophy had started to rule in the Latin American schools of humanities: it rode on the fascist wave and hid behind the phenomenological veil. (Bunge 2016, p. 105)

The Argentina of Bunge's youth, and beyond, was a society with a conservative and reactionary Catholic church, a comfortable ruling elite, and an authoritarian, proto-fascist government that supported Hitler and maintained diplomatic relations with Germany through to 1944. It gave little support to science or to workers' education or their rights. Neither government nor church supported 'free thinking', much less critical philosophy.

The reactionary religious-cultural-political circumstance of Argentina was perva sive throughout most of Latin America. The USA-supported military dictatorships in Paraguay, Uruguay and Chile, set the common standard for anti-democratic authoritarianism. The Enlightenment's advocacy of the separation of Church and State fell largely on deaf ears of the Latin American religious, political, and economic elites. Contraception was illegal, divorce was impossible (it was only legalised in Argentina in 1987), homosexuality was both a sin and a crime, abortion was illegal, censorship of ideas, books, films, theatre was rife, and on and on. The Church had inordinate influence on education, including on the writing of curricula, the training of teachers, and the appointment of principals. In many state universities, passing 'Thomism 101' was a condition of graduation; it was likely a condition in all Catholic universities in Latin America.

Latin America, of course, had no monopoly on religion-based state reaction. Ireland, the Philippines, Portugal, and Spain had comparable Catholic-committed regimes. And the situation was as reactionary in all countries where Islam dominated; and of course, in the USSR, China, East Europe and elsewhere where Marxist ideology dominated. In such regimes, through to the present, it was very costly for academics to make the kind of critical interventions (speeches, papers, books) that now are barely noticed in the liberal West, that take no courage and have zero career consequences. In Bunge's time, in Argentina, non-appointment, fines, dismissal, or jail were the common costs for liberal and socialist dissent. He paid these prices.

Indicative of Bunge's sense of responsibility to the growth of knowledge is that he has always devoted time and energy to the institutions and activities required for it. Bunge has founded and edited journals and book series; he has founded and contributed to scholarly associations in at least five countries; and he has planned and hosted numerous conferences and research seminars, always along with his own constant research and publishing. All of this 'structural' or academic community work is time-consuming, it does not beget research dollars or promotion, and it detracts from writing and personal time. Few scholars have been prepared to make the required monetary, time and career sacrifices. Bunge has.

Alberto Cordero has given a comprehensive account of the history of philosophy of science in Latin America. Of Bunge's publications, translations, 'community building', and international impact, Cordero says: 'No Latin American philosopher had achieved anything comparable before in cosmopolitan philosophy'. He adds that as a

> citizen of the world, perhaps the most universalist of philosophers in the subcontinent, Bunge is nonetheless very South American (it is hard to imagine him growing up anywhere else but in cosmopolitan Argentina). (Cordero 2016)

1.5 Beginnings

Bunge graduated in physics from La Plata in 1942. In 1943 he started to work on problems of nuclear and atomic physics under the guidance of Guido Beck (1903–1988), an Austrian refugee who had been an assistant of Heisenberg in Leipzig. Beck was the inventor of the layer model of the atomic nucleus, the first to propose the existence of the positron, and pioneered the study of beta decay. Bunge believes Beck might have received the Nobel prize for physics had he been working in North instead of South America (Bunge 2016, p. 77). He does thank Beck for 'teaching me not to allow politics to get in the way of my science' (Bunge 1991a, p. 524). Bunge obtained his PhD in physics in 1952 from La Plata with a dissertation on the kinematics of the relativistic electron. 'My doctoral diploma did me no good, because it was not accompanied by the Peronist party card without which I could not even get a job as a dogcatcher' (Bunge 2016, p. 89). Nevertheless, the thesis was subsequently published as a book (Bunge 1960).

Bunge made his international philosophical debut at the 1956 Inter-American Philosophical Congress in Santiago, Chile. He was then aged 37 years. Willard Van Orman Quine, in his autobiography, mentions attending this congress, and the only thing about the congress that he thought worth recording was:

The star of the philosophical congress was Mario Bunge, an energetic and articulate young Argentinian of broad background and broad, if headstrong, intellectual concerns. He seemed to feel that the burden of bringing South America up to a northern scientific and intellectual level rested on his shoulders. He intervened eloquently in the discussion of almost every paper. (Quine 1985, p. 266)

1.6   Systemism

Bunge is a systemist and argues for the unity, not the disunity, of knowledge; for the need for science, social science, and philosophy to be advanced in partnership; and for science education to convey this seamless, interdependent canvas of human knowledge. For some, Bunge is overly systemic, too precise, and ambitiously inter-connected in his writing. But beyond this stylistic commitment, there is a philosophical commitment to systemism as an ontology, as a view about how the natural and social worlds are constituted. Bunge has developed a philosophical system that can be characterized as: materialist (or naturalist) but emergentist rather than reductionist; systemist rather than either holist or individualist; ratio-empiricist rather than either rationalist or empiricist; science-oriented; and exact, that is, built with the help of logical and mathematical tools rather than depending upon purely verbal articulation.

Bunge's philosophical system is laid out in detail in his monumental eight-volume *Treatise on Basic Philosophy* (1974–1989). Its nine individual books are devoted to semantics (one to meaning, another to interpretation and truth), ontology (one to the basic stuff or 'furniture' of the world, another to systems), epistemology (one to exploring the world, another to understanding it), philosophy of science and technology (one to the formal and physical sciences, another to life science, social science and technology), and ethics. He has applied his systems approach to issues in physics, biology, psychology, social science, technology studies, medicine, legal studies, and science policy.

Bunge points to William Harvey introducing systemism into science (natural philosophy) with his study of the heart as part of a cardiovascular system (*De motu cordis*, 1628); and Newton promoting systemic thinking in his postulation of universal gravitation, which led to his unification of planetary and terrestrial motions, the bringing of the heavens down to earth. Early modern philosophers paid little, if any, attention to this scientific innovation.

His systemism is laid out in the first volume of his *Scientific Research*, titled *The Search for System* (Bunge 1967a); the fourth volume of his *Treatise*, titled *A World of Systems* (Bunge 1979a); and in various articles (Bunge 1977a, c, 1979b, 2000a, 2014b). In 2014, he gave a plenary talk ('Big questions come in bundles, hence they should be tackled systemically') at the Vienna congress of the Society for General Systems Research (Bunge 2014b), and there received the Society's Bertalanffy Award.

From the outset, he has been at pains to distinguish his systemism from holism. He regards all variants of holism as more than just philosophically mistaken and obscurantist; they are politically dangerous as they give comfort to statism (Bunge 2016, p. 252).

1.7 Causation

Bunge's first major book in philosophy was his 1959 book *Causality: The Place of the Causal Principle in Modern Science* (Bunge 1959). The book was recommended to Harvard University Press by Quine and reviewed favourably by the physicist-philosophers Henry Margenau and Victor Lenzen (Bunge 2016, p. 127). The book was an instant success and put Bunge, and Latin American philosophy of science, firmly on the international map. It came out of the philosophical 'left field': it was among the few books ever written by Latin American philosophers of science to receive international recognition and review up to the 1950s. The work was translated and published in German, Hungarian, Italian, Japanese, Polish, Russian and Spanish editions. Twenty years later, a third, revised edition was published as a Dover paperback, *Causality and Modern Science* (Bunge 1979c).

The book was a landmark in the subject. For decades, under the influence of positivism and logical empiricism, philosophers had eschewed all serious investigation of causation as understood and investigated by scientists. Outside of Thomism (Wallace 1972), the Humean picture was widely accepted: there was no causation or necessary connection in nature; there was just regularity to which the mind brought the label 'causation'. In Hume's words: 'Upon the whole, necessity is something, that exists in the mind, not in objects' (Hume 1739/1888, p. 165).

Philosophers brought detailed philosophical analysis and debate to the *consequences* of this position, but rarely questioned its empiricist presuppositions (Sosa 1975). Bunge brought detailed *scientific knowledge* of natural processes into the philosophical analysis of causation (Bunge 1961, 1962, 1982). He mounted informed arguments against Humean empiricist and positivist accounts that made causation 'imaginary'; accounts that replaced real-world causation with correlation; that kept the 'causation' label, but denied it had any ontological reference.

Bunge also argued in detail against popular interpretations of quantum mechanics that supposedly had also consigned causation to the Humean bin. Bunge rejected this because it was fanciful philosophy and displayed great ignorance of science. As he wrote in *Causality*:

> The trend of recent science points neither to the decausation preached by positivism in favor of purely descriptive statements or uniformity, nor a return to traditional pancausalism. Present trends show, rather, a *diversification* of the types of scientific law, alongside of an increasing realization that several categories of determination contribute to the production of every real event. (Bunge 1959, p. 280)

But his work also bears upon contemporary, sophisticated non-empiricist accounts of causation. Rögnvaldur Ingthorsson, in his Chap. 12 contribution to this Festschrift, observes that:

> Proponents of powers-based accounts [of causation] seem not to be aware of Bunge's critique of the Aristotelian view of causation, and therefore arguably continue to build on a flawed conception of causal influence, one that is incompatible with the theories and findings of modern science.

1.8 Theory Analysis

Whilst visiting professor of philosophy and of physics at the University of Delaware (1965–1966) Bunge convened a seminar on the 'Foundations of Physics' (Bunge 1967c). His own opening contribution was titled 'The structure and content of a physical theory' (Bunge 1967d), which in turn began with this statement:

> In analysing a physical theory, we may distinguish at least four aspects of it: the background, the form, the content, and the evidence—if any. By the *background* of a theory we mean the set of its presuppositions. By the *form* or structure, the logico-mathematical formalism quite apart from its reference to physical objects or its empirical support. By the *content* or meaning, that to which the theory is supposed to refer, quite apart either its form or the way the theory is put to the test. And the *evidence* a theory enjoys is of course the set of its empirical and theoretical supporters. (Bunge 1967d, p. 15)

This clear and simple fourfold division of the components of the scientific-philosophical analysis of theory represents what Bunge had been doing for the 20 years leading up to the Delaware Seminar, and what he would continue doing for the following 60 years. He analysed theories in physics, chemistry, biology, psychology, economics, sociology, criminology and more, in terms of their *background*, *form*, *content* and *evidence*.

1.9 Axiomatization

The foregoing is the background to Bunge's persistent concern with the axiomatization of theories, a concern which reappears in one of his recent publications, 'Why axiomatize?' (Bunge 2017b). For Bunge, axiomatization simply becomes part of what scientific theorising is, and it is the philosopher's task to both make this clear and to contribute to it; it is scientific work that philosophers can do.

Bunge maintains that any reasonably clear theory can be axiomatized. If it cannot, then it is not clear; and if it is not clear, then it is a deficient or maybe even useless theory. Vagueness, hunches, instincts, feelings are all part of science; they can influence what research paths to take, what might constitute evidence, and so on, but to the extent that they figure in actual outcomes, or scientific theory, then the theory is flawed. Further:

> Contrary to widespread opinion, axiomatization does not bring rigidity. On the contrary,
> by exhibiting the assumptions explicitly and orderly, axiomatics facilitates correction and
> deepening. (Bunge 1999c, p. 28)

In his *Foundations of Physics* Bunge axiomatized five theories: point-particle and continuum mechanics, classical electrodynamics, Einstein's theory of gravitation, and non-relativistic quantum mechanics (Bunge 1967c). Each was presented in a logically ordered sequence: primitive concepts—defined concepts—postulates—theorems. He recognised, of course, that 'this is an artificial logical reconstruction, very different from the rather messy way theories are invented and developed' (Bunge 2016, p. 196). But it facilitated better understanding of the theory, and a clearer grasp of its philosophical commitments and implications. The structure brings clarity to judgements about the defining features of a theory, and what changes would constitute a different theory, or a 'neo' version of it.

Bunge's account of scientific methodology was elaborated in his two-volume *Scientific Research* (Bunge 1967a, b). There are striking similarities between Bunge's analysis of scientific theory and theory change and Imre Lakatos's account of 'rational reconstruction' in history of science (Lakatos 1971) and Lakatos's separation of 'hard core' and 'protective belt' commitments in scientific research programmes (Lakatos 1970)—these being published some 4 years after Bunge's account. The similarities were not unnoticed by Bunge, and along with other matters, led to a falling out between the two philosophers (Bunge 2016, p. 201).

Clear-headed axiomatization is a prerequisite to the successful 'marriage' of different theories or research programmes; axiomatization makes clear what either side needs to give up, what price is to be paid for the marriage. For example, reducing optics to electromagnetic theory; joining thermodynamics with classical mechanics; synthesising evolutionary theory and genetics; and so on. In all these cases, vagueness and ambiguity are exposed by the effort of axiomatization. Vagueness advances nothing in science, though it can do a great deal in politics, religion, and countless pseudo-sciences.

A contributor to the Delaware Seminar, Paul Bernays, a Swiss logician and mathematician, reflected Bunge's own view of the merits of axiomatization:

> Such a strengthened consciousness is valuable whenever the danger exists that we may be
> deceived by vague terminology, by ambiguous expressions, by premature rationalizations,
> or by taking views for granted which in fact include assumptions. Thus, the distinction
> between inertial mass and gravitational mass makes it clear that their equality is a physical
> law—something which might be overlooked by speaking of mass as the quantity of matter.
> (Bernays 1967, p. 189)

In his 'Why Axiomatize?' (Bunge 2017b), Bunge says axiomatizing consists in subjecting a theory that has been built in an intuitive or heuristic fashion to the following operations:

1. Exactification of intuitive constructs, that is, replacing them with precise ideas, as
   when substituting 'set' for 'collection', 'function' for 'dependence', and 'derivative with
   respect to time' for 'rate of change';

2. Grounding and justification of postulates, bringing hidden assumptions to light—assumptions that, though seemingly self-evident, may prove to be problematic;
3. Deductively ordering a heap of known statements about a given subject. (Bunge 2017b)

1.10 Ontology

Bunge is an ontological realist. He believes there is an external, non-subject-dependent world. Nature pre-dated humans, and presumably will post-date them. Further, both the observable and unobservable entities proposed in mature scientific theories—planets, fault lines, elements, chromosomes, genes, waves, atoms, phlogiston, economic class, intelligence, instincts, and so on, are assumed to exist, and that is why they are postulated. And if on appropriate experimental investigation they are found not to exist, then the theory needs to be rejected, or at least this specific postulation within the theory needs to be abandoned. Ontological realism has been refined, and has taken various forms in contemporary philosophy, with selective realism, structural realism, perspectival realism and entity realism being four competing versions (Agazzi 2017). All versions are in conflict with the equally long tradition of ontological idealism, stated loudly in the present day by constructivists. In the words of one proponent:

> ...For constructivists, observations, objects, events, data, laws, and theory do not exist independently of observers. The lawful and certain nature of natural phenomena are properties of us, those who describe, not of nature, that is described. (Staver 1998, p. 503)

Beyond being an ontological realist, Bunge is a realist of the materialist kind (Bunge 1981c, 2000 Pt.1). This is in contrast to immaterialist realists who countenance the existence of non-material explanatory entities—ghosts, spirits, angels, chi, jinns, ancestors, and so on. Bunge's materialism is science-informed; or more strongly, science-dependent. It is scientific materialism. In a recent paper on 'Gravitational Waves and Space-Time', he writes:

> As long as we confine ourselves to macrophysics, we must admit that the recent detection of gravitational waves suggests the counterintuitive thesis that spacetime is a material entity, so that we must rethink our conceptions of matter and materialism, much as people did when Faraday and Maxwell added the concept of an electromagnetic field to that of a body. (Bunge 2018b)

1.11 Epistemology

Bunge is also a realist in epistemology, meaning that the entities and mechanisms postulated by science not only are supposed to exist, but their properties and characteristics can be known. Further, they can be objectively known. That is, the knowledge sought is not subject dependent; knowledge of the entity does not vary from observer to observer, from one knower to another. There are not different Christian, Hindu, Jewish, Islamic, black, white, American, children's, female knowledges of the supposed scientific entities or unseen mechanisms; there is just knowledge, partial knowledge or ignorance of the entities and mechanisms held by people of various nationalities, beliefs, race, age or gender. Everyone is entitled to their own opinion but not to their own facts; let alone 'alternative' facts.

Assuredly the putative knowledge does not come out of the sky. It is created by concrete individuals and groups in specific historical and cultural circumstances, but the knowledge is not constrained by or limited to those circumstances (Minazzi 2017; Musgrave 1993). It is simply a conceptual mistake to talk about 'Christian Science', 'Islamic Science', 'Hindu Science', 'Indigenous Science', 'Children's Science', or 'National Socialist Science', though these labels are, and have been, widely used. There is simply science, good, bad or indifferent; formulated by Christians, Muslims, Hindus, Native peoples, children, or Nazis. The titles can have a short-hand purpose, and they have a legitimate anthropological function, but a limited philosophical one.

Bunge's position had been well expressed by Pierre Duhem, the French Catholic and positivist philosopher, in a 1915 series of lectures on 'German Science'. At a time when it was not fashionable, prudent or a good career move in France to acknowledge the contributions of German thinkers to the development of science, Duhem did so. He elaborated wonderfully and carefully on the achievements, and limitations, of Boltzmann, Einstein, Gauss, Haeckel, Hertz, Kékulé, Mach, Neumann, Weber and many others. He concludes:

> ... if the national character of an author is perceived in the theories he has created or developed, it is because this character has shaped that by which these theories diverged from their perfect types. It is by its shortcomings, and by its shortcomings alone, that science, distanced from its ideal, becomes the science of this or that nation. ... There is no trace of the English mind [*esprit*] in Newton, nothing of the German in the work of Gauss or Helmholtz. In such works one no longer divines the genius of this or that nation, but only the genius of humanity. (Duhem 1916/1991, p. 80)

This Duhem-Bunge 'universalist' position is rejected by postmodernists, and by most adherents of the 'Sociology of Scientific Knowledge' (SSK), or the 'Edinburgh Strong Programme', traditions. Bunge cautions that:

> In particular, epistemology must take into account that scientific research is just one cultural activity, hence the study of it cannot be isolated from the study of other branches, in particular philosophy and ideology. In short, epistemology, if it is to be realistic (not just realist), must be not only structural but also psychological, sociological, and historical. (Bunge 1981d, p. 116)

Bunge believes—*contra* contemporary individualisms, subjectivisms, idealisms, and relativisms—that science can, and does, give us demonstrably the best knowledge of the natural and social worlds. Not perfect or absolute knowledge but the best available. He writes:

> Contrary to a widespread opinion, scientific realism does not claim that our knowledge of the outer world is accurate: it suffices that such knowledge be partially true, and that some of the falsities in our knowledge can eventually be spotted and corrected, much as we correct our path when navigating in new territory. (Bunge 2006, p. 30)

Further, as well as empirical adequacy, there are theoretical virtues or character-istics that can be used to judge competing theories or knowledge accounts against each other, and so to determine which is currently superior. Such considerations also allow identification of the best knowledge-advancing research programmes. These theoretical virtues include empirical testability, plausibility or consistency with extant knowledge, guidance to causal connections, absence of *ad hoc* elements, and having heuristic value for further research. So, theories are not judged just on empirical adequacy: with sufficient *ad hocness* most competing theories in any domain can be reconciled with empirical evidence; but not all can make testable, novel predictions and have them experimentally confirmed.

For Bunge, such typically scientific knowledge is the only sound basis for moral decision making, social and political reform, and personal flourishing. This affirmation brings the charge of scientism down upon him; a charge he happily pleads guilty to—provided it does not entail agreeing to crass, amateurish or plainly mistaken understandings of science (Bunge 1986b, 2014a). He is consequently a critic, indeed a trenchant one, of social forces and academic movements that diminish, or reject, the intellectual authority of reason and science (Bunge 1996).

1.12 Physics

Before and after being awarded his physics PhD, alone or jointly with his former student Andrés J. Kálnay (1932–2002), Bunge published several articles on a number of problems in quantum mechanics. Their subjects included the total spin of a system of particles, the mass defect of the hydrogen atom, new constants of motion, the quantum Zeno paradox, and the measurement process (Bunge 1944, 1945, 1955a, b, 1956; Bunge and Kálnay 1969). Thirty years ago he wrote that one of his 1955 papers—'A New Picture of the Electron'—was 'my best scientific paper':

> In this paper I introduced a new position coordinate (sometimes called the Feynman-Bunge-Corben operator), proved the existence of six new constants of motion of the relativistic electron, and suggested that this particle has an internal structure. (Bunge 1990b, p. 678)

Between 1966 and 1969 Bunge met and discussed quantum physics with Werner Heisenberg, and later contributed to Heisenberg's *Festschrift* (Bunge 1977d). A point that he makes over and over regarding Heisenberg is that his deservedly famous 'Principle'—$\Delta p \ \Delta x \geq h/4\pi$ (the product of the dispersions of the values of the momentum (hence the velocity) and the position of a microparticle is at least $h/4\pi$, where h is Planck's constant)—is not a principle at all, but it is rather a theorem. It is a derived formula that follows rigorously from the axioms and definitions of quantum mechanics. Because the formula is a theorem, to interpret it correctly one must examine the premises that entail it. Bunge maintains that such an examination shows that the formula is quite general, but not in the way most believe it to be. In particular, it refers neither to macrophysical entities, nor to a particle under observation. It is a law of nature for the microphysical, just as much as Schrödinger's equation, which is the basic formula of non-relativistic quantum mechanics. Thus, the popular name 'Uncertainty Principle' is incorrect. As Bunge notes, uncertainty is a state of mind, and quantum mechanics is not about minds but about physical things, most of which are beyond the experimenter's reach.

Much of his work in theoretical physics is gathered in his *Foundations of Physics* (Bunge 1967c) and *Philosophy of Physics* (Bunge 1973). His contributions to theoretical physics have continued to the present day. In his early Eighties he published 'Velocity Operators and Time-Energy Relations in Relativistic Quantum Mechanics' (Bunge 2003b); in his Nineties he published on the Aharonov-Bohm Effect (Bunge 2015) and on Gravitational Waves (Bunge 2018b).

In a comprehensive collection of studies of Bunge's *Treatise,* Manfred Stöckler, a German physicist and philosopher, correctly remarked that:

> There are two characteristic qualities of Mario Bunge's papers on quantum mechanics. Firstly, more than most philosophers he is intimately acquainted with the details of the theory, both with the theoretical structure and with the practice of its applications. So he corrects numerous philosophical claims which are just misunderstandings of physics. Secondly, more than many other experts in the foundations of physics he looks at quantum mechanics from an explicitly philosophical point of view. So most of his writings about the philosophy of quantum mechanics are guided by his fight against people using quantum mechanics in order to refute realism. (Stöckler 1990, p. 351)

Bunge rejected both the popular Copenhagen indeterminist and Bohm understandings, and proposed his own non-local realist interpretation of quantum mechanics. This keeps the mathematical formalism but modifies the positivist interpretation proposed by Bohr, Heisenberg, Pauli, and Born. For example, Bunge interprets the square of the absolute value of the state function not as the probability of finding the object in question in a unit volume (an intrinsically subjective notion), but rather as the probability that it is within a unit volume (an objective version of the former). Bunge argues that electrons and the like are neither particles nor waves, although they appear as such under special circumstances. Talk of waves and particles is metaphorical, an allusion back to classical notions from which quantum mechanics emerged. Bunge maintains:

> Physics cannot dispense with philosophy, just as the latter does not advance if it ignores physics and the other sciences. In other words, science and sound (i.e., scientific) philosophy overlap partially and consequently they can interact fruitfully. Without philosophy, science loses in depth; and without science philosophy stagnates. (Bunge 2000c, p. 461)

Physicists have acknowledged the impact of Bunge's work. In 1989 the *American Journal of Physics* asked its readers to vote for their favourite papers from the journal, from its founding in 1933 to 1989. In the resulting 1991 list of most memorable papers, alongside classics from Nobel Prize winners and luminaries such as Bridgman, Compton, Dyson, Fermi, Kuhn, Schwinger, Wheeler, and Wigner, was Bunge's (1956) 'Survey of the Interpretations of Quantum Mechanics' (Romer 1991). In 1993, the journal repeated the exercise, asking readers for the most influential papers in the journal's first 60 years. In this list, Bunge's (1966) paper—

'Mach's Critique of Newtonian Mechanics'—took its place alongside his 1956 article (Romer 1993). This recognition by physicists of someone who is at once a philosopher, a physicist, and a social scientist, is uncommon, to say the least.

1.13 Psychology and Philosophy of Mind

Bunge has had a serious interest in psychology since his adolescent years. As he writes in his *Memoirs*: 'Psychology had intrigued me, since at age 16, I read some of Freud's books, which sold for a few cents at subway kiosks' (Bunge 2016, p. 43). At the same time, he read Bertrand Russell's *Problems of Philosophy* (Russell 1912). He quickly surmised that the former was 'psychobabble … and sheer fantasy' (Bunge 2016, p. 43). Through the eight decades he has spent on appraising Freudianism and psychoanalysis, these initial evaluations did not change, they only strengthened.

Bunge has contributed to some first-order issues in psychology, including language acquisition, where amongst other things he rejects Noam Chomsky's account of a neurologically-embedded Universal Grammar, and Chomsky's consequent generative linguistics programme (Bunge 1983, 1984, 1986a, c, 1999b). Bunge has written on methodological issues in psychology, with his arguments being stated in articles (Bunge 1985, 1989, 1990a), and the book *Philosophy of Psychology*, written with Rubén Ardila (Bunge and Ardila 1987).

His philosophy of mind is advanced in a series of papers (Bunge 1977b, 1981a, 1987, 1991b), and two major books: *The Mind-Body Problem* (Bunge 1980) and *Matter and Mind: A Philosophical Inquiry* (Bunge 2010). He oft says that he is 'against brainless psychology and mindless cognitive science' (Bunge 1981a) He is against all dualisms in theory of mind, and advances his emergent materialist, monist theory as the only theory of mind consistent with current scientific knowledge of mental processes and consciousness. In a recent paper, he writes:

> nearly all the important findings in psychology, in particular the localization and interdependence of a number of mental processes, from anxiety to morality, have been so many successes of the psychoneural program. (Bunge 2017c, p. 458)

Having said this, Bunge's theory of mind, along with his theory of everything else, is not reductionist. His commitment to systemism (everything except the universe as a whole belongs to some larger system) prevents all radically reductionist moves. This is elaborated in many places (Bunge 1977e, 1991c), summarised in a chapter titled 'A Pack of Failed Reductionist Projects', where he lays out and critiques physicalism, computationism, linguistic imperialism of both the positivist and hermeneutical variety, sociobiology, evolutionary psychology, psychologism, sociologism and rational-choice theory (Bunge 2003a, pp. 149–167). The roll-call of influential theorists whose different reductionist programmes ('everything is a case of ….') are rejected in this chapter includes John Wheeler, Daniel Dennett, Patricia Churchland, Otto Neurath, Rudolf Carnap, Edward Wilson, Richard Dawkins,

Steven Pinker, Wilhelm Dilthey, George Homans, Lev Vygotsky, Michel Foucault, Bruno Latour, Clifford Geertz, and Sandra Harding.

A sense of all these critiques is given in his closing comments on rational-choice theory:

> Rational-choice theory, is currently in vogue, presumably because it looks scientific in addition to purporting to explain much with little, thus producing the illusion that it unifies all the social sciences around a single postulate. However, it can be shown that rational-choice theory is conceptually fuzzy, empirically groundless, or both. Indeed, when the utility functions in a rational-choice model are not specified, as is generally the case, untestability is added to vagueness. (Bunge 2003a, p. 165)

1.14 Social Science

Bunge believes that the lessons learnt from the hard-won successes of natural science should be applied to social science; that the inquiry template forged by the best of natural science can and should be applied to the social and psychological worlds.[1] The disparate Enlightenment philosophers of the eighteenth century were all committed to this thesis. Condorcet, for example, in his influential *Sketch for a Historical Picture of the Progress of the Human Mind* (Condorcet 1795/1955) wrote:

> The sole foundation for belief in the natural sciences is this idea, that the general laws directing the phenomena of the universe, known or unknown, are necessary and constant— why should this principle be any less true for the development of the intellectual and moral faculties of man than for the operations of nature. (Condorcet, 1795/1955, p. 173)

Bunge concurs.

Bunge regards bad philosophy as the major obstacle to the advance of social science. He sees the philosophical deficiencies as logical, ontological, epistemological, and ethical. The logical flaws are conceptual fuzziness and invalid inference; the ontological culprits are individualism and holism; the epistemological errors are sectoralism or tunnel vision, subjectivism, apriorism, pragmatism and irrationalism (Bunge 1998, p. 452).

For Bunge, there are two major moral lapses that contribute to the backwardness of social science:

> One is the frequent violation of the ethos of science, first ferreted out by Merton (1938). Such violation occurs, in particular, when the universality of scientific knowledge is denied, dogmatism is substituted for 'organized scepticism' ....and rigorous testing, or at least testability is jettisoned. The second moral culprit is the attempt to pass off ideology (left, centre or right) for science in basic research, the pretence of moral or political neutrality when tackling practical issues. (Bunge 1998, p. 453)

---

[1] Two issues of the journal *Philosophy of the Social Sciences* were devoted to appraising the implications of Bunge's systemism for social science research (Pickel 2004). He contributed to each issue (Bunge 2004a, b).

Bunge's systemism implies a relatively seamless move from the science of physics, through psychology to sociology and beyond. He does not shy from the label 'scientism', derogatorily applied by others to such a unified theory of knowledge and family resemblance of research methodology (Bunge 1986b, 2014a). Although he rejects Marxist dialectical materialism as 'either unintelligible, too sketchy to be useful, or just plain false' (Bunge 2016, p. 263), he is sympathetic to historical materialism and praises the work of some Marxist historians, such as Eric Hobsbawm, Edward Thompson and Fernand Braudel. His systems concept of society avoids the well-known problems of individualism and holism (Bunge 1979b, 2000a, b).

Tuukka Kaidesoja, a Finnish philosopher who has written extensively on the philosophy of social science, has provided a detailed appraisal of the parallel work of Roy Bhaskar (founder of the 'Critical Realist' programme in social science) and Mario Bunge, and concludes:

> Roy Bhaskar and Mario Bunge have both developed influential realist philosophies of social science. Both of them use the ontological concept of emergence and advocate a doctrine of emergent materialism in their social ontologies. … I argued that Bunge's perspective on emergence enables one to conceptualize levels of organization in complex systems including social systems, while Bhaskar's account of levels of reality is problematic. (Kaidesoja 2009, p. 318)

1.15 Probability and Bayesian Inference

Consistent with his overall realist programme in ontology and epistemology, Bunge has concerned himself with probability and statistical inference in science (Bunge 1951, 1976, 1981b, 1988, 2003a, 2008). The theory of probability is a branch of pure mathematics. It requires interpretation to be connected to the world; either to understand the world or guide actions in the world. Bunge rejects *subjectivist* interpretations of probability, wherein the probability of a statement is a measure of its credibility, or of the conviction a person has in its truth. He rejects *frequentist* interpretations, wherein probability is the long-run relative frequency of observed events. Instead of either, he argues for a *propensity* (or as he also calls it, *objectivist* or *realist*) interpretation of probability, wherein probability is an objective measure of the possibility that some proposed event or state will occur (Bunge 2006, p. 103). He argues that only the third interpretation is compatible with science.

Bunge places Bayesianism, in any of its many forms, within the subjectivist or personalist interpretation of probability; indeed, he believes Bayesianism now occupies the entire subjectivist domain, with no alternatives. Bunge belongs to the minority of philosophers and scientists unequivocally critical of its use in science, social science and medical research. Anti-Bayesianism is a thread through all his writings in the field.

The statistician Leonard Savage, who brought Bayesianism into mainstream statistics and probability theory, stated that 'probability measures the confidence that a particular individual has in the truth of a particular proposition, for example, the proposition that it will rain tomorrow' (Savage 1954, p. 3). Savage used the term 'personalist' probability and defended such Bayesian accounts against relative frequency or objective accounts of probability (Savage 1964). Michael Shaffer states that in Bayesianism 'probabilities are degrees of belief defined over a complete space of propositions' (Shaffer 2012, p. 117).

Beginning almost 70 years ago, in 'What is Chance?' (Bunge 1951), Bunge has continually criticised the popular, subjectivist interpretations of Bayes' Theorem. He sees the theorem as a 'legitimate piece of basic mathematics, which does not refer to the real world' (Bunge 2008, p. 167), but rejects Bayesianism, in particular the 'mindless application of the theorem' (Bunge 2006, p. 101). Not that it works poorly, rather it simply cannot work as routinely interpreted: 'nobody knows how to go about assigning a probability to scientific laws or to scientific data ... in these fields [science and technology] one assigns probabilities to states and events' (Bunge 1988, p. 216). Elsewhere he states the matter as:

> ... in the Bayesian perspective there is no question of objective randomness, random-ization, random sample, statistical test, or even testability: it is all a game of belief and credence. (Bunge 1999a, p. 81)

Bunge's central objection is to the Bayesian linking of probability to 'credence', 'degree of belief', 'confidence', 'expectations', 'conviction', or any such psycho-logical state. For Bunge, psychological states belong to the subjective domain, and should not, in principle, play a determinative role in scientific evaluation or theory appraisal. There are lots of roles that psychological states and conditions can and do play in science, but they have no role in proper theory evaluation. To the degree that they come into theory evaluation, then it is corrupted and arbitrary science. In one long treatment, Bunge says:

> Bayesian statistics and inductive logic are triply wrong: because they assign probabilities to statements; because they conceive of probabilities as subjective; and because they invoke probabilities in the absence of randomness. Adding arbitrary numbers to any discourse does not advance the search for truth: it is just a disguise of ignorance. (Bunge 2008, p. 177)

He is not alone in his rejection of Bayesianism. Ronald Fisher, the statistician and biologist who was largely responsible for the 'modern synthesis' in biology, regarded Bayes' Theorem as measuring 'merely psychological tendencies, theorems respecting which are useless for scientific purposes' (Fisher 1926/1947, p. 7). Bunge favourably cites Fisher's advocacy of randomization in the formation and distribution of control and experimental groups in natural and social science experimentation, and notes that 'Bayesians do not practice randomization .... for them, chance is only in the eyes of the beholder' (Bunge 2008, p. 173). Bunge consistently develops Fisher's claim that measures of psychological states have no place in properly scientific evaluation of theories or hypotheses. Others have also rejected the Bayesian programme because of its inherent subjectivity (Glymour 1980; Kyburg 1978; Levi 1974; Mayo 2004).

An increasing number of statisticians and researchers have abandoned subjective Bayesianism and moved to objectivist versions of the theory, where assigning a numerical value to the prior is done in strict accordance with empirical evidence and/or rational principles, and so is supposedly inter-subjective, trans-subjective or objective (Franklin 2001; Jaynes 2003; Salmon 1990; Shimony 1970). Wesley Salmon writes:

> I proposed that the problem of prior probabilities be approached in terms of an objective interpretation of probability, in particular, the frequency interpretation. I suggested three sorts of criteria that can be brought to bear in assessing the prior probabilities of hypotheses: formal, material and pragmatic. (Salmon 1990, p. 184)

He does acknowledge that while Bayes' theorem provides a mechanical algorithm, 'the judgements of individual scientists are involved in procuring the values that are to be fed into it' (Salmon 1990, p. 181). In some cases, for instance counter-factual hypotheses, there can be no appeal to evidence in quantifying the prior. This because the hypothesis is about what would happen if states of affairs were different from what they are. This is the common case of abstracted or idealised hypotheses and theories in science; for instance, Galileo's claims about what would be the parabolic motion of projectiles in the absence of the known host of actual impediments (Shaffer 2012, pp. 122–124).

Harold Brown surveys the Bayesian retreat and writes[2]:

> The major point urged, with varying degrees of vigor, is that while the use of appropriate algorithms is an important part of the process of arriving at rational evaluations, it is only a part. Judgement is required in order to choose appropriate algorithms and to govern their intelligent application. An account of reason that omits the central role of judgement in determining the inputs to our algorithms and in determining whether and which algorithms to use will be radically incomplete. (Brown 1994, p. 368)

Of this 'retreat', Bunge might say: 'you should not have gone there in the first place'. An informed account of science would have ruled out a priori any flirtation with the subjectivist interpretation of probability, much less with bestowing on it the badge of scientificity. For Bunge, throwing around random numbers and utilising them in long calculations is a hallmark of pseudoscience (Bunge 2008). He lists eight errors with Bayesianism, and concludes:

> Verisimilitude and credibility are often equated with probability. ... This conflation of an epistemological category (verisimilitude), a psychological one (credibility), and an ontological one (probability) is a root of the subjectivist or Bayesian theory. (Bunge 2003a, p. 226)

Bunge develops his 'probability as propensity' account in a number of places (Bunge 1988, pp. 222–226, 1999c, p. 107–108), and in doing so rejects the frequentist alternative to subjective Bayesianism. His central argument is that:

> ... contrary to the frequency view, probability is *not a collective* or ensemble property, i.e., a property of the entire set F, but a property of every *individual* member of F, namely its propensity to happen. ... while each probability function Pr is a property of the ensemble F, its values $Pr(x)$ are properties of the members of F. (Bunge 1988, p. 223)

---

[2]There is a huge literature on this topic. See at least Brown (1994), Corfield and Williamson (2001), Mayo and Spanos (2010), and Swinburne (2002).

He recognises that frequencies can be *indicators* of probability but they are not the probability of an event or episode. The latter is an ontological claim about the event; frequency is not. He writes:

> ... probabilities are theoretical whereas frequencies are empirical (observed or measured). So much so that, unlike probabilities, frequencies depend not only on the sample size (relative to the total population) but also on the sampling method. (Bunge 1999c, p. 108)

Of his own view, he writes:

> In short, the propensity interpretation of probability is consistent with the standard theory of probability and with scientific practice, as well as with a realist epistemology and possibilist ontology. Hence, it solves the old tension between rationality and the reality of chance. None of its rivals has these virtues. (Bunge 1988, p. 226)

1.16 Enlightenment Project

The unifying thread of Bunge's life and research is the constant and vigorous advancement of the Enlightenment project, and criticism of cultural and academic movements that deny or devalue the core principles of the project: naturalism; the search for objective, trans-personal, non-subjective truth; the universality of science; the value of rationality; and respect for individuals. His commitment to the Enlightenment project began in his early 1920s when he was Secretary General of the *Federación Argentina de Sociedades Populares de Educación*. During this time, he wrote his first book, *Temas de Educación Popular* (Bunge 1943), dealing with the principles and practice of popular (workers') education. This was an example of his *practice* of Enlightenment principles; he has no time for un-practised principles, even Enlightenment ones.

Bunge condenses the historical Enlightenment ideology into ten principles:

1. Trust in reason.
2. Rejection of myth, superstition, and generally groundless belief or dogma.
3. Free inquiry and secularism.
4. Naturalism, in particular materialism, as opposed to supernaturalism.
5. Scientism or the adoption of the scientific approach to the study of society as well as nature.
6. Utilitarianism in ethics, as opposed to both religious morality and secular deontologism.
7. Respect for praxis, especially craftmanship and industry.
8. Modernism, progressivism, and trust in the future.
9. Individualism together with libertarianism, egalitarianism (to some degree or other), and political democracy (though not yet for women or slaves).
10. Universalism or cosmopolitanism, for example, human rights and education. (Bunge 1999a, p. 131)

Through his entire scholarly life, he refines and defends each of the foregoing Enlightenment commitments. But he is not uncritical or blinded. In an essay— 'Counter-Enlightenment in Contemporary Social Studies'—he states:

The Enlightenment gave us most of the basic values of contemporary civilized life, such as trust in reason, the passion for free inquiry, and egalitarianism. Of course, the Enlightenment did not do everything for us: no single social movement can do everything for posterity— there is no end to history. For instance, the Enlightenment did not foresee the abuses of industrialization, it failed to stress the need for peace, it exaggerated individualism, it extolled competition at the expense of cooperation, it did not go far enough in social reform, and it did not care much for women or for the underdeveloped peoples. However, the Enlightenment did perfect, praise, and diffuse the main conceptual and moral tools for advancing beyond itself. (Bunge 1994, p. 40)

1.17 Education

Missing from Bunge's above ten Enlightenment 'commandments' is Education. This should take its place on the list as an 11th principle. All the eighteenth century English, French and German founders of the Enlightenment were advocates of a new, different and revitalised education. They saw education as essential for the reformation of current society; and for the more radical thinkers, the creation of a new society. Locke, Priestley, Rousseau, Kant, all wrote works on education (Parry 2007). They established the Enlightenment education tradition whose modern contributors have been Ernst Mach, Thomas Huxley, Frederick Westaway, John Dewey, Philipp Frank, Herbert Feigl, and Gerald Holton (Matthews 2015, chap. 2).

The tradition is characterised by a commitment to the growth of knowledge of the natural and social worlds, the responsibility of the state for the education of all citizens, the extension of knowledge by both formal and informal education, and the utilisation of knowledge for the amelioration of social problems and the betterment of life. These cognitive and applied goals are shared with the Liberal education tradition, and both traditions might broadly be contrasted to utilitarian (whether State or personal) and progressivist movements in education.

Bunge contributes to this tradition, with his very first publication being *Temas de educación popular—Issues in popular education*—(Bunge 1943). He despairs of a great deal of counter-Enlightenment education. He writes of many University Faculties of Arts that:

Here you will meet another world, one where falsities and lies are tolerated, nay manufactured in industrial quantities. The unwary student may take courses in all manner of nonsense and falsity. Here some professors are hired, promoted, or given power for teaching that reason is worthless, empirical evidence unnecessary, objective truth non-existent, basic science a tool of either capitalist or male domination, or the like. ... This is a place where students can earn credits for learning old and new superstitions of nearly all kinds, and where they can unlearn to write, so as to sound like phenomenologists, existentialists, deconstructionists, ethnomethodologists, or psychoanalysts. (Bunge 1996, p. 108)

One educational case he did address in detail was the responsibility of science teachers in dealing with the inevitable 'conflicts' between scientific accounts of the world (its origins, biological evolution, explanations of sickness and healing, natural disasters, historical events, and so on) and alternate 'authoritative' cultural or religious accounts of the same things (creation stories, special creations, divine vengeance, efficacy of prayer, miracles, Chosen People, etc.). Martin Mahner and Bunge contributed a long article on this subject to the journal *Science & Education*, saying that ontologically, metaphysically, and epistemologically, the rival claims of science and religion were inconsistent, and minimally students need to be told this (Mahner and Bunge 1996a).

Mahner and Bunge's many pages of detailed arguments added up to a rejection of the popular, non-confrontational, widely-embraced, 'Non-Overlapping Magisteria Argument' (NOMA) of Stephen Jay Gould (1999), which has become the almost universal default position in science education. The exceptions being, on the one hand, those adherents to religious, cultural, or ideological positions who maintain that such systems of belief can correct specific claims of science; and rationalists who believe the reverse. The arguments of Mahner and Bunge were responded to by educators, theologians, and philosophers; the authors replied (Mahner and Bunge 1996b). Bunge returns to the criticism of NOMA in his *Political Philosophy: Fact, Fiction and Vision* (Bunge 2009).

In 1929 a popular text used for the preparation of English science teachers was published. The author, F.W. Westaway (1864–1946) shared many of Bunge's pre-occupations: he was trained as a scientist. He wrote on scientific method (Westaway 1919/1937), on the history of science (Westaway 1934), on the responsibility, or otherwise, of science for the exaggerated and sophisticated carnage of the Second World War (Westaway 1942), and he was His Majesty's Inspector for Science in English Schools (Brock and Jenkins 2014). On the opening page of his 1929 textbook Westaway characterised a successful school science teacher as one who:

> knows his own subject … is widely read in other branches of science … knows how to teach … is able to express himself lucidly … is skilful in manipulation … is resourceful both at the demonstration table and in the laboratory … is a logician to his finger-tips … is something of a philosopher … is so far an historian that he can sit down with a crowd of [students] and talk to them about the personal equations, the lives, and the work of such geniuses as Galileo, Newton, Faraday and Darwin. More than this he is an enthusiast, full of faith in his own particular work. (Westaway 1929, p. 3)

Bunge embodies Westaway's characterisation of a successful science teacher. He takes for granted Westaway's ideal and is puzzled that anyone would not.

1.18   The Festschrift

This Festschrift of 40 essays and a comprehensive all-languages bibliography, amplifies and evaluates Mario Bunge's systemic thinking and writing across the diverse fields to which he has contributed. The sections are:

1. An Academic Vocation	3 essays
2. A Philosophical System	12 essays
3. Physics and Philosophy of Physics	4 essays
4. Cognitive Science and Philosophy of Mind	2 essays
5. Sociology and Social Theory	3 essays
6. Ethics and Political Philosophy	3 essays
7. Biology and Philosophy of Biology	3 essays
8. Mathematics	3 essays
9. Education	2 essays
10. Varia	3 essays
11. Bibliography	

The hope is that the collection will suitably celebrate Mario Bunge's long life; do justice to his intellectual labours, both by elaborating them and pointing to deficiencies and problems to which they give rise; and bring to the attention of students, teachers, and researchers the commendable, Enlightenment-affirming example of wide and serious scholarship in opposition to obscurantism and pseudoscience, and for the service of human betterment, that Bunge so well represents.

Bernulf Kanitscheider (1939–2017), a German philosopher of science, some 35 years ago wrote:

Few extraordinary personalities have the chance to decisively shape the intellectual geography of a scientific epoch. Mario Augusto Bunge belongs to the small circle of important philosophers of science whose works have already become landmarks in the spiritual landscape of world philosophy. (Kanitscheider 1984, p.viii, cited by Heinz Droste in this *Festschrift*)

The contributions in this Festschrift by scholars from a dozen different disciplines and a dozen different countries, give reason for such high, though optimistic, valuation of Mario Bunge's work. Contributors, readers, and all associated with production of the Festschrift wish the philosophical scientist and the scientifically-driven philosopher well for his centenary birthday.

References

Agassi, J., & Cohen, R. S. (Eds.). (1982). *Scientific philosophy today: Essays in honor of Mario Bunge*. Dordrecht: Reidel Publishing Company.

Agazzi, E. (2017). The conceptual knots of the realism debate. In E. Agazzi (Ed.), *Varieties of scientific realism: Objectivity and truth in science* (pp. 1–18). Dordrecht: Springer.

Bernays, P. (1967). Scope and limits of axiomatics. In M. Bunge (Ed.), *Delaware seminar in the foundations of physics* (pp. 188–191). New York: Springer.

Brock, W. H., & Jenkins, E. W. (2014). Frederick W. Westaway and science education: An endless quest. In M. R. Matthews (Ed.), *International handbook of research in history, philosophy and science teaching* (pp. 2359–2382). Dordrecht: Springer.

Brown, H. I. (1994). Reason, judgement and Bayes' law. *Philosophy of Science, 61*, 351–369.

Bunge, M. (1943). *Temas de educación popular*. Buenos Aires: El Ateneo.

Bunge, M. (1944). A new representation of types of nuclear forces. *Physical Review, 65*, 249.

Bunge, M. (1945). Neutron-proton scattering at 8.8 and 13 MeV. *Nature, 156*, 301.

Bunge, M. (1951). What is chance? *Science and Society, 15*, 209–231.

Bunge, M. (1955a). A picture of the electron. *Nuovo Cimento,*. ser. X, *1*, 977–985.

Bunge, M. (1955b). Strife about complementarity. *British Journal for the Philosophy of Science, 6*, 141–154.

Bunge, M. (1956). Survey of the interpretations of quantum mechanics. *American Journal of Physics, 24*, 272–286.

Bunge, M. (1959). *Causality: The place of the causal principle in modern science*. Cambridge, MA: Harvard University Press. Third Revised Edition, Dover Publications, New York, 1979.

Bunge, M. (1960). *Cinemática del electrón relativista*. Tucumán: Universidad Nacional de Tucumán.

Bunge, M. (1961). Causality, chance and law. *American Scientist, 49*, 432–448.

Bunge, M. (1962). Causality: A rejoinder. *Philosophy of Science, 29*, 306–317.

Bunge, M. (1966). 'Mach's critique of Newtonian mechanics. *The American Journal of Physics, 34*, 585–596. Reproduced in J. Blackmore (ed.), *Ernst Mach—A Deeper Look*, Kluwer Academic Publishers, Dordrecht, (1992), pp. 243-261.

Bunge, M. (1967a). *Scientific research 1, the search for system*. New York: Springer. Republished as *Philosophy of Science*, Vol 1, Transaction Publishers, New Brunswick, (1998).Transaction Publishers, New Brunswick.

Bunge, M. (1967b). *Scientific research 2, the search for truth*. New York: Springer. Republished as *Philosophy of Science*, vol 2Transaction Publishers, New Brunswick, (1998).

Bunge, M. (1967c). *Foundations of physics*. New York: Springer.

Bunge, M. (1967d). The structure and content of physical theory. In M. Bunge (Ed.), *Delaware seminar in the foundations of physics* (pp. 15–27). New York: Springer-Verlag.

Bunge, M. (1973). *The philosophy of physics*. Dordrecht: Reidel.

Bunge, M. (1976). Possibility and probability. In W. Harper & C. Hooker (Eds.), *Foundations of probability theory, statistical inference, and statistical theories of science* (Vol. III, pp. 17–33). Dordrecht: Reidel.

Bunge, M. (1977a). General systems and holism. *General Systems, 12*, 87–90.

Bunge, M. (1977b). Emergence and the mind. *Neuroscience, 2*, 501–509.

Bunge, M. (1977c). The GST challenge to classical philosophies of science. *International Journal of General Systems, 4*, 329–376.

Bunge, M. (1977d). The interpretation of Heisenberg's inequalities. In H. Pfeiffer (Ed.), *Denken und Umdenken: zu Werk und Wirkung von Werner Heisenberg* (pp. 146–156). Munich: Piper.

Bunge, M. (1977e). Levels and reduction. *American Journal of Physiology, 2*, 75–82.

Bunge, M. (1979a). *Treatise on basic philosophy* (A World of Systems) (Vol. 4). Dordrecht: Reidel.

Bunge, M. (1979b). A systems concept of society: Beyond individualism and holism. *Theory and Decision, 10*, 13–30.

Bunge, M. (1979c). *Causality and modern science*. New York: Dover Publications. (Third edition of Bunge (1959)).

Bunge, M. (1980). *The mind-body problem*. Oxford: Pergamon Press.

Bunge, M. (1981a). From mindless neuroscience and brainless psychology to neuropsychology. *Annals of Theoretical Psychology, 3*, 115–133.

Bunge, M. (1981b). Four concepts of probability. *Applied Mathematical Modelling, 5*, 306–312.

Bunge, M. (1981c). *Scientific materialism*. Dordrecht: Reidel.

Bunge, M. (1981d). Popper's unworldly world 3. In his *Scientific materialism*, reproduced in M. Mahner (Ed.), *Scientific realism: Selected essays of Mario Bunge* (pp. 103–117). Amherst: Prometheus Books.

Bunge, M. (1982). The revival of causality. In G. Floistad (Ed.), *Contemporary philosophy* (Vol. 2, pp. 133–155). The Hague: Martinus Nijhoff. Reproduced in Martin Mahner (ed.) *Selected Essays of Mario Bunge*, (2001), Prometheus Books, Amherst, pp.57–74.

Bunge, M. (1983). *Lingüística y filosofía*. Madrid: Ariel.

Bunge, M. (1984). Philosophical problems in linguistics. *Erkenntnis, 21*, 107–173.

Bunge, M. (1985). Types of psychological explanation. In J. McGough (Ed.), *Contemporary psychology: Biological processes and theoretical issues* (pp. 489–501). Amsterdam: North Holland.

Bunge, M. (1986a). A philosopher looks at the current debate on language acquisition. In I. Gopnik & M. Gopnik (Eds.), *From models to modules* (pp. 229–239). Norwood: Ablex Publs. Co.

Bunge, M. (1986b). In defence of realism and scientism. *Annals of Theoretical Psychology, 4*, 23–26.

Bunge, M. (1986c). *Philosophical problems in linguistics*. Tokyo: Seishin-Shobo.

Bunge, M. (1987). Ten philosophies of mind in search of a scientific sponsor. In *Proceedings of the 11th International Wittgenstein Symposium* (pp. 285–293). Wien: Hölder-Pichler-Tempsky.

Bunge, M. (1988). Two faces and three masks of probability. In E. Agazzi (Ed.), *Probability in the sciences* (pp. 27–49). Dordrecht: Kluwer. Reproduced in Martin Mahner (ed.) *Selected essays of Mario Bunge*, (2001), Prometheus Books, Amherst, pp. 211–226.

Bunge, M. (1989). What kind of discipline is psychology: Autonomous or dependent, humanistic or scientific, biological or sociological? *New Ideas in Psychology, 8*, 121–137.

Bunge, M. (1990a). The nature and place of psychology: A reply to Panksepp, Mayer, Royce, and Cellerier and Ducret. *New Ideas in Psychology, 8*, 176–188.

Bunge, M. (1990b). Instant autobiography. In P. Weingartner & G. J. W. Dorn (Eds.), *Studies on Mario Bunge's treatise'* (pp. 677–681). Amsterdam: Rodopi.

Bunge, M. (1991a). A critical examination of the new sociology of science: Part 1. *Philosophy of the Social Sciences, 21*(4), 524–560.

Bunge, M. (1991b). A philosophical perspective on the mind-body problem. *Proceedings of the American Philosophical Society, 135*, 513–523.

Bunge, M. (1991c). The power and limits of reduction. In E. Agazzi (Ed.), *The problem of reductionism in science* (pp. 31–49). Dordrecht: Kluwer.

Bunge, M. (1994). Counter-enlightenment in contemporary social studies. In P. Kurtz & T. J. Madigan (Eds.), *Challenges to the enlightenment: In defense of reason and science* (pp. 25–42). Buffalo: Prometheus Books.

Bunge, M. (1996). In praise of intolerance to charlatanism in academia. In P. R. Gross, N. Levitt, & M. W. Lewis (Eds.), *The flight from science and reason* (pp. 96–115). New York: New York Academy of Sciences.

Bunge, M. (1998). *Social science under debate: A philosophical perspective*. Toronto: University of Toronto Press.

Bunge, M. (1999a). *The sociology-philosophy connection*. New Brunswick: Transaction Publishers.

Bunge, M. (1999b). Linguistics and philosophy. In H. E. Wiegand (Ed.), *Sprache und Sprachen in der Wissenschaften* (pp. 269–293). Berlin/New York: Walter de Gruyter.

Bunge, M. (1999c). *Dictionary of philosophy*. Amherst: Prometheus Books.

Bunge, M. (1999e). Philosophy sociologized: Review of Randall Collins the sociology of philosophies: A global theory of intellectual change. *Contemporary Sociology, 28*(3), 280–281.

Bunge, M. (2000a). Systemism: The alternative to individualism and holism. *The Journal of Socio-Economics, 29*, 147–157.

Bunge, M. (2000b). Ten modes of individualism —None of which works—And their alternatives. *Philosophy of the Social Sciences, 30*, 384–406.

Bunge, M. (2000c). Energy: Between physics and metaphysics. *Science & Education, 9*(5), 457–461.

Bunge, M. (2003a). *Emergence and convergence: Qualitative novelty and the unity of knowledge*. Toronto: University of Toronto Press.

Bunge, M. (2003b). Velocity operators and time-energy relations in relativistic quantum mechanics. *International Journal of Theoretical Physics, 42*, 135–142.

Bunge, M. (2003c). Philosophy of science and technology: A personal report. In G. Fløistad (Ed.), *Philosophy of Latin America* (pp. 245–272). Dordrecht: Kluwer Academic Publishers.

Bunge, M. (2004a). How does it work? The search for explanatory mechanisms. *Philosophy of the Social Sciences, 34*, 182–210.

Bunge, M. (2004b). Clarifying some misunderstandings about social systems and their mechanisms. *Philosophy of the Social Sciences, 34*, 371–381.

Bunge, M. (2006). *Chasing reality: Strife over realism.* Toronto: University of Toronto Press.

Bunge, M. (2008). Bayesianism: Science or pseudoscience? *International Review of Victimology, 15*, 165–178.

Bunge, M. (2009). *Political philosophy: Fact, fiction, and vision.* New Brunswick: Transactions Publishers.

Bunge, M. (2010). *Matter and mind: A philosophical inquiry.* Dordrecht: Springer.

Bunge, M. (2014a). In defence of scientism. *Free Inquiry, 35*(1), 24–28.

Bunge, M. (2014b). Big questions come in bundles, hence they should be tackled systemically. *International Journal of Health Services, 44*(4), 835–844.

Bunge, M. (2015). Does the Aharonov-Bohm effect occur? *Foundations of Science, 20*(2), 129–133.

Bunge, M. (2016). *Between two worlds: Memoirs of a philosopher-scientist.* Dordrecht: Springer.

Bunge, M. (2017a). *Doing science in the light of philosophy.* Singapore: World Scientific.

Bunge, M. (2017b). Why axiomatize? *Foundations of Science, 22*, 695–707.

Bunge, M. (2017c). Evaluating scientific research projects: The units of science in the making. *Foundations of Science, 22*, 455–469.

Bunge, M. (2018a). *From a scientific point of view.* Newcastle: Cambridge Scholars Publications.

Bunge, M. (2018b). Gravitational waves and space-time. *Foundations of Science, 23*, 399–403.

Bunge, M. (2019). Inverse problems. *Foundations of Science,* 1–43.

Bunge, M., & Ardila, R. (1987). *Philosophy of psychology.* New York: Springer-Verlag.

Bunge, M., & Kálnay. (1969). A covariant position operator for the relativistic electron. *Progress of Theoretical Physics, 42*, 1445–1459.

Collins, R. (1998). *The sociology of philosophies: A global theory of intellectual change.* Cambridge, MA: Harvard University Press.

Condorcet, N. (1795/1955). *Sketch for a historical picture of the progress of the human mind* (J. Barraclough, Trans.), New York: Noonday Press.

Cordero, A. (2016). Philosophy of science in Latin America. In *Stanford encyclopedia of philosophy.* https://plato.stanford.edu/entries/phil-science-latin-america/

Corfield, D., & Williamson, J. (Eds.). (2001). *Foundations of Bayesianism.* Dordrrecht: Kluwer.

Denegri, G. M. (2014). *Elogio de la Sabiduria.* Ensayos en Homenaje a Mario Bunge en su 95° Aniversario, EUDEBA, Universidad de Buenos Aires.

Denegri, G. M., & Martinez, G. (2000). *Tópicos actuales en filosofía de la ciencia. Homenaje a Mario Bunge en su 80 Aniversario.* Universidad Nacional de Mar del Plata. Editorial Martín.

Duhem, P. (1916/1991). *German science* (J. Lyon & S. L. Jaki, Trans.). La Salle: Open Court Publishers.

Fisher, R. A. (1926/1947). *The design of experiments* (4th edn). Edinburgh: Oliver & Boyd.

Franklin, J. (2001). Resurrecting logical probability. *Erkenntnis, 55*, 277–305.

Glymour, C. (1980). *Theory and evidence.* Princeton: Princeton University Press.

Gould, S. J. (1999). *Rock of ages: Science and religion in the fullness of life.* New York: Ballantine Books.

Haack, S. (2016). *Scientism and its discontents.* London: Rounded Globe Publishers.

Hume, D. (1739/1888). *A treatise of human nature: Being an attempt to introduce the experimental method of reasoning into moral subjects.* Oxford: Clarendon Press.

Jaynes, E. T. (2003). In G. Larry Bretthorst (Ed.), *Probability theory: The logic of science.* Cambridge, UK: Cambridge University Press.

Kaidesoja, T. (2009). Bhaskar and Bunge on social emergence. *Journal for the Theory of Social Behaviour, 39*(3), 300–322.

Kanitscheider, B. (1984). Introduction to the German translation of Bunge's 'The Mind-Body Problem'. In M. Bunge (Ed.), *Das Leib-Seele-Problem. Ein psychobiologischer Versuch* (pp. VIII–VXII). Tübingen: J.C.B. Mohr Verlag (Paul Siebeck Verlag).

Kyburg, H. E., Jr. (1978). Subjective probability: Criticisms, reflections and problems. *Journal of Philosophical Logic, 7*, 157–180.

Lakatos, I. (1970). Falsification and the methodology of scientific research programmes. In I. Lakatos & A. Musgrave (Eds.), *Criticism and the growth of knowledge* (pp. 91–196). Cambridge, UK: Cambridge University Press.

Lakatos, I. (1971). History of science and its rational reconstructions. In R. C. Buck & R. S. Cohen (Eds.), *Boston studies in the philosophy of science* (Vol. 8, pp. 91–135). Dordrecht: Springer.

Levi, I. (1974). Indeterminate probabilities. *Journal of Philosophy, 71*, 391–418.

Mahner, M., & Bunge, M. (1996a). Is religious education compatible with science education? *Science & Education, 5*, 101–123.

Mahner, M., & Bunge, M. (1996b). The incompatibility of science and religion sustained: A reply to our critics. *Science & Education, 5*, 189–199.

Matthews, M. R. (Ed.). (2003). *Mario Bunge and Quantum Theory. Science & Education,* 12(5–6).

Matthews, M. R. (Ed.). (2012) *Mario Bunge, systematic philosophy and science education. Science & Education,* 21(10).

Matthews, M. R. (2015). *Science teaching: The contribution of history and philosophy of science: 20th anniversary revised and enlarged edition.* New York: Routledge.

Mayo, D. (2004). An error-statistical philosophy of evidence. In M. Taper & S. Lele (Eds.), *The nature of scientific evidence: Statistical, philosophical and empirical considerations* (pp. 79–118). Chicago: University of Chicago Press.

Mayo, D. G., & Spanos, A. (Eds.). (2010). *Error and inference: Recent exchanges on experimental reasoning, reliability, and the objectivity and rationality of science.* Cambridge, UK: Cambridge University Press.

Minazzi, F. (2017). The epistemological problem of the objectivity of knowledge. In E. Agazzi (Ed.), *Varieties of scientific realism: Objectivity and truth in science* (pp. 177–206). Dordrecht: Springer.

Musgrave, A. (1993). *Common sense, science and scepticism.* New York: Cambridge University Press.

Parry, G. (2007). Education and the reproduction of the enlightenment. In M. Fitzpatrick, P. Jones, C. Knellwolf & I. McCalman (Eds.), *The Enlightenment World* (pp. 217–233), London: Routledge.

Pickel, A. (Ed.). (2004) Systems and mechanisms: A symposium on Mario Bunge's philosophy of social science. *Philosophy of the Social Sciences 34*(2,3), 169–210.

Quine, W. V. O. (1985). *The time of my life: An autobiography.* Cambridge, MA: Bradford Books.

Romer, R. H. (1991). Editorial: Memorable papers from the American Journal of Physics, 1933–1990. *American Journal Physics, 59*(3), 201–207. http://wcb.mit.edu/rhprice/www/articles/MemorablePapers.pdf

Romer, R. H. (1993). Sixty years of the American Journal of Physics—More memorable papers. *American Journal of Physics, 61*(2), 103–106. http://web.mit.edu/rhprice/www/articles/MoreMemorablePapers.pdf

Russell, B. (1912). *The problems of philosophy.* Oxford: Oxford University Press.

Salmon, W. C. (1990). Rationality and objectivity in Science, or Tom Bayes meets Tom Kuhn. In C. Wade Savage (Ed.), *Scientific theories: Minnesota studies in the philosophy of science volume XIV* (pp. 175–204). Minneapolis: University of Minnesota Press.

Savage, L. J. (1954). *The foundations of statistics.* New York: Wiley.

Savage, L. J. (1964). The foundations of statistics reconsidered. In H. E. Kyburg Jr. & H. E. Smokler (Eds.), *Studies in subjective probability* (pp. 173–188). New York: Wiley.

Shaffer, M. J. (2012). *Counterfactuals and scientific realism.* New York: Palgrave Macmillan.

Shimony, A. (1970). Scientific inference. In R. G. Colodny (Ed.), *The nature and function of scientific theories* (pp. 79–172). Pittsburgh: University of Pittsburgh Press.

Sosa, E. (Ed.). (1975). *Causation and conditionals.* Oxford: Oxford University Press.

Staver, J. (1998). Constructivism: Sound theory for explicating the practice of science and science teaching. *Journal of Research in Science Teaching, 35*(5), 501–520.

Stöckler, M. (1990). Realism and classicism, or something more? Some comments on Mario Bunge's philosophy of quantum mechanics. In P. Weingartner & G. J. W. Dorn (Eds.), *Studies on Mario Bunge's treatise* (pp. 351–363). Amsterdam: Rodopi.

Swinburne, R. G. (Ed.). (2002). *Bayes theorem*. Oxford: Oxford University Press.

Wallace, W. A. (1972). *Causalilty and scientific explanation*. Ann Arbor: University of Michigan Press.

Weingartner, P., & Dorn, G. J. W. (Eds.). (1990). *Studies on Mario Bunge's treatise*. Amsterdam: Rodopi.

Westaway, F. W. (1919/1937). *Scientific method: Its philosophical basis and its modes of application* (5th edn). New York: Hillman-Curl.

Westaway, F. W. (1929). *Science teaching*. London: Blackie and Son.

Westaway, F. W. (1934). *The endless quest: 3000 years of science*. London: Blackie & Son.

Westaway, F. W. (1942). *Science in the dock: Guilty or not guilty?* London: Blackie & Son Ltd.

Part I
An Academic Vocation

Chapter 2
Mario Bunge: Argentine's Universal Thinker

Guillermo M. Denegri

Abstract I utilise and develop the *wisdom-longevity-books* triad for praising Mario Bunge, Argentine thinker and intellectual of universal citizenship. I had the immense privilege of knowing another extraordinary Argentine whose figure becomes even more colossal with the passing of time, Jorge Luis Borges. I also attended and listened to three of his magisterial conferences. I am convinced that Borges and Bunge are the pride of Argentines, as well as being the most lucid minds of the twentieth century. Bunge has published more than 550 papers in different journals of international prestige, in the most varied areas. He has written over 140 books and they have been translated into many languages, in addition to being permanently reprinted. Mario has published without interruption during 80 years of his long and prolific life, whose 100th birthday is celebrated on September 21 of the year 2019.

I wish to highlight the triad of wisdom-longevity-books as a tribute to this juggernaut of knowledge and longevity that is Mario Bunge. I am convinced that Mario's stature becomes even greater as the years go by, and that his youthful 100 years is the most eloquent demonstration that life can be lived intensively for its whole duration. I am firmly convinced that there is a natural impulse in which knowledge and the constant training of our neurons are indications of a longer and richer life (understood, of course, as quality of life). An example of this is the prolific life of Rita Levi-Montalcini (Italian, Nobel Prize in Medicine of 1986 for the discovery of the Nerve Growth Factor – NGF) who died at 103 years of age, demonstrating that it is possible to improve, every day of our lives, the creative potential of our brains, fostering new and novel nerve connections that enable a better knowledge of the world as well as a fulfilled life.

G. M. Denegri (✉)
Institute of Production, Health and Environment Research (IIPROSAM); Faculty of Exact and Natural Sciences, National University of Mar del Plata (CONICET), Mar del Plata, Argentina
e-mail: gdenegri@mdp.edu.ar

© Springer Nature Switzerland AG 2019
M. R. Matthews (ed.), *Mario Bunge: A Centenary Festschrift*,
https://doi.org/10.1007/978-3-030-16673-1_2

In fact, Mario's life has been like that, and it still is. He has not stopped working for not even a single day, wagering for knowledge, for the hard work that is learning itself, for not letting himself fall for the trickeries of false prophets disguised as intellectual luminaries, and above all for generating new knowledge and reflections that are transmitted in a clear and precise language. As we say in Argentina, "Gardel (1890–1935) is singing better every day." We can also say, unquestionably, that "Bunge is thinking better every day, and even more so, he is writing better every day."

Mario's written production is overwhelming in both quantity and quality. His first published work is from 1939, titled "Introducción al estudio de los grandes pensadores".[1] He was barely 20 years old. Eighty years later, he has published a total of 537 works in many journals of international prestige and in the most varied of fields.[2] What commands attention is the diversity of topics studied and the solvency and intellectual seriousness with which they are dealt. Additionally, there are his books, over 140 with translations into many different languages as well as permanent reprintings.[3]

But there is a datum which I want to especially emphasize in order to have an even better comprehension of Mario's oeuvre, and it is the fact that from the total of his 537 published works, only 12 articles have been co-authored (2%), which means that 98% of his papers have been written by himself alone. Regarding his books, he has published 73 of them, and there is one currently in press, which if we take into account the translations into various languages plus the re-editions, amount to a total of 148. Only three of his 74 books are co-authored (4%). These data are not minor if we take into account the current way of producing scientific and philosophical works, where we usually find many authors without knowing clearly what level of contribution corresponds to each of them in a given publication.

Bunge's work is noteworthy not just for its quantity and quality, but also the path that he has traced in the construction of a novel and creative approach to philosophy as scientific philosophy (Denegri 2000, 2014). It is materialist and emergentist in ontology, realist in epistemology, and systemist in structure.

I want to share an event that was, for me, one of the most vivid and emotional that I have had the great fortune of enjoying. Mario visited Mar del Plata in the year 2000 to give lectures on "Sociology of Science", to the Permanent Seminar of Biophilosophy in the Faculty of Exact and Natural Sciences of the UNMdP. Among the multiple activities that he engaged in during his 1-week stay (interviews in radio and television, talks directed at children from municipal schools, etc.) one of them stood

---

[1] Introducción al estudio de los grandes pensadores (Introduction to the study of the great thinkers). *Conferencias* (Buenos Aires) III: 105–109, 124–126, 1939.

[2] These range from *Nature, Brit. Jrn. Phy. Sc.* and *Am. Jrn. Physics* to *Synthese, Mind, Rev. Metaphysics,* and *Intern. Jrn. of Quantum Chem.,* and more than 30 other widely respected peer-reviewed journals. The complete list of his all-language publications has been prepared by Marc Silberstein and is included in this anthology.

[3] The translations include: French, Spanish, German, Italian, Polish, Hungarian, Russian, Romanian, Chinese, Japanese, and Portuguese.

out. This was the talk he had with the late Dr. Miguel Eduardo Jörg (1909–2002), my cherished and beloved friend, outstanding Argentine scientist in public health during the twentieth century, the last of the living disciples of Dr. Salvador Mazza (1886–1946), co-discoverer of the Chagas disease (Denegri and Sardella 2000).

Bunge and Jörg did not know each other prior to that encounter, so we got together in a bar located in the center of Mar del Plata in order to make the formal presentation and to exchange opinions, anecdotes, stories, and also to organize a talk whose subject matter was yet to be defined. Don Miguel was 90 years old at that time and Mario was 80, they rapidly got along and at a certain moment we noticed that they were talking in German . . . while laughing about something that we could not understand since we did not speak the language. We quickly made arrangements for a dual conference to be held at the Municipal Library "Osvaldo Soriano" (MPGP). We defined the details and the advertising, and when it took place, the room was filled with people, some of them were even sitting on the floor. It was perhaps one of the most important events involving these two intellectual giants, where lucidity, sympathy, knowledge and the gift of location were the main pillars of an afternoon and evening that will never be forgotten. They gave us a clear lesson on the cultivation of knowledge and wisdom which makes of us better people, whole and free, and they showed us that the passing of the years is not a disadvantage. To the contrary, the brain works better every day if we have used it, worked with it, developed it, and pushed it to its limits. Don Miguel Jörg died a few years later at almost 93 years of age, with lucidity and an intact memory.

For those of us who have "left behind the bribery of the Heavens", in the words of Bernard Shaw, a fulfilling life on Earth together with a hunger for knowledge is one of the alternatives that can give us moments of full happiness and above all the opportunity to empower that extraordinary mechanism that is our brain, which can, every day, no matter how many years pass by, generate new and creative nervous connections which, I have no doubt, influence decisively on other organic functions, improving them. In order for this to happen, it is necessary to teach our children in an intellectually stimulating environment and with the firm conviction that only constant hard work can lead us to a good port. Nowadays we frequently witness a pedagogical relativism, influenced on many occasions by postmodern pseudo-schools of thought, which fail to acknowledge something so elementary as teaching others the value of permanent effort, or doing hard intellectual work.

Argentine's first Nobel Prize in Sciences (Medicine and Physiology, 1947) Bernardo Houssay (1887–1971), said something along the following lines in his book *La investigación científica*: "I frequently hear many parents that say . . . 'my son is very smart, too bad he is so lazy' . . . I say to them: if they were smart they would realize that in order to achieve anything you have to work hard and correctly." And this, which is so true, is exemplified in Mario Bunge not only by analyzing his rich intellectual trajectory as a philosopher-scientist, but also in his constant and on occasions solitary struggle against pseudoscience and postmodern posturing, which despise science and accuse it of (almost) every evil that there is. Studying the sciences is a tough and sinuous path, which requires constant effort and sacrifice, with more defeats than victories. Not everyone is willing to even face such a path.

Populist charlatans have seen Mario Bunge as the scientistic enemy that must be defeated, deriding him with fallacious arguments and unintelligible discourses. And to the degree that Bunge has made 'conceptual clarity' his continuing offensive weapon, together with rock-solid arguments, he has received in reply a barrage of epitaphs that are supposedly insulting (such as *scientificist, materialist, realist*, and so on). Bunge's harshest critics have spoken terribly of him, believing that they could confuse and discourage their rival. But this has only made him stronger, and his intellectual coherence has earned him the respect and admiration of thousands who read his articles and books worldwide. On this point, I remember when many years ago (1989), while we were finishing our philosophy courses with my great friend Marta Crivos we sent Mario a monograph that we had written for the course on Logic II, which dealt with the implications of the Duhem-Quine thesis for logic. Shortly afterwards, we received a very friendly letter which, nonetheless, demolished our monograph. We read with great attention Mario's implacable criticisms, ending with this gentle wisdom: "I believe that you two are following a trend. Remember that in philosophy, fashion rarely matches truth."

Mario Bunge's name has been familiar to me since high school (1970–1974), in the Almafuerte Secondary Institute of my beloved town. I seem to recall that one of my excellent professors mentioned him in one of his classes, and I rapidly got access to a book that still accompanies me and that I still use frequently in my Introduction to Biology classes, of the Licenciatura and Professorship in Biology, and in the doctoral courses in sciences of the FCEyN-UNMdP. I'm referring to *Science: its Method and Philosophy* (Bunge 1963) which has as much validity today as when it was published more than half a century ago.

I own a copy of Mario's book *Epistemology* (2nd ed), which he gifted to me with a warm dedicatory note, "For Guillermo, scientist and philosopher, with friendship and gratitude. MB." In the Preface he says:

> In this edition I have introduced some additions and corrections, almost all of them minor, to the first edition of 1980. Since then I have learned much, but I am still a confessed and convicted realist, scientificist, materialist and systemist. The anti-scientificist counter-revolution started by Thomas S. Kuhn (1922–1996) and Paul K. Feyerabend (1924–1994) has not pierced my armor". (Bunge 1997)

This demonstrates Bunge's intimate convictions and the rational argumentation that he has upheld throughout the years, in a philosophical position that has had him as a precursor, defender and representative at a worldwide scale, valiantly and clearly confronting all of the positions that defend the sociologicism-constructivism-relativism interpretation of scientific activity. One of the reasons Bunge gives for why so many intellectuals defend and propagandize (and indoctrinate their students in the Humanities Faculties and sometimes in those of Sciences) those ideas, is due to a supine ignorance of any science whatsoever, or perhaps because they have never studied any of them, let alone having practiced them.

A dimension of Mario Bunge's personality that should be emphasized is his ethical attitude towards life and his great generosity. I would define him as a good person and as a worthy representative of the democratic socialism that he inherited from his father Augusto Bunge (sanitarian physician, sociologist, legislator, professor, journalist and poet). He has always been willing to help others, a happy coincidence with Rita Levi-Montalcini, with his hand reaching out in order to give advice with fraternity and solidarity. His teaching in academia and in life have forged the spirit and talent of many generations of men and women who surely have become better after receiving and learning the lessons of this unmatchable master.

References

Bunge, M. (1963/1960). *La ciencia, su método y su filosofía* (2nd enlarged edition). Buenos Aires : Siglo Veinte.

Bunge, M. (1997). *Epistemologia. Curso de Actualización.* México: Siglo Veintiuno Editores.

Denegri, G. (2000). Hacia un entendimiento fructífero entre científicos y filósofos de la ciencia: un acuerdo civilizado sin exabruptos. In Denegri, & Martínez (Ed.), *Tópicos actuales en filosofía de la ciencia. Homenaje a Mario Bunge en su 80° aniversario* (pp. 79–96). Editorial Martin-Universidad Nacional de Mar del Plata.

Denegri, G. (2014). *Elogio de la Sabiduria. Ensayos en Honor a Mario Bunge en su 95 ° aniversario.* Editorial de la Universidad de Buenos Aires.

Denegri, G., & Sardella, N. (2000). *Elogio de la Integridad. Conversaciones con Miguel Eduardo Jörg.* Editorial Martin.

Chapter 3
Mario Bunge in the Complex Argentina of the 1940s–1960s

Eduardo L. Ortiz

Abstract Towards the end of the decade of 1940 Mario Bunge began to be acknowledged, in Argentina's intellectual circles, as a young theoretical physicist seriously interested in philosophy, and also, deeply committed to social and intellectual advance. Mario came from a family that moved to Argentina in the early nineteenth century, soon after Independence, and acquired a dominant financial position through its activity in the international grain exports business. In the late 1930s there was, in Argentina, a small group of scientists and philosophers interested in exploring the boundaries of their disciplines. Among them Julio Rey Pastor, who had a profound influence in the development of mathematics there; the philosopher Francisco Romero, and the historian of science Aldo Mieli. In 1955 there was a new military coup; its leaders, with more experience in dealing with vocal university students, allowed for some changes to take place at the universities; some were definitely positive. In turn, these changes made possible a creative period that extended from 1956 to 1966, partly under military, partly under obedient civilian rule. However, it ended in disaster, with massive emigration of scientists and intellectuals in 1966. It was in that decade that Mario Bunge was reinstated as a physics professor at the Faculty of Science of the University of Buenos Aires. However, he soon moved to the Philosophy Faculty of the same university, as a full professor of Philosophy of Science. There he began the difficult task of updating the teaching of his subject to more recent and advanced standards, with reference to more modern authors and methodologies. Bunge's activities, mainly in the 1950s, just before he left his country, and their impact in Argentina's intellectual life, are the subject of this paper.

E. L. Ortiz (✉)
Mathematics Department, Imperial College London, London, UK
e-mail: e.ortiz@imperial.ac.uk

© Springer Nature Switzerland AG 2019
M. R. Matthews (ed.), *Mario Bunge: A Centenary Festschrift*,
https://doi.org/10.1007/978-3-030-16673-1_3

3.1   Introduction

Mario Bunge comes from a family that moved to Argentina in the early nineteenth century, soon after independence, and later acquired a dominant financial position through its activity in the international grain export business. He was not, however, the first member of that family to leave a mark on the intellectual life of his native country. In different historical periods, several of the Bunge family distinguished themselves in various fields of activity: literature, sociology, medicine, philosophy, and politics.

Neither was Mario the first to show concern and solidarity with the pains of his country, particularly with the most vulnerable members of society. His own father, a distinguished medical man, represented the Socialist Party in Parliament and, as such, is remembered for promoting the first national medical insurance. His parents' large home, *El Ombú*, in Florida, outside Buenos Aires, became a point of reference for writers and intellectuals from the beginning of the twentieth century, particularly for those with progressive views. In time, this tradition was continued, initially in the same house, by Mario.

3.2   A Workers' University and a Journal of Philosophy

In the 1940s, Mario Bunge began to be acknowledged in Argentina's intellectual circles as a young physicist seriously interested in philosophy, and also, deeply committed to intellectual advance as well as to social and political change. From the late 1930s to the early 1940s, in parallel with his studies at the University of La Plata, Bunge, just 20 years old, participated in several interesting cultural enterprises. The first of them concerned a special and then neglected area of public education.

In 1917, at a time of consolidation of the Mexican Revolution, the educator and revolutionary Vicente Lombardo Toledano (1894–1968) created the *Popular University of Mexico*[1]; 5 years later, he founded a *Night School for Workers*[2] and in 1936, was responsible for the foundation of a *Mexican Workers' University*.[3]

This last institution caught the attention of young Bunge, who apparently established epistolary contact with the Mexican educator and later attempted to translate his ideas to the realities of Argentina. In 1938, he opened an Argentine Workers' University in Buenos Aires: the *Universidad Obrera Argentina*, or *UOA*. The *UOA* began its activities at a time when there was concern about technical education in Argentina. Some educators perceived the need to update and diversify technical education, and others suggested opening new channels to satisfy the

---

[1] Universidad Popular Mexicana.
[2] Escuela Nacional Preparatoria Nocturna para trabajadores.
[3] Universidad Obrera de México.

expanded personnel demands of local industry. The latter needed to be able to produce and compete at a more sophisticated level than ever before.

The armed forces which, as we shall see through this chapter, played a clearly political role in the life of mid-twentieth century Argentina and were not absent from these discussions. They perceived that a new world war, which since the mid-1930s seemed inevitable, would severely limit access to their traditional European suppliers of armaments. This added an extra urgency to the need for a renovation of local industry and, clearly, of technical education. An important technical report, on Industrial Mobilization, produced by one of the most enlightened military men of that decade, General Manuel Savio (1892–1948), is a clear example of these concerns (Ortiz 1996).

At the time, there were specific national schools in Argentina for technical education which were an integral part of the secondary school education system. They ran as an alternative to the baccalaureate, commerce studies, training for admission to Military or Naval Colleges, or training to become a primary school teacher.

However, those technical schools did not adequately attend to the question of remedial adult education. That is, to help re-educate a large population of workers who had been incorporated into industrial production lacking technical instruction. During this period, others proposed the creation of a National Council for Technical Education and to offer some form of training to adults.

It is likely that the section of the population targeted by the *UOA* consisted of semi-skilled workers, with social ambitions, who wished to improve their training and also their social options. The *UOA* offered them introductory courses on a number of topics of contemporary interest, such as basic mechanics, electricity and radio, chemistry, accounting and economics, as well as some topics of mathematics relevant to commerce or to technical applications.

The authorities of the *UOA* had a wide humanistic perspective and, while attempting to improve the technical competitiveness of their students, they also made efforts to widen the horizons of their interests. To do that, in addition to practical, professional courses with a definite technical focus, they attempted to introduce their students to a wider world of culture.

To achieve this last objective, they created a *Scientific Institute*,[4] inside the *UOA*, whose responsibility was to open up the cultural outlook of their students. That Institute even had a *Philosophical Seminar*, which printed monographs written by some of its teachers. In addition, the *UOA*'s authorities – or rather Bunge – invited some progressive intellectual personalities to offer special non-technical lectures, and also courses, to their students. One of those invited was Bunge's friend, Dr. Arturo Frondizi (1908–1995), a progressive lawyer who, later (1958–1962), became President of Argentina. Sadly, this was precisely at the time Bunge found it prudent to leave the country

---

[4] *Instituto Científico de la Universidad Obrera Argentina.*

Interesting as the initiative no doubt was, the fact is that the Argentine private version of the *Mexican Workers' University* didn't enjoy the substantial national support the Mexican government could offer to its own institution. So, the *UOA* became unsustainable and finally had to close its doors in 1943. Its demise was probably accelerated by the political climate created after a military *coup d'état*, the second in 13 years, toppled the elected national government on 4 June 1943.

However, the need for an institution such as the *UOA* was still standing and, 5 years later, the new national government that indirectly emerged from that *coup d'état* created a National Workers' University: the *Universidad Nacional Obrera*.[5] Similarities with the *UOA* no doubt exist, but the new official institution had a radically different philosophy. It was focused on the training of the numerous factory technicians that Argentine industry badly needed – and needed soon – to take full advantage of the present weakness in the European manufacturing exports market. For the new official organisation, the humanist side was not, then, a top priority.

3.3   The Institute for the History and Philosophy of Science

In those years there was already a serious interest among some of Argentina's leading scholars in the study of the connections between philosophy and the sciences. In 1933, on the initiative of mathematician and science historian Julio Rey Pastor (1888–1962), a highly selective *Argentine Group on the History of Science* was established in Buenos Aires. It was the local branch of Paris' *Académie Internationale d'Histoire des Sciences*.

Rey Pastor, whose name will appear more than once in this paper, was a Spanish mathematician who, after graduating, was sent to Germany. There, he trained in modern function theory under David Hilbert (1862–1943) and other leading mathematicians of the Göttingen school. A few years after his return to Spain, in 1917, he was invited to lecture in Argentina and later, in the early 1920s, married and settled down permanently there. In the Spanish-speaking world, Rey Pastor gave mathematics a new modern outlook (Ortiz 2011b).

Five years later, in 1938, an *Institute for the History and Philosophy of Science*[6] was created at the Santa Fe[7] branch of the Universidad del Litoral, north-east of Buenos Aires. Mathematician, José Babini (1897–1984),[8] one of Rey Pastor's first students in Argentina, was a professor there. As his mentor, he was also interested in the history of science and became a pivotal connection with that institute.

In 1939, the above-mentioned *Institute* acquired, as its director, the Italian science historian Aldo Mieli (1879–1950), who had been one of the founders of the

---

[5]From 1958, it came to be known as *Universidad Tecnológica Nacional*.

[6]*Instituto de Historia y Filosofía de la Ciencia*.

[7]North-east of Buenos Aires.

[8]On Babini see Ortiz and Pyenson (1984) and the references given there.

Paris *Académie Internationale d'Histoire des Sciences*. Mieli was forced to emigrate from France on account of war and Rey Pastor helped him move to Argentina.

Young Bunge established a contact with the *Institute*, most probably through Babini, a wise and generous teacher. No doubt, Mieli and Babini must have detected in him a potential science historian. In fact, during this period Bunge produced two interesting studies: one on Newton (Bunge 1943a), which was printed by the *Seminario de Filosofía* of the *UOA*'s *Instituto Científico*; the other, on Maxwell (Bunge 1943b), was printed by the Universidad del Litoral.

These works were favorably received at home and also abroad. Locally, theoretical physicist Guido Beck (1903–1988), who would become Bunge's PhD supervisor, reviewed them (Beck 1944b) in the pages of the journal of the *Argentine Mathematical Union*.[9] A further review appeared in *Nature*, in London.

Bunge's lecture at the *Institute* may have been one of its last activities, as following the aforementioned military *coup d'état* of June 1943, both Mieli and Babini were removed from their chairs at the university. The *Institute for the History and Philosophy of Science* disappeared with them; sadly, Mieli's valuable personal library, which he brought from Europe and deposited at the *Institute* after his arrival in Argentina, suffered serious neglect and pilfering.

However, if work on the subject had officially died, it continued in the hands of a group of specialists. By the end of 1945[10] it was sufficiently strong for Rey Pastor to organize the 'First Colloquium on the History and Philosophy of Science' in Buenos Aires (Valentinuzzi 1946).

3.4 *Minerva, Revista Continental de Filosofía* in Buenos Aires of 1944

In the same period as the *UOA* began to find it difficult to survive, 1943–1944, Bunge turned his attention to a more congenial and complex cultural enterprise: editing a journal of research on philosophy which he called: *Minerva, Revista Continental de Filosofía*. It claimed to be the only journal in Latin America "exclusively devoted to philosophy", and its ambition was nothing other than the modernisation of that discipline in the region.

Because of this second, cultural agenda, *Minerva* can be regarded as a distant relative of a seminal Argentine journal of the recent past: *Revista de Filosofía*, a journal that began to be published under José Ingenieros (1877–1925), one of Argentina's leading intellectuals of his time, from 1915. After his untimely death, *Revista de Filosofía* continued under his disciple Anibal Ponce (1898–1938) but had to close its pages in 1928.

---

[9]*Revista de la Unión Matemática Argentina;* we will refer to it as *RUMA*.

[10]On 18–20 September 1945.

However, the scope of these two journals was quite different. While attending to questions of philosophy, Ingenieros' journal had a much wider fan of interests as its full name suggests: *Revista de Filosofía. Cultura, Ciencias, Educación.* Besides, in addition to these concerns, *Revista de Filosofía* also had a wide programmatic cultural objective; it included the consolidation of a progressive scientific tradition which Ingenieros maintained existed and had evolved in Argentina from the end of Colonial times; this is discussed in his seminal paper (Ingenieros 1914).

In addition, *Revista de Filosofía* supported a further effort: the edition of a series of books, called *La Cultura Argentina,* produced at very affordable prices and aimed at the promotion of classics of Argentine culture or surveys of particular areas of it. The editors of *Minerva* produced a series of pamphlets, called *Cuadernos* (Fløistad 2003, p. 62), aimed at enlarging the discussion of special topics, but did not have the time to develop in that direction.

Yet, there is a further and important link between *Revista de Filosofía* and *Minerva.* Both began to be published at times of isolation from Europe; the former appeared in the mid-1910s, at the beginning of the World War I; the latter in 1944, well into the World War II, when traditional contacts with intellectual Europe had diminished to the point of extinction.

However, there is a significant difference between these two historical periods. While the two world wars almost wiped out short visits from European academics, reaffirming isolation, in *Minerva*'s time Argentina's cultural world was being enriched by a small, but not insignificant, number of intellectuals who had been arriving from Europe and settling down in the country for around a decade. They had initially moved because of political persecutions and, later, due to a war that was more global than ever before.

Some of these new arrivals started to join the locals in their cultural enterprises, adding their efforts and their collaboration to the one already provided by an exceptional group of intellectuals who had recently arrived from Spain, escaping after the Civil War, and had settled down in different parts of Argentina. This enriched intellectual element, more difficult to discern in *Revista de Filosofía*, can easily be detected in *Minerva* to the point of making us wonder if the enterprise would have been at all possible without them. We shall return to this point later, when we discuss *Minerva*'s group of close collaborators.

3.5   The Short Life of *Minerva*

Minerva was an attempt to modernise local philosophy practice and rid it of older, dominant influences such as that of José Ortega y Gasset (1883–1955) and other philosophers of that generation, not necessarily Spanish-speaking. The journal attempted to introduce the work of more modern and perhaps more relevant schools into local philosophical thought and to discuss their works critically.

Minerva opened with a declaration of war on positivism, as well as on its more recent reformulations. In its *'Presentation'* it announced that it was inspired by the goddess Minerva and that "it was armed and in combat: armed of reason and in combat for reason and against *irrationalism*".

3.6 *Minerva's* Editors

Its first issue dates from May to June 1944 and gives the name of Bunge as its director, assisted by four board members: Rodolfo Mondolfo (1877–1976), Simón M. Neuschlosz (1893–1950),[11] Isidoro Flaumbaum and Hernán Rodríguez Campoamor.

The first two board members were European specialists who had recently moved to Argentina and were teaching at national universities outside Buenos Aires. Mondolfo was an Italian anti-fascist émigré, a respected specialist in Karl Marx (1818–1883) who had taught philosophy at the University of Bologna until forced to emigrate. In Argentina, he joined the University of Córdoba and in 1948 moved to the University of Tucumán, in central Argentina.

Neuschlosz was a Hungarian émigré who, since at least 1935, taught biophysics at the Faculty of Medicine of the Rosario branch of the Universidad del Litoral. In addition, he had a serious interest in philosophy and was highly regarded in local circles; he was a representative of science in *Minerva*. Fragments of his Argentine correspondence are preserved in the correspondence of other philosophers; one of them is the leading anti-positivist Argentine philosopher Francisco Romero (1891–1962), whose correspondence has been discussed in Aranda (2012).

As much as Mondolfo, Neuschlosz had an early interest in Étienne Bonnot de Condillac (1714–1780), a philosopher whose ideas had entered Argentina in the very early nineteenth century together with those of Antoine Desttut de Tracy (1754–1826). Neuschlosz' books, (see Neuschlosz 1937, 1939, 1942a, b), were read with interest by those in Argentina interested in the philosophy of science; particularly in the philosophy of physics. Several of his books were published by respected local publishing houses. One of the latter, Editorial Losada, a leading publisher from Buenos Aires, included some of Neuschlosz' works in its then-respected book series on the 'Theory and History of Science'.

Neuschlosz was also a frequent contributor, particularly in 1937–1939, to the prestigious *Colegio Libre de Estudios Superiores* (*CLES*), in Buenos Aires. This was an independent institution of advanced studies, which later acquired branches in several Argentine cities, including Rosario. Neuschlosz' series of lectures at the *CLES* included topics on the history and philosophy of science discussing, in particular, philosophical questions posed by contemporary physics, particularly by relativity and quantum mechanics. Some of his lectures were later published in the *CLES*'s journal, *Cursos y Conferencias*. The other two editors of *Minerva*,

---

[11]The first date is uncertain, referring to his designation at the Universidad del Litoral, (*Boletín Oficial* 1935, p. 532) gives 1893 as his date of birth.

Rodríguez and Flaumbaum, were young students, at the beginning of their careers as Bunge then was; both of them were students at the Faculty of Philosophy.

Flaumbaum came from a German refugee family and was, and remained, active in left-wing politics. All three were good linguists: Flaumbaum would later translate into Spanish some of Antonio Gramschi's (1891–1937) works. Rodríguez, in addition to being Bunge's collaborator, translated into Spanish several of his books, including *Causality* (Bunge 1959, 1961). Later, he put to good use his linguistic and other skills, working for UNESCO in positions of responsibility in Paris.

Bunge, on his part, had also been active in translations; in collaboration with his father he translated Friedrich Engels's (1820–1895) *Dialectics of Nature* into Spanish. This book was published in a collection on *Dialectical Materialism* directed by Dr. Emilio Troise (1885–1976).

Troise was a respected intellectual who after years of concern with racism, became the founder and first president of Argentina's *Committee against Racism and Antisemitism.*[12] The writer José Luís Borges (1899–1986) was his secretary there (Actas 1938). In addition to its campaigns, the *Committee* tried hard to assist foreign intellectuals who wished to find refuge in Argentina, but were discriminated against on grounds of race or political views. In addition to being a friend of the Bunge family, father and son, Troise had correspondence with at least two members of *Minerva*'s editors: Mondolfo and Neuschlosz (Archivo CeDinCI, 'Emilio Troise: Correspondence'). It would not be surprising if Troise had, at some point, assisted some of the later, as he did with many entering Argentina, as both were Jewish and anti-fascist.

3.7 The Work of *Minerva*

As we have already indicated, *Minerva* included the word *Continental* in its title and, in fact, managed to attract contributions from the few active philosophers working in Latin America at the time; however, those from Argentina, or resident there, are predominant in *Minerva*'s index. The journal paid contributors a stipend, probably as high as US$100, which was unprecedented for Argentina at the time.

Contributors to *Minerva* included Rey Pastor, who discussed fictionalism in the philosophy of mathematics[13] (Rey Pastor 1988); Francisco Romero, who considered the problem of the conception of the world, and Risieri Frondizi (1910–1983),[14] a young philosopher with a good understanding of current trends among contemporary philosophers from Latin American and the US. In *Minerva*, Frondizi offered an overview of contemporary Latin American philosophy.

---

[12] *Comité Contra el Racismo y el Antisemitismo de Argentina.*

[13] Bunge has indicated to me, in conversation, that he regarded Rey Pastor's contribution to *Minerva* as one of the most worthy in the collection.

[14] A brother of the already mentioned Arturo Frondizi.

At the time of *Minerva*, in 1944, the range of Argentine journals with some philosophical content in them included at least six other periodicals; we follow the general list of philosophy journals compiled by Tarcus (2007). At least five of them had an important advantage over *Minerva*: they received some form of official sponsorship to cover publication costs; the sixth was not supported by the state, but by the religious Order of Jesus.

Two of these journals were published in Buenos Aires: *Logos,* supported by the Philosophy and Literature Faculty of Buenos Aires University, was edited by the well-known philosopher Coriolano Alberini (1886–1960), a senior professor at that institution, and later by Ángel Battistessa (1902–1993). The journal, as its director, closely followed contemporary trends of German philosophy. At the time, the students' union of that same institution was publishing a 'second series' of its journal, *Verbum.* Although it cannot be strictly described as a *research* publication in the field of philosophy, *Verbum* was an interesting and informative journal addressed, mainly, to university students of philosophy and literature.

Four more philosophical journals were printed outside Buenos Aires. *Philosophia* was published, from 1940, by the new Institute of Philosophy, at the Universidad de Cuyo, in the Andean city of Mendoza; its director was the young philosopher Diego Pró (1915–2000). *Estudios de Filosofía,* was published by the Centre for the Humanities and Philosophical Studies[15] of the University of Cordoba from 1943, and *Humanidades* was edited by the Humanities and Education Faculty at the University of La Plata. Finally, *Ciencia y Fe,* was edited by the Faculty of Philosophy and Theology of the Colegio Máximo de San Miguel, an institute of advanced religious education run by the Jesuit Order. The Colegio Máximo was a part of a large complex situated near Buenos Aires, which included a Cosmic Physics Observatory. Its main interests were science and theology; more marginally, philosophy itself. Pope Francis (1936-) studied at that Colegio Máximo.

Ciencia y Fe published a detailed and objective piece on *Minerva*, quoting extensively from the journal's statements rather than offering their own reading. It applauded *Minerva*'s determination – as the goddess from which it derived its name – to begin a "combat against *irrationalism*", which *Ciencia y Fe* thought was a timely reaction against the so-called philosophical irrationalism (Ciencia y Fe 1942).

It also listed the works contained in the first of *Minerva*'s issues, which were, Mondolfo's 'The Philosophy of Giordano Bruno'; Bunge's 'What is Epistemology'; Neuschlosz' 'Irrationalism in Contemporary Physics'; Flaumbaum's 'Meister Eckart and Martin Heidegger' and Rodríguez' 'Conflict of Life and Death in Antonio Machado'.

Sadly, *Minerva*'s efforts to renovate local philosophical practice were thwarted by financial difficulties which forced it to close before reaching a full 1st year.

---

[15] *Centro de Estudios de Filosofía y Humanidades.*

3.8   Completing a Doctorate in Theoretical Physics in the Argentina of the Mid-1940s

In addition to his work as *Minerva*'s creator, manager and contributor, the year 1944 was also crucial for Bunge as a physics student. Having finished his university course at La Plata, he faced a most difficult task: in order to start working for his PhD, he needed to find a supervisor.

Matters were made even more complex by the fact that he wished to work in the field of theoretical physics, possibly in quantum mechanics. An expert, if there was then any one in that field in the country, prepared to also generously offer his time to a potential PhD student, was even rarer to find. This was not just a matter of selfishness: at the time, a chair was not a full-time job, so scientists – as well as scholars in other fields of research – had to accumulate positions to build a reasonable salary. This was the local version of the French *cumul* system, perhaps more brutal than the original.

Researchers not only taught in more than one institution but, in pursuit of enough jobs to make a decent salary, some had to move from town to town. One of them, mathematician Dr. Juan Carlos Vignaux (1893–1984), a specialist who had contributed with fine work on function theory to leading foreign journals, has recounted that, for him, the railway coach was his study (Ortiz 2011b, pp. 187–88).

Bunge targeted the aforementioned Guido Beck, a remarkable foreign specialist who had migrated to Argentina in mid-1943, escaping from the Second World War. He had done so through a very convoluted itinerary. But before coming to Bunge's contacts with Beck, we need to have a quick look at the state of physics, and more particularly, of theoretical physics research, in Argentina in the early 1940s.

3.9   Theoretical Physics in the Argentina of the Mid-1940s

There is a close connection between the development of a fairly strong mathematical school in Argentina, initially under the tutelage of Julio Rey Pastor, and the emergence, in the early 1930s, of a group of theoretical physicists with a good mathematical training (Ortiz 2011a).

One of the first of Rey Pastor's students who later moved to mathematical physics was the Uruguayan-born Félix Cernuschi (1907–1999). In the mid-1930s, after graduating in Buenos Aires, he was sent to England where he gained a PhD in theoretical physics at the University of Cambridge. On his return to Argentina, Cernuschi decided against remaining in the Rio de La Plata area and accepted a position at the University of Tucumán. This was a good decision: Tucumán had long advocated for a decentralisation of higher education and, at the time, was benefitting from a resurgence of spiritual life in the provinces, and making a serious effort to attract local scholars, as well as European émigrés.

The leading Italian mathematician Alessandro Terracini (1889–1968) had already moved to Tucumán, from Torino, and some time later, in 1945, the Hungarian science historian Desiderio Papp (1896–1993) also moved there. Together with Terracini, Cernuschi started to publish, in Tucumán, the first Argentine journal exclusively centred on research in mathematics and theoretical physics: its name was *Revista de Matemáticas y Física Teórica*. It appeared in 1940 and managed to attract contributions from leading theoretical physicists, among them, Albert Einstein (1879–1955), and from world-class mathematicians, such as Élie Cartan (1869–1951), Tulio Levi-Civita (1873–1941), Guido Fubini (1879–1943), and others.

During those years, in addition to researchers from the University of La Plata's Physics Institute, where there had been physics research almost continuously since the beginning of the twentieth century, there was a small group of researchers at the Faculty of Science,[16] in Buenos Aires. Its leaders were Teófilo Isnardi (1890–1966) and José B. Collo (1887–1962), both members of the first group of physicists that qualified at the then new University of La Plata's Physics Institute around 1910. They were trained there by German physicists, and later sent to Germany for further studies.

There were two other small groups of researchers: one under physicist Enrique Gaviola (1900–1989) at the National Astronomical Observatory, an institution founded in the 1870s by US astronomers hired by the Argentine government, and, of course, in Tucumán where Cernuschi had joined forces with the veteran German physicist, José Würschmidt (1886–1950).[17] Gaviola, also trained in Germany and later in the United States, had been attracted to the Observatory, initially as its vice-director and later as director, with the idea of helping the Observatory's specialists to become more familiar with modern techniques of experimental physics; possibly, adapting them to research in the fields of astronomy and astrophysics.

In 1942, Gaviola and his research group decided to invite their Argentine colleagues to participate in a meeting, exclusively devoted to physics and astronomy, held at the Córdoba Observatory. This was probably the first significant meeting of scientists interested in modern physics in Argentina; it was named *Pequeña Reunión de Astronomía y Física*, and took place in early July 1942.

In addition to the locals, some foreign scientists, who were then visiting Argentina, attended that meeting. One of them was George D. Birkhoff (1884–1944), professor of mathematics and head of the mathematics department at Harvard University. When Birkhoff learned of the Córdoba meeting he contacted Cernuschi, whom he thought highly of[18] and, through him, established a contact with Gaviola.

---

[16] At the time that institution was called the *Faculty of Engineering*, an Engineering School created in the nineteenth century; it had service sections on chemistry, physics, mathematics and natural science. Later, towards the end of the nineteenth century, these sections were allowed to offer degrees in their own fields.

[17] On Würschmidt's scientific personality see Gaviola and Beck (1946).

[18] Later, Birkhoff invited him to Harvard University; later Cerncuschi moved to Princeton and to other leading centres. Much later, in the mid-1950s, he returned to Argentina as a professor

Birkhoff expressed an interest in taking part in that gathering to meet, personally, other local exact scientists. At the meeting, Birkhoff lectured on his alternative formulation of relativity theory.

Birkhoff's visit to Argentina was part of a tour of extended visits to mathematical centres in Mexico, Peru, Chile, Argentina and Uruguay. He viewed his trip as a contribution to Roosevelt's efforts on inter-American friendship and collaboration: the so-called *Good Neighbours Policy* (Ortiz 2003).

There was another reason for Birkhoff's desire to deepen his understanding of the state of the exact sciences in Latin America: he was compiling a report on that subject, which he later submitted to the American institutions sponsoring his trip: the Guggenheim Foundation and, to a much lesser extent, the State Department. His *Report* was later put to good use by the American Mathematical Society, which started a cooperation agreement with the Argentine research groups of Rey Pastor and Cernuschi.

3.10   Guido Beck Arrives in Argentina, from Portugal

As a follow-up to Gaviola's grand meeting of personalities in Córdoba, a group of physicists working at the Córdoba Observatory decided to invite, again to Córdoba, a smaller group of colleagues, working in different Argentine institutions, to a more intimate second meeting where they may report on their recent research work. Those invited were physicists working in Buenos Aires, Córdoba, La Plata and Tucumán.

For the management of this second meeting a new provisional structure was created, which was informally called *Núcleo de Física Teórica*. A number of well-known Argentine senior physicists, such as Collo, Gaviola, Isnardi, Enrique Loedel Palumbo (1901–1962) and others, joined in the initiative. The gathering took place in Córdoba between 27 and 28 November 1943. Contributions from Gaviola's research group dominated that meeting, where six reports or communications were presented.

The Bohemian theoretical physicist Dr. Guido Beck, then 40 years old, who had entered Argentina only a few months earlier, in May 1943, was credited with having been instrumental in the consolidation of that physicists' *Núcleo*. In a note published in *Revista de la Unión Matemática Argentina* (Núcleo de Física 1944a, p. 31), they refer to it as the meeting "organized by Professor Doctor Guido Beck (Córdoba) with the purpose of stimulating studies on the modern directions of theoretical Physics".[19]

---

of physics, head of the physics department and dean at Buenos Aires University's Engineering Faculty.

[19]"[O]rganizadas por el profesor Doctor Guido Beck (Córdoba) con el fin de estimular los estudios sobre la moderna orientación de la física".

Beck arrived in Argentina after a number of temporary exiles in many different European countries. He has been singled out by Lord Beverage, in his well-known book on displaced intellectuals, as the scientist who visited the largest number of countries in exile: 16 in his case (Beveridge 1959, pp. 106–107). After stays as a visiting professor in the United States and in the Soviet Union, and fellowships in several European centres, Beck was in France when the Second World War broke out (Havas 1995). As a foreigner from an enemy country, he was sent to a camp, mainly for people of German origin: Camp de St Antoin.[20]

However, with the help of Portuguese scientists, particularly of physicists Ruy Luís Gomes (1905–1984), Manuel Valadares (1904–1982) and mathematician António A. Monteiro (1907–1980), all committed anti-fascists he had met earlier in France, Beck managed to get permission to immigrate to Portugal then, nominally, a neutral country. He stayed as a refugee in Portugal for about one year, but the Portuguese authorities had some concerns about allowing him to communicate freely with university teachers and students. So, they allowed him entry into the country, but put restrictions on his movements. He was housed in a comfortable pension, in a small Portuguese town, but outside the main university centres.

Nevertheless, Beck established contacts with Portuguese universities, but not on a regular basis. Despite these obstacles, his isolation was not complete and he managed to have a very positive influence on the development of theoretical physics in Portugal. His scientific personality, and his generosity, attracted a small group of students who, encouraged by Ruy Gomes and Valadares, began visiting him, individually, to discuss their research work. At least one Portuguese student gained his PhD fully under Beck's guidance.

Yet, fearing that world war would, one way or the other, involve Portugal, or that the political situation there may deteriorate even further, Beck started to plan for a further exile, possibly in South America, Peru was a possibility he explored at some stage. He also approached the Institute of Physics, in La Plata, which was then going through a difficult time: his letter went unanswered. Finally, his friends in Portugal established a contact with Rey Pastor.

In parallel, through a letter Gaviola wrote in April 1942 to the eminent physicist James Franck (1882–1964), then in Chicago (Archives Emergency Committee 1942, Files 'Franck'), in response to Franck's concerns with Beck's situation, we now know that the chance of him being hired in Argentina had begun to be discussed about a year before his arrival there. Gaviola wrote to Beck, assuming he was at the University of Coimbra, but didn't get a reply. However, again with Franck's help, he finally established a contact and offered Beck a position at the National Astronomical Observatory, Córdoba. His Portuguese friends helped Beck to get the necessary papers and, after nearly a year of exchanges, Beck finally arrived in Argentina.

However, Beck's date of arrival was definitely unlucky, and suggested Argentina would not be the end of his travels. He debarked on 11 May, and some 3 weeks later,

---

[20] Also called *Camp de la Viscose*.

on 4 June, as already said, there was a military *coup d'état* in Argentina. Beck's arrival in Argentina did not go unnoticed by the Argentine security services. He has a file in the *Servicio de Informaciones del Estado* (*SIDE*), the élite intelligence department in those days, which gives accurate details on Gaviola's efforts, full details on Beck's documentation, a list of his friends abroad, his movements in Argentina and, even, his first address in Buenos Aires (Archivo SIDE 1943).

It is likely that the physicist and later eminent writer, Ernesto Sábato (1911–2011), a committed anti-fascist who had studied under Irène Curie in Paris, was also a link between Beck's friends abroad and those in Argentina. In any case, Sábato made contact with Beck immediately after his arrival in Buenos Aires, before he moved to Córdoba, his final destination. It is possible that at this stage, in late May or June 1943, Bunge met him, through Sábato. Apparently, they visited him together, in May or June 1943, at his Buenos Aires first lodging (Callao 181, according to his meticulous *SIDE* file). Possibly, soon after this first meeting, Bunge started to have regular discussions with Beck.

3.11   The Consolidation of the Argentine Physicists' Community

Let us now return for a moment to the *Núcleo* meetings. The second one[21] took place at the Physics Institute, Buenos Aires, 6 months later, on 12–13 April 1944. The number of reports or communications was approximately double the number submitted to the first meeting, suggesting some progress. In addition, participants now covered a wider geographical area, which included groups in, at least, Buenos Aires, Córdoba, La Plata, Rosario and Tucumán: all national universities were now represented.

The following (third) meeting of the *Núcleo* took place at the Physics Institute, La Plata, 4 months later, on 27–29 August 1944. The number of communications was roughly the same as in the last one, suggesting the size of the physicists' group had reached its limits. At this stage, the apparent success of these meetings induced Beck, Gaviola, Loedel Palumbo and other, younger, physicists to establish a more permanent organisation, which was later called *Asociación Física Argentina* (*AFA*).[22]

AFA was created on 27 August 1944, at a meeting held at a neutral territory, a tea room in the city of La Plata, where some 25 researchers gathered. Over half of them were students, either at the end of their studies or, for a few, working towards

---

[21]Which was also the third gathering of Argentina's physicists, taking into account *Pequeña Reunión* of 1942.

[22]Before *AFA* was organised there were already two informal, but active, groups of physics students: *Agrupación de Estudiantes de Física* in La Plata and *Núcleo de Estudiantes de Física* in Buenos Aires.

their PhD. Clearly, most of them were local students from La Plata: Mario Bunge was one of them.

Gaviola was elected *AFA*'s first president, and Loedel Palumbo, Beck and Ernesto E. Galloni (1906–1987) were designated 'local secretaries' for La Plata, Córdoba and Buenos Aires respectively (Núcleo de Física 1944b).

3.12   Guido Beck and the Formation of Young Research Scientists

Beck figured significantly in Bunge's hazardous search for a PhD supervisor. One year after his arrival in Argentina, Beck read a most interesting paper at the second meeting of the *Núcleo*, on 12 April 1944. In it, he candidly offered his views on the current state of research on theoretical physics in Argentina. Politely, Beck didn't discuss the teachers' generation, his colleagues,[23] restricting his views to members of the current generation of young local research students interested in theoretical physics. It interests us since Mario Bunge was one of them.

Beck also made comparisons between the competitive field of *international* scientific research and the current conditions in Argentina, giving us, today, a unique personal overview of the then-current characteristics of physics research in Argentina through the eyes of an experienced foreign expert. The text of his lecture was later published and is available in *Revista de la Unión Matemática Argentina* (Beck 1944a).

For Beck, physics, and more particularly, theoretical physics, was an area that, in those years, was experiencing a significant increase in the number of researchers, worldwide. In the case of Argentina, this observation (for physics in general) seems to be supported by the increase in the number of contributions, meeting after meeting. No doubt, over the last year, and in the field of theoretical physics, Beck's presence in Argentina had been the main reason behind such advances.

Beck indicated in his paper that, while he was in Europe, he had to deal with a large number of students and that all of them, as a rule, had to struggle to do scientific research. That is, in addition to their research, they had to earn their living, get the necessary working facilities, and so on. He noted that "these youngsters took advantage of any chance to do some work, had respect for their work, did not put it in jeopardy and would not abandon it, but for an imperious reason" (Beck 1944a, p. 33); implying, perhaps, that his Argentine students did not do that.

Beck also stated that for the development of a research group (in Argentina) it was "indispensable that a number, even a reduced number, of young students be devoted [to research] without any restriction. Such a group does not yet exist [here]; only a few isolated cases" (Beck 1944a, p. 34). He admitted that a youngster, even

---

[23]But he did so, acidly, in personal correspondence with his friend António Monteiro; see Archivo Monteiro (1946).

if he has qualities for research work, has the right to choose the field of activity that rewards him best; sometimes, with a minimal effort on his part. Beck added that "as life here is very easy, [in Argentina] students have many possibilities. However, in that case, he should abandon research work, which is determined by very different conditions [which are fixed] in other parts of the world". He continued, "if a youngster wishes to compete in research, he must adapt to its intrinsic characteristics and accept all the indispensable sacrifices" (Beck 1944a, p. 34). He clearly meant, to work hard and to be focused on a concrete problem.

He added (Beck 1944a, p. 35) that, in Argentina, he did not find the average talent of his new students to be substantially different from what he had seen in other areas of the world. Later, he would qualify this remark indicating that their mathematics training was very good, but their handling of fundamental concepts of physics was more limited (de Abreu and Candiotti 1989).

Beck insisted that the main problem was not a lack of individual talent, something else was missing: "students do not yet know what they want. One is looking for the simplest way of gaining a doctorate, as if a Dissertation had value in itself, rather than being anything but an introduction to further work. There are others who pretend they wish to work in theoretical Physics, but dream with world's politics, continental philosophy, national history and I don't know what" (Beck 1944a, p. 35). Around the time that this lecture was delivered, May–June 1944, the first issue of *Minerva,* a continental journal of philosophy, began to circulate in Buenos Aires.

3.13 Beck Versus *Minerva*

After managing to be accepted by Beck as a research student, Bunge applied to the *Argentine Association for the Advancement of Science* (*AAPC*) for a research grant, which would allow him to spend 3 months working under Beck in Córdoba, at the Observatory.

In those years, the *AAPC* was the chief Argentine institution for the award of research grants. A decade earlier, the national government had deposited a sum of money in a national bank account and trusted the *AAPC* with the administration of its interests for the award of research subsidies. Later, a discretely administrated financial life support was provided by the Rockefeller Foundation and injected into the *AAPC* to complement its small resources (Ortiz 2015).

The *AAPC* was headed by the Argentine physiologist Bernardo A. Houssay (1887–1971), a future winner of a Nobel Prize in science who had also been a long-standing champion of research and of the establishment of selective full-time positions for research. As funds were very limited, Houssay and other leaders of the *AAPC* applied a very strict policy to the applicants, requesting full-time commitment to research. This last matter had been touched upon by Beck in his lecture and was also the subject of intense discussion in scientific circles in Argentina at the time.

Correspondence between Babini and Bunge, during the period 10 June to 29 September 1944 is preserved[24]; a fragment of it, quoted in Busala and Hurtado de Mendoza (2000, pp. 267–68), indicates that Beck had serious reservations about *Minerva*; or rather, with a piece by Bunge published in it (Bunge 1944d). Beck objected to the tone of Bunge's references to some positivist philosophers and, more specifically, to members of the Vienna Circle. In addition, he disapproved of the use Bunge had made of algebraic operators in a philosophical paper.

These differences caused Beck to discharge himself from the obligation of continuing to be Bunge's supervisor. The withdrawal of support from his supervisor lead to Houssay rejecting Bunge's request for a research grant. More details of these discussions are given in Ramacciotti and Cabrera Fischer (2010), where some of the correspondence exchanged is reproduced. As stated earlier, at the time there were not many alternatives for a graduate student to find a doctoral supervisor in Argentina, not least in a field as arcane then as theoretical physics. The only way out was a mutual reconciliation, which took place after Rey Pastor and Babini negotiated with Beck, on Bunge's behalf.[25]

It must be said that roughly around the time of these exchanges and renunciations, Bunge seems to have been doing his job as a PhD student reasonably well. In 1944, he read two communications on topics of theoretical physics at AFA meetings; they were published locally as short notes (Bunge 1944a, b), but an English version of the last of these notes (on a new representation of nuclear forces) also appeared in *Physical Review* (Bunge 1944c). At the AFA meeting of March–April 1945, Bunge contributed with a communication on neutron-proton scattering (Bunge 1945a), which was also published, as a short piece, by *Nature* (Bunge 1945b). At the following *AFA* meeting, its 6th, in September 1945, Bunge read a paper on resonance phenomena in the diffusion of protons by neutrons (Bunge 1946).

3.14   The Impact on Science, Education and Culture of the 1943 *Coup d'État*

Let us return briefly to the wider contemporary scene of Argentina and to the impact of the new cultural policies that followed the 1943 *coup d'état*. During that year, unofficial conversations took place between a small group of university professors, leading civil servants and military men in which, apparently, each side expressed candidly their views on matters of national interest. Later, a group of university professors (possibly the same group) produced a document inviting the military

---

[24]Fondo José Babini, Universidad de San Martin.

[25]The correspondence between Babini and Bunge is preserved in 'Babini's Correspondence Archive', at the Universidad de San Martin, Buenos Aires; the paper (Busala and Hurtado de Mendoza 2000) uses this source.

authorities, politely, to call for free elections within a reasonable time frame, so that democracy could be restored to the country.

However, once the letter appeared in the press, the response of the military government was not what was expected. It has been argued that, in the meantime, there was a readjustment of the different tendencies that fought for positions within the military Junta. The fact is that extreme right-wing groups gained the upper hand in the areas of education and culture. The latter wished to do away with an education law passed in 1884 which secured compulsory, free and non-religious education at primary level introducing, instead, religious education. In addition, they wished to tighten control on higher levels of education.

The fact is that following publication of that letter, the military Junta dismissed all university professors and civil servants who had signed the document and, in addition, revoked the licence to deal with the state to any professional who had signed the document (Ortiz 1996).

Among those affected by this measure were remarkable personalities from the world of culture as well as leading local scientists: Professor Houssay was one of them. A further, more private outcome of these confrontations was a definite change in the fragile relations between the government and scientists, technologists and other specialists, who were then working in areas that facilitated making the country technologically and industrially more independent. However, as time passed, the certainty that the Axis countries would win the Second World War, as those in power seem to have expected, diminished considerably.

3.15   1946: New Expulsions from Argentine Universities

In the transition from 1944 to 1945, large sectors of the population believed that there was little doubt the Axis powers would not win the war. This brought some changes in the military government's attitudes and, finally, a solution was found to the question of the dismissed teachers: they were reinstated. The discussion now was on the military Junta allowing for fair national elections, and returning quietly to their barracks. This finally happened in early 1946, when national elections were called for.

However, the winner of these elections was General Juan D. Perón (1895–1974), a former member of that Junta. From its installation, the new government entrusted culture and education, again, to extreme right-wing elements, which were still quite powerful. Soon there were new expulsions of university professors and secondary school teachers. In this second instance they were more severe, affecting 1142 positions at university level and some 140 in secondary schools controlled by a university (Avasallamiento de la Universidad *Argentina* 1947; Ortiz 1996, p. 173). As in the previous episode, the one of 1943, about a third of those expelled were in areas of teaching and research related to science, technology and medicine.

My own father was among the many professionals and university professors affected by what was later known as *purifications*. He had specialised in the construction and management of the large national harbours, some of which were very substantial and designed to handle efficiently the very large peaks of activity associated with the grain export season. Others, on the Patagonian coasts, had to be able to withstand powerful waves as well as large tides that created large oscillations at surface level. At the time, with considerable experience in the field, he was the director general of Argentina's National Department of Navigation and Harbors.

Shortly after the depurations, which also affected national departments, my father was invited by United Nations to act as a consultant on the design, or reconfiguration, of several large ports abroad. As such, he designed new harbours in Latin America and participated in discussions on the redesign of European ports that had been affected by the war. This took me away from Argentina for some time.

3.16   My Studies with Bunge in the Early 1950s

On my return, in 1950, after some examinations to complete my secondary school certificate, and pass the University's entrance examinations, I became a student of Mathematics at Buenos Aires University. At the time, we were a very rare species among university students in general.

The name of Mario Bunge was not unknown to me. I had first heard of him from the Rector of my old School, Carlos Federico Ancell (1896–1963), an extremely conservative man in politics, but very open in his views on science; particularly on astronomy which was his real passion. In addition, Ancell had an excellent scientific library and, very generously, had allowed me to use it freely. I understand Ancell was related to Bunge, through his wife Emma Constanza Bunge (b.1894).

At university I soon met Rey Pastor, with whom I shared an interest in mathematics and science history, and later Vignaux, who was then head of a newly founded *Radiotechnical Institute*, a joint venture between the Navy and the university, which offered interesting and modern courses and seminars. The atmosphere of the *Institute* was enhanced by a small group of German scientists who had arrived soon after the end of the war. One of them, Richard Gans (1980–1954), had been in Argentina before, as director of the La Plata Physics Institute from the early 1910s until his return to Germany in 1925. Another interesting scientist in that group was Kurt Fränz (1912–2002), a leading expert in circuit theory and control, a fine mathematical mind.

Vignaux's armed forces connection allowed him to attract a number of local physicists and mathematicians with a minimum of political conditions. In addition, around 1952 there was some relaxation of the rules at the university, so it was now permitted to hire some personnel without the necessary precondition of having to join the Peronist Party beforehand.

At the *Radiotechnical Institute* I became acquainted with one of Viganux's assistants, Dr. Manuel Sadosky (1912–2005), who helped Vignaux with one of his series of lectures. Later, he kindly invited me to his home and it was there where I finally met Bunge personally. Soon after, Bunge invited me to his residence, *El*

Ombú, in Florida, outside Buenos Aires, where he met regularly with a group of most interesting friends: physicists, mathematicians and philosophers. Later some of them became my personal friends. In these meetings, and afterwards in personal visits to his home, I received significant and valuable advice and encouragement from Bunge, which have been of value throughout my life, both personally and professionally.

During this period, Teófilo Isnardi who was, perhaps, the most gifted Argentine theoretical physicist in Argentina at the time, was in charge of courses on that subject at Buenos Aires University. Since his return from postgraduate studies in Germany, over 30 years earlier, Isnardi, together with his old friend Collo, had been valued professors of physics at the Naval School, where they had done an excellent job keeping the teaching of future naval officers right up to date in their own field. This may have been a shield for them: they were not touched by the *purifications* of 1946.

That brief period of calm, around 1952, allowed Professor Isnardi to invite Bunge to become his Assistant in his role as professor of Theoretical Physics. This was a course we students read in the last 2 years of our studies – 4th and 5th years. After passing some extra examinations I jumped a year in my official studies and, in 1952, was able to take one of these courses and, consequently, became an *official* student of Bunge. Traditionally, Assistants' lessons were closely connected to the specific topic the Professor had chosen to discuss in that year. However, Bunge was allowed independence and he chose a topic in electromagnetism: the theory of antennae.

This was a difficult subject, attractive from both a theoretical and a practical viewpoint. From the beginning of his lectures, Bunge tried to teach us to use Maxwell's theory as a tool, asking us to discuss very concrete and often difficult problems which he set for us. This was perhaps the first occasion where we, as students, learned not just a most beautiful theory, but also attempted to *use* it in everyday situations.

3.17 New Political Interference

Sadly, this period didn't last long. Towards the end of 1952 there was a tightening of rules at the university and following that, Isnardi, his assistant, Bunge, professors Rey Pastor, Collo, the distinguished geometer Luis A. Santaló (1911–2001), and several other very good scientists, were dismissed from the Faculty of Science.

Some years later, in an interview, Bunge described his own forced departure in very precise terms:

> The University of Buenos Aires had expelled me for not joining the Peronist Party and for not contributing monthly to Eva Perón's Foundation – two very grave offenses. At the end of 1952, I was fired from the School of Sciences. (Rozzi and Poole 2009, p. 4)

After these expulsions, a group of Bunge's former students visited the *Interventor* (this was the name officially given to a dean compulsorily designated by the authorities) to request a reconsideration of his case. If I remember correctly, I was accompanying two brilliant students: Daniel Amati (1931-) and Alberto Sirling (1930-), who later left Argentina and in time became professors of physics at the University of Rome and New York University, respectively.

The *Interventor* was an engineer by the name of Rioja, a polite man with the air of a visionary. He did not doubt that Dr. Bunge may be a talented young man but, for him, this was not the point. Rhetorically, he asked us why he did not follow the government's line, was he not a Party member? We understood that continuing this discussion would take from his valuable time and, with equal courtesy, said goodbye, leaving him to continue chasing heretics. Sadly, soon after that meeting, he himself was removed from his position. Perhaps, unknown to us, he was also a heretic.

3.18 A New *Coup d'Etat* to Rectify Failures of the Last One

However, the teaching standards at our Faculty, and at the universities in general, improved considerably some 4 years later, around 1956. This was after yet another *coup d'état*, in September 1955, the third since 1930. This one toppled the *elected* government of another military man: General Perón, who was, then, in his second term in office.

After the coup, repression changed signs: it was now directed, more specifically, at *Peronistas*. Particularly to the many who did not agree to change sides or refused to accept the new military government directives. In addition, following a custom firmly established for several decades, repression continued to be applied to the different shades of the political left: an exclusive group among the heretics

By 1955 army leaders had learned from their past experiences. The authorities emerging from that *coup d'état* did not wish to have too much trouble with the university's students. By then they had realised the latter were very vocal, fairly well organised and fond of demonstrating in the streets, which gave the public an impression that the army was not properly in control, and that it could be challenged. In turn, the new designated university authorities had learned not to be too assertive in their dealings with the armed forces. For example, they accepted the suggestion that not all of the formerly *purified* teachers, sacked in different waves by Perón's government from 1946, should be reinstated: a mainly political purification of *purifieds* was necessary.

If we leave aside the violence and economic pressure exerted at the time against some large sectors of the population, often very intense, the acceptance of a selective re-employment of former professors and other concessions made by those tacitly approved by the army to run the university, it is undeniable that, at the same time, many very positive advances were achieved in that period, particularly in science and science education.

Fortunately, Bunge was among those re-admitted. In his own words:

And in 1956, after the fall of Perón, I was reinstated as a Professor there [at the Faculty of Science]. I taught quantum mechanics. (Rozzi and Poole 2009, p. 4)

Another most important achievement was the re-establishment and enhancement of a pre-existing[26] Science and Technology Research Council (*CONICET*),[27] which took up and extended, to a much higher level, the pioneering activities started by the *AAPC*. Professor Houssay, the sage of scientific research, who had remained for several years outside the university, waiting in the circle of the *purified* outsiders, was among those reinstated and was designated head of *CONICET*.

A further important advance in that period was the creation of full-time positions at the university, an old ambition of researchers such as Houssay, Rey Pastor and Gaviola. These positions were established not just at professorial level, but there were also a few at lower levels, which allowed for the intensive training of young researchers. I was fortunate to be in the first group of these young researchers, in 1958.

The creation of a university publishing house, called *EUDEBA*,[28] of which the mathematician and science historian, José Babini, became director, was another important development at the time. One of *EUDEBA*'s many good deeds was the publication, in 1961, of a translation into Spanish of Bunge's 1959 *Causalility*, made by Hernán Rodríguez, the former co-editor of *Minerva*.

3.19 A New Storm Gathers

In the meantime, persons close to the armed forces persistently put the Science Research Council and the universities under a cloud of suspicion. They distrusted those who conducted these institutions as much as some university teachers. Political intolerance permeated deeply and even reached some intellectual circles; some then called it a local form of *McCarthyism*.

In 2016, a privately circulated document, known as *Castaño's Report* (Castaño 1966), produced over 50 years earlier, came to light. It is not, in itself, a significant document in any way, but it illustrates the views of a sector of society which had far more relevance, power and connections than then envisaged. Essentially, *Castaño's Report*, which makes reference to Bunge among many other distinguished intellectuals, is a denunciation of a dissolvent Marxist (atheist, communist, are some of the many synonyms frequently used instead) infiltration at *CONICET* as well as at the

---

[26] *Consejo de Investigaciones Científicas y Técnicas.*

[27] An older version, called *Consejo de Investigaciones Técnicas y Científicas* (note the inversion of words) had been created by Gen. Perón's government on 17 May 1951 by decree Nr. 9695/51.

[28] *Editorial de la Universidad de Buenos Aires.*

top levels of Buenos Aires University. Although plagued by inaccuracies, it reveals the tone and the views of some, perhaps the more extreme, in that period.

These were also years of intense political instability at a national level. The current elected President, Dr. Arturo Frondizi, whom two decades earlier we encountered teaching workers at Bunge's *UOA*, was now waiting to be demoted by the army; this actually happened in 1962. The following elected President would soon be toppled by a further *coup d'état*, in 1966. This last event was followed by an assault on the buildings of some Faculties of Buenos Aires University by troops of an élite armed forces commando unit.

The Faculty of Sciences was one of the institutions most seriously affected, but not the only one. There, troops entered university premises and attacked students, teachers and authorities indiscriminately, before detaining them. It was never clear exactly what the motive for that attack was, if not just crude intimidation.

Unexpectedly for those who ordered this action, one of the teachers affected by the attack on the Faculty of Science was the distinguished American mathematician Warren Ambrose (1914–1995), a professor at MIT. He was then paying an extended visit to the Mathematics Department of the Faculty of Science through a programme supported by UNESCO.

Ambrose did not ask for special consideration when the attack started and was mistreated like anyone else. Later, at a detention centre, when it was his turn to show his documentation, his American passport caused deep embarrassment. The severity of the attack could not, now, be easily dismissed. In addition, immediately after being released, Ambrose sent a letter to the *New York Times* (Ambrose 1966) giving details of the attack, which gave it an international dimension (Argentine Universities 1966).

The total incapacity of the military Junta to grasp the dimension of their actions, their indecisions, their internal squabbling and their failure to act decisively, showed a scene of amazing fragmentation of power, of disorder and of ineptitude. That event precipitated the resignation, in protest, of a large number of university teachers at different levels, totaling over 900.[29]

Physicist José Federico Westerkamp (1917–2014), one of the two professors in his Department to remain in his position, has recalled, in an impartial reference to these events (Westerkamp 1975, pp. 135–36), that the physics department lost 85% of its personnel: "of 17 professors only two remained". Among the auxiliary teachers, the level of resignations was even higher: 95%. The same source recalls that, in physics, "teaching and research activities were, in practical terms, paralyzed" (Westerkamp 1975, p. 136). Similar figures apply to the Mathematics and other Science Departments. Fortunately, some talented graduates continued to emerge.

If the events of 1966 were very dramatic, in so far as precipitating a sudden loss of academic personnel, they were, in no way, unique. A deep sense of instability was

---

[29]A full list of teachers who resigned was only compiled and published in 2016, on the occasion of the 50th anniversary of those events. A list is available at University of Buenos Aires: List of Resignations (1966).

clearly present even *before* the military coup of 1966, which contributed to motivate a brain drain that, despite valuable efforts, has been difficult to contain for a number of years.

I'll refer to one significant case that happened some 3 years *before* the military *coup d'état* of 1966, under the legally elected government of President Frondizi. In a brief autobiographical note written by the Argentine biochemist César Milstein (1927–2002) for The Nobel Organisation: *Nobel Prizes and Laureates of 1984*, he indicated that: "The political persecution of liberal intellectuals and scientists manifested itself as a vendetta against the director of the Institute[30] where I was working. This forced my resignation and return to Cambridge". While at Cambridge, in 1984, Milstein was awarded the Nobel Prize for Medicine.

Another prominent Argentine personality, the physicist and philosopher Mario Bunge, also left for good in those years of *conditional* legality. However, if these events seem then unprecedented, some 10 years later Argentina would suffer a new, and much harder blow through the affair internationally known as the *Desaparecidos* of Argentina.[31]

3.20 Conclusion

Only after an international military operation when badly wrong, with considerable loss of life on both sides, would the armed forces start to gradually lose control of the grand scenario of Argentina's political life. A new, more peaceful and constructive period started then. In it, the legacy of intellectuals of Mario Bunge's dimension began to be more clearly appreciated and absorbed.

References

Actas. (1938). *Actas del Primer Congreso Contra el Racismo y el Antisemtisimo*. Bueno Aires: Consejo Deliberante.

Ambrose, W. (1966). Letter to *The New York Times*, dated August 31, 1966, it was published on September 11.

Aranda, M. M. (2012). Francisco Romero: América en el diálogo epistolar. *Cuyo (Universidad Nacional de Cuyo), 29*, 2. http://www.scielo.org.ar/scielo.php?script=sci_arttext&pid=S1853-31752012000200003.

Archives Emergency Committee. (1942). *Archives of correspondence of the emergency committee in aid of foreign scholars*. New York. Correspondence Franck to Gaviola: Gaviola a Franck, April 15, 1942.

---

[30]The Institute Milstein refers to is a bacteriology research centre known as *Instituto Malbrán*, founded in Buenos Aires following the yellow fever epidemics of the 1870s. At the time, its director was a distinguished scientist, Dr. Ignacio Pirosky (1901–1989), who has written an important document, (Piroski 1986), that discussed the nature and motivations behind the attacks against *Instituto Malbrán*.

[31]See Guest (1990).

Archivo CeDinCI. *Centro de Documentacion e Investigación de la Cultura de Izquierdas en Argentina.* Buenos Aires. Correspondence files Emilio Troise (CeDinCI FA-86): Troise-Mondolfo and Troise-Neustschlosz.

Archivos Monteiro. (1946). Archivos de Correspondencia de Antonio A. Monteiro, Bahía Blanca. Correspondence Beck to Monteiro: 15.03.1946.

Archivos SIDE. (1943). Archivo Reservado del Servicio de Informaciones del Estado. Buenos Aires. File Guido Beck.

Argentine Universities. (1966, September 16). Argentina: Seizure of universities leaves intellectual casualties. *Science, 153*(3742), 1362–1364.

Avasallamiento de la Universidad Argentina. (1947). *Avasallamiento de la Universidad Argentina.* Buenos Aires: Federación de Agrupaciones para la defensa y progreso de la Universidad democrática y autónoma, Buenos Aires.

Beck, G. (1944a). Algunas palabras sobre los trabajos de física teórica. *Revista de la Unión Matemática Argentina, 10*(2), 33–36.

Beck, G. (1944b). Sección Bibliografía. *Revista de la Unión Matemática Argentina, 10*(2), 57–58.

Beveridge, L. W. (1959). *A defence of learning.* Oxford: Oxford University Press.

Boletín Oficial. (1935). *Boletín Oficial de la República Argentina, 1935,* p. 532

Bunge, M. (1943a). *El tricentenario de Newton.* Buenos Aires: Universidad Obrera Argentina, Instituto Científico. Seminario de Filosofía, 8 páginas, 1943.

Bunge, M. (1943b). *Significado físico e histórico de la teoría de Maxwell.* Texto de una conferencia dictada el 21 de junio de 1943 en la Facultad de Química Industrial y Agrícola de la Universidad Nacional del Litoral, en Santa Fe, 16 páginas, Buenos Aires, 1943.

Bunge, M. (1944a). El spin total de un sistema de más de 2 partículas. *Revista de la Unión Matemática Argentina, 10*(1), 13–14.

Bunge, M. (1944b). Una nueva representación de los tipos de fuerza nucleares. *Revista de la Facultad de Ciencias Físicomatemáticas (La Plata), 3,* 221.

Bunge, M. (1944c). A new representation of nuclear forces. *Physical Review, 65,* 249.

Bunge, M. (1944d). Que es la Epistemología? *Minerva, 1.*

Bunge, M. (1945a). Difusión neutrón-protón a 8.8 y 13 MeV (5th Meeting *AFA,* March–April 1945), Comunicaciones. *Revista de la Unión Matemática Argentina, 1946, 11*(3), 103.

Bunge, M. (1945b). Neutron-proton scattering at 8.8 and 13 MeV. *Nature, 156,* 301.

Bunge, M. (1946). Fenómenos de resonancia en la difusión de protones por neutrones. *Revista de la Unión Matemática Argentina, 11*(3), 115.

Bunge, M. (1959). *Causality: The place of the causal principle in modern science.* Cambridge, MA: Harvard University Press. [Third Revised Edition, Dover Publications, New York, 1979].

Bunge, M. (1961). *Causalidad, el principio de causalidad en la ciencia moderna.* Buenos Aires: Eudeba, 1961. (Spanish translation of: *Causality*).

Busala, A., & Hurtado de Mendoza, D. (2000). La Revista *Minerva* (1944–1945), la guerra olvidada. In M. Montserrat (Ed.), *La ciencia en la Argentina entre siglos* (pp. 259–274). Buenos Aires: Manantial.

Castaño, R. (1966). Informe de Ricardo Castaño sobre CONICET. This is part of a collection of documents prepared by Jorge Aliaga.

Cereijido, M. (1990). *La nuca de Houssay: La ciencia argentina entre Billiken y el exilio.* Buenos Aires: FCE.

Ciencia y Fe. (1942). *Ciencia y Fe.* Publicación Trimestral de las Facultades de Filosofía y Teología del Colegio Máximo de San José, (1942)), I, 2, April–June, 1942. In *Actualidades* (pp. 141–142). https://archive.org/stream/cienciayfe12cole/cienciayfe12cole_djvu.txt

de Abreu, A. A., & Candiotti, E. (1989). Interview a Guido Beck. *Ciencia Hoy, 1*(2), 77–80.

Fløistad, G. (2003). *Contemporary philosophy, a new survey. Vol. 8: Philosophy of Latin America.* Dordrecht: Springer.

Gaviola, E., & Beck, G. (1946). José Würschmidt. *Revista de la Unión Matemática Argentina, 9*(4), 158–160.

Guest, I. (1990). *Behind the disappearances.* Philadelphia: University of Pennsylvania Press.

Havas, P. (1995). The life and work of Guido Beck: The European years: 1903–1943. *Anais da Academia Brasileira de Ciencias (The Guido Beck Symposium), 67*(Suppl. 1), 11–36.

Ingenieros, J. (1914). Direcciones Filosóficas de la Cultura Argentina. *Revista de la Universidad de Buenos Aires, 27*, 261–344 (1–83).

Neuschlosz, S. M. (1937). *La Física contemporánea en sus relaciones con la filosofía de la razón pura*. Rosario: Editorial Ruiz.

Neuschlosz, S. M. (1939). *Análisis del conocimiento científico*. Buenos Aires: Losada, (2nd edition in 1944).

Neuschlosz, S. M. (1942a). *El principio de la conservación de la energía y su importancia para el pensamiento contemporáneo*. Rosario: Universidad Nacional del Litoral.

Neuschlosz, S. M. (1942b). *Teoría del Conocimiento* (2nd ed.). Buenos Aires: Losada.

Núcleo de Física. (1944a). Las reuniones del *Núcleo de Física*. *Revista de la Unión Matemática Argentina, 10*(1), 31.

Núcleo de Física. (1944b). La Tercera Reunión del Núcleo de Física. *Revista de la Unión Matemática Argentina, 10*(2), 62–64.

Ortiz, E. L. (1996). Army and science in Argentina: 1850–1959. In P. Forman & J. M. Sánchez Ron (Eds.), *National military establishments and the advancement of science and technology* (pp. 153–184). Dordrecht: Kluwer, 1996, in pp.173–75.

Ortiz, E. L. (2003). La política interamericana de Roosevelt: George D. Birkhoff y la inclusión de América Latina en las redes matemáticas internacionales. *Saber y Tiempo*, Part I, 15, 55–112; Part II, 16, 21–70.

Ortiz, E. L. (2011a). The emergence of theoretical physics in Argentina, mathematics, mathematical physics and theoretical physics, 1900–1950. In L. Brink, & V. Mukhanov (Eds.), *Quarks, strings and the cosmos (Remembering Héctor Rubinstein)* (pp. 13–34). SISSA.

Ortiz, E. L. (2011b). Julio Rey Pastor: su posición en la escuela matemática argentina. *Revista de la Unión Matemática Argentina, 52*(1), 149–194. https://inmabb.criba.edu.ar/revuma/pdf/v52n1/v52n1a16.pdf

Ortiz, E. L. (2015). Maniobras Científicas Ciencia, Investigación y Defensa. In G. A. Visca et al. (Eds.), *La Ingeniería Militar y su contribución al desarrollo Nacional: Perspectiva Histórica* (pp. 17–54). Buenos Aires: Dunken.

Ortiz, E. L., & Pyenson, L. (1984). José Babini, mathematician and historian of science. *Journal of the Spanish Society for the History of Science, 7*(13), 77–98.

Piroski, I. (1986). *1957–1962, Progreso y Destrucción del Instituto Nacional de Microbiología*. Buenos Aires: Eudeba.

Ramacciotti, K. I., & Cabrera Fischer, E. (2010). Un subsidio científico trunco: Mario Bunge y la Asociación Argentina para el Progreso de las Ciencias. *Res Gesta*, (40). http://bibliotecadigital.uca.edu.ar/repositorio/revistas/subsidio-cientifico-trunco-mario-bunge.pdf

Rey Pastor, J. (1988). La Filosofía Ficcionista. In E. L. Ortiz (Ed.), *The Works of Julio Rey Pastor*. London: The Humboldt Society and The International Rey Pastor Centenary Committee. Volumes I–VIII. In Vol. IV, MF 1944 3:1–2.

Rozzi, R., & Poole, A. (2009). The work of Argentinean physicist and philosopher Mario Bunge as an exemplary life for the fruitful integration of philosophy and the sciences. Interview conducted at the University of Montreal on May 6, 2009, revised by Mario Bunge. http://www.cep.unt.edu/papers/bunge-eng.pdf

Tarcus, H. (2007). *Catálogo de Revistas Culturales Argentinas, 1890–2007*. Buenos Aires: CeDInCI. http://www.cedinci.org/PDF/Publicaciones/Catalogos/CCA.pdf

University of Buenos Aires. (1966). List of resignations. http://docplayer.es/75174116-Vlsto-que-el-29-de-julio-de-2016-se-conmemora-el-50-aniversario-de-la-noche-de-10s-bastones-largos-y.html

Valentinuzzi, M. (1946). Crónica: Primer Coloquio de Historia y Filosofía de la Ciencia. *Revista de la Unión Matemática Argentina, 5*, 211–212.

Westerkamp, J. F. (1975). *Evolución de las ciencias en la República Argentina, II: Física*. Buenos Aires: Sociedad Científica Argentina.

Chapter 4
Mario Bunge as a Public Intellectual

Heinz W. Droste

Abstract Mario Bunge is an important philosopher of science. But he does not limit himself to using his "truth-technology" in his particular philosophical discipline. For decades, he has also endeavored to achieve an independent profile as a public intellectual on the basis of his wide-ranging competence. To this end, he authoritatively criticizes authors who market themselves to the public as anti-scientists or as pseudo-scientists. On his home continent of Latin America, Mario Bunge is regarded as a role model because he has achieved international recognition like no other South American philosopher before him.

4.1 Public Intellectuals (PIs)

Is Mario Bunge a "public intellectual"-a "PI"? In the USA, for example, the importance of public intellectuals and their significance has been actively discussed by academics for decades (Etzioni and Bowditch 2006; Posner 2003). These authors share a set of accepted opinions concerning the key attributes of PIs (Etzioni 2006). Accordingly they are characterized by their interest in a wide range of topics, by their tendency to be generalists rather than specialists, and by their tendency not to keep their views to themselves. Especially predestined to play the role of a PI would be people who are "well-travelled and broadly educated men [sic] of letters who (can) speak on a myriad of topics" (Brouwer and Squires 2003). By doing so, they should have the power to raise awareness in key areas of society and present solutions for the endangered welfare of humanity.

Bunge's memoirs (Bunge 2016) show impressively how he meets the agreed criteria of competence, to be considered a PI. Manifestly, he is "well-travelled", as the extensive vacation reports prove. And as a guest professor, he has taught on almost every continent for decades. His philosophical work is unparalleled in terms of the scope and breadth of the questions and problems dealt with, and is exemplary

H. W. Droste (✉)
e-mail: hd@droste-effect.com

© Springer Nature Switzerland AG 2019
M. R. Matthews (ed.), *Mario Bunge: A Centenary Festschrift*,
https://doi.org/10.1007/978-3-030-16673-1_4

in the clarity of the language used, and the quality of argumentation. His detailed knowledge spans the current state of research of a whole spectrum of individual sciences such as physics, medicine, brain physiology, psychology, sociology, political science and economics. He is possibly the only philosophical author who has so far successfully managed to process the intellectual leap of modern physics—especially that of quantum mechanics and the two relativity theories—and to elaborate it as the principle of a comprehensive modern enlightened philosophy. On top of that, as a thinker and author, he uses his in-depth knowledge over and over again to outline solutions to the great current problems of humanity.

For example, in his short article "The Kaya Identity", Bunge demonstrates the "basics" of the prevention of further warming up of our atmosphere and the containment of climate change on just three printed pages in a language understandable even to laymen (Bunge 2012b). On this occasion, he also reveals how the so-called "International Climate Council"—IPCC (Intergovernmental Panel on Climate Change)—unnecessarily complicates the public discussion of the causes of global warming through the careless use of mathematical formalizations.

In another just as easy to understand passage (Bunge 2012c) he deals with the financial crisis in the years 2007 and 2008 and the associated years of decline in international economic activity. He provides convincing arguments that these problems are due to the Western world being theoretically dominated by the erroneous political economy of the time, which could neither predict nor manage economic crises.

Bunge does not conceal his political opinions and, from the public's perspective, draws interesting and illuminating conclusions based on his extensive historical knowledge. When asked about the supposed cultural and religious conflicts between the Arab and the Western world during an interview (Droste 2014a, 2015, p. 177), he argues that this is about material and political interests rather than ideological differences. In one breath, he summarizes facts from the European Middle Ages, facts from the history of Cardinal Richelieu's France, as well as from the history of Iran in the twentieth century.

Mario Bunge clearly has the competence expected of a public intellectual. The next step is to answer the question as to whether his extensive work has enabled him to actually distinguish himself as a PI.

4.2 Public Philosophies Become Public Legends

I confronted Mario Bunge directly with the question. It turned out that in the past he had explicitly given little thought to the idea of being a public intellectual, and that he considered the concept behind it to be ambivalent. The assumption that the concept of "public intellectual" might be ambivalent seems important, which raises the question whether publicity and philosophy are compatible. The broad public associates well-known philosophy with the idea of a consumable cultural product which has the function of providing a worldview and an original approach to understanding human existence. For a publicly viewed philosophy to be understood, it is important that the concept behind it fits in with the widespread thought traditions of the respective population group and makes sense in this context.

The "sense" may be that the considered philosophy provides answers to important questions of life and provides explanations for existential questions—such as the origin of human life, ethical reasons for controversial behaviour—and helps to shape alternative concepts for the human existence. Looking at the function of a philosophy as a resource of sense, we make a discovery—it is not its faithful or literal version that plays a role in the discussion of the audience. Instead, the names of philosophies and names of their authors, as well as selected terms from the respective works, are picked out and are variously associated with ideas and concepts that are not derived from the originals and do not suit them properly. Public "philosophical products" are educational contents that serve individuals. They use them as argumentative templates, as carriers of their thoughts, or as abstract expressive instruments for their own emotions or as argumentative resources they borrow as a substitute for their own train of thought.

An illustrative and well-documented example is the way in which Immanuel Kant's work—in particular, his *Critique of Pure Reason* (Kant 1781/1787; 1976)— was transformed by its public discussion. To this day simple messages and simple concepts are derived and are publicly disseminated from the complicated "critique of reason". For example, getAbstract.com is an online service that promotes the content of textbooks by publishing simply structured abstracts. The publishers of getAbstract have tried to propagate the content of Kant's work—and presented a summary of *The Critique of Pure Reason* that works with a selection of widely used interpretations of Kant's argumentation (getAbstract 2004). In this abstract, its author claims, "Immanuel Kant has triggered a revolution by means of *The Critique of Pure Reason*. The Königsbergian philosopher investigates the foundations of our ability to understand and concludes that it is limited" (translated by HWD).

Since Kant lived in Königsberg most of his life, it is reasonable to call him a "Königsbergian" philosopher. On the whole, however, the interpretation of getAbstract—Kant would have been a kind of thought leader of a constructivist scepticism as well as a doubter of knowledge—has little to do with Kant's intentions. Instead, on the basis of his three critiques, he wanted to define and to secure as inviolable instances of reason a set of concepts—"Ideen"—which, from his point of view, are indispensable to human ethics, such as "human freedom", "the existence of God", etc.[1] In the *Critique of Reason* Kant focused on the unambiguous assurance of insights of reason, not on the foundation of doubts about our ability to understand.

Another widespread "folkloric" interpretation of the *Critique of Pure Reason* points in a completely different direction of thinking. It is popular to assert that Kant's intention in the critique of reason was to justify scientific research and to

---

[1]Accordingly, Kant describes the intention of the *Critique of Pure Reason* in a letter to Christian Garve on September 21, 1798 (Kant 1972, pp. 778).

validate empirical laws by the famous concept of "synthetic propositions a priori" ("synthetische Sätze a priori"). Kant scholars continuously have to point out that this interpretation of the *Critique* points in a completely wrong direction (Hoppe 1969, p. 7; Droste 1985 p. 14). Instead, Kant designed a model of the human "cognitive apparatus" in order to discuss the most general "conditions of possible experiences". The interpretation of the *Critique of Pure Reason* as a means to justify empirical laws is definitely wrong.

It seems to be characteristic that individuals seek ways of living better when they turn to "public philosophies".

A recently published document that shows how this seeking works is the memoirs of former Chancellor and Head of Government of the Federal Republic of Germany, Helmut Schmidt, published shortly before his death in 2015. As a young man during the Third Reich, the later Chancellor of the Federal Republic was obliged early to serve in the army and even had been ordered to the Eastern Front for some time—an experience he had suffered from all his life. In 1980, he met Sir Karl R. Popper in person, whose *The Open Society and its Enemies* he had read before, and with whom he remained friends until Popper's death in 1994. In his memoirs, Schmidt describes that after the war he turned to Kant's philosophy in order to realign himself in situations of uncertainty during his career as a politician: "Kant became my reliable compass in the moral chaos left by the Nazis" (Schmidt 2015—translated by HWD). Kant's philosophy became something like a life-coaching system for him. Schmidt described it as a habit of "anchoring" some Kantian statements in his consciousness. He was guided by the sentence, "Moral action must be based on reason," although he did not claim to be able to derive this quite general statement literally from Kant's work. Schmidt appreciated this idea because he felt that it helped him to take a break during confusing decision-making situations and after that to find better, deliberate solutions.

These examples show how the publication of a philosophical work and its transformation into a public philosophy can have a number of different consequences. Instead of the contents of the "real" critique of reason, a "public legend" is spread, having little in common with what can be found in the original book. The fate of becoming idealized in this way is likely to affect all philosophical works that attract public attention.

To philosophize does not work in public. The faithful dissemination of essentials and content details is impossible in public because of the high level of expertise required. What follows in the best case from a publicly made philosophical work is a certain positive impression, which affects individuals and which might motivate them to study the original texts of a philosophical work in the future. Detailed reading sessions might reveal what the philosophical author actually wrote.

The respective student probably will experience surprises he was unable to prepare for on the basis of the publicly propagated version of the respective philosophy. For example, if he should read the *Physische Geographie* (Kant 1802), where Kant defines a racial hierarchy based on human skin colour: "Humanity achieves its greatest perfection through the white race. The yellow Indians have a comparatively low talent. The Negroes are positioned far lower, and the lowest is a part of the native Americans."[2] (translated by HWD) According to Nina Jablonski, anthropologist and paleobiologist at Pennsylvania State University, Kant was the first author to describe and define the geographical groupings of humans as "races" (Jablonski 2015, pp. 80–1).

As indicated in the cited text he distinguished four races. He characterized them by different skin colours, textures of their hair, skull shapes and further anatomical features. As the most important feature to distinguish them, he claimed their ability to act morally and use their reason. In his *Critique of Pure Reason*, Kant defined a moral standard based on ideas of reason. In his view, these standards are fully applicable only to the "European race", while the rest of humanity, in his opinion, seems to suffer from a deficiency in the use of reason. Kant's work, in this sense, serves as an argumentation guide for racial ideologies—but is hardly to be considered as the infallible "moral compass" which Helmut Schmidt believed he had discovered in it.

In an autobiographical note, Mario Bunge describes as an anecdote how he personally experienced a "surprise effect" that proved to be important for his future. It was Hegel's philosophy, whose "nimbus" first attracted him, and whose reading in French translation led to a strong experience of frustration: "I had swallowed far more than I could digest" (Bunge 2010a, p. 526). This became an experience that led Bunge to be sceptical of "public philosophies" and to see if their public impact might be based solely on the philosophical charlatanry of the pertinent "public intellectual." For "exemplary" cases of such charlatanry, he points to the philosophical concepts behind Dawkins' genetic determinism and Chomsky's nativism. This is indicated by the frequency with which he mentions these authors and reiterates his critical arguments.

Instead of striving for opportunities to have a public impact, Mario Bunge bases his philosophical concept on the knowledge ideal of empirical sciences. He points out that there are proven evaluation criteria that can be summarized in a compact way, such as clarity of the concepts used; content consistency; being proven in the light of relevant empirical facts; coinciding with the mass of already confirmed knowledge; completeness of the scope of considered questions; having the potential to find solutions to these questions as well as the ability to guide future research projects.

Therefore, he criticizes the usual philosophical approaches whose authors do not seek, in his view, to develop general assessment criteria and to comply with them in their publications. From his point of view, in public discourse, philosophies are positively judged or rejected without being based on clear or objective criteria. Instead, intuition, considerations of utility, or feelings are the decisive factors. According to Bunge, this is an unsatisfactory situation. That is why he formulates a guiding principle by means of which we should evaluate philosophical doctrines:

---

[2]"Die Menschheit ist in ihrer größten Vollkommenheit in der Race der Weißen. Die gelben Indianer haben schon ein geringes Talent. Die Neger sind weit tiefer und am tiefsten steht ein Theil der amerikanischen Völkerschaften." (Kant 1802, first section, §4)

"A philosophy is worth what it helps learn, act, conserve our common heritage, and get along with fellow humans" (Bunge 2012a, p. xiv).

To meet this criterion, Bunge has designed his own philosophy as an integrated "truth technology" that includes a complete set of sub-technologies such as ontology, epistemology, methodology, praxeology, value theory and ethics, and political philosophy.

4.3   Public Intellectuals Raising Cain

The interim result of the present analysis is that philosophers and their works can become "public". Philosophies seem to have to fulfil certain conditions in order to function effectively in public. Successful public philosophies should be flexible in their interpretability and usability for a variety of meaningful activities. The lack of expertise of the audience has to be considered. Bunge's concept of truth technology does not seem to be tuned to these conditions.

This raises the question of how certain philosophies succeed in achieving the interpretability and usability required for a broad public impact to become known as a cultural asset and to make their author a public intellectual. Why do most philosophical works not succeed?

An important explanation was found interviewing the German-Canadian political scientist and sociologist Andreas Pickel, who in 2004 had organized a symposium to discuss the social philosophy of Mario Bunge (Pickel 2004). Over the years, Pickel had a regular exchange of ideas with Bunge on questions of the philosophy of science. On the occasion of the 22nd European Meeting on Cybernetics and Systems Research Bunge was awarded the *Bertalanffy Award in Complexity Thinking* in Vienna on April 22, 2014 (Droste 2014b). At this meeting, I asked Pickel how he evaluates the public reception of Bunge's philosophy.

Pickel's assessment was that Bunge's philosophy has not yet achieved the publicity it deserves—despite its scope and relevance to a variety of sciences. As a cause of this failure, he assumes that the Argentine author has not taken care to found his own "school of philosophy". Pickel had observed that this was achieved, for example, in the case of the followers of Karl R. Popper, who would have established a viable circle of "Popperians".

In fact, Popper seems to have received active support from followers during public discussions. One example is the so-called "positivism dispute" ("Positivismusstreit") which occurred in the German-speaking world in the 1960s. This was a dispute about the methodology in the social sciences. The discussion started with a contribution by Popper at a conference of the German Sociological Association (Deutsche Gesellschaft für Soziologie) in October 1961 in Tübingen. As a result of the debate, which was subsequently publicly held in specialized print media and national newspapers, Jürgen Habermas, in particular, appeared as the representative of the Frankfurt School (Frankfurter Schule) in order to attack the position of Popper's critical rationalism. Popper, however, showed no interest in publicly responding to Habermas' contributions.

Popper was asked why he withdrew from the discussion (Grossner 1971). His answer was that he would rather focus on making his own ideas as simple as possible, rather than responding to Habermas' "cruel game" of proclaiming simple and trivial things in presumptuous language (" … Einfaches kompliziert und Triviales schwierig auszudrücken" Grossner 1971, p. 289). For public debate, another representative of critical rationalism was chosen to appear as the defender of Popper's point of view, namely the sociologist and philosopher, Hans Albert. In the following decades, Albert continued the discussion around the positivist dispute and in detail reviewed Habermas' hermeneutic approach as well as his connection to the Christian-religious dogmatics (Albert 1994, pp. 230–62, 2008, p. 92).

By comparison, Bunge doesn't command an intervention force for critical cases of discussion like the Popperians do. In fact, it is doubtful that he would accept any form of proxy. Instead, he prefers to fight his philosophical differences, personally and ruthlessly. "He takes no prisoners," as Andreas Pickel has said.

Bunge takes little account of whether his public discussions are considered sympathetic or are supporting a kind of amiable public image, as exemplified in 1978, when he participated in the 16th World Philosophy Congress from August 27 to September 2 in Düsseldorf. In the middle of the lecture by the Australian brain researcher, and 1963 Nobel Laureate, Sir John Carew Eccles, Bunge stood up to criticize Eccles' concept of the interaction between material brain substance and immaterial consciousness (Bunge 2012a, pp. 22; Droste 2014a). Newspapers reported daily during the Congress. The articles that appeared the following day after Bunge's intervention showed the irritation that was felt by the participants of the event at the Messe-Kongress-Zentrum Düsseldorf.[3] Here the readers could relive the action the next day.

The story went like this: The lecture of a Nobel Laureate of the year 1963, who had written the best-selling book *The Self and its Brain* a year ago, together with the internationally known philosopher Sir Karl R. Popper, was interrupted by a man in an unpleasantly shrill voice, who—by this—disrupted the orderly course of the day.

Regrettably, a number of facts were not explained to the readers of the *Düsseldorfer Zeitung*, such as the background of the "troublemaker", the scientifically proven arguments he put forward at that moment, and the weaknesses of Eccles' body-mind dualism. The newspaper editors focused on the facts of a scandalous disturbance of a renowned, prominent and deserving scientist.

This "Düsseldorf Eccles Scandal" shows how little the scientific and philosophical justification of ideas alone helps to reach public awareness. And the related fact that intellectual competence alone hardly helps intellectuals to achieve the image of a "public philosopher". Public awareness is, to a certain extent, dependent on the deliberate shaping of public opinion, as practised today not only by journalists in mass media but also in the field of so-called public relations.

---

[3]The articles have been documented on microfiche for the city archive of the city of Düsseldorf - Stadtarchiv Düsseldorf, Worringer Strasse 140, 40200 Düsseldorf.

It is an important experience of professional publicity experts in media outlets and in public relations agencies that the use of scandals, in particular, does not hinder the achievement of publicity of people and topics, but instead provides particularly effective support. However, Bunge has not taken advantage of the chance to edit and use "his" scandal in Düsseldorf.

How this is practised successfully has been impressively demonstrated by the already mentioned representative of the Frankfurt School, Jürgen Habermas. Habermas has managed to position himself as an intellectual in the international public sphere most successfully. For example, in 2005, he managed to achieve a seventh place in the 100-person ranking of famous public intellectuals, right behind economist Paul Krugman. This ranking was published in *Foreign Policy* and *Prospect Magazine*. By the standards of this list, Habermas achieves the status of a "top philosopher".

Interestingly enough, Noam Chomsky won first place in the total ranking. Mario Bunge often criticizes Chomsky keenly for the philosophical implications of his linguistic theory. Bunge did not get on this list of "famous" public intellectuals.

How was it possible for Jürgen Habermas to reach this status? According to the standards that Bunge applies to philosophies, Habermas has hardly produced any results:

> The only German philosopher who is well-known outside Germany is Jürgen Habermas, who in my opinion is superficial and long-winded. He has managed to skirt all the important philosophical issues generated by contemporary science, in particular, atomic physics, evolutionary biology, biological psychology, and socioeconomics. His attempt to fuse Hegel, Marx and Freud has not resulted in a coherent system and is not a research project. And his conflation of science, technology and ideology betrays his ignorance of all three. (Droste 2014a, 2015, p. 182)

What is not recognizable at first glance is how Jürgen Habermas skillfully built his reputation career on the basis of public scandals. After completing his studies, he began his career as an editor of the *Frankfurter Allgemeine Zeitung* (FAZ), a daily newspaper widely distributed in Germany. Therefore, he had the opportunity to publish a full-page article on July 25, 1953, in which he scandalized Martin Heidegger and his disclaiming of the Nazi crimes. Habermas did that, although his thinking is strongly influenced by Heidegger's philosophy—during his studies, he was trained by disciples of Heidegger. And today he has changed his strategy. He praises Heidegger publicly as a "great thinker" and even tries to protect his "prophet" from "unjustified discredit" (Habermas 1989). But with the help of his article decades ago, Habermas succeeded in scandalizing Heidegger publicly, a manoeuvre that proved to be an important step to get public attention for himself. This was not an easy thing to do. Heidegger was the most popular philosopher in the German post-war era, even though he had been involved in Nazi politics. The German-American philosopher Walter Arnold Kaufmann witnessed at this time that Heidegger's lectures were so well attended that they had to be transmitted by loudspeaker systems in several lecture halls to satisfy the public interest in his person and his work (Kaufmann 1957).

4.4 Seeking for Applause in the Scientific Community

As pointed out, Andreas Pickel suspects that Mario Bunge did not achieve high public recognition because he did not build something like a philosopher's school. But the reasons might lay deeper. Mario Bunge's concept of public recognition seems to follow a clearly different concept of publicity, in comparison to those journalistic interventions being used to promote the "public philosophers" Habermas and Kant. In this context, the example of the "Düsseldorf Eccles scandal" shows that Bunge obviously does not consider the possible PR effects of his interventions in public discourses.

Bunge does not use "communications campaigns" to make his philosophy known. Significantly, as a sociology of knowledge, he favours the approach of Robert Merton, to which he regularly refers (Bunge 1983, p. 204). As Merton points out in his well-known article on the so-called "Matthew Effect in Science" the successful publicity of an intellectual in the field of science is closely linked to the achievements he has made in his field of research in the past. Merton reveals a paradox of attention here: Scientists who are already well-known because of past achievements are often credited with the merits of others without having achieved the acclaimed results by themselves. Merton points to the Matthew's Gospel: "For unto everyone that hath shall be given, and shall have abundance; but from him that hath not shall be taken away even that which he hath" (Merton 1988, pp. 608–9).

Merton further explains that scientists are looking for a special form of fame that clearly differs from a public celebrity due to paradoxical attribution. He thinks they cleanly distinguish the "gold of scientific fame" from the "brass of popular celebrity". In his article, he mentions economist Paul Samuelson, who, speaking to colleagues, emphasized that scientists do not really desire to be in the "limelight" of the public (Merton 1988, p. 623). Instead, their worthwhile award would be the applause of scientific colleagues.

Bunge's status as a "famous" intellectual can be assessed by examining whether he receives recognition in the sciences. It would be useful, for example, to analyze the so-called "Hall of Fame" of the American Association for the Advancement of Science (AAAS). Here it can be shown that the Argentine philosopher was able to achieve the applause of professional colleagues, the form of fame which Samuelson idealized. This ranking evaluates authors by the frequency with which their works are mentioned in scholarly publications. Bunge's books are quite successful here. His publications are positioned next to the books of prominent scientists such as Richard P. Feynman, physicist and one of the creators of quantum electrodynamics, and Viktor Emil Frankl, psychiatrist and founder of logotherapy.

In contrast, Jürgen Habermas, who, as seen, reaches a top position in a list of international "public intellectuals" assembled by mass media, plays no role in the AAAS Hall of Fame. Its rankings are determined on the basis of scientific qualification and recognition of expertise in scientific communities.

Habermas' books have little success in terms of these criteria. Scholars like the Austrian sociologist Max Haller criticize Habermas because his thinking does not fit into the concept of a "social science as a science of reality" (Haller 2006). From the point of view of Haller—international university lecturer and former president of the Austrian Society for Sociology—Habermas turns away from social reality in order to argue morally instead. Furthermore, Haller criticizes Habermas for failing to support his concepts through scientific explanations of social action or social processes (Haller 2006, pp. 50–51). The Swiss sociologist Peter Max Atteslander calls Habermas' strategy of argumentation "theorizing without research (*empirieloser Theorismus*)" (Atteslander 2006, p. 312). Consequently, Haller favours Bunge's work, when he analyzes the background of sociological terms such as "social system", or "system structure," as in Bunge's *Finding Philosophy in Social Science* (Bunge 1996).

4.5 The Apparent Hopelessness to Answer Big Questions Publicly

Mario Bunge's philosophy has influenced scientists in various fields and has made his works known among scientists. The crucial factor is that he bases his thinking on standards that are common in the natural sciences. As Bunge notes at the end of his memoirs, he regards his books from the standpoint of a scientist as a preliminary work, as part of an ongoing research project—in sharp contrast to, say, the scriptures of a religious sect and its disciples (Bunge 2016, p. 408).

Accordingly, in the past, he had proposed a "B test" with the following "exam questions" to test the value of philosophies. According to it, a good philosophy is:

(a) clear and internally consistent;
(b) compatible with the bulk of the knowledge of its time;
(c) helpful in identifying new interesting philosophical problems;
(d) instrumental in evaluating philosophical ideas;
(e) helpful in clarifying and systematizing philosophical concepts;
(f) instrumental in advancing research both in and out of philosophy;
(g) capable of participating competently, and sometimes constructively, in some of the scientific, moral, or political controversies of its day;
(h) helpful in identifying bunk;
(i) characterized by a low word-to-thought ratio. (Bunge 2003, p. 29)

To comply with this checklist, Bunge designed his work as a comprehensive and integrated truth-technology. By this, he deliberately sets himself apart from most philosophers and their works and gains his individual position as a public intellectual. To achieve this, he emphasizes the perspective of the complete context of problems. His basic assumption is that the great questions of philosophy are closely connected (Bunge 2010b, pp. vii–xii). In his view, this is the reason why these are big problems and not small ones. These do not occur separately or in isolation, but form an intricate series of questions and must be worked on in this overall context in order to be mastered.

According to Bunge, this coping with the entirety of philosophical problems requires clear ideas based on an integrated concept and a systematic approach of processing. Anyone who studies the works of popular philosophical authors from his point of view will quickly realize that pursuing an integrated systematics hardly matters to most of them. These authors obviously favor the mere accumulation of (more or less) profound aphorisms. From Mario Bunge's point of view, the scattered, supposedly intelligent philosophical thoughts presented in this way do not reach the goal of comprehensive philosophical problem clarification. He assumes that these fragmented thoughts are indicative of the fact that writers improvise, rather than develop valuable, coherent, overall concepts.

For Bunge, a valuable philosophy is based on a specific structure. At various points in his work, he explains what he considers to be the decisive pattern (Bunge 2006, pp. 250–82):

Ontology—theory of being, the doctrine of the basic structure of reality.
Epistemology—the doctrine of finding knowledge.
Semantics—the theory of the relation of language and symbols to reality, prerequisites of objective knowledge.
Methodology—theory of methods for the determination of objective knowledge.
Axiology—theory of values and objective criteria for assessing valuations.
Ethics—theory of morality.
Praxeology—theory of effective and moral action.

This peculiarity of his philosophy—its coherence in terms of content and its connection to complex problems—has a decisive effect on the chance of it being known to a broad public. The complexity of Bunge's books overstrains non-philosophical readers and their ambition to think abstractly, making popularization enormously difficult.

In my role as the author of a book, I personally experienced the problem of propagating Bunge's philosophy. This happened about a year before the present analysis was started. At the time I had published a German-language introduction to Bunge's philosophy (Droste 2015). Organized by the publishing house, I got the opportunity to present the book about Bunge's philosophy to a lay audience in spring and summer of 2016 on a lecture tour through southern Germany. The main experience of these lectures was mixed. On the one hand, it turned out that Mario Bunge's philosophy can fascinate an audience. This works even if the listeners have no particular experience in philosophy. However, the presentation and explanation of the overall concept consumed more time than usual for a lecture. I was criticized several times by organizers because the lecture exceeded the allotted time due to inquiries and detailed discussions that followed.

However, the depth and extent of Bunge's thinking making it difficult for laymen to estimate the special value of Bunge's philosophy may also be a major obstacle for the understanding of other target groups. Even scientists who know and appreciate his scientific-theoretical writings in their respective fields might have problems in appreciating the overall concept of his philosophy. For example, a basic problem for "social scientists" on the one hand and "natural scientists", on the other hand, is to understand and to accept Mario Bunge's typical transgression of these two technical "dimensions"—social versus natural.

On this topic, I had talks with two authors who are familiar with Bunge's philosophy and are personally connected with him. As a representative of the social sciences, I selected the aforementioned Andreas Pickel—Professor of Sociology and Political Science at Trent University, Peterborough, Ontario, Canada. On the occasion of a visit to Berlin by the professor, I spoke with him on 10 and 11 September 2015 about the possibility of teaching the theory of science concepts that Bunge recommends for the field of social sciences. For example, Bunge uses the notion of emergence when discussing the concept of social systems. During the discussion of Bunge's "systemic-emergent materialism" with Pickel it became obvious how difficult it is for social scientists to understand how Bunge justifies the differentiation between an epistemological interpretation and an ontological interpretation of emergence (Droste 2011, pp. 157–60). There seems to be a fundamental need for clarification for social scientists on this topic.

Another philosophical insight that is difficult to understand from the perspective of a social scientist is Bunge's concept of subject-independence and materiality of cultural artefacts. It is difficult to explain without causing serious confusion to social scientists, who have often been taught Popper's "three-world doctrine" as their basic philosophical perspective, that within a materialistic ontology there is no room for a "world of the real and objective content of thought" (Droste 2011, pp. 194–7). Bunge speaks in this context of the "myth of subject-independent knowledge" and recommends the use of a moderate fictionalism. Social scientists would have to abandon the usual deep-seated idealistic ideas of the dynamics and effectiveness of cultural phenomena, such as national language, national identities, etc., to understand Bunge's explanation.

Natural scientists—who usually think of themselves as being on a different, a more "scientific", side of understanding the world than social scientists—also encounter problems with Bunge's overall philosophical perspective. This was the result of an interview with Martin Mahner, a doctor in zoology, co-author with Bunge of two books and different articles, whom I interviewed on 17 February 2015 in his office in Roßdorf near Darmstadt.

Bunge had tried to prepare me for the interview with his former postdoctoral candidate. In an email, he wrote, "... the difference between me and Martin Mahner is that he does not believe in the social sciences—he is not only a biologist but also a biologicalist. That's why he calls himself a naturalist"[4] (translated by HWD). The subsequent interview corroborated Bunge's premonition. In the course of the interview, Mahner did not show any interest in a conversation about concepts of

---

[4]"... der Unterschied zwischen mir und Martin Mahner besteht darin, dass er nicht an die Sozialwissenschaften glaubt—er ist nicht nur Biologe, sondern auch ein Biologist. Deshalb bezeichnet er sich als Naturalist."

philosophy of science that mediate between natural sciences and social sciences. Instead, looking back on his postdoctoral stay at McGill University, Mahner criticized Bunge for concentrating on the social sciences at the beginning of the 1990s, rather than focusing on a natural science research field such as biology.

4.6 Intellectuals Playing Roles in Public Dramas

Those who want to achieve the status of a public intellectual are obviously taking risks for themselves and their work. But there are other, related threads. Intellectuals who have ambitions to play a public role have to engage in a social field in which they participate in processes associated with possible risks to an entire society. This was pointed out by US sociologist Amitai Etzioni, who gained extensive experience as "sometime-public intellectual" (Etzioni 2003). He appeared in public as a consultant to US President Jimmy Carter and is regarded as the "spiritual inspirer" of British Prime Minister Tony Blair through his political concept of "communitarianism".

Etzioni reports that his colleagues often ask him for advice because of his successful career as a public intellectual. These intellectuals aspire to become successful PIs too, and therefore ask for insight into Etzioni's know-how. They want specific tips because they dream of being invited by a prestigious national newspaper to publish a major commentary that awakens the powerful people of the nation. Etzioni concedes that he himself was driven by similar visions of public glory. His vision was to write something similar to Émile Zola's famous open letter to the President of the French Republic on January 13, 1898—a full-page article on the title page of the newspaper *L'Aurore*, which initiated the clarification of the famous Dreyfus affair.

Describing this process, Etzioni warns that PIs run the risk of violating their professional standards and thereby forfeiting their credibility as scientists due to these concessions: PIs have to leave aside much of what is important in their everyday scientific work—for example, differentiated reasoning or convincing substantiation of claims through facts—because otherwise, their public contributions would not be "generally understandable enough" and would fit poorly into mass media contents. PIs must speak with an undifferentiated unilateral voice, must either be black or white, argue either as "rightists" or "leftists", fully endorse or fully reject the needs of the market and profit interests, etc., because differentiated interim positions do not fit into the program. Because they are forced to abandon their professional standards when they appear in public, PIs do not receive applause from their own discipline. Instead, they are often criticized from this direction (Posner 2003, pp. 399).

4.7   Public Intellectuals: An Easy Prey in Public Discussions

However, the risks that intellectuals might incur in this regard go much further. Problems are not based solely on the fact that intellectuals jeopardize their own professional fame as Etzioni describes it. Unknowingly, they might contribute to restricting important political decision-making processes within a society.

In their widely acclaimed collaborative book, *Merchants of Doubt*, science historians Naomi Oreskes and Erik Conway describe how an influential circle of intellectuals was "hired" for public performances to prevent important political decisions (Oreskes and Conway 2010): Scientists who had earned a reputation in their earlier careers made use of their prominence to claim publicly that the responsible sciences could not beyond doubt prove facts relevant to the issues of smoking, or warming of the earth's atmosphere.

Oreskes and Conway observe that intellectuals from the empirical sciences have great problems coping with the typical one-sidedness and missing objectiveness in the mass media. Through their commitment to expertise and objectivity, they find themselves in an unpleasant position when it comes to publicly appearing as PIs and contradicting false claims. They feel intimidated and are afraid of being accused of having lost their professional objectivity. Therefore, they tend to stay out of the public discourse when, for example, climate change and its harmful effects are publicly discussed. Privately, they hope for the self-actualizing power of truth.

Scientific researchers see it as their task to find out what is true. But if someone spreads nonsense somewhere in public, they think someone else should take on the role of challenging it, which often does not happen. Due to this attitude, scientists are an easy prey to political and economic managers who deny the facts of the so-called anthropogenic global warming of our atmosphere. Despite fundamental and well-secured findings in climate research, these "climate skeptics" succeed in casting doubt on the human origin of the rapidly advancing climate change without any great contradiction. By persuading some public intellectuals to question well-documented facts, they have achieved a stalemate in the public discussion.

4.8   In the Footsteps of the Radical French Enlightenment: Fighting Against Anti-scientists and Pseudoscientists

Bunge's philosophy is based on the standards of successful empirical sciences. Yet, as an intellectual, he has by no means become an "easy prey" to the described scientific fact-manipulators. Although this is the typical fate of empirical scientists who tend to be unable to defend themselves against the actions of professional hackers of public opinion-formation processes. Instead, Bunge plays the role of a consistent critic of public manipulation. His criticism of distorted interpretations of science is a constantly emerging motif in his books and articles. In a great thematic diversity, he prefers to pick up the favourite thinkers and authors built up by the mass media in order to dismantle them on the basis of his philosophical concepts.

His critique follows a basic strategy that focuses on the discussion of the contributions of intellectuals of two basic orientations. Similar to C. P. Snow in his famous lecture in the spring of 1959 (Snow 1959/1998), Bunge refers to representatives of two antagonistic intellectual cultures and their contributions. On the one hand, the "literary" intellectuals; on the other, intellectuals who appear as beneficiaries of the results of successful modern sciences.

From Bunge's point of view, representatives of both "cultures" are responsible for arguments popularized by the mass media, which misinterpret the results of empirical sciences, jeopardizing the possibility of using important insights to improve the state of humanity (Bunge 2017, pp. 137–59). Bunge invests a great deal of space and critical reasoning in his work in order to distance himself from these two intellectual groups that are generating great public interest. The anti-scientists and anti-realists—intellectuals who want to take modernity and the associated scientific revolution to absurdity. And, secondly, the pseudo-scientific provocateurs and cynics—intellectuals who overestimate the value of certain theories and tend to draw unreflected and unsustainable conclusions from reductionist approaches.

Some examples of "anti-scientists and anti-realists" regularly criticized in his publications are:

- Edmund Husserl and his phenomenology.
- Friedrich Nietzsche and his anti-ethics.
- Thomas S. Kuhn and his concept of scientific revolutions.
- Paul Feyerabend and his anarchist theory of science.
- Jürgen Habermas and his concept of science and technology.
- Martin Heidegger and his fundamental ontology.
- Paul Michel Foucault and his post-structuralism.

Bunge is not afraid to express himself bluntly. For example, in his opinion Heidegger is "one of the most harmful charlatans of his time" (Bunge 2016, p. 218) to be called a *Kulturverbrecher*—a "cultural delinquent" (Bunge 2016, p. 209). Among the popular pseudo-scientific provocations he criticizes are:

- its-from-bits physics with its digital ontology,
- the many-world cosmology,
- Richard Dawkins' genetic determinism, and his meme concept,
- Noam Chomsky's nativism,
- John Carew Eccles and his psychokinetic brain,
- Karl Raimund Popper, his falsificationism and his three-world doctrine,
- Ernst Mach and his phenomenalism.
- Sigmund Freud and his psychoanalysis.

4.9 Conclusion

Mario Bunge has achieved a lot in his career as a philosopher. In the field of scientific theory, he enjoys today the role of an important leader of thought. The German philosopher of science Bernulf Kanitscheider wrote[5]:

> Few extraordinary personalities have the chance to decisively shape the intellectual geography of a scientific epoch. Mario Augusto Bunge belongs to the small circle of important philosophers of science whose works have already become landmarks in the spiritual landscape of world philosophy. (Kanitscheider 1984, p. viii—translated by HWD)

On his home continent of Latin America, Mario Bunge is regarded as a role model because he has achieved international recognition like no other South American philosopher before him. His work is seen as proof that even thinkers who work outside the Anglo-American tradition and are exposed to the most difficult conditions can ascend to take the lead in international discussion. But Bunge does not use these successes to gain popularity among a broad audience or in the eyes of editors in mass media. Instead, he remains unwaveringly on the trail of the radical French Enlightenment. Its authors, too, based on the current state of science and technology of their time had developed a philosophical concept that provoked through profound social criticism.

References

Albert, H. (1994). *Kritik der reinen Hermeneutik: der Antirealismus und das Problem des Verstehens.* Tübingen: Mohr Verlag.
Albert, H. (2008). *Joseph Ratzingers Rettung des Christentums. Beschränkung des Vernunftge-brauchs im Dienste des Glaubens.* Aschaffenburg: Alibri Verlag.
Atteslander, P. M. (2006). *Methoden der empirischen Sozialforschung.* Berlin: Erich Schmidt Verlag.
Brouwer, D. C., & Squires, C. R. (2003). Public intellectuals, public life, and the university. *Argumentation and Advocacy, 39,* 201–213.
Bunge, M. (1983). *Treatise on basic philosophy, Volume 6. Epistemology & methodology II – Understanding the world.* Dordrecht: D. Reidel Publishing Company.
Bunge, M. (1996). *Finding philosophy in social science.* New Haven: Yale University Press.
Bunge, M. (2003). *Dictionary of philosophy.* Amherst: Prometheus Books.
Bunge, M. (2006). *Chasing reality. Strife over realism.* Toronto: University of Toronto Press.
Bunge, M. (2010a). From philosophy to physics, and back. In S. Nuccetelli, O. Schutte, & O. Bueno (Eds.), *A companion to Latin American philosophy* (pp. 525–538). Hoboken: Wiley-Blackwell.
Bunge, M. (2010b). *Matter and mind. A philosophical inquiry.* Dordrecht: Springer.

---

[5]"Wenigen außerordentlichen Persönlichkeiten ist es vergönnt, die intellektuelle Geographie einer wissenschaftlichen Epoche entscheidend mitzugestalten. Mario Augusto Bunge gehört zu dem kleinen Kreis bedeutender Wissenschaftsphilosophen, deren Werke bereits jetzt zu Marksteinen in der geistigen Landschaft der Weltphilosophie geworden sind. " (Kanitscheider1984, p. VIII)

Bunge, M. (2012a). *Evaluating philosophies, Boston studies in the philosophy of science* (Vol. 295). Dordrecht: Springer.

Bunge, M. (2012b). The Kaya identity. In M. Bunge (Ed.), *Evaluating philosophies, Boston studies in the philosophy of science* (Vol. 295, pp. 57–59). Dordrecht: Springer.

Bunge, M. (2012c). Can standard economic theory account for crises? In M. Bunge (Ed.), *Evaluating philosophies, Boston studies in the philosophy of science* (Vol. 295, pp. 77–82). Dordrecht: Springer.

Bunge, M. (2016). *Between two worlds: Memoirs of a philosopher-scientist*. Dordrecht: Springer.

Bunge, M. (2017). *Doing science in the light of philosophy*. Singapore: World Scientific Publishing.

Droste, H. W. (1985). *Die methodologischen Grundlagen der soziologischen Handelstheorie Talcott Parsons' – eine Analyse im Lichte der kritischen Philosophie Kants*. Doctoral thesis, Düsseldorf: Heinrich-Heine-Universität.

Droste, H. W. (2011). *Kommunikation – Planung und Gestaltung öffentlicher Meinung*. Neuss: Pedion-Verlag.

Droste, H. W. (2014a). *The Big Questions come in Bundles! Interviewing Mario Bunge*, https://a-g-i-l.de/mario-bunge-the-big-questions-come-in-bundles-not-one-at-time/

Droste, H. W. (2014b). *Roll over Bertalanffy, tell Parsons and Luhmann the News!, Reportage: System ist nicht gleich System*. https://a-g-i-l.de/roll-over-bertalanffy-tell-parsons-and-luhmann-the-news/

Droste, H. W. (2015). *Turn of the Tide – Gezeitenwechsel. Einführung in Mario Bunges exakte Philosophie*. Aschaffenburg: Alibri Verlag.

Etzioni, A. (2003). *My brother's keeper: A memoir and a message*. Lanham: Rowman & Littlefield Publishers.

Etzioni, A. (2006). Are public intellectuals an endangered species? In A. Etzioni & A. Bowditch (Eds.), *Public intellectuals, an endangered species?* (pp. 1–27). Lanham: Rowman & Littlefield Publishers.

Etzioni, A., & Bowditch, A. (Eds.). (2006). *Public intellectuals, an endangered species*. Lanham: Rowman & Littlefield Publishers.

getAbstract (2004). *Kritik der reinen Vernunft*. Luzern: getAbstract.com

Grossner, C. (1971). Karl R. Popper: Philosophische Selbstinterpretation und Polemik gegen die Dialektiker. In C. Grossner (Ed.), *Verfall der Philosophie – Politik deutscher Philosophen* (pp. 278–289). Reinbek bei Hamburg: Christian Wegner Verlag.

Habermas, J. (1989). Heidegger – Werk und Weltanschauung. In V. Farias (Ed.), *Heidegger und der Nationalsozialismus* (pp. 11–45). Frankfurt am Main: Fischer Verlag.

Haller, M. (2006). *Soziologische Theorie im systematisch-kritischen Vergleich*. Wiesbaden: Verlag für Sozialwissenschaften.

Hoppe, H. (1969). *Kants Theorie der Physik. Eine Untersuchung über das Opus postumum von Kant*. Frankfurt am Main: Vittorio Klostermann Verlag.

Jablonski, N. (2015). Race. In J. Brockman (Ed.), *This idea must die. Scientific theories that are blocking progress* (pp. 80–83). New York: HarperCollins Publishers.

Kanitscheider, B. (1984). Introduction to the German translation of Bunge's 'The Mind-Body Problem'. In M. Bunge (Ed.), *Das Leib-Seele-Problem. Ein psychobiologischer Versuch* (pp. VIII–VXII). Tübingen: J.C.B. Mohr Verlag (Paul Siebeck Verlag).

Kant, I. (1781/1787; 1976). *Kritik der reinen Vernunft*. Hamburg: Meiner Verlag.

Kant, I. (1802). *Physische Geographie* (Zweiter Band). Königsberg: Göbbels und Unzer Verlag.

Kant, I. (1972). *Briefwechsel*. Hamburg: Meiner Verlag.

Kaufmann, W. (1957). Deutscher Geist heute. In *Texte und Zeichen – Eine literarische Zeitschrift* (pp. 633–48). Darmstadt: Hermann Luchterhand Verlag.

Merton, R. K. (1988). *The Matthew effect in science, II – cumulative advantage and the symbolism of intellectual property. Isis – a journal of the history of science society* (pp. 606–623). Chicago: The University of Chicago Press.

Oreskes, N., & Conway, E. M. (2010). *Merchants of doubt. How a handful of scientists obscured the truth on issues from tobacco smoke to global warming*. New York: Bloomsbury Press.

Pickel, A. (2004). Systems and mechanisms: A symposium on Mario Bunge's philosophy of social science. *Philosophy of the Social Sciences, 34*, 169–210; 325–328.

Posner, R. A. (2003). *Public intellectuals. A study of decline*. Harvard: Harvard University Press.

Schmidt, H. (2015). *Was ich noch sagen wollte*. München: Beck Verlag.

Snow, C. P. (1959/1998). *The two cultures*. Cambridge: Cambridge University Press.

Part II
Philosophy

Chapter 5
Mario Bunge's Scientific Approach to Realism

Alberto Cordero

Abstract The first half of this article follows Mario Bunge's early realist moves, his efforts to articulate the achievements of theoretical physics as gains in the quest for objective truth and understanding, particularly in the context of the fights against the idealist and subjectivist interpretations of quantum mechanics that, at least until the mid-1970s, prevailed in physics. Bunge's answers to the problems of quantum mechanics provide a good angle for understanding how his realist positions grew on the "battlefield." The second half discusses Bunge's general conception of the scientific realist stance and confronts it with some current approaches to realism in the mainstream literature.

5.1 Introduction

The last chapter of *Between Two Worlds,* Mario Bunge's autobiographical essay, begins with these words:

> As I put it in my inaugural lecture as the professor of philosophy of science at Buenos Aires University (Bunge 1957), I have tried to philosophize scientifically and approach science philosophically. The philosophical approach to science led me to recast some scientific theories in the axiomatic format, which forces one to focus on the most important concepts and propositions of a field of study, as well as to detect possible sources of trouble. Axiomatize to understand and philosophize to do sound and useful axiomatics. And the scientific approach to philosophical problems has led me to look for both motivation and support in the science of the day. No philosophia perennis for me.
>
> I have criticized views that seemed to me to be utterly wrong, like subjectivism, or harmful, like intuitionism. But I have also attempted to polish nuggets, such as realism, materialism, systemism, and humanism; and turn them from isolated opinions into precise and well-grounded systems (theories). I have also been a militant philosopher rather than a dispassionate commentator, because I believe that philosophy can be beneficial or harmful, and that even apparently neutral and harmless *jeux d'esprit*, such as games in linguistic analysis, are harmful in diverting attention from burning issues. (Bunge 2016, p. 405)

A. Cordero (✉)
Graduate Center and Queens College, City University of New York (CUNY), New York, NY, USA

From the start of his academic career in the 1950s, one of Bunge's most persistent causes has been the articulation and defense of scientific realism, a stance that, along with naturalism, permeates virtually all of his "scientific philosophy." Bunge seeks to integrate and reinforce the various branches of science, epistemology, ontology, metaphysics, semantics, ethics, and social philosophy, groping towards a coherent naturalist proposal. (See in particular Bunge 1977, 1979, 1985). For him, science is the contemporary source of knowledge, and the most reliable basis for action (personal, social, and political).

Scientific realism is an epistemologically and ontologically optimistic position about the theoretical descriptions and explanations best supported by the methods of modern science. From the realist perspective, the theories that achieve scientific acceptance are successful because the world is (at least approximately) as those theories say it is. Accordingly, realists tend to accept as approximately true the descriptions that successful theories offer about entities and processes inaccessible to common observation (e.g., quarks, electrons, protons, atoms, proto-bacteria, social classes).

Bunge's writings and presentations typically display fierce criticism against positions that disparage reason, the objectivity of scientific knowledge, naturalism, the rational search for truth, as well as views that do not respect human beings as individuals. To Bunge, the goal of discerning and articulating the objective facts of reality commits science to realism. (See for example Bunge 1993). The latter, he urges, is the only sensible attitude one can take about science, both theoretically and in practice. The realism he advocates is systemic, just like the arguments he uses to defend it; Bunge's theses gain support from each other and also from components of his broader philosophy.

Systemism is especially apparent in his works on naturalism, emergent ontology, the study of quantum mechanics, relativity, bio-philosophy, the philosophy of mind, cognitive science, social science and economics, as well as his efforts to develop an ethic up to the level of contemporary problems. The major work of Bunge's systemism is his *Treatise on Basic Philosophy*, in eight volumes (1974–1989), the result of a long and vast philosophical adventure that began when Bunge was still a physics graduate at the University of La Plata.

After obtaining his doctorate in 1952, Bunge became a professor of physics and philosophy in Argentina, first at La Plata and then at the University of Buenos Aires until 1966, when he left the country, eventually settling at McGill University in Montreal, Canada. Remembering his early college years, he recounts how shocked he was by the anti-scientific environment that, in his opinion, prevailed in philosophy in Argentinian and Latin American universities in the 1930s. His response was to become mostly a self-taught person. Reading works by philosopher-scientists of the previous generation, in particular, Arthur Eddington and James Jeans, disappointed him because of the subjectivist and anti-realist tone with which they responded to the ontological and epistemological questions surrounding physics, igniting his belligerence as a realist (Bunge 2016, pp. 102–103). Bunge's scientific endeavors have remained loyal to this goal. Regarding the field of physics, for example, he says:

I believe that my main contribution to physics has been my book *Foundations of Physics* (Bunge 1967b), which had a strong philosophical motivation. This was my attempt to prove, not just state, that quantum and relativistic theories are realistic (observer-free) and that their subjectivist (observer-centered) interpretations are illegitimate philosophical grafts. (Bunge 2016, p. 406)

In the late 1940s and early 1950s, the most intense anti-realist fervor within physics occurred in the context of quantum mechanics, a theory that for nearly a century has served as a platform for different idealist positions. Quantum mechanics remains the most fundamental theory of matter, renowned for its extraordinary predictive power but also for its numerous conceptual conundrums. Bunge's answers to the problems of quantum mechanics provide an angle to understand how his realist position operates on the "battlefield" of the contemporary philosophical conversation about science.

5.2 The Case of Quantum Mechanics

Scientifically prosperous, unprecedently fertile and unusually disconcerting, quantum mechanics predicts the behavior of microscopic systems with a level of detail and breadth unparalleled in science. The theory's formalism stabilized more than three-quarters of a century ago,[1] but what the theory says has been from the start a source of ongoing controversy.

In the traditional or "classical" scientific conception of the world, physical objects are dynamically "complete" in that, at every moment, all the magnitudes applicable to them have sharp, dispersion-free, unambiguous values. In particular, they always have position and momentum, angular momentum in every direction, kinetic energy, and so forth. Furthermore, classical physics conceives of observation and measurement as processes aimed at revealing characteristics that the objects studied have before measurement.

Some of our most profound intuitions endorse the classical conception of physical objects. To the surprise of many, however, those intuitions do not fit well with the notion that quantum mechanics might provide complete descriptions of the dynamical properties of material systems. For example, the dynamical state of any quantum object, say an electron, obeys a linear law of evolution that promotes the development of "quantum superpositions" for its properties (situations in which the total state of an object does not presuppose the existence of exact values for one or more of the dynamical properties involved). Such quantum superpositions contradict the classical principle that states that objects always have a complete set of dynamical properties.

The above bafflement is compounded by the way the standard theory of quantum mechanics articulates what happens when a dynamical magnitude is measured on

---

[1] A particularly central reference is von Neumann (1932/1955).

a system that is in a state of superposition with respect to that magnitude. Suppose an electron is in a state of superposition involving, say, widely separated locations. According to the standard narrative, if this electron undergoes a measurement to find out its location, the result will always be a well-defined position. If so, however, where was the electron immediately before that measurement was performed? The electron, the story goes, was in several positions at once, but the measurement process somehow reduced the spectrum of position possibilities to just one. But, how do measurements manage this?

We have a problem here in that the transition from an "unsharp" position state to a sharp position state ("reduction of the quantum state") cannot occur through the theory's linear law of evolution. To account for the measurement process, therefore, it seems indispensable to supplement the dynamics with a second mode of evolution, one of stochastic and non-linear character. What is the structure of this second kind of evolution, and how does it unfold? Those questions are at the heart of the so-called "problem of measurement" in the philosophy of quantum mechanics.

Some physicists and philosophers approach the measurement process by appealing to the participation of "consciousness" (the "observer") in measurement situations, pointing to consciousness as the factor that causes the reduction of the quantum mechanical state. Projects focused on such links have been repeatedly proposed since the early days of the theory. Some of them place the observer as such at the center of fundamental physics (subjectivism); other projects invoke unorthodox physical terms (for example, making consciousness a property that supervenes on certain *components* of the total quantum state on the position-base, as in the so-called "many-mind" interpretations). Bunge has always rejected projects of this type, which he considers ill-conceived or even fraudulent. In his view, quantum mechanics can and should receive an entirely physical interpretation.

The earliest subjectivist responses arose from ideas expressed by Niels Bohr, Werner Heisenberg and other theorists, whose interpretative proposals constitute a variegated stance called the "Copenhagen Interpretation," The most common or "standard" version of this interpretation (SCI) figures prominently in university texts written between the 1950s and the late 1970s. Notoriously, Bunge notes, SCI transforms much of physics into a repository of obscurantist formulas. Nevertheless, for many decades most physicists found the SCI intellectually liberating, making it the dominant interpretation of quantum mechanics until the 1980s.

Complaints against SCI—voiced by Albert Einstein and other theorists in the 1930s—gained strength in the early 1950s. Bunge, initially an SCI supporter, soon joined the critics. In his autobiographical publications, he recounts how, in 1952, David Bohm's writings on hidden variables awoke him from his "dogmatic slumber" (Bunge 1990, 2016). Since then, Bunge has been a radical critic of SCI, whose supporters he repeatedly describes as gullible subjectivists who cannot even practice what they preach (Bunge 2006, section 3.3). Among the ideas associated with SCI that disgust him the most is the claim that the existence of physical objects depends on the "observer," ultimately on the participation of some type of mind (Bunge 2010, ch. 3 and section 7.5). Bunge blames Bohr and his followers for having promoted the idea that, by themselves, electrons and quantum systems in general ("quantons" in Bunge's vocabulary) lack dynamical attributes, ultimately the idea that quantons acquire only those attributes that the experimenter decides to impose on them (Bunge 2006, p. 36). For Bunge, this irreducible inclusion of the "observer" or any other subjectivist element in the description of physical measurements is equivalent to introducing philosophical contraband. A physical system, he insists, should include only physical components. According to Bunge, attributing to the intervention of mind the transition from a quantum superposition of multiple values to a sharp-value state that occurs in measurement processes is dishonest. Such attribution, he thinks, is inconsistent with the way in which quantum mechanics is applied and checked in current science.

Much of what Bunge points out against SCI's anti-realism seems well-deserved. (See for example Bunge 1955, 1967a, b, 1973, 1985, 2006). Explicitly or implicitly, until the late 1970s, virtually all of the major textbooks in physics supported to some degree the subjectivism associated with SCI, which authors did in ways that contradict how their books applied the theory—e.g., in the treatment of dynamic situations such as nuclear reactions. The fervor of the "Copenhagenists" was mostly doctrinal, Bunge notes, since they never offered suitable evidence in favor of the subjectivism they promoted. In his view, the resulting subjectivism was speculative fiction. He, therefore, thought it necessary to denounce the philosophical myopia of most of the followers of SCI, as well as to expose the ideology immersed in the textbooks of the time. From the end of the 1950s on, Bunge has fulfilled that mission, becoming an implacable critic of subjectivist approaches to microphysics in the twentieth century. In this and other respects, objectivists have a debt of gratitude to him.

On the other hand, some of the accusations that Bunge makes against SCI seem unfair (particularly about Bohr's philosophy). Sometimes Bunge gives the impression of not appreciating the positive contributions that SCI has made to contemporary realism, especially Bohr. I would thus like to take this opportunity to comment on two aspects of Bunge's critique: the impact of SCI on the emergence of the most promising objectivist programs in recent works on the philosophy of quantum mechanics, and also the value of "SCI fiction" for the realist project today.

5.2.1 Bohr

In his writings, Bunge repeatedly presents Bohr as a thinker hostile to realism— a positivist, subjectivist, phenomenalist and operationalist.[2] This seems excessive. It is true, however, that many of Bohr's publications lack clarity on the matter. Bohr expressed his ideas with regrettable awkwardness.[3] However, when pressed

---

[2]Representative references of Bunge's ideas about Bohr include, e.g., Bunge (1955, 1967a, 2006, Chapter 2).

[3]Bunge (1955) spells out his main charges of subjectivism against Bohr.

to explain his views, he made it clear that he favored a non-traditional type of realism. In particular, Bohr argued that the interaction between a measurement apparatus and its object, and the physical world in general, are entirely objective, denying that the experimenter could mentally influence the experimental results of quantum physics (Bohr 1958, p. 51). The realism Bohr advocated posited a variety of holism, not subjectivism. In his most articulate presentations, the state of a quanton (say, an electron) and the state of the apparatus used to measure something on it are not quantitatively separable during the measurement process. Only in this sense did Bohr assert that atomic objects lack an independent state. The dependence involved, strictly quantum mechanical, has nothing to do with the "mind" or anyone's "subjectivity." He claimed that the dynamical properties associated with a quanton are relational—relative to the total physical situation (Bohr compared experimental configurations in quantum mechanics to reference systems in the theory of relativity). The resulting holism unveils a quantum world at odds with "classical" ideas about the real and the possible. Bohr's suggestions may be extravagant, but the highly relational world he proposed is independent of the mind as such.

The radical notion that the observer as such plays a privileged role in quantum physics comes mainly from Heisenberg (Howard 2004). It was he who, in the 1940s and 1950s, encouraged subjectivist positions in response to the quantum measurement problem. Some distinguished theorists, among them John von Neumann, Eugene P. Wigner, and John A. Wheeler, welcomed Heisenberg's recommendations with enthusiasm, formulating versions of the measurement-induced reduction of quantum superpositions (Projection Postulate) that involved the intervention of some "conscience." Their respective proposals have been accused of inconsistency with physical practice. However, in quantum mechanics, the "subjectivist option" has proved difficult to eradicate and continues, albeit somewhat diminished, to date. There is no evidence in favor of quantum subjectivism, but the more elaborate formulations of the approach (notably, "the many minds" project[4]) have gained consistency over the years compared with both earlier proposals and current knowledge.

Bunge classifies Bohr as a phenomenalist in at least three respects (Bunge 1973, Ch. 4, 2006, Ch. 2). One focuses on Bohr's interpretation of the Heisenberg relations. According to Bohr, dynamical properties are not intrinsic to individual systems but relational concerning the total physical situation. The second respect corresponds to the idea that measurement creates what is measured—i.e., in general, the results of measurement do not reveal anything that the object studied already had at the start of the operation. The third respect is the positivist view that science seeks to accurately represent only the world we can observe through our unaided natural senses.

---

[4]The March issue of volume 47 (1996) of *The British Journal for the Philosophy of Science* contains a good survey of recent subjectivist approaches.

The first two respects are central aspects of Bohr's idea that quantum mechanics provides a complete description of physical systems. He thought that the quantum state specifies all the dynamical attributes of a system; in his view, dynamical properties with more than one possible value in the particular context at hand (i.e., properties in a state of superposition) are not properly attributable to the system in question. More than a positivist, Bohr was perhaps a confused and confusing Kantian, as Bunge himself acknowledges in some of his writings (Bunge 1973, p. 58). Nor was Bohr an "operationalist." As Don Howard has pointed out:

> [In] stressing the importance of specifying the total experimental context, Bohr was not endorsing operationalism. The quantity "electron momentum" is not defined by a procedure for measuring momentum; instead, one is permitted to speak of the electron's having a definite momentum only in a specified experimental context. (Howard 2004, Sect. 2)

As already said, on these topics the most determined defender of positivism and operationalism was Heisenberg, who did aspire to bring forward a version of quantum mechanics based exclusively on relationships between directly observable magnitudes. His followers are not few, including several prominent Berkeleyians—most outspokenly Mermin (1985)—who think that objects (be they atoms or the Moon) lack dynamical properties when no one is watching them.[5] Bohr did not think that way. In any case, it cannot be denied that even the best versions of SCI have serious deficiencies, in particular:

(a) Poorly articulated notions of holism that turn quantum measurements into processes more unanalyzable and mysterious than there is reason to think they are.
(b) A doctrinal view of the epistemological primacy of classical physics.
(c) An arbitrary conception of the boundary between the worlds ruled by quantum physics and classical physics.
(d) A dogmatic rejection of the possibility of developing a scientific cosmology, since a model of the Universe cannot include a measuring system external to it, as required by quantum mechanics (according to SCI), among other deficiencies.

5.2.2 The Road to Quantum Objectivism

In point of fairness, some cautious presentations of SCI offer radical but insightful ideas, especially regarding aspects such as:

---

[5]This invocation of the Moon is not new (Pais 1979, p. 907). During one of their discussions about quantum mechanics Einstein asked Bohr: "...but, do you really believe the moon only exists when you look at it?" Einstein apparently would not take seriously Bohr's relationist point about dynamical properties.

- The entanglement or "non-separability" that quantum systems develop as they interact with one another.
- The "non-locality" of physical influences within a quantum system.
- The analysis of dynamical properties.
- Bohr's "Principle of Correspondence," which Bunge praises (1967b, Ch. 5).

Although the "canonical" texts associated with SCI deal with interpretation in a disorderly and confusing manner, in fact SCI was the interpretation that most fruitfully and systematically guided the inference of predictions subsequently corroborated in the laboratory. The resulting "theory" is as crudely a predictive instrument as the Ptolemaic model had been about planetary positions. SCI presents a certain level of systematic—if "twisted"—unity, like that of the original Copernican model, which makes SCI an object of philosophical interest. Also, as indicated earlier, not all versions under the SCI label are at odds with realism. Moreover, we should note that, from the 1980s on, many of the critical reactions to SCI assume in some generalized form many ideas initially forged in SCI quarries, especially Bohr's.

The critical reactions to which I refer include the most prominent objectivist and realist projects today. An example, already mentioned, is the mechanical theory Bohr published in 1951, whose explicit objective was to harmoniously integrate into a realist proposal the best ideas of Bohr and Einstein. The leading theoretical physicists of the day rejected Bohr's project almost unanimously. Einstein in particular considered it a frustrating elaboration of SCI, not the innovative approach that in his opinion was needed to free quantum mechanics from the problem of measurement. Despite this lack of acceptance in the upper echelons of physics, Bohm's proposal of hidden variables impressed some talented minds working at the time on the foundations of physics, including Bunge. Initially, Bunge was very sympathetic to the theory, but some aspects disgusted him, leading him to distance himself from the project in the medium term. Bunge lamented, in particular, that Bohm's theory did not eliminate randomness at the fundamental level because it maintained the probabilistic assumptions of standard quantum mechanics instead of deriving them from principles consistent with Einstein's concept of causality. Also, he complained that Bohm's initial proposal did not lead to experimentally corroborative predictions (Bunge 2016, pp. 91–98).

By the end of the 1950s, other lines of generalization and objectivist revision of SCI had begun to emerge, giving rise to very diverse seminal contributions, prominently by Hugh Everett III, G. M. Prosperi, Giancarlo Ghirardi, Murray Gell-Mann, Wojciech H. Zurek and David Deutsch, to name a few. These reinterpretations—in some cases, modifications—of the received quantum theory have served as the basis for the main realist proposals currently in force. To date, the projects generated remain all experimentally inscrutable in practice and, therefore, have the status of speculations. However, although they are not empirically discernible "in practice" for the time being, the proposals describe objectively different worlds and lead to different predictions, something of significant interest to realists.

At a more general level, quantum mechanics continues to fascinate philosophically because projects like the ones highlighted in this paper are seriously transformative. They help both to break down received barriers of the imagination and to specify the limits of empirical underdetermination in current theories. If nothing else, the ongoing projects lead to better identifications of theoretical components best suited for realist commitment in a theory—arguably not a small achievement.

5.2.3 Realism and Fiction in SCI

There is something to characterizing SCI subjectivism as a "science fiction tale". Arguably, however, the Copenhagen story is not entirely fiction. As suggested in the previous section, many SCI ideas have made their way into recent realist-naturalist positions, their scientific character and fallibility. Naturalist positions are open to the possibility of reasoned change, and all their contents are, in principle, revisable—be it regarding the nature of physical reality or how independent of the mind the world investigated by science is. Such issues are not established in advance for all times, regardless of the findings scientists and philosophers make along the way.

SCI cannot philosophically impress because of its prior plausibility (it has none), but because its more careful versions exhibit sufficient coherence to exemplify ways in which physics *might* come to reveal a type of mind-dependence that would ruin the project of naturalist realism from within science. In this sense, SCI has philosophical value—e.g., in the way it unveils the contingent nature of objectivist commitments in physics (and science in general). The point is that the development of quantum mechanics could have gone wrong for realism. It has not, but the moral is that physics could in principle make antirealist stances plausible. If reasoning and evidence ever favored a subjectivist option, that outcome would be supported by the best available knowledge, the same way Bunge's realism draws from the contemporary scientific conception of the world. As long as the goals of science continue to emerge from scientific-naturalist deliberation rather than meta-scientific philosophies or external impositions, neither realism nor objectivism can be immune to what subsequent learning will reveal. The goal of finding facts commits science to corroborable truth, not to realism. In principle, scientific exploration could one day lead to a situation in which idealism and subjectivism (rather than realism and objectivism) become the positions of highest plausibility—which is not to suggest that such a development looks remotely probable today.

Throughout his life, Bunge has been philosophically nurtured by discussions about such themes as quantum mechanics, the mind-independence of the outside world, the intellectual accessibility of aspects of the world beyond the range of ordinary perception, and the perfectibility of knowledge. How does Bunge's contribution to these issues fit with realist responses currently playing a leading role in the philosophy of science?

Bunge's realism is a position forged partly from within the sciences (especially physics and biology), partly from his deep aspiration to keep thought and action harmonious. He agrees that one cannot fully prove or refute the independent reality of the outside world, but in Bunge's view that matters little, because whoever set out to study a process or entity presupposes the existence of numerous other entities and processes—for example, the instruments employed for the task and the environment in which they operate. The remaining sections explore in detail Bunge's realist project.

5.3 Bunge's Realism

In his characterization of scientific realism, Jarrett Leplin lists the following theses, emphasizing that "no majority of which is likely to be endorsed by any avowed realist":

1. The best current scientific theories are at least approximately true.
2. The central terms of the best contemporary theories are genuinely referential.
3. The approximate truth of a scientific theory is sufficient explanation of its predictive success.
4. The (approximate) truth of a scientific theory is the only possible explanation for its predictive success.
5. A scientific theory may be approximately true even if referentially unsuccessful.
6. The history of at least the mature sciences shows a progressive approximation to a true account of the physical world.
7. The theoretical claims of scientific theories are to be read literally, and so read are definitively true or false.
8. Scientific theories make genuinely existential claims.
9. The predictive success of a theory is evidence for the referential success of its central terms.
10. Science aims at a literally true account of the physical world, and its success is to be reckoned by its progress toward achieving this aim. (Leplin 1997, pp. 1–2)

Are there good reasons for taking the foregoing statements seriously, or at least some of them? Critics of scientific realism reject attempts to interpret theories as descriptions of aspects of the world situated beyond the natural range of the senses. For many anti-realists, the aim of scientific theories is, at best, empirical adequacy. In their view, theorists should seek to offer at least one model in which all real phenomena fit well (van Fraassen 1980, p. 12), or achieve theories with high effectiveness in solving problems (Laudan 1986, p. 11).

Bunge is a severe critic of these and other anti-realisms developed in the last quarter of the twentieth century (e.g., phenomenalism and social constructivism). His style of rejection admits few nuances:

> Antirealism is out of step with science and technology, intent as they are on exploring or altering reality. Antirealism is not just wrong; it is utterly destructive because it proclaims a total void: ontological, epistemological, semantic, methodological, axiological, ethical and practical. Such integral nihilism or negativism, reminiscent of Buddhism, discourages not only objective evaluation and rational action but also the exploration of the world. It is at best an academic game. (Bunge 2006, p. 87)

In mainstream contemporary philosophy of science, the project of realism comprises several tasks, in particular, these two: the first is to articulate a clear notion of realism, and the second is to identify which theories or representations merit realist interpretation. The following section offers a synopsis of Bunge's efforts concerning the first task.

5.4 Four Realist Theses

As noted in Sect. 5.1, the realist cause permeates the works of Mario Bunge, whose intellectual perspective—simultaneously philosophical and scientific, materialist and humanist, systemic and emergentist—covers all the central areas of philosophy (ontology, epistemology, semantics, methodology, axiology, moral philosophy and praxeology). This section considers Bunge's realist theses more directly related to science.

5.4.1 The Ontological Thesis

Bunge upholds the existence of a world independent of the mind, external to our thinking and representations (Ontological Thesis). His supporting reasoning on this matter draws from both general considerations as well as some of the special sciences—see, for example, Bunge (1997a, section 3.2, 2006, sections 1.8 and 10.1). One of his most compelling arguments appeals to the epistemic import of error, which he presents as a strong indicator of the existence of a reality independent of our thoughts and social constructions. Subjectivists explain easily why scientists "triumph"—because they build the phenomenal world—but find it excruciatingly difficult to explain the various imbalances between scientific hypotheses and the data provided by observations and experiments.

Another of Bunge's arguments is based on certain assumptions regularly made in scientific disciplines. Biology, for example, explicitly states that, to live and develop, all organisms (including, he stresses, subjectivist philosophers) must take nutrients and energy from sources external to them. Neurologists push in the same direction by unanimously accepting that the brain requires external stimulation to develop. The same trend is apparent in other sciences, he adds, noting as an example that, in the field of history, they assume not only the reality of the past but also presuppose that historical studies neither "change" nor "construct" the past itself.

Also stressed by Bunge is the recognition of invariance as an ontological marker in physics. In the mechanics of Galileo and Newton, frames of reference in relative uniform motion give rise to descriptions of motion that differ only at a "superficial" level. At a deeper level, those frames articulate the forces, masses, and accelerations at play in terms that are invariant to all such frameworks. In physics in general, invariance identifies aspects of the world whose descriptions are the same for everyone. When speaking about the objective existence of concrete things in the world, Bunge uses arguments from the invariance of fundamental laws. For example, Newton's Second Law is invariant to certain transformations of the reference system (in particular, translation, reflection, rotation, and uniform motion). At the dawn of the twentieth century, notes Bunge, Einstein generalized this criterion of objectivity from classical mechanics ("Galilean" invariance), expanding it to the broader and more general scope of Lorentzian invariance as a requirement for attributing objectivity to a magnitude. Since then, physics has been oriented towards frame-invariant theories (See for example Bunge 1967b, Ch. 4).

Like most other realists, Bunge resorts to abductive arguments to justify his proposal, in particular the idea that the achievements of the empirical sciences point to the existence of an external world independent of the mind. Abductive arguments are not conclusive, he acknowledges, underscoring however that their deficiency in this regard is compatible with fallibilist philosophical projects like his. According to Bunge, realism convinces because the success of the sciences fits splendidly with the Ontological Thesis, while the opposite is the case about the anti-realist proposals. Reasons such as these, Bunge argues, should lead us to abandon ontological idealism in all its forms. He insists that it is intellectually counterproductive to erase the distinction between real concrete objects (things) and what happens to them on the one hand, and conceptual objects (ideas, including data, hypotheses, and theories about facts) on the other. Doing so gives rise to uncontrollable confusion and negligence, he worries.

5.4.2 The Epistemological Thesis

Bunge complements the previous proposal with an epistemological thesis made of three major claims (Bunge 1985, section 2.6, 2006, section 10.2):

4.2a. It is possible to know the external world and describe it at least to a certain extent. Through experience, reason, imagination, and criticism, we can access some truths about the outside world and ourselves.

4.2b. While the knowledge we thus acquire often goes beyond the reach of the human senses, it is multiply problematic. In particular, the resulting knowledge is indirect, abstract, incomplete, and fallible.

4.2c. Notwithstanding its imperfections, our knowledge can be improved. Bunge accepts that theories are typically wrong as total, unqualified proposals. In his opinion, however, history shows with equal force that successful scientific theories are not entirely false, and also that they can be improved (Bunge 2006, p. 255). This last statement constitutes the core of Bunge's "meliorism," according to which ideas that are off-the-mark when first presented can be subsequently amended and brought closer to the truth.

Ideas 4.2a and 4.2c oppose skepticism and relativism, while 4.2b distances Bunge's position from both radical empiricism and naive realism. Sub-thesis 4.2b involves five key admissions:

(i) Because knowledge is indirect, theoretical statements gain epistemic force from intricate inferences instead of doing so "directly" from just perception or rational intuition (Bunge 1974, 1998 Chap. 5). Nobody can see viruses without the help of instruments such as the electron microscope. And the physical appearance and biology of our extinct ancestors can only be reconstructed through complex networks of inferences that harmoniously integrate available scientific knowledge, reasoning, imagination, and experience.

(ii) A theory represents a domain of interest in an abstract way, and for this reason, we expect that its models will conform to the intended domain only in an approximate way.

(iii) Idealizations and Simplifications: Scientific theories seek to represent the world, but typically do so by idealizing the complex systems found in nature. For example, theories of the solar system take into account only a fraction of the components of the actual system and then proceed just in terms of gravitational interactions.

(iv) Incompleteness: The vast wealth of information available in nature but not yet noticed by us has epistemological consequences. We build our knowledge of reality based on limited facts and realizations about the objects that interest us. Each scientific research project focuses on only some aspects of the world, selected from an indefinitely wide range of perspectives potentially open to our study. Each discipline looks for matters it considers worthy of representation, explanation, and prediction regarding its particular domain of interest. No discipline provides descriptions of all the possible aspects that could be significant in its field of study. For example, physicists describe the silver atom in terms of basically two quantum numbers (atomic number and mass number) and the energy levels of the system thus constituted. The resulting descriptions ignore abundant aspects of the "total reality" of any concrete silver atom. Indeed, what is left out comprises "most of reality," including the atom's location, the entities around it (e.g., living organisms, rivers, mountains, and so on), the spectra of radio waves around, the sites of silver deposits on Earth, the languages spoken by silver miners today or twenty-seven centuries ago, among numerous other components of "total reality".

(v) Scientific knowledge is fallible due to our inability to prove or disprove conclusively any general statement, so we access a type of knowledge that we cannot establish beyond all possible doubt. This limitation has different sources. One is that inductive inference never leads to certainty. Another source is the set of assumptions under which both the construction and application of theories operate in science—assumptions which typically include metaphysical theories, background hypotheses, and auxiliary hypotheses. The unfading presence of factors of this type ruins the goal of achieving absolutely conclusive verifications or refutations.

5.4.3 The Semantic Thesis

This component of Bunge's realism is framed by the previously stated ontological and epistemological theses. It comprises four interrelated ideas (Bunge 1974, 2006, section 10-3):

4.3a. Some propositions refer to facts (as opposed to only ideas).
4.3b. We can discern the proper ("legitimate") referents of a scientific theory by identifying its fundamental predicates and examining their conceptual connections in order to determine the role those predicates play in the laws of theory.
4.3c. Some factual propositions are approximately true.
4.3d. Any advance towards the truth is susceptible of improvement (semantic complement of the meliorist thesis in 4.2c).

According to Bunge, the referents of a theory are those associated with the variables that figure in its laws. Point 4.3c expresses Bunge's version of truth as correspondence (the general view that truth is a relational property involving a characteristic relation to some portion of reality). Bunge admits that in philosophy the usual notion of correspondence remains vague, incomplete, and awaiting substantial work. In particular, the correspondence approach is unsatisfactory in the case of negative and general propositions, it does not leave room for partial truths, and it does not take into account the importance of what Bunge calls "systemicity"—the integration and coordination between propositions. In Bunge's opinion, however, these problems can be corrected. His specific proposal disagrees with dominant approaches in that it allows propositions to lack inherent truth values. Given a proposition, he suggests, its truth value begins to emerge only after it is tested, and that value may change in the course of the relevant investigation. It is not clear how well this proposal fits with the usual conception of correspondence, or how free it is of the problems associated with the coherentist theory of truth. This part of Bunge's project is still in a preliminary phase.

5.4.4 Methodological Thesis

The fourth facet of Bunge's realism I am highlighting focuses on methodology and comprises at least three proposals: (a) methodological scientism, (b) Bunge's version of the requirement that theories must allow for empirical testing, and (c) a mechanistic agenda for scientific explanation (Bunge 1997b, 2001, Ch. 2, 2006, section 10.4).

4.4a. Scientism asserts that the general methods developed by science to acquire knowledge provide the most effective available exploration strategy at our disposal. The methods of science—whose main use is given in the development and evaluation of theories—use reason, experience, and imagination. The assessment of theories, observes Bunge, does not only resort to induction but its development is mediated by the imagination of the scientists involved, limited by previous learning and current methodological guidelines. For these reasons, he emphasizes, it is necessary to take precautions. From the beginning of its formulation, a scientific hypothesis must pass through rigorous tests to check if it is satisfactory regarding consistency, compatibility with broader scientific knowledge (external adequacy), and empirical refutability (external control).

4.4b. A theoretical proposal should lead to distinctive predictions, and it should be possible to subject at least some of those predictions to demanding empirical corroboration. Given that, in principle, the results of any empirical test draw from all the assumptions and auxiliary hypotheses that contribute to the predictions in question, testing a theory affects its claims "globally" rather than just at the level of localized assertions. Contrasting a theory with nature requires, therefore, a substantial capacity for judgment on the part of experienced scientists (Bunge 1997a, Sect. 3.7).

4.4c. According to Bunge, we cannot be satisfied with merely phenomenological hypotheses of the "black box" type (i.e., structures that do not go beyond articulating correlations between observable phenomena). Good methodology, Bunge insists, presses for further exploration, prompting us to search the world for regularities at deeper levels that provide illuminating explanations of the discovered regularities—ideally "mechanical" ones. In Bunge's writings, the term "mechanical" refers to entities and processes of any kind invoked to explain regularities (Bunge 2006, Ch. 5). In a mechanical explanation, the postulates can be mechanical in either the traditional sense or a generalized sense (in electrical, chemical, cellular, organismic, ecological, psychological, economic, or cultural terms). A hypothesis can emerge as a purely phenomenological claim (for example, that populations evolve), and be subsequently complemented by "mechanical" explanations (for example, regarding general processes by which populations evolve—such as mutation, natural selection, and reproductive isolation).

The theses outlined above are continuous with the rest of Bunge's philosophy. An unambiguous supporter of the Enlightenment project, Bunge's position in other areas fits accordingly. Thus, on moral values, he is also a strong realist, as he is in other fields of practical philosophy in general. Bunge accepts that there are values whose objectivity comes from their roots in biological and social needs, such as health, security, knowledge, and privacy. This aspect of Bunge's realism lies beyond the scope of this article.

5.5 Distance from the Mainstream

Bunge has not shown much interest in the mainstream debate on realism, which is perhaps unfortunate as some recent realist approaches seemingly bring resources for carrying out some aspects of his project. Thus, responding to Larry Laudan's challenge that theories are usually false, since the 1990s a variegated host of "selective realists" (notably, Philip Kitcher, Jarrett Leplin, Stathis Psillos, Juha Saatsi, and Peter Vickers, to name a few) have been working to identify theory-parts and partial narratives that get the story right. Which parts are those? The history of mature science is rich in warnings against excessive optimism about the way theory-parts receive confirmational weight in scientific practice. Nonetheless, selective realism seems both a promising approach and one arguably optimal for advancing of Mario Bunge's realist agenda.

Bunge vehemently rejects some mainstream programs. A case in point is the structuralist view of theories propounded by Patrick Suppes, Joseph Sneed, Wolfgang Stegmuller and their collaborators. In the 1970s he dismissed this line, blaming its practitioners of ignoring the semantic side of scientific theories. In his review of Stegmuller's *The structure and dynamics of theories*, Bunge objected to the author's identification of theories with uninterpreted (abstract) theories, and also to the way structuralists define theory applications as models (in the model-theoretic sense) of them (Bunge 1976). The models considered by model theorists, he admonished, are unrelated to the theoretical models devised by scientists and technologists, which are specific theories, such as that of the simple pendulum. Then (as now) Bunge thus took the model-theoretic approach as *"the fruit of an equivocation, as would be regarding ring theory as dealing with wedding bands, onion rings, and the like"* (Bunge 2017, p. 71). Instead, he has advocated a "dual axiomatics" approach one that accompanies every key mathematical concept with a semantic hypothesis specifying its reference and sketching its sense (Bunge 1967b, c).

Meanwhile, structuralists with broader allegiances have been busy trying to improve the approach, but Bunge does not seem to have considered their work. Cases in point include varieties developed in Britain, most notably by John Worrall and his collaborators (structural realism) and by Steven French and James Ladyman (ontic structural realism). I think these developments matter because they go some way in the reformative directions advocated by Bunge. To Bunge, alas, the whole model-theoretic approach seems no good—an obsolete philosophy, at best, that:

> had it appeared two and a half centuries earlier, it might have shed some light on the discussions among Newtonians, Cartesians, and Leibnizians, that Newton's *Principia* had provoked in 1687. (Bunge 2017, p. 77)

5.6 Conclusion

As suggested at the end of Sect. 5.3, the mainstream contemporary project of scientific realism faces two primary tasks. The first is to articulate a clear notion of realism. Mario Bunge has contributed a great deal to advance this part of the project, with enormous benefit.

The second task is to specify which theories (or parts of them) merit realist interpretation (See for example Cordero 2001, 2003, 2011, 2012, 2016, and references therein). In Bunge's work, this second job remains pending in some important respects. His recommendations are for the most part sensible but perhaps a bit too vague. For instance, Bunge stresses that to contrast a theory with nature one must have "experience and judgment." This recommendation is, of course, well taken, but the real challenge consists in specifying reliable ways to achieve the needed form and level of experience and judgment. We need to know, for example, what criteria we can use to determine, at least roughly, truth content distinctly attributable to a successful theory (as opposed to the array of background theories that converge in efforts to apply it). Issues of detail such as these reveal urgent tasks for contemporary realists, given the acceptance (shared by Bunge) that theories generally err as total constructs and, furthermore, empirical tests do not directly point to the differentially truthful statements implicated in a particular empirical test of a theory.

The realism project that Bunge articulates seems, therefore, to have some major issues still pending. One way to continue the project could be, for example, to apply in detail Bunge's concept or realism to crucial problems in the contemporary debate between scientific realists and anti-realist. Cases in point include truth content in superseded theories that showed remarkable success in their heyday (e.g., Fresnel's theory and Kirchhoff's diffraction theory) as well as empirically successful current theories and ideas marred by interpretive problems (e.g., the multiplicity of seemingly incompatible theoretical models in nuclear physics, the nature of quantum systems, and the quantum state), to mention just a few topics of interest from just physics.

Meanwhile, however, I think there is no doubt that Mario Bunge will continue to make valuable contributions in this and other areas of the realist project, responding with honesty and clarity to the enigmas posed by the most intellectually challenging fundamental theories of our time.

References

Bunge, M. (1955). Strife about complementary. *British Journal for the Philosophy of Science, 6,* 1–12.

Bunge, M. (1957). Filosofar científicamente y encarar la ciencia filosóficamente. *Ciencia e Investigación, 13,* 244–257.

Bunge, M. (1958). *Essays 1932–1957 on atomic physics and human knowledge*, reprinted as *The philosophical writings of Niels Bohr, Vol. II.* Woodbridge (1987). Ox Bow Press.

Bunge, M. (1967a). The turn of the tide. In M. A. Bunge (Ed.), *Quantum theory and reality* (pp. 1–6). New York: Springer.

Bunge, M. (1967b). *Foundations of physics.* Berlin/Heidelberg/New York: Springer.

Bunge, M. (1967c). The structure and content of a physical theory. In M. Bunge (Ed.), *Delaware seminar in the foundations of physics* (pp. 15–27). Berlin/Heidelberg/New York: Springer.

Bunge, M. (1973). *Philosophy of physics.* Dordrecht: Reidel.

Bunge, M. (1974). *Treatise on basic philosophy. Vol., Semantics: Sense and reference.* Dordrecht: Reidel.

Bunge, M. (1976). Review of Wolfgang Stegmüller's the structure and dynamics of theories. *Mathematical Reviews, 55*, 333.

Bunge, M. (1977). *Treatise on basic philosophy. Vol. III, ontology: The furniture of the world.* Dordrecht: Reidel.

Bunge, M. (1979). *Treatise on basic philosophy. Vol. II, ontology II: A world of systems.* Dordrecht: Reidel.

Bunge, M. (1985). *Treatise on basic philosophy. Vol. VII, epistemology and methodology III: Philosophy of science and technology* (Part I (Formal and physical sciences)). Dordrecht: Reidel.

Bunge, M. (1990). Instant autobiography. In Weingartner and Dorn (1990) (pp. 677–684).

Bunge, M. (1993). Realism and antirealism in social science. *Theory and Decision, 35*, 207–235.

Bunge, M. (1997a). *Foundations of biophilosophy* (with M. Mahner). Berlin: Springer.

Bunge, M. (1997b). Mechanism and explanation. *Philosophy of the Social Sciences, 27*, 410–465.

Bunge, M. (1998). *Philosophy of science. Vol I: From problem to theory.* New Brunswick: Transaction Publishers, Chapters 6, 7 and 8.

Bunge, M. A. (2001). In M. Mahner (Ed.), *Scientific realism. Selected essays of Mario Bunge.* New York: Prometheus Books.

Bunge, M. (2006). *Chasing reality.* Toronto: University of Toronto Press, Scholarly Publishing Division.

Bunge, M. (2010). *Matter and mind* (Boston studies in the philosophy of science, Vol. 287). Dordrecht: Springer.

Bunge, M. (2016). *Between two worlds: Memoirs of a philosopher-scientist.* Cham: Springer.

Bunge, M. (2017). *Doing science-in the light of philosophy.* Hackensack: World Scientific.

Cordero, A. (2001). Realism and under determination: Some clues from the practices-up. *Philosophy of Science, 68S*, 301–312.

Cordero, A. (2003). Understanding quantum physics. *Science & Education, 12*, 503–511.

Cordero, A. (2011). Scientific realism and the *divide et impera* strategy: The ether saga revisited. *Philosophy of Science, 78*, 1120–1130.

Cordero, A. (2012). Mario Bunge's scientific realism. *Science & Education, 21*, 1319–1435.

Cordero, A. (2016). Retention, truth-content and selective realism. In E. Agazzi (Ed.), *Scientific realism: Objectivity and truth in science* (pp. 245–256). Cham: Springer Nature.

Howard, D. (2004). Who invented the 'Copenhagen interpretation'? A study in mythology. *Philosophy of Science, 71*, 669–682.

Laudan, L. (1977/1986). *Progress and its problems: Toward a theory of scientific growth.* London: Routledge and Kegan Paul.

Leplin, J. (1997). *A novel defense of scientific realism.* New York: Oxford University Press.

Mermin, N. D. (1985). Is the moon there when nobody looks? Reality and the quantum theory. *Physics Today, 38*, 38–47.

Pais, A. (1979). Einstein and the quantum theory. *Reviews of Modern Physics, 51*, 863–914.

van Fraassen, B. C. (1980). *The scientific image.* Oxford: Clarendon Press.

Von Neumann, J. (1932/1955). *The mathematical foundations of quantum mechanics.* Princeton: Princeton University Press.

Weingartner, P., & Dorn, G. J. W. (1990). *Studies on Mario Bunge's treatise.* Amsterdam: Rodopi.

Chapter 6
Contrasting Materialisms: Engelsian Dialectical and Bunge's Emergentist Realism

Pierre Deleporte

Abstract Despite the fact that dialectical materialism (hereafter DM) raises little interest from contemporary philosophers, Mario Bunge has provided a thorough criticism of DM. This ontology has some merits, being evolutionist (material things change by themselves) and scientific. But the specific 'dialectic laws' for material processes, as viewed by Bunge, are at best vague and of limited scope, and at worst incompatible with materialism, or plainly unintelligible. This analysis can be developed by considering the original dialectic laws in Engels' writings, and also the criticisms and tentative revisions of DM by proponents of dialectics themselves. In this respect, the little-known work of Robert Havemann must be taken into account. This Marxist scientist and philosopher was a severe critic of dogmatic DM and a strong supporter of the scientific method. Nevertheless, he was unable to escape some of the worst obscurities of dialectical thinking. The existence of a modern emergentist, exact and scientist philosophical materialism, exemplified by the clear and large-scoped Bungean synthesis, now makes any excessive concern for DM notions a superfluous, if not even detrimental, intellectual detour.

6.1 Introduction

Philosophical materialism comes in a variety of kinds, among which is *Dialectical Materialism* (hereafter: DM). Once fashionable in academic circles, usually in connection with Marxism, presently the concern for DM has greatly decreased, even among socialist intellectuals. Nevertheless, the philosophical fundamentals of DM have been thoroughly criticized by Mario Bunge, notably in his *Scientific Materialism* (Bunge 1981, pp. 41–63). In a nutshell, Bunge considers the dialectical materialist ontology as partly plausible but not original, and partly original but wrong or obscure to the point of being incomprehensible. Bunge's main point of view on DM will be examined below, with complementary elements of analysis.

P. Deleporte (✉)
Biology Department, Université de Rennes 1, UMR6552 CNRS EthoS, Rennes, France

© Springer Nature Switzerland AG 2019
M. R. Matthews (ed.), *Mario Bunge: A Centenary Festschrift*,
https://doi.org/10.1007/978-3-030-16673-1_6

101

Historical Materialism will not be discussed in detail: Bunge considers Historical Materialism as an important achievement of Karl Marx, the master social critic and "great economist" of his time (Bunge 2012, p. 83). Focus will be put on the core notions of DM as initially propounded by Friedrich Engels in his dialectical laws (Engels 1975, 2013).

Engels attributes to Karl Marx the merit of having promoted Hegelian dialectics in a scientific materialist world view:

> It is Marx's merit (. . .) to have been the first to value again the forgotten dialectic method, its links as well as its differences with Hegelian dialectics, and to have in the same time applied this method, in *Das Capital*, to the facts of an empirical science, the Political Economy. (Engels 1975, p. 53)

But, given that Engels' *Dialectics of Nature* and *Anti-Dühring* both support all of Marx's views and deal with a much larger variety of scientific questions, it is efficient to give priority to the analysis of these texts. For a fair evaluation of DM, it is also important to consider recent critics by declared proponents of dialectics. In this respect, the important but little known work of Robert Havemann, an East German scientist, philosopher and political dissident, will be taken into consideration.

6.2   Bunge on Dialectical Materialism

Bunge (1981) characterized DM ontology by five dialectical principles, on the basis of his own compilation of main authors in this field. He anticipates possible reproaches of inexactness or incompleteness of his report by challenging his critics to elaborate their own manual of dialectics – which, to my knowledge, did not exist in 1981, and seemingly very few proponents of DM are presently willing to elaborate such a manual in a concise and intelligible form. Under the title *What is dialectics?* Ollman, a declared contemporary proponent of Marxist philosophy, even reduces his definition of dialectics to: "dialectics in itself explains nothing, proves nothing, predicts nothing and is the cause of nothing"; he presents dialectics as a mere method of analysis focusing on the processes of change, combined with an ideological posture: "dialectics is by essence critical and revolutionary" and "it forces us to examine where the society comes from and where it is going to" (Ollman 2003, pp. 21–35).

According to Bunge, the five principles of DM are:

1. Everything has an opposite.
2. Every object is inherently contradictory, i.e. constituted by mutually opposing components and aspects.
3. Every change is the outcome of the tension or struggle of opposites, whether within the system in question or among different systems.
4. Development is a helix every level of which contains, and at the same time negates, the previous rung.

5. Every quantitative change ends up in some qualitative change and every new quality has its own new mode of quantitative change (Bunge 1981, p. 42).

Bunge considers only the first three principles as peculiar to dialectics, but says they are obscure and of limited scope rather than universal. Also, no dialectical *ontology* can serve to base any possible dialectical *logic*, given that logic concerns concepts while ontology concerns material things. Bunge's consequent materialism admits no overlap between the class of concepts and the class of concrete objects: nothing is both material and conceptual. Anyway, unlike archaic thinking in terms of *struggles of opposites*, contemporaneous understanding of change does not require a particular logic, not even dialectical.

More precisely, the main Bungean comments and critics may be summarized as follows (Bunge 1981, pp. 44–55):

1. First principle: the notion of anti-things is unclear in DM, and that of anti-properties as well. The mere absence of something cannot oppose it, nor can its whole environment; some things can destroy other things, but they are not necessarily unique antagonists. And the combination of two material things into a third one is a synthesis but it produces an altogether different thing, when polymerization is a combination of equal rather than opposite entities.

   Otherwise, Aristotle's dialectics about *Categoriae* concerned properties, not concrete things. In this domain, negating that an object has a property is a conceptual operation, not a concrete one. It makes sense that some properties can balance or neutralize some other properties, but such cases are not universal, and such counteracting properties may not be unique. The fact that some things can oppose other ones in some respects is rather trivial and not specifically dialectical. But anyway a truly materialist ontology must distinguish between predicates and properties of concrete things. Negative predicates make sense as concepts, but concrete things have only positive properties, hence negative predicates cannot represent properties of concrete things, i.e.: "Negation is a conceptual operation without an ontic counterpart" (Bunge 1981, p. 47).

2. Second principle: again, the notion of opposite is unclear in DM writings. Some material systems show some contradictory properties, but not all material things, hence thesis 2 cannot hold universally. Some systems are polar in some respects, but it is over-simplistic to treat all systems and all processes as such. Polarity is rather a way of thinking about reality than an ontic property of the material world.

3. Third principle: only some changes in some material objects and systems result from conflict, and scientific theories of competition of different kinds are much richer than a simple notion of conflict. Obviously not all changes are due to conflict, not even in human societies where cooperation occurs as well. Even the human cognitive development is not uniquely driven by contradiction and criticism: creation necessarily precedes contradiction, and cooperation contributes to creation.

4. Fourth principle: the notion of *dialectical negation* is no more clear concerning the dialectical view of a helicoidal or spiral form of development (*Aufhebung*,

sublation, or negation of the negation to reach a state that is superior to the initial one). Scientific theories of evolutionary processes are richer than such preconceived polar views. The Thomist catastrophe theory of morphogenesis through conflict concerns geometric objects, not material ones, while the ontology of catastrophe theorists is Platonic, i.e. the view that form pre-exists matter and rules over it. The Bungean acceptance of principle 4 as stated above does not impose universal *progress* in the processes of change: *stagnation* and *regression* occur as well, whatever the definitions of these three notions.

5. Fifth principle: the classic dialectical formulation *conversion of quality into quantity* is unintelligible, because concepts don't evolve and no qualitative property gets properly transformed into a quantitative property, or conversely, for any natural process. The Bungean re-formulation of *principle 5* above escapes this ambiguity. Simply, in every process some new property may happen to emerge, which undergoes its own mode of quantitative variation. Bunge considers *principle 5* as a plausible hypothesis, although not presently demonstrated as a theorem in an elaborated theory of change.

Along with the main DM principles, it is worthwhile to consider the *11th Thesis on Feuerbach* (Marx 1888). This is not a DM law properly, rather a fundamental statement of DM-inspired Marxism, and it bears on the status of philosophy regarding action (pragmatism). This *Thesis 11* reads: "The philosophers have only interpreted the world, in various ways; [but] the point is to change it" (the word *but* is an amendment introduced by Engels). Bunge's interpretation is twofold: "This statement is ambiguous: it may be read either as the claim that praxis trumps theory, or as a call to social action without social science" (Bunge 2012, p. 90).

Bunge did not analyze in detail the diverse interpretations of DM laws provided by a plethora of dialectical materialists and Marxist writers; and understandably so. Philosophically important writings should develop in-depth criticism and revision of DM, something its faithful followers do not do, any more than the superficial ideological opponents to materialism and Marxism. Nevertheless, the five principles of DM mentioned above being clearly Bunge's own synthesis and understanding of the topic, it is useful for the debate to take into consideration at least the main laws or principles of DM as initially formulated by Engels.

6.3   Engels' Dialectical Laws

I draw these laws from the classic references *Dialectics of Nature* (hereafter DoN; Engels 1975) and *Anti-Dühring* (hereafter A-D; Engels 2013). Despite having a larger philosophical scope than A-D, DoN is a mere series of Engels' unpublished notes compiled by the Marx-Engels-Lenin Institute of USSR, so A-D could be viewed as a more elaborated version of Engels' philosophy; hence both texts are worth taking into account for a fair evaluation.

In the chapter "Dialectics" in DoN, Engels mentions three essential dialectical laws: "the law of passing from quantity to quality, and conversely; the law of reciprocal penetration of opposites; the law of negation of the negation" (Engels 1975, pp. 69–70). Engels criticizes Hegel for idealistically developing these laws as pure laws of thought, while he himself wants to deduce these laws from nature. Incidentally, he states, "We don't need here to write a manual of dialectics" (Engels 1975, p. 70). Given the lasting absence of a concise manual of dialectics, it seems that most of DM philosophers understood either: *we still don't need to write such a manual*, or *this is an impossible task*.

In another short paragraph also devoted to dialectics, Engels mentions the fundamental laws this way:

> The so-called *objective* dialectics rules the whole nature, and the so-called subjective dialectics, the dialectical thought, just reflects the reign, in the whole nature, of the movement by opposition of opposites who, by their constant conflict and their final conversion into one another or into superior forms, determines precisely the life of nature. (Engels 1975, p. 213)

Here is added the notion of *conversion* of opposites (sometimes called 'polar' opposites in DM) into one another, as an alternative to sublation to a 'higher' state. These laws are not only presented as universal laws of nature, but also as *reflected* by dialectical thought, which is the Hegelian *reflection* theory of knowledge taken the reverse way: for Engels, thought reflects the laws of nature rather than nature reflects the laws of thought. The same view is also expressed when Engels briefly evokes the reciprocal penetration of opposites:

> In Hegel himself, this is mystical, since categories appear there as preexisting and the dialectics of the real world as their pure reflection. In reality, it is the reverse: dialectics in the head is but the reflection of the forms of movement of the real world. (Engels 1975, p. 204)

But in the views of Hegel and Engels as well, the general laws of change are considered as similar in concepts and in nature via a reflection principle, or process.

A still shorter and ambiguous note sheds little light on the first dialectical law, or even on the dialectical formulation of it: "Conversion of quantity into quality = "mechanical" conception of the world, a quantitative change of the world modifies the quality. This is what these gentlemen never suspected!" (Engels 1975, p. 215). Here, two formulations with quite different meanings are used without explanation: the enigmatic 'conversion' of a property into another one, contrasted with the plain modification of a property caused by some change otherwise. Effectively, this can hardly be 'suspected' from common sense.

In A-D, Engels devotes a special chapter to just two main dialectical laws, namely the first and third laws mentioned above: *quality and quantity* and *negation of the negation*, clearly integrating *sublation* in the latter, i.e. the notion that not only development proceeds by successive 'negations', but the negated negation presents a so-called higher state of development than the initial one. As already mentioned above, these two DM laws correspond to the DM principles considered by Bunge as the more plausible ones, at the cost of a reformulation in sensible terms and a narrowing of their scope.

6.4 Engels on Quality and Quantity

In DoN, below the listing of the three dialectical laws, only the first law is exposed in detail. This *quantity-quality* law is re-formulated this way:

> [In nature] qualitative changes can only occur by quantitative addition or removal of matter or movement (energy, to say it so). (. . .) It is thus impossible to change the quality of any body without addition or removal of some matter or movement, i.e. without quantitative modification of the body in question. Under this form, the mysterious proposition of Hegel appears not only quite rational, but even rather obvious. (Engels 1975, p. 70)

Strikingly enough, here Engels does not use the mysterious Hegelian formulation for this law: *transformation of quantity into quality*. In fact, in the previous formulation used by Engels: "the law of passing from quantity to quality, and conversely", *passing* is an extremely vague statement. What is passing in the system at stake? A given property supposed to transform itself from a quantitative to a qualitative nature, or changes in different properties of the system? Engels' explanation cited above seems to mean simply that some quantitative change in a system is the cause of a qualitative modification in some aspect of the system; hence a 'rather obvious' but quite trivial law with nothing specifically dialectical, because instead of the enigmatic Hegelian *self-transformation* or *conversion* of a given property into something else, Engels describes the mere effect of some change on some other property.

Engels gives examples supposedly illustrating the first law, all in the same vein: a change induces a modification. An amusing example is as follows (Engels 1975, p. 73). If you 'add or remove', to say it so, some atoms to the molecules of alcohol you drink, namely ethyl alcohol versus amyl alcohol (this is the gradual *quantitative* change) then either you get simply dizzy or you suffer of a strong headache on top of it (this is the *qualitative* jump according to the law). The problem in this case is not only that the quantitative change may as well be considered as a qualitative one (the two molecules of alcohol are quite different objects), but the so-called quantitative chemical change in the beverage versus the sudden qualitative jump in brain state concern entirely different properties of the material system *drunk person*. A couple of atoms are not a state of drunkenness, and nothing of the former converts into anything of the latter. Such a transformation means nothing, whether in terms of concepts or of material processes. But then, if we just retain the general notion that any change in a system induces some modification of some of its properties otherwise, be it progressively or by jumps, this is a trivial notion non exclusive to the DM view.

Engels considers that the first law has a universal scope: "[...] a general law of the evolution of nature, society and thought under its universally valuable form" (Engels 1975, p. 74). He anticipates that his opponents would finally consider that this law is trivial and pretend that they always had used it unknowingly, like Molière's theatrical [for theater] character *Monsieur Jourdain* who discovered that he talked in prose without even knowing the notion. This could seem obvious if we consider only the rather trivial formulations and illustrative examples of Engels, but this time he uses not only his previous formulation *passage from quantity to quality*, but also the mysterious Hegelian formula: "quantity and quality convert into one another" (Engels 1975, p. 74). No explanation is given for this shift from the vague formulation *passage*, to the mysterious dialectical one *converts into*. Is it just a will to celebrate the supposed genius of Hegel by using the obscure Hegelian formulation? Anyway, this shift to the *conversion* formulation instead of a simple evolutionary (or causal) one raises a serious problem of logic: the two notions are not equivalent.

In A-D, Engels criticizes the following position of Dühring:

> [...] a transition is made [...] in spite of all quantitative gradations, only through a qualitative leap, of which we can say that it is infinitely different from the mere gradation of one and the same property. (Engels 2013, p. 71)

To argue against Dühring, Engels takes the example of freezing or boiling water to illustrate the dialectical quantity-quality law. Engels writes, explicitly referring to Hegel's *nodal line of measure relations*:

> [...] the purely quantitative increase gives rise to a qualitative leap; for example, in the case of heated or cooled water, where boiling-point and freezing-points are the nodes at which [...] the leap to a new state of aggregation takes place, and where consequently quantity is transformed into quality. (Engels 2013, p. 71)

In another chapter he comments this "best known example": "the merely quantitative change of temperature brings about a qualitative change in the condition of the water" (Engels 2013, p. 167), and he generalizes his point this way (with reference to Marx's *Capital*):

> [...] innumerable cases in which quantitative change alters the quality, and also qualitative change alters the quantity, of the things under consideration; in which therefore, to use the expression so hated by Herr Dühring, quantity is transformed into quality and vice-versa. (Engels 2013, p. 167)

Dühring's position as cited above seems clear and sensible, if understood as the effect of some gradual change on another property. The notion of a gradual quantitative change in 'one and the same' property is quite different from the notion of a qualitative leap induced otherwise. Now, some of Engels' formulations cited above are fairly compatible with this position, namely the quantitative increase *gives rise to* a qualitative leap, and the quantitative change *brings about* a qualitative change (in the condition of the heated water). But in addition to these non-problematical – if rather trivial – statements, he introduces an alternative dialectical formulation as if it were logically entailed by the previous considerations: *consequently*, or *therefore*, quantity is transformed into quality and vice-versa.

When a quantitative change occurs in a property of a system (here its temperature, i.e. some energy is added or withdrawn) and this change 'gives rise to' or 'brings about' a novel property in the system (here its state of aggregation), then the concerned change is a cause of this modification, but in this process nothing ever *transformed* into something else properly, hence the dialectical re-formulation by

Engels is incomprehensible. *Heat* is simply not *aggregation*, and no given quantity of some element has ever been *transformed into* some other property of the system.

The dialectical notion of quality-quantity *transformation* only obscures the description of processes that are otherwise perfectly intelligible in terms owing nothing to any specific dialectical concept. Turning the notion of 'is the cause of something' into 'transforms itself into something else' does not make sense, while maintaining the ordinary description in terms of causality would leave DM with no original contribution to the understanding of the general processes of change.

6.5 Engels on Negation of the Negation

There is no chapter specifically dedicated to the law of negation of the negation and sublation in DoN, but this point is developed in A-D, in defense of Marx's *Capital*. Dühring qualifies the notion of negation of the negation as a piece of Hegelian word juggling that cannot support Marx's view on capitalist social evolution (Engels 2013, p. 171). Engels replies that there is no Marxist a priori but an observation of the evolution of the forms of property, from social to individual and then social again:

> Thus, by characterizing the process as the negation of the negation, Marx does not intend to prove that the process was historically necessary. On the contrary: only after he has proved from history that in fact the process has partially already occurred, and partially must occur in the future, he in addition characterizes it as a process which develops in accordance with a definite dialectical law. (Engels 2013, p. 171)

Hence the DM approach would consist of empirical observation and post hoc qualification as a dialectical evolutionary process, but this would be no demonstration of the general value of this dialectical law. It would seem to 'work' in some cases, but this is no proof of a universal scope, except through mere speculative induction (tentative generalization from some empirical observations). At least this approach by post-hoc qualification would have no predictive value. Even the interpretation of the observation itself raises problems. Why would some form of individual property *negate* a social property, any way we consider it? The more so when Engels himself states that the process of change may have *partially occurred*, which indicates that the forms of property may come in a variety of graduated states.

Engels' assertion *"and must partially occur in the future"* reveals another difficulty. This is no more a post-hoc qualification of an observed phenomenon, but a prediction according to a law, which is thus given the status of a necessity. This dialectical law would have been *proved from history* by Marx. Even a moderately trained DM analyst should notice the logical contradiction inside the above citation of Engels: mere post-hoc commentary and inductive speculation, or a proved predictive law? This is not to deny the important contribution of Marx to the study of social processes, just to question the lawful and universal status of *negation of the negation*, or lack of. A mature scientific law of change should try to explain the processes at work: how and in what circumstances should some process make a material system alternatively shift between two 'polar' states? Mere empiricism and post-hoc qualification will not do. Obviously, not all processes follow such an evolutionary pattern. And again, a law of restricted scope would not be a dialectical exclusivity. Sometimes yes, sometimes not is a trivial notion.

In A-D, Engels exposes his view on the dialectical evolution of philosophy:

> The old materialism was therefore negated by idealism. But in the course of the further development of philosophy, idealism, too, became untenable and was negated by modern materialism. This modern materialism, the negation of the negation, is not the mere re-establishment of the old [. . .]. It is no longer a philosophy at all, but simply a world outlook which has to establish its validity and be applied not in a science of sciences standing apart, but in the real sciences. Philosophy is therefore "sublated" here, that is "both overcome and preserved"; overcome as regards its form, and preserved as regards its real content. (Engels 2013, p. 181)

Taken literally, only idealists could conceive of philosophy in itself as something existing and undergoing an evolution. The unambiguously materialist Bungean position is that ideas have only 'conceptual existence' (a useful materialist notion coined by Bunge) in being imagined by material brains, thus only thinking beings are undergoing changes, not concepts that are not existing in themselves, hence not susceptible to evolve by themselves (Bunge 1981). Also Engel's clause – *no longer a philosophy... but simply a world outlook* – is a mysterious statement, given that a world outlook or world view is a philosophical concept properly, i.e. an ontology. By definition, at least a philosophical materialism of some kind is a necessary minimal ontology in DM thinking.

Even if we generously take Engels' statements above as mere awkward formulations in Hegelian style of otherwise sound notions, has the dialectical sublation principle any universal bearing for philosophy? In the previously cited passage, Engels chooses carefully ad hoc examples to satisfy his demonstration. But what is *the* philosophy after all? A simple check of the diversity of philosophical concepts defended by a variety of contemporaneous thinkers shows that materialism is far from having replaced idealistic thinking throughout the world, or the contrary. Another careful choice could as well support e.g. the pseudo-demonstration of the persistence of different forms of idealism, the continuing invention and development of some of them, or the complete extinction of others. The apparent universality of dialectic laws rests on careful choice of confirming examples while ignoring contrary evidence, as clearly denounced by Bunge.

Concerning negation of the negation and sublation, Engels overtly explains how he proceeds:

> I must not only negate, but also sublate the negation. I must therefore so arrange the first negation that the second remains or becomes possible. How? This depends on the particular nature of each individual case. If I grind a grain of barley [. . .] I have carried out the first part of the action but I have made the second part impossible. (Engels 2013, p. 184)

Here Engels evokes the classic example of the grain of barley, about which he also writes: "the grain...germinates; the grain as such ceases to exist, it is negated [. . .] the plant... the negation of the grain...grows...the stalk dies, [the plant] is in its turn negated." (Engels 2013, pp. 177–178). Here germination is 'negation' and death is also 'negation', where a biologist would simply identify a process of growth and the end of life – just another illustration of the general fuzziness of dialectical concepts such as *negation*, if not just play on words. But here Engels raises a difficulty. If the grain is ground (I presume that it would thus be dialectically 'negated' into flour and bran), the process of negation of the negation cannot be presented as a sublation. A further 'negation' (e.g. into bread or whatever) would not lead the system back to a previous grain state at a higher level. The same impossibility would ensue from the death of the seed via natural processes of predation or parasitism, not to mention the physiological fate of an ingested piece of bread. And here is almost naively exposed a way to 'arrange' negations, which is an element of dialectical method. Or is it just art and craftiness? A careful choice of illustrative examples may give an impression of universality of the dialectical process. Ignoring or at least overlooking part of the processes at work in the real world is a strong philosophical position after all. Turned in Bungean terms, "Polarity resides in our thinking about reality rather than in the world itself" (Bunge 1981, p. 50).

6.6   The Eleventh Thesis on Feuerbach

Concerning the *11th Thesis on Feuerbach* (Marx 1888), a first problem is that it is extremely sketchy, and with little obvious connection with the first ten theses. As mentioned above, Bunge's interpretation is twofold: *praxis trumps theory*, or *promoting social action without social science*. But other interpretations are possible, e.g. a philosophical method (an injunction to put one's philosophy to the test of concrete experience, through our action on the material world); or ethics (an injunction not to be content with describing and explaining the world, but also engage into action for changing the world). In other words, philosophers, quit your ivory tower and play an active part in trying to improve the condition of humankind, in addition to your useful philosophical reflections.

Neither of the two latter interpretations means that pragmatism should entirely *replace* philosophical reflection, nor that blind practice should *replace* theory. But anyway, these two possible interpretations would not make *Marx's 11th* thesis a principle exclusive to DM.

Now, another matter is what has been made of this famous thesis, and Bunge could be right in considering that abandonment of reflection in favor of pragmatic action guided by authoritarian dogmas has frequently been the case in DM-inspired political movements and governments. From inside the Marxist camp, the scientist and philosopher Robert Havemann courageously argued this way.

6.7   Robert Havemann, a Marxist Critic of Authoritarian DM

Bunge observed that there were practically no philosophical critics of DM from inside the DM and Marxist camps: "True, there have been a few feeble attempts to free dialectics from dogmatism, but they have retained the same 'spirit' – imprecision, superficiality, unwarranted claims of universality, and disregard for counterexamples" (Bunge 1981, p. 63). At that time, Bunge notably ignored the remarkable work of the East German Marxist, scientist and philosopher Robert Havemann (1910–1982) that was published from a series of conferences under the title *Dialectics without dogma? Science and world view* (Havemann 1964). Havemann overtly criticized the official DM philosophy in his country for being dogmatic and anti-scientific, so his work deserves a specific examination here.

There are some parallelisms in the philosophical trajectories of Bunge and Havemann. Both are scientists (mainly in physics, and chemistry, respectively), as well as philosophers and philosophers of science, and both have been influenced by DM philosophy in some way. But while dialectics was a rather 'dissident' philosophy in Bunge's Argentina, it was the official philosophy imposed by the authoritarian regime in Havemann's Stalinist and post-Stalinist East Germany. Once an important member of the upper political and academic circles, Havemann became a leader of the 'dissident' opposition to the regime during the 'De-Stalinization' period. His political and ethical vision of socialism required democracy, a complete liberty of thought and expression, and a 'socialist moral'. At first, his unorthodox political and philosophical writings were published, but later on they were censored, and Havemann paid for his bold criticisms with the price of his career and liberty. He nevertheless maintained his Marxist and anti-Stalinist socialist convictions, refusing to emigrate to 'western' countries despite lasting pressures. He was posthumously 'rehabilitated' by the East German communist party (SED) a short 3 weeks after the fall of the Berlin wall.

As underlined by Bunge (1981), DM is not entirely messy. This philosophy shows some worthy qualities, even if they are not original. Notably, Havemann is a scientist and evolutionary materialist (explicitly following Engels in these respects). He views material things as changing continuously by themselves, and he stands firmly in defense of science against idealism and pseudo-science, valuing the scientific method as the best way to investigate and explain material structures and processes. The impetus for this critic of the official views on science in his country was his observation, as an eminent scientific researcher, that the official dogmas go against scientific progress being made in advanced fields like quantum physics or genetics. His core position is that science trumps ideology and philosophical preconceptions, including the way DM laws are used, which he considers as a distortion of true materialist dialectics. His interpretation of the *11th Thesis* on Feuerbach nearly boils down to valuing the scientific method:

> ...the living philosophy...that is at work and proves its efficiency in sciences...is not a universal system but a world view...it evolves with our knowledge...not content with interpreting, it acts and resolves problems by tackling them concretely. (Havemann 1964, p. 139)

Havemann rejects the possibility of building a genuine dialectical logic, comparable to the great system of contemporaneous 'formal' logic (mathematics): "I think however that this is not possible. It is certainly not by accident that, up to now, all the tentatives made this way have completely failed" (Havemann 1964, p. 135). Bunge should approve: "In fact, no one has ever proposed any clear dialectical rules of formation or of inference" (Bunge 2003, p. 74).

Havemann also argues unambiguously that it is absurd to develop a formalist theoretical system of dialectics and impose it on the real world:

> Hegel has discredited his philosophy in the eyes of the scientists. (...) Cut from reality, it [dialectics] becomes incoherent controversies, taking the form of amazing, absurd, meaningless contradictions. Such a dialectic is not even materialist. (Havemann 1964, pp. 135–136)

He agrees with Engels that all scientists are, even unconsciously, influenced by diverse more or less sound philosophies, and that their theoretical work is itself of a philosophical nature, but Havemann observes that dialectical materialism has had nearly no influence on science:

> *Anti-Dühring* is a pamphlet targeting a certain Eugen Dühring...it was repulsive for most uninterested people. (...) as was the *Materialism and Empiriocriticism* of Lenin... largely ignored outside the Russian workers movement. Anyway, it was not addressed to theoreticians of science, nor to physicists, nor to biologists, etc. (Havemann 1964, p. 12)

And:

> One cannot resolve scientific questions by using a manual of dialectics. If it was possible, if such a method was correct, efficient, and valuable, then the scientists would have used for a long time such a satisfying means. (Havemann 1964, p. 17)

This goes very close to denying any use for dialectics in science. Otherwise Havemann explains the scientific method in quite classical terms, insisting on the fact that scientific theories remain open to refutation in the face of new observations of natural phenomena – he qualifies this attitude of 'dialectic', of course. Others would merely call it open mindedness.

In DM, nature is viewed as a material system, as already stated by Engels: "All the nature that is accessible to us constitutes a system, a coherent set of bodies (...) they act upon one another" (Engels 1975, p. 76), and Engels seems to be also philosophically systemist concerning the natural sciences, as could be interpreted from this short fragment: "*The classification of sciences*, every one of which analyses a particular form of movement or a series of related forms of movement and passing from one to the other..." (Engels 1975, p. 254). Otherwise Havemann also underlines that macrophysical and microphysical objects undergo different kinds of processes, the latter presenting some inherent indeterminism, while there is no possibility of requiring hidden variables to explain such indeterminism in quantum processes. He considers that the appeal to hidden variables proceeds from a classic mechanistic philosophy, while the acceptance of the peculiarities of quantum physics should be favored by a truly dialectic attitude. Others would simply call it scientific realism.

Avoiding the vague dialectical notion of 'movement', Bunge gives a clearer general explanation for both the interconnections and differences among the natural sciences, through his systemist-emergentist ontology. At every level of organization, the particular properties of a material system emerge through interactions between its components (moderate reductionism). Bunge makes such conceptual continuity among sciences dealing with 'neighboring' levels of organization a criterion for distinguishing the true natural sciences (interconnected) from the conceptually isolated pseudo-sciences (Bunge 1981).

But a notable difference between the DM and Bungean views is that the latter neatly separates mathematics from the natural sciences, while the dialectical laws are supposed to apply the same way to conceptual systems (like mathematics) and material things. For Bunge, a key notion of a consistent monist materialism is that no object is at the same time material and conceptual, hence there is no reason, and little chance, that the laws of material change should also happen to apply to ideas and conceptual systems. Notably, mathematics per se does not concern the material world (this is the *fictionist* view of mathematics); mathematics merely provides rules for sophisticated coherent reasoning, while every instance of applied mathematics requires a statement of what the mathematical objects refer to in the material world (Bunge 1981).

We saw how far Havemann pushed his criticism of DM, but he did not escape completely some serious dialectical flaws. His views on science and philosophy could boil down to anti-dogmatism, scientific open mindedness, and acceptance of the results of science rather than imposing philosophical preconceptions on the interpretation of the material world. Along with Engels, he considers that this attitude leads to the 'end of philosophy'. Anyway, his views are philosophical after all, including a materialist ontology to begin with, and also the valuation of science, which leads at best to the end of idealism, not philosophy in general.

Nevertheless, things are not that simple with Havemann, who claims to be a convinced proponent of dialectical thinking. He views Hegel as a 'great dialectician', unfortunately suffering from not being a materialist, and thus unable to aid the science of his time, while:

> The dialectical logic is the one contained in the things themselves, that we can only discover in the objects, in reality, and not in our heads. . . . We can effectively grasp dialectics only in its concrete reality. (Havemann 1964, pp. 135–136)

Here we learn that some logic can stand in material objects, ready to be discovered, and that a concept like dialectics has the property of 'concrete reality', which is clearly at odds with materialism. But even admitting this for the sake of the argument, if the discovery of the 'dialectic logic' requires an intimate knowledge of material things, how could Hegel have become a great dialectician through a purely idealistic approach? Are dialecticians so fond of contradictions that they deliberately introduce some incoherence in their own arguments? More charitably, Havemann could be considered as a victim of East German education and indoctrination.

A possible excuse for Havemann's appreciating obscure aphorisms of Hegel is that he had been trained by dialectically twisted teachers professing the official dogma in an oppressive totalitarian society, presenting Hegelian-inspired dialectics as the acme of intellectual skills. By contrast, Bunge is adamantly against obscure philosophy, which he considers as the frequent mask of intellectual imposture: "Philosophy should enlighten, not obscure" (Bunge 2003, p. 5). Bunge was not aware of the work of Havemann when he wrote, "there have been a few feeble attempts to free dialectics from dogmatism" (Bunge 1981, p. 63). Havemann's courageous work has been a rather strong attempt, not a feeble one, given the political conditions of his time. Nevertheless, it shows exactly the limits of such enterprises as already underlined by Bunge, who presently can state, "I admire the integrity of Havemann, not his philosophy" (Bunge 2017, personal communication).

6.8   Conclusion: Why Bother with Dialectical Materialism?

From personal experience, I can understand the appeal for DM to a young person interested in humanism and social progress. Bunge, as a teenage product of humanist and socialist education, was temporarily impressed by DM, up to the point of "seeing dialectics everywhere" (Bunge 2017, personal communication). But, unlike many intellectuals of Marxist obedience, he rapidly objected to this philosophy from a rationalist, realist and scientist point of view (Bunge 2016, pp. 43–44). Later on, he saw that DM ontology tends to hinder scientific progress (Bunge 1981), a serious matter for a philosopher highly concerned by the influence of philosophy on scientific progress (Bunge 2012, pp. 7–9).

Because they stick unshakably to their traditional jargon, obscure formulations, and fuzzy concepts, dialectical philosophers tend to deprive themselves from possible constructive debate with other materialist philosophers and scientists. But it is up to them to try and clarify their positions. Whatever the merits of Havemann's internal critique of some utterly dogmatic and fossilized aspects of DM, the analysis of his philosophy tends to confirm the following view of Bunge:

> A system of dogmas cannot be much improved by partial reformations or revisions: it calls for a fundamental revolution touching upon principles as well as upon the very way basic principles have to be proposed, systematized, tested and discussed. In this regard dialecticians have proved to be conservative, not revolutionary. (Bunge 1981, p. 63)

Havemann, whatever his merits, proved able to perform only part of the task. Notably, he missed the key materialist notion of 'conceptual existence' (Bunge 1981), to surmount an apparent paradox, which is a lasting difficulty for many materialist philosophers. Concepts 'exist' in some way even though they are not material. Material and conceptual existence are properties of different things (real objects versus imaginary ones), hence they are not "opposites", and this distinction owes nothing to a 'dialectical' analysis. Not all differences are contradictions. Otherwise, the readily different nature of material things (existing) and concepts (non-existing by themselves) makes it highly improbable that similar laws of change should rule both the evolutionary processes in material systems and the pseudo 'evolution' of concepts. Concepts are not evolving properly, they are merely imagined through neural processes in a changing brain. The reflection theory is thus unrealistic, whatever the way you may conceive it: Hegelian style from concept to matter, or Engelsian from matter to concept.

Havemann boldly denounced the negative consequences of dogmatically imposed DM on the development of science in the Eastern Germany and Soviet Union of his time. Arguably, this is much less of a concern by now, given the political evolution in the ex-Eastern Bloc countries. But presently, the People's Republic of China remains under influence of a politically enforced DM world view, with lasting detrimental influence on the development of philosophy of science in that country, despite some positive evolution in academic circles (Guo 2014). Hence there are still significant stakes in criticizing DM beyond mere academic studies in the history of ideas. Remarkably enough, in 2011 Bunge was invited to give conferences in several Chinese universities, including to the Communist Party's School of Marxism. There he exposed his main criticisms of authoritarian socialism and of the flaws of DM, notably:

> (. . .) it is high time to ask which components of Marxism have become obsolete, which ones might be salvaged, and how these might be updated to meet the requirements of contemporary science and philosophy. (. . .) Contrary to the conventional criticisms of Marxism, mine are offered in a constructive spirit: my goal is to separate the superseded chaff from the grain, which I take to be materialism, realism, and the ideal of social justice. (. . .) Marxist ontology and epistemology are in deep trouble because they have remained stagnant or worse due to their attachment to dialectics. But there is hope in materialism, for it is the tacit ontology of science and technology, just as realism is their tacit epistemology. (. . .) we need philosophies of values, right conduct, and political action, admitting the existence of universal values, such as welfare and solidarity, as well as the norm of justice enshrined in the motto of the First International: No duties without rights, and no rights without duties. All the norms of practical philosophy should be based on science and technology, and they should be regarded as testable in practice: here, as elsewhere in matters of fact, we should stick to realism and shun dogmatism. (Bunge 2011, unpublished lectures)

Whatever the merits of DM in the nineteenth century, at the present time it appears ineffective and even detrimental to try to base a materialist philosophy on DM notions, while attempting to make sense of them and to revise the most utterly erroneous or fuzzy ones. We have now access to a modern materialist, emergentist, exact and scientific philosophical system, of which Mario Bunge elaborated a clear version with an exceptionally large scope, while carefully avoiding to fix it in dogmatism; this makes DM investigations presently look, to say the least, as a superfluous detour.

References

Bunge, M. (1981). *Scientific materialism*. Dordrecht: Reidel.
Bunge, M. (2003). *Philosophical dictionary*. Amherst: Prometheus Books.
Bunge, M. (2012). *Evaluating philosophies*. Dordrecht: Springer.
Bunge, M. (2016). *Between two worlds: Memoirs of a philosopher-scientist*. Dordrecht: Springer.

Engels, F. (1876/1975). *Dialectique de la nature*. Paris: Éditions Sociales.
Engels, F. (1878/2013). *Anti-dühring & the role of force in history*. Montreuil: Éditions Science Marxiste Eds.
Guo, Y. (2014). The philosophy of science and technology in China: Political and ideological influences. *Science & Education, 23*(9), 1835–1844.
Havemann, R. (1964). *Dialektik ohne Dogma? Naturwissenschaft und Weltanschauung*. Reibek: Rowohlt Tascheubuch Verlag.
Marx, K. (1888). *Theses on Feuerbach*. Marx/Engels internet archive. https://www.marxists.org/archive/marx/works/1845/theses/theses.htm
Ollman, B. (2003). *La dialectique mise en œuvre*. Paris: Syllepse.

Chapter 7
Quantifiers and Conceptual Existence

María Manzano and Manuel Crescencio Moreno

Abstract This chapter examines Bunge's distinction between the logical concept of existence and the ontological one. We introduce a new conceptual existence predicate in an intensional environment that depends on the evaluation world. So that we can investigate restricted areas (worlds) where the different kinds of concepts might exist. We hope this new predicate would encompass Bunge's philosophical position which he designates as conceptualist and fictional materialism. The basic hybridization (adding nominals and @ operators) acts as a bridge between intensions and extensions because @ works as a useful rigidifier. In hybrid logic, the accessibility relation and many properties this relation might have can be easily expressed in the formal language. The initial hypothesis is that hybridization and intensionality can serve as unifying tools in the areas involved in this research; namely, Logic, Philosophy of Science and Linguistics.

7.1 Introduction

In this paper we will revise María Manzano's paper, "Formalización en Teoría de Tipos del predicado de existencia conceptual de Mario Bunge" (Manzano 1985), using the machinery of Intensional Hybrid Type Theory. The point of departure is again Bunge's distinction between the logical concept of existence and the ontological one, but now the logic employed is even more powerful than type theory. The new conceptual existence predicate in an intensional environment depends on the evaluation world. So that we can investigate restricted areas (worlds) where

This research has been possible thanks to the research projects sustained by Ministerio de Economía y Competitividad of Spain with reference FFI2013-47126-P and FFI2017-82554-P.

M. Manzano (✉) · M. C. Moreno
Department of Philosophy, University of Salamanca, Salamanca, Spain
e-mail: mara@usal.es; manuelcrescencio@usal.es

© Springer Nature Switzerland AG 2019
M. R. Matthews (ed.), *Mario Bunge: A Centenary Festschrift*,
https://doi.org/10.1007/978-3-030-16673-1_7

117

the different kinds of concepts might exist. We hope this new predicate would encompass Bunge's philosophical position which he designates as conceptualist and fictional materialism. The basic hybridization (adding nominals and @ operators) acts as a bridge between intensions and extensions because @ works as a useful rigidifier. In hybrid logic, the accessibility relation and many properties this relation might have can be easily expressed in the formal language. The initial hypothesis is that hybridization and intensionality can serve as unifying tools in the areas involved in this research; namely, Logic, Philosophy of Science and Linguistics.

7.2 Historical Background

Bunge's claim that "traditional philosophers have usually defended that existence is a property (or a predicate)" while "modern logicians have stated that existence is not a predicate but a quantifier, i.e., the existential quantifier ∃" (Bunge 1980, p.61), can be evaluated as accurate provided that quantifiers are not considered to be predicates at all. Nevertheless, from Frege on, "existence" has also been regarded as a predicate, but a predicate referring to a second-order property, which turns the predicate of existence into a second-order predicate.[1]

The difference between traditional philosophers and modern logicians concerning existence lies in the order or level where existence belongs to. On the one hand, traditional philosophers have preferred to analyze existence as a first-order predicate, on the other hand, modern logicians have opted for putting existence up a little higher as a second-order predicate. First-order properties can be applied to individuals, as in "Socrates is wise"; second-order properties, however, are said of first-order properties, as positive is said of wisdom, for example, in "Wisdom is positive".

Much has been said about existence in the course of the history of philosophy. Traditionally, philosophers have not found any special difficulty in predicating existence as a property of things. Although different nuances can be found in the treatment of existence by Aristotle, Avicenna, Aquinas or Descartes, they all agree in assigning a predicative use to existence. The first criticisms concerning this predicative use draw back to the eighteenth century. Unlike previous philosophers, Hume and Kant concentrated their efforts in showing how existence "makes no addition" to the mental conception of an object: "A hundred real thalers do not contain the least coin more than a hundred possible thalers"(Kant 1787/1929, B627). Kant therefore concludes with the denial of existence as a *real* predicate, given that it adds nothing to the idea of a particular thing.[2]

---

[1]Nowadays, to discriminate *object language* from *metalanguage* we make a distinction between predicates (of the object language) and properties of the metalanguage.

[2]In spite of this conclusion, Kant allows existence to be a logical predicate for "anything we please can be made to serve as a logical predicate"(Kant 1787/1929, B626).

For his part, Frege moved away from the traditional philosophers by saying that existence is not a first-order predicate but a second-order predicate. In Frege's words: "Affirmation of existence is in fact nothing but the denial of the number nought" (Frege 1884/1950, p.65). Therefore existence is applied to (first-order) properties and not to individuals, and the attribution of existence to a given first-order property amounts to saying that there is at least one individual which instantiates the given property. Existence and instantiation are in consequence equivalent notions for Frege. Sentences of natural language like "Socrates exists" should be analyzed not as if there were a property of existence attributed to the individual Socrates, but as

$$\exists x (x = \text{Socrates})$$

which can be translated as "there is at least one individual which is identical with Socrates" or, much better, "there is at least one individual from which can be said that has the property of being Socrates":

$$\exists x \, \text{IsSocrates}(x)$$

In both cases, the second-order predicate of existence is represented by means of the existential quantifier ∃ which can be translated roughly as "there is". In the first case, the predicate of non-emptiness "there is at least one thing" is said of the predicate "being identical with Socrates". In the second case, "there is at least one thing" is predicated of the expression "being Socrates". Notwithstanding, in both cases nothing is predicated of Socrates.

For Bunge, the problem of the predicate of existence can be solved by "distinguishing two concepts that the modern logicians have confused: the logical concept *something* and the ontological *exists*" (Bunge 1980, p. 61). Consequently, it seems that Bunge is in favor of combining together the traditional and the Fregean views on the existence predicate, defending that existence can act not only as a second-order predicate (as in Frege) but also as a first-order predicate (like the philosophers preceding Hume and Kant). Existence would be then a logical predicate as well as an ontological predicate. In the end, both predicative uses should be perceived as complementary and not as mutually exclusive.

In the next section we will use the language of type theory, inspired on Russell's hierarchy of types, to formulate Bunge's distinction. Let us see how Russell presents his ontology in an easy way:

A term or individual is any object which is not a range. This is the lowest type of object ... the objects of daily life, persons, tables, chairs, apples, etc. are classes as one. (A person is a class of psychical existents, the others are classes of material points, with perhaps some reference to secondary qualities.) These objects, therefore, are of the same type as simple individuals ... Individuals are the only objects of which numbers cannot be significantly asserted.

The next type consists of ranges or classes of individuals …
The next type after classes of individuals consists of classes of classes of individuals.
Such are, for example, associations of clubs; the members of such associations, the clubs,
are themselves classes of individuals. It will be convenient to speak of classes only where
we have classes of individuals, of classes of classes only where we have classes of classes
of individuals, and so on (Russell 1903, §497).

The important issue is not whether existence is a predicate or not, but to be able to add some properties (like to be real) to the predicate we are defining. We will show that the relevant distinction is not the order of the predicate, as both the quantifier and the ontological existence can be defined as second order predicates, but that this treatment allows to add some other properties to the object we are qualifying as existent.

7.3   Bunge's Existence Predicate in Type Theory

In 1985 María Manzano published *Formalización en teoría de tipos del predicado de existencia de Mario Bunge* (see Manzano 1985), where she developed, in the framework of type theory, some of the distinctions Mario Bunge made in Chapter 3 of Bunge (1980), untitled *Naturaleza de los objetos conceptuales*. Manzano agrees to distinguish the logical concept of the existence, as expressed with the quantifier, from the ontological one. By using the powerful language of type theory, the existence predicate is formalized in the system $\mathfrak{TT}$ of Church (1940) and Henkin (1950) as well as in the system $\mathfrak{ET}$ of equational type theory (Henkin 1963; Andrews 1963).

Prior to that formalization, we will see how quantification is defined in these languages. The version of type theory used in both cases is the one denominated *simple type theory*, where we quantify over a hierarchy of universes using a variety of variables to mark the distinction between the different levels.

7.3.1   The Hierarchy of Types

The hierarchy of types and the language used to talk about it, use type symbols to distinguish the different levels that constitute the types. In particular.

$$\mathsf{TYPES} := e \mid t \mid \langle a, b \rangle$$

where $a, b \in \mathsf{TYPES}$.

Intuitively, e marks the type of individuals and t denotes the type of truth values, $\langle a, b \rangle$ denotes the type of functions from elements of type b to elements of type a. The hierarchy of types $\langle D_a \rangle_{a \in \mathsf{TYPES}}$ is built from a non-empty universe of individuals, $D_e \neq \varnothing$, and the two truth values universe $D_t = \{T, F\}$. At level $\langle a, b \rangle$ the universe $D_{\langle a,b \rangle}$ contains functions from D_b to D_a.[3]

The interesting aspect of this organization is that the categories of concepts that Bunge distinguishes in his work can be properly placed at different levels by using the classification created to avoid the classical well known paradoxes whose root Russell identifies with the self-reference or reflexiveness.[4]

> In all the above contradictions (which are merely selections from an indefinite number) there is a common characteristic, which we may describe as self-reference or reflexiveness. The remark of Epimenides must include itself in its own scope. If all classes, provided they are not members of themselves, are members of w, this must also apply to w; and similarly for the analogous relational contradiction ... Thus all our contradictions have in common the assumption of a totality such that, if it were legitimate, it would at once be enlarged by new members defined in terms of itself. (Russell 1908, pp. 224–225).

7.3.2 The Formal Languages $\mathfrak{TT}$ and $\mathfrak{CT}$

The language is also free of other paradoxes originated on the lack of distinction between formal language and metalanguage, as the notions of denotation and truth in a structure are defined according to Tarski's patron.

The formal language $\mathfrak{TT}$ used to talk about the hierarchy of types includes the very relevant lambda operator λ as well as variables of all types, plus some constants. Intuitively $(F_{\langle a,b \rangle} A_b)$ denotes the value of the function denoted by $F_{\langle a,b \rangle}$ at the argument denoted by A_b, both denotations in the corresponding level of the hierarchy $\langle D_a \rangle_{a \in \text{TYPES}}$. Also, $\langle \lambda x_b M_a \rangle$ denotes the function whose value at any element of type b is the denotation of M_a (which generally depends on x_b). If M_a is of type t, $(\lambda x_b M_t)$ is a predicate representing all the elements of type b that verify M_t. $\mathfrak{TT}$ includes the usual connectives ($\neg, \vee, \wedge, \rightarrow, \leftrightarrow$) as constants of types $\langle t, t \rangle$ or $\langle \langle t, t \rangle, t \rangle$ as well as the predicates Π and Σ. In fact, quantifiers in $\mathfrak{TT}$ are introduced by definition using these two constants

$$\forall x_a \varphi_t :=_{Df} \Pi_{\langle t, \langle t, a \rangle \rangle} \langle \lambda x_a \varphi_t \rangle$$

$$\exists x_a \varphi_t :=_{Df} \Sigma_{\langle t, \langle t, a \rangle \rangle} \langle \lambda x_a \varphi_t \rangle$$

This is so because $\Pi_{\langle t, \langle t, a \rangle \rangle}$ is a predicate of predicates and when we apply it to the lambda predicate $\langle \lambda x_a \varphi_t \rangle$, the resulting formula is saying that the lambda

---

[3]In the standard hierarchy, each $\mathcal{D}_{\langle a,b \rangle}$ contains the whole $\mathcal{D}_a^{\mathcal{D}_b}$ while in the general models, $\mathcal{D}_{\langle a,b \rangle}$ is a subset of $\mathcal{D}_a^{\mathcal{D}_b}$ closed under definability. General models were first introduced by Henkin (1950) to prove the completeness of type theory.

[4]For a longer explanation, see section 1.2: "Paradoxes and their solution in4 type theory" in Manzano (1996, pp. 182–186).

predicate is universal; i.e., φ_t is true for all the elements of domain D_a. On the other hand, $\Sigma_{\langle t, \langle t,a \rangle \rangle} \langle \lambda x_a \varphi_t \rangle$ says that the predicate $\langle \lambda x_a \varphi_t \rangle$ applies with truth to some elements of D_a.

The language $\mathfrak{E}\mathfrak{T}$ contains only λ and equality, $\equiv$, as primitives and the quantifiers as well as the connectives are introduced by definition.[5] In particular,

$$\forall x_a \varphi_t := \langle \lambda x_a \varphi_t \rangle \equiv \langle \lambda x_a (x_a \equiv x_a) \rangle$$

$$\exists x_a \varphi_t :=_{Df} \neg \forall x_a \neg \varphi_t$$

Equality is interpreted as identity and so $\langle \lambda x_a (x_a \equiv x_a) \rangle$ is a predicate that is true for all members of D_a. Therefore, to say that $\langle \lambda x_a \varphi_t \rangle$ is equal to $\langle \lambda x_a (x_a \equiv x_a) \rangle$ is a way to express that φ_t is true for all the elements of D_a.

So, in $\mathfrak{T}\mathfrak{T}$ and in $\mathfrak{E}\mathfrak{T}$ the quantifiers are introduced by definition, using the lambda operator and some logical predicate constants (Π, Σ and $\equiv$). Therefore, the existential quantifier can be considered as a second order predicate. According to Bunge's view the existential quantifier has no existential content and the same happens here in $\mathfrak{T}\mathfrak{T}$ and $\mathfrak{E}\mathfrak{T}$ as the symbols used are only logical constants. As we will see in the next section, the ontological existence is also a predicate. The important point is that we need non logical constants as well as the logical ones to define this ontological existence.

7.3.3 Ontological Existence

Bunge distinguishes two kinds of existence predicates that encode the desired ontological existence with existential content: one for conceptual existence and another for physical (or real) existence.

In the first place, using the characteristic function χ_A on a set A, Bunge express "x exists at A" by saying that the value of x under this function is 1, while "x does not exist at A" is to say that the characteristic function gives the value 0. The predicate of relative or contextual existence E_A is defined as a function from A to the set of propositions, and its value for a particular x is just $\chi_A(x)$. So, when x is an object, to say that "x exists conceptually" is the same as "there is a non-empty set C of constructs to which C belongs to"; namely, $E_C(x)$ for a non-empty set of constructs. In a similar way he expresses that "x exists physically".

The binary predicate of existence "in ... exists ..." has an easy formalization in type theory because the whole theory of types was built using lambda operator as its basic tool, and so Manzano proposed to use the lambda term

$$\langle \lambda X_{\langle t,a \rangle} \langle \lambda x_a \left(X_{\langle t,a \rangle} x_a \right) \rangle \rangle$$

---

[5]The definition was first done by Henkin (1963) and improved by Andrews (1963).

for it. This expression can be understood as the membership binary predicate between objects of any type a and $\langle t, a \rangle$, and when applied to a predicate $R_{\langle t,a \rangle}$, we get

$$\left(\left\langle \lambda X_{\langle t,a \rangle} \left\langle \lambda x_a \left(X_{\langle t,a \rangle} x_a \right) \right\rangle \right\rangle R_{\langle t,a \rangle} \right)$$

expressing "in $R_{\langle t,a \rangle}$ exists...". Finally, "in $R_{\langle t,a \rangle}$ exists A_a" is formalized as

$$\left(\left(\left\langle \lambda X_{\langle t,a \rangle} \left\langle \lambda x_a \left(X_{\langle t,a \rangle} x_a \right) \right\rangle \right\rangle R_{\langle t,a \rangle} \right) A_a \right)$$

From this point on, Bunge distinguishes two concepts of existence; namely, the physical existence and the conceptual one. For an object x Bunge establishes the following[6]:

> If x is an object, then
>
> a. x exists conceptually $=_{df}$ There is a non-empty set C of constructs such that $E_C(x)$;
> b. x exists physically $=_{df}$ There is a non-empty set P of physical objects such that $E_P(x)$.
> (Bunge 1980, p. 62).

In Manzano (1985) the constants $C_{\langle t,a \rangle}$ of type $\langle t, a \rangle$ were used to say that "... is a construct" and so "... is a non-empty set of constructs" is defined by

$$\left\langle \lambda X_{\langle t,a \rangle} (\exists x_a \left(X_{\langle t,a \rangle} x_a \right) \wedge \forall x_a (X_{\langle t,a \rangle} x_a \rightarrow C_{\langle t,a \rangle} x_a)) \right\rangle$$

where the quantifiers $\exists x_a$ and $\forall x_a$ can as well be introduced with the logical predicates Σ and Π (or λ and $\equiv$). Finally, "... exists conceptually" is defined by

$$\langle \lambda y_a \exists Z_{\langle t,a \rangle} ((\langle \lambda X_{\langle t,a \rangle} (\exists x_a \left(X_{\langle t,a \rangle} x_a \right) \wedge \forall x_a (X_{\langle t,a \rangle} x_a \rightarrow C_{\langle t,a \rangle} x_a)) \rangle Z_{\langle t,a \rangle})$$
$$\wedge (\langle \lambda X_{\langle t,a \rangle} \langle \lambda x_a \left(X_{\langle t,a \rangle} x_a \right) \rangle \rangle Z_{\langle t,a \rangle}) y_a)\rangle$$

The whole predicate, when applied to a particular A_a to express that "A_a exists conceptually" gives a long formula as formalization

$$(\langle \lambda y_a \exists Z_{\langle t,a \rangle} ((\langle \lambda X_{\langle t,a \rangle} (\exists x_a \left(X_{\langle t,a \rangle} x_a \right) \wedge \forall x_a (X_{\langle t,a \rangle} x_a \rightarrow C_{\langle t,a \rangle} x_a)) \rangle Z_{\langle t,a \rangle})$$
$$\wedge (\langle \lambda X_{\langle t,a \rangle} \langle \lambda x_a \left(X_{\langle t,a \rangle} x_a \right) \rangle \rangle Z_{\langle t,a \rangle}) y_a) \rangle A_a)$$

which in lambda calculus is proven equivalent to the formula $C_{\langle t,a \rangle} A_a$.

---

[6]This is our translation of the Spanish original:

"Si x es un objeto, entonces:

a] x existe conceptualmente $=_{df}$ Algún conjunto no vacío C de constructos es tal que $E_C(x)$;
b] x existe físicamente $=_{df}$ Algún conjunto no vacío F de entes físicos es tal que $E_F(x)$." (Bunge 1980, p. 62).

The result is not surprising as to exist conceptually is the same as being a construct in Bunge's ontology. The goal was to analyze the sorts of existence already mentioned and to identify the logical constants used, and whether or not we need non logical ones as well.

So far we have seen how to formalize in type theory Bunge's distinctions, now we will show that the structures we use in type theory to interpret our formulas are adequate to place Bunge's concepts, defined as the basic units of what he calls constructs.

7.3.4   Concepts and Constructs

But, what are the constructs? Bunge classifies them into concepts (as the basic units used to build propositions), propositions (which are true or false), contexts (sets of propositions with common referents) and theories (contexts closed under logical operations). Concepts themselves are classified into individuals, sets and relations.

Let us see how all these categories are refined in type theory. In the first place, we clearly distinguish between the structures we want to talk about and the language used to do it. We have presented already the formalization of the desired existence predicates, now we will see where in our structures the basic concepts lay.

The structures we talk about in type theory include different quantification domains corresponding to the different types of variables. In the hierarchy of types we have functions of all types $\langle a, b \rangle$. In particular, at the elementary level, in D_e we have individuals, in $D_{\langle t,e \rangle}$ we have sets of individuals (presented as characteristic functions) and in $D_{\langle t, \langle t,e \rangle \rangle}$ we have sets of sets of individuals. Binary relations on individuals are in $D_{\langle \langle t,e \rangle, e \rangle}$ while unary functions on individuals are in $D_{\langle e,e \rangle}$, etc.

Propositions, contexts and theories are represented here as sentences of the language and as sets of sentences with certain properties. In logic, an axiomatic theory with axioms in Γ is the set of its logical consequences, CON

$$CON(\Gamma) = \{\varphi \in SENT(\mathfrak{TT}) : \Gamma \vdash \varphi\}$$

What about existent objects? Shall we divide the set D_e of individuals into physical and conceptual objects, $D_e = \mathcal{P}_e \cup \mathcal{C}_e$, in such a way that $\mathcal{P}_e \cap \mathcal{C}_e = \varnothing$ and assign physical existence just to the elements of $\mathcal{P}_e$? What about higher levels in the hierarchy? What about other types of conceptual objects? In her article of 1985 Manzano wrote:

> I investigate what conceptual objects are. I reach the conclusion that it is better to study a restricted area each time, where existence could even be assigned in different degrees. For instance, in set theory —like in *Animal Farm* of Orwell— every set exists but some "exist more" than others. Of course, in relating degrees of existence to degrees of definability I am not following Bunge. (Manzano 1985, p. 513).

In the same volume Bunge replied with a note: "¿Grados de existencia o grados de abstracción?" (Bunge 1985), where he argues that Manzano's distinction measures degrees of abstraction but not degrees of reality.

In what follows we will introduce another system of type theory where for each type b, the domain of quantification D_b is not unique.

7.4 Existence in Hybrid Logic

Bunge has considered that there is a clear distinction between the existence predicate and the existential quantifier, more precisely the "particularizer or undetermined quantifier" (Bunge 1980, p. 63) as he proposes to rename $\exists$. From the point of view of logic there is no clear opposition between these alternatives. The main reason being that we have tools to deal with the quantifier as a predicate. In fact, it is possible to give a coherent account of both predicates: the quantifier and the existence. As we saw in the previous section, in type theory we can treat existence always as a predicate, the important point is whether or not we need non-logical constants to define it.

In this section we will explore another possibility, we want to incorporate in the formal system the tools needed to define the ontological existence predicate using only logical constants.

How to do it? If we start from the assumption of accepting a varying domain model, we can define a predicate of existence (real or conceptual) EXISTS by means of the quantifier $\exists$, as the following formula shows $\langle \lambda x \exists y (y = x) \rangle$.

$$\text{EXISTS}(\tau) := \langle \lambda x \exists y (y = x) \rangle(\tau)$$

Now we wonder what is the logical system to be used. The main logical systems where we allow different domains of quantification are many-sorted logic and modal logic.

In the first case, the logical language and the structures used to interpret its expressions are conceived as many-sorted; that is, different sets of variables range over the different sorts of domains. For instance, we can use x, y, z, etc. for variables of "real" sort and variables α, β, γ, etc. for variables of "conceptual" sort. Therefore, the variables used in the previous formula are telling you what the domain of quantification is, and so what kind of existence you are predicating of the term τ. In lambda calculus this formula

$$\langle \lambda x \exists y (y = x) \rangle(\tau)$$

is equivalent to $\exists y (y = \tau)$, so either of them can be used.

Let us explore the second option, that of modal logic. In the preface of his inspiring book, *Types, Tableaus, and Gödel's God*, Fitting declares: "[add] the usual box and diamond to the syntax, and possible worlds to the semantics. It is now that

choices must be made, since quantified modal logic is not a thing, but a multitude."
(Fitting 2002, pp. XII–XIII).

In this logic, the usual concept of *truth in a structure* is replaced by the concept of *truth in a world of a structure*. Since a modal model contains many possible worlds, we should decide whether the domain of discourse should be fixed, or be allowed to vary from one world to another. However, there is only one sort of variables to refer to individuals, what makes the difference is the evaluation world.

In modal first-order logic, we can choose between varying domains and constant domain models.

In the first case,

> the domain of the first-order quantifier is allowed to vary form world to world. This interpretation of a quantifier is called *actualist quantification* since a quantifier range over individuals that actually exist, that is, individuals that exist in the actual world. (Braüner 2011, p. 131).

In the following section we will see that in a varying domain model the previous predicate EXISTS is enough to define the ontological existence.

In the second case, "a different interpretation of a quantifier is obtained if it is required that the domain is constant from world to world. This is called *possibilist quantification* since the quantifier in this case ranges over individuals that possibly exist" (Braüner 2011, p. 131).

In case we prefer to start with modal logic with constant domains, we can introduce a constant existence predicate E whose denotation varies with the different worlds. This predicate has to be a non-logical primitive constant predicate and it is not definable by means of the existential quantifier. E is true at any world w of the things that actually exist at w.

These two possibilities in modal logic offer us two choices concerning existence: it can be defined through the existential quantifier in varying domain models; or it can be a first-order predicate constant in constant domain models.

Which choice is better?

Fitting prefers the last one, as far as we know, for pedagogical reasons: "However, either an actualist or a possibilist approach can simulate the other. I opt for a possibilist approach, with an explicit existence predicate, because it is technically simpler" (Fitting 2002, p.XIII).

The final verdict on the very nature of existence is decided when you choose which logical system you prefer to use.

7.4.1 Existence in Hybrid First-Order Logic

In the previous Sect. 7.3.4 we considered the situation where the universe of individuals D_e is divided into two disjoint non-empty sets of physical and conceptual objects ($\mathcal{P}_e$ and $\mathcal{C}_e$). The first-order quantifiers ($\exists x_e$ and $\forall x_e$) run over the whole D_e and so neither the physical existence nor the conceptual existence can be formalized in pure logical expressions. We have to introduce non-logical predicates ($P_{\langle t,e\rangle}$ and $C_{\langle t,e\rangle}$) to formulate the desired concepts.

However, in modal logic we can have different evaluation worlds (say w_1 and w_2) with different universes, such as $\mathcal{P}_e$ and $\mathcal{C}_e$, and quantification can be restricted to a particular domain. In this logic a formula φ can be true at world w_1 but false at world w_2. In symbols, $[[\varphi]]^{\mathcal{M},w_1,g} = T$ but $[[\varphi]]^{\mathcal{M},w_2,g} = F$, where $\mathcal{M}$ is a modal structure, w_1 and w_2 are different worlds in $\mathcal{M}$, and g is an assignment.

The vocabulary of first-order modal logic might include individual constants as well as predicates and functional constants, and their interpretation at $\mathcal{M}$ could vary from world to world; they are defined as functions from worlds to objects. Such functions are called *intensions*.

Some authors call *individual concepts* to the intensions of individual constants whose extensional denotation is not a constant value as it could change from world to world. So, even for atomic formulas such as Mc the value at w_1 could be different from the value at w_2, depending on the values of M and c at w_1 and w_2.

For quantified formulas, the interpretation at different worlds assume different domains and so we can have $[[\forall x_e\varphi]]^{\mathcal{M},w_1,g} = T$ but $[[\forall x_e\varphi]]^{\mathcal{M},w_2,g} = F$ even in the case where the non-logical constants appearing in formula φ have rigid interpretation[7]; that is so because in the first case the domain of quantification is $\mathcal{P}_e$ while in the second it is $\mathcal{C}_e$.

This is good news because we can use the existential quantifier to say that the object named by a rigid term τ exists (physically or conceptually). $\exists x_e(x_e \equiv \tau)$ can be used with this purpose as the formula could be true at w_1 but false at w_2. In the first case the denotation of the term τ must belong to $\mathcal{P}_e$.

However, even though the formula $\exists x_e(x_e \equiv \tau)$ receives different interpretations at different evaluation worlds, there is nothing in the formula revealing where the formula is true or false. This is the weak point of orthodox modal logic: the worlds only appear in the Kripke semantics but there is nothing in the language to refer to them. This lack of expressiveness was first solved by Patrick Blackburn, when he introduced hybrid logic (for surveys of hybrid logic, see Blackburn 2000; Areces and Ten Cate 2007).

Hybrid logic is an extension of modal logic in which it is possible to name worlds using special atomic formulas called *nominals*. Nominals are true at a unique world in any model; that is, a nominal i names the world it is true at.

Once nominals have been introduced, it becomes natural to make a further extension: to add modalities of the form $@_i$, where i is a nominal, and to interpret formulas of the form $@_i\varphi$ as asserting that φ is true at the unique world named by i. The $@_i$ operator can take as arguments not merely formulas but other expressions too. In this approach, $@_i\tau$ denotes the individual that the term τ denotes at the world named by i. To put it another way: $@_i\tau$ is a new term that rigidly designates what τ denotes at the i-world.

---

[7]By a "rigid interpretation" we understand one that does not change from world to world.

Let us have i and j as nominals to name the world w_1 of physical objects and the world w_2 of conceptual objects, respectively. The formula $@_i\varphi$ says that φ is true at the world of physical objects and $@_j\varphi$ says the same for the world of conceptual objects. In this logic, $@_i \exists x_e(x_e \equiv \tau)$ can be used to express that the object named by the term τ exists at the world named by i; the reason being that this formula is rigid and so either

$$[[@_i \exists x_e(x_e \equiv \tau)]]^{\mathcal{M},w_1,g} = T = [[@_i \exists x_e(x_e \equiv \tau)]]^{\mathcal{M},w_2,g}$$

or

$$[[@_i \exists x_e(x_e \equiv \tau)]]^{\mathcal{M},w_1,g} = F = [[@_i \exists x_e(x_e \equiv \tau)]]^{\mathcal{M},w_2,g}.$$

Therefore, in hybrid first-order logic we do not need the existence predicate constant E to say that an object exists either physically or conceptually; we can use the quantifier, the equality and the $@_i$ operator, we do not need specific non-logical predicate constants.

However, there are important issues that are not solved in hybrid first-order logic. In the first place, we wanted a hierarchy of domains, not just a domain of individuals and we wanted to quantify over all the levels in the hierarchy as we do in type theory.

The first question to ask is, are these higher-order domains also different from world to world or rather the requirement of having different universes corresponding to the different worlds only apply at the elementary level? Does it make sense an actualist quantification?

Another interesting question is, do we allow partial functions as elements of universes $D_{\langle a,b \rangle}$?

7.4.2   Hybrid Partial Type Theory with Various Domains

Our goal is to develop a formal system to integrate the machinery of hybrid logic into a type theory in which partial functions could be reasoned about naturally. In what follows we are going to justify that decision on the basis that they are in accordance with Bunge's view.

7.4.2.1   Actualist Quantification Semantics Justification

There are two points in Bunge (1980) that allow us to consider that existence, not only of individuals but also of first-order properties and higher-order properties, can vary from the domain of a possible world to the domain of another.

In the first case, Bunge remarks that "existence statements are responsible",[8] hence, in formal sciences and in factual sciences we are not allowed to lose a minute coming up with postulates and conjectures about the useless existence of insubstantial objects. However, what can be an insubstantial object in one world can become a valuable object in another. A property such as "yellow" can be worthless in a world where mathematical objects are taken into account, while it can be of the utmost importance in a world where biological objects, such as plants, are considered.

In the second case, Bunge qualifies his "conceptology" as conceptualist, fictionalist and materialist.[9] By materialist Bunge understands that the concepts, which exist as fictitious objects, are invented by beings with flesh and blood. Accordingly, all the abstract objects (first-order and higher-order properties included) that a human being can conceive are limited and determined by the experiences felt by a human being in his flesh. It is possible, then, to postulate the existence of abstract entities or concepts that vary from one world to another, given that the concepts that are conceivable at a given world (and therefore existing at this world) are not necessarily equal with the concepts imaginable by a rational non-human being at another world.

We maintain therefore that a varying domain model is not only helpful in order to deal with the existence of individuals, but also with the existence of given properties. Properties of every order can be thought as being world dependent.

7.4.2.2   Partial Functions Justification

Bunge warns us about the danger of violating the division of the objects into things and constructs. If this is done, the construct produced has to be considered as ontologically not well formed. "The attribution of conceptual properties to physical objects, and of physical properties (or chemical, biological or social) to conceptual objects belongs to the category of ontologically not well formed objects." (Bunge 1980, p. 58).

In order to avoid the formation of ontologically not well formed objects, partial functions can be used. Partial functions do not have values for every element in the domain but only for some of them, and hence, functions can have undefined values for some arguments.

---

[8]"Tanto en ciencias formales como en ciencias fácticas las afirmaciones de existencia son responsables: se tiene algún motivo razonable y no se pierde el tiempo inventando postulados o conjeturas de existencia de objetos ociosos que no desempeñan función alguna tales como mundos posibles." (Bunge 1980, p. 64).

[9]"En este trabajo exploraremos una alternativa, que llamaremos *materialismo conceptualista y ficcionista*" (Bunge 1980, p. 54).

It does not have any sense to talk about the mechanic, electric, chemical, mental or social state of number 2, and much less of its possible changes of state: it does not happen anything to number 2 and it will not happen anything to it. (Bunge 1980, p. 59)

7.4.3   The Formal Language $\mathfrak{HPTT}$

We want to develop a formal system to integrate the machinery of hybrid logic into a type theory in which partial functions could be reasoned about naturally. However, the use of partial functions leads to nondenoting terms because any argument outside the domain of the function produces this kind of terms.

How do we treat them? There are several possibilities, as Farmer (1990) pointed out. The first one is to treat "nondenoting expressions as non-well-formed terms". The main problem is that there is nothing in the term itself telling that it lacks denotation prior to the interpretation of the term in a particular context (structure). The second possibility mentioned is to consider "functions represented as relations". Even though, from a strictly mathematical point of view n-ary functions can easily be represented as n+1 relations, Farmer criticizes this choice by saying: "Reasoning about functions strictly using relations is neither natural nor efficient, since function application must be represented in a verbose, indirect fashion." (Farmer 1990, p. 1272).

The two final choices of Farmer proposals are:

- "Partial valuation for terms and formulas".
- "Partial evaluation for terms but total evaluation for formulas".

We agree with him in choosing the last one on the basis that we believe it is in accordance with the principles of both the physical and the conceptual worlds that a term denotes only if all its subterms denote, but atomic formulas including non denoting terms are false.

One of the task we want to solve in the sequel is just the answer to the question: How to define a semantics of hybrid type theory encoding these general principles?

The first distinction we need to introduce concerns the set of TYPES that is now divided into two kinds: The set TYPES_e of types of kind e is the subset of TYPES defined by

$$\mathsf{TYPES_e} := e \mid \langle a, b \rangle$$

where $a \in \mathsf{TYPES}_e$, $b \in \mathsf{TYPES}$. The set TYPES_t of types of kind t is the subset of TYPES defined by

$$\mathsf{TYPES_t} := t \mid \langle a, b \rangle$$

where $a \in \mathsf{TYPES_t}$, $b \in \mathsf{TYPES}$. Obviously, $\mathsf{TYPES}_e \cup \mathsf{TYPES}_t = \mathsf{TYPES}$.

As far as language $\mathfrak{H}\mathfrak{P}\mathfrak{I}\mathfrak{T}$ is concerned, we add to type theory the nominals, the @ operator and the modal operators $\square$ and $\lozenge$.

The structures used to interpret our formulas are pairs $\mathcal{M} = \langle \mathcal{S}, \mathsf{I} \rangle$ such that

$$S = \big\langle \langle D_a \rangle_{a \in \mathsf{TYPES}}, W, R, \langle D_{wa} \rangle_{wa \in W \times \mathsf{TYPES}} \big\rangle$$

is a skeleton and I is a denotation function.

In Areces et al. (2014) we presented a system where the domains of quantification are independent of the world, there is a unique D_b at each level b.

The novelties now are that our hierarchy is a partial one, and that the quantification domains are all local; namely, the quantification is actualistic.

In our case, I is an intensional interpretation; in particular, for any constant C_a of the language, $\mathsf{I}(C_a)$ is a function from W to D_a. That is, we allow the constant C_a to denote different extensional elements of D_a at different worlds in W. We can even allow constants without denotation; in that case I should be a partial function.

In the definition of $\langle D_a \rangle_{a \in \mathsf{TYPES}}$ as a partial type hierarchy, we distinguish types $\langle a, b \rangle \in \mathsf{TYPES}_e$ and $\langle a, b \rangle \in \mathsf{TYPES}_t$.

- In the first case $D_{\langle a,b \rangle}$ is a subset of the set $PF(D_a, D_b)$ of partial (and total) functions from D_b into D_a

$$D_{\langle a,b \rangle} \subseteq PF(D_a, D_b) = \{ f \mid f : D_b \rightsquigarrow D_a \}$$

(the symbol $\rightsquigarrow$ stands for partial functions); that is, the domain of any function $f \in D_{\langle a,b \rangle}$ could be a proper subset of D_b.

- In the second case, $D_{\langle a,b \rangle}$ is a subset of the set $D_a^{D_b}$ of all total functions from D_b into D_a. Levels D_e and D_t are, as usual, (1) a non-empty set, (2) the set $\{T, F\}$, respectively.

W is a non-empty set of worlds, R is a binary relation on W.

We have also decided to have different worlds at all levels of the hierarchy and so we define $\langle D_{wa} \rangle_{wa \in W \times \mathsf{TYPES}}$ as the local domains in the hierarchy. In the first place, $D_{wt} = \{T, F\}$ for all the worlds w. However, D_{we} is just a subset of D_e. To set $D_{w\langle ab \rangle}$ we distinguish between kind e and kind t. In the first case, functions in $f \in D_{\langle a,b \rangle}$ are partial, and the restriction is that the domain of $f \in D_{w\langle a,b \rangle}$ must be a subset of D_{wb}. In the second case, functions in $D_{w\langle a,b \rangle}$ are total but they give the special value F_a for all the elements that are not in D_{wb}. This value F_a is a generalization of the false value F and it is recursively defined by saying (1) $F_t = F$ and (2) $F_{\langle a,b \rangle}(\theta) = F_a$ for any $\theta \in D_b$.

The set ME_a of **meaningful expressions of type** a consists of the basic and complex expressions of type a. Basic expressions are nominals, non-logical constants and variables

$$i \in \mathsf{ME}_t \mid c_{m,a} \in \mathsf{ME}_a \mid v_{n,a} \in \mathsf{ME}_a$$

Complex Expressions are recursively generated as follows:

$$\gamma\beta \in \mathsf{ME}_a \mid \langle \lambda u_b \alpha \rangle \in \mathsf{ME}_{\langle a,b \rangle} \mid @_i \alpha \in \mathsf{ME}_a$$

$$\{\alpha \equiv \delta,\ \neg\varphi,\ (\varphi \wedge \psi),\ \exists u_a \varphi,\ \Box\varphi,\ \mathsf{E}\varphi\} \subseteq \mathsf{ME}_t$$

where $\gamma \in \mathsf{ME}_{\langle a,b \rangle}$, $\beta \in \mathsf{ME}_b$, $\alpha \in \mathsf{ME}_a$ and u_b is a variable of any type b and $i \in \mathsf{NOM}$; α and δ are both in ME_a, φ and ψ are in ME_t.

The semantics of expressions of kind e and t are very similar to each other in Areces et al. (2014) because no partial functions are included in the hierarchy. Therefore, for any expression A_b (of any type b), any world w and any assignment g, $[[A_b]]^{\mathcal{M},w,g} \in \mathsf{D}_b$. In Fitting (2002) the hierarchy of types includes no functions at all because there functions are represented as relations. However, intensional types are present all over.

In the present system, the semantics of expressions of kind e and t are significantly different from each other: expressions of kind e may lack denotation but formulas always denote classical values true or false. Therefore, $[[A_b]]^{\mathcal{M},w,g} \in \mathsf{D}_b$ for any expression of kind t ($b \in \mathsf{TYPE}_t$), but for kind e ($b \in \mathsf{TYPE}_e$) $[[A_b]]^{\mathcal{M},w,g}$ could be undefined.

In particular, expressions of kind e may denote partial functions and the application $(F_{\langle a,b \rangle} A_b)$ of kind e has a denotation if and only if both $F_{\langle a,b \rangle}$ and A_b have a denotation and the denotation $F_{\langle a,b \rangle}$ is defined at the denotation of A_b. However, an application $(F_{\langle a,b \rangle} A_b)$ always denote for expressions of kind t. In particular, $(F_{\langle t,b \rangle} A_b)$ is false if A_b is non-denoting.

Once we have defined the semantics, we wonder whether denotation is again just another semantical distinction unable to be formalized in the object language. We know from the start that non-denoting expressions could be well-formed expressions and so the question is about definability. Fortunately, we can say that A_a denotes, $D(A_a)$, using the lambda predicate $\langle \lambda x_a(x_a \equiv x_a) \rangle$ because $\langle \lambda x_a(x_a \equiv x_a) \rangle (A_a)$ is always true unless A_a is non-denoting. We can also define the existence predicate as well as the non-existence predicate in pure logical terms using λ and $\equiv$.

$$D(A_a) := \langle \lambda x_a(x_a \equiv x_a) \rangle (A_a)$$

$$\mathsf{EXISTS}(A_a) := \langle \lambda x_a \exists y_a(y_a \equiv x_a) \rangle (A_a)$$

$$\overline{\mathsf{EXISTS}}(A_a) := \langle \lambda x_a \neg \exists y_a(y_a \equiv x_a) \rangle (A_a)$$

When we interpret these formulas in a world $w \in W$ of a structure $\langle \mathcal{M}, g \rangle$:

- The formula $D(A_a)$ means that the expression A_a denotes at the world w under the interpretation $\langle \mathcal{M}, g \rangle$.
- The formula $\mathsf{EXISTS}(A_a)$ means that the expression A_a not only denotes but its denotation belongs to the domain D_{wa}; i.e., $[[\mathsf{EXISTS}(A_a)]]^{\mathcal{M},w,g} = T$ iff $[[A_a]]^{\mathcal{M},w,g} \in \mathsf{D}_{wa}$.

- The formula $\overline{\text{EXISTS}}(A_a)$ means that the expression A_a denotes but is not in the local domain of the world w under the interpretation $\langle \mathcal{M}, g \rangle$; i.e., $[[\overline{\text{EXISTS}}(A_a)]]^{\mathcal{M},w,g} = T$ iff $[[A_a]]^{\mathcal{M},w,g} \in \mathsf{D}_a - \mathsf{D}_{wa}$.

In fact, for any world $w \in W$ and structure $\langle \mathcal{M}, g \rangle$ we can see that

$$[[D(A_a) \leftrightarrow \text{EXISTS}(A_a) \vee \overline{\text{EXISTS}}(A_a)]]^{\mathcal{M},w,g} = T$$

We concluded Sect. 7.4.1 on hybrid first-order logic by saying that to express that an object exists either physically or conceptually we can use the quantifier, the equality and the $@_i$ operator, we do not need specific non-logical predicate constants as the $@_i$ operator is forcing the formula to be evaluated at i-world. This is also the case in our system of hybrid partial type theory, $\mathfrak{H}\mathfrak{P}\mathfrak{T}\mathfrak{T}$.

7.5 Existence in Intensional Hybrid Type Theory

In the previous system $\mathfrak{H}\mathfrak{P}\mathfrak{T}\mathfrak{T}$ we allow our constants to receive intensional interpretations, but intensions as such are only semantical devices as they lack a proper place in the hierarchy of types $\langle D_a \rangle_{a \in \text{TYPES}}$ as well as in the object language. In particular, our variables always take extensional values and intensional predication is forbidden. In hybrid type theory, predication is always done of the extensionalized denotations of the terms, but not of the intensions themselves. We can say that "the temperature is 20 Celsius degrees", but not that "the temperature is rising". That is, even thought $I(M)$ is a function from W to $D_{\langle t,e \rangle}$ and $I(c)$ is a function from W to D_e, neither $I(M)$ or $I(c)$ are in the hierarchy, and when we evaluate Mc at world w we only use the extensional values $I(M)(w)$ and $I(c)(w)$.

Shall we consider a new hierarchy $\langle D_a \rangle_{a \in \text{TYPES}}$ with intensional types as proper members? The answer is positive, as in Fitting (2002):

> Next, we must go up the ladder of higher types. Doing so extensionally, as in classical logic, means we take subsets of the ground-level domain, subsets of these, and so on. Going up intensionally, as Montague did, means we introduce functions from possible worlds to sets of ground-level objects, functions from possible worlds to sets of such things, and so on. What is presented here mixes the two notions-both extensional and intensional objects are present. (Fitting 2002, p. XIII)

Intensional hybrid type theory is an extension of hybrid type theory with certain intensional machinery; namely, intensional quantifiers and intensional predication. In this new environment, we are considering that "the temperature" is an intensional term, and therefore it can be interpreted as a function from possible worlds to a number of degrees. Sometimes we want to use "the temperature" to refer to a concrete number of degrees, this would be the extensional use of the intension, but we sometimes prefer to refer to the very concept of "temperature", in this case we would need to have a function as denotation. When we say that "the temperature is rising", what increases is not a determinate number of degrees, in fact, "20" is not rising; what really rises is "the temperature", which designates a function from times to numbers. Moreover, it is possible to introduce also an extensionalizing operator, ⌊, which allow us to obtain from the function denoted by the intension at the evaluation world, the value of this function at this particular world. There is therefore a different formalization of "the temperature" (t):

1. As an intensional term which has been extensionalized in "the temperature is 20 Celsius degrees": $\lfloor (t) = 20$.
2. As a proper intensional term in "the temperature is rising": $R(t)$, where R stands for "is rising".

7.5.1   The Formal Language 𝕴𝕳𝕻𝕿𝕿

Our Intensional Hybrid Partial Type Theory will not only count with an intensional interpretation at the semantic level for the meaningful expressions of the language, but it will have also intensional variables, terms and predicates as proper expressions of the language.

We can extend our previous language of 𝕳𝕻𝕿𝕿 and incorporate intensions. Firstly, we modify our definition of the set of TYPES, which is now divided in three kinds. The set TYPES_e of types of kind e, the set TYPES_t of types of kind t and the new set TYPES_s of intensional types, defined by:

$$\mathsf{TYPES}_s := \langle a, s \rangle$$

where $a \in \mathsf{TYPES}$. Note, however, that s is not itself a type, but it is used to form functional types.

The set of TYPES is thus defined as follows:

$$\mathsf{TYPES} = \mathsf{TYPES}_e \cup \mathsf{TYPES}_t \cup \mathsf{TYPES}_s$$

Once intensional expressions of type $\langle a, s \rangle$ are introduced in the object language, we can allow intensional predicates of type $\langle t, \langle a, s \rangle \rangle$ and so $A_{\langle t, \langle a,s \rangle \rangle} B_{\langle a,s \rangle}$ is a well-formed formula. As we have intensional variables, $\forall X_{\langle a,s \rangle} A_{\langle t, \langle a,s \rangle \rangle} X_{\langle a,s \rangle}$ is well-formed as well.

What are the necessary changes to accommodate these new features of our system?

In Sect. 7.4.3, the structures used to interpret our formulas are pairs $\mathcal{M} = \langle \mathcal{S}, \mathsf{I} \rangle$ such that

$$\mathcal{S} = \langle \langle \mathsf{D}_a \rangle_{a \in \mathsf{TYPES}}, W, R, \langle \mathsf{D}_{wa} \rangle_{wa \in W \times \mathsf{TYPES}} \rangle$$

is a skeleton and I is a denotation function. Now we will make some changes.

First, we need to include to the hierarchy $\langle D_a \rangle_{a \in \text{TYPES}}$ the types $\langle a, s \rangle$. The type of intensions $D_{\langle a,s \rangle}$ is a subset of the set $PF(D_a, W)$ of partial (and total) functions from W into D_a; that is, the domain of any function $\mathbf{f} \in D_{\langle a,s \rangle}$ could be a proper subset of W. Moreover, this domain of intension quantification is not required to be the set of all functions (partial and total) from W to D_a, that is:

$$D_{\langle a,s \rangle} \subseteq \{ \mathbf{f} \mid \mathbf{f} : W \rightsquigarrow D_a \}$$

The restriction in $D_{\langle a,s \rangle}$ is based on some coherence requirements that intensions should fulfill, and not everything mathematically definable has to meet a coherent condition for being a meaningful intension.

Second, I is an intensional interpretation; in particular, for any constant C_a of the language, $I(C_a)$ is a function from W to D_a. In particular, $I(C_{\langle a,s \rangle})$ is a function from W to $D_{\langle a,s \rangle}$. We can even allow constants without denotation; in that case, I should be a partial function.

Once intensional terms have been introduced, we can broaden the scope of hybrid operators of the form $@_i$ to qualify not only formulas and expressions of kinds e and t, but also expressions of any other type of the new kind s. Hence, we can build from the intensional expression A of type $\langle a, s \rangle$, a new expression $@_i A$ of type $\langle a, s \rangle$. The difference between A and $@_i A$ is that A can denote a different function depending on the world of evaluation, while $@_i A$ denotes the same function at each world; namely the function denoted at the i-world. $@_i$ acts as a rigidifier when applied to A, and gives us the function denoted by A at the world named by i. Thanks to the operator $@_i$, which rigidifies the expression A of type $\langle a, s \rangle$, the new term $@_i A$ (of type $\langle a, s \rangle$ as well) has the same denotation at every possible world w. Moreover, it is possible to introduce also the extensionalizing operator, $\downarrow$, which allows us to obtain not only the function denoted by the intension at a given world, but also the value of this function at the evaluation world. The new expression $\downarrow (@_i A)$ is of type a; and when we interpret $\downarrow (@_i A)$ at the world w, we obtain the value at the world w of the function denoted by $@_i A$ at the i-world.

Denotation issues can also be raised by expressions of the form $@_i \tau$ provided that the intensional term τ does not have any denotation at the world named by i. One way of dealing with this problem is to allow the denotation function I to be partial. Terms can denote or can not denote depending on whether I is defined or not at a given world.

Let us illustrate this with an example. Suppose that τ is the intensional term "the highest executive office in France".[10] When it is interpreted at a world it can denote different functions, for example, in the seventeenth century it denotes the king of France, while in the twentieth century it denotes the president of the Republic of France. If we want to fix the value of τ to be the king of France, assuming that i is a name for the seventeenth century, $@_i \tau$ should be used.

---

[10]The example is based on Tichy (1979).

Moreover, it is possible to introduce also an extensionalizing operator, $\downarrow$, which allow us to obtain not only the function denoted by the intension at a given world, but also the value of this function at a given world. The extensional operator would provide us the concrete individual which is the king of France or the president of the Republic of France at a given world; we will use the expression $\downarrow(@_i\tau)$ to name him.

7.5.2 Existence Revisited

We remember how Bunge has considered that there is a clear distinction between the existence predicate and the existential quantifier, $\exists$. From the point of view of the logical systems we have presented so far there is no clear conclusion about what is really existence. Our formal languages have tools for dealing with existence as a predicate and also as a quantifier. In fact, it is possible to give a coherent account of both alternatives.

In the first place, remember that in $\mathfrak{TT}$ the existential quantifier, $\exists$, was defined by means of a predicate:

$$\exists x_a \varphi_t :=_{Df} \Sigma_{\langle t, \langle t, a \rangle \rangle} \langle \lambda x_a \varphi_t \rangle$$

This definition reveals the very nature of the existential quantifier, which according to Frege, should be intended as a second-order predicate or a predicate of predicates. The ontological existential predicate was also defined in $\mathfrak{TT}$.

Alternatively, in $\mathfrak{HPTT}$, assuming a semantics with various domains, the predicate of existence, EXISTS, can be defined by means of the quantifier $\exists$:

$$\mathsf{EXISTS}(A_a) := \langle \lambda x_a \exists y_a (y_a \equiv x_a) \rangle (A_a)$$

Note that in case a is type e, the predicate of existence would behave as a first-order predicate applicable to individuals.

If we had preferred to have a constant domain model in $\mathfrak{HPTT}$, we should have introduced a primitive existence predicate E. A predicate that is not definable by means of the existential quantifier and demands a new specific non-logical constant. Anyway, E is true at any world w of the things that actually exists at w.

The introduction of intensions in the language of our Intensional Hybrid Partial Type Theory ($\mathfrak{IHPTT}$) opens a new possibility concerning existence which has not been taken into account previously. It is related with considering existence as a predicate of intensions.

In the previous sections, we have been analyzing existence as a predicate either of individual objects or of predicates, both extensionally intended. Mostly, predication has been centered in extensional terms, and intensional predication has not received the attention it merits.

In our $\mathfrak{IHPTT}$, existence can also be predicated of intensions, and we should expand our previous definition to include terms of type $\langle a, s \rangle$:

$$\text{EXISTS}(A_{\langle a,s \rangle}) := \langle \lambda x_{\langle a,s \rangle} \exists y_{\langle a,s \rangle} (y_{\langle a,s \rangle} \equiv x_{\langle a,s \rangle}) \rangle (A_{\langle a,s \rangle})$$

7.6 Conclusion

Intensions were introduced as a way of dealing with the *senses* of the expressions of the language. These *senses* are considered to belong to the realm of concepts. In fact, for Church, a *concept of x* is anything which is capable of being the sense of a name of x.[11] Therefore, *conceptual existence*, in Bunge's words, should be applied properly to intensions. In fact, Bunge's *constructs* can be explained successfully by means of intensions. Bunge's *concepts* would be our intensional terms. Bunge's *propositions* would be intensions of type $\langle t, s \rangle$, i.e., functions from possible worlds to truth values. Finally, Bunge's *contexts*, as they are explained in terms of concepts and propositions, can also be described with intensions.

When we apply existence to intensions we are predicating existence to concepts, and that is not a trivial issue. Physical theories are based on concepts, and hence the assumption that some concepts exist and others do not raises a crucial question. Do an absolute time and space exist? Does phlogiston exist? Do gravitons really exist? Do the vibrating strings of superstring theory exist?

In fact, when we explain scientifically the world what we are really doing is to indicate what concepts exist and what concepts do not.

Acknowledgements To Mario Bunge, with gratitude.

References

Andrews, P. (1963). A reduction of the axioms for the theory of propositional types. *Fundamenta Mathematicae, 52*, 345–350.

Areces, C., & Ten Cate, B. (2007). Hybrid logic. In *Handbook of modal logic* (pp. 821–868). New York: Elsevier.

Areces, C., Blackburn, P., Huertas, A., & Manzano, M. (2014). Completeness in hybrid type theory. *Journal of Philosophical Logic, 43*(2–3), 209–238.

---

[11] "In order to describe what the members of each type are to be, it will be convenient to introduce the term *concept* in a sense which is entirely different from that of Frege's *Begriff*, but which corresponds approximately to the use of the word by Russell and others in the phrase 'class concept' and rather closely to the recent use of the word by Carnap, in *Meaning and Necessity*. Namely anything which is capable of being the sense of a name of x is called a *concept of x*." (Church 1951, p. 11).

Blackburn, P. (2000). Representation, reasoning and relational structures: A hybrid logic manifesto. *Logic Journal of the IGPL, 8*, 339–365.

Braüner, T. (2008). Adding intensional machinery to hybrid logic. *Journal of Logic and Computation, 18*(4), 631–648.

Braüner, T. (2011). *Hybrid logic and its proof-theory*. Dordrecht/Heidelberg/London/New York: Springer.

Bunge, M. (1980). *Epistemología*. Barcelona: Ariel.

Bunge, M. (1985). ¿Grados de existencia o de abstracción? *Theoria. Segunda época, 2*, 547–549.

Church, A. (1940). A formulation of the simple theory of types. *The Journal of Symbolic Logic, 5*, 56–68.

Church, A. (1951). A formulation of the logic of sense and denotation. In P. Henle, H. M. Kallen, & S. K. Langer (Eds.), *Structure, method and meaning: Essays in honor of Henry M. Sheffer* (pp. 3–24). New York: The Liberal Arts Press.

Farmer, W. M. (1990). A partial functions version of Church's simple theory of types. *The Journal of Symbolic Logic, 55*(3), 1269–1291.

Fitting, M. (2002). *Types, Tableaus, and Gödel's God*. Dordrecht/Boston/London: Kluwer Academic Publishers.

Fitting, M., & Mendelsohn, R. L. (1998). *First-order modal logic*. Dordrecht/Boston/London: Kluwer Academic Publishers.

Frege, G. (1884/1950). *The foundations of arithmetic*. Oxford: Blackwell.

Henkin, L. (1950). Completeness in the theory of types. *The Journal of Symbolic Logic, 15*, 81–91.

Henkin, L. (1955). The nominalistic interpretation of mathematical language. *Bulletin de la Société Mathématique de Belgique, 7*, 137–141.

Henkin, L. (1963). A theory of propositional types. *Fundamenta Mathematicae, 52*, 323–344. (Henkin, L. (1964). Errata. *Fundamenta Mathematicae, 53*, 119).

Kant, I. (1787/1929). *Critique of pure reason*. London: Macmillan.

Manzano, M. (1985). Formalización en teoría de tipos del predicado de existencia de Mario Bunge. *Theoria. Segunda época, 2*, 513–534.

Manzano, M. (1996). *Extensions of first order logic*. Cambridge: Cambridge University Press.

Manzano, M., Martins, M. A., & Huertas, A. (2014). A semantics for equational hybrid propositional type theory. *Bulletin of the Section of Logic, 43*(3–4), 121–138.

Russell, B. (1903). *The principles of mathematics*. Cambridge: Cambridge University Press.

Russell, B. (1908). Mathematical logic as based on the theory of types. *American Journal of Mathematics, 30*(3), 222–262.

Tichý, P. (1979). Existence and god. *The Journal of Philosophy, 76*(8), 403–420.

Chapter 8
Truth in the Post-Truth Era: Evaluating the Theories of Truth with a Table of Contingency

Dominique Raynaud

Abstract Notwithstanding what the champions of post-truth may think, truth remains a key concept of scientific research. In this contribution, several theories of factual truth are evaluated with the help of a simple and straightforward tool, i.e., a logical table of contingency. Its application shows that most contemporary disconcerting theories of truth have critical flaws, and that, albeit imperfect, truth-as-correspondence is still the best theory. Further criticism of correspondence as *adaequatio*, compliance and isomorphism suggests that *correspondence-as-mapping* constitutes a more acceptable theory. This contribution provides a definition and examples in the empirical sciences.

8.1 Introduction

Let us be glad and rejoice: the post-truth era has come! In 2016, post-truth became *Oxford Dictionary*'s word of the year. The editors pointed out that the term is to be understood as "belonging to a time in which the specified concept [truth] has become unimportant or irrelevant". So unimportant, indeed, that when I came to evoke the concept of truth before an audience of social scientists,[1] a participant raised the question:

> I was born in 1969, in the era of the Internet. It no longer mattters whether Neil Armstrong set foot on the Moon or if the Apollo 11 Mission was shot in-studio. Frankly, I don't see the point of having a debate on the concept of truth.

---

[1]Different social scientists were invited at this encounter that took place on June 18, 2015, at PACTE, Grenoble. Among them, several STS enthusiasts created a "LatouringClub".

D. Raynaud (✉)
Université Grenoble Alpes, Grenoble, France
e-mail: dominique.raynaud@univ-grenoble-alpes.fr

© Springer Nature Switzerland AG 2019
M. R. Matthews (ed.), *Mario Bunge: A Centenary Festschrift*,
https://doi.org/10.1007/978-3-030-16673-1_8

Should the Apollo 11 Mission and Moon Hoax[2] be placed on an equal footing? Only those who are immersed in the media crazed-world can believe it.

Apollo (Apache Point Observatory Lunar Laser-Ranging Operation) was mainly intended to improve the gravitational theory of the Moon through Earth-Moon millimetric measurements. The main scientific instrument of the mission was the LRRR Laser retroreflector, to be placed on the lunar ground to reflect Earth laser shots. Since the laser determination of the Earth-Moon distance was devised by French astronomer Jean Rösch, with the help of Alain Orszag, the first shots were made from the 1 m telescope of Pic-du-Midi Observatory in the nights of 5 and 6 December 1970. The follow-up of this research showed that the Earth and Moon moved away from about 3.8 cm per year (Rösch and Orszag 1968; Rösch et al. 1970). By chance, I served as astronomer trainee at Pic-du-Midi in 1979–1980 when the observatory was still headed by Jean Rösch, and I could even see the traces of assembly of the laser telemeter on the 1 m telescope.

Most importantly, social scientists should understand that giving credence to the post-truth doctrine comes down to negating the social component of science and sawing the branch on which social scientists themselves sit, i.e., people. And for this reason: Totalling the scientists and engineers who conceived and analyzed the Earth-Moon telemetric datasets, those who conceived and analyzed the records of the passive seismograph PSEP, those who conceived and analyzed the collection of lunar samples, and all those who worked at the Kennedy Space Center at the time, there were about 2000 persons involved into the program—omitting those who worked on the Lunar Reconnaissance Orbiter Mission, which supplied images of the Apollo 11 Mission in the Sea of Tranquility in 2012.

The conspiracy thesis implies that thousands people were prey to collective self-delusion. This amounts to saying that the hoax thesis can be supported only if the collective functioning of science is negated—a quite surprising assumption from social scientists.

Truth has caused a lot of philosophical ink to flow during the twentieth century so that there exists today a very broad spectrum of theories, the most famous of which being the correspondence, coherence, pragmatist, deflationary-minimalist, eliminativist, and subalternation (to usefulness, power or social consensus) theories. From the outset, this "non-overlapping magisteria" range of solutions is quite surprising: there is one world, and we need one theory of truth. Therefore it is advisable to adopt Mario Bunge's suggestion to evaluate these theories:

> To evaluate a science or a scientific theory scientists use a battery of objective criteria accepted by almost all investigators [. . .] Nothing like that happens in philosophy: usually philosophical doctrines are accepted or rejected, fully or in part, without resorting to any clear and objective criteria. The evaluation of philosophical doctrines tends to be intuitive, utilitarian, or even emotive. (Bunge 2012b, p. xiii)

---

[2]The first conspiracy theory dates back to the 1970s: Kaysing and Reid (1976).

This paper paints a big picture of the problem—not details.[3] The meaning of (approximate) truth in the empirical sciences (both observational and experimental) will be discussed from the philosophy/sociology of science perspectives. And then reviewed pursuant to the same plan: first the exposition, then criticism.

Plan: The following section (Sect. 8.2) is about the "table of contingency," the tool with which the different theories of truth will be assessed. The contribution will review the eliminative (Sect. 8.3), subalternative to usefulness, power or social consensus (Sect. 8.4), and coherentist (Sect. 8.5) theories of truth. Truth-as-correspondence will be discussed in more detail (Sect. 8.6). Taking the opportunity that this "theory is still a research project after two and a half millennia" (Bunge 2012b, p. 66), *adaequatio* and isomorphism will be assessed and rejected in favour of a new thesis, i.e., the correspondence-as-mapping theory of truth.

8.2 The Logical Table of Contingency

Definitional problems arise in all sciences. As an introduction, it is worth recalling that "to define something is to state a set of conditions $\{C_1, C_2 \ldots C_N\}$, which are both individually necessary and collectively sufficient" (Cozic 2010). These conditions are individually necessary, because if one condition is lacking, the case falls outside the definition; there are collectively sufficient, because if all conditions are met, the case does meet the definition.

The choice of the conditions belongs to the one who builds the definition, and that choice usually results from qualitative evaluation. As several people may evaluate differently the same issue, it is not uncommon that multiple definitions of the same concept be given. Such be-fruitful and multiply motto is not admittable about truth.

Instead it is relevant to assess the strength of the conditions involved in the definition of a concept. In a previous paper, I have introduced the "logical table of contingency" as an extension of both the "bilateral diagram" (xy, $x\overline{y}$, $\overline{x}y$, $\overline{x}\overline{y}$) (Carroll 1897, pp. 22–38) and the statistical "contingency table" that measures the dependence of characters on a two-way array (Raynaud 2017b).

The logical contingency table proceeds in much the same manner as the statistics table. However it only uses qualitative data (presence vs. absence) and its purpose is to test an alleged relationship, rather than establishing its existence.

	Property	¬ Property
Concept	1	3
¬ Concept	4	2

---

[3]Only a few references will be given not to accumulate too much literature. Each theory has collected a hundred of scholarly articles. The reader will find useful pointers to literature in Changeux 2003, Engel 2002, Kirkham 1992, and *Stanford Encyclopedia of Philosophy* online.

When one gets to the point of using a contingency table to test dependencies between a concept C and its defining condition or property P, four cases come up. The two diagonal cells (1, 2) are filled: there is *equivalence* between the condition and the concept. Three cells are filled (1, 2, 3) and the cell which comes in addition to the main diagonal is at the top: the condition *over-determines* the concept under consideration. Three cells are filled (1, 2, 4) and the cell which comes in addition to the main diagonal is at the bottom: the proposed condition *under-determines* the concept. The four cells (1, 2, 3, 4) are filled: there is *independence* between the concept and the proposed defining condition.

These four cases can be illustrated by examples in evolutionary biology, where species are defined by autapomorphies and higher-level taxa by synapomorphies. Both are irreversible innovative characters (Lecointre and Le Guyader 2006, p. 8). Characters are more or less fit to the taxa. (1) Autapomorphies and synapomorphies are intrinsically evolutionary because they stem from the reform of systematics. (2) Despite the tables deal with one character at a time, the list of characters is open because a taxonomy can rely on various morphological, anatomical, embryological and molecular data—hence the interest of gauging their relevance.

Cells (1, 2) are filled when the taxon [M: Mammalia] and character [G: mammary glands] are intersected on the table. Even if we think hard, the two cells [¬GM] [G¬M] are left blank. As only the two diagonal cells are filled, there is *equivalence* between mammary glands and mammals: mammary glands properly define mammals.

Cells (1, 2, 4) are filled when the taxon [H: Hominidae] and character [N: no tail] are intersected on a contingency table. As Gibbons (*Hylobates agilis*) are not Hominidae but lack a tail, there is an extra-cell filled at the bottom of the table: the absence of tail under-determines the phylum Hominidae.

Cells (1, 2, 3) are filled when the taxon [M: Mammalia] and character [P: placenta] are intersected on a table of contingency. Since Metatherians, such as the Kangaroo (*Macropus rufus*), are mammals deprived of placenta, the cell 3, at the top right of the table, is filled. Therefore the placenta over-determines Mammalia. This situation is symmetrical to the previous situation.

Cells (1, 2, 3, 4) are filled, for example when the taxon [H: Hominidae] and character [B: bipedalism] are intersected on the table: Orangutans (*Pongo pygmaeus*), though not bipeds, are Hominidae; Ostriches (*Struthio camelus*), though not Hominidae, are true bipeds. As four cells are filled, bipedalism does not properly define Hominidae (probably because bipedalism is a function that can be achieved by a number of species, including by homoplasy).

Although over-simplified, the previous examples inform us that a contingency table can be applied as soon as a test of dependence of conditions, properties or characters is intended—even if their statistical frequencies are unknown. The table of contingency allows to weigh the conditions that come within a definition.

8.3   The Elimination Theory of Truth

8.3.1   Exposition

In the 1970s, philosophers and sociologists of science have abandoned the classical theories of truth—in particular the correspondence theory that "truth is what corresponds to reality"—in favour of new ideas. Their approaches can be divided into two classes: either eliminate or subalternate truth. As doubts were increasing about the old notion of truth, some have claimed it should be retired completely from the study of science.

Radical criticism is the outright elimination of the concept of truth, which is considered unnecessary to science.

This is the position of most relativists and constructivists, such as Bloor, Collins, Latour and many others. One can study the conception of truth of these authors without going into details, because most expressed themselves clearly on the subject.

Within this anthology, one should cite David Bloor: "This poses a problem about the notion of truth, for why not abandon it altogether?" (1991, p. 40). Latour supported several theses. The first one is just an extension of Bloor's idea: the sociologist of science should refrain from invoking the difference between truth and falsehood: "Sociological explanation should be impartial with respect to truth or falsity [. . .] Our argument is that the implicit (or explicit) adoption of a truth value alters the form of explanatory account which is produced" (Latour and Woolgar 1986, p. 149). This position would be incorporated in Actor Network Theory (hereafter noted as ANT), which aims at showing that truth and falsehood, human and non-human, and the like, are just social-linguistic constructions. Sociology would be unable to use these notions to show how they are constructed—otherwise there would be circular reasoning. This is reflected in the "Third Rule of Method," (a parody of Descartes) acknowledging the disappearance of truth: "Since the settlement of a controversy is the cause of Nature's representation, not the consequence, we can never use the outcome—Nature—to explain how and why a controversy has been settled" (Latour 1987, p. 258).

8.3.2   Criticism

Whoever ignores Nature has no way of knowing which theory is true or false. A bizarre consequence of the ANT is that all theories are equal. So any scientific debate could be rewritten in favor of the losers. For example, the debate over the theory of evolution could be rewritten in favour of Wilberforce against Darwin; the controversy on the capacity of water molecules to remember what has previously been dissolved, water memory, in favour of Jacques Benveniste. This strategy has no bounds. Latour himself applied these ideas to the reconstruction of the spontaneous generation debate in favor of Pouchet. He explicitly declares: "Winners do not have to be protected by the historian, but only the vanquished, to which we shall give, in a sense, a second chance before the tribunal of history" (Latour 1989, p. 430). Truth is no more than the result of collective self-delusion.

First Objection The elimination of truth is flawed because many authors confuse "truth" as a standard regulating the production of propositions, and "truths" (in plural) that is "true propositions" that comply with the standard of truth. The sentence: "All innovative work runs the risk of modifying this tradition, and the definition of rationality which it carries" (Stengers and Schlanger 1989, p. 62) means that rationality would change in history at the pace of scientific discoveries. Each "tradition" would elicit its own truths (propositions)—hence truth (as a standard). However "truth" and "true propositions" are different. Whatever the case, any standard is more robust than the products that are supposed to comply with it. Even if there are constant advances in logic, logical rules are stable: the structure of the syllogism ($p \subset q \subset r$) or modus ponens ($p \supset q, p \vdash q$) have never changed over time, and are as valid today as they were in the time of Aristotle. The confusion between "truth" and "true proposition" is just a trick to claiming that universal concept is, in fact, relative.

Second Objection Relativists and constructivists need to convince the audience that they are right. This forces them to implicitly resort to truth. The reading of texts is piquant. For example, Latour asks to set aside the "true" and "false," but nevertheless writes: "Organisation, taste, protocol, bureaucracy, minimisation of risks are not common technical words to describe a chip. *This is true*, however, only once the chip is a black box sold to consumers [. . .] *The same is true* for the endorphin assay when the dissenter, losing his temper, accuses the Professor of fabricating fact [. . .] It is crucial for us, laypeople who want to understand technoscience, to decide which version *is right* [explaining the nature vs. the networks] [. . .] Napoleon won because he had the power and the others obeyed. *Exactly the same is true* of the relations between the handful of scientists and millions of others [. . .] In taking the asymmetric stand, *it is true* that the tiny size of scientific networks was ignored" (Latour 1987, pp. 5, 73, 97, 119, 196).

Waiving truth while continuing to use it is an inconsistency that is facing the "retortion argument": If truth is an illusion, why should I believe the one that ensures me it does not exist? This is an old argument (Aristotle, *Metaphysics*, Book 4, 1008b). The strong thesis that urges us to abandon truth has a fatal flaw, because whoever argues that "truth does not exist" cannot prove that he is right. Therefore, there is no reason to trust him.

8.4   The Subalternation Theories of Truth

Other philosophers and sociologists of science, realizing that it was too radical to eliminate truth, tried to make it disappear otherwise, i.e., by subordinating it to other concepts.

Subalternation theories of truth consist of subordinating truth to a concept considered more fundamental than truth.

This theory is available in three main variants, namely: the subordination to usefulness, to power and to social consensus.

8.4.1   Subalternation of Truth to Usefulness

8.4.1.1   Exposition

The subordination of truth to usefulness has been first expressed by Francis Bacon: "Truth and therefore utility are the very same thing" (Bacon 1620, aphorism cxxiv). This idea was then developed by pragmatism—as long as the word refers to James' doctrine, because Peirce, who was a strong supporter of the correspondence, said: "So with scientific research. Different minds may set out with the most antagonistic views, but the progress of investigation carries them by a force outside of themselves to one and the same conclusion [. . .] What we mean by the truth, and the object represented in this opinion is the real" (Peirce 1878, pp. 138–9). James sharply departed from this view: "Truth is what gives us the possible maximum sum of satisfactions [. . .] You can say of it then either that 'it is useful because it is true' or that it is 'true because it is useful.' Both these phrases mean exactly the same thing" (James 1907, pp. 204–5). By reducing truth to action, James' theory is opportunistic. Truth is no longer "what is," but anything that can serve someone's purposes or interests.

8.4.1.2   Criticism

Let us build a contingency table of truth vs. usefulness:

	Usefulness	¬ Usefulness
Truth	Strength of materials	Nucleosynthesis
¬ Truth		Physiognomy

[UT] Many theories are both true and useful. This is the case of the strength of materials inaugurated by the study of elementary forces by Hooke, Newton,

Leibniz, Bernoulli and others. The first synthesis would appear in 1818, in Henri Navier's lessons at the École Polytechnique of Paris;

[¬U¬T] There are also many theories both false and useless—or even harmful—like physiognomy (Lavater) and morphopsychology (Corman) that became tools for the groundless detection of criminals, as served as a basis for the racial theories of the twentieth century;

[¬UT] Some true theories are devoid of any utility. This is the case of stellar nucleosynthesis (i.e., the study of the formation and evolution of chemical elements in stars, a science inaugurated by an article of Burbidge et al. (1957);

[U¬T] Finally, there can be no false-useful theory, because any false theory fails when it is aimed at solving a practical problem. Think again of the strength of materials (in connection to buildings), or aerodynamics (in connection to planes).

As the extra-cell filled is at the top of the contingency table, utility over-determines truth. This opposes the utilitarian and pragmatist view that "truth and utility are the very same thing." In fact, they are not.

8.4.2 Subalternation of Truth to Power

8.4.2.1 Exposition

Constructivists frequently refer to the "political economy of truth," a term coined by Michel Foucault.[4] Let us examine this idea, which was advocated in *Vérité et pouvoir*: "Every society has its regime of truth, its 'general politics' of truth; that is to say, the types of discourse which the society accepts and makes it work as true; the mechanisms and instances which enable us to distinguish between true and false statements" (Foucault 1977, p. 25). Foucault's thesis is that true and false depend on power: "True is the name of what the most powerful imposes upon the weakest."

This position was then endorsed by sociologists of science: "Once it is realized that scientists' actions are oriented toward the agonistic field, there is little to be gained by maintaining the distinction between the 'politics' of science and 'truth'" (Latour and Woolgar 1986, p. 237). Again: "By analyzing the allies and resources that settle a controversy we understand everything that there is to understand in technoscience" (Latour 1987, p. 97). Again: "There is no great divide between minds [i.e., between true and false ideas], but only networks, longer or shorter than others" (Latour 1989, p. 428).

In science as elsewhere, true and false would be labels derived from some power games that are established between the actors. Whenever there is a scientific discussion, a network outweighs the other, not because he holds the truth, but because it has more numerous and more powerful allies. In this version, advocated

---

[4]Note that Foucault is mentioned six times in *Laboratory Life* (Latour and Woolgar 1986, pp. 107, 184, 229, 251, 258 and 265).

by Latour, Knorr Cetina and others scholars, the notion of truth becomes an ancillary concept of power.

8.4.2.2 Criticism

Let us create a table of contingency truth vs. power:

	Power	¬ Power
Truth	Vernalization	*Conics*
¬ Truth	$\pi = 3.2$	Radiesthesia

[PT] Some true theories have benefited from political backing—at times without much discernment. In the USSR, Lysenkoism became a state doctrine between 1935 and 1956. Lysenkoism, which defended the inheritance of acquired characters against genetics was completely false, but it was promoting "vernalization" which has proved to be correct;

[¬P¬T] Some false theories have never been politically supported. As far as I know, no government or party actively supported radiesthesia (dowsing pendulum);

[¬PT] There is no piece of evidence establishing—or even suggesting—that Apollonius' *Conics* or Desargues' projective geometry received any political support;

[P¬T] Sometimes a false theory is supported politically, as was the case of the theory of inheritance of acquired characters in the Soviet Union under Stalin. But the most amazing case is that of Edward J. Goodwin, a physician who thought he had found the solution of squaring the circle. He introduced a bill in the Indiana House of Representatives, with a view to fixing the value of π. On February 5, 1897, the House voted unanimously (67 vs. 0) a law fixing the value of π to 3.2, even if Lindemann had shown in 1882 that π was a transcendental number! Fortunately, this insane project was stopped by mathematician Clarence A. Waldo just before it was passed by the Senate (Beckmann 1982, pp. 175–7).

As the four cells of the contingency table are filled, power says nothing about truth. The subalternation of truth to power is delusive.

8.4.3 Subalternation of Truth to Social Consensus

8.4.3.1 Exposition

Other sociologists have attempted to reduce truth to social consensus. In science, the state of the art merely reflects the knowledge of the best-informed scientists of the moment, so truth would be reducible to intersubjective agreement within the scientific community.

This interpretation has been advocated on several occasions: "The [sociologist, like any scientist] will naturally have preferences and these will typically coincide with those of others in his locality. The words 'true' and 'false' provide the idiom in which those evaluations are expressed" (Barnes and Bloor 1982, p. 27). Here, the truth is subalternated to collective opinion... More recently, it has been claimed that "Talking about the objectivity of science is actually [...] a way of naming the intersubjective agreement of the members of a scientific community" (Gingras 2013, p. 48). The latter author does not support this idea on his own: he refers to Popper's *Open Society and Its Enemies*. Being mentioned, the source can be checked through: "Scientific objectivity can be described as the inter-subjectivity of scientific method [...] Two aspects of the method of the natural sciences are of importance [...] First, there is something approaching *free criticism* [...] Secondly, scientists try to avoid talking at cross-purposes [...] In the natural sciences this is achieved by recognizing *experience* as the impartial arbiter of their controversies" (Popper 1947, p. 217). In reality, Popper does not support the idea that truth is subalternated to social consensus. He admits the social dimension of scientific work (the collective control in science) but critical thinking and experimentation remain the main criterions for truth. Therefore Gingras expands on Popper's argument. The control and regulation of scientific activity—the fact that we do follow rules—does not necessarily implies "truth = inter-subjectivity."

8.4.3.2 Criticism

As previously, a contingency table truth vs. consensus is built:

	Consensus	¬ Consensus
Truth	Refraction	Relativity
¬ Truth	Geocentrism	N-rays

[CT] In optics, the longstanding laws of refraction are true and consensual;

[¬C¬T] Some contested scientific hypotheses, such as the "N-rays" (Blondlot 1904), the "gravitational magnetism" (Blackett 1947) or "cold fusion" (Fleis-chmann and Pons 1989) never reached a social consensus;

[¬CT] History of science also reports on cases where a solid theory has generated much skepticism. This was the case of the "continental drift" (Wegener 1915), a theory which was only accepted in the 1960s with the advancement of plate tectonics. This also happened with Relativity (1905) still fought in the 1930s, whereas it had received decisive evidence through the constancy of the speed of light (De Sitter 1913), explanation for the advance of Mercury's perihelion (Einstein 1915) and deflection of light in a gravitational field (Dyson et al.

1920). In 1922, when Louis de Broglie invited Einstein to expose his theory in Paris, Emile Picard, then the Permanent Secretary of the Academy of Sciences, simply forbade his lecture. About 10 years later, an assembly of authors led by Erich Ruckhaber published *Hundert Autoren gegen Einstein*, a pamphlet against Einstein and his Relativity, to which Einstein merely replied: "If I were wrong, one would have been enough!"

[C¬T] There exist false consensual statements. Albeit untrue, geocentrism and epicycles (Ptolemy), fixism (Cuvier), phlogiston (Stahl), spontaneous generation (Pouchet) have aroused strong memberships in their time. So do pseudo-sciences today. Homeopathy (Hahneman), iridology (von Peczely), chromatherapy (Agrapart), psychogenealogy (Ancelin Schutzenberger) and quantum therapy (Kaplan) are popular, albeit groundless.

As the four cells of the table are filled, truth is not only a matter of intersubjective agreement: men are able to convince themselves from the true and false with the same ease. The subalternation argument is rather strange because its flaw has long been pointed out, e.g., in *The Progress of Reason in the Search for Truth*: "The experience must convince that the multitude of opponents proves nothing against the truth of any proposition" (Helvetius 1775, p. 61; also Bunge 1974, p. 127).

As these characters are independent, social consensus is absolutely silent about truth.

8.5   The Coherence Theory of Truth

8.5.1   *Exposition*

Based on the argument that reality cannot be accessed directly, some scholars have tried to define truth without referring to external reality. They have argued that truth is the agreement between our propositions and our representations of reality: "Truth is what is logically consistent" (coherence). The outside world is left out, and the non-contradiction test applies to propositions only. This position was defended by several authors, among which the English idealist Francis H. Bradley:

> "Prof. Stout denies, I understand, that coherence will work as a test of truth in the case of facts due to sensible perception and memory [. . .] Mr. Russell again [. . .] I contend that this test works satisfactorily, and that no other test will work" (Bradley 1914, p. 202).

Albeit with a few differences, this was also the opinion of Otto Neurath:

> "If a statement is made, it is to be confronted with the totality of existing statements. If it agrees with them, it is joined to them; if it does not agree, it is called 'untrue' [*unwahr*] and rejected [. . .] there is no other notion of truth [*Wahrheitsbegriff*]" (Neurath in Sarkar 1996, p. 53).

Truth-as-coherence is to say that if some theory is flawless, it is true; if it is contradictory, it is false.

8.5.2   Criticism

The main objection against the coherence theory is that by Russell: a non-contradiction test is fundamentally inadequate to define truth. There always may be a set of false propositions that nevertheless successfully pass the test of non-contradiction. Let be a set of true propositions (about the real world). Now consider the negation of all propositions (in a dream). This new set of propositions is logically consistent. However, both sets cannot be true at the same time. Therefore, one is false. Hence, the non-contradiction test cannot determine the truth. It is the kind of objections Russell has in mind, when he claims that the coherence theory is flawed. Russell illustrated the case by a "good novel" (Russell 1910, p. 156–7), then by "life as a long dream": "It is possible that life is one long dream [. . .] but although such a view does not seem inconsistent with known facts, there is no reason to prefer it to the common-sense view, according to which people and things do really exist [. . .] Thus coherence as the definition of truth fails because there is no proof that there can be only one coherent system" (Russell 1912, p. 191–2).

This can be represented in the following table: truth vs. coherence:

	Coherence	¬ Coherence
Truth	Scientific theories	
¬ Truth	Good novel	Delirium

[CT] Many sets of propositions are both logically consistent and true, as most scientific theories well supported by data;

[¬C¬T] Inconsistent sets of propositions are false, such as the speech produced during the alcohol-induced delirium;

[C¬T] Some sets of propositions, though false, may be coherent (Russell's good novel);

[¬CT] There is no set of true-inconsistent propositions.

The extra-cell filled is at the bottom of the contingency table; therefore coherence under-determines truth. As any set of propositions can be both false and consistent, truth-as-coherence is not, in itself, an acceptable theory.

However Russell does not fully rejects coherence: he only rejects the claim that it could provide a good—self-sufficient—definition of truth. He declares: "For the above two reasons, coherence cannot be accepted as giving the meaning of truth, though it is often a most important test of truth after a certain amount of truth has become known" (Russell 1912, p. 193). This declaration is just the result of the under-determination relationship between coherence and truth: all that is true is logically consistent; but all that is logically consistent is not necessarily true. Coherence can be claimed only if it is certain that there is at least one true proposition in the set of propositions under consideration. This is the reason why only a few authors have supported a self-sufficient coherence theory. Most have chosen instead a combination of correspondence and coherence.

Descartes, for example, advocates both correspondence and coherence, and explicitly accepts the theory of truth as "adequatio rei et intellectus" in a letter to Mersenne dated October 16, 1640 (Descartes 1898, p. 597). Even an idealist and avowed proponent of coherentism, such as Francis Bradley, admitted not being able to support a pure coherentism: "Truth, to be true, must be true of something, and this something itself is not truth" (Bradley 1914, p. 325). In recent years, Bunge has similarly advocated for a "synthetic theory of truth" (Bunge 1974, pp. 93–96).

All this suggests that the coherence theory is ancillary to some other theory, but with theories T1–T4 now discarded, there remains just one possibility: that the coherence theory be ancillary to the correspondence theory.

8.6 The Correspondence Theory of Truth

8.6.1 Exposition

The correspondence theory tells us that "truth is what corresponds to reality." Here the core notion is the correspondence between a proposition and a fact. Correspondence is no doubt the oldest theory of truth: it was supported by Aristotle, the Scholastics authors, up to the present: "To say that 'what is' is not, or that 'what is not' is, is false; but to say that 'what is' is, and 'what is not' is not, is true" (Aristotle, *Metaphysics* 4, 1011b); "The truth or falsehood of a belief always depends upon something which lies outside the belief itself [. . .] Truth consists in some form of correspondence between belief and fact" (Russell 1912, p. 189). "I accept the commonsense theory (defended and refined by Alfred Tarski) that truth is correspondence with the facts (or with reality); or, more precisely, that a theory is true if and only if it corresponds to the facts" (Popper 1972, p. 44). Correspondence is also the common theory of truth in everyday life. If I say "Berlin had a sunny day on 4 June 2016 at 4 PM" and it was sunny, it is true; otherwise it is false.

Among the theories of truth, I will make no special room to the deflationary-redundancy theory, which is just a surrogate of the correspondence theory.[5]

---

[5]Among the main proponents of the deflationary position, the German logician Gottlob Frege declares: "The sentence 'I smell the scent of violets' has the same content as the sentence 'It is true that I smell the scent of violets'" (1956, p. 293). We also read in Tarski: "'It is snowing' is true if and only if it is snowing" (1933, p. 157), an expression that Wittgenstein reduced to: "'p' is true $= p$" (2010, §136). Frege, Tarski and Wittgenstein all play with the quotation marks around the sentence, but they do it in very different ways. The redundancy theory claims that the word "true" can be abandoned without any loss of meaning. This is untrue, because the disquotation works only if the hidden condition "if and only if" is met, that is, if some "material adequacy condition" *binds* propositions and real facts (Tarski 1933). Otherwise, this theory loses

8.6.2   Criticism

Correspondence is facing Frege's objection: the correspondence "can only be perfect if the corresponding things coincide and are, therefore, not distinct things at all" (Frege 1956, p. 291). Frege considers that the authenticity of a 20-mark banknote can be established by comparing it with an authentic banknote, but not with a 20-mark gold piece, because there is no strict correspondence in this case. Frege waives the correspondence theory "for it is absolutely essential that the reality be distinct from the idea. But then there can be no complete correspondence [. . .] So the attempt to explain truth as correspondence collapses" (Frege 1956, p. 291).[6]

Frege's objection against the correspondence theory is twofold: (1) strict correspondence is illusory; (2) only things belonging to the same class can be compared. Both objections resurfaced in later works. Frege's second argument was revived by Neurath. When it is said that truth is an agreement between propositions and empirical facts, correspondence connects physical and mental objects, which belong to two different orders of reality: "*Thus statements are always compared with statements*, certainly not with some 'reality,' nor with 'things'" (Neurath in Sarkar 1996, p. 53). The argument has been repeated by Lakatos: "Propositions can only be derived from other propositions, they cannot be derived from facts" (1978, p. 16). So we return to Frege's argument: if establishing truth is to compare different things, things cannot be in complete correspondence.

If we accept to weigh the previous criticisms and sift through the theories of truth, we note that Frege-Neurath's objection against correspondence is the weakest one. Let us see why and how it can be rebutted.

Frege-Neurath's criticism is twofold. It is claimed that "correspondence can only be perfect if the corresponding things coincide" and "are not distinct at all" (Frege 1956, p. 291). Correspondence is thus seen as a mere identity ("not distinct at all") and this is the ground for the argument that no proposition can be compared with facts. The view that correspondence is reducible to identity is baseless. To correspond has basically four meanings:

1. to be related, or associated by natural or logical links;
2. to be in conformity, or compliance, with;
3. to be in relationship of equivalence, by nature or function; and finally
4. to be identical.

---

any difference between an unfounded opinion and a proven fact. The redundancy theory thus comes down to the correspondence theory with implicit elements.

[6]Frege does not promote any other theory to replace the truth-as-correspondence: "Every other attempt to define truth collapses too [. . .] the content of the word 'true' is unique and indefinable" (Frege 1956, p. 291). Truth is a primitive concept.

8.6.3 Correspondence-as-Adaequatio

Identity is just one possible meaning of correspondence. Ancient theories of truth are worth reading in this respect. In *Disputed Questions on Truth*, Aquinas introduces the idea that truth adds something to the being: truth is not defined by identity, but by some relationship of compliance between "being" and "true." This is where (erroneously quoting Isaac Israeli), he gives the definition of truth as "*adaequatio rei et intellectus*" (Aquinas 1952, q. 1, a. 1). How should we precisely translate this? Du Cange's *Glossarium mediae et infimae latinitatis* explains the verb "*adaequari*" as "*aestimari, diligenter discuti, inspici*" a phrase that we could translate as "to be assessed, carefully examined, inspected" (Du Cange 1887, 1, col. 68c). *Adaequatio* thus places emphasis on assessing the facts: correspondence is the correctness of a comparison, not the identity of things being compared. Hence, "*adaequatio rei et intellectus*" may be rendered as "correspondence-as-compliance between the thing and the knowledge [of the thing]." Two facts are worth mentioning:

1. Compliance does not require the facts and statements to be identical.
2. There is no need that propositions and facts belong to the same class: the facts can belong to the outside world, when the statements are parts of the mind.

This correction nullifies Frege-Neurath's argument, yet despite the above discussion of *adaequatio*, correspondence is still unclear. "The trouble with the correspondence theory is that its key idea, correspondence, is just not made adequately clear" (Haack 1978, p. 92), "A proposition is factually true if it fits the facts it refers to. But what does 'fit' (or 'match,' or 'correspond to' mean?" (Bunge 2004, p. 239; 2012a, p. 66). Consequently the concept of correspondence requires further clarification.

8.6.4 Correspondence-as-Isomorphism

When it is said that a physical theory corresponds to the facts, there is less than an identity, but more than just a "compatibility" between the two. How could this compliance be described?

1. A theory has no physicality. It is an abstract representation. It does not represent the facts themselves, but the relationships between them: it is a structural correspondence;
2. A theory is a kind of representation of reality: it corresponds to it, if its relationships are identical in some respect with the real relationships it describes.

While commenting on such matters, contemporary philosophers have argued that the correspondence theory of truth could reduce to structural isomorphism which is "a shared structure of form, between a belief and a wordly state of affairs"

(Clark 2016, p. 248). Expressions of this view have been given by many authors (Haack 1978, p. 92, Kirkham 1992, p. 119, Elsby 2015, pp. 65–68). Harold N. Lee (1965) has provided a crystal-clear conception of truth-as-isomorphism even though the term "isomorphism" never appears in his text. He uses the phrase "one-to-one correspondence" (Lee 1965, pp. 102, 106 and 108), which implies a bijective isomorphism between a proposition and the real world.

Isomorphism asks for too much because of its own definition. A function f: $X \rightarrow Y$ is called an isomorphism, or one-to-one correspondence between X and Y, if f is a bijection with $f(x) = y$ and $x = f^{-1}(y)$. Albeit intuitively clear, isomorphism is facing many objections. Bunge (1974, pp. 93–94) highlighted some of them. There are others:

1. No statement can encompass the wide range of aspects that can be discovered in the real world. Every statement implies "filtering" (I will touch on that shortly, Sect. 7.4).
2. No doubt that men create theories that represent reality, but the symmetrical claim that reality represents a theory is spurious. As new theories are discovered daily, at any time, some part of reality lacks its theory. There can be no one-to-one correspondence: "to comply with" is not a bijection (Fig. 8.1) but an injection (Fig. 8.2), where C denotes the part of reality which still lacks a description. This is why isomorphism fails.

Fig. 8.1 A bijection

Fig. 8.2 An injection

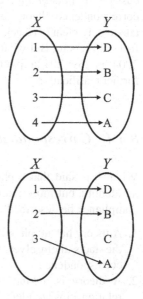

8.6.5 Correspondence-as-Mapping

As an "injection" is defined as a "one-to-one mapping," isomorphism could preferably be replaced by "mapping," a word that both refers to the "matching" of objects and to the "map" of, and differing from the territory (Korzybski 1933). Now, as any theory is finite, the correspondence-as-mapping is finite too. It is not established between all propositions and the whole reality, but just between a set of propositions and a part of reality.

To clarify what a "part" is, it is worth looking at the theory of observation. The first ideas on the matter were expressed by mathematician Robert Vallée (1951). The key concept is that an observation is an active process. It is reflected by operator $\mathcal{O}: \xi \mapsto \mathcal{O}(\xi) = \eta$. This operator accounts for both the correspondence between a proposition and real fact, and its possible distortion through observation. In other words, there is only partial correspondence between propositions and reality. The observation process is accounted for by two main sub-classes of operators, which can be as numerous as necessary:

Sampling. Vallée calls the first-class operators "field operators" $A: \xi \mapsto A(\xi)$. Let us rename it sampling, which is a much easier term (parcelling, or framing would work as well). This class includes spatial framing $A(x, y, z)$ with the limits of any observation field, and time slicing $A(t)$, as is the case with stroboscopic observations. Reality is observed through a "window" containing just some part of it—and it is generally assumed that what is observed within the frame is largely independent of what happens outside. In short, sampling means that the observer never observes the whole universe, but only a parcel of it. Sampling can be seen as an extension if the idea of "selective description" (Popper 1957).

Filtering. Vallée calls the second-class operators "filtering operators" $Z: \xi \mapsto Z(\xi)$. Filtering is immediately understandable (aspect would work as well). This class includes all output/input transfer functions, spatial $Z(x, y, z)$ as well as temporal $Z(t)$. This process is both passive and active. On the one hand, data are naturally filtered by sense channels: only a small part of the electromagnetic spectrum (400–700 nm) is sensed by human vision. On the other hand, any observation yields information that is not of much interest: the scientist must select data, e.g., observe a phenomenon in a given wavelength. Filtering echoes several classical themes such as Leibniz's "perspective multiplication" (Leibniz 1714; Poincaré 1902; von Bertalanffy 1968) and Mach's "red glass" (Titchener 1922; Hahn 1929).

Sampling operators and filtering operators combine in the form $Z(A(\xi)) = ZA(\xi)$, without generally commuting with each other: $ZA \neq AZ$.

While observation limits reality to some aspect of it $\mathcal{O}: \xi \mapsto \mathcal{O}(\xi) = \eta$, scientific discovery proceeds in reverse order from some aspects of reality to reality, and in the best case, it goes from the known to the unknown (more on this in Raynaud 2017a). Scientific discovery is an inverse transfer function $\mathcal{O}^{-1}: \eta \mapsto \mathcal{O}^{-1}(\eta) = \xi*$.[7]

These notions can be best understood with the help of an example, which I take from solar astronomy. Astronomers make a difference between quiescent vs. eruptive prominences which raise above the photosphere: the first ones are long-term phenomena, while the latter can disappear in a few minutes (for the classification of prominences, see Zirin 1988).

Sampling. The distinction between quiescent and eruptive solar prominences is analytic in nature. It is nonetheless usual to 'isolate' the prominences even if they do appear within the same region of the Sun. Plate 8.A shows an eruptive prominence 200.000 km high, which developed between 19:29 and 20:44 UT on 6 July 2013. We note at its base a 25.000 km high platform arch, which did not change in shape throughout the period considered. The discussion of the 'eruptive prominence' presupposes its independence from the platform arch standing at its base. Data is sampled.

Filtering. Since the end of the nineteenth century, solar astronomers are aware that what is seen of the Sun can be misleading. So they generate variations in the wavelength to get different views of the event under study. As the Sun is mainly made of hydrogen, a favourite line of observation is the red H-alpha line (at the wavelength of 6563 Ångström). Since then, other wavelengths have been explored, such as that of the ionized Calcium II K (3934 Å), of the ionized Carbon C IV (1600 Å), of the ionized Helium He II (304 Å) and of the highly ionized Fe IX/X (171 Å). All filtered images displayed on Plate 8.B record different facets of the Sun, and it comes as no surprise that visible light opens a very narrow window on what is the Sun. For example, the magnetic loops occuring in the upper transition region and corona aptly recorded by Fe IX/X lines, are invisible to the eye with white light. Data is filtered.

For a statement to be true, it must be assigned spatio-temporal coordinates (sampling) and some kind of data (filtering). For example, the statement "There is a spot on the Sun" is neither true or false, until it is explicitly referred to some *Fkc* event[8] of heliographic coordinates 14.13, 6.97 that occurred on 4 Jul 2013 at 14:59 UT (sampling) and was seen in visible light (filtering) (See Plate 8.B-1).

---

[7] Arguably, the result of the operation $\mathcal{O}^{-1}(\eta)$ is just a part of ξ, hence the notation $\xi*$.

[8] Sunspots are labelled according to McIntosh (1990) classification. In *Fkc*, *F* means "an elongated bipolar sunspot group with penumbra on both ends," *k* "large, assymetric", *c* "compact".

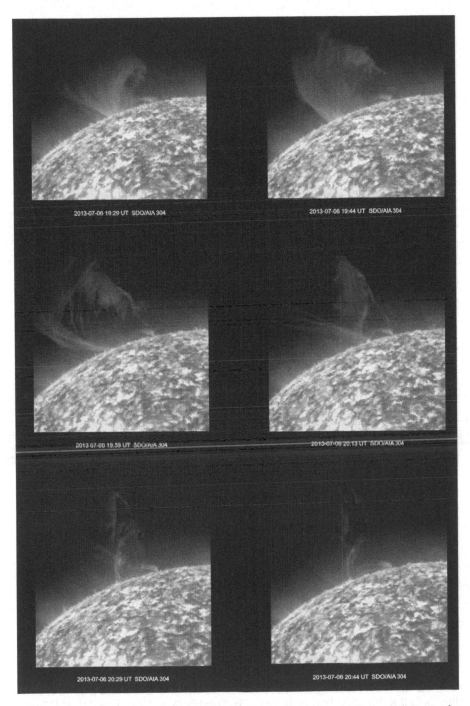

Plate 8.A Six images of the Sun showing the intermingling of an eruptive prominence and a platform arch at 19:29 (Fig. 1), 19:44 (Fig. 2), 19:59 (Fig. 3), 20:13 (Fig. 4), 20:29 (Fig. 5) and 20:40 UT (Fig.6), on 6 July 2013. Courtesy of NASA/SDO and the AIA, EVE, and HMI science teams

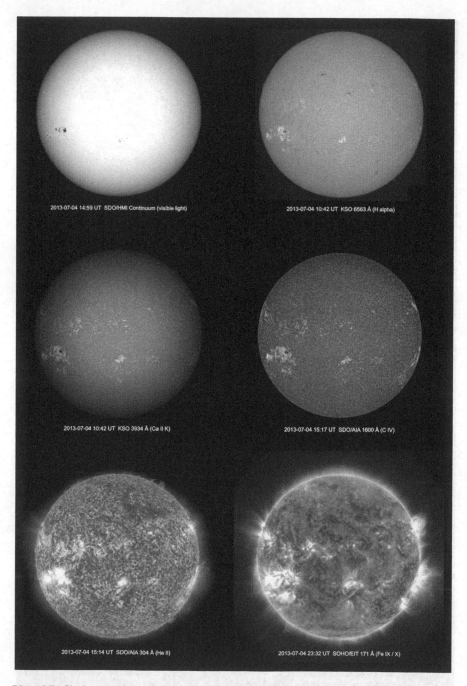

Plate 8.B Six images showing the Sun in different wavelengths: visible light (Fig. 1), H alpha (Fig. 2), Ca II K (Fig. 3), C IV (Fig. 4), He II (Fig. 5) and Fe IX/X (Fig.6) around 10:42 UT, on 4 July 2013. Courtesy of NASA/SDO and the AIA, EVE, and HMI science teams

8.7   Conclusion

The current best theory of truth is correspondence-as-mapping. It can be specified as follows:

The Cpp* correspondence-as-mapping theory of truth is the thesis that proposition p* corresponds to fact p, iff there exists a one-to-one mapping $p^* \mapsto p$ involving specific sampling $A_0 \ldots A_i$ and filtering $Z_0 \ldots Z_j$ operations.

Some tests can be made to check whether this specification works in general. Frege's 20-mark banknote will be taken along several cases in the empirical sciences, alternating factual and theoretical statements.

Table 8.1 suggests that the correspondence-as-mapping theory of truth is suited to any field of knowledge. Once sampling A and filtering Z are properly specified,

Table 8.1 Testing the correspondence-as-mapping theory of truth

Science	p^* proposition	C mapping	p fact	A sampling	Z filtering
Philosophy	This is a 20-mark banknote		A 20-mark piece of gold	Used in 1918 Germany	From the standpoint of its monetary value
Astronomy (fact.)	There is a sunspot on the Sun		The SDO/HMI observation	Of the western region of the Sun seen on 4 Jul 2013 at 14:59 UT	In the H-alpha line
Physics (theo.)	Maxwell's theory of electromagnetic waves		The loop-antenna and spark-gap experiments made	By Hertz (1887) in Berlin	With a frequency 100 MHz
Chemistry (fact.)	Element sodium exists	Is true if it corresponds to . . .	The observation of a new substance made	By Sir Humphrey Davy (1807) in London	By electrolysis of soda (sodium hydroxide)
Biology (theo.)	The theory of evolution		Geospiza's beak variation	Observed by Abzhanov et al. (2004)	As a function of Bmp-4 gene
Archaeology (fact.)	Donnart report (2012)		The archaeological excavations	On the site of Champ-Durand, 3400–3000 BE	Regarding lithic macro-outillage
Sociology (theo.)	Durkheim (1897) social theory of suicide		Statistical data	In Europe in the years 1875–1880	About divorce, religion and suicide

correspondence-as-mapping provides an acceptable definition of factual truth. However, as with any theory, correspondence-as-mapping is open to further improvement.

Acknowledgements Plates A and B: Courtesy of NASA/SDO and AIA, EVE, and HMI science teams. Courtesy of SOHO/EIT consortium. SOHO is an ESA/NASA project of international cooperation.

References

Abzhanov, A., Protas, M., Grant, B. M., et al. (2004). Bmp-4 and morphological variation of beaks in Darwin's finches. *Science, 305*, 1462–1465.

Aquinas, T. (1952). *Quaestiones disputatae de veritate, questions 1–9* (R. W. Mulligan, S.J., Trans.). Chicago: Henry Regnery Company.

Aristotle (1933). *Aristotle in 23 volumes: Metaphysics* (H. Tredennick, Trans.). Cambridge, MA: Harvard University Press, Vols. 17–18.

Bacon, F. (1620). *The new organum, or true directions concerning the interpretation of nature.* London: Typographium Regium.

Barnes, B., & Bloor, D. (1982). Relativism, rationalism and sociology of knowledge. In M. Hollis & S. Lukes (Eds.), *Rationality and relativism* (pp. 21–47). Oxford: Blackwell.

Blackett, P. M. S. (1947, May 17). The magnetic field of massive rotating bodies. *Nature, 159*, 658–666.

Beckmann, P. (1982). *A history of Pi.* Boulder: Golem Press.

Blondlot, R. (1904). Sur une nouvelle espèce de rayons N. *Comptes rendus de l'Académie des Sciences, 138*, 545–547.

Bloor, D. (1991). *Knowledge and social imagery.* Chicago: The University of Chicago Press. First ed1976.

Bradley, F. H. (1914). *Essays on truth and reality.* Oxford: Clarendon Press.

Bunge, M. (1974). Truth. In *Treatise on Basic Philosophy, II: Interpretation and Truth* (pp. 81–132). Dordrecht: Reidel.

Bunge, M. (2004). The centrality of truth. In E. Agazzi (Ed.), *Right, wrong and science* (pp. 233–241). Amsterdam: Rodopi.

Bunge, M. (2012a). The correspondence theory of truth. *Semiotica, 188*, 65–76.

Bunge, M. (2012b). *Evaluating philosophies.* New York: Springer.

Burbidge, E. M., Burbidge, G. R., Fowler, W. A., & Hoyle, F. (1957). Synthesis of elements in stars. *Review of Modern Physics, 29*, 547–650.

Carroll, L. (1897). *Symbolic logic, 1. Elementary.* London: Macmillan and Co.

Changeux, J. P. (Ed.). (2003). *La Vérité dans les sciences.* Paris: O. Jacob.

Clark, K. J. e. (2016). *The Blackwell companion to naturalism.* Chichester: Wiley.

Cozic, M. (2010). *Philosophie de la connaissance, 1. La Définition.* Paris: Université Paris Val-de-Marne. Retrieved September 10, 2016, http://mikael.cozic.free.fr/connaissance0910/connaissance-0910-definition1-np.pdf

Davy, H. (1808). The Bakerian lecture: On some new phenomena of chemical changes produced by electricity. *Philosophical Transactions, 98*, 1–144.

Descartes, R. (1898). Descartes à Mersenne, 16 octobre 1639. In C. H. Adam & P. Tannery (Eds.), *Oeuvres II: Correspondance, mars 1638 à décembre 1639 (110 à 180)* (pp. 587–599). Paris: Cerf.

De Sitter, W. (1913). Ein astronomischer Beweis für die Konstanz der Lichtgeschwindigkeit. *Physik-alische Zeitschrift, 14*, 429.

Donnart, K. (2012). Le macro-outillage lithique. In R. Joussaume (Ed.), *L'enceinte néolithique de Champ-Durand à Nieul-sur-l'Autise (Vendée)* (pp. 443–482). Chauvigny: Association des publications chauvinoises.

Du Cange, C. (1887). *Glossarium mediae et infimae latinitatis . . .* (10 volumes). Niort: L. Favre.

Durkheim, E. (1897). *Le Suicide: Étude de sociologie.* Paris: F. Alcan.

Dyson, F. W., Eddington, A. S., & Davidson, C. R. (1920). A determination of the deflection of light by the Sun's gravitational field, from observations made at the solar eclipse of May 29, 1919. *Philosophical Transactions of the Royal Society A, 220,* 291–333.

Einstein, A. (1915). Explanation of the perihelion motion of mercury from the general theory of relativity. English transl. In A. J. Knox, M. J. Klein, R. Schulmann (Eds.), *The collected papers of Albert Einstein* (Vol. 6, pp. 234–242). Princeton: Princeton University Press, 1996.

Elsby, C. (2015). Aristotle's correspondence theory of truth and what does not exist. *Logic and Logical Philosophy, 25,* 57–72.

Engel, P. (2002). *Truth.* Bucks: Acumen.

Fleischmann, M., & Pons, S. (1989). Electrochemically induced nuclear fusion of deuterium. *Journal of Electroanalytical Chemistry, 261,* 301–308.

Foucault, M. (1977). Vérité et pouvoir, entretien avec M. Foucault. *L'Arc, 70,* 16–26. Reprinted in *Dits et Ecrits II* (pp. 140–160). Paris: Quarto, 2001.

Frege, G. (1918). Der Gedanke. Eine Logische Untersuchung. *Beiträge zur Philosophie des deutschen Idealismus, 1,* 58–77. English transl. Thought: A Logical Inquiry. *Mind,* 1956, 65, 289–311.

Gingras, Y. (2013). *Sociology of science.* Paris: Presses universitaires de France.

Haack, S. (1978). *Philosophy of logics.* Cambridge: Cambridge University Press.

Haack, S. (1996). Concern for truth: What it means, why it matters. *Annals of the New York Academy of Sciences, 775,* 57–63.

Hahn, H. (1929). Superfluous entities or Occam's razor. Engl. transl. In H. Hahn & B. McGuinness (Ed.), *Empiricism, logic, and mathematics* (pp. 1–19). Dordrecht: Reidel, 1980.

Helvetius, C. A. (1775). *Les Progrès de la raison dans la recherche du vrai.* London: no name.

James, W. (1907). Pragmatism's conception of truth. *Journal of Philosophy, Psychology and Scientific Methods, 4,* 141–155.

Kaysing, B., & Reid, R. (1976). *We never went to the moon: America's Thirty Billion Dollar Swindle.* Pomeroy: Health Research Books.

Kirkham, R. L. (1992). *Theories of truth.* Cambridge, MA: MIT Press.

Korzybski, A. (1933). *Science and sanity: An introduction to non-Aristotelian systems and general semantics.* Brooklyn: Institute of General Semantics.

Lakatos, I. (1978). In J. Worrall & G. Currie (Eds.), *The methodology of scientific research programmes* (Philosophical Papers Volume 1). Cambridge: Cambridge University Press.

Latour, B. (1987). *Science in action. How to follow scientists and engineers through society.* Cambridge, MA: Harvard University Press.

Latour, B. (1989). Pasteur et Pouchet: hétérogenèse de l'histoire des sciences. In M. Serres (Ed.), *Éléments d'histoire des sciences* (pp. 423–445). Paris: Bordas.

Latour, B., & Woolgar, S. (1986). *Laboratory life. The construction of scientific facts.* Princeton: Princeton University Press. First ed. 1979.

Lecointre, G., & Le Guyader, H. (2006). *The tree of life: A phylogenetic classification.* Cambridge, MA: The Belknap Press of Harvard University Press.

Lee, H. N. (1965). A fitting theory of truth. In E. G. Ballard et al. (Eds.), *The problem of truth* (pp. 93–110). New Orleans: Tulane University.

Leibniz, G. W. (1881). *La Monadologie* (1714) (E. Boutroux, Ed.). Paris: C. Delagrave.

McIntosh, P. S. (1990). The classification of sunspot groups. *Solar Physics, 125,* 251–267.

Neurath, O. (1996). Physicalism. In S. Sarkar (Ed.), *Logical empiricism at its peak: Schlick, Carnap, and Neurath* (pp. 74–79 [52–57]). New York: Garland.

Peirce, C. S. (1878). How to make our ideas clear. *Popular Science Monthly, 12,* 286–302. Reprinted In C. J. W. Kloese (Ed.), *Writings of Charles S. Peirce. A Chronological Edition* (Vol. 3, pp. 257–276). Bloomington: Indiana University Press, 1986.

Poincaré, H. (1902). *La Science et l'Hypothèse*. Paris: Flammarion.

Popper, K. R. (1947). *The open society and its enemies, 2. The high tide of prophecy: Hegel, Marx, and the aftermath*. London: Routledge & K. Paul.

Popper, K. R. (1957). *The poverty of historicism*. London: Routledge.

Popper, K. R. (1972). *Objective knowledge: An evolutionary approach*. Oxford: Clarendon Press.

Raynaud, D. (2015). *Scientific controversies. A socio-historical perspective on the advancement of science*. New Brunswick: Transaction Publishers; 2nd ed. New York: Routledge, 2017.

Raynaud, D. (2017a). Qu'est-ce que la science: inférer l'inconnu à partir du connu. In *Qu'est-ce que la science pour vous?* (pp. 215–220). Paris, Editions matériologiques.

Raynaud, D. (2017b). Inside the ghetto: Using a table of contingency and cladistic methods for definitional purposes. *Bulletin of Sociological Methodology, 133*, 5–28.

Rösch, J., & Orszag, A. (1968). Les applications de mesure de distance Terre-Lune par laser. *Bulletin Astronomique, 3*, 453–459.

Rösch, J., Orszag, A., et al. (1970). *Premiers échos lumineux sur la Lune obtenus par le télémètre laser du Pic du Midi* (Vol. 270, p. 1637). Comptes rendus de l'Académie des Sciences.

Ruckhaber, E. (Ed.). (1931). *Hundert Autoren gegen Einstein*. Leipzig: R. Voigtländers Verlag.

Russell, B. (1910). *Philosophical essays*. London: Longmans & Green.

Russell, B. (1912). *The problems of philosophy*. London: Williams & Norgate. *Stanford Encyclopedia of Philosophy Online*. http://plato.stanford.edu. Accessed 10 Aug 2017.

Stengers, I., & Schlanger, J. (1989). *Les Concepts scientifiques: invention et pouvoir*. Paris: Editions La Découverte.

Tarski, A. (1933). The concept of truth in formalized languages. English transl. In J. Cocoran (Ed.), *Logic, semantics, metamathematics* (pp. 152–278). Indianapolis: Hackett, 1983.

Titchener, E. B. (1922). Mach's lectures on psychophysics. *American Journal of Psychology, 33*(2), 213–222.

Vallée, R. (1951). Sur deux classes d'opérateurs d'observation. *Comptes-Rendus de l'Académie des Sciences, 233*, 1350–1351.

von Bertalanffy, L. (1968). *General system theory. Foundations, development, applications*. New York: G. Braziller.

Wegener, A. (1915). *Die Entstehung der Kontinente und Ozeane*. Braunschweig: F. Vieweg & Sohn.

Wittgenstein, L. (2010). *Philosophische Untersuchungen*. English transl. by G. E. M. Anscombe, P. M. S. Hacker, & J. Schulte, *Philosophical investigations*. Oxford: Wiley-Blackwell.

Zirin, H. (1988). *The astrophysics of the sun*. Cambridge: Cambridge University Press.

Chapter 9
Is Simplicity a Myth? Mach and Bunge on the Principle of Parsimony

Íñigo Ongay de Felipe

Abstract This chapter examines the contrast between Ernst Mach's and Mario Bunge's accounts of parsimony as a principle in theory evaluation. The point is argued that simplicity, central as it undoubtedly is to Mach's philosophy, does not constitute a proper epistemic virtue within his framework. In his 1963 book *The Myth of Simplicity,* Mario Bunge endorses a very different position. Bunge begins by noticing that *simplicity is not that simple.* Indeed, the notion can be accorded a plurality of distinct meanings whether they are semantic, logical, epistemic, ontological, syntactic, or even aesthetic. Thus, doubts do inevitably arise regarding what sort of simplicity is accorded epistemic importance. Unlike Ernst Mach, Mario Bunge adopts a critical realist position and so he argues that scientific theories aim at representing the world. Importantly however, Bunge does not mean to claim that the principle of parsimony is devoid of any relevance. Nonetheless, replacing this modest and transitional usage of the principle with a *cult of simplicity* coupled with the ungrounded metaphysical assumption that the world is intrinsically simple (which Bunge calls *simplicism*) conflicts with other heuristic goals of science. I will show that Bunge's analysis deserves credit for reminding us that research customarily unveils the multiple -fold inner complexity of reality and so it is always good advice to recall that our best theories are to be kept simple enough but not *simpler.*

9.1 Introduction

It is quite a paradox that scientific simplicity represents an epistemic virtue with many avatars. Ever since the Middle Ages, the idea that *other things being equal,* simpler theories or hypotheses are to be preferred over more complex ones has been fleshed out in a rather complex plurality of ways. One possible angle from which this paradox may be tackled is to notice that while there seems to be general consensus

Í. O. de Felipe (✉)
Instituto de Estudios de Ocio, Universidad de Deusto, Bilbao, Spain

© Springer Nature Switzerland AG 2019
M. R. Matthews (ed.), *Mario Bunge: A Centenary Festschrift,*
https://doi.org/10.1007/978-3-030-16673-1_9

to the effect that simplicity is epistemically important, a similar level of agreement is lacking when it comes to specifying what it means for a given theory to be simpler than others. While some philosophers would want to understand parsimony in terms of ontological conservativeness, others prefer to frame it as referring to a range of properties including explanatory unification, the use of the principle of the common cause, the number of adjustable parameters a theory contains, or even as a primarily aesthetic virtue to be taken into account in theory evaluation. A separate albeit somewhat related issue is that there exist also an equally considerable variety of accounts of why simplicity, however conceived of, should matter at all. This is of course not to say that the principle of parsimony lacks a general justification, as the problem is rather about that very many different justifications do exist and that they are often incompatible with, or at least clearly different from, one another.

This chapter examines two contrasting philosophical positions about the principle of simplicity as an epistemic value. First, the chapter considers Ernst Mach's approach to what he terms 'the principle of economy of thought' in science to show that ironical as it may seem at first sight, Mach does not really consider parsimony as virtue in theory evaluation. Rather, as will be shown, what he argues is that theories that aren't simple enough cannot even count as scientific in character. There is a reason for this: Mach's conception of parsimony is grounded on an anti-realist view of science, which denies that the aim of scientific theories is to construct a true-like representation of the world.

Secondly, Mario Bunge's treatment of the principle of simplicity in *Is Simplicity a Myth?* (Bunge 1963) is discussed. Different from Mach, Bunge does affirm that there are cases where simplicity is epistemically significant. Indeed, he argues that whereas it is perfectly sensible to prefer simpler theories to more complex alternatives in some scientific contexts, such cautionary and prudent usage of parsimony does not authorize a general cult of simplicity (something which Bunge terms *simplicism*) as there are other important epistemic virtues to consider also. It will be shown that part of the reason for this discrepancy lies in the fact that while Mach endorses a roughly instrumentalist approach, Bunge in turn adopts a critical-realist interpretation of the goals of science.

Finally, the chapter will explore an array of more recent ideas from John Dupre (1993), Hasok Chang (2012), Kyle Stanford (2006) and Gustavo Bueno (2013) on realism and scientific pluralism to conclude that there are some positive points to be commended both in Bunge's prudential endorsement of simplicity as an epistemic value and his critical reservation about an unrestricted reliance on the principle of parsimony.

9.2 Mach, Simplicity and Naturalism

Ernst Mach is famous for having advocated a highly controversial version of the principle of scientific simplicity, which he terms the *principle of economy of thought* (*Das Prinzip der* Denkokönomie). According to this view of parsimony, it is the goal of science to provide the most economical conception of nature compatible with human experience and able to allow for new precise predictions about what future experience will be like. Here is how Mach formulates his position in the introduction to his book *The Science of Mechanics:*

> Economy of communication and of apprehension is of the very essence of science. Herein lies its pacificatory, its enlightening, its refining element. Herein, too, we possess an unerring guide to the historical origin of science. In the beginning, all economy had in immediate view the satisfaction simply of bodily wants. With the artisan, and still more so with the investigator, the concisest and simplest possible knowledge of a given province of natural phenomena – a knowledge that is attained with the least intellectual expenditure – naturally becomes in itself an economical aim; but though it was at first a means to an end, when the mental motives connected therewith are once developed and demand their satisfaction, all thought of its original purpose, the personal need disappears.
>
>   To find, then, what remains unaltered in the phenomena of nature, to discover the elements thereof and the mode of their interconnection and interdependence – this is the business of science. It endeavors by comprehensive and thorough description, to make the waiting for new experiences unnecessary; it seeks to save us the trouble of experimentation by making use, for example, of the known interdependence of phenomena, according to which if one kind of event occurs, we may be sure beforehand that a certain other event will occur. (Mach 1919, p. 6)

In his *Principles of the Theory of Heat,* Mach endorses the same account of the goals of science and adds some more nuance about the sort of economy he has in mind:

> The aim of scientific economy is to provide us with a picture of the world as complete as possible – connected, unitary, calm and not materially disturbed with new occurrences: in short, a world picture of the greatest possible stability. The nearer science approaches this aim, the more capable it will be of controlling the disturbances of practical life, and thus serving the purpose out of which its first germs were developed. (Mach 1986, p. 336)

It is worthwhile to note that such a conception of simplicity led Mach to embrace a unificationist approach to science regarding the relationship between physics and psychology. Famously, this is an aspect of Mach's thought that in due time would prove powerfully appealing to the neo-positivists of the Vienna Circle (Frank 1970). It is also very significant that Mach's ideas about the principle of economy of thought go hand in hand with the anti-metaphysical remarks with which he starts off his book *The Analysis of Sensations* (Mach 1914). To capture with exactitude what Mach is thinking about when shying away from metaphysics and defending the unity of science, we need to pause to consider the following: for Mach *common sense realisms* represents a metaphysical view of nature, if there is one, and the non-metaphysical alternative he envisions is phenomenalist in character.

As it is generally known this side of Mach's philosophy was critically scrutinized by some of its commentators—from Lenin to R. Musil—although one does well to notice that, contrary to Lenin's famous interpretation of his stance Mach's phenomenalism does not constitute an attempt to reduce the physical to the psychical. On the contrary, what Mach tries to show is that the only way possible to unify both realms is by postulating a neutral substratum to be used as a *tertium quid*. This is how Mach describes the situation in a way which inevitably reminds the reader of the ontological view of reality which will subsequently go by the label of "neutral monism" (Banks 2003, 2010, 2014).

Thus, perceptions, presentations, volitions and emotions, in short the whole inner and outer world are put together in combinations of varying evanescence and permanence, out of a small number of homogeneous elements. Usually, these elements are called sensations. But as vestiges of a one-sided theory inhere in that term, we prefer to speak simply of elements as we have already done. The aim of all research is to ascertain the mode of connection of these elements. (Mach 1914, p. 22)

Even if Mach autobiographically attributed the end of his naïve realism to the influence of Kant's philosophy at the age of 18 years (Blackmore 1972; Blackmore et al. 2001), to sort out the kind of an idealist conception Mach would later want to part ways with, one needs to avoid reading too much into such encounter with Kant's *Prolegomena*. In fact, Mach's doctrine of the elements is in a way closer to Carnap's later ontological neutrality towards the divide between realism and instrumentalism in philosophy of science than to the postulation of an actively transcendental ego and a correspondingly mysterious *Ding an sich*. In a footnote in *The Analysis of Sensations,* Mach describes this ontological division between the ego and the real world as a metaphysical dilemma to avoid rigorously, thereby taking distance from the influence of both Kant's transcendental idealism and Berkeley's theist variety of phenomenalism insofar as they are both based on the divide between the *outside and the inside*:

Shall I once again state the difference in a word? Berkeley regards the "elements" as conditioned by an unknown cause external to them (God); accordingly Kant in order to appear as a sober realist, invents the "thing-in-itself": whereas in the view which I advocate, a dependence of the "elements" on one another is theoretically and practically all that is required. It seems to me that in the interpretation of Kant, this very natural and psychologically intelligible fear of being considered fantastic has not been sufficiently taken into account. It is only from this point of view that we can understand how, while holding that only those concepts had meaning and value which were applicable to a possible experience, he could posit a thing in itself, of which no experience is conceivable. Over against the particular sensation, the plain man and the man of science both set the thing as the presentational complex of all the experiences, whether remembered or still expected which are connected with the sensation in question; and this procedure is extremely shrewd. But for anyone who has assimilated Kant's way of thinking, it becomes meaningless at the limits of experience. (Mach 1914, pp. 361–362)

Make no mistake in this regard: According to Mach's standpoint in ontology the postulation of an external thing-in-itself is as metaphysical a move as the very notion of *ego* both Kant and Berkeley had taken for granted. There is no subjective idealism to Mach's view as he sustains that the notion of ego as well as that of external physical substance can be safely explained away by means of the functional – *nota bene*: not causal – relationships between the elements as he aptly put it in *Knowledge and Error*. In dissolving the psychological ego into a plurality of elements, Mach is without doubt adopting an anti-dualist tenet. Such doubly reductive stance is one Mach admitted to having inherited from Lichtenberg. It interesting to see that some scholars (Baatz 1992) and Mach himself in a famous footnote to the *Analysis of Sensations* (Mach 1914, p. 356) identify this view as akin to the doctrine of non-duality in Buddhism.

Nonetheless, even if he shows no desire to endorse idealism, there is an important respect in which Mach *is* an anti-realist. This position resonates in many of the very controversial views Mach endorsed over the course of his career such as his famously phenomenalist interpretation of the theoretical terms of thermodynamics, his fierce opposition to the reality of atoms against Max Plank or his reservations about Einstein's theory of relativity. It is tempting to see this recurrent hostility towards any realist interpretation of the status of unobservable entities as involving a view of science akin to that of Bas van Fraassen's later constructive empiricism (1980).

Indeed, even if Mach's and Van Fraassen's positions present important differences (Mach to begin with makes no claim about what acceptance of a scientific theory amounts to) they still keep a vague anti-realistic *family resemblance* with each other. Even if not of the sort van Fraassen advocates for, it is fair to say that there is a good pinch of empiricism to such dimension of Mach's philosophy as has been recognized both by Mach himself and by many of his critics. It is well to note in this direction that while the Austrian novelist and engineer Robert Musil in his famous dissertation on *Mach's Theories* argues that Mach's *Humean* attack on the notion of an externally existing material substance is self-defeating when it comes to providing a sound conception of science (Musil 1982), Mach himself doesn't shy away from acknowledging his proximity to certain ideas of David Hume:

> By studying the physiology of the senses, and by reading Herbart, I then arrived to views akin to those of Hume though at that time I was still unacquainted with Hume himself. To this very day I cannot help regarding Berkeley and Hume as far more logically consistent thinkers than Kant. (Mach 1914, p. 368)

Mach's dismissal both of material substances and psychologically standing substantial egos has another source aside from Hume's radical empiricism. This further line of attack stems from the principle of economy of thought, where it is argued that avoiding representationalism in epistemology by way of an immanentist reduction of the physical and the mental to an array of sensation-like elements, constitutes a more parsimonious account of knowledge in comparison to realism and idealism alike.

It has been noted that Mach's characterization of the aim of science involves the usage of this version of the idea of economy. Now, this is not the same with arguing that simpler theories are better because they convey a more accurate representation of the reality outside. One way to highlight the difference is by taking notice that according to Mach's view of science it is essential and not accidental to a scientific theory to treat the experience at hand parsimoniously. It might seem that appealing to the difference between essential and accidental properties doesn't cohere with Mach's declaredly anti-metaphysics inclinations. If so, here's an alternative illustration of the same thought: while simplicity is undoubtedly central to science, it is yet worth elucidating the reasons behind such centrality.

The idea is sometimes advanced that parsimony, used with caution, is a relevant indication of a theory being true. Mach never explicitly discusses this view but it is arguable that the approach to the relationship between science and the notion of economy of thought he endorses, stands in dire contrast with this usage of parsimony. In his doctrine simplicity is just not an indication of a theory offering an adequate representation of the world outside but is rather what makes any theory count as scientific in the first place. It just happens that given the purported economic goal of scientific research as a form of intellectual endeavor, an overly complex theory would fail to meet such an aim and therefore its scientific character, as different from its adequacy, could be safely discarded altogether.

In other words, Mach's treatment of the principle of economy of thought suggests that simplicity is a demarcation criterion instead of an epistemic virtue. If so, the comparison between degrees of parsimony of different opposing conceptual systems wouldn't be a tool to be used when deciding which of our scientific theories are truth-like. Truth is absolutely beside the point here as the aim of science is to deliver economy of intellectual labor in bringing about prediction about new empirical experience.

This being said however, the reader is probably left with the question why anyone would consider it sensible to assign such a purpose to science at the first place. Ernst Mach gives an answer to this question when he writes at the very start of his *Knowledge and Error* that:

> Lower animals living under simple, constant and favourable conditions adapt themselves to immediate circumstances through their innate reflexes. This usually suffices to maintain individual and species for a suitable period. An animal can withstand more intricate and less stable conditions only if it can adapt to a wide range of spatial and temporal surroundings. This requires a farsightedness in space and time, which is met first by more perfect sense organs, and with mounting demands by a development in the life of imagination. Indeed an organism that possesses memory has wider spatial and temporal surroundings in its mental field of vision than it could reach through its senses. It perceives, as it were, even those regions that adjoin the directly visible, seeing the approach of prey or foe before any sense organ announces them. What guarantees to primitive man a measure of advantage over his animal fellows is doubtless only the strength of his individual memory, which is gradually reinforced by the communicated memory of forbears and tribe. Likewise, what essentially marks progress in civilization is that noticeably wider regions of space and time are drawn within the scope of human attention. With the partial relief that a rising civilization affords, to begin with through division of labour, development of trades and so on, the individual's imaginative life is focused on a smaller range of facts and gains in strength, while that of society as a whole does not lose in scope. Gradually the activity of thinking thus invigorated may itself become a calling. Scientific thought arises out of popular thought, and so completes the continuous series of biological development that begins with the first simple manifestation of life. (Mach 1976, p. 1)

Interpreted in this roughly evolutionary fashion science and scientific thought, whatever the degree of cognitive sophistication we may still accord to them, constitute one of the ways in which humans, as biological organisms, adapt themselves to a dauntingly variable set of circumstances. Here's the thought: If the environment a biological individual inhabits grows in mutability and unpredictability so as to overwhelm individual and collective memory with its constant variations, then it becomes a good evolutionary strategy to save mental energy by keeping conceptual schemata about facts as simple as they can possibly get at least insofar as that simplification does not bring a parallel decrease in terms of heuristic and predictive powers. Mach believes that it is the aim of science to make this adaptive parsimony of cognition possible by replacing first hand experience with an organized web of functional relationships between elements. In this sense, if science is a successful adaptation to the environment, that adaptive success can only be explained in terms of the degree of economy it exhibits as opposed to the more casual set of observations marshaled by the layman.

It is interesting to observe that if this is the role of the principle of economy of thought as Mach conceives it, then his account of human cognition in general and of science in particular is not too far from a very thriving albeit doctrinally multifold epistemological program which has been embraced by a variety of authors working in multiple philosophical and scientific domains (from Konrad Lorenz or Donald T. Campbell to Ronald Giere, Patricia Churchland, Dan Dennett, or Philip Kitcher). Callebaut and Stotz (1998) have made use of the label "Evolutionary Epistemology" understood as the "naturalistic explanation of cognition". It is easy to see though that taken at that level of generality, such a summary characterization is still susceptible to receive a virtually massive number of interpretations. Perhaps as a result of Quine's influential *Two dogmas of Empiricism* it has become quotidian to insist on the need for the philosophy of empirical science to *go naturalized*. It is commonly admitted also that precisely due to the lack of contact with actual scientific practice, which has plagued the discipline for many decades, developments within the philosophy of science go routinely ignored by scientists themselves. *No reason for desperation though*, the honest philosopher should muster with relief: the solution is available at hand as many in the field would agree, too. It is a solution that involves the naturalistic transformation of the discipline so that it can at last be made capable of capturing the interest of those folks working in the arenas of (hard) science.

So broadly construed, very few in the domain of science studies (maybe with the very particular exception of the social-constructivists but it is generally recognized that they, too, face many a pitfall of their own), would challenge the point even if it's also true that not so many seem to have a coherent and unambiguous characterization of what *naturalism* as a new *lingua franca* of the philosophy of science involves precisely. Most of the self- described *naturalists* of the day would feel content with Quine's rather vague prescription that Philosophy be continuous with the empirical science(s) without noticing that this epistemological *recipe*, at least the way it stands in Quine's seminal piece (Quine 1969) is hardly a properly developed thesis. Quine's indication rather represents a program to be worked out at length in ways that remain unclear in his original formulation.

Given this proliferation of naturalism(s) it might seem perhaps relatively un-informative to add the name of Ernst Mach to the list of authors who in one way or another have defended different versions of such an epistemic program. However, this characterization of Mach's account is not entirely inconsequential either. If it is the key point of science to provide a systematically succinct account of facts that, while bringing about satisfaction to human beings' needs in their never ending struggle for survival and self-preservation, operates by replacing other more tentative and unconscious forms of cognitive adaptation to the world (Mach 1976, p.

361), then, this undoubtedly naturalistic vision of the status of scientific knowledge, as well as of its biological and evolutionary origins, seems to entail a displacement of truth as an epistemic value to be taken into account when ascertaining which theories, whether scientific or unscientific, are to be held.

Under this framework, parsimony is important not because it provides a sign of truth or likelihood but purely on the grounds of its alleged superior adaptive value. What is more, if truth were the goal, then science would be a battle lost before it's even begun, for our experience is inexhaustible even if by way of a series of strong idealizations and simplifications it is always possible to adapt our thought to the regularities observed in the empirical world. Notice that this is not the same as saying that Mach gives up the aspiration that the laws of science may perhaps correspond to certain degrees of approximation to the real regularities of the natural world. On the contrary, Mach sustains no skepticism in this regard as he significantly finds reason in his ontological monism to believe that the laws of the part (namely, the laws of psychology) cannot be a very long distance away from the laws of the whole.

Although this vague aspiration may very well ultimately persist in the name of Ernst Mach's doctrine of the unification of science and nature, that doesn't mean that within such a naturalistic view the concept of truth bears any prominence either in explaining the origins and goals of scientific theories or in characterizing what makes them peculiar in comparison to other more casual knowledge claims. Mach gives a reason showing why this is so when he highlights that both truth and error flow from the same psychological sources and that it is adaptive success alone and not representative accuracy that tells one from the other. It is a familiar (and a highly controversial) idea often associated with the strong program in sociology of science that the explanatory factors used to account for true and false knowledge claims alike should be kept impartial and symmetrical with regards to the truth or falsehood of the claims to be explained. The same demand applies, too, to Mach's naturalistic view of the biological origins of science.

9.3   Bunge: Why Ockham's Razor Should Be Used with Caution

In his book *The Myth of Simplicity* Mario Bunge (1963) advances a critical account of the role of the principle of parsimony in scientific reasoning and theory evaluation which while giving an explanation of why simplicity may matter at times as an epistemological criteria of value, offers also a criticism of what the author terms *the cult of simplicism* insofar as oversimplification is a dangerous tendency both in science and in philosophy. It is worthwhile to see that if according to Quine's characterization, naturalism consists in the desideratum that philosophy be continuous with the sciences, Bunge's model of *metascientific elucidation* would indeed fit the standard well, as it is presented as a form of *scientific philosophy.*

Now, different from Mach's treatment of the topic of parsimony, which is comparatively much more general in character, Bunge proposes a classification of the very different meanings the rule of simplicity may be accorded. The point of the analysis here is to show that one of the problems with parsimony is that far from representing one single principle giving univocal advice to the scientist when constructing or evaluating theories or hypotheses, simplicity covers up a promiscuous array of things so that there is more than one way in which a theory can be simple. The matter gets even more complicated if you take into consideration the fact that sometimes these multiple forms of simplicity do demand incompatible things from a theory.

It is one thing to pretend to keep a theory logically simple; it is another to search for extra-logical parsimony, be it semantical, epistemological or pragmatic in character. Bunge says that there are good reasons to ascertain that logical simplicity can hardly be a true objective of science as the *simplest logical* version of a theory would involve the conjunction of the observational proposition in its field. That construct however wouldn't be a theory at all because it would lack both explanatory and predictive power as no further proposition can be logically deduced from such a conjunction.

Epistemological simplicity (ES) is in turn much less idle. In Bunge's reconstruction ES amounts to the economy of transcendent terms, in other words the closeness of a given theory to sense observation and correspondingly the relative paucity of theoretical and abstract terms. This is exactly the type of simplicity that phenomenalists often make reference to when they say that simplest theories are to be preferred. Bunge argues nonetheless that contrary to what authors like Mach or Russell would have thought, this sort of theoretical succinctness comes at a cost:

> Predominantly phenomenalist languages—to the extent to which they are possible—achieve epistemological simplicity, or triviality, at the cost of both syntactical complexity and epistemological shallowness: it takes longer to say less in phenomenalist languages. The latter are not economical, but just poor. (Bunge 1963, p. 72)

Bunge contends that in contrast to the requirements of epistemological paucity, there are cases where preference should be given to theories formulated in a physicalist language about unobservable references. Granted that they may be more complex at least if their degree of complexity is measured by the presence of theoretical terms, but it is precisely this complexity that allows for a number of epistemological virtues which are also inescapably important if scientific theories should be made tractable. He writes:

> Predominantly physicalist languages, in which theoretical constructs occur, such as "electric charge" and "surplus value", allow for a greater logical cohesiveness and compactness, a deeper understanding, and an easier refutation (they take more risks than phenomenalist constructions, which ultimately inform only about our subjective feelings). Besides, physical-object languages enable us to distinguish the subject from the object, appearance from reality, the finite known from the infinite unknown, and so on. Finally, they render communication possible (a strictly phenomenalist language would be private), and they are interesting. (Bunge 1963, p. 73)

This line of thinking about the perils of epistemological simplicity does not involve a more general plea for the proliferation of theoretical constructs in science for their own sake. Bunge's point here is prudent. Let us grant that the complexity of the reality outside sometimes calls for a shrewd introduction of theoretical terms whatever their remoteness from sense-data might be, but with that in mind, it is also good to notice that unqualifiedly multiplying scientific constructs is as damaging as restricting them without a good explanatory reason. The rule here is neither unrestrictive epistemological paucity nor excessive theoretical largesse but a more modest usage of Ockham's razor which, while demanding that theoretical entities be avoided whenever they are in vain, allows for the introduction of theoretical constructs if demonstrably required by the intricacies of the real processes to be explained.

Although Bunge associates epistemic simplicism to Mach's formulation of the principle of economy of thought, Mach's own version of phenomenalism, different perhaps from other varieties of the same philosophy, includes no commitment to any radical prohibition of the use of theoretical terms in mechanics. However, there is a further variety of extra-logical simplicity in Bunge's reconstruction, which fits Mach's position better. *Psychological simplicity* is a sort of pragmatic simplicity consisting in the ease of understanding of a given theory. Bunge connects this pragmatic version of the principle of parsimony to Mach's thesis that the goal of science is to make possible the minimization of expenditure of intellectual labor. It is worth noticing that even though Bunge does not dispute that there might be reasons to recommend keeping scientific theories psychologically simple within certain limits, he also insists that it is just a mistake (and one attributable to Mach's philosophy of science) to overdo the importance of psychological economy.

There are two reasons why an excessive reliance on psychological simplicism goes astray. First, it is not always clear what this type of parsimony means in practice as different theoretical constructions may seem easier or harder to grasp to different people relative to empirically variable educational factors. Secondly, there are cases where more complicated theories are demonstrably better, so contrary to Mach's thesis it is not the ultimate objective of science to restrict the spending of psychological work. Bunge argues:

> This is the type of simplicity that seems to be envisaged by the partisans of the economy of thought and particularly by Mach, who described scientific research as a business (Geschäft) pursued for the sake of saving thought. Psychological simplicity is desirable both for practical (e.g., didactic and heuristic) purposes: if preliminary theoretical models are to be set up, details must be brushed aside and easily understandable notions ("intuitive" ideas) must be seized upon. The obvious must be tried first, if only to dispose of it early: this is a well-known rule of intellectual work.
>
> However, it should be borne in mind that (a) psychological simplicity is culturally and educationally conditioned, i.e., is not an intrinsic property of sign systems; (b) the deliberate neglect of a given factor should be always justified; (c) we must be prepared to sacrifice psychological economy to depth and accuracy whenever the former becomes insufficient. For, whatever science is, it is certainly not a *business* whose concern is to save experience with the minimum expenditure of work. (Bunge 1963, pp. 77–78)

9.4 Realism and Simplicity: Why Parsimony (Sometimes) Matters

Mario Bunge's quarrel with Ernst Mach ultimately revolves around their respective views on the true nature of the aim of science. In this connection, while Mach's doctrine may be roughly considered as *instrumentalist* in character in that it denies truth any bearing upon the definition of what science is essentially about, Bunge's account in contrast is a *critically realist* one (see for example Bunge 1973, 1979, 2012). He formulates this realist thesis by arguing that against whatever Mach had wanted to say, far from limiting itself to the mere description or inventory of perceptual data, the business of science does not consist in the instrumentalist purpose of saving intellectual work but rather in the construction of a more or less adequate picture of real processes out there. This does not mean that we can ever get a direct representation of a real physical (or chemical or biological or psychological) process which maps univocally onto it, for scientific theories always refer to the target system by way of a web of theoretical models which pictures it in a somewhat impoverished fashion.

Whatever the case, let us not forget, Bunge demurs, that "the goal of research is not to restrict itself but to attain truth" (Bunge 1963, p. 84). With such a representation of the objective of science in mind, he points out that as long as epistemological simplicism is concerned, a phenomenalist interpretation of science, however simple it may seem at first sight when compared to a realistically physicalist alternative, threatens to make science nonsensical:

> Of course, the assumption that there are physical objects somehow corresponding to my perception *is* risky. But without assuming this and many other hypotheses there is no science. (Besides, the assumption that my perceptions are never caused by external objects is even more risky—and barren). Epistemological simplicism, the injunction to eschew the unfamiliar and abstract in order to diminish the danger of error, is an invitation to remain ignorant. "Safety first", a good slogan for driving, is lethal to the venturesome spirit of scientific research. The possibility of obtaining new and deeper and more accurate knowledge is drastically cut down by the allied requirements of simplicism, fundamentalism and infallibilism. In science, just as in everyday life, "Nothing venture, nothing win", as Pauli once said." (Bunge 1963, p. 87)

In any event, if the goal of science is to attain a truth-like representation of the real processes of nature, what verdict does this entail on the usefulness of simplicity in general as an epistemic criterion? Bunge's meta-scientific elucidation has two reasons to challenge the general value of parsimony. Firstly, as has been seen already, the principle of parsimony is multiple-fold: rather than a single and unified *razor,* simplicity can be represented as a multi-edged and *not–so-simple Swiss knife.* What makes things inescapably complicated in this respect is that each of the edges of the knife brings about different and incompatible advice on what is simple and what is complex. The logical or mathematical formulation of a physical theory may be condensed in ways that make that theory epistemologically convoluted in terms of the multiplicity of theoretical terms it involves. Conversely, when the epistemic succinctness in the sense just explained increases, the syntactic complexity of the theory in question may very easily grow to a point in which the theory itself becomes heuristically and pragmatically intractable. What about psychological simplicity? Well, as Bunge shows convincingly, that is a subjective value is entirely dependent on contingent and variable educational and cultural circumstances.

Bunge's second reason for thinking that the principle of simplicity, whatever the form it may adopt, is not an unrestricted epistemic value as simplicism takes it to be, is that good theory construction has many other requirements. Indeed, scientists tend to want their theories to exhibit a number of virtues other than parsimony. A good theory is supposed to be explanatorily and predictively powerful, fertile, original, empirically interpretable, linguistically exact, syntactically correct and externally consistent with the bulk of accepted knowledge. It is clear that not all these standards are epistemically created equal as scientists are routinely prepared to under-emphasize some of them for the sake of the others. It is at any rate, clear also that an overuse of the multiple edges of Ockham's razor may at times jeopardize many of the rest of the legitimate requirements of science. Simplicity if anything is a rule to be used in conjunction with other principles of hypothesis-evaluation but not something to be taken in isolation:

> It is consequently false that, as the intuitivist and simplicist rule has it, "We ultimately prefer the simplest hypothesis compatible with the available empirical information". We normally prefer the hypothesis, whether simple or complex, compatible with (a) the available empirical evidence *and* (b) the bulk of knowledge *and* (c) the period's metascientific principles and tenets. (Bunge 1963, p. 95)

Now, having said this, it is interesting to consider that Bunge's debunking of simplicism does not go so far as to denying that simplicity (in any number of its many faces) may be a desirable feature of a scientific theory in a plurality of contexts. It is good to consider that a judicious degree of syntactic, logical or even epistemic succinctness may, when other requirements of science are not ignored, make a theory mathematically tractable, more fruitful explanatory-wise or easier to falsify or to confirm. Certainly there is no doubt that these are all very desirable attributes for a scientific construct to exhibit.

Bunge thus grants that there is nothing wrong with simplicity when taken in moderation. Yet, the question remains that the task of science is not to search parsimony for its own sake but rather to attain relatively adequate models of real physical processes:

> The second reason for the devaluation of simplicism is the enormous growth of scientific theory and the accompanying realization that the task of the theoretician is not merely to describe and summarize experience in the most economical way, but to build theoretical models of bits of reality, and to test such images by means of logic, further theoretical constructions, empirical data and metascientific rules. Such a constructive work certainly *involves* the neglect of complexities, but it does not *aim* at disregarding them. Rather, a desideratum of every new theory is to account for something that has been overlooked in previous ways. (Bunge 1963, p. 112)

This paragraph reads as a clearly anti-Machian declaration of scientific realism. Nevertheless, how does it resonate in relation to the validity of the principle(s) of simplicity as a mark of truth in science? Bunge claims that whatever other justifications may obtain to desire to make a theory as simple as possible, when taken at face value there is a *prima facie* reason for remaining skeptical about simplicity as a symptom or an indication of truth. In fact it is if anything safer to assume that *simplicity is rather the seal of falsehood.* Here's the idea in a nutshell: The simpler the assumptions of a given theory might be, the more unrealistic the theory will be shown with regards to its reference.

In order to develop this contention, there are some epistemological qualifications to be drawn here. If science proceeds by constructing models of natural processes then we should always keep in mind that such constructions of scientific models involve by their very nature the neglect of very many dimensions of worldly systems. If that is so, it also entails that given the amount of abstraction and idealization theoretical models invariably imply (Psillos 2011), there is at least something we can know a-priori. Taken literally, all our best theories are probably false—or more optimistically put, only partially correct—as they apply exactly to schematic models and only inexactly and in approximation to the references of those models (Bunge 1963, p. 112). Even if this feature of scientific representation may at first sight be taken as vindicating anti-realism, it really doesn't do so in the mind of Bunge and others (see Giere 2009; Hesse 1963; McMullin 1984), because for all their inevitable deficiencies, models are still partially *like* certain aspects of the system to be covered. Those systems however are known to be far more complex than their approximations and then simplicity on the part of a theoretical model can be used as an indication that the theory which carries it is incomplete and defective and will be superseded in due time by another more accurate representation of the system.

What epistemic good does the rule of simplicity have to offer? Bunge's answer to this question is entirely in line with his realist thesis about the aim of science and sounds like a sensible alarm to the perils of simplicism. After all, what seems to follow from his analysis is that there is no value whatsoever to the principle of parsimony as long as simplicity, with any of its many faces, is taken in isolation as a symptom of truth. In contrast, Bunge argues, if combined with other requirements of science-construction, the benefits certain forms of simplicity can bring about albeit modest in nature are not negligible. Whatever the case, a legitimate use of the razor(s) should not ignore the fact that when soundly used, simplicity does not constitute an independent rule of scientific validity but rather a derivative version of a much more general prescription: *avoid the pomp of speculative ideas whenever they are not really called for.*

Truth, however difficult the elucidation of this concept may be, is the central target of scientific research; hence to truth all other desiderata, including certain partial simplicities, should be subordinated. Owing to the incompatibility of the unqualified demands for economy in every respect with most symptoms of truth, simplicity should neither be regarded as mandatory nor counted as an independent criterion on a par with others, let alone above others.

Rightly understood, the rule of simplicity, far from being independent, falls under the general norm: *Do not hold arbitrary (ungrounded and untestable) beliefs.* In the light of this basic rule, an arbitrary simplification is as sinful as an arbitrary complication (Bunge 1963, p. 113).

9.5 Simplicity and Scientific Pluralism: Why Less Is Not Always Better

In his 1865-piece *An Examination of Sir William Hamilton's Philosophy*, John Stuart Mill raises an argument in favor of the principle of parsimony. Mill says that contrary to many of its usual interpretations, parsimony does not depend on any ontological view of nature as being fundamentally simple because it takes its force from a more general prescription. Whatever the ontological intricacies of the world itself, science should always avoid pointless speculation:

> The "Law of Parsimony" needs no such support; it rests on no assumption respecting the ways or proceedings of Nature. It is a purely logical precept; a case of the broad practical principle, not to believe anything of which there is no evidence. When we have no direct knowledge of the matter of fact, and not reason for believing it except that it would account for another matter of fact, all reason for admitting it is at an end when the fact requiring explanation can be explained from known causes. The assumption of a superfluous cause is a belief without evidence; as if we were to suppose that a man who was killed by falling over a precipice, must have taken poison as well. (Mill 1865, pp. 467–468)

As Sober (2015) has rightly put it, Mill's 'conception of simplicity' is in fact a variety of the general rule of evidentialism not to hold beliefs in the absence of evidence for them. It is conceivable that many a commentator would express doubts as to whether this evidentialist proposal deserves to be called parsimony proper. After all, Mill suggests that simplicity has no role of its own except insofar as it represents a particular case of the broader rule against ungrounded conjectures in science. As it has been shown already, Mill's view on this matter aptly captures the essence of Mario Bunge's modest defense of simplicity. It is also safe to assume that considering his empiricist background, Ernst Mach wouldn't necessarily feel at odds with the anti-speculative spirit of this view of the Ockham's razor even though he clearly had other more substantive uses of the role of parsimony in science.

If this all seems a bit too little for such a distinguished principle, then think of what a stronger alternative would take. One regular way of justifying the idea that simplicity is a mark of truth is to assert that Nature itself abhors complexity. However, as both Mill and Bunge show, this sort of an ontological justification involves a series of metaphysical commitments with many pitfalls of their own. In light of that there are reasons to commend Bunge's strategy of keeping things modest when defending why parsimony may at times be a heuristically advisable policy.

Quite apart from the uncertainties of the ontological view of Nature as inherently simple, there is a second aspect to commend to Bunge's account of simplicity. The history of science shows eloquently that there are instances where unrestricted parsimony *is not a good idea*. Present day theoretical biology for instance shows a particularly thriving state of development in which ideas about non-genetic inheritance processes, epigenetic systems and developmental flexibility make manifest exactly where some of the central tenets of the modern synthesis in evolutionary biology about the allegedly principled impossibility of soft inheritance go astray. It just happens that things in evolution are not *that* simple as an unduly extensive use of parsimony could easily lead us to believe. In general, it is good to note that research customarily unveils the manifold inner complexity of real processes.

Examples could be multiplied with regards to very many successful contemporary theories in physics (quantum mechanics and the theory of relativity), chemistry (chemical physics and atomic theory), psychology (cognitive psychology or neuropsychology), compared to their seemingly more parsimonious counter-parts of the old days. Whatever the case and no matter which discipline we decide to look at, the lesson to be learned here has the same invariable flavor. Regardless the understandable appeal of parsimony, it is well not to ignore that modern science, at some very real levels, is a *complex* area of intellectual endeavor.

To put this discussion in a more contemporary context it is interesting to notice that both Stanford's (2006) thesis about the inevitability of unconceived alternatives to any of our pet theories, and Hasok Chang's (2012) advocacy for the benefits of scientific pluralism, help cast light on the reasons why plurality and the proliferation of systems of knowledge may often times be way better for scientific growth than the grim sobriety a strict adherence to parsimony can bring about. Nothing of this involves of course a call for ungrounded fantasizing in science. On the contrary, it would be a mistake to assume that pluralism and proliferation is tantamount to a declaration of epistemological relativism and scientific irresponsibility. Chang has it right in this respect when he notes that pluralism should not be confused with abdication of judgment (Chang 2012). It is precisely here that the principle of simplicity, rightly understood, has still a proper role to play. In discussing law-like statements Bunge calls the attention of the reader both to the undesirable presence of inscrutable-predicates in them and the heuristically healthy services of Ockham razor when it comes to shaving such dispensable ontological inflation:

> (...) propositions denoting entities the existence of which cannot be ascertained by any empirical means – such as *élan vital* or *id*- so that they should be shaven with Occam's razor. Ex: expressions in which the terms "absolute velocity of bodies" or "libido energy" occur. Although inscrutable predicates have often been introduced in the preliminary stages of sciences (remember "phlogiston", "caloric" and "ether") and will probably continue to be, they are finally expelled from scientific discourse or are elucidated in terms of scrutable predicates (e.g., via law statements), because they lead to the formation of untestable assumptions. (Bunge 1963, p. 166)

This is a good anti-speculative point which indicates what a good use of the razor would be like. Yet, here again Bunge is right to warn that Ockham's razor as many other razors, is a hazardous item to be handed with caution; an undoubtedly good point to take if scientific growth is not to be unduly *shaved* in the name of economy and restriction.

There is however a further angle to be tackled in relation to scientific pluralism and the dangers of simplicism. The type of realism endorsed by authors such as Dupré (1993), Galison and Stump (1996) or Bueno (2013), to name but a few, takes inspiration from scientific plurality to point to a conclusion ontologically more ambitious than that of Chang or Stanford. This would be the idea in a nutshell: For all the calls to cautionary simplicism, perhaps the best way of thinking about the actual plurality and disunity within various scientific fields is to take it as a face value as an intrinsic indicator that the world itself is irreducibly divided also. Discussing this further ramification of the topic clearly exceeds the scope of this paper but suffice it to say that if sound this ontological point would fly in the face of one of the very few aspects Mach and Bunge share in common, namely the idea that nature is unified (perhaps *systemically unified*, vis Bunge 1979), and the aspiration that science, too, should be so (maybe *methodologically so*, vis Bunge 1973).

9.6 Conclusion

It may seem as a paradox that a philosopher who puts so much emphasis on the role of simplicity of thought in science as Ernst Mach does, doesn't himself propose the idea that parsimony is an epistemic virtue to be considered in theory evaluation. Perhaps it is indeed a paradox, however the upshot of my analysis indicates that this is precisely the case. To see why this is so it is helpful to consider the following feature of Mach's thought. Science is essentially all about ordering sense data by using functional relationships so that prediction of new experience is made possible. The adaptive dividend science so produces is huge and goes hand with hand with a correlative minimization of the amount of time and effort invested by the layman. Notice that such characterization of science leaves aside the notion of truth as though it was a metaphysical burden to get rid of. To say that the principle of simplicity is an epistemic rule to follow in the scientific search of truth, would be just totally beside the point under a framework like this.

Bunge's realist philosophy includes a strong rejection of such a conjunction of naturalism and phenomenalism. In sharp contrast to what Bunge takes to be the mistakes of Mach (see his 1966), Bunge with a pinch of realism and another bit of semantics refrains from identifying the aim of science with the economy of thought. There is another more ironical dimension to his line of thinking. Again different from Mach, Bunge's realist account of scientific theories is able to accommodate a weak vindication of the heuristic validity of some versions of simplicity at least so long as parsimony works as a defensive barrier against the proliferation of inscrutable speculation. There is however another proviso to be taken seriously here. When used without restriction, simplicity may easily become simplicism, which constitutes a relevant danger jeopardizing the growth of science.

By using some more contemporary ideas from current philosophical debates on realism and pluralism, this chapter has shown that Bunge's siren call about the perils involved in the cult of simplicity is a point worth making. The connection between his and Mach's ideas on the unity of science and the view of the ontological (dis) unity of the world advanced by present-day philosophers of science is a worthwhile task to explore but also one that clearly escapes the scope of this chapter.

References

Baatz, U. (1992). Ernst Mach- the scientist as a Buddhist? In J. T. Blackmore (Ed.), *Ernst Mach – a deeper look. Documents and new perspectives* (pp. 183–199). Dordrecht: Springer.

Banks, E. C. (2003). *Ernst Mach's world elements. A study in natural philosophy*. Dordrecht: Kluwer Academic Publishers.

Banks, E. C. (2010). Neutral monism reconsidered. *Philosophical Psychology, 23*(2), 173–187.

Banks, E. C. (2014). *The realistic empiricism of James, Mach and Russell. Neutral monism reconceived*. Cambridge: Cambridge University Press.

Blackmore, J. T. (1972). *Ernst Mach. His work, life and influence*. Berkeley/Los Angeles/London: University of California Press.

Blackmore, J. T., Itagaki, R., & Tanaka, S. (Eds.). (2001). *Ernst Mach's Vienna 1895–1930. Or phenomenalism as philosophy of science*. Dordrecht: Springer.

Bueno, G. (2013). *Sciences as categorial closure*. Oviedo: Pentalfa.

Bunge, M. (1963). *The myth of simplicity*. Englewood Cliffs: Prentice-Hall, Englewood Cliffs.

Bunge, M. (1966). Mach's critique of Newtonian mechanics. *American Journal of Physics, 34*(7), 585–596.

Bunge, M. (Ed.). (1973). *The methodological unity of science*. Dordrecht/Boston: Reidel Publishing Company.

Bunge, M. (1979). *Treatise on basic philosophy. Vol 4. Ontology II: A world of systems*. Dordrecht/Boston: D. Reidel Publishing Company.

Bunge, M. (2012). *Evaluating philosophies* (Boston studies in the philosophy of science, Vol. 295). Dordrecht: Springer.

Callebaut, W., & Stotz, K. (1998). Lean evolutionary epistemology. *Evolution and Cognition, 4*(1), 11–36.

Chang, H. (2012). *Is water H₂0? Evidence, realism and pluralism*. Dordrecht: Springer.

Dupre, J. (1993). *The disorder of things. Metaphysical foundations of the disunity of science*. Cambridge, MA: Harvard University Press.

Frank, P. (1970). Ernst Mach and the unity of science. In R. S. Cohen & R. J. Seeger (Eds.), *Ernst Mach. Physicist and philosopher* (pp. 235–244). Dordrecht: Springer.

Galison, P., & Stump, D. J. (Eds.). (1996). *The disunity of science: Boundaries, contexts and power*. Stanford: Stanford University Press.

Giere, R. N. (2009). Why scientific models should not be regarded as works of fiction? In M. Suárez (Ed.), *Fictions in science. Philosophical essays on modeling and idealisation* (pp. 248–259). London: Routledge.

Hesse, M. (1963). *Models and analogies in science*. New York/London: Sheed and Ward.

Mach, E. (1914). *The analysis of sensations and the relation of the physical to the psychical*. Chicago/London: The Open Court Publishing Co.

Mach, E. (1919). *The science of mechanics. A critical and historical account of its development*. Chicago/London: The Open Court Publishing Co.

Mach, E. (1976). *Knowledge and error. Sketches on the psychology of enquiry*. Dordrecht: Springer.

Mach, E. (1986). *Principles of the theory of heat: Historically and critically elucidated*. Dordrecht: Springer.

McMullin, E. (1984). A case for scientific realism. In J. Leplin (Ed.), *Scientific realism* (pp. 8–40). Berkeley: University of California Press.

Mill, J. S. (1865). *An examination of Sir William Hamilton's philosophy and of the principal questions discussed in his writings*. London: Longmans Green and Co.

Musil, R. (1982). *On Mach's theories*. Washington, DC: The Catholic University of America Press.

Psillos, S. (2011). Living with the abstract. Realism and models. *Synthese, 180*(1), 3–17.

Quine, W. V. O. (1969). Epistemology naturalised. In *Ontological relativity and other essays*. New York: Columbia University Press.

Sober, E. (2015). *Ockham's razors. A user's manual*. Cambridge: Cambridge University Press.

Stanford, K. (2006). *Exceeding our grasp. Science, history and the problem of unconceived alternatives*. Oxford: Oxford University Press.

Van Fraassen, B. C. (1980). *The scientific image*. Oxford: Clarendon Press.

Chapter 10
Quantitative Epistemology

Nicholas Rescher

Abstract This chapter follows Mario Bunge's unswerving dedication to the idea that philosophical deliberations should be precise in formulation and cogent in substantiation. In deliberating about the range of human knowledge from a quantitative point of view it emerges that three very different ranges of consideration have to be addressed. The range of what we human individuals can actually and overtly know is bound to consist of a finite number of items. And given the recursive nature of language the range of what is knowable—i.e. propositionally formulated truth—is at most denumerably infinite. But the range of what is theoretically knowable—the manifold of actual facts—is going to be transdenumerably large, if only due to the role of real-valued parameters. A proper heed to this quantitative disparity has interesting implications for the status of our knowledge in a world where we must deal with digital conceptualization of an analogue reality.

I have always admired Mario Bunge's unswerving dedication to the idea that philosophical deliberations should be precise in formulation and cogent in substantiation. The present deliberations will hopefully proceed in the spirit of these aspirations. These deliberations do not address the epistemology of mathematics; rather, they will deal with the mathematics of epistemology (Rescher 2004, 2010). Their starting-point lies in the consideration that our factual knowledge—our knowledge that something-or-other is the case—has to be formulated symbolically, by means of language (broadly speaking). And language as we deal with it is articulated recursively, developed from a finite vocabulary by a finite number of basic rules of combination. This means that our symbolically articulable claims—and thereby our accessible knowledge—are theoretically at most enumerable and practically actually finite in scope. And in actuality no more than a finite number of facts have

N. Rescher (✉)
Philosophy Department, University of Pittsburgh, Pittsburgh, PA, USA
e-mail: rescher@pitt.edu

© Springer Nature Switzerland AG 2019
M. R. Matthews (ed.), *Mario Bunge: A Centenary Festschrift*,
https://doi.org/10.1007/978-3-030-16673-1_10

ever been expressed and accordingly be known. As G. W. Leibniz already noted in the seventeenth century, the manifold of *known information* is destined to be finite.[1]

This cannot, however, be said of the manifold of *fact*. For there is good reason to think that this manifold is not merely infinite but even transdenumerably infinite in scope. For consider:

1. The mathematical realm embodies collections of items which, like the real numbers, are transdenumerably infinite in extent.
2. If (as it seems) the physical realm involves various descriptive parameters that vary along a continuous scale, this too points to a transdenumerably diverse body of information. For once we are dealing with a feature subject to continuous (real-number) parametrization, a trans-countable manifold of factuality emerges.
3. Even if the number of basic objects within in a theoretical domain or in the natural world is only enumerable in quantity, still the number of compound and complex objects engendered by viewing these in combination is going to be transdenumerably large. (This follows from Cantor's theorem that the cardinal number of subsets of a given set is always greater than that set's own cardinality).

All of this means is that the scope of what is known is going to be quantitatively far smaller than the totality of fact, so that we finite knowers will have to come to terms with a vast volume of unknown fact.

To be sure, at this point the temptation to deliberate as follows arises. Recall that there are two very different ways of presenting and considering a body of information. One is by way of an inventory: one simply provides a register in inventory of the facts at issue. And of course given this way in which we have to proceed here, this will always be something of finite scope and thus limited. But there is another, altogether different way afforded by the monumental discovery of the ancient Greeks—the way of axiomatization. Here one indeed starts out with a finite register of basic facts. But on the pathway of inferential derivation can amplify this potentially ad infinitum. And thus a modest register of basic facts can embody a potentially infinite range of fact by way of implicit logical derivation rather than explicit textual articulation.

But this well-intentioned idea will not achieve its present purpose. For one thing, lies in the logic of the situation that no inventory can ever ensure its own completeness. For to do so we would have to be able to move inferentially from $(\forall x \in I)Fx$ to $(\forall x)Fx$, and such a step is logically impracticable without begging questions about the nature of I.

However, the non-axiomatizability of the truth at large is also rooted in fundamentally logical considerations. For suppose that we endeavor to axiomatize the truth. We would then have to set out a group of axiomatic truths

$$t_1, t_2, t_3, \ldots t_n$$

[1]For details see the author's "Leibniz's Quantitative Epistemology." *Studia Leibnitiana*, vol. 36 (2004), pp. 210–231.

that purports to provide an inferential basis for deriving what is true:

$$(\forall x)\left(T \vdash x \to T(x)\right) \tag{10.1}$$

where

$T = t_1, \& \ t_2, \& \ t_3, \& \ \ldots \& \ t_n,$
$T(x)$ has it that "X is true," and
$\vdash$ represents logical derivability

Given the truth of all those t_i, this contention is unproblematic. But of course when what is additionally needed for overall axiomatization is:

$$(\forall x)\left(T(x) \to T \vdash x\right) \tag{10.2}$$

This of course nowise follows from (10.1) itself And is—by contrast—decidedly problematic in respect of substantiation. For while (1) rather modestly claims that what follows from T is true, (2) far more ambitiously claims that anything and everything that is true must be a T -consequence. It is difficult indeed to see how this could be.

To be sure, if we could conjure with truths in suitably enterprising ways then (10.2) could perhaps be secured. And in particular if among the components of T we could insert *the conjunction of all truths whatsoever* we would be home free. But this sort of thing is just not practicable. For truth (unlike fact) is simply not conjunctively totalizable. The mega-conjunction of all facts is still a fact. But truths (unlike facts) are symbolism (linguistic) formulations that must be finitely articulated. And accordingly simply is no prospect of combining *all* truths into a single megatruth.

The upshot of these considerations is that the manifold of fact will be deductively incomplete: any supposed axiomatization of it is bound to leave something out.

The cognitive ideal of course, is to know "the truth, the whole truth, and nothing but the truth." Here the radical sceptic is concerned to deny that we can ever know the truth of things; as he sees it the realization of certifiably true knowledge is beyond us. And the moderate sceptic is concerned to deny that we can be know "nothing but the truth"; as he sees it our information always contains an admixture of misinformation and does so in a way that we cannot securely distinguish the one from the others. But neither of these modes of scepticism has been at issue in the present deliberations which have no quarrel with the idea of cognizable truth. They are addressed only at the third element of that dictum, by rejecting the idea that we can ever know "the whole truth." Accordingly these deliberations do not of themselves make for cognitive scepticism of some sort, but only for cognitive modesty.

Quantitative disparity brings to the fore some crucial considerations regarding knowability—considerations which the analogy of musical chairs helps to bring into clearer view. For the question "Are all truths knowable?" is akin to the question "Are all persons seatable?" And here consider that not all persons are even candidates for seatability since the room will be too small to admit more than a handful of players, and among these candidates it is inevitable that some players will ultimately be unseated when the music stops since limited number of chairs imposes limits on seatability. Accordingly, we will have to distinguish between the number of people that are *effectively* seatable (i.e., can ultimately be seated), the number of them that can *possibly* be seated through being participants in the game, and the number of them that are *theoretically* seatable as potential players. And this situation will carry over by analogy to the cognitive case, where we must distinguish between

- the actually achievable body of knowledge as a finite subset of the manifold of truths.
- the potentially realizable body of knowledge (as the manifold of truths at large) which is denumerably infinite in scope, and
- the theoretically available knowledge relative to the vast, non-denumerably extensive body of fact.

However, one very important point needs to be made. The situation that the range of fact vastly outruns the reach of knowability does not however, in and of itself mean that there are untenable facts. To see that this is so, consider the game of Musical Chairs where there are more players than seats. The circumstance that *not all players can possibly be seated* does not mean that there is any one player who is in principle unseatable. And so this is in prospect for fact knowability as well. However the situation is not quite as easy as this might make it seem. For there is at least one fact—viz. the megafact consisting of all facts taken together—whose unknowability is inherent in the numerical disparity at issue.[2]

We are thus carried back to the fundamentals of Leibnizian epistemology. We humans are finitely discursive thinkers who have to conduct their cognitive business with truths framed within the discursive instrumentalities of recursively (combinationally) articulated language. But the world of actual fact is more ramified and complex than that. We conduct our cognitive affairs by digitally discrete means in an analogue world of analytic continuities. And so our comprehension of the realm of fact is bound to be imperfect and to some extent conjectural.

References

Grim, P. (1991). *The incomplete universe*. Cambridge MA: MIT Press.
Rescher, N. (2004). Leibniz's quantitative epistemology. *Studia Leibnatiana, 36*, 210–231.
Rescher, N. (2010). *Epistemetrics*. Cambridge: Cambridge University Press.

---

[2]It is, to be sure, possible that as my friend Patrick Grim stressed the idea of a totality of fact—the fact of all facts—in epistemology will encounter the same theoretical difficulties as the idea of a set of all sets in mathematics (Grim 1991).

Chapter 11
Mario Bunge on Causality: Some Key Insights and Their Leibnizian Precedents

Richard T. W. Arthur

Abstract Mario Bunge wrote his classic *Causality and Modern Science* more than 60 years ago, and a third revised edition was published by Dover in 1979. With its impressive scope and historical perspective it was a long way ahead of its time. But many of its insights still have not been sufficiently appreciated by physicists and philosophers alike. These include Bunge's distinction between causation and other types of determination, his critique of the still-dominant Humean accounts of causality as leaving out the productive aspect of determination, his critique of the conflation of determination with predictability, his insistence on the fictional character of isolated causal chains, and his demonstration that "causal connectability" depends only on the principle of retarded action, not causation. It will be argued that there is a (perhaps surprising) degree of agreement between his views on determinism and causality and those of Leibniz, and a comparison between the two thinkers' views is used to throw further light on these matters.

11.1 Introduction

One of Mario Bunge's main assets is his keen eye for *Idols of the Theatre*, Francis Bacon's term for views that are tacitly adopted just because they are widely accepted in the theatre of learning. He goes by the maxim: if a doctrine makes no sense to you, be brave enough to say so! Of course, Bunge has generally taken this expedient as licensing him to ignore huge swathes of work by his contemporaries who have, unfortunately, responded in kind. Thus, for instance, one can look in vain for a reference to his fine book on causality in the SEP entry (by Jonathan Schaffer) on 'The Metaphysics of Causation', which summarizes a huge variety of writing on causation over the last few decades, including contributions by Armstrong, Bennett, Cartwright, Castañeda, Davidson, Dowe, Dummett, Lewis, Mackie, Mellor, Quine,

R. T. W. Arthur (✉)
Department of Philosophy, McMaster University, Hamilton, ON, Canada
e-mail: rarthur@mcmaster.ca

© Springer Nature Switzerland AG 2019
M. R. Matthews (ed.), *Mario Bunge: A Centenary Festschrift*,
https://doi.org/10.1007/978-3-030-16673-1_11

Russell and Van Fraassen (Schaffer 2016).[1] Bunge wrote his classic *Causality and Modern Science* more than 60 years ago, and the 1963 second edition was republished as a Dover book in 1979.[2] With its impressive scope and historical perspective it was already a long way ahead of its time. But, it will be argued here, that is no less true of it today.

This paper draws attention to some of the insights in those pages that have still not been sufficiently appreciated by physicists and philosophers alike. These include Bunge's distinction between causation and other types of determination, his critique of the still-dominant Humean accounts of causality as leaving out the productive aspect of determination, his critique of the conflation of determination with predictability, his insistence on the fictional character of isolated causal chains, and his demonstration that "causal connectibility" depends only on the principle of retarded action, not causation. The paper will proceed by stressing the (perhaps surprising) extent of agreement between his views on determinism and causality and those of Gottfried Wilhelm Leibniz, using this consilience in their views to expand upon some of Bunge's most seminal insights.

On a superficial reading, it might seem perverse to take Leibniz, of all philosophers, as a foil for expounding Bunge's views. If a monad produces all its states out of its own store, without any interaction with things external to it, what possible role could causation have in his philosophy? As a matter of fact, the answer to this is that it played a large role in his philosophy. Leibniz originated the first causal theories of time, appealed to the identification of causes as the way to break the "Equivalence of Hypotheses" about what is really moving (in the face of the relativity of motion), and identified as the founding principle of his dynamics the Principle of Equipollence, according to which "the full cause is equipollent to the entire effect" (Arthur 2015)—a principle which Leibniz credited with having led him to his discovery of the conservation of energy.[3] These considerations already indicate that Leibniz did not equate causation with the wider category of determination, that he saw the principle of causality as a valuable principle in scientific epistemology, and that he conceived causation as a physical action, all in agreement with Bunge. We will see further specific agreements in what follows.

Granted, Leibniz was an opponent of the materialist doctrines of his contemporaries, Thomas Hobbes and John Toland, and defended a Christian and Creationist account of the universe. Given this, one might expect that Bunge's uncompromising advocacy of scientific materialism would lead him to be dismissive of the views of the German polymath. He is indeed critical of several of Leibniz' doctrines, such as his argument for the necessity of a first cause, the hypothesis of unlimited universal interconnection, and a tendency to confuse cause with reason.[4] But Bunge does not

---

[1] One could say the same about Carl Hoefer's article on causal determinism in the same encyclopedia (Hoefer 2016).

[2] The Preface to the First Edition is signed "M. B., Buenos Aires, December 1956".

[3] See Bunge 1979, pp. 238, 59 and 229.

[4] See Bunge 1979, p. 214.

allow such deprecations to blind him to Leibniz's positive contributions, such as his commitment to the genetic principle, the self-activity of matter, and the non-productive nature of time (contrary to Hermann Cohen's later misappropriation of Leibniz's views).[5]

Moreover, several others of Bunge's estimable insights are foreshadowed by Leibniz, especially in connection with the former's criticisms of the empiricist and romantic traditions. The most notable disagreement concerns their assessments of the place of teleology in science. Here Leibniz's views are more defensible and in accord with modern science than Bunge allows. Given all this, a comparison of Bunge's views on causality with Leibniz's will make for an interesting way to assess some of the former's insights.

11.2   Laplacian Determinism

A good place to start is with what is usually taken to be the classic statement of causal determinism, that given by Pierre Simon de Laplace in the introduction to his *Théorie analytique des probabilités* in 1812:

> We ought, then, to envisage the present state of the universe as the effect of its earlier state and as the cause of the state that is to follow. An intelligence that could, for a given instant, know all the forces by which nature is animated and as well as the relative situations of the beings that compose it, would comprehend in the same formula the motions of the greatest bodies in the universe and those of the lightest atom, provided it were sufficiently vast to submit all the data to analysis; to it, nothing would be uncertain, and the future, like the past, would be present to its eyes. The human mind, in the perfection that it has been able to give to astronomy, affords a feeble outline of such an intelligence. (Pierre Simon de Laplace 1812, Introduction, pp. ii–iii; RTWA translation)

Much has been written about the implications of such determinism. It has been said to entail the necessity of all that happens, thus precluding free will, and making contingency only another name for human ignorance (as in the philosophy of Spinoza). Laplacian determinism, equated with predictability, has been held to rule out genuine novelty, becoming and even temporal succession. Causal processes have been interpreted as those following necessarily according to Newton's Laws, so that, in Whitrow's words, given a complete specification of the initial conditions, "the future history of the universe pre-exists logically in the present" (Whitrow 1961, p. 295). In the judgement of Milič Čapek (expounding and agreeing with Henri Bergson), Laplacian determinism entails that "'Future' is merely a label given to the unknown part of reality which coexists with our present moment in a sense similar to that in which distant scenery coexists with our limited visual field." Consequently, Čapek asserts, we are left with two and only two choices: "either real succession

---

[5]See Bunge 1979, pp. 138, 178, 195 and 76, respectively.

with an element of real contingency, or complete determinism with the total absence not only of possibilities, but of succession as well" (Čapek 1971, p. 111).

All these alleged implications are convincingly contested by Bunge in his *Causality* book. The claim that determinism is incompatible with free will, made already in the seventeenth century by Isaac Newton and Samuel Clarke, is repeated by modern authors who see the indeterminism of quantum theory as a salvation. But this rests on a notion of the freedom of the will as its being *unconditioned*, whereas the decision of the agent must be one of the determining conditions if it is to be a free will worth defending. It also confuses causal determinism with necessitarianism. For the claim that every effect must have a cause does not entail that the cause necessarily has its intended effect: causation is contingent. Causation, moreover, is only one of several species of determination: even the indeterminism characteristic of certain quantum processes involves statistical determination, allowing for an approximate determinism to be regained "in the large" (i.e. for processes whose action is very large compared with Planck's quantum of action).

The identification of determinism with predictability, moreover, conflates an ontological doctrine with an epistemological one: lack of predictability may be evidence for the failure of a particular causal hypothesis, or for the failure of the principle of causality to apply universally, but it does not refute lawful determination as such. Indeed, successful prediction by non-causal laws already shows the fallacy of identifying predictability with causation. As for (qualitative) novelty, our inability to predict it, far from indicating the failure of determinism, "only suggests that the law of emergence of the new feature under consideration has not yet been found" (Bunge 1979, p. 331).

In what follows these objections will be discussed in greater detail, beginning with the most extreme claim, that determinism precludes becoming and temporal succession, and then proceeding to the claims that it entails a strict necessitarian worldview that leaves no room for freedom of the will, contingency or novelty. This will be followed by a discussion of causal processes, and, finally, the status of final causes in modern science.

11.3   The Genetic Principle

Does determinism really entail a strict necessitarian worldview that leaves no room for contingency or time's passing? On the basis of the above passage, Laplace is often credited with having been the first to clearly articulate such a view. His image of an [originally, the] all-knowing and inconceivably vast intellect has been dubbed "Laplace's Demon", perhaps with Descartes's and Maxwell's rather different demons in mind. But as Laplace himself makes perfectly clear just prior to the quoted passage, the omniscient intelligence he alludes to there is simply the divine being to which Leibniz had earlier imputed the same powers. Actual events, Laplace claims, have a linkage with preceding ones founded on Leibniz's *principle of sufficient reason*, whereby "a thing cannot begin to exist without a cause that produces it" (Laplace 1812, p. ii).

For Leibniz the principle of sufficient reason is based on the premise that God would create nothing without a reason, so that there would be determinate reasons for everything that happens in the world he decides to create. If an agent acts freely, God would know the motives determining the action, as well as everything else that must be in place in order for the action to be performed, and would create everything accordingly. But, Leibniz insisted, creation is not causation; in order for the actions to be ascribed to the agent and not to God, the agent would have to be equipped with the means to produce its own actions, according to a law of succession that is intrinsic to that agent. Moreover, these actions would have to be consistent with everything else, so that each enduring thing in the world (each substance) would likewise have to have a law of production of its states known beforehand to God, so that it would pass from its present state into its future states in accordance with its own law and in harmony with the states of all other substances.

We may presume that such a theological foundation for determinism would not be conducive to Laplace's way of thinking, given what he is alleged to have told Napoleon: God is "a hypothesis I do not need in my philosophy". But this did not prevent him from taking for granted that there is always a sufficient reason why an event occurs, a set of circumstances sufficient for its production, even if that is opaque to us. Without needing to appeal to a divine guarantor he could still make the assumption that every event is produced in a determinate way out of previous events, even if we do not know the law of production. In fact, this is the assumption on which his entire theory of probability proceeds, according to which the appeal to probabilities is necessitated by lack of knowledge on our part.

There is therefore nothing in the conception of determinism advocated by Leibniz and Laplace that precludes the reality of succession or becoming: in fact, for them determinism requires both. A process is simply a sequence of events produced out of prior events according to some principle or principles of determination. The order in which these events come out of events in their past is the order of succession, an inherently asymmetric ordering. Without becoming there is no process. But deterministic theories certainly represent processes. So, the idea that there is any incompatibility between deterministic theories and the reality of succession or becoming is simply false. This principle that events come to be out of prior events, evident in both Leibniz and Laplace, has been identified by Mario Bunge as one of two principles that are constitutive of determinism in general (along with the *principle of lawfulness*, which we will come to presently). He calls it: 'the *genetic principle*, or principle of productivity, according to which nothing can arise out of nothing or pass into nothing' (Bunge 1979, pp. 25–26).

That the genetic principle has not been widely recognized is probably due mainly to the prevalence of the Humean view of causation in the "Theatre" of philosophy of science. For this view denies precisely any production of the effect by the cause, reducing causality to constant conjunction rooted in past experience. As Bunge writes, in this tradition, where "experience is the sole source of legitimate knowledge, ... the category of causation as a genetic link was a dispensable figment of abstract thought" (Bunge 1979, p. 89).

For Leibniz too, although experience is a *source* of knowledge—in fact, a necessary source for our knowledge of existing things—reality is not *constituted* by ideas or impressions, as it is for Locke and the empiricists. Instead, according to his doctrine of expression, "One thing is said to *express* another when the relationships that hold in it correspond to the relationships of the thing that is to be expressed. ... [S]olely from a consideration of the relationships of the expression we can arrive at a knowledge of the corresponding properties of the thing to be expressed."[6]

This presupposes that the relations we ascribe to things, especially, for instance, as embodied in complex mathematical equations, are not solely relations of ideas, but instead are intended to model relations among the things modelled. In his advocacy of such a realist view of representation, Leibniz was clearly more perceptive than Hume. Constant conjunctions do not constitute causal relationships, they are simply evidence for them. In fact, constant conjunctions are useless unless they are explicable in terms of laws; and these are established hypothetico-deductively. Certainty is a complete red herring: you can postulate that these laws necessarily hold, i.e. hold with certainty, *cæteris paribus*; but this is not the same as it being necessary that they should hold, as Leibniz well understood.[7] Bunge acknowledges Leibniz's clarity on this point:

> In other words, the law *statements* are contingent truths (in Leibniz's sense of the term) because they lack the certainty that is supposed to characterize analytic statements (necessary truths). This does not mean, of course, that the laws themselves (the objective patterns of being and becoming) are contingent. (Bunge 1979, p. 239)

(We shall come back to the questions of contingency and lawfulness below).

Of course, there must be some restrictions on the Ionians' genetic principle as stated by Bunge above. As Leibniz pointed out, states or accidents arise where there were none before. This is unobjectionable, since they are modifications of something already existing, so they do not arise out of nothing. But for Leibniz the "something already existing" is a *substance*, something that cannot naturally not be. It can only come into being or perish through a miracle. Bunge is dismissive of such substance-accident ontology, although he allows merit in Leibniz's (Spinozan) characterization of substance as something that acts.

There is a genuine difference in ontology here, though, which will limit the scope of agreement between Leibniz and Bunge on causation. Superficially, they

---

[6]Leibniz, from an unpublished manuscript 1677 (A VI 4, 1370). All quotations from Leibniz are given in my translation from the Akademie edition or the Gerhardt ones, although I have also referenced the corresponding page numbers to available English translations such as Loemker's (Leibniz 1976) where possible.

[7]In his *New Essays* Leibniz writes "however many instances confirm a general truth, they do not suffice to establish its universal necessity" (A VI 6, 49/Leibniz 1981, p. 49) and "when a new situation appears similar to its predecessor, it is expected to have the same concomitant features as before, as though things were linked in reality just because their images are linked in the memory" (A VI 6, 51/Leibniz 1981, p. 51).

appeal to the genetic principle to criticize their contemporaries in similar ways. Bunge used it to criticize not only interpretations of the Big Bang as a creation from nothing, but also the Steady State cosmology advocated by Fred Hoyle (and afterwards adopted by Burbidge et al.),[8] according to which hydrogen atoms are produced throughout spacetime from nothing. The latter cosmological hypothesis, Bunge writes in response to his critics, "should not be invoked against the genetic principle: if we wish to build a scientific ontology we should not employ *ad hoc* conjectures conflicting with well established laws of conservation" (Bunge 1979, p. 354). "The concept of emergence out of nothing is characteristically theological or magical—even if clothed in mathematical form" (Bunge 1979, p. 24).

Similarly, Leibniz appealed to the genetic principle in criticism of his contemporaries' notion of continuous creation. In the seventeenth century context, this meant a continuous *re*-creation: many of Descartes's followers, and others such as Erhard Weigel and Johann Christoph Sturm, held that God had to continually recreate the world from nothing at each moment of its existence.[9] This, Leibniz objected, is an unwarranted intrusion of the supernatural into explanations of the natural. These thinkers, moreover, were persuaded to their view by the consideration that one created substance (whether mental or physical) acting on another is unintelligible. They therefore denied any causation in the created world. An apparent instance of causation, such as my sitting down in a chair as a consequence of having decided to do this, is for them explained by the fact that God creates the succession of qualities and positions of the parts of my body and the cushion on the chair, *on the occasion of* my decision: he creates the successive physical states in conformity with one another and in conformity with the physical laws he has established.

Leibniz followed the Occasionalists in holding that the action of one created substance on another is unintelligible. But he objected to their attributing all causation to God: this would undermine the autonomy of action of all creatures, as well as free will in humans. His solution was to relocate action in individual substances, and to redefine causes as "concomitant requisites". Leibniz believed it nonetheless possible to save orthodox accounts of one thing truly causing another to move by recasting causation in terms of there being more reason to attribute the action to one thing than another. One thing acts on another or causes it to move if and only if the reason for the change is more clearly represented in it than in the thing acted upon, that is, if its being held to instigate the motion is the most intelligible hypothesis. This enabled him to hold that "*every individual created substance exercises physical action and passion on all others*", while still insisting

---

[8]See e.g. Hoyle (1952) and Hoyle et al. (1995). "The basic theory underlying the quasi-steady state cosmological model". Proceedings of the Royal Society A. 448: 191.

[9]Leibniz, *Theodicy* §384; GP VI 343/Leibniz 1985, p. 355.

that "in rigour, *no created substance exercises metaphysical action or influence on any other*".[10]

This means that the genetic principle has a different scope for Leibniz than it does for Bunge. According to Leibniz, each substance produces its own states out of its own store, whereas for Bunge states are produced in a given thing by its interactions with other things. Nevertheless, since for Leibniz it is the substance that produces the states, it is no less true for him than for Bunge that "States are not causes, but simply antecedents of later states" (Bunge 1979, p. 71). The consequent states follow from the antecedent ones according to the "law of the series" characterizing the substance, but their cause is not just this antecedent state but also all the states of other substances required to cooperate in the production of the effect: these are the "concomitant requisites". So a cause for Leibniz is not something producing its effect in isolation from other things, and this again agrees with Bunge's account.

11.4   The Principle of Lawfulness: Free Will and Self-Determination

Now let us proceed to the second principle that Bunge identifies as constitutive of determinism in general. This is the *principle of lawfulness*, according to which "nothing happens in an unconditional and altogether irregular way—in short, in a lawless, arbitrary manner" (Bunge 1979, p. 26). A consideration of this principle will help lay to rest the claim that general determinism leaves no room for freedom of the will, contingency or novelty. We can illustrate this by reference to Leibniz's defense of the compatibility of determinism with free will, which provides an excellent example of the application of the principle of lawfulness. The same example will also throw light on the distinction of determinism from causation.

Immediately prior to the famous passage from the *Théorie analytique des probabilités* quoted above, Laplace notes that according to Leibniz, all events are subject to this principle, including contingent events such as actions of an agent with free will:

> This axiom, known under the name *principle of sufficient reason*, extends to even the most indifferent actions. The freest will cannot be without a determining motive, giving birth to it; for if all the circumstances of two situations [*positions*] were exactly the same, then if the will acted in the one case but refrained from acting in the other, its choice would be an effect without a cause. It would be, says Leibniz, the blind chance of the Epicureans. (Laplace 1812, p. ii)

Here Laplace is correctly describing the position Leibniz had taken in his celebrated controversy with Samuel Clarke. For Newton rejected Leibniz's assertion of

---

[10]These claims are made, with Leibniz's own emphasis, at the beginning of consecutive paragraphs in an unpublished manuscript, *Principia logico-metaphysica* (A VI 4, 1646–7/Leibniz 1976, p. 269), probably drawn up in 1689 for his Italian colleagues.

determinism, claiming it to be contrary to the possibility of free will, and on this Samuel Clarke took Newton's side. In likening the will to "a balance where reasons and inclinations take the place of weights" (GP VII 359/Leibniz 1976, p. 680), Clarke charged, Leibniz was reducing the world to a mere clockwork mechanism, subject to fatal necessity.

According to Clarke (and Newton), on the other hand, the mind, being active, has a "self-motive principle" by means of which it can choose in the absence of determining causes. The mind, Clarke asserts, may have reasons for acting even "when there may be no possible reason to determine one particular way of doing the thing rather than another" (Fifth Reply, §§1–20/GP VII 422). To Leibniz this seemed an obvious self-contradiction. If the mind has reasons for its choice, then there must be a possible reason for it to choose one alternative as better than the rest. The mind "acts by virtue of its motives, which are its dispositions to act" (Fifth Paper, §15, GP VII 392/Leibniz 1976, p. 698). To imagine that the mind can act in defiance of its own motives is "to divide the mind from its motives, as if they were outside the mind as the weight is distinct from the balance, and as if the mind had, besides its motives, other dispositions to act by virtue of which it could reject or accept the motives" (392/698). Here Leibniz seems to be perfectly correct. If there is no reason for the mind to determine to do one thing rather than another, then it does not have sufficient reason to act. It is only if the mind does have a sufficient reason for its course of action that one would want to count the act as freely chosen. So, Leibniz concludes, having a sufficient reason is required for freedom of choice, not contrary to it.

On this view, agents are free to act to the extent that they are the authors of their own actions: they must be autonomous agents. The deciding motive of the individual agent, together with all the other conditions necessary for the action, would then constitute its full cause. Since the agent's autonomous participation is itself a necessary condition for the act to take place, there is no question of the act being externally compelled just because it is fully determined. An act does not have to be consciously willed in order to be free; and freedom is not absence of self-determination. Again, Bunge agrees: "Freedom, in the general sense in which it is here understood, need not be conscious; and it is not an undetermined remainder, an arbitrary, lawless residue, but consists in the lawful self-determination of existents on whatever level of reality" (Bunge 1979, p. 182). So, we see that being determined is not the same as being caused to happen.

As Leibniz sagely points out, to suppose that an event's being determined means that it will happen no matter what you do, is a fallacy. It is: "what the Ancients called the 'Lazy Argument', a sophism that has troubled people in almost every age" (GP VI 30/Leibniz 1985, p. 54). Leibniz relates the apocryphal story about Zeno of Citium, the founder of the Stoics, who was about to flog his slave for stealing. The slave, aware of the Stoics' doctrine that everything is necessitated, protests: "But, master, I was fated to steal!" "And I," replied Zeno, "am fated to flog you for it!" The fallacy, Leibniz explains, is as follows:

It is false that the event happens whatever one may do: it will happen because one does what leads to it; and if the event is written, the cause that will make it happen is written too. Thus the connection of effects and causes, so far from establishing the doctrine of a necessity detrimental to conduct, serves to destroy it. (GP VI 33/Leibniz 1985, p. 57)

Bunge agrees: "general determinism does not recognize anything unconditional and, as a consequence, it does not entail any inevitability other than what results from the lawful concurrence and interplay of processes—among which the human conscious conduct may eventually intervene" (Bunge 1979, p. 104). Every event must have a cause, but a cause does not necessarily produce its intended effect: "A given set of causes may be hindered for the production of their otherwise normal effects, by the interposition of other causes" (Bunge 1979, p. 103). In order for a cause to produce its effect, other conditions must concur. Zeno's slave cannot validly claim that he was necessitated to steal when his own decision to steal was one of the principal conditions making up the cause of his act of theft; had he decided otherwise, or had any of the other conditions necessary for the theft been unfulfilled, no theft would have taken place.

So far we have seen that there is nothing in determinism to preclude succession or becoming, and that freedom of the will, far from being opposed to it, requires determination in order to deliver a conception of freedom worth defending. Causation, we saw, requires many necessary conditions to be in place in order for the effect to follow, and the potential failure of any of these conditions to be fulfilled is what makes for the contingency of the effect's following from the cause. Still, it may be objected, this analysis takes for granted the notion of contingency. But how can there be genuine contingency in a deterministic world?

11.5 Contingency, Lawfulness and Predictability

Contingency, as we saw above, is supposed by critics such as Čapek and Whitrow to be incompatible with Laplacian determinism. One reason for this is their understanding of scientific laws as necessitating events. Another is their equation of lawfulness with predictability. If with the aid of the relevant scientific laws future states can be predicted with certainty from prior states, so it is alleged, then they are logically bound to happen.

Here I am going to leave aside Bunge's astute observation that scientific laws are not necessarily causal. As he points out, in addition to statistical determination, there is also the functional determination typical of states linked by equations. Not only are initial states not causes of succeeding states, as noted above, such functional relationships are usually reciprocal, contrary to the unidirectionality assumed for causation. But even if we understand Laplacian determinism as strictly causal, the claim that it is incompatible with contingency involves the two other conflations, that of determination with necessitation, and that of lawfulness with predictability. Again, a comparison with Leibniz will be illuminating.

Leibniz was adamant that being determined is not synonymous with being necessitated, where the notion of necessity in question is *logical* necessity. This, it should be observed, is the very notion of necessity appealed to by Whitrow in his claim quoted above, "the future history of the universe pre-exists logically in the present" (Whitrow 1961, p. 295). But something is logically necessary, Leibniz argued, if its not being the case entails a contradiction. This is not true of any existing state of affairs, he reasoned, since in order to come about an infinity of particular conditions must be satisfied. This depends on Leibniz's conception of the sufficient reason for an existing thing as consisting in all its requisites—a doctrine which he owed to Hobbes[11]—and his doctrines that a requisite of any existing thing "can only be conceived through others", and thus through an actual infinity of requisites. Correspondingly, a demonstration that one existing state of affairs logically follows from another "involves infinitely many reasons, but in such a way that there is always something that remains for which we must again give some reason".[12] But given Leibniz's doctrine that all demonstrations must involve only a *finite* number of steps, this means it is impossible to demonstrate a contingent truth, since such a demonstration would require an infinite analysis.

This may make it sound as though Leibniz is denying determinism. If it is impossible to demonstrate a contingent truth, then how could there be such truths? And correlatively, if there is an infinitely complex chain of intermediate states between any two states, how can they be related in a deterministic way? To these questions Leibniz answers that, despite the fact that "there is never a full demonstration, the reason for the truth nevertheless always subsists, although it can be perfectly understood only by God, who alone can go through the infinite series in a single mental thrust" (A VI 4, 1650/Leibniz 1989, p. 28). God knows truths through intuition; he does not have to know them through a step-by-step demonstration. This is consistent with Leibniz's claim that the complete concept of each individual substance is known only by God; only he knows the "law of the series" of each substance's individual states, as well as how these reflect the states of coexisting substances at each moment of its existence. "That there should be such a persistent law, which involves the future states of that which we conceive to be the same, is exactly what I say constitutes it as the same substance" (GP II 264/Leibniz 1976, p. 535).

But what about *our* knowledge of laws relating the states of things we experience? Here Leibniz is far from denying that we can make judgements perceiving "the objective patterns of being and becoming" among phenomena to which Bunge refers. In fact, he concurs completely with Bunge's second constitutive principle of general determinism stated above, the *principle of lawfulness*, according to which "nothing happens in an unconditional and altogether irregular way—in short, in a

---

[11]Leibniz explicitly acknowledges this debt to Hobbes in his *Theodicy* (GP VI 389/Leibniz 1985, pp. 394–95).

[12]Leibniz, "On the Origin of Contingent Truths", (A VI 4, 1662/Leibniz 1989, p. 99). For an account of Leibniz's theory of contingency, see Arthur (2014).

lawless, arbitrary manner" (Bunge 1979, p. 26). In his *Discours de métaphysique* Leibniz writes, "not only does nothing happen in the world that is absolutely irregular, but also we can't even imagine such a thing" (A VI 4, 1537/Leibniz 1976, p. 306). In support, he argued as follows:

> Suppose, for example that someone puts a number of completely haphazard dots on a piece of paper, as do people who practise the ridiculous art of geomancy. I say that it is possible to find a geometrical line whose notion is constant and uniform according to a certain rule, such that the line passes through all the points, and in the same order in which they were drawn. ... [Likewise,] there is no face whose contours are not part of a geometrical line, and which could not be drawn in a single line by some rule-governed motion. (*Discours* §6, A VI 4, 1537–38/Leibniz 1976, p. 306)

Of course, Leibniz granted, if the line is very complex, it will appear irregular, even if it is the result of a divinely imposed order. "Therefore what counts as extraordinary is only so with respect to some particular order established among created beings. For as regards the universal order, everything conforms to it." (A VI 4, 1538/Leibniz 1976, p. 306) Here Leibniz is distinguishing lawfulness as it corresponds to known physical laws, from the lawfulness corresponding to the divine order instituted by God. The particular laws we discern in the behaviour of things are hypothesized by us to be the case. They involve abstraction from all the complexities of things as they actually exist, but may nevertheless contain truths about the relations of things. But since we are not omniscient, we will never know for sure whether we have correctly abstracted the relevant features and related them rightly. In this he would concur with Bunge: "All general propositions with a factual content are more or less probable hypotheses, but ... universal law statements are not all *equally* probable; some of them are almost certainly true and others almost certainly false" (Bunge 1979, p. 323).

Thus, contingency for Leibniz is born of the infinite, both the infinitude of the connection between one event and another, which must traverse all the infinite complexity of the chain of events between them, and also the infinitude of the information needed to completely specify the state of any existing thing. But just as we are not prevented from identifying individuals by some finite subset of properties of all the infinite properties defining them, so we can abstract from these complexities and articulate laws connecting them and governing their behaviour. Such laws will nonetheless be contingent: there is no logical necessity that the one event will follow another according to this law, just because we have abstracted from all the other relevant factors, and have assumed that they will not affect the law's holding true—an assumption that may, of course, have greater or lesser empirical and theoretical support.

In this respect, Leibniz's position is more sophisticated than Laplace's, at least, as it has traditionally been interpreted. For the latter assumes as a theoretical posit knowledge of all the initial conditions of the universe, but then also assumes that a given set of laws in principle accessible to the human mind, will govern all subsequent development. Indeed, classically these were assumed to be Newton's laws. For Leibniz, by contrast, God's knowledge of the "laws of the series" for all existing things, together with his omniscient grasp of initial conditions, allows for all subsequent events to be determined. But such laws governing individual substances are not accessible to us; we must remain content with a partial knowledge of initial conditions and laws that we hypothesize as correctly describing the objective pattern on some level of abstraction. These laws, according to Leibniz, are contingent on the optimal design of the universe God chooses to create; but they are not logically necessary. What this example shows, then, is that one can commit to an ontological determinism, according to which all events follow from preceding events in a lawful way, without being committed to the view that anything that happens according to the laws we discover involves any logical necessitation.

The same analysis illustrates Bunge's point that the identification of determinism with predictability conflates an ontological doctrine with an epistemological one: lack of predictability may be evidence for the failure of a particular causal hypothesis, or for the failure of the principle of causality to apply universally, but it does not refute lawful determination as such. "Very few facts in the concrete world are predictable with near certainty, and none in all detail, because scientific forecast is based on the knowledge of laws and of specific information regarding singular facts, neither of which is ever complete and exact" (Bunge 1979, pp. 330–331).

11.6   Causal Processes and the Fiction of Isolated Causal Chains

So far, discussion has been about determinism in general. Not much has been said about causal processes in general, and their distinction from merely deterministic processes.

A *causal process*, as conceived by today's physicists, is an isolated process tracing a path through space and time—a process that could go from one point in spacetime to another without exceeding the speed of light, for example. This is particularly prominent in discussions of special relativity, where events that are connectible by slower-than-light physical processes are usually termed *causally connectible*. The significance of this is that if such a process can extend from an event a to an event b, then a is (absolutely) before b. Such a relation is invariant, and does not depend on the reference frame chose to represent the process in space and time coordinates. It may be seen as a modern incarnation of Leibniz's causal theory of time, according to which "If one thing is the cause of another, and they are not able to exist at the same time, the cause is *earlier,* the effect is *later*. Also, earlier is whatever is simultaneous with the earlier."[13] Alfred Robb and others showed how this feature can be taken as definitional, and from this assumption together with minimal symmetry requirements and the hypothesis of a maximum signal velocity, the Lorentz equations of special relativity may be derived. This delivers a causal

---

[13] Leibniz (1923-) A VI 4, 568); see Arthur (2016b) for an exposition of Leibniz's causal theory of time.

theory of spacetime, which may thus be regarded as an update of Leibniz's causal theory of time.[14]

Now it is perhaps understandable that physicists should call events that are connectible by physical processes *causally* connectible, in that in order for something at one such point to cause something to happen at the second point by physical means there would have to be an influence, or some chain of influences, propagated from the first point to the second, and it would need to traverse some such path slower than light. But as Bunge has observed, such time-like-separated events need not be connected by a *causal* process at all. The operant principle here is the Principle of Retarded Action, according to which any physical process must take a certain finite quantity of time. There is no necessity for this process to be a casual one:

> The principle of retarded action is independent of the causal principle, and, whenever it is postulated, it entails a restriction on the possible genetic connectivity of the physical level of reality . . . [M]oreover, the mere statement of retarded action, far from being essentially committed to causality, is consistent with noncausal categories of determination. (Bunge 1979, pp. 67–68)

For example, a non-causal process of becoming will still be consistent with the principle of retarded action. In fact, an inertial motion is a perfect example of such a non-causal motion, as Bunge observes (Bunge 1979, pp. 110–111). And on the other hand, instances of causation may be held to be instantaneous, as in Newtonian gravitation theory. So "causal connectibility" is actually a misnomer. The term is probably too well established to be changed now, but as Roberto Torretti has argued elsewhere (Torretti 2007), such terminological infelicities may well breed conceptual confusion.

Bunge has made other keen and insufficiently appreciated observations regarding causal processes: first regarding the *abstraction* involved in the very idea of a causal process, and second with regard to the *isolation* necessary to the existence of a causal chain. Regarding the first he notes that "in order that a process may be regarded as causal, either one causal factor or one of the consequences must be singled out of a whole constellation of determiners" (Bunge 1979, p. 127). This he calls a "causal paradox": the very applicability of causal reasoning requires that certain factors are *not* causally connected to the phenomena under investigation to any significant degree, contrary to the idea that everything is causally interlinked.

This paradox is exacerbated if we are concerned with a chain of influences connecting two events, in which case we are talking not about one causal process, but a *causal chain*. For if other things interfered with such a causal chain along the way, this chain of influences might not have its effect. So the chain must be an isolated process.

Such causal isolation is, of course, a fiction. "The main ground why causal chains can at best work as rough approximations for short periods of time is that they assume a fictitious *isolation* of the process in question from the remaining processes" (Bunge 1979, p. 127). Such complete isolation, of course, can never

---

[14]See Winnie (1977) for details.

occur in fact, because "actually an infinity of neglected factors—Galileo's *cause accidentarie* or *cagioni secondarie*—are constantly impinging on the main stream" (Bunge 1979, p. 130). Here Bunge approvingly quotes J. D. Bernal: "chance variations or side reactions are always taking place. These never completely cancel each other out, and there remains an accumulation which sooner or later provides a trend in a different direction from that of the original system" (Bernal 1949, p. 31 in Bunge 1979, p. 131).

So, we have a paradox: "the fiction of an isolated 'causal chain' will work to the extent to which such an isolation takes pace" (Bunge 1979, p. 130). Yet this fiction is a methodological requirement for the application of causal ideas to reality: "The isolation of a system from its surroundings, of a thing or process from its context, of a quality from the complex of interdependent qualities to which it belongs—such 'abstractions', in short, are indispensable not only for the applicability of causal ideas but for any research, whether empirical or theoretical" (Bunge 1979, p. 129). So, on the one hand, "The picture of linear causal chains is ontologically defective because it singles out a more or less imaginary line of development in a whole concrete stream" (Bunge 1979, p. 132); and, on the other, "such a linear character of causation is not altogether fictitious; it does work in definite respects and in limited domains. Causal chains are, in short, a rough model of real becoming" (Bunge 1979, p. 147).

If we compare these considerations to Leibniz's philosophy, we can see again the surprising consilience between Bunge's philosophy of science and that of the German polymath. For this very contrast between the inherent complexity of reality as it exists *in concreto*, and the necessity of making abstractions from that complexity in order to have any knowledge of it, is right at the heart of Leibniz's philosophy, as evident from the following passage from his *New Essays on Human Understanding*:

> if we thought in earnest that the things we do not apperceive are not there in the soul or in the body, we should fail in philosophy as in politics, by neglecting το μικρόν, imperceptible changes; whereas an abstraction is not an error, provided we know that what we are ignoring is really there. This is the use made of abstractions by mathematicians when they speak of the perfect lines they ask us to consider, and of uniform motions and other regular effects, although matter (that is to say the mixture of the effects of the surrounding infinite) is always providing some exception. We proceed in this way so as to distinguish the various considerations from one another, and, as far as is possible, to reduce effects to their reasons, and foresee some of their consequences. For the more careful we are to neglect no consideration which we can subject to rules, the more closely does practice correspond to theory. (A VI 6, 57/Leibniz 1981, p. 57)

11.7 Final Causation: The Status of Teleology

One repeated theme in Bunge's book is the disparagement of final causation. A representative passage is the following:

To say that in behaving the way they do physical objects move "with the purpose of" minimizing or conserving the intensity of a given quality is not too different from asserting that things happen as they do "in order that" the laws of nature may be satisfied. Extremum principles are no more indicative of end-seeking behaviour than any other physical laws, and the association of integrals with purpose or design belongs in the same class of confusion of dimensions of language as the association of differential equations with [efficient] causation. (Bunge 1979, pp. 83–84)

Bunge sees the appeal to final causes on the basis of extremal principles in physics (expressed in integral equations) as the mirror image of the appeal to differential equations as modelling efficient causes, a mistaken notion of efficient causation that he has already subjected to refutation: "Differential equations are as little the carriers of efficient causation as integral equations are the carriers of final causation" (Bunge 1979, p. 87). In criticizing the notion of differential equations as models of efficient causation, Bunge notes the non-operative character of physical states, and also that equations model functional relationships, which are mostly not causal. These criticisms, it seems, are directed in particular against Max Planck, who in his *Religion und Naturwissenschaft* had claimed that "extremum principles are the mathematical translation of—purpose" (Bunge 1979, p. 82). In this regard, Bunge commends Voltaire's satire of Maupertuis's interpretation of the Principle of Least Action in his *Histoire du Docteur Akakia* (Bunge 1979, p. 83).

Again, a brief comparison with Leibniz's thought will prove illuminating. For, on the one hand, there are good grounds for crediting Leibniz with the origination of the Principle of Least Action, as a natural concomitant to his invention of differential equations and his pioneering of the method of minimization and maximization of relevant quantities to solve physical problems.[15] On the other hand, his work in what we would call biological science was in many ways far ahead of his contemporaries', as is only now being appreciated, and in it he recommended the use of final cause reasoning.[16]

It was Leibniz who introduced the term "organism" as a substantive term to describe each of the substantial individuals that he took to constitute the created world, each of them an embodied monad, and a living creature. He characterized the body of such an animal as a machine, though one whose functions, in contrast to artificial machines, were not externally imposed. These functions include self-motion, the ability to sustain itself in existence by extracting energy from its environment through nutrition and converting this to a mechanical equivalent, and also the ability to reproduce other creatures of the same kind. Although all of this could in principle be described mechanically, in default of knowledge of the requisite mechanisms, an explanation appealing to the functions of the various organs would serve to advance biological knowledge.

Bunge follows Descartes and Spinoza in wishing to extirpate teleological reasoning from science. "Needless to say," he writes, "goal-directed structures,

---

[15] See in particular Goldenbaum (2016) for a defence of Leibniz's claim to originating the principle.

[16] See in particular Duchesneau (2010) and Smith (2011). See also Arthur (2016a) for a discussion of the relevance of Leibniz's views to contemporary biology and astrobiology.

function and behaviors need not be purposefully planned by anybody" (Bunge 1979, p. 19). Leibniz would agree. Purposive behaviour, according to him, is not exclusive to animals capable of rational thought, although only rational agents will be aware of their behaviour and able to reflect on its possible future consequences in order to guide them more effectively towards the ends they are seeking. But all organisms need to be able to sense sources of food or danger (they need at least rudimentary perception), and need to be motivated to endeavour towards the first and away from the second (they cannot act without appetition).[17] On Leibniz's account, this requires the information necessary for this behaviour to exist in them, so that the states or perceptions of the organism follow one another (deterministically but contingently) according to an in-built programme ("the law of the series"), but always in perfect conformity with the motions of its own body and with the bodies and motions among which it is situated.

For non-living, inanimate processes, on the other hand, a third kind of teleology is operative. Here a process such as that of a light ray is end-directed, but involves no reference to any intentions or knowledge (however rudimentary) on the part of the systems involved. As Leibniz showed, one can derive the laws of reflection and refraction by assuming the light ray will take an optimal or "most determinate" path, a path that will be either a maximum or a minimum. But as he wrote in 1678–1679, "it is not the ray itself, but the founding nature of optical laws that is endowed with cognition, and foresees what is best and most fitting" (A VI 4, 1405/Leibniz 2001, p. 257).

These three distinct manifestations of teleology correspond fairly well with three of the five species of teleology identified by the distinguished biologist and philosopher, Ernst Mayr: respectively, (1) "animal behaviour that is clearly purposive, revealing careful panning" (Mayr 2004, p. 57), (2) the teleonomic processes or behaviours that "are most conspicuous in the behaviour of organisms" and owe their goal-directedness to an internal programme (Mayr 2004, p. 51), and (3) the teleomatic processes that are goal directed, but "automatically achieved" as the result of the operation of natural laws (Mayr 2004, p. 50)—what Peirce called "finious processes" (Pierce 1958 CP v. 7. p. 471). Moreover, Mayr's claim that, contrary to the long tradition of rejection of final causes in science initiated by Descartes, "there is no conflict between teleonomy and [efficient] causality" (Mayr 2004, p. 52), was already anticipated by Leibniz, as noted above.

The fourth type of teleology identified by Mayr is that appealed to in reasoning about "adaptive features". This is the kind of reasoning parodied by Voltaire in his satire *Candide*, where every adapted feature is supposed to have been divinely selected for the purpose it happens to serve: "Observe", says Pangloss, "that the nose is formed for spectacles, and therefore we come to wear spectacles; the legs are visibly designed for stockings, and therefore we come to wear stockings!" (Voltaire

---

[17]"There is reason to think that there are infinitely many souls, or more generally of primitive entelechies, possessing something analogous to perception and appetite, and that all of them are and forever remain substantial forms of bodies" (*New Essays*; A VI 6, 318/Leibniz 1981, p. 318).

2009, p. 48). Such reasoning abounds in Noël Antoine Pluche's *Spectacle de la Nature*,[18] but not in Leibniz, for whom optimal design is not necessarily to be found in any given individual process, event or substance taken by itself, but is a feature of the world taken as a whole (and over its entire history).

It is this idea of a world designed so as to increase towards progress or perfection that is the fifth species of teleology identified by Mayr, cosmic teleology. Here Leibniz's views can be seen to be in direct conflict with the Darwinian programme. The very idea that design could be the product of natural selection, premised on variation in genetic heritage and the contingent survival of individuals remaining after the least fit among them have died out, was at complete odds with his theological standpoint.

The fact remains, however, that the teleonomic processes common among even rudimentary living organisms are entirely consistent with efficient-causal explanations, as Leibniz claimed, and do not involve any appeal to conscious intentions; but they do involve an appeal to function, and this is perfectly legitimate, especially where no efficient-causal explanations are possible. Secondly, teleomatic processes, such as those governed by the principle of least action in physics, are goal-directed in an innocuous sense compatible with Bunge's strictures, but nonetheless extremely useful in scientific explanations where efficient causal explanations cannot be brought to bear.

11.8   Conclusion

It has been argued that some of Bunge's chief insights, perhaps still insufficiently appreciated, are the following: the distinction of causation from other types of determination, his critique of the still-dominant Humean accounts of causality as leaving out the productive aspect of determination, his critique of the conflation of determination with predictability, and his insistence on the fictional character of isolated causal chains. These insights have been discussed and explicated through a comparison with the views of Leibniz, with which they have considerable common ground, despite the latter's rejection of the intelligibility of the action of one substance on another. In particular, Leibniz's views were used to illuminate and fortify Bunge's claims of the compatibility of determinism with free will, the distinction of lawfulness from predictability, the idealization inherent in the very idea of a causal chain, and to clarify the status of final causes in physics and biology.

---

[18]See the excerpts given by Palmer in his critical edition of Voltaire (2009).

References

Arthur, R. T. W. (2014). *Leibniz*. Cambridge: Polity Press.

Arthur, R. T. W. (2015). The relativity of motion as a motivation for Leibnizian substantial forms (chapter 10). In A. Nita (Ed.), *Leibniz's Metaphysics and Adoption of Substantial Forms* (pp. 143–160). Dordrecht: Springer.

Arthur, R. T. W. (2016a). Leibniz, organic matter and astrobiology. In L. Strickland, E. Vynckier, & J. Weckend (Eds.), *Tercentenary Essays on The Philosophy and Science of G. W. Leibniz* (pp. 87–107). New York: Routledge.

Arthur, R. T. W. (2016b). Leibniz's causal theory of time revisited. *Leibniz Review, 26*, 151–178.

Bernal, J. D. (1949). *The Freedom of Necessity*. London: Routledge and Kegan Paul.

Bunge, M. (1979). *Causality and Modern Science* (3rd revised edition; originally published 1959, Harvard University Press). New York: Dover.

Čapek, M. (1971). *Bergson and Modern Physics*. Dordrecht: D. Reidel.

Duchesneau, F. (2010). *Leibniz: le vivant et l'organisme*. Paris: Vrin.

Gerhardt, C. I. (1875–1890). *Der Philosophische Schriften von Gottfried Wilhelm Leibniz*. (Berlin: Weidmann; reprint ed. Hildesheim: Olms, 1960), 7 vols; cited by volume and page, e.g. GP II 268.

Goldenbaum, U. (2016). *Ein gefälschter Leibnizbrief? Plädoyer für seine Authentizität*. Hannover: Wehrahn.

Hoefer, C. (2016). Causal determinism. In E. N. Zalta (Ed.), *The Stanford Encyclopedia of Philosophy*. URL = https://plato.stanford.edu/archives/spr2016/entries/determinism-causal/

Hoyle, F. (1952). *The Nature of the Universe*. London: Blackwell.

Hoyle, F., Burbidge, G., & Narlikar, J. V. (1995). The basic theory underlying the quasi-steady state cosmological model. *Proceedings of the Royal Society A, 448*, 191.

Laplace, P. (1812). *Théorie analytique des probabilités*. Paris: Ve. Courcier.

Leibniz, G. W. (1923). *Sämtliche Schriften und Briefe* (Akademie der Wissenschaften der DDR, Ed.). Darmstadt: Akademie-Verlag; cited by series, volume and page, e.g. A VI 2, 229.

Leibniz, G. W. (1976). *Philosophical Papers and Letters* (2nd ed., Ed. and Trans. L. Loemker). Dordrecht: D. Reidel.

Leibniz, G. W. (1981). *New Essays on Human Understanding* (Ed. and Trans. P. Remnant & J. Bennett). Cambridge: Cambridge University Press.

Leibniz, G. W. (1985). *Theodicy: Essays on the Goodness of God, the Freedom of Man, and the Origin of Evil* (Trans. E. M. Huggard of *Essais de Théodicée sur la bonté de Dieu, la liberté de l'homme et l'origine du mal*, 1710). La Salle: Open Court.

Leibniz, G. W. (1989). *Philosophical Essays* (Ed. and Trans. R. Ariew & D. Garber). Indianapolis: Hackett.

Leibniz, G. W. (2001). *The Labyrinth of the Continuum: Writings of 1672 to 1686* (Selected, edited and translated, with an introductory essay, by R. T. W. Arthur). New Haven: Yale University Press.

Mayr, E. (2004). *What makes Biology Unique? Considerations on the autonomy of a scientific discipline*. Cambridge: Cambridge University Press.

Peirce, C. S. (1958). In A. W. Burks (Ed.), *Collected papers* (Vol. 7). Cambridge, MA: Harvard University Press.

Schaffer, J. (2016). The metaphysics of causation. In E. N. Zalta (Ed.), *The Stanford Encyclopedia of Philosophy*. URL = https://plato.stanford.edu/archives/fall2016/entries/causation-metaphysics/

Smith, J. E. H. (2011). *Divine Machines: Leibniz and the Science of Life*. Princeton: Princeton University Press.

Torretti, R. (2007). 'Rod contraction' and 'clock retardation': Two harmless misnomers? In F. Minazzi (Ed.), *Filosofia, scienza e bioetica nel dibattito contemporaneo: Studi internazionali in onore di Evandro Agazzi* (pp. 1029–1039). Roma: Istituto Poligrafico e Zecca dello Stato.

Voltaire. (2009). *Candide; or, All for the Best*. With excerpts from Pope, Voltaire, Rousseau, Bayle, Leibniz, Pluche and Boswell. Ed. with introductory essay by E. Palmer. Peterborough: Broadview Press.

Whitrow, G. J. (1961). *The Natural Philosophy of Time*. CA: Harper. London: Nelson.

Winnie, J. (1977). The causal theory of space-time. In J. Earman, C. Glymour, & J. Stachel (Eds.), *Foundations of Space-Time Theories* (pp. 134–205). Wisconsin: Minnesota University Press.

Chapter 12
Mario Bunge and the Current Revival of Causal Realism

Rögnvaldur D. Ingthorsson

Abstract Mario Bunge's *Causality and Modern Science* is arguably one of the best treatments of the causal realist tradition ever to have been written, one that defends the place of causality as a category in the conceptual framework of modern science. And yet in the current revival of causal realism in contemporary metaphysics, there is very little awareness of Bunge's work. This paper seeks to remedy this, by highlighting one particular criticism Bunge levels at the Aristotelian view of causation, and illustrating its relevance for contemporary powers-based accounts. Roughly, the Aristotelian view depicts interactions between objects as involving a *unidirectional* exertion of influence of one object upon another. This idea of unidirectional action is central to the Aristotelian distinction between active and passive powers, and its corresponding distinction between active and passive objects. As Bunge points out, modern physics does not recognise the existence of any unidirectional actions at all; all influence comes in the form of reciprocal action, or interaction. If this is right, all notions deriving from or influenced by the idea of unidirectional actions—such as the concept of *mutual manifestation* and *reciprocal disposition partners*—risk being false by the same measure. Bunge drew the conclusion that the Aristotelian view is ontologically inadequate, but still advocated its use as the most useful approximation available in science. He considered, but ultimately rejected the possibility of a modified view of causation built on reciprocal action, because, in his view, it couldn't account for the productivity of causation. Bunge's critique of this particular aspect of the Aristotelian view cannot be overlooked in contemporary metaphysics, but it is possible to construe a modified view of causation that takes the reciprocity of interactions seriously without loss of productivity.

This paper is based on research in the project "Scientific Essentialism: Modernising the Aristotelian View", funded 2015–2018 by *Riksbankens Jubileumsfond: Swedish Foundation for Humanities and Social Sciences*, Grant-ID: P14-0822:1.

R. D. Ingthorsson (✉)
Department of Philosophy, Lund University, Lund, Sweden
e-mail: rognvaldur.ingthorsson@fil.lu.se

12.1   Introduction

In another contribution to this volume, Ingvar Johansson expresses puzzlement about the lack of awareness among philosophers of science about Mario Bunge's views about mechanisms and approximate truth. In this paper the focus is on another puzzle, the lack of awareness among contemporary metaphysicians about Bunge's views on causality. The particular concern is with those engaged in the revival of Aristotelian causal realism, now in the form of what is typically labelled *powers-based accounts of causation*. Proponents of powers-based accounts seem not to be aware of Bunge's critique of the Aristotelian view of causation, and therefore arguably continue to build on a flawed conception of causal influence, one that is incompatible with the theories and findings of modern science.

The twenty-first century has seen a dramatic revival of realist approaches in metaphysics. An important part of this revival is the re-examination of the relevance of Aristotelian metaphysics, often, but not always, with the aim of construing a scientifically informed account of causation.[1] A focal point in this re-examination is the notion of powers—the idea that natural properties are best understood as determinate ways an object can affect other objects and be affected by them— and how this notion might help to elucidate causation, as well as other notions not discussed here, such as laws of nature, natural kinds, and agency.

While twenty-first century realists reach back to twentieth century thinkers like Bhaskar (1975), Harré and Madden (1975), as well as Cartwright (1989), they hardly ever mention Mario Bunge's *Causality and Modern Science* (1959/1979).[2] Notable exceptions are Johansson (1989) and Ingthorsson (2002). This is unfortunate. Bunge is one of the very few to have noted that the Aristotelian view of causality—which Bunge calls the 'strict doctrine of causality'—is in one particular respect falsified by the theories and findings of empirical science. Roughly, the Aristotelian view depicts interactions between powerful particulars as involving a *unidirectional* exertion of influence of one object (the *Agent*) upon another (the *Patient*) which then results in a change in the Patient. Consequently, a cause on the Aristotelian view consists in an action of an Agent upon a Patient, and an effect is the change produced in the Patient. Modern science, on the other hand, does not recognise the existence of any unidirectional actions at all. It insists that all influence comes in the form of perfectly reciprocal action, or *interaction*, between two objects that subsequently both suffer a change.[3] Indeed, Bunge drew the conclusion that the Aristotelian

---

[1] See, for instance, Ellis (2001), Ingthorsson (2002, 2007), Molnar (2003), Heil (2003, 2012), Lowe (2006), Bird (2007), Martin (2008), Marmodoro (2007, 2017), Mumford and Anjum (2011) and Jacobs (2011).

[2] First published 1959 as *Causality: The Place of the Causal Principle in Modern Science*, but references in this paper are to the 1979 edition.

[3] The fact that all interactions are perfectly reciprocal is universally acknowledged in the sciences, and is impressed on students very early in their education. To verify, consult any undergraduate textbook in physics, such as Resnick et al. (2002).

distinction between active and passive powers, and between Agents and Patients, is ontologically inadequate. If this is right, then in so far as modern powers-based accounts are influenced by the Aristotelian ideas about unidirectional influence, they risk being false by the same measure.

Bunge still advocated the Aristotelian view as the most anthropocentrically useful approximation of causality available in science. He considered, but dismissed, the possibility that causality could be modelled instead on the basis of reciprocal action. I think Bunge's critique of the Aristotelian view is conclusive and cannot be overlooked in contemporary metaphysics. On the other hand, I think Bunge is wrong to dismiss the possibility of a conception of causality as a phenomenon grounded in reciprocal action. In my previous works (Ingthorsson 2002, 2007), I have tried to modify the Aristotelian view to accommodate the reciprocity of interactions and to arrive at an ontologically adequate characterisation of causation that retains most of the characteristics Bunge thought were essential to causality, notably production and asymmetry. However, in that early work no effort was made to relate to the development of neo-Aristotelian powers-based accounts. This will be done here.

12.2   The Strict Doctrine of Causality, or the Aristotelian View

Causal realism—in its most general form—is the view that causation is an objective feature of the world and not only a feature of our thinking (see, for instance, Bunge 1979, ch. 1.1.2; Price 2001, p. 106; Esfeld 2011). There are many philosophers who endorse causal realism in that general sense, but fewer who explicitly attempt to characterise it in greater detail. Those who do attempt such a characterisation tend to include some stance on most (if not all) of the following characteristics: *necessity, production, efficiency, uniformity,* and *process.*[4] The affinity that their characterisations have with the older Aristotelian tradition is obvious, even if not always explicit.

Modern characterisations of causal realism where the connection to the Aristotelian view is made explicit have appeared regularly throughout the twentieth century, for instance in the works of Johnson (1924), Bunge (1979, ch. 2), Emmet (1985), Johansson (1989, ch. 12), and Dilworth (1996). But these works appear before the current Aristotelian revival and for some reason are rarely mentioned in the current literature on powers-based causation. The causal realist view that emerges in these writings is that new states of affairs are brought into existence when an already existing material body, or complex of bodies, changes due to an external influence without which the change would not have occurred and the new state of affairs never have existed; a view that echoes the spirit of Aristotle's claim

---

[4]See, for instance, Bunge (1979, ch. 1), Ingthorsson (2002), Huemer and Kovitz (2003), Chakravartty (2005), Marmodoro (2007), Esfeld (2011) and Mumford and Anjum (2011).

that "everything that comes to be comes to be by the agency of something and from something and comes to be something" (*Metaphysics*, bk.7, 7, 1032a13). Typically, the external influence, or cause, is depicted in terms of an *extrinsic motive agent* (or simply *Agent*), which is an object possessing the power to influence other objects and which actually does *act* upon another object, an object sometimes referred to by the term *Patient* since it lacks a similar active power but is instead able to passively receive the influence exerted by an Agent and change in some particular way. Accordingly, a *cause* is the action of some object upon another object, and an *effect* is the change produced in the object acted upon; or, in other words, a cause is the exertion of influence by an Agent upon a Patient and an effect is the resulting change in the Patient.

Embedded in the presuppositional depth-structure of this view are three fundamental convictions about natural reality that continue to be implicitly assumed but are seldom brought out explicitly. Among the very few to explicitly discuss them are Bunge and Craig Dilworth. First is the belief that nothing comes into being out of nothing or passes into nothing. Dilworth calls it the *principle of (the perpetuity of) substance* (1996, p. 53), while Bunge prefers the term *genetic principle* (1979, p. 24). The latter label captures very well the core idea at play here, notably that everything has a *natural origin*; everything comes to be out of something already existing and due to the influence of something existing.

Second is the conviction that causal changes come about as the result of an influence being exerted by something on something else. I call it the *principle of action*, which is my own term for what I consider to be the proper causal realist take on what is usually called the *causal principle* (Bunge 1979, p. 26), or *the principle of causality* (Dilworth 1996, p. 57). The most watered-down form of the causal principle is 'there is a cause for everything', which leaves it entirely open what can count as a cause; even Humean regularity can fit the bill. However, in the causal realist tradition it is without doubt the exertion of influence that is meant to distinguish causal from non-causal changes of various kind, say, inertial motion or spontaneous decay (for discussion of other kind of determinations, see Bunge 1979, p. 17ff).

Third is the conviction that the world changes in regular and determinate ways. This is often taken to be equivalent to the belief that the world behaves in accordance to natural laws, which is why Bunge calls it the *principle of lawfulness* (Bunge 1979, p. 26), but others call it *the principle of the uniformity of nature* (Dilworth 1996, p. 55). The principle does not entail a commitment to the existence of some kind of abstract entities, *the laws*, which somehow rule the behaviour of concrete entities. The principle is equally compatible with the idea—explicitly incorporated into the Aristotelian view—that objects behave in what Bunge calls 'immanent patterns of being and becoming' in virtue of their intrinsic and universal properties, patterns that can be expressed in the highly generalised form of law-statements (Bunge 1979, p. 249).

In other words, the principle is equally compatible with immanent realism and powers-based metaphysics, as it is with Platonism about laws. Indeed, when Stephen Mumford insists there are no laws, what he means to say is that whatever uniformity there is in nature (and he believes it to be imperfect and therefore strictly speaking lawless) it is determined by the intrinsic powers of the particulars and not by general laws that directly rule the behaviour of the particulars, or indirectly by determining the causal role of the powers that the particulars bear (Mumford 2005). Anyhow, these three principles are part of the metaphysical framework of the Aristotelian view.

There are three salient features of the Aristotelian view that I would like to particularly emphasize. First, that it depicts *causal influence* as something that is exerted between substances, i.e. by an Agent on a Patient. In other words, the Aristotelian view depicts actions as occurring between *persistent objects*, not between *events* or *states*. This is contrary to what is the received view today, notably that causal influence is a relation between events or states. On the Aristotelian view, the states and events are always outcomes of an interaction between objects. As I have pointed out before (Ingthorsson 2002) this means that on the Aristotelian view, causal relations (between cause and effect) are not themselves productive, but are the products of something more fundamental, notably the interaction between powerful particulars. Second, that the Aristotelian view depicts the exertion of influence as being *unidirectional*; it goes from Agent to Patient. Third, that it does not depict effects as the product of the action of the agent alone, but of the sum total of material, formal and efficient causes; the effect is depicted as the product of the way two or more material objects act on each other in virtue of their powers to produce a change in those very objects. In Hobbes' words, "the efficient and material causes are both but partial causes, or parts of that cause, which in the next precedent article I called an entire cause" (1656, ch. IX, 4). It is this interaction between Agent and Patient that I identify with a *process of production*, and which we should consider to be the proper referent of the term 'causation'.

12.3 Bunge's Critique of the Aristotelian View

It is not feasible to recount here in full Bunge's discussion of the faults of the Aristotelian view. Interested readers should consult *Causality and Modern Science*, in particular Chap. 6. But the gist of it is that modern science flatly contradicts the Aristotelian assumption that influence is exerted unidirectionally between objects; and insists instead that all influence comes in the form of reciprocal action. The most accessible treatment of the concept of reciprocal action in physics is the one found in classical mechanics, and yet it requires something of a shift of perspective to get a full grasp of the concept. The claim 'for every action an equal and opposite reaction' is familiar enough, but it is often misunderstood to mean that effects react back on the cause, making cause and effect interdependent, or as saying that cause and effect are simultaneous (which is arguably the idea Kant labours with in the second analogy of experience; Kant 1787, A189/B233).

The assumption is that 'action' and 'reaction' denote two different phenomena, of which the action has some form of ontological priority over the reaction; i.e.

that the action provokes the reaction. This is such a common misunderstanding that every textbook on classical mechanics has to address this issue to imprint on the novice student a different, and correct, understanding.

We are not talking about the sense in which two lovers reciprocate feelings, nor the way we talk about human communication as reciprocal, nor the mutuality involved in the joint contribution of an active and passive power in the production of a change in the patient. No, action and reaction, or force and counterforce as they are also called, are reciprocal in the sense that they occur simultaneously in opposite directions, are of equal magnitude and of the same kind, and that therefore we really cannot objectively identify either of them as ontologically prior to the other; we are therefore as Heinrich Hertz observes "free to consider either of them as the force or the counterforce" (Hertz 1956, p. 185).

The correct understanding of reciprocal action is that it isn't composed of two different kinds of actions, of which one gives rise to or provokes the other, but of mutual influence of the same kind occurring simultaneously between two objects. Bunge sums the point up by saying that "physical action and reaction are, then, two aspects of a single phenomenon of reciprocal action"(1979, p. 153). In an interaction, so understood, neither side of the interaction has priority, and the terms 'action' and 'reaction' really are labels that can be used arbitrarily for either side. The point comes out very nicely in the following passage by James Clerk Maxwell:

> The mutual action between two portions of matter receives different names according to the aspect under which it is studied, and this aspect depends on the extent of the material system which forms the subject of our attention. If we take into account the whole phenomenon of the action between the two portions of matter, we call it Stress [. . .] But if [. . .] we confine our attention to one of the portions of matter, we see, as it were, only one side of the transaction—namely, that which affects the portion of matter under our consideration—and we call this aspect of the phenomenon, with reference to its effect, an External Force acting on that portion of matter. The other aspect of the stress is called the Reaction on the other portion of matter. (Maxwell 1877, pp. 26–7)

The exact sense of the reciprocity of interactions is perhaps most clearly expressed in Newton's third law of motion which says that the force by which object 1 acts on object 2 is equal to the oppositely directed force by which object 2 acts on object 1 ($F_{1on2} = -F_{2on1}$). Taking into account that the designations 'F' and '$-F$', as well as 'object 1' and 'object 2' are assigned arbitrarily to the forces and objects involved, then the proper understanding of the reciprocity of interaction is that the influence exerted mutually between two objects is always of the same quantity, of the same kind, and occurs *simultaneously* in opposite directions, which is why we are unable to assign *causal* priority to one or the other. The only thing we can do, as Maxwell observed, is to assign *explanatory* priority, but then only on the basis of our subjective interests. In an interaction between a baseball and a window, we are as a general rule more concerned about what happens to the window than what happens to the ball. We therefore attend to the influence of the ball on the window, and the resulting change in the window, conveniently neglecting the influence exerted by the window on the ball and the resulting change in the state of motion of the ball.

Bunge is one of the very few to have noted that the fact that interactions are in reality perfectly reciprocal in this sense, is a serious threat to the Aristotelian view. It entails that "the polarization of interaction into cause and effect, and the correlative polarization of interacting objects into agents and patients, is ontologically inadequate" (Bunge 1979, pp. 170–1). The reciprocity between interacting objects shows that there are no strictly passive substances, i.e. substances who only receive influence but do not themselves influence other things, nor are there substances who influence other things without being themselves affected in any way. Let us now examine how Bunge's finding affects contemporary powers-based accounts of causation.

12.4 Powers-Based Accounts of Causality in the Twenty-First Century

Powers-based accounts come in many different forms, but there is a growing consensus around a couple of core ideas. One idea, often traced back to Martin (1997) even though it has also been expressed by Bhaskar (1975, p. 99) and Johansson (1989. p. 164), is that individual powers never really bring about anything on their own, but only together with other powers. The term *reciprocal dispositional partners* (Martin 1997, pp. 201–02) is now the favoured label used to denote the powers that jointly produce an outcome, and the term *mutual manifestation* is used to denote the outcome so produced (Martin 1997, p. 202).

Despite widespread use of the terms *mutual* and *reciprocal*, the common assumption is still that interactions are unidirectional in the traditional Aristotelian way, i.e. that an active power of one object will partner up with a passive power of another object to jointly bring about a change in the object acted upon: a brick with the power to break will act upon a fragile window to break the window, and water with the power to dissolve will act upon soluble salt to dissolve the salt. Powers, on this view, operate mutually and/or reciprocally only in the sense that dispositional partners are equally important in bringing about the manifestation.

The neo-Aristotelian manner of talking about reciprocal disposition partners as pairs of active and passive powers that jointly bring about a mutual manifestation fits very naturally to our everyday conceptual scheme. We tend to think about what we can (and want to) achieve with our actions, and we rarely are interested in both sides of physical interactions. However, there is a clash with the conceptual framework of science. The clash comes out clearly if we compare the everyday and scientific understanding of the favourite examples used by friends of powers-based accounts, say, 'water dissolves salt'. On the powers-based, and everyday interpretation, water does something to salt that salt does not do to water; water dissolves salt, but salt does not dissolve water. However, is it correct to attribute to water as a unified body a simple power to being able to dissolve salt? If we ask science, the relevant power that makes water into a 'universal solvent' is actually the polarity of *individual* H_2O molecules. It is this polarity that allows an H_2O molecule to bind strongly to either the positively charged sodium ion (Na^+), or the negatively charged chloride ion (Cl^-), and separate them from each other. Water is able to dissolve salt only as long as individual H_2O molecules are able to bond with the components of individual NaCl molecules. In solid form, i.e. as ice, the water molecules are strongly bound to each other and therefore do not interact as easily with NaCl. However, if you sprinkle salt on ice, it will to some extent melt the ice (depending on the temperature), and the resulting saline does not freeze as easily. Arguably, the effect salt has on ice is akin to the effect water has on salt; it dissolves it. The NaCl molecules interact with the H_2O molecules in such as way to break the bonds between H_2O molecules and create bonds with Na^+ and Cl^- instead, which in turn reduces the tendency of the H_2O molecules to bond with each other to form solid ice.

The main point is that what appears in our everyday lives as something a certain amount of liquid does to solid grains of salt, without any apparent reciprocity, appears under the microscope of science as an aggregate of a multitude of perfectly reciprocal interactions between powerful particulars. At the very least, there is here an issue that hardly anybody has paid attention to, much because Mario Bunge's work on causality has for some reason come to be neglected (but see Johansson 1989; Ingthorsson 2002).

12.5 Powerful Particulars or Just Powers?

Although contemporary powers-based accounts of causation have an obvious affinity with the older Aristotelian tradition, they also are clearly different. For one, they exclude final causes, which is actually a break with older tradition that was already established by Hobbes (Hobbes 1656, ch. X, 7). Another difference is that today the objects bearing the powers play a less prominent role than they used to do. The assumption is that it is the *powers* that are causally efficacious, not the *objects* that have the powers, so much so that the object is all but neglected. As a result, contemporary neo-Aristotelians do not often talk in terms of interactions between objects. Notable exceptions to this trend are Ingthorsson (2002) and Marmodoro (2007, 2017). Instead the focus is entirely on powers jointly contributing to the manifestation of some or other outcome. So, for instance, the vectorial model suggested by Mumford and Anjum represents causation with the help of the conceptual construct of a 'quality space' in which various powers are represented as being present at a time and place and as mutually manifesting something, regardless of the objects that have them (Mumford and Anjum 2011, ch. 2.3ff). The powers are presented as horizontal arrows emanating from a common vertical line, each pointing in opposite directions towards one or other outcome, F or G (Mumford and Anjum 2011, ch. 2.3). There is nothing in the model that represents the objects that have the powers.

Even though the particulars bearing the powers disappear in the background, contemporary powers-based accounts still typically operate with the distinction between active and passive powers. On that view, causation is a phenomenon where active and passive powers jointly bring about a mutual manifestation; the heat of the fire and the ability of the hand to be heated results in the heating of the hand; the power of water to dissolve combines with the power of salt to dissolve to mutually manifest the dissolution of the salt. Indeed, the resulting image naturally gives rise to the *problem of fit*, notably why a particular active power only combines with some particular passive power to manifest an outcome; why don't they combine any which way (Williams 2010)?

Now, the shift in focus towards the causal primacy of the powers themselves, and away from the substances that have the powers, may in fact be traced back to Hobbes, but without making the object explanatorily redundant. He makes a point of stating that it isn't bodies *qua* bodies, that exert an influence on other bodies, but bodies *qua* so and so empowered (1656, ch. IX, 3). The fire doesn't heat because it is a body, but because it is hot. And yet Hobbes clearly depicts the power as something belonging to the object; it is the object that has the power to do to another object, and not the power that has the power to do to another power. Indeed, in the following statement from Locke it becomes abundantly clear that although objects act on each other in virtue of the powers, it is still the objects and not the powers that act:

> Thus we say, fire has a power to melt gold, i.e. to destroy the consistency of its sensible parts, and consequently its hardness, and make it fluid; and gold has a power to be melted: that the Sun has a power to blanch wax, and wax a power to be blanched by the Sun, whereby the yellowness in destroyed, and whiteness made to exist in its room. (Locke 1690, bk. II, ch. xxi, 1)

To be explicit, the power to melt gold doesn't change the quality of hardness to become the quality of fluid; it is the fire that changes the gold from hard to fluid. The lesson to be learnt is that it is quite possible to accept that a body may not cause *qua* body but *qua* so and so empowered and yet retain the idea that it is the body that exert the influence on other objects. Indeed, this is how Hobbes conceived of it:

> A body is said to work upon or *act*, that is to say, do something to another body, when it either generates or destroys some accident in it: and the body in which an accident is generated or destroyed is said to suffer, that is, to have something done to it by another body; as when one body by putting forwards another body generates motion in it, it is called an AGENT; and the body in which motion is so generated, is called the PATIENT; so fire that warms the hand is the *Agent*, and the hand, which is warmed, is the *Patient*. That accident, which is generated in the Patient, is called the EFFECT. (Hobbes 1656, ch. IX, 1)

We have then, on the one hand, the older account of physical causation in terms of interactions between powerful particulars, but one in which it is still assumed that influence is not reciprocal but unidirectional, and that in such interactions the active power of the Agent together with the passive power of the Patient jointly produce a change in the Patient. On the other hand, we have the more modern accounts, which merely talk about powers jointly manifesting some outcome, that outcome typically being the obtaining of some or other power.[5]

The main problem with a model of causation in terms of mutual manifestations without particulars is that it fails to connect explicitly to questions about the proper subject of change and to issues about persistence. To be sure, it represents different powers obtaining at different times, but says nothing about the object to which they belong, and which would then have changed through the interaction. Powers-based accounts that take objects rather than powers as the entities that act and are acted upon, have a definitive explanatory advantage in the field of metaphysics more generally.

12.6 Bunge's Defence of Aristotelianism as a Useful Approximation

While Bunge is one of the very few philosophers who have acknowledged and illustrated the ontological significance of the fact that physical interactions are reciprocal, he still resisted the possibility of constructing causality on the model of reciprocal action. On the one hand he thinks it is often possible to treat one side of an interaction as if it was the primary cause, without introducing significant errors, and on the other he thinks that a conception of causation in terms of reciprocal action would distort the concept of causation beyond recognisability. To be more precise, he thinks it would lead to a conception of causation in which we no longer have any asymmetry between cause and effect and in which causes do not produce effects. In total he has four objections:

 (i) Things are typically in a state of flux, and therefore actions are (usually) temporally prior to the reaction (a ball is in motion prior to its collision with a window).
 (ii) Interactions are often so asymmetric that the reaction (and corresponding effect) can be quantitatively neglected (a stone falling to Earth has a negligible effect on the Earth's state of motion, even though in principle they act reciprocally on each other).
(iii) Some processes, e.g. the spontaneous decay of particles, cannot be described as interactions (for instance the decay of a pion into a muon and neutrino).
(iv) If actions do not produce reactions, then interaction does not involve productivity.

---

[5] Ann Whittle has recently argued in favour of what she calls substance causation (2016), to which Andrei Buckareff has replied arguing in favour of powers causation (2017). Unfortunately, neither of them relate to Hobbes or Locke. I think their arguments reveal to some extent an assumption that there has to be a choice between the two, while Locke and Hobbes seem to advocate a view in which it is the unity of object/power that is the active ingredient of causation.

I have elsewhere dealt with these objections in detail (Ingthorsson 2002). Here I will mainly focus on (i) and (iv), and only briefly mention that (ii) isn't really an objection to the reciprocity of interactions, but rather a defence of the anthropocentric value of the Aristotelian view despite its shortcomings. The argument is that in many cases it involves no error at all to think of interactions as unidirectional even though they really are reciprocal, e.g. the interaction between a stone and the whole Earth. As Bunge writes:

> Only if one of the masses is much smaller than the other (for example, a stone as compared with the whole Earth), can the greater mass be regarded as the *cause* of the acceleration of the smaller one, and the reaction of the latter's motion be *quantitatively* neglected [. . . i]n some cases this involves no error at all. (Bunge 1979, p. 150–51)

I have two worries about (i). First, to treat the motion of the brick prior to the collision as an 'action' is arguably to confuse two distinct phenomena that are both called 'action' in classical mechanics: (i) the sum of the kinetic energy of an object over a period of time whether it is interacting with anything, and (ii) exertion of influence of an object a on an object b. As Hertz observed, to use the term 'action' for the first phenomenon is likely to create confusion:

> the name 'action' for the integral in the text has often been condemned as unsuitable [. . .] these names suggest conceptions which have nothing to do with the objects they denote [*in mechanics, RI*]. It is difficult to see how the summation of the energies existing at different times could yield anything else than a quantity for calculation [. . .]. (Hertz 1956, p. 228)

Arguably, before the ball actually comes into contact with a window, it does not 'act' in the second sense. That kind of action only begins on contact and then as one side of an interaction.

Second, modern physics denies there are absolute states of motion or rest. The window is therefore just as much in a state of motion as the brick, and should therefore also count as being in a state of 'acting' (in the first sense) prior to their collision. Since everything is thus in flux prior to an interaction, flux does not give causal priority to anything in particular.

Bunge's fourth objection—that interactions do not involve production—is the most serious objection, but one that can be dealt with in a way that Bunge did not anticipate. Bunge's objection is built on the assumption that to model causation on reciprocal action commits us to treat 'action' as cause and 'reaction' as effect, and thus to construe causes and effects as standing in a perfectly symmetric and non-productive relation to each other: "Let us agree to call interactionism, or functionalism, the view according to which causes and effects must be treated on the same footing, in a symmetrical way excluding both predominant aspects and definitely genetic, hence irreversible, connections" (1979, p. 162). This view, he claims, may be regarded as a "hasty extrapolation of the mechanical principle of the equality of the action and the reaction" (1979, p. 162). I agree that what he calls 'interactionism' is a hasty extrapolation of this kind, but it is not how I propose to construe causation on the basis of reciprocal action (for a more detailed account, see Ingthorsson 2002).

There is no relation of one-sided existential dependence between action and reaction, rather a mutual dependence. But this only shows that reactions *should not* be construed as effects, and actions *should not* be construed as causes. The route to go is to accept that the Aristotelian view has been wrong to characterise causation in terms of Agents acting externally and unidirectionally on Patients. Instead we should accept, like Hobbes (1656, Chap. IX, 4) that the only thing that does *efficiently* produce change is the interaction as a whole and that the entity that suffers a change is the compound whole of interacting things. In other words, interactions (not actions) are causes producing a certain outcome, which then counts as the effect. The relation between action and reaction is symmetrical, yes, but it does not follow from this that the relation between the interaction and the change it produces is symmetrical and non-productive. This is a possibility that Bunge does not consider.

12.7   Conclusion

The lesson to be learnt at the end of all this is that Bunge's critique of the Aristotelian view undermines the validity of central notions in contemporary powers-based accounts of causation. If interactions are genuinely reciprocal in the manner laid out above, then we should not conceive of powers-based causation in terms of active and passive powers jointly producing a mutual manifestation, one that only encompasses the change produced in the object possessing the passive power. That is a conception that arguably is anthropocentrically biased in the manner described by Maxwell. However, I have also argued that we should not think of Bunge's critique as a decisive refutation of the possibility of an account of causation in terms of reciprocal action between powerful particulars. Such interactions, as I have described them, *can* be conceived to involve *production* due to the *exertion of influence* (i.e. reciprocal action) such that the producer and product hold a relation of *one-sided existential dependence*, although this requires that we combine these components in a somewhat different way than before. It also fits certain widely accepted metaphysical principles, i.e. the genetic principle and the principle of lawfulness, but requires that the principle of action be substituted for what may be called the principle of reciprocity. This, I believe is however only a minor modification, and which serves to correct the Aristotelian view in light of advances in our best sciences.

References

Aristotle. (1924). *Aristotle's Metaphysics*. W. D. Ross (ed.). Oxford: Clarendon Press.
Bhaskar, R. (1975). *A realist theory of science*. London: Verso.
Bird, A. (2007). *Nature's metaphysics*. Oxford: Clarendon Press.

Buckareff, A. A. (2017). A critique of substance causation. *Philosophia, 45*(3), 1019–1026.

Bunge, M. (1979). *Causality and modern science.* Cambridge, MA: Harvard University Press.

Cartwright, N. (1989). *Nature's capacities and their measurement.* Oxford: Oxford University Press.

Chakravartty, A. (2005). Causal realism: Events and processes. *Erkenntnis, 63*(1), 7–31.

Dilworth, C. (1996). *The metaphysics of science.* Dordrecht: Klüwer Academic Publishers.

Ellis, B. (2001). *Scientific Essentialism.* Cambridge: Cambridge University Press.

Emmet, D. (1985). *The effectiveness of causes.* Albany: SUNY Press.

Esfeld, M. (2011). Causal realism. In D. Dieks, W. J. González, S. Hartman, & M. Stöltzner (Eds.), *Probabilities, laws, and structures* (pp. 157–168). Dordrecht: Springer.

Harré, R., & Madden, E. H. (1975). *Causal powers: A theory of natural necessity.* Oxford: Basil Blackwell.

Heil, J. (2003). *From an ontological point of view.* Oxford: Oxford University Press.

Hertz, H. (1956). *The principles of mechanics.* New York: Dover Publications.

Hobbes, T. (1656). Elements of philosophy concerning body. In W. Molesworth (Ed.), *The English works of Thomas Hobbes of Malmesbury* (p. 1839). London: Bohn.

Huemer, M., & Kovitz, B. (2003). Causation as simultaneous and continuous. *The Philosophical Quarterly, 53*, 556–565.

Ingthorsson, R. D. (2002). Causal production as interaction. *Metaphysica, 3*(1), 87–119.

Ingthorsson, R. D. (2007). Is there a problem of action at a temporal distance? *SATS–Northern European Journal of Philosophy, 8*(1), 138–154.

Jacobs, J. (2011). "Powerful Qualities, Not Pure Powers". *The Monist* 94(1): 81–102.

Johansson, I. (1989). *Ontological investigations: An inquiry into the categories of nature, man, and society.* New York: Routledge.

Johnson, W. E. (1924). *Logic, part III: The logical foundations of science.* Cambridge: Cambridge University Press.

Kant I. (1787/1965). *Critique of pure reason.* (N. K. Smith, Trans.). New York: St. Martin's Press.

Locke, J. (1690). *An Essay Concerning Human Understanding* (abridged and edited by K. P. Winkler). Indianapolis: Hackett, 1996.

Lowe, E. (2006). *The four category ontology.* Oxford: Oxford University Press.

Marmodoro, A. (2007). The union of cause and effect in Aristotle: Physics 3.3. *Oxford Studies in Ancient Philosophy, 32*, 205–232.

Marmodoro, A. (2017). Aristotelian powers at work: Reciprocity without symmetry in causation. In J. Jacobs (Ed.), *Causal powers* (pp. 57–76). Oxford: Oxford University Press.

Martin, C. B. (1997). On the need for properties: The road to pythagoreanism and back. *Synthese, 112*, 193–231.

Martin, C. B. (2008). *Mind in nature.* Oxford: Oxford University Press.

Maxwell, J. C. (1877). *Matter and motion.* New York: Dover.

Molnar, G. (2003). *Powers: A study in metaphysics.* Oxford: Oxford University Press.

Mumford, S. (2005). Laws and lawlessness. *Synthese, 144*(3), 397–413.

Mumford, S., & Anjum, R. L. (2011). *Getting causes from powers.* Oxford: Oxford University Press.

Price, H. (2001). Causation in the special sciences: The case for pragmatism. In M. C. Galavotti, P. Suppes, & D. Costantini (Eds.), *Stochastic causality* (pp. 103–121). Stanford: CSLI Publications.

Resnick, R., Halliday, D., & Krane, K. (2002). *Physics* (5th ed.). New York: Wiley.

Whittle, A. (2016). A defence of substance causation. *Journal of the American Philosophical Association, 2*(1), 1–20.

Williams, N. (2010). Puzzling powers: The problem of fit. In A. Marmodoro (Ed.), *The metaphysics of powers: Their grounding and their manifestations* (pp. 84–105). New York: Routledge.

Chapter 13
Systemic Thinking

Evandro Agazzi

Abstract Modern natural science followed Galileo's proposals regarding ontology, epistemology and methodology, limiting investigation to a few measurable properties of bodies by the adoption of the experimental method. Force appeared as a specialization of the traditional concept of efficient cause within the new science of Mechanics, which was soon able to incorporate practically all branches of physics. The concept of system had been introduced into scientific vocabulary in the seventeenth century and the limitations of the mechanistic approach in physics emerging by the end of the nineteenth century prompted Bertalannfy to deeply re-elaborate it as System Theory. This permitted him to complement the analytic outlook with a rigorous characterization of the notion of organized totalities. This could be applied to life sciences and several other scientific domains, allowing rigorous treatment of traditional concepts such as finality, and new concepts like complexity and interdisciplinarity.

13.1 Biographical Details

The cordial friendship and mutual estimation that have linked Mario Bunge and me for half a century were nourished by the fact that we shared several aspects of our respective philosophical views, especially in the philosophy of science, independently of our different formation and cultural background. I made the personal acquaintance with Mario at a meeting of the International Academy for Philosophy of Science that took place in Amsterdam in September 1967, when I was working on my book on the philosophy of physics (*Temi e problemi di filosofia della fisica*). I had just read his volume *Foundations of Physics* (1967) and we had a very fruitful exchange of ideas and considered with interest the considerable affinities of our approaches. A feeling of mutual sympathy immediately arose, and a relation of cordial friendship began that has lasted till now. We have had a

E. Agazzi (✉)
Interdisciplinary Center for Bioethics, Panamerican University, Mexico City, Mexico

regular correspondence and several opportunities of personal meeting, especially on the occasion of the conferences of the International Academy for Philosophy of Science (of which Bunge has been a member since 1965 and over which I presided from 1978 to 2016) but also on the occasion of other scientific and philosophical conferences all over the world. During these decades our philosophical positions have lost that proximity that they had when our common field of interest was essentially the philosophy of physics. For example, Bunge has advanced strong criticism not only towards several pseudo-sciences (which I largely share), but also regarding certain schools of contemporary philosophy, from existentialism, to phenomenology, to hermeneutics that I find much less pertinent. Still, there is a substantial part of Bunge's philosophical outlook on which we have kept a basic agreement, and which we have applied in different but complementary directions, that is 'systemism' or, to put it better, a systemic way of thinking.

Just to indicate the most salient points of our philosophical agreement I can mention, first, the *realist* conception of science (both in the ontological and epistemological sense, that is, in the sense that science aims at knowing a reality that *exists* independently of scientific investigation, as well as in the sense that science succeeds in this *cognitive* endeavour, though without attaining the epistemic – not epistemological – privilege of absolute certainty). This view is based on the attribution of equal importance to the empirical and the theoretical sides of scientific investigation, which, in particular, means recognizing a genuine cognitive purport to theories in science and vindicating the preeminent role of thinking over simple observation.

Therefore, we both reject that subjectivist view of science that was often advocated by certain philosophical interpretations of quantum mechanics in the first half of the twentieth century and has reappeared in certain more recent 'postmodern' philosophical doctrines. Therefore, for both of us science is an *objective knowledge* in a sense that is partially in keeping with the Popperian meaning of this expression, but also different in other respects. Regarding the nature of scientific objectivity, I may point out a certain difference between Bunge's approach and mine: I have spent much time reflecting in order to explain how knowledge (which undoubtedly originates in an act of first-person subjective apprehension of reality) can acquire an intersubjective status, and I have found the solution in the fact that the intersubjective agreement relies not on the identity of the way of *conceiving* a notion but in the way of *using* it, this use being bound to certain *operations* that are different from the linguistic and conceptual tools adopted for the construction of the scientific statements and theories. In this I have found an aspect of truth in the doctrine of operationalism which, on the contrary, has been discarded by Bunge (who charged it with the mistake of attributing to operations a semantic function instead of a methodological one).

The pivotal role attributed by me to operations (in particular owing to their 'referential' function and their being criteria for immediate truth) is the major factor that distinguishes Bunge's scientific realism from mine. This difference, however, does not lessen a second common feature of our philosophy of science, that is, the fact of having passed through a serious confrontation with logical empiricism that has allowed us to overcome its limitations while preserving its merits. The limitation we both have overcome is the radical empiricism, that restricts to sense experience the whole cognitive purport of science and reduces logic to the role of a formal machinery for transforming linguistic expressions without increasing the volume of knowledge proper. Analytic philosophy of science (that has represented for several decades the main stream in philosophy of science) has remained prisoner of that framework and for this reason has been unable to solve its internal difficulties.

Bunge and I have retained, so to speak, the 'methodological lesson' of that approach by always attributing to a duly qualified experience its proper place and also by making abundant use of formalizations and formal tools in the construction of the discourse. This is why we can be qualified, in a certain sense, as analytic philosophers and this concerns the respect for *rigor* that can be found in our scholarly production. Mario has coined a happy terminology for denoting this intellectual attitude, speaking of *exact philosophy* in a general sense and meaning by this a style of philosophizing in which clear definitions of concepts, explicit declaration of premises and cogent logical arguments are applied to the study of a great variety of issues.

One can say that this ideal of an exact philosophy mirrors the epistemological structure of the 'exact sciences', and this is true, but does not entail that Bunge's philosophical interests have been confined to the philosophy of such sciences. Quite the contrary, the extraordinary range of his production covers not only the reflection on many sciences that are not usually classified among the 'exact sciences', but concerns also several domains that are not bound to science. The most eloquent confirmation of this range is certainly his *Treatise on Basic Philosophy*, a work of eight volumes in nine parts published across 15 years (1974–1989), which encompasses the most fundamental and 'classical' branches of philosophy (from ontology to semantics, epistemology, methodology, ethics) and many specific applications to particular domains (especially to those represented by different natural and social sciences).

Nevertheless, Bunge has not proposed a unique model of science and one can say that the concept of science is, for him, 'analogical' in the classical sense of this term, that is, that it applies partly in the same way and partly in different ways to the different sciences, that are recognized concretely in their historical realization. This is another point at which Bunge and I converge, since I too defend an analogical concept of science, though proposing certain precise criteria for differentiating the various sciences. Not withstanding this relaxed, but realistic, account of science, Bunge has developed a strong criticism against the 'pseudo-sciences', a criticism that is usually pertinent and has, in a certain sense, the advantage of not being dependent on a preconceived abstract model of scientificity.

Mario and I were becoming seriously interested in general system theory (GST) at the beginning of the 1970s without being aware of this common interest. At that time GST was the object of a very radical controversy between enthusiastic supporters and fierce enemies, the first being attracted by the fact that GST was offering a legitimacy to concepts such as ordered totality, global unity, goal oriented processes, specific function, multilevel realities, and emergent properties, which are frequently and profitably used in several sciences, from biology to psychology, sociology and other 'human' sciences. The enemies rejected such concepts, considering them to be vague, imprecise, and belonging to the superficial level of common sense language that should be banned from the rigorous discourse of science. This attitude was in keeping with the positivistically inspired scientific culture still predominant in the first half of the twentieth century.

I was convinced that a correct understanding of the issue had to avoid both extreme positions, and in 1975 organized a conference of the International Academy for Philosophy of Science, devoted to the study of systems. In 1978 I published as editor a volume in Italian containing most of the papers presented at that conference and a few additional contributions: it was a rather rich display of chapters authored by prestigious specialists and presenting the general features of system theory and several of its applications to the sciences and to philosophy. Just one year later Bunge published Volume IV of his *Treatise on Basics Philosophy* with the title *A World of Systems* (1979) that contained, in particular, a full-fledged axiomatization of GST according to the most rigorous formal methodology adopted in mathematical logic and in the foundational studies of various disciplines. That book can be considered as the final demonstration of the genuine scientific character of system theory that objectively supported and encouraged the broad adoption of its schemes in different fields of investigation, not only in the work of Mario Bunge, but also at a much larger scale.

13.2 The Systemic Point of View

At a distance of nearly four decades from that historical achievement the systemic way of thinking has shown itself as the most proper tool for understanding complexity and investigating complex realities, and stimulates reflections capable of revisiting classical philosophical concepts, basic metaphysical and ontological principles, the deepest sense of fundamental developments in the history of science, the critical appraisal of merits and confronting the limitations of many present research programs in various fields. Considering the richness of these applications one might wonder at the rather recent appearance of the systemic approach (and not without oppositions) on our cultural horizon. The reason for this fact may emerge by considering the difference in meaning of two cognate adjectives, "systematic" and "systemic". "Systematic" is primarily understood as the property of a discourse, of a written or oral presentation regarding a certain subject matter according to an *order* showing a logical structure, a global unity and completeness, and even a certain architectonic elegance. By extension, this adjective can be applied to a procedure following a well determinate order of steps with the view of attaining a certain result. "Systemic", on the other hand, is a recent term introduced as a shortening of "system-theoretic", that is, in order to qualify a concept, a principle, a model according to what they are meant in *system-theory*.

The difference is not negligible and is comparable, for example, with that existing between two meanings of "physical": when we say "a stone is a physical object" we mean that it is some material thing, different, for instance, from a legal prescription, a wish, a dream, a proposal; when we say "this is a physical law" we mean that the statement we are presenting is expressed in a particular scientific discipline that we call "Physics". Here a spontaneous remark surfaces: if we qualify something as 'systematic' we mean that it has the form, or better, that it is an example, of a *system*, and when we speak of "system-theory" we again make reference to *systems*. Hence, we remain with the question, "What are systems, what is a system"?

The answer to this question is not simple, not because this term circulates only in specialized sophisticated languages, but because, on the contrary, it has acquired a very large community of applications in a variety of contexts. So, for example, in ordinary language we speak of linguistic system, legal system, political system, economic system, bureaucratic system, productive system, industrial system, energetic system, railway system, metric system, and so on. In addition, within specialized disciplines one finds the mention of several systems, like muscular system, nervous system, endocrinal system and many other systems in biology, or numerical system, Boolean system, Euclidean system, equation system in mathematics, elastic system, gaseous system isolated system in physics, not to speak of the many systems that are considered in chemistry, crystallography, astronomy, geology, geography and, in the domain of humanities the capital notion of philosophical system.

Such a variety of applications may at first produce the reaction of considering the notion of system as endowed with a vague and confused meaning belonging to ordinary language that can only give rise to ambiguities and misunderstandings, and must be overcome and replaced by a rigorous and technical treatment. This impression is wrong, because this generalized use rather testifies that it belongs to *common sense*, that is, to that complex set of basic concepts and principles that make possible our understanding of reality and whose meaning is, therefore, not ambiguous, but *analogical*, that is, such that it must be applied partly in the same way and partly in different ways to different kinds of reality (as we have already noted above regarding the concept of science). If this is the situation, the stimulating task is that of making explicit that common semantic core that underlies the analogical use and makes it possible.

This core can be described in a rather intuitive way by saying, for example, that a system is an entity constituted by parts that are linked by mutual relations, making up a complex ordered unity which is endowed with its own individuality in the sense that its characteristic properties and functioning are different from those of its constituent parts though depending on them to a certain extent. In a shorter way we could perhaps say that we mean by a system an ordered totality of interrelated parts whose characteristics depend both on the characteristics of the parts and on the web of their interconnections.

GST can be considered as an effort to make explicit and precise this conceptual core, by sometimes revisiting concepts and principles that philosophy has already defined and analyzed in the past but have been neglected or abandoned for several historical reasons, especially as a consequence of the predominance attained by the conceptual framework of certain successful sciences in Western culture. This remark suggests a hypothesis that we shall check later on, that is, that the notion of system has entered Western culture coming from the sciences, and that this fact has contributed to the dismissal of certain traditional philosophical concepts. If this hypothesis is correct, it appears natural that GST was born rather recently in the domain of science (more precisely, in the domain of biology with the work of Bertalanffy) and has then known conspicuous developments many of which are of a philosophical nature and have implied a certain recovering of traditional philosophical concepts.

It would make no sense to summarize here the basic ideas of GST. It will be sufficient to say that initial hints contained in Bertalanffy's early writings of 1934–1937 fostered developments in research and teaching activity in Vienna, Canada and the United States, attracting collaborators from different domains and leading, in particular, to the foundation in 1958 of the Society for General System Research. The crystallization, so to speak, of this intense and scattered work was the publication of Bertalanffy's book *General System Theory* in 1968, presenting GST as a recognized solid scientific discipline. Bunge's volume *A World of Systems* (Bunge 1979), which appeared about 10 years later, appears as the crowning of this theory, consisting in something like a 'systematic' presentation of system theory, more or less in the same spirit as Hilbert's *Foundations of Geometry* had been an original systematization of the whole body of elementary geometry.

No exaggeration is contained in this appraisal. Indeed, the first chapter of Mario's volume displays a detailed analysis of all aspects characterizing systems presented through an abundant use of formal and symbolic expressions accompanied by useful clarifications formulated in plain ordinary language and supported by all sorts of diagrams and figures. All this provides the solid ground and the inventory of conceptual and methodological tools that are then used in the following chapters for 'exemplifying' the notion of system in a few fundamental domains where it plays a particularly relevant role, such as the chemical, the biochemical, the domain of life and functions, the phenomenon of evolution, the study of mind and that of society with its interconnections of subsystems and supersystems. In the course of this presentation many conceptual clarifications of a general nature are introduced that complement what has been offered in the first chapter, so that the result is really a systemic worldview, that is, the conception of the world as an integrated coherent system of systems.

This general conception is actually a philosophical outlook that one could call metaphysical, though Mario would perhaps feel allergic to this terminology. He explicitly presents his position (which he calls "systemism") as intermediate between two erroneous extremes, "atomism" and "holism". The weakness of atomism resides in that it ignores the relevance of properties and especially relations, without which it is impossible to distinguish a simple "aggregate" from a "system". The weakness of holism resides (according to Bunge) in its pretention that the knowledge of the whole must precede and make possible the knowledge of the parts. Systemism avoids both mistakes by recognizing that the whole 'results' from the correlation of its parts and at the same time has influence on their functioning. Bunge does not offer explicit examples of holistic doctrines in this sense, and his rejection of holism is more clearly hinted at when he says that holism "is anti-analytic and therefore anti-scientific" (Bunge 1979, p. 41), but this is, again, a rather peculiar way of describing holism.

I personally prefer to consider holism a correct overcoming (due especially to system theory) of the atomistic perspective dominant in science, not in the sense that knowledge of the whole is a precondition for the knowledge of the parts, but in the sense that the 'point of view of the whole' is indispensable in any adequate cognition, both because a global *Gestalt*, a global 'point of view', provides the conceptual space in which a given investigation takes place, and because the items of information acquired in that investigation must be organized in a unity (which can usually modify in different degrees the original *Gestalt*). This is however, rather a terminological than a substantial difference with respect to Bunge.

More significant is the difference in our positions regarding *reductionism*, which I reject rather strongly whereas Bung accepts what hc qualifies as "weak reductionism" (for which he describes the conditions). We had the opportunity of comparing our views at a conference on "The problem of reductionism in science" that I organized at the Polytechnic School of Zurich in 1990 as President of the Swiss Society of Logic and Philosophy of Science. At that meeting Bunge made a presentation on "The power and limits of reductionism" and my talk was on "Reductionism as a negation of scientific spirit". Our views were only *prima facie* irreducibly opposite since Mario expressed a positive judgment on "micro-reductionism" essentially on the ground of historical considerations. He pointed out that increasing efforts to interpret a great variety of physical phenomena within the framework of classical mechanics (that was in a way a form of reductionism) had worked as a very powerful propeller of modern science, but at the same time he recognized its lack of synthetic power as a limit of reductionism and also recognised the fact that the world is made of systems.

My opposition to reductionism was based on my fundamental view regarding the specificity of any single science due to its being a discourse on reality made from a certain specific point of view, equipped with specific operational criteria of reference and truth. This not only stresses that reality consists of a world of systems, but also stresses the need for these systems to keep their identity while at the same time being strongly interrelated. Nevertheless, I admit that every science must be "methodologically reductionist" in the sense of trying to interpret and explain as much as possible of reality using its tools, while at the same time recognizing that there are aspects of reality that lie outside of its domain of reference. Considering these details, one can see that our positions were not so strongly opposed, which was apparently confirmed when I published the proceedings of that conference (in which Mario's paper (Bunge 1991) and mine appeared) as a book that was subsequently included in Reidel's series "Episteme", edited by Bunge.

In my opinion one reason for our consonance on several matters is a fundamental philosophical attitude that we share, that is, the attribution of the character of reality to properties and relations. This marks our distance from positivism and is the key to our adoption of a systemic way of thinking not only when we explicitly speak of systems, but also when our discourse does not mention them. Of course, this approach is more evident when we actually apply GST in a determinate field, as Mario has done several times and I have exemplified by proposing a system-theoretic approach to the delicate problem of the ethics of science, in which the disciplinary autonomy and freedom of science is harmonized with its duty of being ethically responsible in its practice (Agazzi 2014, chap. 9.2). An indirect presence of the systemic way of thinking, however, in the treatment of particular issues can also be noted as, for example, in my treatment of interdisciplinarity as well as in many of Mario's contributions to different issues.

It would not make much sense to go on with this mention of similarities and differences that Mario and I present in our common adhesion to the systemic way of thinking. Therefore, I prefer to pass to a short survey regarding the history of the notion of system. This can offer us interesting clarifications as, for example, emerge from the fact that the first appearance of the very notion of system occurred in modernity and in the scientific domains, and was later applied especially to philosophy. Also the usefulness of this notion in order to understand the birth of the evolutionary perspective and then of the different evolution theories will be noted, as well as the reasons that favoured the exclusion of 'final causes' and the adoption of the mechanistic outlook in the natural sciences at the beginning of modernity. The reasons for the adoption of the machine as the privileged model for scientific understanding, the historical reasons for the imperialism of the analytic method in science and outside science, the fascination and equivocations of reductionism, the fruitfulness of a (correctly understood) 'holistic' approach to problems and, finally, the fundamental clarifications that the systemic approach offers in the correct understanding of complexity and interdisciplinarity will also be touched upon.

The mention of so many items could suggest the impression that I advance a pretention of completeness. On the contrary, in what follows I will simply offer a survey of some important steps in the evolution of the notion of system, being conscious that other important details are overlooked. Things that one does not speak about, however, are always infinite, and the only reasonable aspiration is that what is being said is relevant and correct.

13.3   Moments of the History of the Notion of System

13.3.1   The System of the World

The concept of system (or, better, the term "system") is so widely used in our linguistic contexts – as we have already noted – that one is inclined to assume that it has belonged in our academic vocabulary from times immemorial; but it is likely that we will find its first irruption on the stage of Western culture in the title of the most famous work of Galileo, the *Dialogue Concerning the Two Chief World Systems* published in 1632, in which the reasons supporting, respectively the

Ptolemaic and the Copernican astronomic theories are compared. This occurrence, however, was not new, if we consider that already in the conclusion of his *Sidereus Nuncius* (1610) Galileo mentions as a commonly used notion that of the Copernican system and promises to later present his own system.

Pushed by our curiosity, we might consult certain very authoritative dictionaries of those times, such as the *Lexicon philosophicum* by Rudolf Goclenius published in 1613, in which several concepts occurring in Galileo's writings and belonging to the Scholastic ontological and epistemological doctrines are present, but the item "systema" does not occur. We can go then to a celebrated much later dictionary whose scope is apparently not limited to philosophy, that is the famous *Lexicon totius latinitatis* by Egidio Forcellini (1771) and here we find a couple of lines devoted to "systema", which is presented as a pedagogic device for making easier the presentation of "scientific" discourses (where science, however, is understood in its classical general sense of sound knowledge and not in the sense of the modern natural sciences), and the only author cited for the occurrence of this term is Martianus Capella, a scholar active in Cartago in the fifth century who cultivated encyclopaedic interests. It is possible that more accurate historical investigation brings about some additional information, but what we have mentioned seems sufficient to maintain that the concept of system entered the background of Western culture in modernity coming from the domain of the natural sciences.

The obvious consequence of this circumstance was that the domain of application of this concept was that studied by the new natural sciences, that is, the domain of the physical bodies considered in its generality, more or less in the sense of 'external world'. This is why the new science felt itself charged with the task of providing also a *system of the world* (or a 'worldsystem' for brevity) and we actually find this expression as the title of the concluding book of Newton's *Philosophiae Naturalis Principia Mathematica* (1687) that reads *De mundi systemate* ("On the system of the world").

This phrase became quickly standard and we find it, for example, in the title of Laplace's work *Exposition of the World System* (1796). The 'World' considered by Newton and Laplace is actually what we now call the planetary solar system, which is far from being the total system of the universe. It is true, however, that already with Laplace the proposal was that of formulating a *cosmological theory*, in which a certain hypothesis about the original state of the universe could explain the present structure of the visible world, that was, at those times, practically the planetary system. This cosmological meaning is mirrored in the famous work by Pierre Duhem, *The System of the World. A History of the Cosmological Doctrines from Plato to Copernicus* published in ten volumes between 1913 and 1959.

Today this expression is no longer used in the context of contemporary cosmological theories, and not even in physics (a new meaning has been attached to it starting in the 1970s for denoting the complex structure of the present world economy).

13.3.2 The System of Nature

Nevertheless, a similar concept soon emerged in the context of another natural science, biology, and is present in the title of one of the most famous works of this discipline, Charles Linné's *Systema Naturae* which has known, between 1735 and 1768, ten successive enlarged and revised editions during the life of its author. The interesting fact is that the term "world" is replaced by "Nature", not because the intention was that of enlarging the perspective in order to include also the living beings near the physical bodies, but rather with the intention of taking into special consideration precisely that specific wide domain, and this is already an evidence of the analogical sense of "system".

The novelty, however, is more significant if one considers the methodological structure of this work in comparison with the treatises of physics: as is well known, the most important contribution credited to Linné is the introduction of the *binomial nomenclature* for the classification of living species which, with certain improvements, has remained in use until today. Leaving aside the historical question that this method was partially anticipated two centuries before him by the brothers Bahuin, what must be stressed is the intellectual outlook that guided Linné's generalized adoption of this criterion.

For two centuries the naturalists had been looking for a 'natural criterion' for the classification of the living beings, and none had proved satisfactory. Linné's idea was that a natural classification should reflect that logical order that exists in Nature due to the fact of its being the expression of the supreme intelligence of the Creator God. In other words, the order of living creatures had to be a *logical order* for which formal logic had provided a well-known scheme in the ancient 'Porfiry tree' regarding the hierarchic disposition of genera and species. According to this view, the entirety of Nature was conceived as a kind of mosaic in which the position of any single piece is strictly determined following a design constituted by a web of logical relations, in which each piece occupies 'its' proper place. This is the fundamental worldview of *fixism* that was never abandoned by Linné, in spite of his admission (in his later years) to have personally ascertained occasionally the appearance of new species.

If we compare his 'System of Nature' with the 'System of the World" proposed by the physicists we can recognize that the latter was in a certain sense more significant, since the systemic architecture of the World was conceived as consisting in causal links expressed in terms of natural laws and forces, and not simply through the fragile spider web of a conceptual order, and we may also add that Linné's system has only a *descriptive* aim and purport, whereas the physicists' systems had an *explanatory* aim.

13.3.3 Philosophical System

All this is true, but we cannot ignore the historical background of that time, in which a prominent thinker such as Spinoza could formulate in the second book of his *Ethica More Geometrico Demonstrata* the aphorism *ordo et connexio idearum idem est ac ordo et connexio rerum* ("the order and connection of the ideas is identical with the order and connection of things"). That statement can seem dogmatic today, but it fascinated the representatives of the romantic 'Philosophy of Nature' at the end of the eighteenth century and also the thinkers of the German transcendental idealism of the nineteenth century. That fascination was produced by that impression of intellectual rigor, systematicity and architectural elegance that transpires in Spinoza's work and was easily taken as a warranty of speculative soundness, as opposed to that 'rhapsodic' way of thinking (to use Kant's term *rhapsodistisch*) that marks the style of those scattered reflections that ignore the need of strong logical links.

In Kant's work this kind of appraisal is explicitly made, and it is significant that the term "system" widely occurs in order to express the satisfaction of these requirements. This explains how the phrase "philosophical system" has become customary for denoting the whole complex of the speculation of a single thinker, independently of the fact the he uses or not this denomination to this end (like does, for example, Schelling in his *System of the Transcendental Idealism*). The generalized use of the notion of system by philosophers, however, is a consequence of the fact that this concept had gradually penetrated the vocabulary of cultivated people at large, particularly in the milieu of the Enlightenment. This is testified, for example, by the fact that Diderot, in the article "Encyclopédie" of the famous *Encyclopédie*, uses the concept of system in order to explain the meaning of the encyclopedic approach as such, qualifying it as the program of bringing to a coherent unity, based on mutual correlations, a disparate variety of notions and items of knowledge, and the analogy he uses is that of the global structure and functioning of a machine.

13.4 The Evolutionary Point of View

The overcoming of the Linnean fixism, however, did not occur as a consequence of epistemological criticism but due to the gradual penetration of the category of *historicity* (which had been applied traditionally only to human actions, enterprises and realizations) also into the realm of nature. For this approach it is appropriate to mention a particularly significant scholar, George Luis Buffon, whose famous *Natural History, General and Particular* (of which he published 36 volumes during his life between 1749 and 1788) contains, in particular, a part with the title *Epochs of the Earth* which is considered as the inaugural text of geology. Since it has always been well-established that the properties and functions of living beings are in strict interdependence with their natural environment, it was almost automatic to derive from the recognition that the Earth (that is, the natural environment) has undergone changes, that also the forms of life have undergone a similar change.

Buffon and Linné were intellectual adversaries on the cultural stage of the eighteenth century, but the final acceptance of Buffon's geological history (that in particular favoured the interpretation of the fossils as relics of extinct past living species) meant a dismissal of the Linnean fixism in its literal form, which Linné himself had expressed in the statement *tot numeramus species quot a principio creavit Infinitum Ens* ("we count as many species as were created from the beginning by the Infinite Being").

Nevertheless, a mitigated form of fixism was compatible with the admission that life has had a history, and this was actually realized in the doctrine of George Cuvier. This great biologist, founder of paleontology and comparative anatomy, proposed a classification of the animals based on the analogies and differences of their fundamental functional and morphological structure that were strictly coupled by the "correlation principle". This was actually a systemic view that had the same character of rigidity as the Linnean one, the difference being that anatomic and functional correlations had replaced the logical correlations of Linné; and Cuvier did not admit gradual changes and transformations. The way of making compatible this view with the history of living species was "catastrophism": the thesis that transitions from one geological era to another had consisted in terrible cataclysms that radically changed the surface of the Earth and produced the sudden violent extinction of all the forms of life (some samples of which have remained petrified in the form of fossils). Life began to develop again in the new natural environment, taking forms totally different from the preceding ones, resulting in full and rich varieties of new species mutually linked according to the correlation principle. This new situation then remained unchanged until the next catastrophe.

In the same years Jean-Baptiste Lamarck was patiently working at a description and classification of the almost unexplored domain of the invertebrates whose results he published in several books the most important of which are the seven volumes of his *Natural History of the Invertebrates* published between 1815 and 1822. This wide study had brought him to focus on the different degrees of *complexity* that appear in the animals and which can be organized according to a suitable order. This idea gives rise to the conception of a great classification of the living beings based on a general order of complexity that is objectively present in nature, satisfying in such a way the need for a 'natural classification' that was generally felt by the naturalists.

The major novelty, however, was that such an order was not born, so to speak, at the origin of life, but was the result of a gradual *historical transformation*. In Lamarck's *Zoological Philosophy* (1809), he presents his views regarding the nature of life: this rises from non-living matter by spontaneous generation in the form of a few very simple living organisms which have the capability of *adapting* themselves to their environment. This adaptation consists in the performance of certain functions that in turn entail also modifications in the form and structure of the organism. These small modifications are transmitted to the offspring and accumulate slowly and gradually from generation to generation and eventually, between the times of geological transformation, this phenomenon results in the existence of new species.

This doctrine is the first appearance of the *evolutionist* theory of life, whose conceptual core consists in the thesis that the present species are the descendants of less numerous and less complex original species. Besides this basic statement, Lamarck also offered the first *evolution theory*, that is, the first theory in which are proposed the ways or procedures thanks to which evolution has occurred, and they are – as we have seen – the intrinsic adaptability of living matter to the environment and the hereditary transmission of the acquired characteristics to the descendants. We can say that with Lamarck the world of life is really conceived as a system in which the fundamental order is a *genetic order* of increasing complexity.

Lamarck's doctrine was the subject of strong debates during the five decades that separated its appearance from the publication of Charles Darwin's *The Origin of Species by Means of Natural Selection* (1859), and during those years the evolutionary approach had gained increased acceptance among naturalists. This favoured the greater fortune of Darwin's work, though it is anti-historical and superficial to consider him as the 'founder' of evolutionism, as so often occurs. If we compare the two doctrines, we must recognize that Lamarck's approach is more 'systemic' because full weight is attributed to the internal functioning and dynamics of living organisms, as well as to their interaction with the environment, from both of which factors is produced the genetic gradual transformation of the species. In Darwin's perspective, interaction between environment and living organisms plays no real role, because the organisms are not modified by the environment, and even less are internally pushed to 'adapt' to the environment. Natural selection simply consists in the contingent fact that, among the individuals of a certain group living in a particular environment, a minority can exist of those who share a particular characteristic. If that characteristic favours their survival when a shortage of the living resources occurs they (and their descendants that share the same exceptional characteristic), will survive and reproduce, finally become the only population living in that environment.

One may wonder why Darwin's doctrine finally attained a much larger acceptance than Lamarck's more 'biologically flavoured' doctrine and the fact that Lamarck had taken for granted that the *acquired characters* that an individual attains through its adaptation to the environment can be *inherited* by its descendants, may represent an objective reason for this. This conviction, however, had no empirical confirmation and was even blocked by a tenet almost universally accepted in the biology of the late nineteenth century, that is, the doctrine of the strong distinction between the 'somatic plasma' and the 'germinal plasma' in the living organism, the first being constituted by the cells of the body, and the second by those of the gametes. Only changes affecting the germinal plasma can be inherited, and since the changes produced in an organism by interaction with its environment affect only its body (the phenotype) and not its 'genotype', they cannot be transmitted to the descendants. This has remained a persistent difficulty that has hindered the acceptance of Lamarck's ideas and has been overcome only by the research in modern genetics, that have shown that environmental actions can affect the genome and, therefore, the corresponding phenotypic modifications are transmissible to the descendants.

In addition to this objective difficulty, however, another more philosophical reason was acting against Lamarckism, that is, the fact that Darwin's theory of evolution was much in keeping with the *mechanistic* outlook dominating science in the positivistic culture of that time, whereas Lamarck's conception of life was reminiscent of traditional notions with a 'metaphysical' flavour. For this reason, the new systemic approach that had been introduced in the domain of the life sciences (or of Nature), consisting of open, adaptive and dynamic systems, had to wait several decades before becoming the inspirer of General System Theory in the 1930s.

13.5   The Issue of Finalism

Modern natural science was born – as we have seen – according to certain ontological, epistemological and methodological restrictions that had been proposed by Galileo and accepted by Newton and their followers. Among these restrictions one, in particular, was more implicit than explicit but of paramount importance, and regarded the concept of *cause*.

The 'principle of causality' is one of the most fundamental metaphysical principles that can be formulated, in its simplest form, as the statement that *every change has a cause*. This principle is so fundamental that it can be considered as an indispensable condition for understanding reality and, as such, universally admitted. However, the concept of *cause* is far from being univocally understood and very many meanings of this concept have been proposed in the history of philosophy. The one that is probably the most common in ordinary language is that which was called an "efficient cause" in the philosophical tradition. This corresponds to the idea of something that *produces* something else as its *effect,* and whose most familiar examples are the physical actions that bring about new objects, or certain observable processes.

Common language, however, has no difficulty in accepting as meaningful, for example, the discourse of Socrates in Plato's *Phaedo*, when he explains that the *cause* of his coming and remaining in prison (waiting for his capital execution) were not his legs, bones and muscles, (which could serve equally well to run away) but his desire to obey his city's laws. That physical situation had an immaterial cause. This depends on the fact that the Greek word *aitìa* (that is commonly translated as "cause") had a polysemous sense, which we might better express through the notion of "the reason for which". In this way we can easily understand Aristotle's doctrine of the 'four causes' (formal, material, efficient and final), of which only the efficient means the 'production' of an effect, whereas the others concern the 'reasons' for which something occurred, these reasons being the presence of a material substratum, the 'form' or *essence* of the entity concerned, and a *goal* orienting the process. This goal can be either the *aim* or purpose pursued by an external operator performing the process, or a *pattern* inscribed in the internal essence of the entity, a kind of *design* presiding over its development; and also over its way of behaving in different circumstances.

A generalized rejection of the final causes can be found at the beginning of modern philosophy, both in empiricist philosophers like Francis Bacon, and in rationalist thinkers like Descartes and Spinoza. No wonder, therefore, that it enters also the new natural science, especially considering that Galileo had explicitly excluded from the objectives of this science the investigation of the *essence* of the physical bodies, and the final cause was precisely meant to reside in their essence. The same attitude is explicitly adopted by Newton who, in the *Scholium generale* of his *Principia*, after having admitted that he had been unable to uncover the *cause* of gravitation, declares that he will not try to "imagine hypotheses", by postulating "hidden causes", like those that the Scholastic tradition was accustomed to locate in the substantial forms of things.

This kind of reasoning has nothing to do with a refusal of finalism in Nature that, according to certain authors of our time, would be a subtle improper tool for admitting the interference of religion into science by requiring the existence of an intelligent omnipotent God as the cause of the marvellous order or design present in Nature. Indeed, the Newtonian statement *hypotheses non fingo* appears in the conclusion of the *Scholium generale* in which ample space is given to a series of *theological* considerations according to which only the existence of such a supernatural spiritual Creator can account for the global order of the world, while the impossibility of uncovering the cause of gravitation is linked with the impossibility of natural science (called "experimental philosophy") to bypass the external properties of things and penetrate their intimate essence (which is an *epistemological* reason).

As a matter of fact, the mechanically interpreted order of Nature has remained for a long while one of the fundamental arguments for the existence of God as its cause, even for anti-religious thinkers like Voltaire, and Darwin's evolution theory (in which no finalism is present) was considered by him as the more compatible with divine creation (as he says in the final lines of the *Origin of Species* that have remained until the last edition):

> There is a grandeur in this view of life, with its several powers, having been originally breathed by the Creator into a few forms or into one; and that, whilst this planet has gone circling on according to the fixed law of gravity, from so simple a beginning endless forms most beautiful and most wonderful have been, and are being evolved. (Darwin 1859/1958, p. 450)

It lies outside the scope of this paper to investigate the deep epistemological reasons of this exclusion of finalism from science, which has received due attention in the speculation of Kant. He conceived causality according to the strict deterministic character of the efficient cause in the *Critique of Pure Reason,* but in the *Critique of Judgement* he formulated the distinction between "determinant judgment" (which is that expressing cognition in a proper sense, and in particular the cognition acquired in science) from the "reflecting judgment", which depends not on intellect but on sentiment, and distinguishes two kinds of reflecting judgments, the aesthetic and the *teleological.*

The first concerns the appreciation of beauty, whereas teleological judgments ascribe ends or purposes to natural things, or characterize them in purposive or functional terms. Therefore, the finalistic point of view is not discarded or ignored, but Kant claims that it is an *a priori* principle of reflecting judgment that nature is "purposive for our cognitive faculties" or "purposive for judgment." We cannot assert that nature is, as a matter of objective fact, purposive for our cognitive faculties, but it is a condition of the exercise of reflecting judgment that we assume Nature's purposiveness for our cognitive faculties. In short, teleology does not have the cognitive objective purport of efficient causality, but we consider natural beings and processes "as if" they were purposive even for knowing them in strict intellectual form, and this is especially useful in the case of living organisms and for understanding the ordered architecture of Nature. These brief and approximate hints at Kant's doctrine explain why his doctrine is revisited today in the philosophy of biology, but at the same time because it clearly contains several elements of the systemic point of view.

13.6   The Model of the Machine

There is, however, a more concrete reason for the imperialism acquired by the efficient cause in the natural sciences and it consists in the fact that the 'specialization' of this concept is represented by the notion of *force* that produces the *change of motion* (and not motion itself, which is considered primitive in physics, needs no explanation for its existence and enjoys a 'principle of conservation' like matter). The object of motion is matter, whose 'specialization' in natural science is *mass*. In such a way matter, motion and force were the fundamental concepts of the new-born natural science and force was conceived as something acting on the material bodies *from the outside* and, therefore, independently from the particular nature of the material entity concerned. This was perfectly in keeping with the Galilean 'revolutionary' proposal to ignore the specific "intimate essence" of natural entities (and the equivalent views expressed be Newton and their followers).

In such a way all natural beings could be interpreted, at least to a certain extent, as mechanical *machines*, including living beings, and this was sketchily done by Galileo himself and developed by some of his indirect disciples like Borelli. As is well known, Descartes has presented in his *Treatise on Man* the first detailed interpretations of the human corporeal constituent as a complicated machine, while considering the spiritual constituent (the thinking capability) as a separate "substance". This 'dualistic' conception advanced with the aim of securing to spiritual metaphysics and theology their autonomous legitimate domain, was not shared by those authors, like La Mettrie, who, adhering to a materialist metaphysics, proposed a portrayal of the *Machie Man* (1748) in which the interpretation of the totality of the human characteristics was expressed in mechanical terms.

There are essentially two reasons for the growing success of this machine-modelling. One is historical: at the time of the Renaissance several admirable amusing realizations were produced by very skilful craftsmen who constructed 'automata' capable of imitating the form and function of certain animals (like the famous duck of Vaucanson in which 400 mobile parts interacted) or could produce astonishing plays of fountains and other hydraulic effects in the gardens of royal palaces. This was the path of 'simulation' that still has a notable force today. The second reason is more intellectual: nothing is mysterious in a machine but everything clearly depends on the appropriate disposition of several constituent parts and the effect of natural laws acting on these parts. Its constitution and functioning are known, understood and explained before its effective construction, because they correspond to a project that has been elaborated by a conscious application of empirical knowledge and scientific theories.

From this correct view a spontaneous consequence seems to follow: if we want to understand any natural entity or process, and we are able to represent it as a particular machine, we have the impression of having understood it completely, or at least 'essentially', considering irrelevant those aspects that we have been unable to include in the machine-model. Precisely this spontaneous attitude is the core of *reductionism*, that consists in the cognitive position of considering irrelevant all the features that cannot be encompassed in the pre-selected approach.

The watch was the paradigmatic example of such machines, also because it brings together the features of autonomy, precision and miniaturization of its internal mechanisms, stimulating the idea (explicitly expressed by Descartes) that if humans are able to realize such machines, much more has this capability Nature, that can realize these mechanisms in its most concealed minute parts. The combination of this idea with the general Christian worldview (that was all-pervasive at that time in Western culture) that Nature is the creation of an infinitely intelligent God, produced the conception of the "Watchmaker God" widely accepted in the seventeenth and eighteenth century, which is present in less pictorial terms in Newton's *Scholium generale*, in Voltaire's *Treatise of Metaphysics* and in several other thinkers of the Enlightenment and could offer an intellectual blend of deterministic efficient causality (acting at the level of single realities and processes) with a global finalism organizing 'from the outside' (that is, as an action of God) the complexity of the whole of Nature.

This general view was common also to thinkers, such as Leibniz, who replaced the atomistic view of a Nature dominated by the rigid determinism of the efficient causes by a more flexible conception of the elementary constituents of reality being "monads" endowed with different kinds of representational capabilities whose interactions can be explained by the principle of "sufficient reason", and whose global coexistence in the real world is granted by the "pre-established harmony" impressed by the wisdom of God, such that the real world is "the best of all possible worlds".

Leibniz' conception was particularly open to the understanding of living and intelligent beings but could not rely on a sufficient development of the life sciences at that time. Therefore, the mechanistic view could quickly spread on the whole domain of physics, gradually 'absorbing' into mechanics domains such as those of acoustics, optics and theory of heath that were traditionally distinct on the basis of the respective sensorial access to their specific phenomena.

Outside physics, however, new sciences rapidly began to acquire their epistemological independence, by determining their empirical methodologies and theoretical frameworks. This is particularly clear in the case of biology, where the cellular theory recognized a certain 'elementary' constituent of every living being (namely the cell) that in addition was already a complex *organism* (so that "unicellular organisms" could be envisaged in which the cell is not a 'part' in a proper sense). Moreover, life was characterized by a series of specific characteristics not common to non-living entities, and it became more and more clear that any living being is an organism in the more precise sense that its constituent parts have specific structure and functions and are mutually interrelated, so that the whole organism has its own properties that are not shared by its parts, though depending also on them and their relations.

The evidence of these specificities produced a doctrinal dispute between "mechanists" (who maintained that the mechanical interpretation of the phenomena of life was sufficient to explain them) and "vitalists" that denied such sufficiency. They were right, but could not clearly triumph in the dispute because they pretended to explain the specific properties of the living beings by the presence of a "vital principle" or "vital force" of which they were unable to provide any empirical evidence. The vitalists remained still partially prisoners of the mechanistic epistemological framework that required a special 'force' to causally explain, as its effects, certain observable phenomena. However, the real reason for difference had to be looked for in the direction of functions, relations, structures, unity, and this did not happen before the birth of system theory.

This is confirmed by the fact that the mechanistic reduction had already shown its shortcomings in the domain of physics itself before the end of the nineteenth century. Maxwell's electromagnetic theory of light and general theory of the electromagnetic field had challenged the most brilliant physicists to propose mechanical models of this field but they never attained success. Therefore, optics and electromagnetism could not be 'absorbed' into mechanics. The same occurred with thermodynamics, owing to the incapability of obtaining a satisfactory explanation of the second principle of thermodynamics from the purely mechanical kinetic theory of gases (and of matter in general). Even the much more drastic crisis that occurred for the so-called "classical mechanics" through the advent of Quantum Mechanics and Relativity Theory did not produce a completely radical change, as might be attested by the fact that we still speak of quantum and relativistic *mechanics*.

13.7 System Theory

One can correctly point out that Bertalanffy's study, in which he presents the first seeds of system theory, focuses on the inadequacy of the second principle of thermodynamics for the explanation of the phenomenon of biological growth of individual organisms (*Investigations on the laws of growth,* 1934). This principle, however, had been often criticized in physics. The novelty of Bertalanffy's approach is the consideration that it applies to closed systems, whereas living organisms are *open systems.* Therefore, the problem is not that of 'criticizing' the second principle of thermodynamics (that is right under its specific hypothesis) but to recognize that it is not fully *pertinent* in the case of living organisms because they are not *systems* of the kind envisaged by the said principle. Therefore, the issue is that of making a pertinent investigation regarding the different kinds of systems and possibly their common features.

In particular, the question of 'lawlikeness' was important, because the privilege of the physical closed systems was that of being regulated by deterministic 'natural laws', permitting predictions and experimental tests, while nothing comparable appeared possible for living systems. Nevertheless, it is also evident that living organisms are able to preserve a certain identity underlying their continuous change, that they realize and tend to keep a *steady state* which is different from the simple *equilibrium* (be it the mechanic or the thermodynamic one). These are among the best-known characteristics that are studied in General System Theory (GST). They suggest that, due to their difference with regard to the conceptual tools usually admitted in the sciences, they can offer the opportunity of revisiting certain other more general philosophic concepts that are appreciated within the systemic way of thinking.

In other words, GST, that is born in the field of *science,* can help us to recover the intellectual importance of philosophical concepts that had been marginalized as a historical consequence of the advent of modern science in the Renaissance.

13.7.1 Holism

The ontology of GST consists of a web of single *totalities* each one of them being individually characterized by its own internal structure and proper functions. In order to appreciate the novelty of this ontology it is sufficient to compare it with that of another great foundational theory in mathematics, that is, set theory. In set theory only one relation is primitive, that of membership of the elements in the set, but the elements are in a certain sense all equivalent, since they have no property. Moreover, they have no internal structure, not even being linked by particular relations but they can be arbitrarily aggregated in sets, subsets and supersets.

On the contrary, the primitive constituents in system theory are systems, each having its specific characteristics and internal structure, and they do not simply 'belong' to the global system, but are mutually interrelated with the other systems and are not 'elements' but 'subsystems' of the global system, according to a net of relations that allow the global system to have certain properties and perform certain functions.

Due to this fact, every system is at the same time 'simple' (in the sense that it is well determined in what it is, independently of its relations with other systems) but also 'complex' (as far as it has an internal structure, constituted by a web of relations among its subsystems, from which its own specific properties depend). It is neither arbitrary nor difficult to recognize in what we have just said the classical notion of *substantial form*, that was precisely the ontological principle expressing the fact that any entity is what it is due to a particular organization of its constituent parts whose status was qualified as *matter* (not because they are simply 'raw material', but because they belong to a lower level of organization). After these precisions it should be clear that "holism" is here understood as the appreciation of the 'point of view of the whole' as opposed to "atomism", and has nothing to do with the notion of holism that Bunge rejects.

13.7.2 Complexity

Complexity is included in the holistic perspective: the properties of a system are the result of the correlations among the subsystems that constitute it, and also of the relations it has with its environment. Modern science, on the contrary, had followed Galileo's proposals not only in the exclusion of the investigation of the intimate essence of things, but also in the practice of studying an isolated phenomenon concerning one single property by trying to create an artificial situation in which all possible 'disturbances' were eliminated. This is the basis of the experimental method that has given a tremendous impulse to the natural sciences and has permitted the establishment of numberless *physical laws* of a strictly deterministic type, from which exact predictions can be inferred. All this represents the merits of the *analytic method*.

Nevertheless, already at the end of the nineteenth century the limitations of this approach appeared in connection with the awareness of the impossibility of adopting this model for the treatment of *complex systems*. Non-linearity and several forms of 'indeterminism' are too well known to be recalled here. Therefore, the *synthetic approach* has emerged not at variance with, but as complementary to, the analytic approach, and has produced a wide investigation of complexity that is strictly cognate with GST. This situation has promoted important philosophical discussions regarding the meaning of natural *laws*, and the applicability of this concept also in other domains – like psychology, sociology, economics – as well as a deeper analysis of the notions of determinism and causality, that is, of fundamental ontological and epistemological issues. All this occurs apart from more specific problems of the philosophy of science, like, for instance, the proposal of admitting as 'explanation' of phenomena 'mechanisms' that describe 'how' they occur, rather than 'why' they occur: or the legitimacy of speaking of laws for single phenomena, just to mention a few examples.

13.7.3   Finalism

In system theory the concept of finality could receive a sense purified of any psychological flavour linking it with the intention or purpose set down by a subject (a meaning that, however, is perfectly legitimate in the study of human actions). This objective meaning of finality simply reflects the condition for qualifying something as a system: that it is an ordered totality of parts, endowed with properties that objectively contribute, thanks to a precise order of relations and correlations (and not to another one), to the existence of properties and functions of the global system. This is actually the classical notion of "final cause", which expresses the specific way in which a certain entity behaves because it has a specific nature. If we prefer, we could say that the final cause expresses the dynamic aspect of the nature of an entity. This type of causality can be amplified also to include the super-systems of a particular system, and in such a way can concern even the universe, as it was the case with Aristotle's doctrine of the "Immobile motor": this acts as supreme final cause and not as an efficient cause. It is due to the Judeo-Christian doctrine of God's creation that this was also seen as efficient cause, and this – as we have already noted – produced the diffidence of certain contemporary authors against the admission of final causes in science.

GST offers a conceptualization of finalism or "teleology" that is neutral and not entailing *per se* any "theological" consequence. However, this does not prevent one from taking this finalism as an objective feature present in the world that allows proposing *specific philosophical arguments* for proving the existence of God, whose force must be judged according to philosophical criteria.

It may be noted, in addition, that the notion of *propensity* introduced by Popper and taken up by recent scholars for the explanation of several phenomena in the natural and especially in the human sciences, is a rather patent recovery of the concept of final cause.

13.7.4   Interdiscipinarity

Already Bertalanffy had pointed out that the systemic approach can be applied in different domains and this idea was strongly reinforced when the notion of an 'open' system was extended not only to the existence of exchanges of matter and energy with the environment, but also of exchanges of *information*. Concepts such as feedback, regulation and self-regulation, together with all models elaborated in cybernetics, could be used for a significant improvement of the description of the interactions within systems and between systems and environment.

This means that the concept of system is *transdisciplinary*, that is, that it can be profitably used in different disciplines. The systemic approach, however, is equally important in every *interdisciplinary* research that is in the treatment of *complex problems*. By complex problem we do not mean a 'difficult' problem, but one in which different *aspects* of an issue must be taken into consideration. In these cases, the best strategy is that of making explicit the differences and specificity of the disciplines that can approach each aspect, with their specific criteria of investigation, of testing, of making arguments, and then to make the effort of making a certain translation and especially of finding correlations between these disciplinary results. The 'global' result will not be, and must not be, a 'unique' portrayal of the reality investigated (obtained by *reduction* to a single allegedly 'fundamental' discipline) but a multifaceted portrayal in which the contribution of every discipline can be appreciated because it 'contributes' to a better understanding *of the whole*.

13.8   Conclusion

Considering the enormous quantity of complex problems that are surfacing in our contemporary world, and which will increase in number and complexity in the coming future, we can conclude that a generalized adoption of a systemic way of thinking will be the more suitable intellectual attitude to be promoted in our societies. And this, for at least the past 50 years, has been a fundamental, and exhaustively elaborated, contention of Mario Bunge.

References

Agazzi, E. (2014). *Scientific objectivity and its contexts*. Dordrecht: Springer.
Bunge, M. (1979). *Treatise on basic philosophy. Vol. 4: A world of systems*. Dordrecht: Reidel.
Bunge, M. (1991). The power and limits of reduction. In E. Agazzi (Ed.), *The problem of reductionism in science* (pp. 31–49). Dordrecht/Boston: Kluwer.
Darwin, C. (1859/1958). *The origin of species*. New York: Mentor Books.

Chapter 14
Mechanism Models as Necessary Truths

Ingvar Johansson

Abstract The paper argues that there is a fruitful analogy to be made between classic pre-analytic Euclidean geometry and a certain kind of mechanism models, called *ideal mechanisms*. Both supply necessary truths. Bunge is of the opinion that pure mathematics is about fictions, but that mathematics nonetheless is useful in science and technology because we can go "to reality through fictions." Similarly, the paper claims that ideal mechanisms are useful because we can go to real mechanisms through the fictions of ideal mechanisms. The view put forward takes it for granted that two important distinctions concerned with the classification of fictions can be made. One is between ideal and non-ideal fictions, and the other between social and non-social fictions. Pure numbers, purely geometric figures, and ideal mechanisms are claimed to be ideal and social fictions.

For decades, I have been amazed by the way many seemingly erudite philosophers of science completely neglect or are completely unaware of the views of Mario Bunge. Once in 2014 and once in 2015, however, I became truly shocked by this negligence. I had optimistically thought that as soon as mechanisms came to be discussed, then, surely, Bunge's views and reflections must be considered by those interested. But no. Neither in the overview article "A Field Guide to Mechanisms" (Anderson 2014) nor in *Stanford Encyclopedia of Philosophy*'s entry "Mechanisms in Science" (Craver and Tabery 2015) is Bunge's name mentioned. That is, to take all doubts away, not mentioned at all! Leaving this highly remarkable fact as a future case study for the sociology of philosophy to explain, let me present his and my thoughts about mechanisms.

I. Johansson (✉)
Department of Historical, Philosophical and Religious Studies, Umeå University, Umeå, Sweden
e-mail: ingvar.johansson@umu.se

© Springer Nature Switzerland AG 2019
M. R. Matthews (ed.), *Mario Bunge: A Centenary Festschrift*,
https://doi.org/10.1007/978-3-030-16673-1_14

14.1   Bunge on Mechanisms and Models of Mechanisms

From the paper "Phenomenological Theories" (Bunge 1964) and onwards, Bunge
has stressed a distinction between black box theories and representational translucid
box theories. The essential difference between them is located in the concept of
mechanism: "In other words, a 'mechanism' linking I [input data] to O [output data]
is wanted in the translucid box approach" (Bunge 1964, p. 239).[1] Nowadays, Bunge
defines mechanisms as follows:

> *Definition 1.5* A *mechanism* is a set of processes in a system, such that they bring about
> or prevent some change – either the emergence of a property or another process – in the
> system as a whole. (Bunge 2003, p. 20)

Two of his claims about mechanisms are: "There is no method, let alone logic, for
conjecturing mechanisms" and "the covering-law model [of explanation] fails to
capture the concept of explanation used in the sciences, because it does not involve
the notion of a mechanism" (Bunge 2006, pp. 138 and 139). I agree.

Mechanisms must be kept distinct from *models* of mechanisms. "Mechanisms are
processes in concrete (material) systems," but models are "conceptual and semiotic
systems" (Bunge 2006, p. 129)[2] – and:

> The simplest sketch or model of a material system σ is the list of its composition,
> environment, structure, and mechanism, or
>
> $$\mu\,(\sigma) = <C\,(\sigma)\,,\,E\,(\sigma)\,,\,S\,(\sigma)\,,\,M\,(\sigma)>.$$
>
> Here, $C(\sigma)$ denotes the set of parts of σ; $E(\sigma)$ the collection of environmental items that
> act on σ or are acted upon by σ; $S(\sigma)$ the structure, or set of bonds or ties that hold the
> components of σ together, as well as those that link it to its environment; and $M(\sigma)$ the
> mechanisms, or characteristic processes, that make σ what it is and the peculiar ways it
> changes. (Bunge 2006, p. 126)

Only one of these four model-elements, $M(\sigma)$, models a mechanism. Bunge has not
tried to explicate the notion of $M(\sigma)$ as such. He has rested content with presenting
a number of mechanisms put forward in the natural and the social sciences. Late in
life, he has also brought in examples from medicine (Bunge 2013).[3] In this chapter

---

[1] Apart from Bunge (1964), see also in particular (Bunge 1997, 2004, 2006, Ch. 5). In a "Personal
Postscript" in Bunge (1997), he presents "a brief account of my struggle with the concepts of
mechanism and mechanismic explanations" (ibid. p. 458). For me, Bunge (1964) was a seminal
paper. The notion of mechanism has ever since the end of the 1960s played an important role in my
philosophical endeavors. In Johansson (2004 [1989], Ch. 14) it is central. Subchapter 14.3 has the
heading "Mechanisms and their parts." The views about mechanisms stressed in the present paper
were first outlined in Johansson (1997).

[2] In many contexts the expression 'material model' is used, but, as far as I know, Bunge uses it only
on one occasion (Bunge 1967, p. 146). I take him to mean that material systems become material
models only when being part of "conceptual and semiotic systems."

[3] A co-author and I have, partly influenced by Bunge (see footnote 1), ever since the 1990s stressed
the importance of distinguishing in medicine between *mechanism knowledge* and *correlation*

I will show that the mechanism concept is of such a character that we ought to distinguish between mechanism models that do and do not contain necessary truths, respectively.

This does not mean that I am interested in what exactly can be meant by necessary truths. In contemporary philosophy, the term is often avoided in favor of terms such as (depending on philosopher) 'formal truths', 'analytic truths', 'tautological truths', 'logical truths', 'conceptual truths', 'true in all possible worlds', or something else. What is important to me is the analogy between classic geometry-thinking and the kind of mechanism-thinking that I will present and defend. By Euclidean geometry I mean geometry as it was before Descartes invented the coordinate system. By means of this, large parts of geometry became amenable to arithmetic and algebraic treatment. That is, analytic geometry came into being; but I will be concerned only with classic pre-analytic geometry.

By the expression 'necessary truths' I mean statements that are true in the same sense – whatever that is – in which Euclid's five Postulates and five Common Notions (his axioms) except the fifth postulate (the so-called parallel axiom) are still regarded as in some sense being necessary truths (Euclid 2017, p. 2). When from such axioms theorems are validly deduced, the theorems are of course necessary truths, too.[4]

Let me take two examples, and at the same time (in contradistinction to Euclid) make the necessity claim explicit. Euclid's first Postulate says that, necessarily, between any two points there is a straight line, and his first Common Notion says that, necessarily, things which are equal to the same thing are also equal to one another. It is statements like these that I claim my forthcoming proposed mechanism axioms are analogous to. From a subjective point of view, a necessary truth says that this is the way the things talked about have to be; they cannot possibly be different.

When using the term 'proposition', I will use it in Bunge's sense of a man-made construct and non-Platonic entity (Bunge 1974b, pp. 85–86). A necessary truth can then be represented by the form 'it is true that, necessarily, p', where p is a variable for propositions. Factual (contingent) truths can be given several forms. I will use only the strictly singular and the strictly universal, i.e., 'it is true that, at time t and place x, p' and 'it is true that, always, p', respectively.

---

knowledge. See Johansson and Lynøe (2008, Ch. 6); we did earlier put forward the distinction in similar but much smaller books in Swedish in 1992 and 1997, and in Danish 1999.

[4]What today are called theorems, Euclid calls *propositions*. In my discussion I disregard the fact that Euclid's first three Postulates are not statements that describe a state of affairs as existing; they state that it is always possible to *draw*, *produce*, and *describe*, respectively, a certain kind of geometric figure.

14.2   The Cogwheel Mechanism

My first example of a mechanism is the cogwheel mechanism. In models of systems that have cogwheels as their characteristic mechanism, $C(\sigma)$ denotes a number of cogwheels with their axles, $E(\sigma)$ denotes something that makes at least one of the cogwheels rotate, and $S(\sigma)$ denotes the structure which keep the wheels in place and somewhere makes the cogs of one wheel fit into those of at least one other. But what does the very mechanism model $M(\sigma)$ denote? According to Bunge, it denotes "a set of processes in a system, such that they bring about or prevent some change."

Let us think of a mechanism consisting of two connected circular cogwheels whose axles are paralleled (Fig. 14.1).

When the wheels are equally large, I will label it *the simple cogwheel mechanism*, and call the two wheels W_1 and W_2, respectively. When nothing else is said, it is taken for granted that something external to W_1 makes W_1 rotate, and that W_2 moves because of W_1. One single rotating cogwheel cannot be a mechanism; it constitutes merely a process. As Bunge says: "Every mechanism is a process, but the converse is false" (1997, p. 416). A possible practical function of this mechanism is to change rotation direction from one axle to another.

As just indicated, when the wheels in the simple cogwheel mechanism rotate, they rotate in opposite directions. The question I will discuss is whether or not this is a necessary truth. Since, normally, the mechanism in question is part of a material system, the first-hand answer seems to be negative. Why? Because no descriptions of a material system can contain necessary truths. But let us take a closer look at the issue; to start with, let us memorize these two factual hypotheses:

The singular hypothesis H_S: At place x and time t, when in the simple cogwheel
   mechanism wheel W_1 makes a single turn in one direction, wheel W_2 makes a
   single turn in the other direction.
The universal hypothesis H_U: Always, when in the simple cogwheel mechanism
   wheel W_1 makes a single turn in one direction, wheel W_2 makes a single turn in
   the other direction.

So far, nothing has been said about the material of the cogwheels. It has been abstracted away, even though all real cogwheels must be made of some kind of

Fig. 14.1 Circular
cogwheels

material and have a mass. Now, assume as a first specified case, that wheel W_1 is made of steel and W_2 of fragile glass. In all probability, if W_1 starts to rotate, it will crush a number of cogs on W_2, and when W_1 has made a whole turn, W_2 has not. That is, H_S is in this case false, and the universal hypothesis H_U is falsified.

Secondly, assume that the glass wheel (W_2) is exchanged for a very soft wool wheel. This time the cogs of W_2 will not be crushed, only bent. However, the result is similar to the glass case. When W_1 has made a whole turn, W_2 has in all probability not. Therefore, even in this case H_S is false and H_U falsified. Moreover, there are with such wheels even more conspicuous ways of falsifying H_U. Assume that the wheels W_1 and W_2 are by different force sources made to rotate in the *same* direction. Is this possible? Yes, since the steel cogs will bend the wool cogs.

Thirdly, assume that both the wheels are made of steel. Now, at last, H_S (with properly chosen values for x and t) becomes true. However, we are not allowed to extrapolate and claim that H_U is true as well, which would falsely mean that the process could go on for an infinite number of rotations. In the long run, due to wear and tear, the mechanism will start to malfunction and not rotate properly; and in the very long run, due to rust and other kinds of decay, it will stop function at all. That is, H_U will in the end become falsified even by the pure steel wheel mechanism.

Assume, fourthly, that both the cogwheels are *absolutely rigid bodies*, i.e., they are bodies that:

(i) cannot possibly change shape
(ii) cannot possibly change size
(iii) cannot possibly lose parts or take up new parts
(iv) cannot possibly be destroyed.

By definition, no parts of such wheels can be broken (like the cogs of the glass wheel) or bent (like cogs of the wool wheel), and the mechanism process is not susceptible to wear and tear (as steel wheels are). In fact, absolutely rigid bodies are analogous to the absolutely non-changing two-dimensional figures (triangles, squares, circles, etc.) and three-dimensional solids (pyramids, cubes, spheres, etc.) of Euclidean geometry. Often, such Euclidean objects are called *ideal figures* (plane geometry) and *ideal solids* (solid geometry), respectively; and I will follow suit.

The essential difference between the notions of *ideal solid* and *absolutely rigid body* is that the former notion does not bring in temporal duration, whereas duration is essential to the definition of the latter. To be rigid is defined as not being changeable during a certain time period. Euclidean solids are *atemporal* objects, while rigid bodies are *temporal* objects. An absolutely rigid body is a solid that, necessarily, endures without changing anything else than spatial position. I will also call it an *ideal body*. An ideal figure has a two-dimensional shape, an ideal solid has a three-dimensional spatial shape, but an ideal body has a shape that is extended not only in three spatial dimensions but also in a temporal dimension; i.e., an ideal body can be ascribed a four-dimensional shape.

The generic notion of *ideal objects* will for a while simply be taken for granted, but in later sections the first part, *ideal*, will be explicated. If I were not allowed to insert this term in the necessity-claims I will put forward, I guess they would look extremely odd. But the second part, *object*, is a primitive and undefined notion in the wide sense that Bunge speaks of objects (Bunge 1974a, p. 26). Both property bearers and properties are objects in this sense, and I will say no more about it.

The statement A $_{CM}$1 below I regard as a first axiom (A) in what I would like to call a *mechanism geometry* for cogwheel mechanisms (CM). In none of the axioms and theorems that follow are any operating forces mentioned; otherwise my term 'geometry' would be wide off the mark. Here is the first necessity-claim:

(A_{CM}1) Necessarily, two directly connected ideal circular cogwheels do when rotating rotate in opposite directions.

This implies that if the rotational forces in a real corresponding mechanism with very rigid cogwheels are such that the wheels are made to rotate in the *same* direction, then this can nonetheless not happen. Instead, something, whatever it is, must happen to the force sources or what links these to the cogwheels.

If we change the mechanism by making one of the circular wheels larger than the other, then we can state another axiom:

(A_{CM}2) Necessarily, if two directly connected ideal circular cogwheels are not equally large, then the one with the shortest diameter must when rotating rotate faster than the other.

Extending the first mechanism to three circular cogwheels we can derive the two theorems (T) below. I regard them as being analogous to, for example, Euclid's theorem that the three interior angles of a triangle are equal to two right angles (Euclid 2017, p. 24; proposition 32). Here they are:

(T_{CM}1) Necessarily, in a linear chain of three connected ideal cogwheels, the third rotates in the same direction as the first.[5]

(T_{CM}2) Necessarily, three mutually and directly connected ideal cogwheels cannot rotate.[6]

Extending the mechanisms to non-circular cogwheels we can find quite a number of non-possibility axioms of the following kind:

($A_{CM}n$) Necessarily, a circular ideal cogwheel cannot form a rotational mechanism together with an ideal "cog-square."

For illustrations of non-circular cogwheels that *can* constitute a mechanism, see Fig. 14.2.

Just like Euclid's axioms and theorems, the ones that I have presented do not contain numerical variables for real numbers. This is not an accidental manoeuver of mine. Since the axioms are concerned with rotating *shapes*, and this property

---

[5] According to A_{CM}1, if W_1 rotates clockwise, W_2 rotates anti-clockwise, and if W_2 rotates anti-clockwise, W_3 rotates clockwise just as W_1 does.

[6] According to A_{CM}1, if all the three wheels are mutually connected and rotating, then one of the wheels must be able simultaneously to rotate in two opposite directions, which is a contradiction.

Fig. 14.2 Non-circular cogwheels

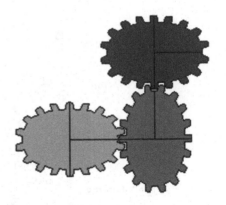

dimension has not yet been quantified the way property dimensions such as length, mass, and energy have, they cannot at the moment be given the form of mathematical equations.[7] This fact is not in conflict with Bunge's views about mechanisms; according to him, mechanisms can be modeled without equations (Bunge 1997, p. 423).

My simple examples of cogwheel mechanisms can easily be complemented by much more complex mechanisms with ideal bodies of both circular and non-circular shapes. Mechanical clocks are the first examples that come to mind (Fig. 14.3). However, for the purposes of this paper, I need not discuss complex mechanisms. My present aim is only to defend this minimal thesis: *certain mechanism models contain necessary truths*.

I will, however, in the next two sections show that ideal bodies are not the only kind of ideal objects that can figure in mechanism models.

14.3 Archimedes' Water Screw Mechanism

Let us take a look at Archimedes' famous water screw (Fig. 14.4). Normally, it is presented as a device for moving water, and it consists of ordinary rigid bodies that do not let water through. When the screw is made to rotate clockwise, the water moves from the lower basin to the higher. But what is here true of water is true of most liquids, so in what follows I will call it Archimedes' liquid screw mechanism.

The same kind of thought variations of the material that I made in relation to the cogwheel mechanism, can also be made in relation to Archimedes' screw mechanism. I divide the mechanism into three parts: the cylinder, the screw, and

---

[7]In fact, I have the much stronger view that the property dimension of shape cannot be quantified. See Johansson (2011), where a proof to this effect is given. So far, no one has been able to find a mistake in the proof, which relies on transfinite mathematics.

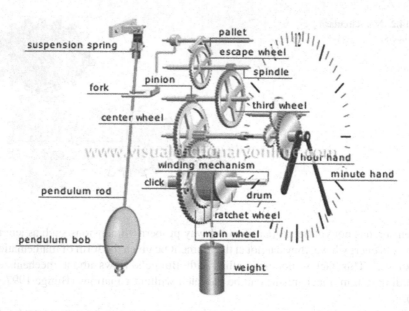

Fig. 14.3 Clock mechanism

Fig. 14.4 Archimedes' screw

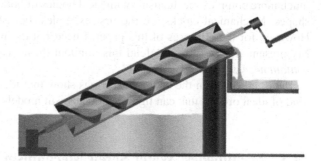

the liquid. When within the screw-in-the-cylinder, the liquid takes on the shape of a spiral formed pillar.

First case: the cylinder and the screw are made of a weak and porous material that lets many liquids through, and the liquid is water. In this case, not all of the water will be moved from the lower to the higher place; partly because the screw may break, and partly because some of the water will move through the cylinder into the outside of the mechanism. That is, in this case the mechanism doesn't function properly even though part of the liquid will be translocated.

Second case: the cylinder and the screw are made of steel, but the liquid has such a high viscosity (= resistance to shape deformation) that it is incapable of taking on the spiral shape it has to have in order to fill the volume between the screw and the cylinder. That is, neither in this case does the mechanism function properly, even though part of the liquid will be translocated.

Third case: the cylinder and the screw are again made of steel, but the liquid has now such an extremely low viscosity that some of it falls apart into droplets when it is to enter the screw. Again, the mechanism doesn't function properly, even though part of the liquid will be translocated.

Now let us see whether we can find objects that are analogous to the ideal cogwheels, i.e., objects which make Archimedes' mechanism function properly by necessity.

The first change to make is quite obvious: let us regard both the cylinder and the screw as being absolutely rigid (ideal) bodies without holes. But, in fact, a similar change can be introduced in relation to the liquid-volume that is meant to be moved through the cylinder. I will alternately call this idealization an *absolutely malleable liquid-volume* and an *ideal liquid-volume*.

Liquid-volumes are often regarded as being the same and as having the same volume when they are poured from one kind of container into another; for instance, from a pot to a bottle, or from a bottle to a glass. The liquid-volume is regarded the same in spite of the radical shape changes it undergoes in the process. By stipulation, I now define an absolutely malleable (ideal) liquid-volume as a liquid-volume that:

(i)   can take on any shape
(ii) cannot possibly change volume (size)
(iii) cannot possibly lose parts or take up new parts
(iv) cannot possibly be destroyed.

The defining characteristic (iv) secures that the individual liquid-volume in question has an enduring individual (numerical) identity during the processes it partakes in. If we regard liquids as aggregates of molecules, then neither statement (iii) nor (iv) can possibly be true. That is, however, beside the point, since it is an ideal object that is defined.

Characteristic (i) introduces a feature that cannot be accepted if ideal objects are regarded as Platonic objects; by definition, Platonic objects are atemporal, and so by definition non-changeable. However, when ideal objects are regarded as fictions in the way I will propose and defend in later sections, then even temporal ideal objects become acceptable.

If we conceive of Archimedes' liquid screw mechanism as consisting of absolutely rigid (ideal) bodies that move an absolutely malleable (ideal) liquid-volume, we have in thought constructed what adequately can be called Archimedes' *ideal* liquid screw mechanism (AM). About this we can truly state this axiom (A):

(A_{AM}1) Necessarily, when the ideal screw is rotating in the ideal cylinder, the ideal liquid-volume moves from one of the possible positions to the other.

Note that in the axiom there is no mention of a lower and a higher place between which the liquid is moved. The reason is that such talk would bring in associations to gravitational forces that make the liquid of itself move downwards. But all the objects involved are assumed to be as massless, colorless, and outside all causal factors just as the figures and solids of Euclidean geometry are.

14.4   The Piston Mechanism

The steam engine was central to the first industrial revolution, and the internal combustion engine to the second. Common to both kinds of engine is that they contain – and essentially so – a piston mechanism, see Fig. 14.5. The piston itself is contained in a cylinder in which it can move back and forth. There is a gas in the cylinder, and when it expands (in the internal combustion engine because the gas is ignited and explodes) the piston moves. It is only this movement that, in the light of the preceding sections, I will say some brief words about.

Of course, whether a real piston mechanism will function or not depends on the materials of the cylinder and the piston, and on what the gas is like. This time, in contradistinction to the cogwheel and the liquid screw mechanisms, I will immediately ask whether in relation to the mechanism it is possible to introduce ideal objects that give rise to a necessary truth.

To begin with, the cylinder and the piston can be regarded as absolutely rigid (ideal) bodies without holes, and I postulate that the piston without friction can freely move inside the cylinder. But what about the gas? Let me introduce still another idealization. By stipulation, I define an *absolutely gaseous (ideal) gas-substance* as an object that:

 (i)   can take on any shape
 (ii) can take on any volume (size)
(iii) cannot possibly lose parts or take up new parts
(iv)   cannot possibly be destroyed.

As in the ideal liquid-volume case, characteristic (iv) secures that an individual ideal gas-substance has an enduring identity during the processes in which it partakes. And, again as with liquid-volumes, if we regard gases as aggregates of molecules, then the statements (iii) and (iv) cannot be true. However, the gas-substance defined is an ideal object whose parts are left outside the definition. This means that ideal gas-substances cannot be identified with aggregates of particles (e.g. molecules), which, in turn, means that the notion of an *ideal gas-substance* must not be conflated with the notion of an *ideal gas* (such a gas is constituted by perfectly elastic collisions between particles).

Fig. 14.5 Piston mechanism

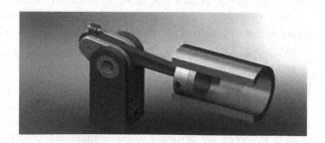

When an ideal gas-substance expands in the ideal cylinder-with-piston mentioned, then, necessarily, the piston moves in the cylinder. The process can be given the form of a piston mechanism (PM) axiom (A):

(A_{PM}1) Necessarily, when the ideal gas-substance expands in the ideal cylinder, the ideal piston moves as long the gas-substance expands.

From an epistemological point of view, the necessary truths A_{CM}1, A_{AM}1, and A_{PM}1 are on a par; and they are trivially true. In the first case there are only two ideal bodies, in the second there are two ideal bodies plus an ideal liquid-volume, and in the third there are two ideal bodies plus an ideal gas-substance. Since neither ideal liquid-volumes nor ideal gas-substances have an enduring or changing shape of their own, they must be combined with ideal bodies in order to be able to have a specific function within a specific mechanism. By combining absolutely rigid bodies, absolutely malleable liquid-volumes, and absolutely gaseous gas-substances, one can create models of many ideal mechanisms; models that, like the ones presented, can contain necessary truths.

The definitions of ideal bodies, liquid-volumes, and gas-substances given do not allow them to be transformed into one another. But, of course, transformations of real bodies, liquids, and gases into each other are central to many real mechanisms. The purpose of this paper, however, I remind the reader, is not to discuss and understand mechanisms in general, only to defend the minimal thesis that certain mechanism-models contain necessary truths.

Now time has come to explicate the generic notion of ideal objects that I am using.

14.5 To Reality Through Fiction

The heading above is stolen from a heading by Bunge (2006, Ch. 8), and the reason behind my "theft" will become clear some paragraphs below.

Mostly, Bunge says (as already quoted): "Mechanisms are processes in concrete (material) systems." In his *Definition 1.5* (as also already quoted), on the other hand, he is more general and says only: "A *mechanism* is a set of processes in a system." This definition allows one to speak not only of concrete mechanisms, but also of *ideal mechanisms*. In the spatiotemporal world, I take it for granted, there are no absolutely rigid (non-subatomic) bodies, absolutely malleable liquid-volumes, or absolutely gaseous gas-substances. This is the reason why the necessary truths I have presented are about ideal objects. In what way then does the naturalist Bunge conceive of ideal objects?

Throughout the history of philosophy, pure mathematical numbers have been reckoned prototypical ideal objects, and as such also reckoned prototypical Platonic objects; that is, being regarded as existing non-spatially and non-temporally in a non-changing realm distinct from our common spatiotemporal universe and from our minds. By definition, no naturalist can accept Platonic objects, so what is

Bunge's view of pure numbers? He claims they are fictions of a certain kind. I quote:

> Still, I submit that fictionism, while utterly false regarding factual science, is quite true concerning pure mathematics. [...] Consequently, the concept of existence occurring in mathematical existence theorems is radically different from that of real or material existence. [...] In short, mathematicians, like abstract painters, writers of fantastic literature, "abstract" (or rather uniconic) painters, and creators of animated cartoons deal in fictions. To put it into blasphemous terms: ontologically, Donald Duck is the equal of the most sophisticated nonlinear differential equations, for both exist exclusively in some minds. (Bunge 2006, p. 192)

My own views on pure numbers are similar, even though they differ somewhat. These differences, however, are of no importance for this paper, and will not be discussed.[8] I will at the moment rest content with using Bunge's views on numbers as a foil for my discussion of ideal mechanisms. One of my points about the latter is analogous to Bunge's about numbers. I claim that *we can go to real mechanisms through fictional mechanisms*.[9] Or, in other words, we can sometimes obtain a kind of knowledge of features of real spatiotemporal mechanisms by means of necessary truths about ideal fictional mechanisms. How I think this works will be explained in a later section; first more words about numbers.

In the world we perceive, it is impossible to point at pure numbers such as 1, 2, 3, 4, 5, etc. We can only point at their symbols, i.e., the corresponding numerals. I have just used the first five numerals of the decimal system (which has ten as its base); if I had used the binary system (having the base two) I should have written "pure numbers such as 1, 10, 11, 100, 101, etc." One and the same number may in number systems with different bases be represented by different numerals. Before mathematics was invented, there were not even numerals to point at. Conclusion: numbers are not mind-independent spatiotemporal objects.

For thoroughgoing naturalists, like Bunge and myself, who find both nominalism, psychologism, and many-worlds thinking incoherent, talk of pure numbers must be regarded either as complete nonsense or as being about fictional objects of some kind; that is, be about man-made objects without being reducible to mental states. But to regard mathematics as nonsense is of course out of the question. Therefore, the only option left is to regard pure mathematical numbers as being fictional objects of some sort. Alternately, Bunge also calls *fictional* objects *abstract* or *conceptual* objects: "mathematics [. . .] deals exclusively with ideal (or abstract or conceptual) objects" (Bunge 2006, p. 28).

In conformity with Bunge's views, I will now introduce two distinctions. First, Bunge's general bipartite distinction between *real objects* and *conceptual objects*

---

[8]My views are put forward in Johansson (2013, 2015). The way I differ from Bunge is presented in sect.2.2 of the first paper.

[9]There have in this decade been some discussions about whether certain scientific models should be regarded as fictions or as analogous to fictions See for instance Contessa (2010), Frigg (2010), Toon (2010), Morrison (2015), and Bueno et al. (2018). But in none of these papers and books is my thesis that certain models can contain necessary truths put forward.

(Bunge 2006, p. 27). Real objects exist in real space and time, but purely conceptual objects exist only in fictional discourses. The latter objects have as such no *real* spatiotemporal existence, even though the fictional discourse in which they exist may allow a fictional spatiotemporal framework; normally, fictional novels tell a narrative that unfolds in both space and time.

Second, I will divide conceptual objects into two sub-kinds, ideal and non-ideal (e.g. literary) fictions. According to Bunge, real, ideal, and literary objects can when referred to in propositions figure in three corresponding kinds of truths: *factual*, *formal*, and *artistic* truths, respectively (Bunge 2006, p. 194).

Bunge claims that "Logic, philosophical semantics, and mathematics are formal sciences," and that as such they can contain formal truths (Bunge 2006, p. 193). If I were to use this terminology, then I would have to say that the mechanism axioms I have presented are formal truths. However, in an obvious sense they have contents. Therefore, I have chosen to call them necessary truths in spite of the fact that they must be reckoned formal truths in Bunge's conceptual apparatus.

The distinction between ideal and non-ideal fictions is of utmost importance in order to understand Bunge's (and my) conception of the ontological status of numbers. Of course, if both pure numbers and fictional figures in novels and cartoons are fictions, then there must be some radical and important difference between them. As repeatedly pointed out in the philosophical discussions of fictions in literature and cartoons, conceptual and pictorial objects like Donald Duck are *ontologically indeterminate*. That is, as long as the creators of Donald Duck do not, for example, tell exactly how old he is, what his weight is, etc. – and that by so-called conversational implicatures such descriptions do not follow from what has been explicitly said – there simply are no corresponding facts of the matter. The natural numbers, on the other hand, cannot possibly have such ontological spots of indeterminacy.[10] No mathematician can freely add a feature to the posited natural numbers the way the creators of Donald Duck can freely fill in what hitherto has been a spot of indeterminacy.

This difference between ideal and non-ideal fictions does by no means imply that number theory cannot contain unsolvable problems. Trivially, where there are spots of ontological indeterminacy there are unsolvable epistemological problems; but the converse is not true.

Bunge says as follows: "Indeed, it [his kind of fictionism] distinguishes between mathematical fictions on the one hand and myths, fairy tales, theological specula-tions, abstract paintings, parapsychological and psychoanalytic phantasies, as well as many-worlds philosophical theories, on the other" (Bunge 2006, p. 193). After having asked "how does the fundamental theorem of algebra differ from the claim that Superman can fly or Mickey Mouse can speak?," he lists 12 points in which he regards the difference to manifest itself (Bunge 2006, p. 204), but for the purpose of this paper they need not be discussed.

---

[10]The expression 'spots of indeterminacy', which I like very much, is taken from Roman Ingarden; see Johansson (2010, p. 94, 2013, pp. 29–30).

Briefly put, Bunge's view that we can go "to reality through [mathematical] fiction" can be reached by an inference of the following kind:

(P1) Modern physics gives us knowledge of mind-independent reality
(P2) In much of modern physics, mathematics is an indispensable part of the content
(P3) Pure numbers are fictions

Hence:

(C) We can go to reality through fictions.

I have no qualms about the inference. However, the view is in need of a credible description of what such pure-numbers-to-reality transitions are like. One problem can be stated thus: natural laws are *not* necessary truths, but mustn't applications of mathematical necessary truths give rise to new necessary truths? Bunge does not say much about the transition, but I will outline an answer that I find both reasonable and as not being contradicted by anything I know Bunge has claimed.

Look at the following simple true arithmetic statement: 'necessarily, $1 + 1 + 1 = 3$'. Since the numbers are not to be found in spatiotemporal reality, there must be some intermediary level that is at work before we can state, for example: 'necessarily, 1 tomato + 1 tomato + 1 tomato = 3 tomatoes'.

In order to get a concrete case, let us think of a child called Adam, who has just learned to count and add a little. His parents want to profit from his new abilities, and they ask him to go to the nearby shop and buy three tomatoes. The following scenario then unfolds.

In the shop, Adam goes to a shelf full of tomatoes and takes first one tomato and puts it in his bag, then a second, and so at last a third. When doing this he says to himself: "one tomato, plus one tomato, plus one tomato; now I have three tomatoes." If we symbolize this real-world adding by the symbol '$+^r$' and the real-world result by '$=^r$', then we can represent Adam's procedure by the expression '1 tomato $+^r$ 1 tomato $+^r$ 1 tomato $=^r$ 3 tomatoes'. This *physical addition* is in contradistinction to the corresponding arithmetic addition not a necessary truth, which will soon become evident.[11] When Adam comes home, and proudly wants to show that he has been able to do what was expected of him, it turns out that on his way home his arms have been too movable; the bag has bumped into various walls and posts. In the bag there is now only a certain amount of tomato porridge; no distinct tomatoes can be discerned.

In the preceding sections I have introduced non-mathematical idealizations, and such a move can be made even here. By stipulation, I now define an *ideal tomato* as a tomato that:

 (i) can take on some (but not all possible) different shapes
 (ii) can take on some (but not all possible) different sizes

---

[11] The notions of *physical addition* and *physical sum* can be found in Bunge (1967, pp. 199–202). I have earlier discussed physical and pre-mathematical addition in Johansson (1996).

(iii) can lose some parts and take up some new parts

(iv) cannot possibly be destroyed (and so lose its individual identity).

So defined, the fourth characteristic of an ideal tomato makes it, like Plato's atemporal tomato-idea, an indestructible tomato, but unlike Plato's idea it exists in a temporal framework. Nonetheless, as can be seen from the corresponding definitions, ideal tomatoes differ radically from absolutely rigid bodies; the former can change, the latter cannot. If Adam had been able to pick ideal tomatoes and put them into his bag, then, surely, he would have brought home three distinct tomatoes independently of what his bag would have bumped against during the walk.

As far as I can see, the best way to understand the transition from the purely arithmetic expression '$1 + 1 + 1 = 3$', which represents a necessary truth, to the expression '1 tomato $+^r$ 1 tomato $+^r$ 1 tomato $=^r$ 3 tomatoes', which at best represents a contingent truth, is to introduce *additions of ideal objects*. Instead of distinguishing only between, on the one hand, pure arithmetic numbers and arithmetic addition, and on the other real-world tomatoes and physical addition, we should posit intermediate strata and additions of objects within these. I think there are four kinds of proposition strata and three kinds of between-strata relations to be taken into account:

(d) Necessarily, $1 + 1 + 1 = 3$

(c) Necessarily, 1 indestructible object $+1$ indestructible object $+1$ indestructible object = 3 indestructible objects

(b) Necessarily, 1 ideal tomato $+1$ ideal tomato $+1$ ideal tomato = 3 ideal tomatoes

(a) Contingently, 1 tomato $+^r$ 1 tomato $+^r$ 1 tomato = ^{r} 3 tomatoes

Top-down, the relation between the strata (d) and (c) is a straightforward case of *application*; if you apply a necessary truth, the result should also be, as here, a necessary truth. Bottom-up, the relation can be seen as a kind of *abstraction*; the indestructible objects are abstracted away in the sense of being disregarded. There are interesting things to discuss about both this top-down and this bottom-up relation, but I will not delve into these.[12] The relation between stratum (b) and stratum (c) is that of logical *subsumption*; an ideal tomato is simply by the definitional characteristic (iv) made into a kind of indestructible object.

The most interesting question for this paper is this: How can there be a relation between the necessary truth (b) and the factual truth (a)? My answer is: because real objects can be *approximations* of ideal objects. The term 'approximation' is meant in a wide and informal sense. Ideal tomatoes are absolutely indestructible, but all real tomatoes can be destroyed. In this sense, there is an ontological gap between them. Nonetheless, different real tomatoes can be ranked as in a certain situation being more or less easily destroyable, and also as in and of themselves being more or less prone to decay. Therefore, despite the ontological gap, certain

---

[12]I discuss them in Johansson (2015, sects. 6 and 7), and I think they point towards what might be called a property view of the natural numbers.

real tomatoes can be more similar to the ideal tomato than some others are, i.e., be better approximations of the ideal tomato than the others.

The view I have presented must not be conflated with the view that we regard some real tomatoes *as if* they *are* ideal tomatoes. When I count real tomatoes, I do not look upon them as if they are indestructible; I look upon them as for some time probably enduring, i.e., for some time probably approximating ideal tomatoes.

My explanation of what it looks like to go from purely arithmetic (ideal and fictional) addition to physical addition can easily be extended from the noun 'tomato' to all so-called count nouns. Now, whether this explanation is right or wrong, I will take it for granted when later I will look at the relationship between ideal and real *mechanisms*. In the next section, I will in passing say some words about approximations between factual theories that contingently are true or false. They cast some further light on the informal approximation notion I rely on.

14.6 Interlude: Bunge and Popper on Approximations to Factual Truths

In the twentieth century, it was quite common among physics teachers to say things such as these: (i) from the perspective of special relativity, Newtonian mechanics still gives rise to *approximately true* predictions for bodies whose velocity is small compared to the velocity of light, and (ii) from the perspective of Newtonian mechanics, Galileo's law for falling bodies still gives rise to *approximately true* predictions for bodies falling near to the earth. In spite of this, mainstream analytic-philosophical philosophy of science was very skeptical to notions such as 'approximate truths', 'partial truths', 'degrees of truth', and 'truthlikeness'. Bunge and Popper are the two outstanding great dissidents from this orthodoxy, if I may call them so. They think, as I do, that a notion of approximation between factual propositions/theories is not only meaningful, it is also needed in order to understand and further the development of science. I quote Bunge:

> Contrary to a widespread opinion, scientific realism does not claim that our knowledge of the outer world is accurate: it suffices that such knowledge be *partially true*, and that some of the falsities in our knowledge can eventually be spotted and corrected. (Bunge 2006, p. 30; italics added)

That a factual statement is partially true means, that when it is compared to a factual statement that is (for the moment) regarded as literally true, it (i) is literally false, but (ii) has nonetheless quite a similarity to the true statement.

Looking back at the philosophy of science since the 1960s, I have the impression that Bunge's conception of approximate truth did not give rise to such a widespread discussion that Karl Popper's for some decades did. My explanation is this.

In Bunge's book *Scientific Research*, the general notion of partial truth is rather informal: "It is only some logicians who still oppose the very idea of partial truth, as a result of which we continue using an intuitive, presystematic concept of partial truth" (Bunge 1967, p. 301). Later, in his *Treatise of Basic Philosophy*, he gives an implicit definition of the formal features of 'degrees of truth' by means of a number of postulates (Bunge 1974b, Ch. 8.3.2-3).

Popper, on the other hand, did for a long time (probably from the end of the 1950s to the end of the 1970s) wrongly believe that much more could be done. During this period, he claimed that the notion of approximate truth could be given an explicit definition and well-defined measure, too. This claim gave rise to wide discussions. Most philosophers of science at the time thought that without a precise explication the expression 'approximate truth' would be meaningless. When, at last, consensus was reached that such a definition or measure is impossible to construct, most of those involved in the discussion dropped the whole notion. But not Popper. He still defended the importance of retaining the original informal notion. For this history, and a further elaboration and defense of the informal notion of approximate or partial factual truths, see my (Johansson 2017).

In the earlier sections, I have without any arguments taken an informal notion of approximation for granted, but the arguments needed can be found in the paper just mentioned. In the sections below, I will continue to use the same informal notion. I think the notions 'approximate truths' and 'objects approximating each other' imply each other. If there are statements that are approximately true, they must contain some concepts whose referents are only approximations of these concepts; and if there are concepts whose real-object referents are only approximations of certain concepts, then statements containing these concepts can be created.

Let me add that I regard the notion of approximate truth as a *complement* to the simple binary opposition between being true and being false. The latter opposition is needed in order to understand what a valid deductive inference is, the former notion is needed in order to understand what an empirical-scientific development is. I end this section with another quotation from Bunge:

> And yet, paradoxically, no one seems to have produced a detailed, true, and generally accepted theory of objective and partial truth. We only have a few insights into the nature of factual truth, the way its knowledge emerges from tests, and the confluence of truths attained in different fields. (Bunge 2003, p. 249)

14.7 To Real Mechanisms through Ideal

Models and descriptions of real mechanisms can only be contingently true, but models of ideal mechanisms can, as I hope to have shown, contain necessary truths. How, then, can the latter kinds of models shed light on real mechanisms? My answer aligns of course with my views on how arithmetic truths can be applied to ordinary kinds of objects. Bunge says:

> models are seldom if ever completely accurate, if only because they invite more or less brutal simplifications, such as pretending that a metallic surface is smooth, a crystal has no impurities [etc.]. These are all fictions. However, they are all stylizations rather than wild

> fantasies. Hence, introducing and using them to account for real existents *does not commit us to fictionism.* (Bunge 2006, pp. 189–190; italics added)

Here, unusually, I think Bunge moves to fast. Whereas he claims that fictionism is true for mathematics, he can in the quotation easily be interpreted as denying that fictionism can ever be true for models. In my opinion, what should be said is this:

> Fictionism is as true for *ideal* mechanisms as it is for arithmetic addition, and it is as false for *real* mechanisms as it is for physical addition.

That Bunge himself doesn't say so is, I guess, due to the fact that he never has thought about the possibility of fictional but nonetheless ideal mechanisms.

Before proceeding, let's go back to Plato. Even he can be said to have worked with a relation of approximation. He distinguished between a realm of ideas and the spatiotemporal sensible world in which we normally live. The spatiotemporal objects, he claimed, do *participate more or less* in corresponding objects in the world of ideas. When this assumption is accepted, it becomes natural to say that spatiotemporal objects are better or worse approximations of the posited ideas or abstract objects. Euclidean geometry affords a good example. In the idea of the circle the circumference is exactly the diameter multiplied by π, but in all spatiotemporal circles this is only approximately the case. Similarly, in the idea of a triangle the sum of the interior angles is exactly $180°$, but in all spatiotemporal triangles this is only approximately so. Actual macro- and mesoscopic spatiotemporal circles and triangles can only more or less approximate the ideal figures in the realm of ideas (I leave it for physics to decide about the shapes of microscopic particles).

On the approach I am defending, there is no problem in handling real everyday spatiotemporal circles and triangles the way I have analyzed additions of spatiotemporal tomatoes. Now, however, the strata (c) and (d) are not needed. The schema below is enough:

(b) Necessarily, in *ideal* triangles, first angle + second angle + third angle $= 180°$
(a) Contingently, in *real* triangles, first angle $+^r$ second angle $+^r$ third angle $=^r$ $180°$

With this in mind, let us look at mechanisms. Even though there are no absolutely rigid bodies, absolutely malleable liquid-volumes or absolutely gaseous gas-substances in the spatiotemporal world, there are rigid bodies, malleable liquids, and gaseous substances. Moreover, such rigidity, malleability, and gaseousness can take degrees. Therefore, these properties can more or less approximate the corresponding absolute property. For instance, the steel wheel in the cogwheel example is closer to absolute rigidity than the wool wheel and the glass wheel are. In the mechanism examples presented, there is no mathematics, so there is no analogy to be made with the four-strata arithmetic tomato example, but the two-strata Euclidean triangle example affords the analogy searched for:

(b) Necessarily, two directly connected *ideal* circular cogwheels do when rotating rotate in opposite directions.

(a) Contingently, two directly connected *real* circular cogwheels do when rotating rotate in opposite directions.

Since real cogwheels can approximate ideal cogwheels, there is, when approximation obtains, a relation between the strata (a) and (b). Let it be noted, that since the ideal wheel is a man-made fiction, the approximation relation is here a relation between something real and something fictional. The relation is an internal relation in the sense that *if* both the relata exist, then, necessarily, the approximation relation obtains; it then comes so to speak for free. Compare the relation between these two spots: ● ·. *Given* the existence of both, necessarily, the spot to the left is larger than the one to the right, i.e., the relation *larger than* is an internal relation in the sense meant.

Of course, as there can be approximation relations between *fictional ideal* and real mechanisms, there can also be such relations between *fictional non-ideal* mechanisms and real mechanisms. An inventor can first invent in thought either an ideal mechanism or a non-ideal mechanism, and then later try to create in the world an approximately similar mechanism. I have in this paper chosen to stress fictional ideal mechanisms, but in technology, probably, fictional non-ideal mechanisms are of more importance.

What has just been said about mechanisms in relation to inventors, is also applicable to researchers. A researcher can first in thought conjecture either an ideal mechanism or a non-ideal mechanism, and then later try to test whether the world contains an approximately similar real mechanism. This is true in relation to both non-living nature, living nature, and social reality.

I have made no attempt to relate my views on ideal mechanisms to the general discussion about idealizations in the philosophy of science, but I would like to make a brief remark in relation to Max Weber's famous notion of *ideal types* (or ideal-type models).

Weber was both a philosopher of science and one of the founding fathers of sociology, and he used the notion of ideal types in order to defend the view that sociology can be a science. In the light of my distinction between ideal and non-ideal fictional mechanisms, all his ideal types should probably be classified as non-ideal fictions, but it might also be argued that he simply never considered such a distinction. Be this as it may, in my opinion both the ideal and the non-ideal mechanisms I have spoken of can function the way Weber intended his ideal types to function. His view was that many social patterns and structures can only be grasped conceptually by being seen through the artificial lens of an ideal type.[13] I think the same is true of many real natural and technological mechanisms. In order to be able conceptually to communicate about many of them, we need ideal types.

---

[13]For a comprehensive discussion of Weber's ideal types, see von Schelting (1934). He finds distinct kinds of ideal types in Weber. Along one dimension, von Schelting divides ideal types into "causal-real" and "non-causal ideal," and along another into "generalizing" and "individualizing."

From the Weberian perspective hinted at, ideal mechanisms are ideal types that can function the way mathematics and classic pre-analytic Euclidean geometry function when applied to real-world phenomena.

14.8 Abstract Objects as Fictions

From Plato and ancient Neoplatonism, via a number of medieval Islamic and European philosophers, up until Frege and much of contemporary world-wide analytic philosophy, abstract objects are regarded as having a necessary existence. Hereby, they are also ascribed a mind-independent existence. To me, it is a mystery how contemporary philosophers can have such a view of the abstract objects they work with, be it numbers, sets, propositions, or all of these, and at the same time call themselves naturalists, which many of them do. Normally, being a naturalist means thinking that there is a mind-independent reality with spatiotemporal objects, but that there is no reality such as a God, Platonic objects, or Kantian transcendental faculties. For a naturalist, abstract objects cannot have a necessary existence; they must either be regarded as being directly immanent in the mind-independent spatiotemporal world, or as being man-made (or perhaps animal-made) constructions. And I think that our constructions of abstract objects have such a character that they deserve to be called fictions, even though this semantic option implies a widening of the traditional notion.

As I have already said, a distinction has to be made between ideal fictions and the non-ideal fictions in novels; numbers, sets, and propositions do not belong to the extension of the traditional notion of fiction. However, I will only stress this fact, not say more about it, even though there are many details to be discussed.

Traditionally, the term 'fiction' is connected both to what we experience in dreams and to what we are presented to in novels. When we stress dreams, all fictions easily look purely personal, but the fictions in novels are normally regarded as intersubjective social constructions. From the point of view of mind-independent nature, as well as from the point of view of transcendent realms, all fictions, both personal and social, are in a straightforward sense just fictions. However, from the point of view of social reality, social fictions are simultaneously also real. That is, they belong for some time to some social reality. The expression 'socially real fictions' is not a contradiction in terms.

As a matter of fact, we can in many conversations and writings identify and re-identify fictional objects of various sorts; both temporally across the mental states of ourselves and inter-subjectively across the mental states of ourselves and others. This should be taken as a fact that philosophical investigations should elucidate and explain, not as something that philosophy can show to be an illusion.

Therefore, in this paper, I have taken the existence of fictions as social facts for granted.[14]

Since a social fiction is in some sense individually the same in all the mental states in which it exists, it cannot be reducible to a class of mental states, even though it is the case that it would not exist at all without mental states directed at it. Should they then be called subjective or objective objects? About the ideal fictions of mathematics, Bunge says:

> the statements in pure mathematics are not ontologically objective: they do not refer to the real world. But of course they are not subjective either: they do not report on the speaker's state of mind. Thus, they are neither objective nor subjective in an ontological sense, even though they are impersonal and asocial [in the sense of being socially neutral]. (Bunge 2006, p. 190)

I have earlier quoted Bunge saying that "paradoxically, no one seems to have produced a detailed, true, and generally accepted theory of objective and partial truth." In analogy with this, I would now like to say: paradoxically, no one seems to have produced a detailed, true, and generally accepted theory of what is neither objective nor subjective in an ontological sense.[15] This notwithstanding the fact that, in the last decades, social ontology has become more or less a philosophical discipline of its own. There, however, ideal objects are never discussed.

Since I believe in the existence of certain kinds of non-empty necessary truths, and since I am a Bunge-like anti-Platonist and anti-Kantian naturalist, I will end by putting forward a question that despite my naturalism has a Kantian ring. Kant took the existence of synthetic a priori-truths for granted, and asked how they are possible. In my opinion, we must nowadays take it for granted that abstract objects can be given a place in a naturalist framework. Therefore this question: *how are abstract objects, e.g. numbers and ideal mechanisms, possible?*

Acknowledgements For useful comments on a preliminary version, I would like to thank Rögnvaldur Ingthorsson, Ingemar Nordin, and Christer Svennerlind.

---

[14]My analysis of how it can be possible for different persons to refer to the same individual fiction is presented in Johansson (2010). It differs from Bunge's, but it also differs from "the pretence theory of fiction" and "the make-believe theory of fiction." Both the latter have been used in discussions of fictional models in science; see Frigg (2010) for the first and Toon (2010) for the second.

[15]In the sixth section, I briefly compare Bunge's and Popper's conceptions of approximate truths. A comparison between Bunge's view of what is "neither objective nor subjective in an ontological sense" and Popper's view of his so-called "world 3" would lay bare a further striking similarity between them.

References

Anderson, H. (2014). A field guide to mechanisms: Part I and part II. *Philosophy Compass, 4*, 274–297.

Bueno, B., Darby, G., French, S., & Rickles, D. (Eds.). (2018). *Thinking about science, reflecting on art*. Abingdon: Routledge.

Bunge, M. (1964). Phenomenological theories. In M. Bunge (Ed.), *The critical approach to science and philosophy* (pp. 234–254). New York: Free Press.

Bunge, M. (1967). *Scientific research II. The search for truth*. Berlin: Springer.

Bunge, M. (1974a). *Treatise on basic philosophy* (Vol. 1). Dordrecht: Reidel.

Bunge, M. (1974b). *Treatise on basic philosophy* (Vol. 2). Dordrecht: Reidel.

Bunge, M. (1997). Mechanism and explanation. *Philosophy of the Social Sciences, 27*, 410–465.

Bunge, M. (2003). *Emergence and convergence: Qualitative novelty and the unity of knowledge*. Toronto: University of Toronto Press.

Bunge, M. (2004). How does it work? The search for explanatory mechanisms. *Philosophy of the Social Sciences, 34*, 182–210.

Bunge, M. (2006). *Chasing reality: Strife over realism*. Toronto: University of Toronto Press.

Bunge, M. (2013). *Medical philosophy: Conceptual issues in medicine*. Hackensack: World Scientific Publishing.

Contessa, G. (2010). Scientific models and fictional objects. *Synthese, 172*, 215–229.

Craver, C. & Tabery, J. (2015). Mechanisms in science. *Stanford Encyclopedia of Philosophy*. https://plato.stanford.edu/entries/science-mechanisms/. Accessed 11 Jan 2018.

Euclid. (2017). *Euclid's elements – All thirteen books in one volume (The Heath translation; Ed. D. Densmore)*. Santa Fe: Green Lion Press.

Frigg, R. (2010). Models and fiction. *Synthese, 172*, 251–268.

Johansson, I. (1996). Physical addition. In R. Poli & P. Simons (Eds.), *Formal ontology* (pp. 277–288). Dordrecht: Kluwer Academic Publishers.

Johansson, I. (1997). The unnoticed regional ontology of mechanisms. *Axiomathes, 8*, 411–428.

Johansson, I. (2004 [1989]). *Ontological investigations. An inquiry into the categories of nature, man and society*. Frankfurt: Ontos Verlag.

Johansson, I. (2010). Fictions and the spatiotemporal world – In the light of Ingarden. *Polish Journal of Philosophy, 4*, 81–103.

Johansson, I. (2011). Shape is a non-quantifiable physical dimension. In J. Hastings, O. Kutz, M. Bhatt, S. Borgo (Eds.), *Shapes1.0, Proceedings of the First Interdisciplinary Workshop on Shapes*. Karlsruhe. http://ceur-ws.org/Vol-812/invited1.pdf. Accessed 11 Jan 2018.

Johansson, I. (2013). The ideal as real and as purely intentional – Ingarden based reflections. *Semiotica, 194*, 21–37.

Johansson, I. (2015). Collections as one-and-many – On the nature of numbers. In S. Lapointe (Ed.), *Themes from ontology, mind, and logic. Present and past. Essays in honour of Peter Simons* (pp. 17–58). Leiden: Brill Rodopoi.

Johansson, I. (2017). In defense of the notion of Truthlikeness. *Journal for General Philosophy of Science, 28*, 59–69.

Johansson, I., & Lynøe, N. (2008). *Medicine & philosophy. A twenty-first century introduction*. Frankfurt: Ontos Verlag.

Morrison, M. (2015). *Reconstructing reality. Models, mathematics, and simulations*. Oxford: Oxford University Press.

von Schelting, A. (1934). *Max Weber's Wissenschaftslehre*. Tübingen: Mohr.

Toon, A. (2010). The ontology of theoretical modelling: Models as make-believe. *Synthese, 172*, 301–315.

Chapter 15
Bunge *contra* Popper

Joseph Agassi and Nimrod Bar-Am

Abstract Most of our colleagues are either dogmatists or justificationists. This makes friendship with them a delicate matter: one constantly faces the dilemma of either doing them the (closed society) curtesy of overlooking their faults, or offering them the (open society) service of readiness to criticize their opinions. Bunge is one of the few who make both friendship and criticism easy: he avoids both dogmas and justifications.

15.1 Introduction

Mario Bunge and Joseph Agassi are lifelong friends. They met in London, ages ago, when Agassi was a young startup, and Bunge was already an established academic of world renown (although they have less than a decade between them). Bunge, like a true member of the open society, approached Agassi and they instantly broke into a friendly debate. A friendship was struck. It, and the lively debate between them, continues to this day. Most of our colleagues are either dogmatists or justificationists. This makes friendship with them a delicate matter: one constantly faces the dilemma of either doing them the (closed society) curtesy of overlooking their faults, or offering them the (open society) service of readiness to criticize their opinions. Bunge is one of the few who make both friendship and criticism easy: he avoids both dogmas and justifications.

J. Agassi (✉)
Tel Aviv University, Tel Aviv, Israel

York University, Toronto, ON, Canada
e-mail: agass@post.tau.ac.il

N. Bar-Am
Sapir Academic College of The Negev, Israel

© Springer Nature Switzerland AG 2019
M. R. Matthews (ed.), *Mario Bunge: A Centenary Festschrift*,
https://doi.org/10.1007/978-3-030-16673-1_15

15.2 Dogmatism and Skepticism

Traditionally, those who avoid both dogmatic adherence to their views and efforts to justify them are considered skeptics. This is fine, as etymologically the word means 'searchers'; but it is confusing since the most famous skeptics are the Pyrrhonists and they aspired to avoid both dogmatic adherence to their opinions and efforts to justify them by declaring, quite dogmatically, that they advocate no opinion. Everyday conduct betrays holding opinions, as even the apocryphal anecdotes about Pyrrho illustrate. Mario Bunge, on the other hand, is quick to express an opinion: it is hard to find a good question on which he has not published a detailed and enlightening discussion,[1] and it is impossible to find a publication of his in which he advocated no opinion. Indeed, his advocacy is always clear and forceful, yet never dogmatic. For this he deserves Brownie points regardless of agreement or disagreement with him. For, one way or another it is hard if not impossible to find an issue whose debate he has not advanced by his decisive interventions.

It is our custom to consider as a separate (and rare) breed all those who belong to neither the dogmatist nor the justificationist clans. Einstein and Popper, Chief among them are Russell (at least in spirit, if not always in philosophy)[2]. Of course, some philosophers have made significant contributions despite being dogmatists or justificationists. This is particularly so for those who are disposed (as we are) to view as philosophers all those who have made contributions that are philosophically significant even those that they did not make as philosophical. Some of our friends find it an excessive burden to add to the philosophical agenda information that comes from different fields of research. This is amusing: unless they are particularly ambitious, academics have a very light burden and they have no pressing obligation to study or to update their knowledge. Teaching in Israeli Universities is up-to-date in mathematics and in physics but not in philosophy: the philosophy they teach is at least a century out-of-date. (We ignore here the just protest that Bunge has launched against the popular advocacy of irrational philosophies and of inhumane ones; somehow, those are quickest to enter cutting-edge ultramodern reference lists). This is not surprising; the surprise is that Bunge was asked to return to his university after his retirement to continue teaching up-to-date philosophy despite all opposition.

---

[1] See Bunge (2003, 2011) for the wide range and the diversity of his ideas.

[2] Russell objected to Popper's anti-inductivism. He said, we distinguish between the ravings of a mad person and an Einstein. And whatever Popper says is the advantage of the one over the other may count as his principle of induction. Popper admitted that this is so, and that it makes sense, yet he found this kind of preference significantly different from justification, since it is possibly erroneous, hopefully given to improvement, and should apply well also to our preferences of scientific theories which have already been openly acknowledged as refuted, over the madman's ravings.

15.3   Scientific Philosophy

By 'up-to-date philosophy' we mean philosophy done by philosophers who take science seriously; for example, Polanyi, Quine and Bunge. Of course, this raises the problem of demarcation of science: what theory is scientific? For example, why do we declare astrology a pseudo-science? The answer is obvious: it is too arbitrary. This is somewhat circular: we try to have our opinions non-arbitrary by following science and we declare what we deem non-science arbitrary. Yet this need not alarm us. Anyone who cares even a little about astrology should know that the Babylonian constellations differ from the Hindu and the Chinese, and that astronomy declares them non-existent. Two stars adjacent in a constellation can be unimaginably distant from each other but look adjacent as they appear within a small angle from each other. In other words, we have fairly good ways to judge arbitrariness, and one reason scientific ideas are usually so appealing is that they are usually by far the least arbitrary among the available ideas, and even obviously so.

Of course, one may wish to seek criteria for non-arbitrariness nevertheless. One obvious criterion is simplicity. What is simplicity? Few have struggled with this problem.[3] Bunge is by far the richest of them, and there is no thinker who has come close to the wealth of ideas that his essay on simplicity exhibits. Since it is impractical to survey Bunge's tremendous output, we will mention here only a very small number of items from three essays. Such a choice may elicit the suspicion that it is arbitrary. This may be correct; our intention is never to convince readers. We hope that readers will find the following comments sufficiently enlightening to serve as illustrations.

Philosophers whose output is neither justificationist nor arbitrary are rare because such a feat requires delicate balancing skills, a balance between two intolerable extremes. Such a feat keeps one on one's toes, ready to view opinions critically and seek ever newer interesting explanations. Most intellectuals find this attitude flippant and thus unbecoming, or else too burdensome a permanent task. They take themselves too seriously. Some measure of flippancy is essential for the search for ideas, newer ideas that may hopefully serve as better explanations; and the task of always seeking improvements is more a delight than a burden. However serious our peers are, they declare that they sponsor the idea of intellectual progress, that they advocate constant discovery; they usually go even further, and declare that they support the idea of the inductive method for this very reason. This program is hardly possible to entertain on and off: it soon becomes a chronic illness.[4]

Now, the dream of a method for discovery is ancient. Its revival opened the scientific revolution. Its great expression was in *Novum Organum* of 1620 of Francis Bacon. It is a book written in prophetic aphorisms that fired hope in ever so many

---

[3] For a rich list of references see Bunge (1961a, 120–149); see also the 12 chapters in Bunge (1963).

[4] Sadly, Ludwig Wittgenstein found the search for philosophical explanation irksome, Wittgenstein (1953, §§109, 116, and 255).

individuals who decided to become researchers with almost no training.[5] That this theory of induction is a pipe dream, is easy to see; except that when so many researchers express faith in it and when science is so powerful, it is not that simple to swim against the strong scientific current. The right metaphor is perhaps not swimming; it is that of tight-rope walking.

15.4 Bacon's Example

For the tight rope, light foot is essential. We need some flippancy to help us rid ourselves of some intellectual ballast. We owe a debt of gratitude to Bacon for his having encouraged us to jettison the ballast of scholasticism unceremoniously. Unfortunately, he was too serious about induction. And so now we need to rid ourselves of the ballast of inductive excess information. Here Mario Bunge excels. He is amazingly highly informed about many fields of knowledge. He has contributed to the technically cumbersome fields of theoretical physics and of logic and more. Yet in his many publications, he is always terse and tidy with almost no ballast: every item of information that he cites plays a distinct role in his discussion, and with no delay.

When we rid ourselves of some ballast, we cease pretending that there is assurance of success in research. As it turns out, a reason for the steady progress of science superior to inductivism is simply our immeasurable ignorance. It allows for the hope that whatever we touch upon is in great need of improvement and for the hope that the required progress will not stop right now with no reason. We may seek the reason for our past progress and examine it with the hope to continue with research with no pretense of assurance of any success: research is not inductive. What then is the right scientific method? The answer that comes easy to mind is that of a sieve; invent as many ideas as you can and use a sieve to reduce their number fast and capture the hopefully more valuable nuggets among them. Of course, we have criteria of value, and these may be erroneous. We reduce this risk by using our urgent problems as the criteria. We then try to view recent improvements in a pretty straightforward manner and we find gratifying any new solution to an interesting problem, even if on a second or third thought we give it up. This may count as scientific progress despite all its shortcomings. It is famous as trial and error.

Removal of as much ballast as possible is, thus, part and parcel of scientific method as we present it here. However, maintaining a light foot is a tightrope-walk in itself. Is it a vicious circle again? One popular manner of carefree production of ideas and solutions is *brainstorming*. It has partial results (especially when participants lack education for light-foot) and it is often not cost-effective. It contrasts sharply with Bunge's writings that have a surprisingly high rate of interesting ideas per page. How does he manage this? What enables him to avoid

---

[5]For more details of the story see Agassi (2012).

the ballast that so often makes brainstorming inefficient? For this question we need some idea of what is the ballast, or, more specifically, what is the ballast that tends to impede scientific research as well as scientific philosophy?

15.5   Freud and Confirmation

The ballast in science, Popper said, is the high rate of confirmation of worthless ideas. Bunge offers examples in his paper on simplicity: psychoanalysis has too much confirmation, most of which he rightly judges useless. Although our view of psychoanalysis is more favorable than that of Bunge, we agree with him: the writings of Freud are full of pointless confirmations. Popper made that same point a few decades earlier. Consider for example Freud's report in *Psychopathology of Everyday Life* of him noticing a woman signing with her maiden name. He suspects at once that she is experiencing marital troubles. And, he adds triumphantly, within a few months she was divorced. He considers the divorce confirmation, and even emphatically so; yet the absence of it would not count as infirmation, and even obviously so. We should not consider confirmation of a theory the results of any test of it that cannot possibility refute it.

This suggestion, nowadays ascribed to Popper, perhaps properly belongs to William Whewell. In Whewell's view, however, confirmations may amount to verification, namely to immunity from all conceivable future refutation. Consequently, he found the refutation of Newton's optics disturbing. He recognized it as a slippery slope that implies the possible undermining of Newton's theory of gravity, which seemed to him a catastrophe. He claimed to have invalidated that threat by showing that, unlike Newton's theory of gravity, Newton's optics never underwent a proper testing procedure. He also deemed refuted theories unscientific. This renders Kepler's laws unscientific. He discovered this fact but refused to consider it seriously. Popper's demarcation criterion renders Kepler's laws scientific, but his demand for corroboration renders many of Kepler's terrific (testable, refuted, yet uncorroborated) conjectures unscientific.[6] Pity.

Popper thus reduced the ballast of scientific theories considerably, but not enough: all of Kepler's ideas are refutable but only some of them are remembered as significant. Which ones? Popper did not mind that so many hypotheses are ignored. But for the distinction between Kepler's rejected hypotheses and his famous laws we need more. That shows how inadequate is Popper's view that the scientific consists of all refutable hypotheses. Agassi's proposal, borrowed from Popper's own lecture courses, is to view as empirical all testable hypotheses and as scientific all empirical explanations, thus viewing the scientific a proper subset of the empirical (with much technology as empirical and dependent on science but not scientific).

---

[6]Popper (1963, Chapter 10, note 31).

15.6 Karl Popper

Now we are discussing ballast. And the suggestion before us is that only explanatory hypotheses are not ballast. This however is insufficient: in research, ideas turn up that are not explanatory, yet they signify as potential explanations. Popper's methodology expressly ignores potential explanations as he had his hands full in discussing actual ones. The gratitude that Popper deserves should not stop us from discussing potential explanations. They appear in the discussion of Henri Poincaré of the idea that proper confirmation comes with proper tests. In his classic, breathtaking *Science and Hypothesis* (Poincaré 1905) he says: we can make the law of conservation of energy precise by adding to it a list of all the possible kinds of energy. And then we may risk its refutation by the discovery of a new kind of energy. Or else, we can leave it vague by adding a list of the possible kinds of energy without claiming that it is complete. This will prevent its refutation by the discovery of a new kind of energy, but at a cost: if a theory is not empirically infirmable then it is also not empirically confirmable.[7]

Popper made this observation of Whewell and Poincaré into a pillar of the philosophy of science: no empirical infirmability, no empirical confirmability. Bunge fully endorses this great idea.[8] This is why we consider him a Popper ally. Of course, he has criticisms of Popper. (So do we). Nevertheless, we respect his preference for being considered not a Popper ally and even anti-Popper.[9] So be it. Suffice it that he agrees with Popper on the refutability of scientific theories[10] and on the idea that confirmation with no refutability is sheer ballast.

15.7 Bunge's Criticism of Popper

On three items, Bunge sharply criticizes Popper: on confirmations, on social institutions and on the mind-body problem. Let us say a few words on these.

On confirmations, Bunge says they are important. Popper never denied that. Bunge scarcely explained this: he found it redundant, everyone knows that scientists value confirmation; so does the Nobel committee.[11] Now, the Nobel committee is no authority: its view on science is notoriously old-fashioned and its awards are notoriously much influenced by (regrettably often inevitable) power-struggles. Much information on confirmations that goes around is simply inaccurate public

---

[7]Bunge (1961b, p. 279) offers a competing view: he prefers science with confirmations even at the cost of refutations.

[8]Bunge (1961a).

[9]Bunge (2017, pp. 3–12).

[10]Bunge (1967, p. 342). Bunge allows scientific theories to include untestable hypotheses. He claims that in this he disagreed with Popper (Bunge 1961a,b 136, note 25); this is a slip of his pen.

[11]Bunge (2017).

relations.[12] A staple example is the great 1934 meson theory of Hideki Yukawa. It found tremendous confirmation in the discovery of mesons; it was also infirmed as his assessment of their mass was wide of the mark. The standard concealment of this error of his seems respectful, but it is a shamefaced and uncalled-for admission of guilt; worse, it conceals the brilliant reasoning of his interesting research, thereby rendering the discovery miraculous.

Quite generally, new ideas are usually improvements upon older ones and so their possible confirmation has to be the outcome of the crucial experiment between the old and the new. Hence, it is simultaneously infirmation and confirmation. Why do we prefer to view it as confirmation? Because of the traditional cult of success. Even though the success is often temporary and the refutation is often final (from both the logical and the historical perspectives). The cardinal question of methodology is, how is theoretical learning from experience possible? As Bunge does not succumb to the myth of induction, we may expect him to offer an answer to it—as an alternative to Popper's or in agreement with it. The crux of Popper's answer is the hypothesis that theoretical learning from experience is by successful refutation. This answer is so outlandish that Carl G. Hempel and Hillary Putnam, for two conspicuous examples, have ascribed to Popper the opposite answer: they ascribed to Popper the view that theoretical learning from experience is by failed refutation. There are other forms of learning, such as the learning of mathematics that is decidedly different from the learning from experience.

There is also learning by confirmation; it is different from learning from refutation. Learning from refutation is less problematic in that it does not suffer from Hume's critique of induction. This is very clear. Bunge dislikes what he calls "the Hume cult".[13] This is understandable: as he never assumed that science is inductive, he did not find it troublesome; in this sense, he rightly does not see great value in Hume's critique of it. In another sense, the widespread of inductivism does render it significant. Similarly, as he recognizes many kinds of learning, he does not share Popper's idea of negative learning (from experience). The idea that science is a prolonged Socratic dialogue seems exciting even if the concept of Socratic dialogue is in need of a new look and of some tweaking.

The second difference between Bunge and Popper concerns social institutions. Agassi has called Popper's view 'institutional individualism', claiming that Popper has synthesized the strong aspects of both traditional individualism and traditional collectivism by recognizing collectives as significant objects that influences the behavior of individuals—while dismissing the claim that they have their own rationality: only individuals can have motives, not collectives. Species, nations, tribes, and social institutions do not; not even the lovely society for the prevention of cruelty to animals. Like Popper, Bunge too is concerned with the roles collectives play in determining the behavior of individuals and opposes the holism and idealism

---

[12] Agassi (1996) Reprinted in Agassi (2003, pp. 152–163).

[13] Bunge (2017). There is a slip of the pen here: he overlooks Hume's assertion that as Newton's theory survived the test by foreigners, it will never be refuted (Hume 1742, p. 122).

associated with traditional collectivism. He calls his philosophy "systemism". It is difficult to imagine what difference there is between Popper's institutional individualism and Bunge's systemism. Too much hinges on whether Bunge would allow levels of existence into his ontology. Clearly, he admits institutions as artificial, and thus in some sense as real: they do influence the behavior of individuals. But are they as real as individuals? (Are individuals as real as quarks?) To say that they do (as Bunge sometimes says[14]), to admit one and only one sense of "existence" brings about a host of problems of methodological, logical and ontological nature. Sometimes Bunge takes the idea of a universe of discourse sufficiently seriously to allow for different senses of existence, depending on the different universes of discourse. This raises many new problems, but it drastically dispenses with the old ones that lead many a logician to idealism. It seems Bunge owes it to himself to discuss diverse kinds of existence and dispense with his odd idea that mathematical entities are all fictitious. The great Quine said numbers exist as tables and chairs do. Bunge said they exist but fictitiously. There is much room for in-between positions.[15]

The same goes for the third of Bunge's differences with Popper, namely his critique of Popper's writings on the Mind-Body problem, specifically his postulation of World 3. This is not to defend Popper: Bunge's claim that Popper ignores brain physiology is just; it seems to us that the trailblazing researches of Alexander R. Luria deserve the attention of philosophers who study the mind-body problem, and Bunge is a pioneer in this (as is Anne-Lise Christensen). Also, the terminology of Frege, of C. S. Peirce and of Popper may be objectionable, and it may allude to Plato's metaphysics. But terminology is of no significance. What is important is that living things are emergent systems that are not reducible to material systems, yet Bunge is right in insisting on his version of materialism (that differs from classical materialism in quite a few significant ways).

Popper's system is likewise materialist. He co-authored a book with John Eccles in which the aim of Eccles was to reconcile his following of Catholic dogma with some results of his exciting studies in neuro-physiology. Popper's cooperation with both Eccles and Konrad Lorenz are hardly honorable, as he compromised his opinions in each of these joint volumes for the sake of friendship and acceptability. This, however, is no evidence that Bunge disagrees with Popper on the two chief points: for all we know, living systems are evolved material systems but are not reducible to them—they have souls—and their souls are mortal. To maintain a sense of proportion, we should agree that all those who share this view are more partners than opponents.

It is not that we regard agreement as better than dissent; both agreement and dissent are possibly intelligent but not necessarily so. Both can occur in the closed society and in the open society. It is easier to express agreement—by simply

---

[14]Bunge (2009, pp. 64, 82–3, 90, 290).

[15]For more details see Lejewski (1954–1955). See also Bochenski (1990) and Agassi (1990), as well as Bunge's "Replies", all in Weingarten and Dorn (1990).

repeating known arguments—than to develop intelligent dissent. And often received opinions are the better ones, and then it is easier to agree with them than to develop criticism of them. The popular (foolish) demand that criticism should be "constructive", namely, that critics should offer alternatives to the ideas that they criticize, renders it ever so much harder to criticize than otherwise. Thus, it is very hard to criticize general relativity but ever so much harder to offer a viable alternative to it. It is therefore wiser to invite any criticism of any extant theory, since it may then stimulate others to offer an interesting alternative to the received view. In other words, intelligent criticism is always constructive: knowing where one stands should be better than living in illusion.

15.8 Conclusion

It is useful to contrast the philosophies of Popper and Bunge. It is especially beneath dignity to defend a philosophy come what may. Nevertheless, we need some sense of proportion. Seeing that Popper and Bunge are generally allies, in comparison with most philosophers around, we may then go into detail and try to contrast their views as best we can, starting with the most important disagreement. It is advisable, however, not to forget the general agreement, and to have the disagreement as the next task on the current philosophical agenda.

References

Agassi, J. (1990). Ontology and its discontent. In Weingartner & Dorn (1990) (pp. 105–122).

Agassi, J. (1996). Towards honest public relations of science. In S. Amsterdamski (Ed.), *The significance of Popper's thought, Poznań studies in the philosophy of the sciences and the humanities* (Vol. 49, pp. 39–57). Reprinted in Agassi, J. (2003), pp. 152–163.

Agassi, J. (2003). *Science and culture.* Dordrecht: Kluwer.

Agassi, J. (2012). *The very idea of modern science.* Cham: Springer.

Amsterdamski, S. (1996). *The significance of Popper's thought: Proceedings of the 'Karl popper 1902–1994. Conference, Poznań Studies in the Philosophy of the Sciences and the Humanities* (p. 49).

Bochenski, J. (1990). On the system. In Weingartner & Dorn (1990) (pp. 99–104).

Bar-Am, N., & Gattei S. (Eds.). (2017). *Encouraging openness: Essays for Joseph Agassi on the occasion of his 90th birthday* (Boston studies in the philosophy and history of science). Cham: Springer.

Bunge, M. (1961a). The weight of simplicity in the construction and assaying of scientific theories. *Philosophy of Science, 28,* 120–149.

Bunge, M. (1961b). Kinds and criteria of scientific laws. *Philosophy of Science, 28,* 260–281.

Bunge, M. (1963). *The myth of simplicity.* Englewood Cliffs: Prentice-Hall.

Bunge, M. (1967). *Scientific research II: The search for truth.* Dordrecht: Springer.

Bunge, M. (2003). *Philosophical dictionary.* Amherst: Prometheus Books.

Bunge, M. (2009). *Political philosophy: Fact, fiction and vision.* New Brunswick: Transaction Publications.

Bunge, M. (2011). *100 ideas*. Penguin Random House Grupo Editorial Argentina.
Bunge, M. (2017). Why don't scientists respect philosophers? In Br-Am & Gattei (2017) (pp. 3–12).
Hume, D. (1742). *Essays, moral, political, and literary*. Strand/Edinburgh: A. Millar and A. Kincaid & A. Donaldson.
Lejewski, C. (1954–1955). Logic and existence. *British Journal for the Philosophy of Science, 5*, 104–119.
Poincaré, H. (1905). *Science and hypothesis*. New York: The Walter Scott Publishing Co.
Popper, K. (1963). *Conjectures and refutations*. London: Routledge.
Weingartner, P., & Dorn, G. (1990). *Studies in Bunge's treatise*. Atlanta: Rodopi.
Wittgenstein, L. (2001/1953). *Philosophical investigations*. Oxford: Blackwell.

Chapter 16
Bunge is Correct About Positivism, but less so about Marxism and Hermeneutics

Alberto Cupani

Abstract Bunge is critical of three traditional philosophical trends: Positivism, Marxism and Hermeneutics. In relation to the first, he disapproves of its empiricisms, its nominalism and its phenomenalism, as well as its rejection of traditional ontological and axiological questions. In Marxism, Bunge mainly criticizes the ambitions of Dialectics and the thesis that ideas are socially determined. In relation to Hermeneutics, Bunge is convinced that it is radically failed for confusing the knowledge of social events with the interpretation of texts. In this article I show that although Bunge's criticisms of Positivism are fair, his judgement of the two other doctrines is not totally so.

Bunge is critical of three traditional philosophical trends: Positivism, Marxism and Hermeneutics. In relation to the first, he disapproves of its empiricism, its nominalism and its phenomenalism, as well as its rejection of traditional ontological and axiological questions. In Marxism, Bunge mainly criticizes the ambitions of Dialectics and the thesis that ideas are socially determined. In relation to Hermeneutics, Bunge is convinced that it is radically failed for confusing the knowledge of social events with the interpretation of texts. In this article I show that although Bunge's criticisms of Positivism are fair, his judgement of the two other doctrines is not totally so.

16.1 Positivism

There is a tendency, above all in Latin America, to denominate as "positivism" the exaltation of the theoretical and practical value of scientific research according to the model of the natural sciences. In this sense, the Bungean system is probably

A. Cupani (✉)
Department of Philosophy, Federal University at Santa Catarina, Florianópolis, Brazil

M. R. Matthews (ed.), *Mario Bunge: A Centenary Festschrift*,
https://doi.org/10.1007/978-3-030-16673-1_16

the best illustration of "positivism".[1] Nevertheless, historical Positivism, whether in the sense of a philosophical school of the nineteenth century (Auguste Comte, John Stuart Mill, Herbert Spencer, among others), or in the sense of an intellectual movement of the first decades of the twentieth century (Neopositivism or Logical Empiricism of the "Vienna Circle": Rudolph Carnap, Otto Neurath, Moritz Schlick, etc.) did not consist only in the exaltation of scientific knowledge. Scholars of historical Positivism such as Kolakowski (1981) and analysts such as Giedymin (1975) and Stockmann (1983) indicate that empiricism, phenomenalism and nominalism are characteristic of all forms of Positivism.

Bunge explicitly challenges these theoretical positions. He does not believe that the information provided by the senses is the exclusive basis of our knowledge of the world (on the contrary, he constantly emphasizes the role of theory). Bunge denies that our knowledge, particularly scientific knowledge, is limited to appearances (phenomena). He argues that true knowledge is explanatory, and is developed by means of (realist) conjectures about the "internal mechanism" of production of phenomenal aspects of real events. He thus equally rejects nominalism, because for him constructs (concepts, propositions, theories), although fabricated by the human mind, correspond – symbolically and approximately – to traits of reality.

The historical positivists also tended to be inductivists. But Bunge, although criticizing Popper's anti-inductivism, defends the hypothetical-deductive interpretation of scientific research. And as much as he shares the trust of positivists and neopositivists in the methodological unity of science, he challenges the mere imitation, by new disciplines, of the procedural forms of the older sciences (particularly, physicalism, encouraged by some neopositivists). For all of these reasons, Bunge's epistemology is not positivist *sensu stricto*.

Moreover, historical Positivism was also, declaredly and programmatically, antimetaphysical. Ontological questions, as well as the queries and presuppositions that transcend any possible experience, were disdained by the positivists as obsolete or insoluble concerns, or even as pseudoquestions. As is known, Bunge believes this position to be wrong at least concerning principles. Although certain questions (such as the "body-soul problem") may lack meaning in the way that they were traditionally formulated, and certain doctrines (such as spiritualism, which Bunge considers to be a manifestation of animism) are marked, in his opinion, by the immaturity of the human mind, the great majority of the problems raised by Western metaphysics (epistemological, ontological and axiological problems) can and should be reformulated according to the scientific approach. And unlike the old and new positivists, for Bunge this scientific stand presupposes ontological convictions (specifically realist ones, as is also well known).

On the other hand, Bunge does not share the axiological "emotivism" of certain neopositivists and defends the possibility (and the need) to justify objective values,

---

[1] I examine "positivism" (more specifically scientific naturalism) in various articles, mainly in my book *A Crítica do Positivismo e o futuro da Filosofia* (1985). Bunge's "positivism" is analyzed in Cupani (1991).

especially in the moral field. In this sense, he is equally opposed to a trait considered to be characteristic of historical Positivism by scholars of this movement: the denial of the cognitive character of value judgments and of the formulation of norms, as well as the separation of knowledge of reality from its evaluation. It is well known that Bunge limits to pure or basic science the abstention of valuing its object, recognizing evaluations (and social commitments) as inherent to applied science and to technology.[2]

For Bunge, Positivism and Neopositivism were fundamentally worthy of esteem, as philosophies that valued science, although they did not offer a correct image of it. One of the merits of Neopositivism, in particular, would have been its effort to attain a clear and rigorous way of thinking, and to have converted symbolic logic into an instrument of philosophy. But, in Bunge's opinion, the neopositivists (due to Ludwig Wittgenstein's influence on them) were exaggeratedly dedicated to the analysis of scientific language, increasingly neglecting real science, its evolution and its great transformations in the twentieth century. As a result, they cultivated an artificial philosophy of science, dedicated to formal problems that interested the philosophers, although not scientists very much.[3]

16.2 Marxism

In regard to Marxism, as is known, it encompasses both a cosmovision (so-called "Dialectical Materialism") and a theory of the socio-historical evolution of humanity ("Historical Materialism"). According to the first, everything that exists is material, and matter is in constant evolutive transformation produced by intra-conflicts ("contradictions") operating within every real entity. This peculiar mechanism of universal evolution is called "dialectical" and can be understood as a function of some grand laws formulated by Fredrick Engels (*Dialectics of Nature*, 1873–1883) though inspired by Georg W. Hegel. According to Marxism, there is a peculiar way of conducting scientific research, the "dialectical method", which allows understanding the transformations that occur in natural or social reality, based on dialectical laws. The most important of these are the principles:

1. That everything that exists includes its own "negation" (a law denominated at times "the unity and struggle of opposites");
2. That reality evolves by overcoming its "contradictions", and.
3. That quantitative transformations produce, over time, sudden qualitative transformations.

---

[2] On the other hand, pure science is for him linked to values in regard to its *ethos*.

[3] Bunge gives as an example of this dedication to problems that are of little use to scientists the attempt by authors such as Reichenbach, Carnap and Bar-Hillel to elucidate, in terms of probability, the philosophical concepts inherent to science, such as truth, simplicity, confirmation, etc. (Bunge 1980, Chap. 1).

In turn, a fundamental thesis of historical materialism is that human society, at any moment of its development, is constituted by a material "infrastructure" (the mode of production specific to the time, which includes both the economic system as well as the social relations that arise from it) and a political-ideological "superstructure" (the state, law, morality, etc., including science), which is determined by the first. Another basic thesis of historical materialism maintains that class struggles are the ultimate "motor" of history. For Marxism – always, obviously, in a summarized characterization of it – men can know reality of which they are part through information provided by their senses, with ideas, when verified in praxis (that is in action), being a "reflection" of the reality. At the same time, Marxism calls attention to the social conditioning of knowledge, in such a way that the nature of the ideas (even in abstract or technical issues) cannot be completely understood without linking them to their social context and its conflicts.

Marxism is seen by Bunge as a philosophy that defends correct ontological and epistemological theses, but in an incorrect manner, and that is mistaken in its understanding of socio-historic issues.[4] Bunge praises Marxism for maintaining (against spiritualist and subjectivist philosophies) that reality is material and dynamic; and that it can be gradually known, although he rejects Dialectics, a doctrine which for him is "half confused and half false". From this doctrine comes the "dialectical method"; something which for Bunge is non-existent as an alternative to usual scientific methodology.

In relation to the alleged dialectical laws, Bunge affirms that they contain incorrect affirmations (propositions, but not real things can be "contradictory", for example) and/or affirmations that cannot be verified, particularly the principle of the unity and universal struggle of opposites.[5] Bunge observes:

> Certainly *some* systems are formed by things or processes that are opposed to each other in *some* senses. For example, an atom is composed of a positively charged nucleus and surrounded by a set of negatively charged electrons. But in these conditions no qualitative change is produced: in a stationary state the atom does not change. For there to be some qualitative change the atom should no longer be a "unity of opposites": it should lose or gain an electron or nucleon. That is, "contradiction", far from being a source of any qualitative change, as Dialectics affirms, in this case is a guarantee of stability. Another example: the current [1985] international political equilibrium is based on a parity of forces. At the time in which one of the two superpowers exceeds the other in nuclear weaponry, it can produce a qualitative change, as in a final war. In general, a "unity of opposites" does not always exist, and when it does exist, it is not a condition for qualitative transformation, but at times, of equilibrium (Bunge 1985, p.164, emphasis in the original, AC translation.)

For Bunge, the main problem of Dialectics consists in that it expresses and stimulates a confused manner of thinking by denominating as "contradictory" what

---

[4]See especially Bunge (1985, Chap. 12, 1987, Chap. 6, 1996, Chap. 11.5).

[5]For Bunge, the supposed principle does not apply either to things ("it is easy to show the existence of simple material objects, such as the electron, the neutrino and the photon") or for properties ("it is not true that all things are simultaneously small and large, valuable and without value, etc."), as for processes ("it is not true that all things heat at the same time as they cool, that all goods that become cheap also become expensive, etc.") (Bunge 1985, pp. 163–164, AC translation).

is, strictly, opposed; or as "opposite", what is merely different. In general, calling any change "dialectical" is for Bunge an indication of not knowing how to describe it precisely.

Regarding epistemological issues, Bunge criticizes the empiricist and pragmatic aspects of Marxism, rejecting the thesis that all concepts, even abstract ones such as mathematics, are based on sensory experience. Bunge, as is known, defends the *sui generis* nature of constructs and the autonomy of formal disciplines. He also does not approve of the Marxist thesis that the criteria of truth is practice, or of the metaphor of reflection as a characterization of the epistemological status of ideas (better defined as symbolic representations of the degree to which one is able to understand reality). For Bunge, intellectual clarity requires differentiating between *applying* an idea and *verifying it*, between the eventual effectiveness of our representations of the real and its truth. Thus:

> Practice does not establish the *truth* of any proposal, but only the *effectiveness* of rules or prescriptions for action. For example, the marvels of Egyptian or Roman engineering do not demonstrate the truth of the rough Egyptian or Roman understanding of physics, as the initial success of Nazism does not prove the truth of the myth of racial superiority of the Germans.

> Professional or political practice does not prove anything more than the effectiveness of the rules employed by the people who apply them. Praxis is realized in conditions that are not controlled experimentally, so that their success or failure can be attributed to an accumulation of factors that escape control [...] (Bunge 1985, p. 168, emphasis in the original, AC translation).

Beyond empiricism and pragmatism, Bunge criticizes Marxist sociologism, expressed in the thesis that knowledge "reflects" social circumstances. For Bunge, it is not possible to totally explain the content of scientific ideas by the social conditions that favor or impede them, although these conditions should also be considered in order to understand in a realistic manner the production of knowledge. Nevertheless, the latter Bungean criticism is only one aspect of his criticism of Historic Materialism. He also rejects the Marxist distinction between "real infrastructure" and "ideal superstructure" of society, arguing that the determinism of the infrastructure is proved false by countless cases in which cultural or political factors stimulated economic and social changes, and that the Marxist model, in addition to improperly emphasizing a "primary motor" which is social implies an involuntary dualism, because it recognizes as material *sensu stricto*, only the economic dimension of society.[6]

Moreover, historical materialism involves another contradiction by simultaneously affirming that the infrastructure determines all of social life and that the class struggle (a mode of political struggle, for Bunge) "moves" history. For Bunge, explaining the dynamics of history requires that, beyond the role of class

---

[6]Bunge believes that a similar contradiction is found in Marxist expressions such as that of the brain being the "material base" of the mental events, which would suggest that the mental events are not material, contrary to universal materialism (Bunge 1985, p. 166).

conflicts, the role of ethnic and religious conflicts be recognized as well. Also, it is necessary to admit the existence of problems that concern all social classes, such as environmental degradation, global overpopulation and the discrimination of women. In any case, for Bunge, Marxism underestimates the role of collaboration in social structuring and social change, unilaterally emphasizing the importance of conflict.

For all of these reasons, Bunge defends the systemic model he himself formulated as more appropriate for understanding human society and its evaluation. According to his model, social life encompasses three interactive subsystems: the economic, the political, and the cultural; and two necessary processes: cooperation and conflict (Bunge 1979, p. 208).

In summary, for Bunge, Marxism is a philosophy that did not know how to properly develop some of its correct intuitions. He does not deny Marx's genius or the importance of his description and criticism of nineteenth century capitalism. Nor does he deny the value of the theses about the influence of economic conditionings and of the class struggle, with the condition that they are relativized. As a philosophy, or as a social doctrine, Marxism, in Bunge's opinion, neutered its initial scientific potential due to the predominant dogmatism and scholasticism that is always found among Marxists. In his opinion, they limit themselves to repeating, illustrating or interpreting the theses of the founders (Marx, Engels, Lenin) instead of revising these theses, submitting them to authentic testings (instead of proclaiming their truth) and contributing to their improvement.

16.3   Hermeneutics

The third philosophical trend in relation to which it is worth situating Bunge's thought is the hermeneutical approach in the human sciences (Bleicher 1990). This trend historically corresponds to the affirmation, which dates back to the nineteenth century, of the *sui generis* character of human phenomena (or at least, of a certain type of researching of human events). According to Hermeneutics, human phenomena require an effort at *comprehension* (*Verstehen*) of human expressions, actions and productions, instead of an *explanation* of them, as in the natural sciences, a thesis defended by Wilhelm Dilthey (*Einleitung in die Geisteswissenschaften*, 1883). This demand is similar to the exegesis of biblical, literary, historic, legal, philosophical texts, which since the eighteenth century had associated the understanding of human life to the *interpretation* of documents produced by human beings. It received a new foundation by the phenomenological philosophy of Edmund Husserl and the existential philosophy of Martin Heidegger.

Husserl's phenomenology conceived philosophy as a reflection on the consciousness that "constitutes" the world of daily experience, giving "meaning" to "phenomena" (literally, "what appears", what is "given" to consciousness) (Husserl 1962). Husserl's work inspired a new type of sociology, represented by the writings of Alfred Schutz, who systematically analyzed the "significant construction of the social world" according to the title of his main work (Schutz 1932/1972). This tradition culminated in the book by Peter Berger and Thomas Luckmann *The Social Construction of Reality* (1966/1972). They inaugurated a sociological approach in which social life is entirely made possible and structured by meanings produced and maintained by social interactions, such as those experienced by the consciousness of the agents. All of this was consistent with Martin Heidegger's (*Sein und Zeit*, 1927) thesis that interpreting and understanding are not occasional activities of human beings, but basic modes of their specific existence, which is thus inherently significant or signifying (Heidegger 1927/1967, p. 142).

The influence of Phenomenology in the social sciences combined with that of Analytical Philosophy, particularly with the practice of the analysis of vulgar language as was encouraged by Ludwig Wittgenstein in his so-called "second phase" (*Philosophical Investigations*, 1953). Wittgenstein called attention to the various "games" (uses, functions) of language, which manifest different "forms of life", thus inspiring an anthropological approach addressed at understanding the rules that make human practices "significant" (see Winch 1958). In general, it can be affirmed that the growing interest in language that characterized a good part of the philosophy of the twentieth and twenty-first centuries, and epistemology in particular, had as a result a way of understanding the human sciences (a way often mentioned as the "linguistic turn" or the "textual revolution" of the sciences) which is based on the presumption that human phenomena are like texts to be deciphered (see Dallmayr and MacCarthy 1977).[7]

For Bunge, this whole tradition in the human sciences is radically mistaken, beginning with the use – which he finds incorrect – of the notions of "meaning" and "interpretation".[8] He argues that this involves *semiotic* categories (that is, those relative to signs), improperly applied to facts or events. *Stricto sensu*, only a construct or its symbolization could have a "meaning", and not extralinguistic entities (actions for example) which, therefore, cannot be "read" or interpreted". For Bunge, this categorical confusion results from not distinguishing, in the various modalities of the hermeneutical approach, between real things and the discourses through which we refer to them. This is ultimately caused by an idealist-subjectivist attitude, both ontological as well as epistemological. For Bunge, as is well known, this position is incompatible with effective scientific practice.

Nevertheless, one could suppose that Hermeneutics appeals to an expanded concept of "interpretation" which could be applied both to words and to social signs (such as gestures) and to natural "signs" (such as dark clouds that "indicate" rain). But Bunge objects that this type of definition of the term "interpretation" can only be the result (and the cause) of confusion.

---

[7]Moreover, this general theoretical trend admits a range of modalities, from the defense that it is possible to understand actions as if they were texts (Paul Ricoeur) to the conclusion that "there is nothing outside of text" (Jacques Derrida), passing through the theses that "man is a self-interpretive animal" (Charles Taylor) and that culture is "an assemblage of texts" (Clifford Geertz) (see Bunge 1996, pp. 290–291).

[8]For this critique see: Bunge (1980, Chap. 11, 1995, Chap. 9, 1996, Chap. 11).

Artificial signs, such as words and traffic signals, are interpreted with the help of explicit conventions; social signals are "interpreted" with the help of more or less explicit generalizations, and natural "signs" are "interpreted" according to natural laws. (Bunge 1996, p. 292)

For Bunge, what Hermeneutics indiscriminately denominates as the "interpretation" of human actions, is in reality the *formulation of hypotheses*, relative to the ideas, feelings or intentions of human beings,[9] which should be conveniently submitted to testing, giving origin to conclusions that are always provisory, as in any form of science. The hermeneutical approach, in direct proportion to its phenomenological inheritance, is inclined, not only to believe that it can and should "interpret" actions, but also to excessively trust in the "interpretations" obtained in this way.

In addition to confusedly characterizing the task of human sciences, Hermeneutics makes another more serious mistake, according to Bunge. To a realist vision of the social world, this world is certainly constituted by the "meanings" that human beings attribute to their actions, deeds and words (and to their congeners, contemporary or past). But it is also, and fundamentally, constituted by those actions, deeds and words understood as *objective* entities and processes, of a physical-biological basis (in addition to which they also have their own psychical, social and cultural characteristics). Of course, the peculiar structure and the laws that "govern" the evolution of social processes must be researched by social scientists (sociologists, anthropologists, historians, linguists and others) without ignoring the "meanings" that human beings attribute to their social existence. This would be equivalent to a now outdated behaviorism. Nevertheless, they should also not *reduce* their research to the genesis of these "meanings", a genesis that is misunderstood when it is conceived as based only upon the subjectivity of the agents. Above all, social scientists cannot claim that the "meaning" is equivalent to the *existence* of social processes. In sum:

> Normal people distinguish between signs, such as words and their referents. Moreover, they do not attribute syntactic, semantic or phonetic properties to things such as stars, people and societies, for the simple reason that we cannot read them, write them or interpret them. For this reason, we empirically study and construct conceptual models of them, instead of consulting dictionaries and grammar books [to understand them]. Certainly, in this process, we use or produce texts, but only as registers of the empiric and conceptual operations relative to our objects of study – stars, people or whatever. Even in literate societies, the social facts, such as those that imply kinship, work, commerce and political power, are not texts, nor [are they] "like" texts. Therefore, the linguists, hermeneutics and literary critics are not prepared to study social life. (Bunge 1996, p. 291)[10]

---

[9]For the formulation of these hypotheses, Bunge admits that *Verstehen* can help, but not in the form of "putting oneself in the place of the Other" (Dilthey), but in the form of looking for the (typical) motives of the Other's actions (as proposed by Max Weber). In any case, *Verstehen* does not excuse us from having to prove our hypotheses (Bunge 1996, pp. 154–155).

[10]These alleged confusions lead Bunge to frequently affirm that Hermeneutics, as well as Phenomenology and Existentialism, are irrational theoretical positions. He considers "Postmodernism" analogously (see Bunge 1995, Chaps. 9 and 10).

Moreover:

> Social facts are not bunches of symbols (...) our "interpretation" of the actions of others is not the sole determinant of our own actions: we are motivated or inhibited by a quantity of material conditions (natural and social) and by the way that we "interpret" the actions of others (Bunge 1996, p. 289).

The "hermenêutical analysis" should, therefore, be confined to its natural place, that is, the understanding of texts in the strict sense. Even so, Bunge considers it necessary to distinguish three uses of hermeneutical practice: to gain an understanding of the original meaning of a text, to evaluate actions in the light of a certain code or doctrine (the standard procedure in Law), or to turn to the content of texts to understand or modify present events. In the first case, the attempt is valid and can be fertile, although it is not free of risks, as forcing the author to think like his interpreter. The value of the second modality of Hermeneutics depends on the validity of the code or doctrine that is used as a reference. And this validity must be appreciated in comparison to other standards, which can lead to modifying the code or the doctrine. The third use of Hermeneutics is nevertheless the most problematic for Bunge. It involves explaining present reality by means of texts produced by different people, in circumstances that are also different from the current ones. That practice involves a distortion, whether of the text, or of the current events, or of both things, because otherwise the events of today could not adapt to the content of the text (Bunge 1983, pp. 221–222).

16.4 Conclusion

Bunge's argument in relation to Positivism *sensu stricto* is very clear, and helps in not confusing Positivism with Scientism, as I observed in another work (Cupani 1991). Moreover, I believe that Bunge's portrait of the pretensions and limitations of Positivism and Neopositivism, particularly when confronted with effective scientific practice, is correct and convincingly drafted. Nothing different could be expected from a thinker with Bunge's education as a physicist. His criticism of the positivist rejection of traditional philosophical problems (in particular, the ontological and axiological ones) has the unquestionable merit of reviving a range of questions that human beings seem not to be able to avoid, theoretically and practically. His revival shows that these questions can be resolved in a manner compatible with the level of the critical spirit that humanity has currently attained. In this way the solutions to philosophical problems can be useful to other human activities, beginning with science. For these reasons (and no less importantly), it is possible to establish that the scientificist Bunge is not a positivist, as at times it is heard commented in a critical tone.

But, if the Bungean criticism of Positivism deserves applause, his criticism of Marxism warrants some reservations. As is evident from Bunge's presentation of the Marxist theses, he considers them in their most generic form, and therefore superficially. He seems to have in mind the so-called "vulgar Marxism", that is, of the enunciations into which Marxist ideas, seen as a doctrine, are generally condensed, above all in ideological debates. In this superficial plane, Bunge's critical observations are not difficult to share, and are not new. Thus, for example (and to limit myself to two theses), the doubts in relation to Dialectics, understood either as a research method or as a set of grand cosmic laws, have accompanied Marxism since its origin, and were even raised by Marxists thinkers. And the question of the relationship between the social "infrastructure" and the "superstructure" was recognized as complex by most Marxist scholars.[11] Meanwhile, Bunge correctly affirms (although he is not innovative here either) that the defenders of Marxism generally adopt a dogmatic attitude in relation to the theses they defend, and that this attitude harms the evolution of the theory. Nevertheless, in contrast, it is clear that Bunge does not make any effort to understand the theoretical and social *peculiarity* of Marxism. At this other level of analysis, the value of his criticism is doubtful.

Concerning the theoretical peculiarity of Marxism, it results from its Hegelian roots, specifically, the idea of Dialectics. Well, it is impossible to understand the latter (even not agreeing with it) without provisionally suspending the conviction that the bivalent formal logic constitutes the key to all valid thinking, as Bunge maintains. Only under this condition do notions such as "(real) contradiction", "unity of opposites", etc. make sense. In Bunge's analysis, Dialectics can be nothing else than how it appears to him: as a set of confused affirmations. Confused, and to the degree to which they are "clarified", false, as for example in the composition of the atom brandished by Bunge.

By the way, the example would unlikely be accepted by a partisan of Dialectics as illustrative of what Dialectics actually consists of. The same could be said of Bunge's criticism of Marxism's criteria of truth, which is raised by Bunge to equal conditions of pragmatist criteria. Here, the failure is in not perceiving that for Marxism truth is not decided in *practice*, but in "praxis", a notion that consists in the "synthesis" of theory and practice, as is known. These notions, regardless of their value, appear to be beyond Bunge's comprehension.

Regarding the social peculiarity of Marxism, it resides in the constant remission of social ideas, institutions and practices to a base constituted by the conflicts of a society that does not cease to be divided into classes of confronting interests. Without this examination, the analysis of any social event is subject, for Marxism, to the danger of "alienation", that is, to not perceive its true nature, regardless of how "objective" one believes himself to be. It appears to me that in this case Bunge also does not know how to admit, at least experimentally, the Marxist approach, and it is symptomatic of this failure that he considers class struggle as a *political* conflict

---

[11]It should be noted that the use of the expression "ideal superstructure" by Bunge to identify a supposed contradiction, is not found as a rule in Marxist writers. They commonly use the expression "superstructure" to designate political elements (state and law) and ideological ones (morality, religion, education, art, philosophy, science, etc.), to which are *not* attributed an "ideal" existence, according to the famous principle that "consciousness can never be something other than the consciousness [of concrete men]" (Marx and Engels 1977, p. 37).

when, formally, for Marxism, the political aspect of human confrontations derives from their basically *social* character. Without recognizing this, the merely political consideration is naïve and deceiving.

Similar reflections inspire Bunge's position in relation to Hermeneutics. He is not able to assume, not even as an exercise, the intellectual attitude in which the questions conceptualized by this philosophical trend make sense. Here, the main reason for Bunge's difficulty is found in his immediate rejection of all forms of subjectivist Idealism, allied to the severity with which he semantically analyzes the concepts used by Hermeneutics. Nevertheless, it is necessary to observe that Phenomenology and Heidegger's Existential Analysis, which decisively influenced twentieth century Hermeneutics, had the purpose of avoiding or going beyond the realism-idealism dichotomy (as can be appreciated in Husserl's concept of *Lebenswelt* and *in-der-Welt-sein* by Heidegger). At the same time, and for the same reason, both philosophies aim to clarify (each with its own approach, through the description and interpretation of personal experiences) the way in which our conviction arises that an "objective" world "confronts" what we consider our "self". In other words, it involves becoming aware of what is commonly not perceived: our sense of "reality".[12] In this task, the activity of consciousness reveals itself in turn as essentially significant or signifying, "a donor of meaning", as is often read in phenomenological texts, and the world winds up as a vast system of meanings. On the other hand, Phenomenology explores the various modalities that consciousness can adopt, mainly the passage of daily life with its pragmatic motivations to *theory* and *objective* knowledge.[13]

All of this is achieved based on the typical procedure of Phenomenology: the "placement between parenthesis" or "suspension" (*epokhé*) of the "natural attitude" (Husserl). This means, of certainties that are part of our spontaneous daily consciousness, our common sense that includes – it should be recognized – the conviction that scientific knowledge is superior to vulgar knowledge for understanding what we denominate as the facts of the world and their causes. To see what Phenomenology intends to show, it is essential to practice this "suspension", which is not equivalent – it is important to note – to *denying* what remains provisionally "suspended" (in particular, the existence of a world of facts researched by science), because the purpose of Phenomenology is to arrive at the understanding of how the thus revised convictions are "constituted" (produced). The phenomenological "suspension", *naturally* (or that is, from the perspective of one who is not disposed to abandon the "natural attitude"), does not make sense to Bunge, nor could it. Logically, the thesis of a phenomenological sociology, about the "signifying

---

[12]The proposals of both Husserlian Phenomenology and Heidegger's Existential Analytics are certainly much more complex. Here I limit myself to the aspect that is of interest in relation to Bunge's criticism.

[13]For readers uninformed about Phenomenology and Hermeneutics I recommend Bleicher (1990) and Dartigues (1973).

construction" of the social world and the need to "interpret-it" also could not.[14] Hermeneutical terminology can only be imprecise or inadequate for Bunge,[15] who seems incapable of understanding it in its own terms and above all, of perceiving the usefulness that hermeneutical studies can have for understanding *the very realistic and scientific attitude* that he defends.

Analogously to the case of the pretensions of the sociology of scientific knowledge (which appears similar in certain versions with Phenomenology and Hermeneutics), I believe that in the hermeneutical analysis one can recognize the exploration of an effective aspect of all scientific activity. I refer to the constant *interpretation* (it seems difficult to substitute the word) of experience implied by human action, and the degree to which the presuppositions of any activity (its object, the questions and proposals referred to it, the form of social exchange that is exercised, etc.) depend on that interpretation. And it is worth repeating, this is not equivalent to denying the value of the interpreted activity.[16]

However, although I thus find Bunge's criticism of Hermeneutics quite limited, I believe that his defense of researching social phenomenon as objective events is opportune in face of attempts to reduce the social sciences to the hermeneutical approach, something that also does not seem acceptable to me. If such an approach is necessary, or at least convenient for avoiding an extreme objectivism, "positivist" social research regards the concern of not falling into an also extreme subjectivism.[17] Naturally, there is a risk here of perpetuating a dialog of the deaf among the partisans of each approach, and for this reason, to conclude I mention the importance of the attempts to integrate these apparently antagonistic perspectives, as Habermas did in his *Theorie des kommunikativen Handelns* (1981).

References

Berger, P., & Luckmann, T. (1966/1972). *La Construcción Social de la Realidad* (The Social Construction of Reality, 1966). Buenos Aires: Amorrortu.

Bleicher, J. (1990/1980). *Contemporary hermeneutics*. London/New York: Routledge.

Bunge, M. (1979). *Treatise on basic philosophy, tomo 4: Ontology II: A world of systems*. Dordrecht: Reidel.

Bunge, M. (1980). *Epistemologia* (*Epistemología*, 1980). São Paulo: Queiros-Edusp.

---

[14] I offer a detailed view of the limitations of Bunge's criticisms of the sociology of knowledge in Cupani (2000).

[15] And in general, for everyone that defends "positivism", as I showed in Cupani (1986). Moreover, his bias in favor of science prevents Bunge from appreciating the analysis of vulgar language as is practiced since Wittgenstein, because Bunge is always suspicious of the (in his understanding) mistaken assimilation of the vulgar and scientific use of terms.

[16] See Heelan (1983) for the specific case of the natural sciences.

[17] Bunge criticizes, pointing to that danger, Winch's thesis according to which social relations are "expressions of ideas about reality" (Winch 1958) commenting that, in that case "social relations must be a kind of internal relations" (Bunge 1996, p. 288).

Bunge, M. (1983). *Treatise on basic philosophy, v. 5: Exploring the world*. Dordrecht: Reidel.
Bunge, M. (1985). *Seudociencia e Ideología*. Madrid: Alianza.
Bunge, M. (1987). *Vistas y Entrevistas*. Buenos Aires: Siglo Veinte.
Bunge, M. (1995). *Sistemas Sociales y Filosofía*. Buenos Aires: Sudamericana.
Bunge, M. (1996). *Finding philosophy in social science*. New Haven: Yale University Press.
Cupani, A. (1985). *A crítica do Positivismo e o futuro da Filosofia*. Florianópolis: Ed. da Universidade Federal de Santa Catarina, Brasil.
Cupani, A. (1986). A Hermenêutica ante o Positivismo. *Manuscrito (Brazil), IX*(1), 75–100.
Cupani, A. (1991). A filosofia da ciência de Mario Bunge e a questão do «positivismo». *Manuscrito (Brazil), XIV*(2), 113–142.
Cupani, A. (2000). Realismo científico: el desafío de la sociología de la ciencia. *Revista Adef (Buenos Aires), XV*(1), 29–40.
Dallmayr, F. R., & MacCarthy, T. (Eds.). (1977). *Understanding and social inquiry*. Notre Dame: University of Notre Dame Press.
Dartigues, A. (1973). *O que é a Fenomenologia? (Qu'est-ce que la phénomenologie?*, 1973). Rio de Janeiro: Livraria Eldorado Tijuca Ltda.
Giedymin, J. (1975). Antipositivism in contemporary philosophy of social sciences and humanities. *British Journal for the Philosophy of Science, 26*, 275–301.
Habermas, J. (1981). *Theorie des kommunikativen Handelns*. Frankfurt a.M.: Suhrkamp Verlag.
Heelan, P. (1983). Natural science as a hermeneutic of instrumentation. *Philosophy of Science, 50*, 181–204.
Heidegger, M. (1967, orig. 1927). *Sein und Zeit*. Tübingen: Max Niemeyer Verlag.
Husserl, E. (1962). *Ideas relativas a una fenomenología pura y una filosofía fenomenológica (Ideen . . .*, 1913). México: Fondo de Cultura Económica.
Kolakowski, L. (1981). *La Filosofía Positivista (Panstwowe Wydawnictwo Naukowe*, 1966). Madrid: Cátedra.
Marx, K., & Engels, F. (1977). *A ideologia alemã (Die deutsche Ideologie*, 1973). São Paulo: Grijalbo.
Schutz, A. (1972). *Fenomenología del Mundo Social (Der sinnhafte Aufbau der sozialen Welt*, 1932). Buenos Aires: Paidós.
Stockmann, N. (1983). *Antipositivist theories of the sciences*. Dordrecht: Reidel.
Winch, P. (1958). *The idea of a social science and its relation to philosophy*. London: Routledge & Kegan Paul.

Part III
Physics and Philosophy of Physics

Chapter 17
Physics and Philosophy of Physics in the Work of Mario Bunge

Gustavo E. Romero

Abstract This brief review of Mario Bunge's research on physics begins with an analysis of his masterpiece *Foundations of Physics*, and then it discusses his other contributions to the philosophy of physics. Following that is a summary of his more recent reactions to scientific discoveries in physics and a discussion of his position about non-locality in quantum mechanics, as well as his changing opinions on the nature of spacetime. The paper ends with a brief assessment of Bunge's legacy concerning the foundations of physics.

17.1 Introduction

Before embarking on a more formal examination of Bunge's work in philosophy, physics and the overlap of both, I would like to indulge in some personal narrative and reflect on the depth of influence which Mario Bunge has had on my career, as an example of how he has been a central figure in the work of so many others over many decades. Given the purpose of this paper and the context of its inclusion in a larger work recognizing the breadth and depth of his footprint across many fields, such a diversion does not seem misplaced. Following this background story, there is a brief summary of some of the major points in Mario's own early life and academic career. This provides some context to the subsequent analysis of his works and his Cartesian approach to the development of a fundamental re-construction of the theory and practice of scientific methods.

Mario Bunge began academic studies in physics in March 1938. His doctoral thesis was presented at the University of La Plata in his native Argentina and was supervised by Guido Beck, a distinguished theoretical physicist (Bunge 1960). By 1945 Bunge had published papers on nuclear physics in *Physical Review* and *Nature*, two top international research journals. His involvement in philosophy and quixotic

G. E. Romero (✉)
Instituto Argentino de Radioastronomía (IAR, CCT La Plata, CONICET-CIC), Buenos Aires, Argentina

© Springer Nature Switzerland AG 2019
M. R. Matthews (ed.), *Mario Bunge: A Centenary Festschrift*,
https://doi.org/10.1007/978-3-030-16673-1_17

enterprises such as the philosophical journal *Minerva* and the Argentine Workers University, of which he was founder and main driving force, resulted in long delays in his academic career. Bunge worked as a teaching assistant in physics courses at the University of La Plata until, after the fall of the Peronist government, he could obtain an appointment as Professor of Physics. Soon afterwards, he also gained positions in physics and philosophy at the University of Buenos Aires. From 1958 and until he left the country in 1963 he was Full Professor of Philosophy at the University of Buenos Aires. During this lengthy formative period Bunge assimilated research habits, was in direct contact with all major theories of modern physics and started to shape his unique philosophical views. His experiences as a research scientist and his involvement in the controversies in physics during this period, in particular those related to the interpretation of quantum mechanics, were key in the formulation of his philosophical project, which was later expressed in his *Treatise on Basic Philosophy* (1974–1989) and a series of books that followed. The basic outline of his philosophy was already present in his *Foundations of Physics* (Bunge 1967a), a superb work of clarification of all major theories in this field of science. This chapter discusses how physics and philosophy interrelate in Bunge's formative years and how this interrelation leads to a new approach to philosophy.

In his *Other Inquisitions*, the Argentine writer Jorge Luis Borges evokes a phrase (perhaps apocryphal) of William Henry Hudson, where Hudson says that many times in his life he undertook the study of metaphysics, but happiness always interrupted him. I confess, not without some melancholy, that few events have interrupted my many years of study and dedication to physics. However, I remember some bright moments of intellectual happiness in a life that may have lacked many things, but not thought. Even today I recall my excitement when I first read Mario Bunge's *Foundations of Physics*. It was my reading of Bunge's *Foundations* that re-ignited my enthusiasm for physics in the late 1980s. This was at a time when local and national politics were had more to do with petty and destructive power struggles than genuine leadership, and this had an adverse effect on researchers aspiring to further their knowledge unhindered by such negative external influences. His book had not been widely acknowledged by philosophers, but he did write extensively and with clarity about physics, using logic and mathematics. It was this rigorous approach which I adopted from the outset in my own pursuit of scientific knowledge.

When Bunge wrote about general relativity or quantum mechanics, he focused on the theories and their development: axiomatized, formalized, and clarified. Logic, mathematics, semantics: all were used in his analysis of a science in which I intended to base my own career. His work was a continuation of Hilbert's program, with knowledge of the formal limitations and carried out in the most rigorous possible way, by an expatriate Argentine who had studied in the same classrooms as me.

I decided to study all of Mario's work, and with my friend Santiago Pérez Bergliaffa we formed a group to read and analyze his books. By 1991 we had already mastered the essentials. It was then that we learned that Bunge was to teach a course on Epistemology at the University of Santa Catarina, in Florianópolis, Brazil. We did not hesitate: we bought a bus ticket (which we paid for with our lean wages) and went to hear Bunge. The course was brilliant and after posing many questions in class, we accepted his cordial invitation to the lobby of his hotel. He encouraged us to analyze the axiomatization of quantum mechanics. When we returned to Argentina, we decided to attack the problem. We discussed the technical issues with Mario, and later with our thesis supervisor. Shortly afterwards we had produced a new axiomatization of non-relativistic quantum mechanics which we sent to Bunge. To our surprise, he liked it and urged us to publish our results (something that had never occurred to us). Soon the paper appeared in the *International Journal of Theoretical Physics* (Perez Bergliaffa et al. 1993).

It would be the first of many articles I would write about philosophy of physics and scientific philosophy, despite having developed my main professional career in astrophysics. It would also be the beginning of a friendship with Mario that continues to the present day. When he presented his *Between Two Worlds: Memoirs of a Philosopher-Scientist* (Bunge 2016) at the University of Buenos Aires, I said publicly that my encounter with Mario Bunge has been one of the main events of my intellectual life. And of all his works, however much I have admired his *Treatise* (Bunge 1974a, b, 1977a, 1979, 1983a, b, c, 1985a, b, 1989), his *Causality* (Bunge 1959), his *Chasing Reality* (Bunge 2006), it was *Foundations of Physics* that had the greatest influence.

17.2   Early Years and First Works on Physics and Its Philosophy

Mario Bunge was born in West Florida, 17 km from the city of Buenos Aires, Argentina, on September 21, 1919. As the son of a doctor (Dr. Augusto Bunge) and socialist congressman of a patrician family, the young Mario always lived in an environment of intellectual freedom and social commitment. His mother of German origin, Marie Müser, had been a nurse with the Red Cross in China. Young Mario was soon attracted to philosophy. In 1936 he began to make more or less systematic readings of philosophical topics, but he soon became convinced that if he wanted to do philosophy seriously he had to first know science thoroughly.

He enrolled in the physics courses at the National University of La Plata in 1938, graduating in 1944, and later received his doctorate in 1952, with a thesis on relativistic electron kinematics. His mentor in physics was Guido Beck (1903–1988), who had been an assistant to Heisenberg in Leipzig.

In 1944 Bunge published his first physics article in the respected journal *The Physical Review* (Bunge 1944b) on the subject of the representation of nuclear forces. It was the first theoretical article on this subject undertaken in Argentina. In that same year he participated in the founding of the Argentine Physical Association (AFA, after its Spanish name, it is a society that still exists and is very active).

But before these scientific activities, Bunge had ventured into philosophy, publishing in 1939 a volume entitled *Introdución al Estudio de los Grandes Pensadores*

(Introduction to the Study of Great Thinkers) (Bunge 1939). In 1944 he founded the philosophy magazine *Minerva*, where articles were published by several of the most respected thinkers in Latin America. Bunge himself wrote several papers for the magazine, which lasted just one year. One of those articles, "What is epistemology?" (Bunge 1944a), is perhaps the first article on modern philosophy of science written in Spanish.

In 1945, in *Nature*, he published a new article on nuclear physics, on the scattering cross-sections between protons and neutrons (Bunge 1945). In 1951 his first philosophy article in English appeared in the journal *Science and Society:* "What is chance?" (Bunge 1951). The article would be the starting point of Bunge's research on causality, which culminated in the publication of his classic book *Causality* (Bunge 1959). In that work, Bunge considered the role of causality in modern physics, and concluded that the world, although it is deterministic, is not strictly causal. Determinism implies legality in the sense that it should follow certain laws. There are no events that are not restricted by natural laws, but those laws are not necessarily causal. The world also admits probabilistic laws.

By that time (the mid-1950s), Bunge began to explore the different interpretations of quantum mechanics and to clarify his own position on this important and controversial issue. During this period, he published two important articles: *A survey of the interpretations of quantum mechanics* (Bunge 1956), and *Strife about complementarity* (Bunge 1955). That time also marks the first contact between Bunge and Karl Popper, who at the time also defended a realistic interpretation (although in his case a statistical one) of quantum mechanics. In the early 1960s Bunge moved first to the USA, then Germany, and finally settled in Canada in 1966, taking a chair in philosophy at McGill University, where he remained till his retirement and beyond.

Bunge produced a large number of works in philosophy of physics, published in various media such as the *American Journal of Physics* (Bunge 1956, 1957, 1961, 1966), *The Monist* (Bunge 1962), *Reviews of Modern Physics* (1967c), and *British Journal for Philosophy of Science* (1967d), among other journals. His essential vision of the great themes of physics would mature in these years and culminate in the publication, in 1967, of his extraordinary *Foundations of Physics* (op. cit.).

17.3 Foundations of Physics

Foundations of Physics (Bunge 1967a) was a pioneering and unique book. Pioneering for its rigorous style both from the philosophical and scientific points of view. Although many times physicists had written about philosophy, and philosophers had written on physics, this time the author was in complete control of both fields of research. It was a unique book also in its scope (it dealt with general methodology in philosophy of science and all the main theories of physics), in its depth, and in its rigor. Bunge presented rigorous axiomatizations of all the theories he discussed.

The book is full of observations and clarifications of many conceptual errors that, given the clarity of Bunge's explanations, become obvious to the reader. It is a very stimulating text in that it provides many ideas and suggestions for further investigation. The second chapter includes an outline of a philosophical research program that Bunge went on to develop between 1974 and 1989: a complete system of scientific philosophy that includes semantics, ontology, epistemology and ethics. It provided the seed of what would later be the monumental *Treatise on Basic Philosophy* (op. cit.), a work of eight volumes in which the originality of Bunge's philosophical system is apparent. The *Treatise* may be the most important philosophical research project of the twentieth century and has its origin in the fact that, when conducting his research for the *Foundations of Physics*, Bunge realized that the underlying philosophy on which the science is based was not adequately developed. That gave rise to his development of a complete system of scientific philosophy, which would become a very influential reference.

After *Foundations of Physics*, Bunge continued to develop different aspects of his philosophy of physics in his *Delaware Seminar* (Bunge 1967e), *Philosophy of Physics* (Bunge 1973), *Controversies in Physics* (1983a, in Spanish), in Volume 7 of the *Treatise*, and in *Mind and Matter* (Bunge 2010). His most recent papers on philosophy of physics include: "Does the Aharonov-Bohm Effect Occur?" (Bunge 2015) and "Gravitational Waves and Spacetime" (Bunge 2018), both published in the journal *Foundations of Science*. Thus, Bunge has a continuous record of more than 70 years of work in both physics and philosophy of physics – a rare achievement.[1]

17.4 Outline of Bunge's Views

The two essential points of Bunge's philosophy of physics are its epistemological realism and ontological materialism. Scientific theories represent a reality that has an objective and independent existence apart from the subject. That does not imply, of course, that in our descriptions of reality we can dispense with a reference system. A description of reality can be perfectly objective and realistic, but it is always relative to a given reference frame.

In the 1960s, Bunge addressed the axiomatization of the major theories of physics, always highlighting the existence of a class of reference frames for each theory. These frames are concrete rather than conceptual, and certainly should not be confused with the coordinate systems we use to specify dimensional locations in physics.[2] Bunge is very careful to point out that mathematics is a system of

---

[1] See his publications (1967b,1967g through to 1970d, and 1977c, 2000, 2002a, 2003, 2012) in the Reference list. And his publications with Kálnay (1969, 1983a, b) and with Maynez (1977).

[2] This confusion is so common among physicists that it can be found even in the young Einstein.

formal abstractions that we use to represent aspects of real systems. Our ideas and representations of reality are mathematical, not reality itself.

Bunge represents the properties of physical systems by means of functions, or sets of functions, and concrete systems by sets. Thus, in general relativity for example, spacetime is represented by a real, 4-dimensional and pseudo-Riemannian manifold (a very particular class of set), and its properties by means of a set of 10 functions defined on the manifold. These functions are called the metric coefficients. The properties of material systems other than spacetime are represented by another set of functions structured in the so-called energy-impulse tensor. Then, the laws of the theory are represented by relations between functions that determine restrictions on the form of these through systems of equations. If the laws are purely local, the equations are differential. In general, the laws of physics are represented by integro-differential equations, allowing for non-local actions.

In this way, the axiomatic structure of the theories is characterized by three sets of axioms: (1) the purely formal ones that fix the mathematical form of the theory, (2) the semantic axioms, which relate certain formal constructs with physical items, and (3) the law statements, which express the restrictions upon the state space of the referents of the theory. Law statements represent natural laws which are the patterns which arise in the occurrence of events in classes of things.

Bunge was the first to point out the importance of making explicit the semantic axioms in order to clarify the questions of interpretation of the theories of physics. To deal accurately with the interpretative aspects, he developed a whole semantic theory in the first two volumes of his *Treatise*.

Foundations of Physics is the most serious attempt that has been made so far to implement something similar to the Hilbert program of axiomatization of physics. The criticisms that some authors have made on the grounds that the program is obsolete due to Gödel's incompleteness theorems are baseless. These criticisms only show, in fact, a complete misunderstanding of the incompleteness theorems. The incompleteness theorems affirm that (1) no formal theory capable of describing arithmetic is at once consistent and complete, and (2) the consistency of a theory cannot be proved within the same theory. In the case of the theories of physics, if they are completely and correctly formalized, there is no danger of inconsistencies since the theories do not aspire to completeness: what we seek in physics is the best possible description of reality, not a theory complete in the formal sense. Nor are we interested in testing the consistency of our theories within the theory itself: every physicist works using appropriate meta-languages, even if he or she does not know it.

On the other hand, much is gained in clarity by axiomatizing a theory, as Bunge explains in detail in his *Philosophy of Physics*. For example, the axiomatization of quantum mechanics shows that the recurrent "observer" of textbooks is nowhere to be found. Quantum mechanics is not "subjective": the world does not exist, as Berkeley thought, because we observe it. It is the other way around: we can observe it because it exists. The observer plays no role in quantum mechanics. Bunge has pointed this out countless times, but even today we continue to hear about the supposed relationship between the consciousness of the observer and the "wave function", or worse, about its collapse.

The wave function of quantum mechanics does not "collapse". How would it collapse if it is a mathematical object, that is, a conceptual creation of the human mind? Countries, nerves, and buildings can collapse, but not the wave function. The wave function is a complex function in an infinite-dimensional space! Bunge showed with his axiomatic procedure that what actually evolves, in any case, is the quantum system in interaction with its environment, but not the wave function (Bunge 1967c, d, f).

The dynamic equation of quantum mechanics refers to physical systems, things such as electrons or photons. The wave function, as postulated by Born, simply gives the probability that the system is in this or that state, when the equation is solved for this or that set of boundary conditions. The relationship with the environment, Bunge correctly points out, must be non-linear, and cannot therefore be described by the wave function that satisfies a linear differential equation. To study the relationship between the quantum system and the environment, a quantum theory of measurement is needed. Long after Bunge wrote *Foundations of Physics*, the theory of decoherence was developed, which shows how quantum properties are rapidly lost when systems of many-particles interact with a complex environment.

17.5 Bunge on Quantum Mechanics

In his books and articles, Bunge has demolished many popular myths about quantum mechanics and its interpretations. He has pointed out, for example, that Heisenberg's inequalities are a theorem, not a principle (Bunge 1977b); and they have nothing to do with observations, since they are deduced from the non-commutativity of quantum operators and the (purely mathematical) inequality of Schwarz.

Bunge has pointed out how the wave function of Schrödinger's famous cat does not exist: what exists is a scribble that some stray people write in its place and to which they attribute supposed quantum properties. It is the same thing if they write $\Psi|_{cat}$ as if they prefer $\Psi|_{Jupiter}$ or $\Psi|_{sausage}$. The symbol Ψ must denote a mathematical expression that satisfies the corresponding equation. In a system as complex as a cat, the quantum properties have disappeared at a level of composition just above that of a molecule. Hence, cats are not in a superposition of quantum states. Even if the master of the cat is Schrödinger!

In later works (for example, *Controversies in Physics* (Bunge 1983a) and *Mind and Matter* (Bunge 2010)) Bunge deals with the EPR paradox, Bell's inequalities, the alleged violation of realism by Aspect's experiments and the like. He also discussed the so-called quantum entanglement. Bunge always offers us the same thing: the most sensible response that corresponds to the real formalism of the theory. He points out, for example, that the refutation of Bell's inequalities does not imply that realism is not valid, but that deterministic theories with hidden variables are incompatible with the assumption of locality. Quantum mechanics is clearly not local. But that does not imply a problem for a realistic interpretation, unless you have a classicist conception of realism, something that certainly is not the case for Bunge.

Perhaps we could say that Bunge's interpretation is the most direct and simple of all the proposals. I call it *the minimalist interpretation of quantum mechanics*. We could say that it is a non-local realistic interpretation with a sui generis ontology, its famous "quanton" postulation. Bunge interprets the operators of quantum mechanics as mathematical objects that represent physical properties of the referents of the theory: these are sui generis quantum systems. These objects do not share the properties of classical systems. Under certain circumstances they can behave as if they were particles and under others as if they were waves, but they are neither waves nor particles, which are classic objects (Bunge 2002b). Much less they are both, particle and wave, at the same time. *Quantons* are quantum, singular objects, with quantum properties different from the classical ones. Spin, for example, is not the intrinsic angular momentum of the quantum system: quantum systems do not rotate. Spin resembles in some respects the intrinsic angular momentum of a classical object, but it is a quantum property. The same happens with the other properties, such as lepton number, isospin, location, velocity, and so forth. The only exception is energy, which is the capability of the system for changing. Energy, for Bunge, is the only truly universal property. It is what all existents share. In the quantum domain, it is a discrete property and, consequently, the *quantons* change discretely.

How is it possible that a composite quantum system, prepared in a certain state, remains in some way bound, even long after it has been separated, even if the separation is such that it makes any direct or causal interaction impossible? What is the origin of quantum entanglement? Before seeing how Bunge responds, certain facts, sometimes overlooked, need to be emphasized: (1) Entanglement is not universal (like gravitational action, say); it occurs only between quantum systems that were prepared in a certain way; (2) It does not involve the propagation of energy at a speed greater than that of light; (3) It does not imply a change of state of the system or a remote action; (4) It cannot be used to transmit information.

Let us examine an example of simple entanglement. Consider a non-polarized light source that emits photons in all directions. Suppose that two photons come out in opposite directions. Their total polarization, by the way they were produced, is null. Each of the photons, however, has an equal and opposite polarization. Quantum mechanics implies that one of them has a probability ½ of having a polarization (+) and a probability also ½ of having the opposite polarization (−).[3] Suppose now that we determine that the polarization is (−). Then, we conclude with probability 1 that the other photon has polarization (+). However, the same theory of quantum mechanics predicts that if the other photon is an independent system, its probability of being in the (+) mode is not 1 but ½. The experiments show that the second photon is always in the state that corresponds to the initial preparation, regardless

---

[3] Photons have only two polarization modes.

of the determination of the state of the first photon. The system is still "connected" in some way. This is the "spooky" action at distance mentioned by Einstein.

Actually, there is no connection, but non-local correlation: once an interlaced state has been formed, the system remains entangled regardless of the spatial separation of the components. Does it contradict common sense? Yes, for sure. And it is not the only thing that contradicts it in the quantum world. This should not surprise us, since common sense has been forged with experiences obtained in a different ontological level, the macroscopic one, which we call "classical".

When we specify the state of the first photon, the state of the second photon is specified, according to the initial preparation of the system. Once an interaction has destroyed the entanglement, the components are separated and there are no more correlations. Note that if we want to use entanglement to transmit information faster than light we will fail: there is no way that at the moment the second polarization is measured the value of the first one is known. That information can only be transmitted at the speed of light, as always. There is also no instant "work" on the second photon performed by the first. There is no change of state, but a *specification* of the state of the second photon that is correlated with that of the first photon. If there is work on the photon, the second detector does it, locally.

Bunge, then, invites us to accept the real world as it is: non-local, legal (the laws represented by the equations of quantum mechanics always apply), and independent of the cognitive subjects: it does not matter whether the second photon is registered by an instrument or interacts naturally. All these processes have been occurring in the universe long before human beings appeared on Earth, and they will continue to occur long after life has disappeared.

Bunge's position has sometimes been compared to that of Sir Karl Popper (see, for instance, Popper 1992). While it is true that Popper, like Bunge, was realistic in his interpretation of quantum mechanics, he thought that the wave function corresponds to a collection of particles and not to individual quantum objects. In that sense, Popper's position was similar to that of American physicist Leslie E. Ballentine, in his book *Quantum Mechanics: A Modern Development* (Ballantine 1998). Popper, on the other hand, was willing to accept that certain quantum phenomena have a relevance in the explanation of consciousness, something also unacceptable to Bunge. Popper supported indeterminism, in the sense that some non-legal events might occur. Bunge, on the contrary, always claimed that the world is fully legal, albeit in case of the quantum realm, the laws can be probabilistic. So, we can say that both thinkers agreed on the realistic interpretations of quantum mechanics, by differed on ontological views of many particular issues.

17.6 Bunge on Fields

Most of the *Foundations* is devoted to classical theories of physics, such as mechanics, electrodynamics, and the general theory of relativity. In Bunge's analysis of these theories, we can already glimpse the general ideas underlying what would later form the core of his ontology, expressed in volumes 3 and 4 of the *Treatise*. Bunge considers that the components of the world are corpuscles and fields. The form in which they associate or compose to form larger entities is different: fields overlap, while particles juxtapose (the classical ones at least).

Bunge's analysis of field theories is particularly illuminating. He clearly separates the mathematical from the physical concept of fields. A mathematical field is a conceptual object that satisfies a theory that incorporates a principle of minimum action and is described by Lagrangian density. A physical field, on the other hand, is a real object. Bunge distinguishes three types of theories of fields in physics: theories of pure, material and mixed fields. The first have as referents physical objects that cannot be eliminated by mere changes of the reference system adopted in the formulation of a model of the theory. The second type represent properties of extended material systems. Examples of pure fields are the electromagnetic and the gravitational fields. Examples of material fields are the velocity field of a fluid and the temperature field of a body. It is not possible to eliminate the electromagnetic or gravitational field globally by changing our description of nature. On the other hand, everything we say about material fields can be expressed in terms of particles, at least in principle. The theories of mixed fields are those that combine pure fields with material ones, such as macro-electromagnetism.

Bunge's analysis of the electromagnetic field is remarkably lucid. Among other issues, he points out that theory does not properly contain the hypothesis that charges are the sources of the field. Strictly speaking, electromagnetic theory is a theory of the interactions between fields and charged particles. The hypothesis that charges are sources of the field must be added to the Maxwell equations in order to discard the advanced contributions (determined by future events) to the solutions of the inhomogeneous Dalambertian equation. This is done by means of the application of the principle of causality. This hypothesis is logical, ontological and epistemologically independent from the rest of the theory.

Something similar happens with the general theory of relativity: the hypothesis that matter is the cause of the curvature of spacetime (gravitational field) is *a posteriori*. Bunge points out other points poorly understood in connection with general relativity. For example, that the difference between gravitational and inertial mass is a mere conjecture, refuted by experience, and hence the identity of "both" masses is not a physical law. In fact, there are no such masses.

Bunge points out the purely heuristic character of the principle of equivalence (which can be derived as a theorem in his axiomatic analysis), and clarifies that the principle of general covariance is a meta-nomological statement (that is, a prescriptive statement for the formulation of basic laws). He also asserts the incompleteness of the theory, the fact that it is not only matter that determines the geometry of spacetime but matter *plus* the initial and contour conditions, and finally that the existence of gravitational waves is compatible with the theory, but it is not a direct theorem of it.

With the discovery of gravitational waves in 2016 (Abbott et al. 2016), Bunge modified his vision of the ontology of general relativity, admitting the identity of spacetime and the gravitational field (Bunge 2018). However, he has maintained his "presentist" position, in the sense that he thinks that only present events exist, and not future and the past events, contrary to what is held by "substantivalists" and most of the realists regarding spacetime, and my own controversy with him in this area (Romero 2012, 2013, 2015, 2017, 2018) points to an ongoing debate.

17.7 Conclusion

Throughout more than 70 years, Mario Bunge has examined and analyzed the foundations of physics. Unlike many other philosophers dedicated to the philosophy of this science, Bunge has also researched in physics and has been a university professor of physics. This has given him a unique insight and depth in his views on this field. Also, in a certain way, it has isolated him from his peers, more inclined to semantic discussions. Most of the philosophy that still appears in academic journals is based on examples that appeal to common sense or adopt a physics that is closer to that of Democritus than to modern science – Ladyman and Ross (2007) provide a critique on the current state of analytic philosophy.

Fifty years after its publication, *Foundations of Physics* continues to be a book of enormous depth. No one who engages with this text will leave empty-handed. I, on re-reading it for this paper, felt its extraordinary impact again, just like in those memorable days of my youth.

Acknowledgements I thank Mario Bunge for more than 25 years of stimulating discussions on physics and philosophy. I am also indebted to Santiago Perez Bergliaffa, Pablo Jacovkis, Daniela Pérez, Luciano Combi, Federico López Armengol, Janou Glaeser, and Héctor Vucetich for valuable discussions, questions, and criticisms. I am grateful with the members of the working group on Mario Bunge's work in Facebook, in particular with Gerardo Primero, Emerson Salinas Caparachin, Manuel Carroza Muro, Nicolás Pérez, Ed Cidd, Silvio Sánchez Mújica, Sergio Riva de Neyra, Gustavo Garay, and Matías Castro. I appreciate the kind hospitality of the University of Barcelona and the Institute of Cosmos Science (ICCUB) and its Head, Prof. Josep M. Paredes. Finally, I am thankful to Micheal Matthews for significantly improving the original manuscript and his superb job as editor of this volume. My research is supported by grant PIP 0338 (CONICET) and grant AYA2016-76012-C3-1P (Ministro de Educación, Cultura y Deporte, España).

References

Abbott, B. P., et al.. [LIGO Collaboration](2016). Observation of gravitational waves from a binary black hole merger. *Physical Review Letters, 116*, 061102.
Ballantine, L. E. (1998). *Quantum mechanics: A modern development*. Singapore: World Scientific.
Bunge, M. (1939). Introducción al estudio de los grandes pensadores. *Conferencias, III*, 105–109. and 124–126.
Bunge, M. (1944a). ¿Qué es la epistemología? *Minerva (Buenos Aires), 1*, 27–43.
Bunge, M. (1944b). A new representation of types of nuclear forces. *Physical Review, 65*, 249.
Bunge, M. (1945). Neutron-proton scattering at 8.8 and 13 MeV. *Nature, 156*, 301.
Bunge, M. (1951). What is chance? *Science and Society, 15*, 209–231.

Bunge, M. (1955). Strife about complementarity. *British Journal for the Philosophy of Science, 6*, 1–12. and 141–154.

Bunge, M. (1956). A survey of the interpretations of quantum mechanics. *American Journal of Physics, 24*, 272–286.

Bunge, M. (1957). Lagrangian formulation and mechanical interpretation. *American Journal of Physics, 25*, 211–218.

Bunge, M. (1959). *Causality: The place of the causal principles in modern science*. Cambridge, MA: Harvard University Press.

Bunge, M. (1960). *Cinemática del electrón relativista*. Tucumán: Universidad Nacional de Tucumán.

Bunge, M. (1961). Laws of physical laws. *American Journal of Physics, 29*, 518–529.

Bunge, M. (1962). Cosmology and magic. *The Monist, 44*, 116–141.

Bunge, M. (1966). Mach's critique of Newtonian mechanics. *American Journal of Physics, 34*, 585–596.

Bunge, M. (1967a). *Foundations of physics*. New York: Springer.

Bunge, M. (1967b). *Scientific research*. Berlin/Heidelberg/Nueva York: Springer.

Bunge, M. (1967c). Physical axiomatic. *Reviews of Modern Physics, 39*, 463–474.

Bunge, M. (1967d). Analogy in quantum mechanics: From insight to nonsense. *British Journal for the Philosophy of Science, 18*, 265–286.

Bunge, M. (1967e). *Delaware seminar in the philosophy of science*. Berlin/Heidelberg/Nueva York: Springer.

Bunge, M. (1967f). The structure and content of a physical theory. In M. Bunge (Ed.), *Delaware seminar in the foundations of physics* (pp. 15–27). Berlin/Heidelberg/New York: Springer.

Bunge, M. (1967g). Quanta y filosofía. *Crítica, 1*(3), 41–64.

Bunge, M. (1968). Physical time: The objective and relational theory. *Philosophy of Science, 35*, 355–388.

Bunge, M. (1969). Corrections to foundations of physics: Correct and incorrect. *Synthese, 19*, 443–452.

Bunge, M. (1970a). Problems concerning inter-theory relations. In P. Weingartner & G. Zecha (Eds.), *Induction, physics and ethics* (pp. 285–315). Dordrecht: Reidel.

Bunge, M. (1970b). The so-called fourth indeterminacy relation. *Canadian Journal of Physics, 48*, 1410–1411.

Bunge, M. (1970c). Virtual processes and virtual particles: Real or fictitious? *International Journal of Theoretical Physics, 3*(6), 507–508.

Bunge, M. (1970d). The arrow of time. *International Journal of Theoretical Physics, 3*(1), 77–83.

Bunge, M. (1973). *Philosophy of physics*. Dordrecht: Reidel.

Bunge, M. (1974a). *Treatise on basic philosophy, vol. 1, sense and reference*. Dordrecht/Boston: Reidel.

Bunge, M. (1974b). *Treatise on basic philosophy, vol. 2, interpretation and truth*. Dordrecht/Boston: Reidel.

Bunge, M. (1977a). *Treatise on basic philosophy, vol. 3, the furniture of the world*. Dordrecht: Reidel.

Bunge, M. (1977b). The interpretation of Heisenbergs inequalities. In H. Pfeiffer (Ed.), *Denken und Umdenken: zu Werk und Wirkung von Werner Heisenberg* (pp. 146–156). Munich: Piper.

Bunge, M. (1977c). Quantum mechanics and measurement. *International Journal of Quantum Chemistry, 12*(Suppl. 1), 1–13.

Bunge, M. (1979). *Treatise on basic philosophy, vol. 4, a world of systems*. Dordrecht/Boston: Reidel.

Bunge, M. (1983a). *Controversias en Física*. Madrid: Tecnos.

Bunge, M. (1983b). *Treatise on basic philosophy, vol. 5, exploring the world*. Dordrecht: Reidel.

Bunge, M. (1983c). *Treatise on basic philosophy, vol. 6, understanding the world*. Dordrecht: Reidel.

Bunge, M. (1985a). *Treatise on basic philosophy, vol. 7, parte I, formal and physical sciences*. Dordrecht: Reidel.

Bunge, M. (1985b). *Treatise on basic philosophy, vol. 7, parte II, life science, social science, and technology.* Dordrecht: Reidel.

Bunge, M. (1989). *Treatise on basic philosophy, vol. 8, the good and the right.* Dordrecht/Boston: Reidel.

Bunge, M. (2000). Energy: Between physics and metaphysics. *Science and Education, 9,* 457–461.

Bunge, M. (2002a). Twenty-five centuries of quantum physics: From Pythagoras to us, and from subjectivism to realism. *Science and Education, 12,* 445–466.

Bunge, M. (2002b). Quantons are quaint but basic and real. *Science and Education, 12,* 587–597.

Bunge, M. (2003). Velocity operators and time energy relations in relativistic quantum mechanics. *International Journal of Theoretical Physics, 42,* 135–142.

Bunge, M. (2006). *Chasing reality: Strife over realism.* Toronto: Toronto University Press.

Bunge, M. (2010). *Matter and mind* (Boston Library in the Philosophy of Science, No. 287). Berlin: Springer.

Bunge, M. (2012). Does quantum physics refute realism, materialism and determinism? *Science and Education, 21,* 1601–1610.

Bunge, M. (2015). Does the Aharonov-Bohm effect occur? *Foundations of Science, 20,* 129–133.

Bunge, M. (2016). *Between two worlds: Memoirs of a philosopher-scientist.* Berlin: Springer.

Bunge, M. (2018). Gravitational waves and spacetime. *Foundations of Science, 23*(2), 405–409.

Bunge, M., & García Maynez, A. (1977). A relational theory of physical space. *International Journal Theoretical Physics, 15,* 961–972.

Bunge, M., & Kálnay, A. J. (1969). A covariant position operator for the relativistic electron. *Progress of Theoretical Physics, 42,* 1445–1459.

Bunge, M., & Kálnay, A. J. (1983a). Solution to two paradoxes in the quantum theory of unstable systems. *Nuovo Cimento, B77,* 1–9.

Bunge, M., & Kálnay, A. J. (1983b). Real successive measurements on unstable quantum systems take nonvanishing time intervals and do not prevent them from decaying. *Nuovo Cimento, B77,* 10–18.

Ladyman, J., & Ross, D. (2007). *Everything must go.* Oxford: Oxford University Press.

Perez Bergliaffa, S. E., Romero, G. E., & Vucetich, H. (1993). Axiomatic foundations of non-relativistic quantum mechanics: A realistic approach. *International Journal of Theoretical Physics, 32,* 1507–1522.

Popper, K. (1992). *Quantum theory and the schism in physics: From the postscript to the logic of scientific discovery.* London/New York: Routledge.

Romero, G. E. (2012). Parmenides reloaded. *Foundations of Science, 17,* 291–299.

Romero, G. E. (2013). From change to spacetime: An Eleatic journey. *Foundations of Science, 18,* 139–148.

Romero, G. E. (2015). Present time. *Foundations of Science, 20,* 135–145.

Romero, G. E. (2017). On the ontology of spacetime: Substantivalism, relationism, eternalism, and emergence. *Foundations of Science, 22,* 141–159.

Romero, G. E. (2018). Mario Bunge on gravitational waves and the existence of spacetime. *Foundations of Science, 23*(2), 405–409.

Chapter 18
Causal Explanations: Are They Possible in Physics?

Andrés Rivadulla

Abstract The existence of causal explanations in science has been an issue of interest in Western philosophy from its very beginnings. That is the reason this work, following an idea of Mario Bunge, makes a historical review of this matter. The modern treatment of this subject takes place since the postulation by Popper and Hempel of the D-N model of scientific explanation, whose viability is scrutinized here from different points of view in the current philosophy of science. The main object of this paper is to present two arguments against the possibility of causal explanations in theoretical physics. The first one concerns the existence, in certain cases, of inter-theoretical incompatibilities, and the second refers to the need to resort, in other cases, to concatenations of laws of different theories and disciplines. The final conclusion will be the defence of a form of theoretical explanation, which follows the Popper-Hempel model, but devoid of any ontological and metaphysical connotations.

18.1 Introduction

Since Plato and Aristotle introduced the idea of causality, a complex network of conforming and opposing positions developed around both the reality of causation and its comprehension, without so far achieving a unanimous position. These issues are identified by Mario Bunge (1982) as the ontological and methodological aspects of causality, respectively.

Complutense Research Group 930174 and Research Project FFI2014-52224-P supported by the Ministry of Economy and Competitiveness of the Spanish Government.

A. Rivadulla (✉)
Department of Logic and Theoretical Philosophy, Complutense University of Madrid, Madrid, Spain
e-mail: arivadulla@filos.ucm.es

© Springer Nature Switzerland AG 2019
M. R. Matthews (ed.), *Mario Bunge: A Centenary Festschrift*,
https://doi.org/10.1007/978-3-030-16673-1_18

303

Already in *Logik der Forschung*, 1935, Karl Popper presented for the first time the logical structure of scientific explanation. Thirty years later Carl Hempel took up the issue and popularized the idea of explanation by deductive subsumption under general laws, which he called deductive-nomological explanation, and claimed that causal explanations are special cases of it.

This article deals precisely with the possibility of causal explanations in theoretical physics. My main contribution will be to present two arguments against this possibility. I will focus *first* on the question of whether deductive subsumption under theoretical principles is sufficient condition to offer a causal explanation of the *explanandum*. In particular: Does the subsumption of Kepler's laws under more general principles offer a causal explanation of planetary motions? My answer will be: definitely not. The reason is that, as there are incompatible gravitational theories that give theoretical explanations of these movements, if such explanations were causal we would have mutually incompatible causal explanations of planetary motions. The *second* argument consists in showing that it is sometimes necessary to resort to long theoretical chains, concatenations of laws, whose existence makes the possibility of a clear and concrete explanation, much less causal, of the event in question fade away. We can find traces of this in Whewell, Lewis and Hanson, among others. The example that I am going to present is the astrophysical explanation of the brightness of the stars.

The final conclusion is then disappointing for the hopes of causal explanation in theoretical science. It is not intended, of course, to affirm that in Nature there is no causality. But theoretical science faces serious difficulties if we assume that its goal is knowledge of causes. Thus, if theoretical science does not guarantee causal knowledge, and, coinciding with Berkeley, perhaps that is not even its aim, would it not be reasonable to give up causal explanations in physics?

In the first two sections I will summarize the history of causality from the ontological and methodological points of view. In the third section I will focus on how the current philosophy of science faces the issue of scientific and causal explanation and their relationships. In the fourth and fifth sections I present my aforementioned arguments about the difficulty of causal explanations in theoretical physics based on: (1) the coexistence of incompatible theories – the most serious threat of scientific realism–, and (2) the necessity of resorting to chains of theoretical laws, the existence of which raises serious doubts about the possibility of causal explanations. In the last section I will argue that the unificationist approach to scientific explanation stumbles before the pitfall of inter-theoretical incompatibility. Finally, in the Conclusion, I will argue in favour of the idea that the goal of theoretical science can only be the search for theoretical explanations, explanations free of any ontological and metaphysical connotations, instead of causal explanations proper.

18.2 Explanation and Causality in the History of Western Philosophy of Science: Optimistic and Sceptical Stances

Mario Bunge identifies the ontological problem of causality with the following questions: "What is causation? What are the relata of the causal relation (things, properties, states, events, mental objects)? What are the characteristics of the causal nexus? Are there causal laws? How do causation and randomness intertwine?" (Bunge 1982, p. 133). The purpose of this section is to present a brief historical review of this ontological problem, from Plato to our days.

For Plato "everything which becomes must of necessity become owing to some Cause" (Plato 1929, 28) and "that which has come into existence must necessarily, as we say, have come into existence by reason of some Cause" (Plato 1929, 29).[1]

Aristotle develops the idea of cause both in *Metaphysics*, where he claims that causes "are that out of which these [things, A.R.] respectively are made" (Aristotle 1985, Book V, Chap. 2) as in *Physics*, Book II, Chap. III, but it is in *Posterior Analytics* where he offers a treatment closer to our interest as philosophers of science. Aristotle expresses the relationship between science and causal knowledge in the following way: "we think that we understand something when we know its explanation" (Aristotle 1975, 94ª20). A translation equally acceptable might be: "We think we know a fact when we know its cause" (Ross 1949, p. 637), and even: "We think that we have science when we know the cause".[2]

In *Posterior Analytics* Aristotle affirms that "What explains why something is coming about (and why it has come about, and why it will be) is the same as what explains why it is the case. (...) E.g. why has an eclipse come about? – Because the earth has come to be in the middle. And it is coming about because it is coming to be there; it will be because it will be in the middle; and it is because it is" (Aristotle 1975, Chap. 12, 10–15).

But Aristotle claims also that "it is evident that what it is and why it is are the same. What is an eclipse? Privation of light from the moon by the earth's screening. Why is there an eclipse? Or Why is the moon eclipsed? Because the light leaves it when the earth screens it" (Aristotle 1975, 90ª15). He maintains that "to know what something is, is the same as to know why it is" (Aristotle 1975, 90ª30) and that "to know what something is and to know the explanation of whether it is are

---

[1]It is not my aim to make here an interpretation of the concept of causality in Plato. To that end, and without any claim of completeness, I refer to Ashbaugh (1988).

[2]This is the translation of Miguel Candel in the Spanish edition of the *Organon* (Aristóteles 1988, p. 412). In any case as Ross recognizes: "In history and in natural science we are attempting to explain events, and an event is to be explained (in Aristotle's view) by reference either to an event that precedes it (an efficient cause) or to one that follows it (a final cause)" (Ross 1949, p. 79, Introduction).

the same; and the account of the fact that something is the explanation" (Aristotle 1975, 93ᵃ5).[3]

Aristotle inaugurates what we might call the *optimistic* tradition in the philosophical thought of the West around the notions cause, causal relation and explanation. In this tradition Francis Bacon (1561–1626) affirms in *Novum Organum* that "true knowledge is that which is deduced from causes" (Bacon 1952, Second Book, II, 2).

Immanuel Kant (1724–1804), in his *Critique of Pure Reason*, 2nd edition of 1787, enunciates the Principle to which the Analogies of Experience are subjected: *Experience is possible only through the representation of a necessary connection of perceptions.* Well, the second analogy is the Law of Causality, according to which *all alterations occur in accordance with the law of the connection of cause and effect.* Kant's conclusion is that "it is only because we subject the sequence of the appearances and thus all alteration to the law of causality that experience itself, i.e., empirical cognition of them, is possible; consequently, they themselves, as objects of experience, are possible only in accordance with this law" (Kant 1998, p. 305). Or: "the principle of the causal relation in the sequence of appearances is valid for all objects of experience (under the condition of succession), since it is itself the ground of the possibility of such an experience" (Kant 1998, p. 312).

For John Herschel the explanation of any new phenomenon is "reference to an immediate producing cause" (Herschel 1830, p. 137). And the causal relation is characterized by: (1) Invariable antecedence of the cause and consequence of the effect; (2) Invariable negation of the effect with absence of the cause; (3) Increase or diminution of the effect with the increased or diminished intensity of the cause; (4) Proportionality of the effect to its cause; and (5) Reversal of the effect with that of the cause (Herschel 1830, p. 145).

John Stuart Mill (1806–1873) proposes in his *System of Logic*, Book III *Of Induction*, Chap. V 'Of the Law of Universal Causation', the Law of Causation, i.e. the law that every consequent has an unconditional invariable antecedent that means a cause, as the universal law of successive phenomena. Moreover, he claims that the notion of cause is the root of the whole theory of Induction: "For every event there exists some combination of objects or events, some given concurrence of circumstances, positive or negative, the occurrence of which is always followed by that phenomenon ... On the universality of this truth depends the possibility of reducing the inductive process to rules" (Mill 1843, p. 214).

---

[3] In this the current philosophy of science clearly disagrees with Aristotle. Hempel for instance claims that "To explain the phenomena in the world of our experience, to answer the question 'why?' rather than only the question 'what?' is one of the foremost objectives of empirical science" (Hempel 1965, p. 245). Lawrence Sklar considers that "To explain, we feel, is to answer the question *why* what occurs, and not just to describe *what*, in fact, does occur" (Sklar 1992, p. 100). And Peter Lipton argues that "The starting point of enquiry into explanation ... is the gap between knowing that something is the case and understanding why it is" (Lipton 2001b, p. 103).

Mill points out that:

> when in the course of this inquiry I speak of the cause of any phenomenon, I do not mean a cause which is not itself a phenomenon, ... the causes with which I concern myself are not *efficient*, but *physical* causes. They are causes in that sense alone in which one physical fact is said to be the cause of another. Of the efficient causes of phenomena, or whether any such causes exist at all, I am not called up to give an opinion. (Mill 1843, p. 213)

Finally, Mill's concept of explanation is: "An individual fact is said to be explained by pointing out its cause, that is, by stating the law or laws of causation of which its production is an instance" (Mill 1843, p. 305).

William Whewell (1794–1866) maintains in his *Philosophy of Inductive Sciences*, that "to proscribe the inquiry into causes would be to annihilate the science" (Whewell 1847, p. 105).

At the end of the nineteenth century, there has been so much philosophical reflection on the concepts of causality and causal explanation in science that the British logician, economist and philosopher of science William Stanley Jevons (1835–1882) did not overlook the debate around these terms. In addition, the stream that we might call *sceptical* began to gain in intensity. For Jevons every explanation is completely unrelated to the idea of cause or causation: "the most important process of explanation consists in showing that an observed fact is one case of a general law or tendency." Or: "Whenever, then, any fact is connected by resemblance, law, theory, or hypothesis, with other facts, it is explained" (Jevons 1958, pp. 533–534). Here Jevons clearly advances the modern notion of theoretical explanation. Jevons maintains against the optimistic causal tradition and against Mill that:

> The words *Cause* and *Causation* have given rise to infinite trouble and obscurity, and have in no slight degree retarded the progress of science. From the time of Aristotle, the work of philosophy has been described as the discovery of the causes of things, and Francis Bacon adopted the notion when he said '*vere scire esse per causas scire.*' Even now it is not uncommonly supposed that the knowledge of causes ... consists, as it were, in getting possession of the keys of nature ... In Mill's *System of Logic* the term *cause* seems to have re-asserted its old noxious power. (Jevons 1958, p. 222)

The Austrian philosopher of science Ernst Mach (1838–1916), Professor of Philosophy of Inductive Sciences at the University of Vienna between 1895 and 1901, proposes the substitution of the idea of causality by that of mathematical function:

> The old traditional conception of causality is of something perfectly rigid: a dose of effect follows on a dose of cause ... The connections of nature are seldom so simple, that in any given case we can point to one cause and one effect. I therefore long ago proposed to replace the conception of cause by the mathematical conception of function. (Mach 1959, p. 89)

And he adds:

> The principal advantage for me of the notion of function over that of cause lies in the fact that the former forces us to greater accuracy of expression, and that it is free from the incompleteness, indefiniteness and one-sidedness of the latter. The notion of cause is, in fact, a primitive and provisional way out of a difficulty. (Mach 1959, p. 92)

As a matter of fact, for Mach "The conception of 'effective cause' and of 'purpose' both have their origin in animistic views, as is quite clearly seen from the scientific attitude of antiquity" (Mach 1959, p. 96).

The climax of the sceptical tradition in nineteenth-century British philosophy of science is reached with the mathematician, biologist and philosopher of science Karl Pearson (1857–1936). In full consonance with the mathematician and philosopher William Kingdon Clifford (1845–1879) – not by chance is Pearson the publisher of Clifford's posthumous book, *The Common Sense of the Exact Sciences* – Karl Pearson claims that "Beyond such discarded fundamentals as 'matter' and 'force' lies still another fetish amidst the inscrutable arcana of even modern science, namely, the category of cause and effect" (Pearson 1911, p. xii). Pearson's epistemological conception is indisputably instrumentalist and full of Machian resonances. In the Preface to the Second Edition of his *Grammar of Science* Karl Pearson maintains that "Step by step men of science are coming to recognise … That all science is description and not explanation" (Pearson 1911, p. xiv), and in the Preface to the Third Edition he still claims that: "Nobody believes now that science *explains* anything; we look upon it as a shorthand description, as an economy of thought" (Pearson 1911, p. xi).[4]

"For science, cause is meaningless" claims Pearson (1911, pp. 128–129). Thus, "The aim of science ceases to be the discovery of 'cause' and 'effect'; in order to predict future experience it seeks out the phenomena which are most highly correlated" (Pearson 1911, pp. 173–174). An opinion which he reiterates later, against Aristotle and all the tradition I have come to call *optimistic*:

> the object of science … is not to explain but to describe by conceptual shorthand our perceptual experience. […] Strong in her power of describing *how* changes take place, science can well afford to neglect the *why*. (Pearson 1911, pp. 302–303)

To finish this first analysis of the notions of causality and explanation in the history of Western scientific philosophy, I refer to the French physicist, historian and philosopher of science Pierre Duhem (1861–1916), widely regarded by many epistemologists as the major driver of contemporary scientific instrumentalism. His instrumentalist positions on what is a physical theory and on the possibility of scientific explanation are of course fully consistent with his antirealist and anti-metaphysical orientation. Following Ernst Mach, several of whose works he mentions explicitly, Duhem (1954, Part I, Chap. II, §2) considers also physical theories as economies of thought: economic representations of experimental laws, and not as explanations thereof. Duhem accurately describes Mach as "One of the

---

[4] According to Mach, science is an economy of thought, a clearly instrumentalist idea. He sums up this concept in the following terms: "The economy of thought, the economical representation of the actual, – this was indicated by me, in summary fashion first in 1871 and 1872, as being the essential task of science, and in 1882 and 1883 I gave considerably enlarged expositions of this idea. As I have shewn elsewhere, this conception, … can be traced back to Adam Smith, and, as P. Volkmann holds, in its beginnings even to Newton. We find the same conception again, …, fully developed in Avenarius (1876)" (Mach 1959, p. 49).

thinkers who have insisted most energetically on the point that physical theories should be regarded as condensed representations and not as explanations" (Duhem 1954, p. 39; French version 1906, p. 59).[5] And he reiterates that "Ernst Mach has defined theoretical physics as an abstract and condensed representation of natural phenomena" (Duhem 1954, p. 53; French p. 82).

Based on Mach, Duhem maintains that "Most often we find that physical theory cannot . . . offer itself as a *certain* explanation of sensible appearances, for it cannot render accessible to the senses the reality it proclaims as residing underneath those appearances" (Duhem 1954, p. 8; French, p. 7). For Duhem, the question of the existence, and, where appropriate, of the nature of a reality other than sensible appearances is something that transcends the experimental method of physics and therefore it is the object of Metaphysics. Thus, Duhem concludes: "*if the aim of physical theories is to explain experimental laws, theoretical physics is not an autonomous science; it is subordinate to metaphysics*" (Duhem 1954, p. 10; French, p. 10).

What, then, is a physical theory for Duhem? His answer is twofold. First, what is not: "*A physical theory is not an explanation.*" Secondly, what it is: "*It is a system of mathematical propositions, deduced from a small number of principles, which aim to represent as simply, as completely, and as exactly as possible a set of experimental laws*" (Duhem 1954, p. 19; French, p. 26). Thus, "a true theory is not a theory which gives an explanation of physical appearances in conformity with reality; it is a theory which represents in a satisfactory manner a group of experimental laws . . . *Agreement with experiment is the sole criterion of truth for a physical theory*" (Duhem 1954, pp. 20–21; French, p. 28).

Needless to say, the content of this section does not exhaust the whole history of the optimistic and sceptical traditions about the possibility of causal explanation in science. The next section shows that, when it comes to analyzing whether or not causes are likely to be known, the debate is both more extensive and dense. The optimistic and sceptical positions reappear, indeed alternating, for centuries throughout history.

18.3   The Debate on the Cognition of Causes in the Western Philosophy of Science

This polemic responds to what Mario Bunge calls the *methodological problem* of causality, namely: "What are the indicators and criteria of causation? How do we recognize a causal link, and how do we test for a causal hypothesis?" (Bunge 1982, p. 133). The purpose of this section is to present, as briefly as possible, the characters and ideas found about the cognition of causes, which the history of the West has produced so far. It is obvious that the attitudes I have come to call *optimistic* and

---

[5]Henceforth, French refers to the first edition of Duhem's book.

sceptical respond to contrary underlying philosophical stances and that presenting them to the full extent exceeds the objective of this section, which obviously is not at all to offer a complete history of Western theory of knowledge.

In the Middle Ages, in the Franciscan School of Oxford, Robert Grosseteste (1175–1253) assumes an optimistic position regarding the knowability of the causes. For him all natural effects are explained by mathematical physics: "omnes enim causae effectuum naturalium habent dari per lineas, angulos et figuras." (indeed, all causes of natural effects have to be given by means of lines, angles and figures) (Quoted from Étienne Gilson 1952, Chap. IX, Sect. I). His disciple Roger Bacon also supports the idea of the acceptance of mathematics for the knowledge of the causes of phenomena: "per potestatem mathematicae sciverunt causas omnium exponere." (based on the power of mathematics, they knew how to set forth the causes of all things.) However, for the founder of the later Franciscan school Duns Scotus (1266–1308) "Nulla demonstratio, quae est ab effectu ad causam, est demonstratio simpliciter." (No demonstration that proceeds from effect to cause is simply a demonstration.) (Cited from Gilson).

Whereas for Thomas Aquinas (1224–1274) the final cause is the cause of all causality, the inductive methods of Grossetest, Scotus and Ockham constitute procedures for the discovery of the underlying causes of observed phenomena. Nicolas d'Autrécourt (1300–1350) denies the possibility of reaching a necessary knowledge of causal relations through these methods, since it cannot be said that a correlation that has been observed has to continue in the future. Both Grossetest and Bacon demanded the experimental test of the principles obtained by induction.

At the beginning of the Modern Era, Andreas Osiander wrote the anonymous preface to the 1543 edition of Copernicus's *De Revolutionibus*, entitled "To the reader concerning the hypotheses of this work". In that preface, referring to astronomy, he claims: "it is sufficiently clear that this art is absolutely and profoundly ignorant of the causes of the apparent irregular movements. And if it constructs and thinks up causes – and it has certainly thought up a good many – nevertheless it does not think them up in order to persuade anyone of their truth but only in order that they may provide a correct basis for calculations" (Copernicus 1993, pp. 505–6). This sentence is considered one of the most characteristically instrumentalist statements in history.

Johannes Kepler maintains, on the contrary, that "the old hypotheses ... do not give the reasons for the number, extent, and time of the retrogressions, and why they agree precisely, as they do, with the positions and mean motion of the Sun." Therefore *"Copernicus's postulates cannot be false, when so reliable an explanation of the appearances – an explanation unknown to the ancients– is given by them, insofar as it is given by them"* (Kepler 1981, pp. 75–77. My emphasis, A.R.).

Francis Bacon (1561–1626) relates scientific discovery to that of causes by asserting that his method, which proceeds by induction and deduction, is "to deduce causes and axioms from effects and experiments; and new effects and experiments from those causes and axioms" (Bacon 1952, First Book, p. 117).

From this moment on, the sceptical and optimistic positions alternate. For instance, George Berkeley (1685–1753) claims the unknowability of causes:

For when we perceive certain ideas of sense constantly followed by other ideas, and we know this is not of our doing, we forthwith attribute power and agency to the ideas themselves and make one the cause of another, than which nothing can be more absurd or unintelligible. (Berkeley 1970, 32)

And in a similar line of thought, David Hume's (1711–1776) analysis of the concept of 'causal relationship' led him to conclude that if a causal relationship involves both a 'constant conjunction' and a 'necessary connection', then it will never be possible to attain any knowledge of a causal relationship. Our sensory impressions never give us necessary connections.

Contradicting these opinions, the British astronomer John Herschel (1792–1871) asserts that the causes are knowable, since experience shows that the real causes of phenomena exist. For example, the cause of the motions of objects is the force acting on them. In mechanics John Herschel maintains that:

To such causes Newton has applied the term *verae causae*; that is causes recognized as having a real existence in nature and not being mere hypotheses or figments of the mind. (Herschel 1830, p. 138)

Vera causa is real cause and knowledge of causes is inductive. Indeed, according to him, "we have been led by induction to the knowledge of the proximate cause of a phenomenon or of a law of nature (Herschel 1830, p. 172).

In a position opposite to Herschel, the father of French Positivism Auguste Comte (1798–1857) maintains in his *Cours de Philosophie Positive*, 1830–1842, Première Leçon, the impossibility of the cognition of the causes: "In the final, the positive state, the mind has given over the vain search after absolute notions, the origin and destination of the universe, and the causes of phenomena, and applies itself to the study of their laws – that is, their invariable relations of succession and resemblance."

According to Comte, in what may be the first version of the notion of theoretical explanation, "What is now understood when we speak of an explanation of facts is simply the establishment of a connection between single phenomena and some general facts, the number of which continually diminishes with the progress of science" (Comte 2000, p. 28; French, pp. 21–22).[6] He insists later:

the first characteristic of the Positive Philosophy is that it regards all phenomena as subjected to invariable natural *Laws*. Our business is, – seeing how vain is any research into what are called *Causes*, whether first or final – to pursue an accurate discovery of these Laws, with a view to reducing them to the smallest possible number. (Comte 2000, p. 31; French, p. 25)

A contrary position to Comte is maintained by John Stuart Mill, who explicitly defends the dependence of the validity of all inductive methods of science "on the assumption that every event ... must have some cause, some antecedent, on the existence of which it is invariably and unconditionally consequent. ... The universality of the law of causation is assumed in them all [the Inductive Methods,

---

[6]French means here the original edition of Comte's *Cours de Philosophie Positive*.

AR]" (Mill 1843: Chap. XXI, § 1, p. 369). All effect therefore follows invariably and unconditionally from its corresponding cause.

In apparent opposition to Mill, William Whewell claims that "our idea of causation, which implies that the event is necessarily connected with the cause, cannot be derived from observation." (Whewell 1847, Part One, Book III, Chap. II, p. 168) But Whewell is a Kantian:

> The idea of cause, like the ideas of space and time, is a part of the *active* powers of the mind. The relation of cause and effect is a relation or condition under which events are apprehended, which relation is not given by observation, but supplied by the mind itself. (Whewell 1847, Part One, Book III, Chap. III, p. 176)

Astronomy is for Whewell paradigmatic for the knowledge of the causes of phenomena. But while he maintains that "the force of gravity causes bodies to move downwards if they are free" (Whewell 1847, Part One, Book III, Chap. IV, p. 179), about the very cause of gravity Whewell mentions Newton, who "In his second Letter to Bentley (1693) he says: 'The cause of gravity, I do not pretend to know, and would take more time to consider of it'"(Whewell 1847, Part One, Book III, Chap. IX, pp. 257–258). In fact, in a letter to Robert Boyle of 1678/1679 on the Cause of Gravitation Newton (1782, Supposition 5, pp. 387–388), proclaimed the ether as the cause of gravity, but in Letter II addressed to Richard Bentley (1662–1742), dated January 17, 1693, Newton is very cautious about the knowledge of the cause of gravity, penning the lines quoted by Whewell: "You sometimes speak of gravity as essential and inherent to matter. Pray do not ascribe that notion to me; for the cause of gravity is what I do not pretend to know, and therefore would take more time to consider of it" (Newton 1782, p. 437). And in Letter III to Bentley on February 25 of the same year Newton expands his idea:

> It is inconceivable, that inanimate brute matter should, without the mediation of something else, which is not material, operate upon, and affect other matter without mutual contact; as it must do, if gravitation, in the sense of *Epicurus*, be essential and inherent in it. And this is one reason, why I desired you would not ascribe innate gravity to me. That gravity should be innate, inherent and essential to matter, . . . , is to me so great an absurdity, that I believe no man who has in philosophical matters a competent faculty of thinking, can ever fall into it. Gravity must be caused by an agent acting constantly according to certain laws; but whether this agent be material or immaterial, I have left to the consideration of my readers. (Newton 1782, p. 438)

Nonetheless, the importance of hypotheses about causes is such that for Whewell "Men cannot contemplate the phenomena without clothing them in terms of some hypothesis, and will not be schooled to suppress the questionings which at every moment rise up within them concerning the causes of the phenomena" (Whewell 1847, Part Two, Book XI, Chap. VII, p. 104–5).

Whewell finds in Newton a support in his commitment to the thesis of the cognition of causes. Indeed, he refers to *Optics*, where in Query 28 Newton maintains that "The main business of natural philosophy is to argue from phaenomena without feigning hypotheses, and to deduce causes from effects, till we come to the First cause, which certainly is not mechanical" (Newton 1782, p. 237). In Query 31, near the end, Newton claimed also that "By this way of analysis we may proceed from compounds to ingredients; and from motions to the forces producing them; and in general, from effects to their causes; and from particular causes to more general ones, till the argument end in the most general" (Newton 1782, p. 263).

Stanley Jevons maintains a position similar to Whewell's interpretation of Newton regarding the use of hypotheses in science. In fact, Jevons states that "Newton said that he did not frame hypotheses, but, in reality, the greater part of the *Principia* is purely hypothetical, endless varieties of causes and laws being imagined which have no counterpart in nature" (Jevons 1958, p. 583).

That said, it must be recognized that for Jevons the use of the term *cause* in science is philosophically dangerous. Incorporated into a tradition that goes back to Berkeley, Jevons is emphatically opposed to the scientific cognition of the causes: "nothing is more unquestionable than that finite experience can never give us certain knowledge of the future, so that either a cause is not an invariable antecedent, or else we can never gain certain knowledge of causes. (...) The word *cause* covers just as much untold meaning as any of the words *substance, matter, thought, existence*" (Jevons 1958, p. 222). He also maintains this position when he affirms that "To us, then, a cause is not to be distinguished from the group of positive or negative conditions which, with more or less probability, precede an event" (Jevons 1958, p. 226).

With this reference to Jevons I close here this journey through the optimistic and sceptical positions on the knowability of causes until the nineteenth century.

18.4 From Prehistory to History: The Debate Around Hempel's D-N Model of Scientific Explanation

In the preceding sections I have presented the debate on the acceptance or rejection of scientific explanation, including causal explanation, in the history of Western theory of knowledge. But nothing has been said about the structure of scientific explanation. It will be up to the philosophy of science of the twentieth century to fully develop this aspect. This was the compromise, first by Einstein, and then by Popper, with the idea of causal scientific explanation, with which the German philosophy of the early twentieth century dissociates itself from Machian and Duhemian instrumentalism and from Berkelyan anti-causalism.

Nonetheless, that scientific explanation takes the form of a deductive argument already has a long tradition in Western philosophy of science. It was John Stuart Mill, who, in the nineteenth century, truly intuited the modern form of the scheme of scientific explanation. Indeed, as Mill claims, "The ascertainment of the empirical laws of phenomena often precedes by a long interval the explanation of those laws by the Deductive Method; and the verification of a deduction usually consists in the comparison of its results with empirical laws previously ascertained" (Mill 1843, Chap. XVI, § 1, "Of Empirical Laws," p. 339).

It is interesting to notice that Popper acknowledges, in his *Intellectual Autobiography*, that Mill had already anticipated his own model of scientific explanation:

> In section 12 of *Logik der Forschung* I discussed what I called 'causal explanation', or deductive explanation, a discussion which had been anticipated, without my being aware of it, by J. S. Mill, though perhaps a bit vaguely (because of his lack of distinction between an initial condition and a universal law). (Popper 1974, p. 93)

Indeed according to Karl Popper "To give a *causal explanation* of an event means to deduce a statement which describes it, using as premises of the deduction one or more universal *laws*, together with certain singular statements, the *initial conditions*."

Popper anticipated Hempel's D-N model of scientific explanation,[7] Hempel (1965, p. 337, note 2) himself acknowledged it reluctantly and Wesley C. Salmon claimed that "Although Popper's *Logik der Forschung* (1935) contains an important anticipation of the D-N model, it does not provide as precise an analysis as was embodied in (Hempel and Oppenheim 1948)" (Salmon 1984, p. 21, note 4). Since, moreover, as Salmon claims, Popper's views on scientific explanation were not widely influential until the English translation of his book appeared, he concludes: "It is for these reasons that I chose 1948, rather than 1935, as the critical point of division between the history and the prehistory of the subject" (Salmon 1984, p. 21, note 4).

It is well known that for Hempel scientific explanations are "deductive arguments whose conclusion is the *explanandum* sentence, E, and whose premise-set, the *explanans*, consists of general laws, L_1, L_2 ... L_r and other statements, C_1, C_2, ..., C_k, which make assertions about particular facts" (Hempel 1966, p. 51). They are called deductive-nomological explanations and "the explanatory information they provide implies the *explanandum* sentence deductively and thus offers logically conclusive grounds why the *explanandum* phenomenon is to be expected" (Hempel 1966, p. 52).

Mario Bunge (1979) takes Hempel's D-N model as a referent in order to develop his own point of view about scientific explanation. The first thing that strikes Bunge is that sentences that begin with the conjunction *because* need not be causal to be scientific "although causal explanation does constitute an important ingredient of scientific explanation in many cases" (Bunge 1979, p. 282). According to Bunge the causal relation is only a relation among events. Bunge claims that "The causal generation of events is lawful rather than capricious. That is, there are causal laws or, at least, laws with a causal range" (Bunge 1982, p. 137). This is the reason why for a scientific explanation to be causal the explanation must 'invoke' a causal law because "it is characteristic of scientific explanation in general to be made by means of laws, some of which have a causal component while others lack it" (Bunge 1979, p. 305).

---

[7]For instance, Popper says: "This concept of explanation (now commonly referred to as the 'deductive-nomological concept of explanation') is further elaborated in the *Logik der Forschung* (1934, 2nd ed., 1966), Sect. 12" (Popper 1979, p. 86, note *2. My own translation, A.R.).

However, as the D-N model focuses exclusively on the logical-deductive relationship between *explanans* and *explanandum*, Bunge complains that this model fails to discern causal from non-causal explanation (1979, *ibid.*). In any case, Bunge recognizes that "Causal laws are only a species of scientific law" (Bunge 1979, p. 248). In conclusion, scientific explanation is not necessarily explanation by causes. This is fully consistent with his claim that

> *The World is not strictly causal although it is determinate*: not all interconnected events are causally related and not all regularities are causal. Causation is just one mode of determinism among several. Hence, determinism should not be construed narrowly as causal determinism. Science is deterministic in a lax sense: it requires only lawfulness (of any kind), and nonmagic. (Bunge 1982, p. 137)

Actually, Bunge's discussion with Hempel about the existence of causal explanations is not as crude as it may seem. Hempel affirms that

> the causal explanation implicitly claims that there are general laws –let us say $L_1, L_2, \ldots,$ L_r– in virtue of which the occurrence of the causal antecedents mentioned in $C_1, C_2, \ldots, C_k$ is a sufficient condition for the occurrence of the *explanandum* event. This relation between causal factors and effect is reflected in our schema (D-N): causal explanation is, at least implicitly, deductive-nomological. (Hempel 1966, p. 349)

Thus, causal explanations are special types of D-N explanations. In short, Bunge and Hempel seem to agree.

Many philosophers of science have discussed the D-N model.[8] Psillos (2002) collects a lot of alleged counter-instances of Hempel's D-N model, particularly the problems of symmetry and irrelevance.

From examples such as the law of the pendulum, Boyle's law and even Kepler's laws, Psillos (2002, p. 230) concludes that the counter-examples leave the Basic Thesis (BT) of Hempel's D-N model unscathed. The reason is that what BT claims is that "all causal explanations of singular events can be captured by the deductive-nomological model" (Psillos 2002, p. 223), i.e., that "if Y is a causal explanation of a singular event, then Y is also a DN explanation of this event" (Psillos 2002, pp. 230–231).

Hempel himself maintained that causal explanations are deductive-nomological, but not that all deductive nomological explanations are causal explanations. Moreover, the counter-examples to Hempel's D-N model mentioned by Psillos are empirical or phenomenological laws, and, according to Salmon, "they do not afford causal explanations of the events subsumed under them" (Salmon 1984, p. 136). From this point of view I claim that it was not necessary to devote so many pages to find alleged counterexamples of Hempel's model on the basis of mere empirical generalizations.

---

[8] See Feyerabend (1962), Friedman (1974), Salmon (1984), Lewis (1986), Kitcher (1981, 1989), Sklar (1992), Lipton (2001a, b), and Psillos (2002).

Lipton, for his part, acknowledges that causation is an asymmetrical relationship: "while causes explain their effects, effects do not explain their causes. The recession of the galaxy explains why its light is red shifted, but the red shift does not explain why the galaxy is receding, even though the red shift may provide essential evidence of the recession" (Lipton 2001a, p. 51). The so-called flagpole-and-shadow case is by no means a serious counter-example of Hempel's D-N model. And Kitcher affirms also that "shadows do not explain the heights of towers because shadow lengths are causally dependent on tower heights" (Kitcher 1989, p. 436). Indeed, it would be absurd to claim that the D-N model allows for the period of a pendulum to cause the pendulum to have the length it does.

Nonetheless, Lipton presents as a problem of the DN model that this "notoriously counts as explanatory the worthless deduction of a law from the conjunction of itself to another unrelated law" (Lipton 2001b, p. 100). He adds that Hempel acknowledged this difficulty but what Hempel really claimed was that even if we consider that the conjunction of Kepler's laws, K, with Boyle's law, B, allows the derivation of K from this conjunction, the "derivation of K from the latter would not be considered as an explanation of the regularities stated in Kepler's laws; rather, it would be viewed as representing, in effect, a pointless 'explanation' of Kepler's laws by themselves." Only "The derivation of Kepler's laws from Newton's laws of motion and of gravitation, ..., would be recognized as a genuine explanation in terms of more comprehensive regularities, or so-called higher-level laws" (Hempel 1965, p. 273, note 33). Obviously, Lipton's objection is a bit frivolous.

Something similar happens with Kitcher who claims that the Hempelian account of theoretical explanation should admit "that we explain laws by deriving them from more fundamental laws", so that "If the causal approach to explanation is to be fully developed, it must provide some way of saying what is meant by the intuitive (but murky) thesis that some laws are more fundamental than others." (Kitcher 1989, p. 428) Now, Hempel does not mention the expression 'fundamental laws' at all. Literally Hempel reads as follows: "The derivation of Kepler's laws from Newton's laws of motion and of gravitation, ..., would be recognized as a genuine explanation in terms of more comprehensive regularities, or so-called higher-level laws" (Hempel 1965, p. 273, note 33). Hempel recognizes indeed that the establishment of clear-cut criteria for the distinction of levels of explanation constitutes as yet an open problem. However, as I will argue later, a theoretical explanation takes place by deductive recovery of the *explanandum* within a broader theoretical framework. And this answers the question.

Against the deductive-nomological model of explanation Lipton (2001b, p. 99) states also that it is unsuitable in the context of inference to the best explanation. But this cannot be taken as a serious reproach. The reason is quite simple: whereas Inference to the Best Explanation is situated in the context of abductive reasoning, the D-N model belongs completely to deductive reasoning.

18.5   Inter-theoretical Incompatibility and the Case of Kepler's Third Law

Einstein affirmed, in a paper on Newton's mechanics, that:

> Kepler's empirical laws of planetary movement, deduced from Tycho Brahe's observations, confronted him [Newton, A.R.], and demanded explanation. These laws gave, it is true, a complete answer to the question of *how* the planets move round the sun ... But these rules do not satisfy the demand for causal explanation. (Einstein 1973, p. 254)

This is the general fact – planetary motion – around which I am going to develop my first argument on the impossibility of causal explanations in physics. But before entering it I will briefly present some elements of the debate around the real or true cause of planetary movements in the philosophy of science in the nineteenth century.

Kepler's laws of planetary motions are empirical laws, and therefore unverified inductions "deduced by Kepler entirely from a comparison by observations with each other, *with no assistance from theory.* These laws ... amply repaid the labour bestowed on them, by affording afterwards *the most conclusive and unanswerable proofs of the Newtonian system."* So wrote Herschel (1830, p. 187. My emphasis, AR). The force of gravity in Newton's theory is assumed by him as a true cause: "This force, then, which we call the *force of gravity,* is a real cause" (Herschel 1830, p. 209).

In Book III "Of Induction", "Miscellaneous Examples of the Explanation of Laws of Nature," §1, Mill asserts that "it was very reasonably deemed an essential requisite of any true theory of the causes of the celestial motions, that it should lead by deduction to Kepler's laws; which, accordingly, the Newtonian theory did" (Mill 1843, pp. 303–304). And in Chap. XIV: "Of the limits to the explanation of laws of nature and of Hypotheses", §4, Mill reiterates: "the true law, that the force varies as the inverse square of the distance ... suggested itself to Newton, in the first instance, as an hypothesis, and was verified by proving that it led deductively to Kepler's laws" (Mill 1843, p. 322). This deduction is for Mill what makes of Newtonian celestial mechanics a theory of the real or true causes of planetary movements.

For his part, William Whewell, in his commentary on Newton's *Rules of Philosophizing* of *Principia Mathematica*, attributes to gravitation the character of *vera causa* as well: "the first Rule is designed to strengthen the inference of gravitation from the celestial phenomena, by describing it as a *vera causa,* a true cause" (Whewell 1847, Part Two, Book XII, Chap. XIII, p. 279). And in Sect. 7 Whewell asks:

> But what do we mean by calling gravity a 'true cause'? How do we learn its reality? Of course, by its effects, with which we are familiar; – by the weight and fall of bodies about us. These strike even the most careless observer. ... Hence, it may be said, this cause is at any rate a true cause, whether it explains the celestial phenomena or not. (Whewell 1847, p. 281)

From an opposite perspective Ernst Mach shows himself very temperate about the possibility of a causal explanation by Newton of planetary movements: "when Newton gives a 'causal explanation' of the planetary motions,..., he is only pointing out or describing facts, which, although by a roundabout path, yet have been reached by observation." Thus: "explanation is nothing more than a description in terms of elements. Every particular case can then be put together out of spatial and temporal elements, the relations between which are described by equations" (Mach 1959, p. 336).

And according to his instrumentalist stance, Karl Pearson also maintains a position antagonistic to that of John Herschel on the relation of the concepts of force and cause:

> we can describe *how* a stone falls to the earth, but not say *why* it does. Thus scientifically the idea of *necessity* in the stages of the sequence – stone in motion, broken window – or the idea of enforcement would disappear; we should have a routine of experience, but an unexplained routine. (Pearson 1911, pp. 116–117 and p. 118)

There are therefore no causal scientific explanations for Pearson. The law of gravitation only describes how the bodies move. To say that they move owing to the force of gravitation "is merely throwing the answer on the beyond of sense-impression –it is the metaphysical method of avoiding saying: We don't know" (Pearson 1911, p. 299). The following quotation closes Pearson's approach to scientific causality and the task of science: "*Why* things move thus becomes an idle question, and *how* things are to be *conceived* as moving the true problem of physical science" (Pearson 1911, p. 352).

Finally, Pierre Duhem claims explicitly that:

> the laws of all celestial phenomena are found condensed in the principle of universal gravitation. Such a condensed representation is not an explanation; the mutual attraction that celestial mechanics imagines between any two parts whatsoever of matter permits us to submit all celestial movements to calculation, but the cause itself of this attraction is not laid bare because of that. (Duhem 1954, p. 47; French edition, pp. 71–72)

Apart from this historical discussion on the real cause of the planetary motions, the derivation of Kepler's Third Law in the context of Newtonian mechanics is very easy. Indeed, let us take a system consisting of two bodies – the sun and any planet – of masses M and m, respectively separated by an average distance r, among which there is a balance between attractive gravitational $F = Mm/r^2$ and centrifugal $F = m \times \omega^2 r$ forces (ω is the average angular velocity of the planet in its circular orbit around the sun). Equating both equations, we obtain $M = \omega^2 r^3$, also known as *Kepler's 1-2-3 law*.

According to Hempel "the principles of Newtonian mechanics … explain certain 'general facts', i.e., empirical uniformities such as Kepler's laws of planetary motion; for the latter can be deduced from the former" (Hempel 1965, pp. 173–174). So, my question is: Does the Newtonian derivation of Kepler's third law provide a causal explanation of planetary motions?

The answer to this question depends in turn on the answers we give to the two following questions: (1) Is Newtonian mechanics (NM) the only theory providing an explanation of Kepler's laws or is there besides NM any other gravitational theory that could also apply as an explanation? (2) In the event that such theories exist, what is the relationship between them?

In relation to the first question the answer is that such an alternative theory exists. It is the general theory of relativity (GRT). Indeed, GRT would have been stillborn if it were not able to allow for the mathematical derivation of Kepler's Third Law, and no self-respecting theoretical physicist could evade the task to provide the deduction of Kepler's third law in the framework of GRT. As a matter of fact, Misner et al. (M-T-W) (1973 25.12, p. 655) claim that Kepler's *1-2-3* law results from the formula for the motion of a test particle in Swarzschild's geometry described by the line element, and Straumann (2004, §3.6) also provides a deduction of Kepler's law.

Now, if the derivation of Kepler's third law in a given theoretical context provided a causal explanation of planetary motions, then the derivation of this law in another theoretical context would also provide a causal explanation of the aforementioned planetary movements. In other words, if the Newtonian explanation of Kepler's law were a causal one, then the relativistic explanation should be causal too. And the same would happen for any rival gravitational theory.

This is a serious problem indeed. But it can be further complicated depending on the answer we give to the previous question 2, namely: what kind of relationship exists between NM and GRT? In the case at hand we find that the 'universal' laws that constitute the premises (*explanans*) of the deductive explanation of Kepler's third law (*explanandum*) refer to causes of totally different nature in NM and GRT: Gravitational forces in the case of NM and the curvature of spacetime in the case of GRT. Both are mutually exclusive. As M-T-W claim: "Gravitation shows up in no way other than in curvature of the geometry, in which the particle moves as free of all 'real' force" (Misner et al. 1973, p. 649). Both theories deny the fundamental postulates of each other. They are incompatible with each other!

Mario Bunge does not appear to be affected by the problem of inter-theoretical incompatibility. He is not a naive realist, but a critical realist. As such, he admits that "in a sense, causation *is* in our minds – but not in our minds *alone*" and that "scientific theories introduce certain concepts which have no factual correlates, such as the components of vector quantities, lagrangians and the average American" (Bunge 1962, pp. 312–313). But what the incompatibility between NM and GRT puts on the table is whether the gravitational force has a real referent. Bunge's position seems to be clearer when he later states that "Every law, whether of physics or sociology, is valid exactly only with respect to the corresponding theoretical model, which disregards complications." And he concludes that "Strict and pure causation is as ideal – yet as useful – as any model can be" (Bunge 1962, p. 315). But this does not solve the problem either. NM and GRT offer theoretically incompatible models of the world in general and of gravitational phenomena in particular. If truths were relative to models, this would only dodge the problem of causality in physics. Or, to get out of this situation, we would have to settle for the idea that causation is only a useful ideal.

In short, we face the following dilemma: Either theoretical physics allows the existence of causal explanations, incompatible with each other, of the same phenomena, or theoretical physics must renounce causal explanations. The answer to this dilemma is very clear to me: the search for causal explanations is not a goal of theoretical physics, and that the concept of causal explanation is dispensable in the philosophy of physics.

18.6 Nomic Chains and Theoretical Explanations

In this section I want to introduce a new element of reflection that will deepen the doubts about the possibility of causal explanations in physics. This is the fact that very often the explanation of a phenomenon requires entering an explanatory return. If the explanation should be causal, then this procedure would require going back to other causes, that is, it should reveal the existence of nomic chains. Since the explanation of an event might imply the concurrence of several explanatory laws, that is, a concatenation of explanations, then, if the explanation should be causal, causation would be distributed among several causes. This circumstance contributes to blur the image of the existence of a clear and distinct efficient causation. Let us approach gradually to this idea from different perspectives.

To begin with, already Whewell (1847), in a Section entitled *Induction of Ulterior Causes*, introduces a sceptical reflection on the discovery of causes:

> The first Induction of a Cause does not close the business of scientific inquiry. Behind proximate causes, there are ulterior causes, perhaps a succession of such. Gravity is the cause of the motions of the planets; but *what is the cause of gravity?* [...] Thus we are referred back from step to step, in the order of causation. ... We make discovery after discovery in the various regions of science; each, it may be, satisfactory, and in itself complete, but none final. Something always remains undone. The last question answered, the answer suggests still another question. (Whewell 1847, Part Two, Book XIII, Chap. X, Sect. 6, p. 434 (My emphasis, A.R.))

In a similar line of thought, David Lewis maintains that the explanation of any event must take into account its *causal history*. Lewis's thesis is that "Any particular event that we might wish to explain stands at the end of a long and complicated causal history" (Lewis 1986, p. 214). Thus *"to explain an event is to provide some information about its causal history"* (Lewis 1986, p. 217). And taking his position therein, Hanson affirms that "Causes certainly are connected with effects; but this is because our theories connect them, not because the world is held together by cosmic glue" (Hanson 1958, p. 64). Cause and effect are only intelligible "against a pattern of theory". In other words: "what we refer to as 'causes' are theory-loaded from beginning to end" (Hanson 1958, p. 54).

Now, if we take seriously Hanson's view that causes are theory-loaded, what should we expect if we tried to apply it to the search for explanations in situations of great physical complexity, for instance when we ask for the cause of stars' brightness?

What is the stars' energy source? The first answer to this question was provided by Helmholtz-Kelvin's contraction theory, which assumed that the stars' power supply is maintained by converting gravitational energy into heat due to the star's gradual contraction. Indeed, in 1854 Hermann von Helmholtz (1821–1894) first raised the idea of solar power generation by gravitational collapse.

Twelve years later Lord Kelvin adhered to Helmholtz's meteoric theory[9]:

> *That some form of the meteoric theory is certainly the true and complete explanation of solar heat can scarcely be doubted.*

And

> *It seems, therefore, on the whole most probable than the sun has not illuminated the earth for 100,000,000 years*, and almost certain that he has not done so for 500,000,000 years. As for the future, we may say, with equal certainty that inhabitants of the earth cannot continue to enjoy the light and heat essential to their life, for many million years longer, *unless sources now unknown to us are prepared in the great storehouse of creation.* (Kelvin 1903, pp. 493–494. My italics, AR)

But the explanation based on the assumption of gravitational contraction is unacceptable. Ostlie and Carroll (1996, p. 329) show that if the stars shine due to the conversion into heat of Newtonian gravitational potential energy, then the current age of the Sun should be 10^7 years, known as the *Kelvin-Helmholtz time scale*. The gravitational collapse cannot be *the cause* of the brightness of stars, and the Helmholtz-Kelvin's contraction theory must be definitely rejected.

Today it seems indisputable that the stars' brightness resides in the nuclear reactions taking place inside stars. But to say that stars shine because of the nuclear reactions inside them opens the way to a lengthy chain of laws, whose intricate connections lead us to have to walk back along the paths of different physical theories and disciplines: Nuclear physics, of course, but also atomic physics and quantum mechanics, and equally classical mechanics, thermodynamics, statistical physics, hydrodynamics, acoustics and the wave theory of electromagnetism, elementary particles physics, theory of relativity, etc. In a more intuitive way: giving a scientific explanation of the question of the brightness of the stars, represents a real monographic course in a high school of physics, if not even a whole degree in astrophysics. After this long period, and only if the information acquired step by step has been well assimilated to make one an astrophysicist, one can understand why the nuclear reactions that occur inside stars are *the reason* why they are bright and produce heat.

How do astrophysicists tackle the explanation of the brightness of the stars? What they really do is to *build a theoretical model* of a star that allows them to get an idea of a star of the so-called *main sequence*. The explanation of stellar brightness astrophysicist offer cannot be expected to be 100% correct, i.e. to be a causal explanation properly. But it's all we can expect. To build such a model of the stellar interior astrophysicists resort to the available empirical and theoretical results, proceeding from all physical theories and disciplines which they accept –

---

[9]This theory "consists in supposing the sun and his heat to have originated in a coalition of smaller bodies, falling together by mutual gravitation, and generating, as they must do according to the great law demonstrated by Joule, an exact equivalent of heat for the motion lost in collision" Kelvin (1903, pp. 493–4).

though they do not necessarily assume to be true – in order to combine them mathematically and to derive a set of interrelated equations. These combinations, needless to say, must scrupulously respect the dimensional analysis. This way of reasoning that anticipates or advances theoretical results not yet available is called *theoretical preduction*.[10]

To 'know' about the structure of the stars amounts to constructing a *theoretical model* of the stellar interiors, assuming some ideal conditions: stars as static symmetric spheres of ideal gases and as black bodies. In its standard form this model consists of five basic differential equations, five gradients: Hydrostatic equilibrium, mass conservation, interior luminosity and the temperature gradient for radiative transport and adiabatic convection. Newtonian mechanics, classical statistics mechanics and quantum physics must be combined in order to obtain the pressure gradient. While for the derivation of the temperature gradient of radiative transfer it is necessary to combine classical and quantum physics, and classical statistical mechanics and thermodynamics of adiabatic processes to obtain the temperature gradient of a monoatomic ideal gas expanding adiabatically (Ostlie and Carroll 1996, Chap. 10).These concatenations of laws, or nomic chains, are those that allow us to get an idea, perhaps approximate, of the processes that result in the stars shining in the sky.

The explanation of stellar brightness is very complex indeed. We cannot approach a star inside, much less probe it physically. But we long for an idea of the physical processes that keep a star in equilibrium, as well as the processes – gravitational and nuclear – that generate energy and of the procedures by which this energy is transported abroad. This allows us to manage theoretically enormously complex physical conditions. To give a complete causal explanation of the brightness of stars decidedly escapes our material and theoretical capabilities. We are definitely left with the only means we have for now: to offer explanations through the construction of theoretical models.

18.7 Explanatory Unification and the Challenge of Inter-theoretical Incompatibility

One of the most serious attempts to date to overcome Popper-Hempel's D-N model of scientific explanation has been a project dating back to Michael Friedman, which involves linking scientific explanation with understanding. The question posed by Friedman is the following:

---

[10]Since this is neither the place nor the occasion to develop the philosophical theory of preduction as a strategy of creativity or innovation in theoretical physics, I refer to Rivadulla (2008, 2010, 2016a) to see more on preductive reasoning.

what is the relation between phenomena in virtue of which one phenomenon can constitute an explanation of another, and what is it about this relation that gives understanding of the explained phenomenon? (Friedman 1974, p. 6)

That is, how is it possible that the explanation of a phenomenon allows us to understand it? Friedman confesses: "I don't see how the philosopher of science can afford to ignore such concepts as 'understanding' and 'intelligibility' when giving a theory of the explanation relation" (Friedman 1974, p. 8). Thus, Friedman reproaches Hempel in that the "D-N theorists have not succeeded in saying what it is about the explanation relation that provides understanding of the world" (Friedman 1974, p. 9). Indeed, "Our theory should somehow connect explanation and understanding – it should tell us what kind of understanding scientific explanations provide and how they provide it. This is where D-N theorists have been particularly negligent" (Friedman 1974, p. 14).

For Friedman (1974, p. 15), the essence of scientific explanation is to *reduce* the number of given independent phenomena and this is what allows us to understand the World. This is called the *unifying effect of scientific theories*. For instance, the great Newtonian synthesis would have unified Kepler's laws of planetary motions and Galileo's laws of falling bodies, thereby reducing the number of independent phenomena and achieving a better understanding of the world. Already Comte in the nineteenth century acknowledged the unifying character of Newtonian theory:

We say that the general phenomena of the universe are explained by it, because it connects under one head the whole immense variety of astronomical facts;... whilst the general itself is a mere extension of one which is perfectly familiar to us...; –the weight of bodies on the surface of the earth. (Comte 2000, p. 31. French edition, pp. 25–26)

To develop the view of explanation as unification is the goal of Kitcher (1981, 1989). In the wake of Friedman, Philip Kitcher maintains that "A theory of explanation should show us *how* scientific explanation advances our understanding" (Kitcher 1981, p. 508). Thus if K is a consistent set of statements formulated in a language L and accepted by a scientific community, to claim that $E(K)$ is the *explanatory store over K* means that $E(K)$ is the set of explanations or derivations that best unifies K (Kitcher 1981, p. 512, 1989, p. 431), i.e., that best systematizes K, since "explanation consists in the systematization of our beliefs" (Kitcher 1989, p. 476).

The touchstone of Kitcher's unificationist approach is his account of scientific change. Kitcher's starting point is the realization that since scientific change may involve changes in belief and changes in language, then:

the principle of explanatory unification ought to be formulated so as to enable us to decide whether it would be reasonable to modify our scientific practice from $<L, K, E(K)>$ to $<L'$, $K', E(K')>$ on grounds of attaining greater unification in our beliefs. (Kitcher 1989, p. 488)

Kitcher's answer is:

I suggest that appeals to explanatory unification can underwrite transitions from $<L, K,$ $E(K)>$ to $<L', K', E(K')>$ only subject to the proviso that the shifts from K to K' and the shifts from L to L' are defensible. This does not require that K and K', L and L' be identical –

... – but, roughly, that there are no strong arguments from the perspective of $<L, K, E(K)>$ against the shifts envisaged. (Kitcher 1989, p. 491)

Well, let's look at the transitional situation between NM and GRT. Indeed, NM is a theory with a great systematizing power because it achieved the synthesis or unification of celestial and terrestrial phenomena. But the unifying power of GRT is even greater. For it not only encompasses the predictive success of NM, but it also accounts for many phenomena for which NM did not offer a satisfactory answer, such as the deviation of light by the sun and the advancement of perihelion of the planets, and even GRT is capable of systematizing absolutely unsuspected phenomena from the Newtonian perspective, such as the gravitational red shift and the existence of gravity waves. In principle, then, there seem to be no arguments against the transition from NM to GRT, but quite the opposite.

Is the transition from Newton to Einstein correctly reflected by Kitcher's scheme as the transition from a unifying scheme $<L, K, E(K)>$ to another unifying scheme $<L', K', E(K')>$? The problem here is that the languages L of NM and L' of GRT are incompatible with each other. Not that they are incommensurable with each other, as Kitcher seems to have in mind, but incompatible. For not only does it happen that some fundamental theoretical entities of NM, as the gravitational force, do not exist in GRT, where it is replaced by the curvature of spacetime, but even the 'underlying' structure of the world changes radically from a three-dimensional Euclidean world to a non-Euclidean four-dimensional world. We can even say that not even K and K' coincide exactly: phenomena such as black holes, gravitational lenses and gravity waves do not belong to the set of statements accepted by the Newtonian community. Thus, if I am right, the transition from NM to GRT is the transition from one conceptual schema $<L, K, E(K)>$ to another incompatible conceptual schema $<L, K, E(K)>$, and this goes well beyond Kitcher's proviso that "that the shifts from K to K' and the shifts from L to L' are defensible". They are of course defensible, but not in the sense that Kitcher may have got in mind.

What then constitutes a true or correct explanation? According to Kitcher, "*correct* explanations are those derivations that appear in the explanatory store in the limit of rational development of scientific practice" (Kitcher 1989, pp. 494, 498). "To explain", concludes Kitcher, p. 500, "is to fit the phenomena into a unified picture insofar as we can. What emerges in the limit of this process is nothing less than the causal structure of the world." This was also Salmon's (1984, p. 19) idea.

But the idyllic image of science that evokes this beautiful sentence responds to an ideal of linear scientific progress which does not correspond at all with the actual development of science. The doubts manifested by James Woodward on Kitcher's contention "that one can begin with the notion of explanatory unification, understood in a way that does not presuppose causal notions, and use it to derive the content of causal judgments" (Woodward 2017, §5.3) are firmly confirmed when we consider that the incompatibility between cohabiting or succeeding theories is a common occurrence in the history of science. Incompatibility takes very different forms: between the theoretical entities of different theories, for example between pre-Copernican and Copernican astronomies; between the fundamental postulates of conflicting theories, for example NM and GRT; between theories with different domains of application, for example, theory of relativity and quantum mechanics; between the underlying metaphysical assumptions, for example orthodox indeterministic quantum mechanics and Bohmian deterministic quantum mechanics; between the underlying mathematical structures, for example linear quantum mechanics and non-linear quantum mechanics, etc. As I say in Rivadulla (2016b, p. 529): "Incompatibility is omnipresent in the realm of theoretical physics."

Unification, yes. But the problem is that the unificationist approach must face the question of whether true or correct explanations are possible when the transitions occur between incompatible theories. This seriously challenges both the idea of a limit of scientific inquiry and that the causal structure of the world can ever be made evident.

18.8 Conclusion

In the field of scientific explanation it seems reasonable to relinquish causal explanations – explanations by reference to efficient causes – on behalf of theoretical explanations – explanations in a given theoretical framework. The difference between the *causal* explanations we give, say, of a solar eclipse and of the brightness of a star, lies 'only' in the variety of theories that in the second case intersect each other and which are necessary ingredients for the 'understanding' of the phenomenon at issue. But this is decisive when it comes to convincing ourselves whether we are offering a causal explanation properly or merely a *theoretical* explanation of the phenomenon under consideration.

The term *theoretical explanation* appears in Hempel indeed and it means *explanation based on theoretical principles*. The importance of this form of explanation is for Hempel that

> a theoretical explanation deepens our understanding for at least two reasons. First, it reveals the different regularities exhibited by a variety of phenomena, … Secondly, …, the generalizations previously accepted as correct statements of empirical regularities will usually appear as approximations only of certain lawlike statements implied by the explanatory theory, and to be very nearly satisfied only within a certain limited range. (Hempel 1965, p. 345)

Further, Hempel claims that when a scientific theory is superseded by another, "then the succeeding theory will generally have a wider explanatory range, including phenomena the earlier theory could not account for" (Hempel 1965, p. 354). Well, I affirm that we give a theoretical explanation of an *explanandum* – whether this is an event, a phenomenological hypothesis, a semi-empirical law, a theoretical law, a model or even a theory itself – simply when we are able to deductively recover it within a broader theoretical framework. This is the point of view that I maintain, from a non-realistic perspective, in Rivadulla (2005, pp. 166–167 and p. 169). A theoretical explanation only shows the deductive relationship between a theoretical context, which serves as *explanans*, and the *explanandum*. Thus, it is an extension of Popper-Hempel's D-N model of scientific explanation, free of any ontological commitment.

As Hempel himself (1965, pp. 173–174) accepts that Newtonian mechanics provides an explanation of Kepler's empirical laws, this explanation is an excellent example of theoretical explanation. For the same reasons the deduction of Kepler's *1-2-3 laws* in the context of the General Theory of Relativity also provides a theoretical explanation of Kepler's third Law. And, of course, the deduction, in the context of Bohr's atomic model, of Balmer's empirical law on the distribution of the spectral lines of Hydrogen also provides a theoretical explanation of the said formula. This is also true of the deduction of Balmer's formula in the context of Schrödinger's wave mechanics (Rivadulla 2005, pp. 169–173) but the same can be said of the deduction of Planck's radiation law in the context of Bose-Einstein's quantum statistical mechanics (Rivadulla 2005, pp. 175–177). The explanation of the deviation of light by the Sun, the gravitational lens effect, and of Mercury's perihelion, offered by the general theory of relativity, is a typical form of theoretical explanation. And, of course, the explanation of the stars' brightness provided by the stellar interior model for main sequence stars is a typical case of theoretical explanation as well. And so on.

Philip Kitcher claims that "Theoretical explanation provides some support for the Hempelian idea that explanation is derivation" (Kitcher 1989, p. 428) and he assumes that the Newtonian derivation of Kepler's laws is a paradigm of theoretical explanation. Of course, Kitcher has to accept that the general theory of relativity also offers a theoretical explanation of Kepler's laws. What Kitcher does not realize is that both theories are incompatible with each other, and that they offer incompatible theoretical explanations of the planetary motions.

Theoretical explanations are, of course, less evocative than the causal ones. The latter rely on the idea that science is an activity capable of entering into direct contact with reality. However, the aim of theoretical science might be more modest, it could only seek to provide tools or instruments for predictive and anticipative, i.e. innovative, approaches to nature, and always with moderate success.

The reasons I have presented in this article, and they alone, lead me to agree with Woodward (2017, §7.1), after having travelled a very different path, that scientific explanation is certainly a topic that is somewhat independent of causation. This is a position Bunge himself could be willing to accept.

References

Aristóteles. (1988). In M. Candel (Ed.), *Tratados de Lógica (Órganon)* (Vol. II). Madrid: Editorial Gredos.

Aristotle. (1975). *Posterior analytics* (J. Barnes, Trans.). Oxford: Clarendon Press.

Aristotle. (1985). In J. Barnes (Ed.), *The complete works* (Vol. II). Princeton: Princeton University Press.

Ashbaugh, A. F. (1988). *Plato's theory of explanation. A study of the cosmological account in the Timaeus*. Albany: State University of New York Press.

Bacon, F. (1952). Novum organum. In F. Bacon (Ed.), *Advancement of learning. Novum Organum. New Atlantis*. William Benton, Publisher. Chicago: Encyclopaedia Britannica. 1. First published in 1620.

Berkeley, G. (1970). *A treatise concerning the principles of human knowledge*. C. M. Turbayne (Ed.), with critical essays. Indianapolis: The Bobbs-Merril Company, Inc. First published in 1710.

Bunge, M. (1962). Causality: A rejoinder. *Philosophy of Science, 29*, 306–317.

Bunge, M. (1979). *Causality and modern science*. New York: Dover Publications. First edition 1959 by Harvard University Press.

Bunge, M. (1982). The revival of causality. In G. Floistad (Ed.), *Contemporary philosophy. A new survey* (Vol. 2, pp. 133–155). The Hague: Martinus Nijhoff. Reprinted in Martin Mahner (ed.), *Selected Essays of Mario Bunge. Scientific Realism*. New York: Prometheus Books 2001.

Clifford, W. K. (1955). *The common sense of the exact sciences*. New York: Dover Publications, 1955. First published in 1885.

Comte, A. (2000). *The positive philosophy of Auguste Comte*. Freely Translated and Condensed by Harriet Martineau. Batoche Books. Kitchener, French original version: *Cours de Philosophie Positive*. Presentation et notes par Michel Serres, François Dagognet, Allal Sinaceur. Paris: Hermann, 1998.

Copernicus, N. (1993). On the revolutions of the heavenly spheres. In M. J. Adler (Ed.), *Great books of the Western World. 15. Ptolemy, Copernicus, Kepler*. Chicago: Encyclopaedia Britannica, 1952, Fourth Printing. First published in 1543.

Duhem, P. (1954). *The aim and structure of physical theory*. Princeton University Press. French original version: *La Théorie Physique. Son objet et sa structure*. Paris: Chevalier & Rivière, 1906.

Einstein, A. (1973). The mechanics of Newton and their influence on the development of theoretical physics. In A. Einstein (Ed.), *Ideas and opinions* (pp. 253–261). London: Souvenir Press Ltd. Originally published in *Die Naturwissenschaften*, vol. 15, 1927.

Feyerabend, P. (1962). Explanation, reduction and empiricism. *Minnesota Studies in the Philosophy of Science, III*, 28–97. Reduced version: Erklärung, Reduktion und Empirismus, Chap. 4 of Paul K. Feyerabend, *Probleme des Empirismus* (pp. 73–125). Braunschweig/Wiesbaden: Vieweg & Sohn, 1981.

Friedman, M. (1974). Explanation and scientific understanding. *Journal of Philosophy, 71*, 5–19.

Gilson, E. (1952). *La Philosophie au Moyen Age* (2nd ed.). Paris: Payot. 1952.

Hanson, N. R. (1958). *Patterns of discovery. An inquiry into the conceptual foundations of science*. Cambridge: Cambridge University Press.

Hempel, C. G. (1965). *Aspects of scientific explanation and other essays in the philosophy of science* (2nd ed.). New York: Free Press.

Hempel, C. G. (1966). *Philosophy of natural science*. Englewood Cliffs: Prentice-Hall, Inc.

Herschel, J. F. W. (1830). *A preliminary discourse on the study of natural philosophy*. Chicago: Chicago University Press, 1987. (Facsimile edition).

Jevons, S. (1958). *The principles of science. A treatise on logic and scientific method*. New York: Dover. First published in 1873.

Kant, I. (1998). *Critique of pure reason* (P. Guyer & A. W. Wood, Trans. & Ed.). Cambridge: Cambridge University Press. Originally published in 1781.

Kelvin, L. (1903). On the age of the sun's heat. Appendix E. In L. Kelvin (Ed.), *Treatise on natural philosophy*. Cambridge: Cambridge University Press. First published in *Macmillan's Magazine*, March 1862.

Kepler, J. (1981). *Mysterium cosmographicum. The secret of the universe* (A. M. Duncan, Trans.). New York: Abaris Books. Originally published in 1596.

Kitcher, P. (1981). Explanatory unification. *Philosophy of Science, 48*, 505–531.

Kitcher, P. (1989). Explanatory unification and the causal structure of the world. In P. Kitcher & W. C. Salmon (Eds.), *Scientific explanation* (Minnesota Studies in the Philosophy of Science) (Vol. XIII). Minneapolis: University of Minnesota Press.

Lewis, D. (1986). Causal explanation. In D. Lewis (Ed.), *Philosophical papers* (Vol. II). Oxford: University Press.

Lipton, P. (2001a). What good is an explanation? In G. Hon & S. Rakover (Eds.), *Explanation. Theoretical approaches and applications* (pp. 43–59). Dordrecht: Kluwer Academic Publishers.

Lipton, P. (2001b). Is explanation a guide to inference? A reply to Wesley C. Salmon. In G. Hon & S. S. Rakover (Eds.), *Explanation: Theoretical approaches and applications* (pp. 93–120). Dordrecht: Kluwer.

Mach, E. (1959). *The analysis of sensations and the relation of the physical to the psychical*. New York: Dover. Originally published in 1886.

Mill, J. S. (1843). *A system of logic ratiocinative and inductive*. London: Longman. 1970.

Misner, C., Thorne, K., & Wheeler, J. A. (1973). *Gravitation*. New York: Freeman and Co.

Newton, I. (1782). *Opera Quae Exstant Omnia*. Tom IV. London. (Samuel Horsley, Facsimile ed.). Stuttgart: Friedrich Frommann Verlag.

Ostlie, D., & Carroll, B. (1996). *Modern stellar astrophysics*. Reading: Addison-Wesley Publication Co., Inc.

Pearson, K. (1911). *The grammar of science* (1st ed. 1892). New York: Meridian Library edition, second printing 1960.

Plato. (1929). *Timaeus*. In *Plato in twelve volumes* (Vol. IX). Cambridge, MA: Harvard University Press.

Popper, K. (1959). *The logic of scientific discovery*. London: Hutchinson, 1959. First German edition, 1935.

Popper, K. R. (1974). Intellectual autobiography. In P. A. Schilpp (Ed.), *The philosophy of Karl Popper* (pp. 3–181). La Salle: Open Court.

Psillos, S. (2002). *Causation & explanation*. Chesham: Acumen Publishing Ltd.

Rivadulla, A. (2005). Theoretical explanations in mathematical physics. In G. Boniolo et al. (Eds.), *The role of mathematics in physical sciences* (pp. 161–178). Dordrecht: Springer.

Rivadulla, A. (2006). Theoretical models and theories in physics. A rejoinder to Karl Popper's picture of science. In I. Jarvie, K. Milford, & D. Miller (Eds.), *Karl Popper. A centenary assessment. vol. III: Science* (pp. 85–96). Aldershot: Ashgate.

Rivadulla, A. (2008). Discovery practices in natural sciences: From analogy to preduction. *Revista de Filosofía, 33*(1), 117–137.

Rivadulla, A. (2010). Complementary strategies in scientific discovery: Abduction and preduction. In M. Bergman, S. Paavola, V. Pietarinen, & H. Rydenfelt (Eds.), *Ideas in action: Proceedings of the applying Peirce conference* (Nordic Studies in Pragmatism 1) (pp. 264–276). Helsinki: Nordic Pragmatism Network.

Rivadulla, A. (2016a). Abduction and beyond. Methodological and computational aspects of creativity in natural sciences. *IFCoLog Journal of Logic and Its Applications, 3*(3), 105–121.

Rivadulla, A. (2016b). Models, representation and incompatibility. A contribution to the epistemological debate on the philosophy of physics. In J. Redmond et al. (Eds.), *Epistemology, knowledge and the impact of interaction* (Logic, Epistemology and the Unity of Science) (Vol. 38, pp. 521–532). Switzerland: Springer.

Ross, W. D. (1949). *Aristotle's prior and posterior analytics*. Oxford: Clarendon Press.

Salmon, W. C. (1984). *Scientific explanation and the causal structure of the world*. Princeton: University Press.

Sklar, L. (1992). *Philosophy of physics*. Oxford: Oxford University Press.

Straumann, N. (2004). *General relativity. With applications to astrophysics*. Berlin: Springer.

Whewell, W. (1847). *The philosophy of the inductive sciences* (Part One and Part Two). London: Frank Cass and Ltd, Second Edition.

Woodward, J. (2017). Scientific explanation. In E. N. Zalta (Ed.), *The Stanford encyclopedia of philosophy* (Spring 2017 ed.). https://plato.stanford.edu/archives/spr2017/entries/scientific-explanation/

Chapter 19
A Realist Analysis of Six Controversial Quantum Issues

Art Hobson

Abstract This essay provides a philosophically realistic analysis of six phenomena central to our current confusion: quantization, field-particle duality, superposition, entanglement, nonlocality, and measurement. As will be shown, these are logically related: Understanding measurement depends on properly understanding nonlocality, entanglement, and superposition; understanding these three depends on properly understanding field-particle duality and quantization. All six will be resolved, based on a realistic interpretation of standard quantum physics. The analysis is internally consistent as well as consistent with the relevant experimental facts. Thus, at least for these issues, QP concurs with the scientific paradigm as it has been known since Copernicus: nature exists on its own and science's goal is to understand nature's operating principles, which are independent of humans.

19.1 Introduction

Quantum physics (QP) is in a scandalous state (van Kampen 2008). Although founded at the dawn of the *preceding* century, and although arguably the most wide-ranging, highly accurate, and economically rewarding scientific theory of all time (Hobson 2017, pp. 7–10), its fundamentals remain in disarray. In particular, there is little consensus about wave-particle duality, superposition, entanglement, nonlocality, and measurement.

This disarray has spawned an astonishing assortment of interpretations and alterations of the theory. A quantum foundations conference polled its 33 expert attendees about their favorite interpretation (Schlosshauer et al. 2013). Nonrealistic interpretations, which claim that QP describes only our knowledge of the microworld rather than the microworld itself, gathered 24 votes. These sub-divided as 14 for the Copenhagen interpretation, 8 for the view that QP is about "information,"

A. Hobson (✉)
Department of Physics, University of Arkansas, Fayetteville, AR, USA
e-mail: ahobson@uark.edu

© Springer Nature Switzerland AG 2019
M. R. Matthews (ed.), *Mario Bunge: A Centenary Festschrift*,
https://doi.org/10.1007/978-3-030-16673-1_19

and 2 for "Bayesianism" according to which quantum states represent *personal* degrees of belief (Fuchs et al. 2014). Six chose the many worlds interpretation, which is realistic in the sense that the trillions of universes supposedly created every time one photon strikes a light-sensitive surface are presumed to be real. The remainder chose "other" or "none." Remarkably missing from the listed options was "standard quantum physics realistically interpreted."

It's not sufficiently recognized that this quantum confusion has spawned a serious social threat. Pseudoscience and related fantasies are rampant, especially in my country (Andersen 2017). Earth cries out for rationality and scientific literacy (Sagan 1995), yet quantum-inspired pseudoscience has become a threat to both. A pseudoscientific movie, *What the bleep do we know,* won several film awards and grossed $10 million in 2004; it argues that we create our own reality through consciousness and QP, and features physicists saying things like "The material world around us is nothing but possible movements of consciousness" (Shermer 2005). A non-technical physics literacy textbook being taught in high schools and universities bears the title *Quantum Enigma: Physics Encounters Consciousness* (Rosenblum and Kuttner 2006). According to the book's dust cover, "Every interpretation of QP encounters consciousness." It's striking that Stenger (1997) and Shermer (2005) have, by coincidence, the same title: *Quantum Quackery.*[1]

This essay provides a philosophically realistic analysis of six phenomena central to our current confusion: quantization, field-particle duality, superposition, entanglement, nonlocality, and measurement. As will be shown, these are logically related: Understanding measurement depends on properly understanding nonlocality, entanglement, and superposition; understanding these three depends on properly understanding field-particle duality and quantization. All six will be resolved, based on a realistic interpretation of standard quantum physics. The analysis is internally consistent as well as consistent with the relevant experimental facts. Thus, at least for these issues, QP concurs with the scientific paradigm as it has been known since Copernicus: nature exists on its own and science's goal is to understand nature's operating principles, which are independent of humans.

Non-realistic interpretations generally regard QP to be merely the study of "what we can say" about the microworld rather than the study of the actual microworld. As Niels Bohr reportedly stated:

> There is no quantum world. There is only an abstract quantum physical description. It is wrong to think that the task of physics is to find out how nature is. Physics concerns what we can say about nature (quoted in French and Kennedy 1985, p. 305).

At least with regard to the analyzed issues, this non-realistic view need not be accepted. It is possible, instead, to regard QP as the study of the general principles of the microworld itself, just as fluid dynamics is the study of fluids and geology is the study of Earth's structure. This conclusion concurs with Mario Bunge's lifetime work and further verifies his conclusions (Bunge 1967, 2003a, b, 2012; Mahner 2001).

---

[1] For further discussion, see Hobson (2013, 2017, pp. 12–13).

19.2   The Quantum

The quantum entered history on 14 December 1900 at a meeting of the Deutsche Physikalische Gesellschaft in Berlin when Max Planck first publicly wrote his famous $E = hf$. We view this today as the equation for the energy (in joules) of a photon having frequency f (in Hertz) where $h = 6.6 \times 10^{-34}$ joule-seconds is Planck's constant to two figures.

It wasn't always thought of that way, certainly not by Planck in 1900. For Planck, $E = hf$ represented the best possible fudge factor. He, like other physicists, had been searching for a theoretical explanation of the electromagnetic (EM) energy radiated at each frequency by a "black body" such as a small opening in an oven. Previous explanations predicted that the energy radiated at short wavelengths (high frequencies) was so enormous it should blind us whenever we look at a fire (Tegmark and Wheeler 2001)! Planck assumed correctly that radiation comes from vibrations of atoms such as those on the inner walls of the oven. Unable to solve the ensuing mathematics, he resorted to the numerical trick of assuming the amount of energy emitted by an atom could not range over a continuum of values but was instead restricted to a discrete set of values 0, ε, 2ε, 3ε, etc., where the increment ε is a small number. Today, we would say he assumed the amount of energy emitted to be "quantized." His plan was to calculate the radiation formula in this simpler approximation, and then find the real result by allowing ε to approach 0 since, as everybody knew, energies can always vary continuously. But, as happens frequently in quantum history, what "everybody knew" turned out to be wrong. When Planck allowed ε to approach 0, he recovered the bizarre result that others had found. It turned out the only way to get agreement with experiment was to allow ε to be the *non-zero* frequency-dependent number hf.

Planck had happened upon a principle deeper and stranger than he could have known: microscopic energies range over a discrete, rather than continuous, set of possible values.

This is the source of quantum jumps. Here's why. If the energy emitted by an atom must be an integral multiple of a positive number ε, then an atom must emit energy in *instantaneous bursts,* rather than continuously over time. That is, the least energy an atom can radiate is ε joules, and *this energy must be emitted at a single instant*. This is because, if ε joules are emitted continuously over some positive time $\tau > 0$, then some fraction of ε joules must be emitted in any shorter time, but *Planck's rule does not allow this*. Thus, there are physical "processes" or, better, "events," involving positive amounts of energy, that happen in "zero"[2] time. Today we call them quantum jumps. Discontinuity and quantum jumps were baked into QP from the beginning, and we should not be disconcerted when they re-appear in atoms, nonlocal phenomena, measurement, etc.

---

[2]But taking into consideration possible fundamental limitations such as the Planck time.

The greatest works of art are surprising. A Beethoven symphony's change of key, a Picasso painting's fractured face, are entirely unexpected yet, once we experience them, we realize they are perfect. The quantum is perhaps nature's ultimate stroke of genius. Quantum physics is the study of these quanta.

Little attention was paid to Planck's idea until 1905, when Albert Einstein reasoned that, if EM energy is emitted in small (in energy but not necessarily in spatial extent) lumps, then it probably travels through space in small lumps; he employed this notion to explain the emission of electrons from metal surfaces struck by radiation of sufficiently high frequency. Classical EM theory was unable to explain this in terms of smoothly extended EM fields, but Planck's energy lumps did the trick. Later, the lumps came to be called "photons." Einstein's contribution was crucial, for it extended the notion of the quantum as a finite small increment of energy to the notion of the quantum as a physical object: an energetically small and highly unified lump of energy. Photons, electrons, and quarks are examples of simple, or fundamental, quanta. Atoms and molecules are also quanta, but they are made of simpler quanta "entangled" into a highly unified "compound" quantum. All quanta are made of "quantized" (i.e. obeying the principles of QP) amounts of various sorts of energy. For photons, electrons, and quarks, the energies arise, respectively, from the EM field, electron-positron field, and strong field, three "quantum fields" that pervade the universe.

Einstein was the first to notice that quantum discontinuity has nonlocal implications. At the Firth Solvay Conference in Brussels in 1927, shortly after Heisenberg and Schrodinger invented quantum theory, Einstein asked the gathering to consider a single electron passing through a small hole in a partition. Schrodinger's equation predicts that its wave function Ψ, after passing through the hole, spreads out broadly over a distant viewing screen which Einstein assumed was spherical so Ψ would reach all of it simultaneously. Yet the electron impacts at only a small place on the screen. Einstein's written version states:

> The scattered wave moving towards [the viewing screen] does not present any preferred direction. If psi-squared were simply considered as the probability that a definite particle is situated at a certain place at a definite instant, it might happen that *one and the same* elementary process would act *at two or more* places of the screen. But the interpretation, according to which psi-squared expresses the probability that *this* particle is situated at a certain place, presupposes a very particular mechanism of action at a distance (Gilder 2008, p. 374).

Einstein's concern was that, if the impact appears at some point x, then the impact must also *not* appear at other points y, so the status of such other points must *instantly* switch from "possible impact point" to "impossible impact point." Instantaneous correlations must therefore exist between x and y, and special relativity does not seem to allow this. This was 5 years prior to the first broad analysis of measurement (von Neumann 1932) and 37 years prior to the earliest real understanding of nonlocality (Bell 1964).

Planck had practical reasons for inventing his fudge factor, but for nature it's not a fudge factor. Why the quantum? It's a question famously asked by John Wheeler (Barrow et al. 2004).

My suggestion would be that nature prefers countable sets, or perhaps even finite sets, over continuous and therefore uncountable sets. There are many more numbers along any single interval of the continuous real line than there are integers. If the universe is constructed out of countable quanta, reality becomes *much* simpler. This suggests that the universe may even be constructed from a *finite* number of quanta, which would be far simpler still.

The countability of quanta might suggest that they are particles. But as will be demonstrated, quanta are not particles and are not necessarily even tiny.

19.3   Is Reality Made of Fields, Particles, or Both?

The notion that everything is made of tiny particles runs through scientific history. Early Greeks such as Democritus of Abdera perceived reality to be made of small material particles moving in empty space. Newton's physics begins from particles obeying Newton's laws, so the motion of complex objects follows from that of their particles. Indeed, Newton writes:

> All these things being considered, it seems probable to me that God in the beginning formed matter in solid, massy, hard, impenetrable, movable particles, of such sizes and figures, and with such other properties, and in such proportion to space, as most conduced to the end for which he formed them (Newton 1998/1704).

The particle notion runs deep. Schrodinger's equation for, say, a moving electron is clearly a "field equation" for a scalar (i.e. number-valued rather than vector-valued) field $\Psi(x, y, z, t)$ entirely analogous to Maxwell's equations for the vector EM field $\mathbf{E}(x, y, z, t)$, $\mathbf{B}(x, y, z, t)$. Nevertheless, the QP founders retained the Newtonian language, speaking consistently of quantum "mechanics" and quantum "particles." Physicists continue to apply the term "particle" to essentially every quantum object, including even the Higgs field which is clearly a universe-filling quantized field. Because language so shapes our perception of reality, I doubt we will transcend our quantum confusion until we adopt more appropriate words. Electrons, photons, and other quanta are not "particles."

The question of fields versus particles is crucial because, once one adopts the particle misconception, most other issues become unfathomable. For example, if quanta are particles separated by empty space, distant nonlocal connections become incomprehensible.

The best example is still the double-slit experiment.[3] The set-up is a beam of monochromatic light or mono-energetic electrons passing through a pair of parallel narrow vertical slits cut into an opaque partition, with a viewing screen beyond the slits. This experiment with light was performed in 1801 by Thomas Young, who found that the viewing screen displayed an interference pattern: many vertical bands

---

[3]The remainder of this section is based on Hobson (2013), which should be consulted for references and details.

of light separated by dark bands. Physicists correctly inferred that light is a wave in a field later identified as the universal EM field. The interference arises from the intersection of two light waves, one from each slit, with "constructive interference" (the light bands) arising where crests meet crests and "destructive interference" (the dark bands) arising where crests meet valleys. Such waves indicate an underlying field that "carries" the waves. Thus, light is a wave in the universal EM field, much as ocean waves are waves in the water. This was the consensus among nineteenth-century physicists, and it's correct today.

The analogous experiment with an electron beam was performed for the first time by Claus Jonsson in 1961. The long-predicted result was an interference pattern just like Young's result! The only significant difference: Jonsson's interference bands were much smaller because typical electron wavelengths are smaller than photon wavelengths.

But most of us were taught that light is a spatially extended nonmaterial wave while electrons are tiny material particles orbiting within atoms. How can electrons be similar to light?

Here's how. In either Young's or Jonsson's experiment, one can dim the beam so extremely that only a single quantum (photon or electron) comes through at a time. Each quantum makes a tiny flash somewhere on the viewing screen. After 10–100 quanta have impacted, and assuming each flash persists indefinitely, we see 10–100 flashes distributed apparently randomly over the entire screen. But by the time 1000 quanta have impacted the screen, we begin to see an interference pattern forming from the distribution of flashes, the way a pointillist painting forms from tiny dots. With more impacts, it becomes clear that both interference patterns result from large numbers of small impacts. The similarity of the two experiments defies the notion that the first arises from waves and the second from particles.

Either experiment can be performed using large numbers of identically-prepared photons or electrons. The interference pattern that spreads over the entire screen then demonstrates quantum randomness: Identical preparations but a variety of outcomes. However, impacts are not entirely random but tend to cluster in regions of constructive interference. Since preparations are identical, *the entire pattern must be carried, i.e. "known," by each quantum.*

This suggests trying the experiment with only one slit open. The interference bands then vanish, replaced by a broad swath of tiny impacts distributed randomly all over the screen.[4] This is perplexing if quanta are particles. If a quantum is a small particle, then by definition it can come through only one slit regardless of whether the other slit is open. How does it "know" whether the other, relatively distant, slit is open? The experiment has been done not only with photons and electrons, but also with neutrons, atoms and molecules of all sorts, and there are even plans to use viruses. It's hard to imagine any long-range force that could inform these quanta as to whether the other slit is open or closed. In fact, the forthright particles advocate

---

[4]This assumes the slit's width is smaller than the wavelength, so the light passing through the slit spreads out into a single broad diffraction band without side fringes.

Richard Feynman said "you will get down the drain into a blind alley" if you think too hard about this (Feynman 1965).

According to the field view, every photon or electron is a wave rippling through a universe-filling field, namely the "quantized" (i.e. obeying quantum rather than classical rules) EM field or positron-electron field. Fields, because they naturally spread, resolve the question that got Feynman down the drain: Each photon or electron has sufficient spatial extension to come through both slits, encoding the double-slit or single-slit pattern. As Paul Dirac put it, "Each photon ... interferes only with itself" (Dirac 1958).

So fields explain the interference, but what about the particle-like impact points? This is the "measurement problem," explained more fully below. Briefly, the quantum approaching the screen extends over the entire interference pattern, and instantaneously "collapses" to microscopic size upon impact. This happens because the quantum "entangles" with the screen's atoms, instantaneously "localizing" the quantum. Most physicists would not phrase it this way. The standard words are that the quantum's "wave function" describes the entire pattern, and the "wave function" then collapses upon impact. But this is mincing words. Why would one invent a new object, the "wavefunction," to explain measurements? The quantum and its wavefunction are not separate objects. The quantum is its wavefunction. Fields are all there is.

This is a good place to clarify a confusion that annoyed John Bell and many others. Bell complained that QP seems "exclusively concerned about 'results of measurement,' and has nothing to say about anything else." Who, he asked, qualifies as the measurer? Is it every living creature? Or must it be a physicist? "The aim remains to understand the world. To restrict quantum mechanics to be exclusively about piddling laboratory operations is to betray the great enterprise. A serious formulation will not exclude the big world outside the laboratory" (Bell 1990).

It's hard to understand why this point wasn't forcefully stated long before 1990. The answer is not difficult. Every quantum measurement involves a measuring device that detects a quantum object and records a macroscopic mark. This happens not only in laboratories, it happens all the time in nature as for example when a cosmic ray strikes and moves a sand grain on Mars. A "quantum measurement" is simply any process by which a quantum phenomenon causes a macroscopic change. It has nothing necessarily to do with humans.

On a similar note, the "Born rule" connecting a quantum's state (or wavefunction) Ψ with the macroscopic world is usually stated this way: "The probability density for finding the quantum at some point x upon measurement is Ψ-mod-squared evaluated at x." The word "finding" is inappropriate because it suggests a human who does the finding, and it suggests the measured quantum is "at" some single point x and is therefore a particle. The quoted statement should be replaced by: "If a quantum interacts with another object, the probability density for the interaction to occur at x is Ψ-mod-squared evaluated at x."

Further linguistic points: Planck's energy lumps are "quanta of the EM field." There are other kinds of quanta: Electrons are quanta of the electron-positron field, quarks are quanta of the strong field, etc. "Quantum physics" is the study of fields that are bundled into highly unified lumps called "quanta."

Standard non-relativistic QP based on Schrodinger's equation is not the best basis for analyzing field-particle duality, because spontaneous quantum energy fluctuations plus relativity's principle of mass-energy equivalence entail that quanta can be created or destroyed. Relativistic QP was invented to deal with such possibilities. But it isn't easy to fit QP into a relativistic framework. In fact, rigorous theorems show that particles cannot exist in a universe that obeys both QP's so-called "unitary" time development and special relativity. The reason is that, even if we generously define a particle as any entity that is "localized" in the sense that it is contained, with 100% probability, within *some* region of finite volume, particles are impossible.

Here's why. If such a particle exists at some specific time such as t = 0, then unitary evolution implies it must *instantly* expand and have a positive probability of interacting an arbitrarily large distance away from its initial localization region at any positive time t > 0. But such instantaneous expansion implies that a particle on Earth at t = 0 could be detected on the moon an arbitrarily short time later, and this violates special relativity's ban on superluminal transport.

Thus, even under a broadly inclusive definition of "particle," quantum particles contradict special relativity. Fields, which fill all space, do not. Louis de Broglie put it bluntly in his 1924 PhD dissertation (a dissertation that delighted Einstein):

> The energy of an electron is spread over all space with a strong concentration in a very small region. That which makes an electron an atom of energy is not its small volume that it occupies in space, I repeat it occupies all space, but the fact that it is undividable, that it constitutes a unit (Baggott 2011, p. 38).

Quantum particles have many other conflicts with relativity. In fact, the only known version of relativistic quantum theory is quantum field theory which is, as its name implies, a theory of fields, not particles (Eberhard and Ross 1989). The Standard Model of high-energy physics, for example, is two quantum field theories known as the electroweak theory and quantum chromodynamics. Non-relativistic quantum theory based on Schrodinger's equation can also be cast as a quantum field theory.

It's well known that quantum field theories predict "vacuum states" having energy but no quanta. This is for the same reason (namely the uncertainty principle) that a mechanical oscillator obeying QP cannot be at rest but must instead have a ground state energy of $hf/2$. The quantum vacuum is quite real: It's the source of exquisitely predicted and verified phenomena such as the Lamb shift, the Casimir effect, and the electron's anomalous magnetic moment. Furthermore, excited states of atoms and other systems could not spontaneously transition to ground states without the quantum vacuum. This quantum vacuum is embarrassing for particle theories. If the universe is made of particles, then what is it that has this energy in the state that has no particles?

Another example of this embarrassment is the Unruh effect. Quantum field theory predicts that an accelerating observer in vacuum observes quanta while a non-accelerating observer of the same vacuum observes no quanta. If particles form the basic reality, how can they be present for the first observer and absent for a second observer of the same region? If fields are basic, this conundrum resolves itself: The first observer's acceleration promotes the second observer's vacuum fluctuations into higher-energy excitations.

Single-quantum nonlocality provides further evidence that the vacuum has properties arising from real fields. In fact, Einstein's remark at the 1927 Solvay Conference, quoted earlier, suggests individual quanta have nonlocal properties, and experimental and theoretical work since 1991 has shown that such properties arise from a single quantum's entanglement with the quantum vacuum (Tan et al. 1991).

In summary, the particle view encounters contradictions at every turn, but everything is consistent under an all-fields view. Particle-like phenomena arise from the discrete, countable nature of quanta, and from a quantum's partial localization upon detection. Picture a single photon, for example, as a spatially-extended cloud big enough to pass through both slits and then to extend over its entire interference region, which collapses into an atom-sized region upon impact.

19.4   Superposition

Since quanta are waves in fields, it's not surprising that they obey a "superposition principle": If a quantum can be in any one of several different states, then it can be in all of them simultaneously. Similarly, two water waves can exist simultaneously in the same pond and can even pass through each other. This doesn't seem odd for water waves, which are just shapes of the pond's surface, but it seems odd for, say, electrons and atoms. It entails that quanta can be in two places simultaneously.

The experiment diagrammed in Fig. 19.1 furnishes a perfect example. The figure shows an interference measuring device or "interferometer." Imagine a monochromatic light beam entering at the lower left. The interferometer comprises two "optical paths" along which the beam can travel. BS1 is a "beam splitter," a device that reflects half the beam upward into path 1 and transmits the other half rightward into path 2. M1 and M2 are mirrors that bring the two beams back

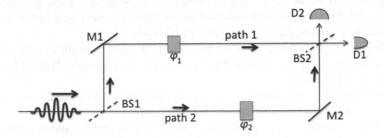

Fig. 19.1 A "Mach-Zehnder" interferometer

together so they cross at a second removable beam splitter BS2. D1 and D2 are light detectors, and ϕ_1 and ϕ_2 are "phase shifters" that can lengthen path 1 and path 2, respectively, by any amount up to one wavelength of the light.

If we remove BS2 and send a light beam through, half the light goes to D1 and half to D2 independently of how we set the phase shifters: evidence that the two half-beams simply pass through each other at the crossing point.

With BS2 inserted, half the light in each path goes to each detector, so the two paths mix together. The amount of light going to each detector varies as the path lengths change. At some phase settings, all the light goes to D1. This demonstrates interference: constructive at D1 and destructive at D2. If we then lengthen, say, path 1 slightly, a fraction of the light goes to D2. Lengthening path 1 by a quarter wavelength results in half the light going to each detector. Lengthening by a half wavelength results in all the light going to D2. The two paths are clearly interfering at the detectors, verifying that light is a wave.

If we dim the beam sufficiently, quantum effects must show up at some point. What happens when a single photon encounters BS1? Planck's rule entails that the photon cannot split, with half of it following each path. We discover, experimentally, that the entire photon instead reflects *and* transmits. It follows both paths! This sounds crazy. How do we know this?

Here's how. When a single photon passes through the interferometer with BS2 absent we find after many "trials" that the entire photon always impacts either D1 or D2, never both, verifying the photon to be a single unified object that doesn't split. The impacts occur with 50–50 random statistics at D1 and D2. It's interesting and significant that the randomness is *perfect*. For example, with a long enough series of trials we find the perfect fraction (1 part in 2^{10}) of ten D1s in a row. Quantum indeterminacy is more perfectly random than any classical device such as a roulette wheel or coin flips (Hobson 2017).

Continuing the explanation: Inserting BS2 and varying the path lengths reveals a surprising result: The statistics at D1/D2 vary depending on the *difference between* the two path lengths, as shown in Fig. 19.2, so the measured outcome responds equally to phase alterations of *either* path. The simplest conclusion is that each photon goes both ways; it is "superposed" along both paths. When the two "copies" of the photon pass through BS2, each copy again goes both ways so the two paths mix and interfere at either detector: If crest meets crest at D1, interference is constructive and we detect a wave (a photon). If crest meets valley, we detect no photon at D1. Figure 19.2 shows how the probabilities of detecting the photon at D1 vary with the difference between the two path lengths.

If photons are particles, similar to peas only smaller, superposition is remarkable. But if a photon is an extended real lump of field, it's expected: The field simply spreads along both paths.

Interferometer experiments have also been performed with material quanta such as atoms, with the same result. Atoms, too, are fields.

Gui-Lu Long and colleagues recently performed an interferometer experiment that amounts to a direct demonstration of quantum realism (Long et al. 2018). The experiment is an extreme version of a "delayed-choice" experiment (Jacques

Fig. 19.2 Outcome of the experiment of Fig. 19.1, evidence that each photon travels both paths

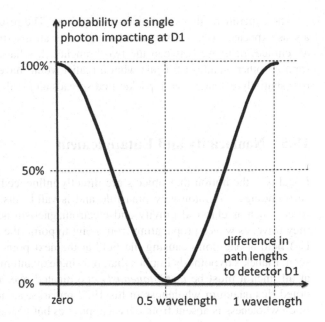

et al. 2007). In Jacques' experiment, a single photon passes through BS1 and *then* a random choice is made to either insert or not insert BS2. Bohr's notions about the complementary dual wave-particle nature of quanta had misled some physicists into thinking a photon might mysteriously retain whichever aspect, wave or particle, it had acquired *before* entering the interferometer, regardless of the last-minute switching of BS2. As one might expect, this proved false. The photon interfered with itself if and only if BS2 was inserted, regardless of its history. A photon is not sometimes a wave and sometimes a particle; it is always a bundled lump of field that spreads whenever it can.

Realistically, a single photon is an EM wave packet stretched out in space. Accordingly, Long proposed to quickly insert (or remove) BS2 *while the photon is passing through the crossing.* Long's ingenious implementation allows him to insert BS2 at any time during the photon's passage through the crossing; his results show 16 data points corresponding to trials made at 16 "delay times" between 0 (when the photon's front end arrives at the crossing) and T (when its back end arrives). If Long had considered only delays of either 0 or T, the experiment would mimic Jacques' experiment, and indeed Long's results recapitulate Jacques' in these cases: phase-dependent interference with BS2 present (delay time of 0), phase-independent random 50–50 outcomes with BS2 absent (delay time of T).

For intermediate delay times, Long and colleagues analyze the four "sub-waves" (the two branches of the superposition, both before and after insertion of BS2) in detail to predict the probability of "particle-like" behavior (phase-independent random outcomes) and "wave-like" behavior (phase dependent interference) at the detectors, as a function of both the delay time and phase difference.

The experiment slices and dices each photon. The results for all four sub-waves are as expected from realistic standard QP, over all insertion times from 0 to T and all choices of phase between the two branches. It's hard to believe that a photon (and all other quanta) isn't just what a realist would have guessed decades ago: A real spatially-extended wave packet that spreads along all available paths.

19.5 Nonlocality and Entanglement

Locality – the notion that objects are directly influenced only by their immediate surroundings – is intuitively plausible and is valid outside of QP. Long-distance forces such as classical gravity and electromagnetism act only "locally" because they travel as waves propagating from point to point, the changing gravitational or EM field at one point causing the field at the next point to change. But as we've seen, Planck's hypothesis implies that, if a single quantum has any spatial extension at all, then it must be *instantaneously* coordinated over its entire extent. Einstein noted this quantum wholeness in his 1927 remarks, as have others (Bohm 1980). Such wholeness is absent from classical physics but pervasive in QP and the source of quantum nonlocality.

Nonlocality is most obvious when two or more quanta are involved. When two quanta interact, their spatially extended states can become "entangled." Theory and experiment show entangled quanta can influence each other instantaneously regardless of their separation distance. Figure 19.3 indicates this conceptually. Assume the black quantum and gray quantum, initially moving rightward and upward respectively, are independent, i.e. alterations in one don't affect the other. They then come together and overlap their spatially extended states. This can "entangle" them so that, as they separate, each retains some portion of the other's state. The unity of each pre-entanglement quantum, implied by Planck's hypothesis, then suggests the entangled quanta form a single unified object.

Both the black and gray quanta now move in two ways simultaneously, in the following superposition:

"Black quantum moves rightward and gray quantum moves upward," superposed with "black quantum moves upward and gray quantum moves rightward."

This two-quantum superposition turns out to be remarkable. As suggested by Fig. 19.3, any alteration of one of the two final objects entails *simultaneous* changes in the other object, *regardless of their separation.* Such instantaneous coordination at a distance is perhaps not surprising when we recall the nonlocal implications of Planck's quantum hypothesis, and that quanta are simply ripples in a field: Two ripples have met and separated so that part of each incoming ripple now appears in both outgoing ripples.

Nonlocality is massively documented. Einstein suggested nonlocal consequences of entanglement (Einstein et al. 1935); Bell found a mathematical criterion for whether entangled systems really behave nonlocally (Bell 1964); Clauser demon-

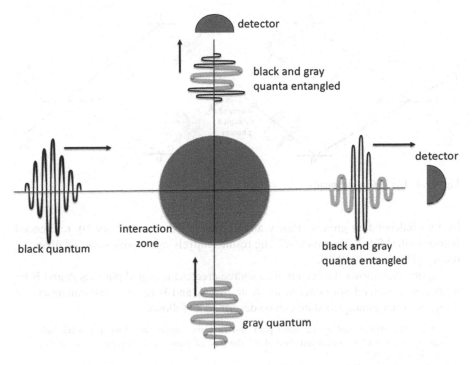

Fig. 19.3 Two quanta can entangle when they meet, a condition that persists even after separation

strated nonlocality in entangled systems (Clauser and Freedman 1972); Aspect demonstrated nonlocal coordination occurs faster than lightspeed (Aspect et al. 1982); and many experiments demonstrate nonlocality across great distances (Pan et al. 2017). Nonlocality raised enormous skepticism, but experiments managed to close all loopholes simultaneously.[5]

The result is that it's now known beyond any reasonable doubt that "local realism" fails, meaning outcomes of measurements on entangled pairs of quanta are not fully determined by real properties (such as quantum states) carried along "locally" by either quantum. Either individual ("local") systems are instantaneously influenced by distant events, or the properties (such as quantum states) of entangled systems don't objectively exist. It's to quantum theory's great credit that it correctly predicts the failure of local realism. But which one fails, locality or realism? The Copenhagen camp claims realism fails, but others (Aspect 2007) argue locality fails.

It's helpful to consider a specific experiment. Of the plethora of nonlocality experiments since 1972, the most pedagogically useful are the experimental investigations of pairs of momentum-entangled photons performed nearly simultaneously

---

[5] See Giustina et al. (2015), Hensen et al. (2015), and Shalm et al. (2015).

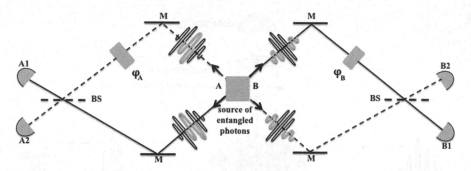

Fig. 19.4 The RTO experiments

by two independent groups (Rarity and Tapster 1990; Ou et al. 1990), referenced below as the "RTO experiments." The results entirely agree with standard quantum theory (Horne et al. 1990).

Figure 19.4 shows the layout. The source creates entangled photons A and B by a process that need not concern us. Note that A and B are the *post*-entanglement photons. Their entangled state can be described as follows:

> "A moves leftward along the solid path and B moves rightward along the solid path," superposed with "A moves leftward along the dashed path and B moves rightward along the dashed path."

More briefly, A and B move away from each other along the solid path and also along the dashed path. A phase shifter is situated on each path.

If A and B were not entangled, the experiment would be simply two back-to-back interferometer experiments, each one like Fig. 19.1 but with BS1 lying inside the source. However, the entanglement changes everything. The single photon of Fig. 19.1 interferes only with itself, but neither photon in Fig. 19.4 interferes with itself! This is a key result and as we'll see below it resolves the measurement problem. Regardless of either phase setting, both detectors (A1/A2 and B1/B2) register perfectly random 50–50 mixtures of outcomes. This means that neither photon has a phase of its own, even though non-entangled photons have phases of their own and can interfere with themselves.

However, the entangled pair *does* have a phase, and a state, of its own. Upon varying the phases we find the highly organized two-photon correlations graphed in Fig. 19.5, showing both photons "know" the phase *difference*. The graph shows the photons' "degree of correlation" for various phase differences, where 360° represents one wavelength difference between the solid and dashed path lengths. A "degree of correlation" of +1 means that, in a long series of trials, the two outcomes are always the same (either both are 1 or both are 2); a correlation of −1 means they are always different; zero correlation means the outcomes are the same on 50%, and different on 50%, of the trials; values between 0 and +1 indicate "same" is more likely than "different." For example, at a 45-degree phase difference the outcomes are the same in 71% of trials. For mathematical definitions, see Hobson (2018).

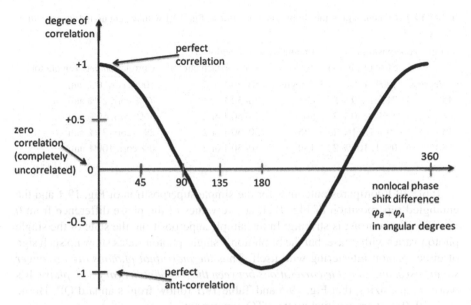

Fig. 19.5 The degree of correlation between RTO's two entangled photons varies with the non-local phase difference $\phi_B - \phi_A$ between the photons

The perfect correlation at phase zero is manifestly nonlocal because of the presence of the beam splitters: A's beam splitter sends photon A *randomly* to A1 or A2, and similarly for B's beam splitter. Nevertheless, each photon "knows" the "choice" of the other photon and correlates its path accordingly. How can A know B's choice? The answer: Entanglement. At zero phase, the two photons form a single object. In fact, it's enlightening to consider the *pair* of photons as a single "bi-photon," an "atom of light" (atoms are also internally entangled) superposed along two bi-photon paths, solid and dashed.

The coordination shown in Fig. 19.5 is shocking, and entirely independent of the photons' separation: The entire bi-photon responds to changes in either phase shifter by adjusting its distant correlations accordingly. In fact, the experiment's violation of Bell's famous inequality shows that any change in phase entails an alteration in the future behavior of *both* photons. For example, if B's phase shifter changes, then both photons' future outcomes become different from what they would have been. Furthermore, the experiments of Aspect and others show this alteration occurs faster than light can travel between the two photons.

Just as we can obtain Fig. 19.2 by varying either phase shifter in Fig. 19.1, we can obtain the correlations of Fig. 19.5 by varying either phase shifter in Fig. 19.4. Thus, just as Fig. 19.2 entails that the single photon really is superposed along both paths, Fig. 19.5 entails that the bi-photon really is superposed along both the solid path and the dashed path of Fig. 19.4. The bi-photon simply spreads along both available paths, because it's a single unified field.

Table 19.1 Comparison of the simple superposition of Fig. 19.1 with the entangled superposition of Fig. 19.4

Simple superposition		Entangled superposition		
ϕ	State of photon	$\phi_B - \phi_A$	State of each photon	Correlation between photons
0 degrees	100% 1, 0% 2	0 degrees	50–50 1 or 2	100% corr, 0% anti
45	71% 1, 29% 2	45	50–50 1 or 2	71% corr, 29% anti
90	50% 1, 50% 2	90	50–50 1 or 2	50% corr, 50% anti
135	29% 1, 71% 2	135	50–50 1 or 2	29% corr, 71% anti
180	0% 1, 100% 2	180	50–50 1 or 2	0% corr, 100% anti

Table 19.1 compares outcomes for the simple superposition of Fig. 19.1 and the entangled superposition of Fig. 19.4, at five values of the phase difference from 0 to 180°. The contrast is striking: In the simple superposition, the state of the single photon varies with phase, but the bi-photon's single-photon states show no such sign of either photon interfering with itself. Thus *the individual photons are no longer superposed, and only the correlations between the two photons vary with phase.* It's worth emphasizing that Fig. 19.5 and Table 19.1 follow from standard QP (Horne et al. 1990) and are verified by the RTO experiments.

It's been amply demonstrated that nonlocal correlations are established faster than light can connect the two measurements. This might seem to violate special relativity. But special relativity prohibits superluminal transfer only for causal signals. Nonlocality is implemented entirely with correlations, and correlations alone cannot transfer any causal effect. In the RTO experiment, for example, A cannot send an instantaneous signal to B by changing A's phase shifter, because nothing changes at B. The only way B can know something changed is by obtaining information about A's outcomes and noting the change of *correlations*. Thus, the bi-photon is instantaneously coordinated across its entire extent without violating special relativity.

We've seen that the RTO experiment is manifestly nonlocal. Nonlocality is also manifest in Planck's hypothesis, and in Einstein's comment at Solvay. Experiments such as RTO only reinforce the centrality of nonlocality to the quantum world. The consensus today is that either locality or realism fails, but there is no reason to abandon five centuries of scientific history by renouncing realism because nonlocality is clearly written all over QP.

A realistic picture of superposition and entanglement emerges. A fundamental quantum such as a photon is a highly unified bundle of field energy carrying one "excitation" (*hf* joules). In Fig. 19.1, a photon spreads along paths 1 and 2. Upon measurement at D1/D2, the photon's reality emerges in its unified nature, its amplitudes (which needn't be 50–50), and its phase. Two entangled photons also form a single highly unified bundle of field energy but carrying two excitations. In Fig. 19.4, a bi-photon spreads out along the solid and dashed paths and is detected at A1/A2 and B1/B2. Although the bi-photon is superposed, neither single photon is superposed. Neither photon has a phase (Hobson 2018), but the bi-photon has a phase and a quantum state as revealed by nonlocal correlations. Upon interaction, the bi-photon behaves as a single unified quantum regardless of its extension. A similar realistic view of superposition and entanglement has been expressed by Perez-Bergliaffa (Perez-Bergliaffa et al. 1996).

19.6   Measurement and Entanglement

Measurements – processes in which a quantum event leads to an irreversible macroscopic change – have long presented difficulties for physicists. As will be shown, the main stumbling block has been entanglement. Entanglement is subtle. It is "*the* characteristic trait of quantum mechanics" (Schrodinger 1935, his emphasis). Today we still have no clear consensus about its nonlocal implications, although recent experiments might be changing this. Thus the continuing conundrum, and the variety of proposed solutions, are not surprising.

As a simple example, consider the single photon of Fig. 19.1, without BS2, as the superposed photon moves along both paths toward D1/D2. As the photon nears the detector, the photon and detector interact to form the following entangled "measurement state":

"Photon on path 1 and D1 registers," superposed with "photon on path 2 and D2 registers."

Most physicists agree with the analysis up to this point, a point reached long ago in the first detailed analysis of measurement (von Neumann 1932).

The conundrum is that this state appears to entail that the macroscopic detector is superposed, and registers both D1 and D2. This would present two problems: (1) there is no definite outcome, so the measurement is useless; (2) such a macroscopic superposition, while physically possible, would be excruciatingly difficult to create and certainly could not be created as easily as this. Schrodinger dramatized the paradox in his famous example involving a radioactive nucleus, with a cat playing the role of detector by dying or living according to whether or not the nucleus decayed. The superposition seems to imply a dead and alive cat.

But we understand entanglement better today than in Schrodinger's day or indeed any other day, and this understanding resolves the supposed paradox (Hobson 2018). The above measurement state is simply a version of RTO's entangled state at zero phase (correlation of +1), with one of the quanta being (or triggering) a macroscopic detector. From Table 19.1, then, it can be seen that there is no paradoxical macroscopic superposition: the state of the photon is a random 50–50 probability of either path 1 or path 2, the state of the detectors is a random 50–50 probability of either D1 or D2, and the two outcomes are perfectly correlated. There is no macroscopic superposition because *neither subsystem is superposed*. In fact, we've seen that neither subsystem has a phase of its own, implying it cannot interfere with itself. Neither subsystem is a "coherent" quantum. Only *correlations between subsystems* are superposed, not *states* of subsystems, and thus there is no paradox. Even if one subsystem is macroscopic, there is no paradox. All this agrees with standard quantum theory (Horne et al. 1990).

This solves the problem of "definite outcomes" also known as "Schrodinger's cat." But it doesn't entirely solve the measurement problem, because a completed measurement must irreversibly record a single outcome. In fact, the mathematical formulation of the measurement process shows that the measurement state is in principle reversible, so it cannot yet be a permanent indicator such as a macroscopic sound wave or visible mark (Hobson 2018).

When the photon entangles with D1/D2 it transfers its energy to a single electron in either D1 or D2. This electron triggers a many-electron "avalanche," creating a detectable electric current in either D1 or D2. Experiments show the photon actually interacts with both detectors, but the triggering happens within just one of them while the other detector experiences only the quantum vacuum (Fuwa et al. 2015). In other words, entanglement with *both* detectors ensures that *one* detector is triggered while the other is *not* triggered, resolving Einstein's question at the 1927 Solvay Conference. The process is necessarily non-local due to the separation between D1 and D2 and the requirement of instantaneous correlation. The indeterminism between D1 and D2 is a fundamental quantum property, just as it is throughout quantum physics.

Concerning irreversibility: the electron avalanche involves a large number of microscopic processes, each of them reversible or "unitary" in principle but collectively so numerous and diverse that the probability of reversal is essentially zero. Such processes are what the second law of thermodynamics is all about. The question of whether such a process is "really" irreversible or simply irreversible "for all practical purposes" has been argued since Ludwig Boltzmann's day (Hobson 1971), and arises in identical fashion even for classical systems. In other words, at this point our task of explaining quantum measurements is finished.

To summarize: Quantum measurement occurs when a superposed microscopic quantum entangles with a macroscopic object. The entanglement is not a paradoxical macroscopic superposition because (1) it is not a superposition of states of any single subsystem but merely a superposition of correlations between states of different subsystems; and (2) each subsystem has no phase and so cannot interfere with itself. The entanglement entails an indeterminate but definite quantum choice of macroscopic outcomes. The macroscopic object can be a laboratory detector or it can be natural, i.e. measurement has nothing necessarily to do with humans. Thus, when put into the proper realistic physical context, the measurement problem is not difficult.

19.7 Conclusion

Properly understood, and as has also been argued by Mario Bunge, the issues of quantization, field-particle duality, superposition, entanglement, nonlocality, and measurement present no barrier to a consistent and realistic interpretation based on standard quantum physics. At least to this extent, quantum physics is consistent with the scientific view as it has been known since Copernicus: nature exists on its own and science's goal is to understand its operating principles, which are independent of humans.

References

Andersen, K. (2017). *Fantasyland: How America went haywire, a 500-year history*. New York: Random House.

Aspect, A. (2007). To be or not to be local. *Nature, 446*, 866–867.

Aspect, A., Dalibard, J., & Roger, G. (1982). Experimental test of Bell's inequalities using time-varying analyzers. *Physical Review Letters, 49*, 1804–1807.

Baggott, J. (2011). *The quantum story* (p. 38). Oxford: Oxford University Press. As translated in English by Kracklauer, A. F., from de Broglie, L. (1924), "Recherches sur la Theorie des Quanta," PhD thesis, Faculty of Science, Paris University.

Barrow, J. D., Davies, P. C. W., & Harper, C. L. (2004). *Science and ultimate reality: Quantum theory, cosmology, and complexity*. Cambridge: Cambridge University Press.

Bell, J. (1964). On the Einstein Podolsky Rosen paradox. *Physics, 1*, 195–200.

Bell, J. (1990). Against "measurement". *Physics World, 1*, 33–40.

Bohm, D. (1980). *Wholeness and the implicate order*. London: Routledge.

Bunge, M. (1967). *Foundations of physics*. New York: Springer.

Bunge, M. (2003a). Twenty-five centuries of quantum physics: From Pythagoras to us, and from subjectivism to realism. *Science & Education, 12*, 445–466.

Bunge, M. (2003b). Quantons are quaint but basic and real, and the quantum theory explains much but not everything: Reply to my commentators. *Science & Education, 12*, 587–597.

Bunge, M. (2012). Does quantum physics refute realism, materialism and determinism? *Science & Education, 21*, 1601–1610.

Clauser, J. F., & Freedman, S. J. (1972). Experimental test of local hidden-variables theories. *Physical Review Letters, 26*, 938–941.

Dirac, P. (1958). *Quantum mechanics* (4th ed.). Oxford: Oxford University Press.

Eberhard, P. H., & Ross, R. R. (1989). Quantum field theory cannot provide faster-than-light communications. *Foundations of Physics Letters, 2*, 127–148.

Einstein, A., Podolsky, B., & Rosen, N. (1935). Can quantum-mechanical description of physical reality be considered complete? *Physical Review, 47*, 777–780.

Feynman, R. (1965). *The character of physical law*. Cambridge, MA: MIT Press.

French, A. P., & Kennedy, P. J. (1985). *Niels Bohr: A centenary volume*. Cambridge, MA: Harvard University Press.

Fuchs, C. A., Mermin, N. D., & Schack, R. (2014). An introduction to QBism with an application to the locality of quantum mechanics. *American Journal of Physics, 82*, 749–754.

Fuwa, M., Takeda, S., Zwierz, M., Wiseman, H. M., & Furusawa, A. (2015). Experimental proof of nonlocal wavefunction collapse for a single particle using homodyne measurements. *Nature Communications, 6*, 6665.

Gilder, L. (2008). *The age of entanglement: When quantum physics was reborn*. New York: Alfred A. Knopf.

Giustina, M., et al. (2015). Significant-loophole-free test of Bell's theorem with entangled photons. *Physical Review Letters, 115*, 250401.

Hensen, B., et al. (2015). Loophole-free Bell inequality violation using electron spins separated by 1.3 kilometers. *Nature, 526*, 682–686.

Hobson, A. (1971). *Concepts in statistical mechanics*. New York: Gordon and Breach, Chp. 5.

Hobson, A. (2013). There are no particles, there are only fields. *American Journal of Physics, 81*, 211–222.

Hobson, A. (2017). *Tales of the quantum: Understanding physics' most fundamental theory.* Oxford: Oxford University Press.

Hobson, A. (2018). Review and suggested resolution of the problem of Schrodinger's cat. *Contemporary Physics, 59,* 16–30. Also available at https://arxiv.org/abs/1711.11082.

Horne, M. A., Shimony, A., & Zeilinger, A. (1990). Introduction to two-particle interferometery. In A. I. Miller (Ed.), *Sixty-two years of uncertainty* (pp. 113–119). New York: Plenum Press.

Jacques, V., Wu, E., Grosshans, F., Treusart, F., Grangier, P., Aspect, A., & Roch, J.-F. (2007). Experimental realization of Wheeler's delayed choice gedanken experiment. *Science, 315,* 966–968.

Long, G. L., Wei, Q., Yang, A., & Li, J.-L. (2018). Realistic interpretation of quantum mechanics and encounter-delayed-choice experiment. *Science China: Physics, Mechanics & Astronomy, 61*(3), 03031.

Mahner, M. (2001). *Scientific realism: Selected essays of Mario Bunge.* Amherst: Prometheus Books.

Newton, I. (1998/1704). *Opticks.* Palo Alto: Octavo. ISBN 1-891788-04-3. Originally published 1704. For quotation, see https://en.wikiquote.org/wiki/Isaac_Newton

Ou, Z. Y., Zou, X. Y., Wang, L. J., & Mandel, L. (1990). Observation of nonlocal interference in separated photon channels. *Physical Review Letters, 65,* 321–324.

Pan, J.-W., et al. (2017). Satellite-based entanglement distribution over 1200 kilometers. *Science, 356,* 1140–1144.

Perez-Bergliaffa, S. E., Romero, G. E., & Vucetich, H. (1996). Axiomatic foundations of quantum mechanics revisited: The case for systems. *International Journal of Theoretical Physics, 35,* 1805–1819.

Rarity, J. G., & Tapster, P. R. (1990). Experimental violation of Bell's inequality based on phase and momentum. *Physical Review Letters, 65,* 2495–2498.

Rosenblum, B., & Kuttner, F. (2006). *Quantum enigma: Physics encounters consciousness.* Oxford: Oxford University Press.

Sagan, C. (1995). *The Demon-haunted world: Science as a candle in the dark.* New York: Random House.

Schlosshauer, M., Kofler, J., & Zeilinger, A. (2013). A snapshot of foundational attitudes toward quantum mechanics. *Studies in History and Philosophy of Modern Physics, 44,* 222–230.

Schrodinger, E. (1935). Discussion of probability relations between separated systems. *Proceedings of the Cambridge Philosophical Society, 31,* 555–563.

Shalm, L. K., et al. (2015). Strong loophole-free test of local realism. *Physical Review Letters, 115,* 250402.

Shermer, M. (2005). Quantum quackery. *Scientific American, 292*(1), 34.

Stenger, M. (1997). Quantum quackery. *Sceptical Inquirer, 21*(1), 37–42.

Tan, S. M., Walls, D. F., & Collett, M. J. (1991). Nonlocality of a single photon. *Physical Review Letters, 66,* 252–255.

Tegmark, M., & Wheeler, J. (2001, February). 2000 years of quantum mysteries. *Scientific American, 284,* 68–75.

van Kampen, N. G. (2008). The scandal of quantum mechanics. *American Journal of Physics, 76,* 989–990.

von Neumann, J. (1932). *Mathematische grundlagen der quantenmechanik.* Berlin: Springer. Translated into English by R. T. Beyer (1955) *Mathematical foundations of quantum mechanics.* Princeton: Princeton University Press.

Chapter 20
On the Legacy of a Notable Quantum Dissident: David Bohm (1917–1992)

Olival Freire Junior

Abstract David Bohm was a long-standing critic of the standard Copenhagen interpretation of quantum mechanics, thus a quantum dissident. He devoted much of his professional time to an alternative interpretation of quantum theory, the causal interpretation. When he suggested it in the early 1950s, research on the foundations of quantum mechanics was viewed with suspicion among physicists, as most of them considered foundational issues already solved by the founders of this physical theory. As time went by, interest in foundational issues widened, indeed it became a regular field for physical research, and Bohm's works went on to be considered more positively, or at least with more tolerance. Bohm's work had made an early and lasting impression on Mario Bunge. However, they did not always follow the same approach to the quantum riddles.

While this paper deals primarily with the work of Bohm, it contributes to our appreciation of Bunge's legacy in this area by establishing significant historical background and context to Bunge's views on Quantum Mechanics and the kinds of resistance the profession tended to offer. In this paper I will examine Bohm's achievements and how these achievements were received by the public, especially other physicists. I address this subject through the lens of scientometric data, mainly the number of citations of papers over time, adding to these data some qualitative information concerning how these papers were seen by other physicists.

20.1 Introduction

Mario Bunge's interest in the philosophy of physics was awoken when he was 17 years old and decided to study physics in order to become a philosopher of science. His graduate work was supervised by the physicist Guido Beck, a European émigré who had gone to Argentina escaping Nazi persecution. His early philosoph-

O. Freire Junior (✉)
Instituto de Fisica, Universidade Federal da Bahia, Salvador, Brazil
e-mail: freirejr@ufba.br

© Springer Nature Switzerland AG 2019
M. R. Matthews (ed.), *Mario Bunge: A Centenary Festschrift*,
https://doi.org/10.1007/978-3-030-16673-1_20

ical concerns were related to the role of chance and necessity in physical theories. Thus, it was no surprise that Bunge was struck by David Bohm's 1952 new interpretation of quantum mechanics recovering a causal description for the physical phenomena. From Buenos Aires, Bunge wrote to Bohm, who was living in exile in São Paulo, commenting on this work. Bohm's reaction to Bunge's letter was: as you are asking too many questions to discuss, why don't you come to São Paulo for a while and we can work together on these issues? Bohm arranged a grant from the Brazilian Conselho Nacional de Desenvolvimento Científico e Tecnológico, an organization of the Brazilian federal government and Bunge spent 1 year in São Paulo. However, the stay was not particularly fruitful and no scientific publications resulted.

Bunge was working on the toughest issue for the causal interpretation, how to deal with the new interpretation of quantum mechanics in the relativistic domain. Bohm, himself, would spend years grappling with this problem. Bunge remained interested in Bohm's causal interpretation for more than a decade, teaching and publishing on the subject (Bunge 1955, 1956a, b). In the mid-1960s, however, while working on the axiomatization of quantum mechanics, Bunge began to think differently (Bunge 1967). The challenge was not to recover determinism in the quantum domain, rather, the issue was how to interpret quantum theory from a realistic perspective.[1] Later Bohm would follow similar philosophical lines. Thus, it seems appropriate to honor Mario Bunge on his 100th birthday by reviewing the legacy of David Bohm, the physicist with whom Bunge first worked on the foundations of physics.

This chapter uses scientometric data to assess the legacy, in other words the influence in physics of the works of David Bohm, one of the notable quantum dissidents.[2] Bohm's earliest work on plasmas and collective coordinates had already made his reputation. However, his most original and heterodox contribution to physics may have been the elaboration of the causal interpretation of quantum physics, published in 1952. This strongly departed from the standard theory in its conceptual and philosophical assumptions but still got the same predictions, thus opening the way for alternative interpretations of quantum mechanics. In the late 1950s, Yakir Aharonov's and Bohm's seminal paper contributed to our understanding of the role of phases and electromagnetic potentials in the quantum domain (Aharonov and Bohm 1959). Later, Bohm put the causal interpretation aside and turned his attention to the highly mathematical approach called implicate order, working with Basil Hiley, looking for the most basic algebraic structures from which quantum theories could emerge. In fact, while abandoning the quest for the recovery of a deterministic view of quantum phenomena, Bohm remained committed to a realistic view of physical theories.

---

[1] Letters from Mario Bunge to the author, 01 Nov 1996 and 12 Feb 1997.

[2] I used the term 'quantum dissidents' for the physicists who did not accept the dominant view according to which the foundations of quantum mechanics did not deserve further investigation as all foundational issues were already solved by the founding fathers of this physical theory. Instead, they thought that these issues "were worth pursuing as part of a professional career in physics." See Freire Junior (2015, p. 2).

In the late 1970s, two students, Chris Dewdney and Chris Philippidis surprised Bohm and Hiley by building diagrams of the paths and potentials from Bohm's early interpretation (Philippidis et al. 1979). Indeed, they used computers to make such diagrams. At that time, interest in the foundations of quantum mechanics was being revived and Bohm spent his last years trying to reconcile his own different approaches to the quantum. This attempt is embedded in Bohm and Hiley's textbook, *The Undivided Universe*, published posthumously (Bohm and Hiley 1993). Bohm's work generated several strands, including Hiley's algebraic work; Bohmian mechanics, led by Detlef Dürr, Sheldon Goldstein, and Nino Zanghi; and attempts to use and test Bohm's ideas in the cosmological domains.

Bohm's personal life was also marked by the vicissitudes of the times. Caught up in the American anticommunist anxiety, he left the US for a job in Brazil, and later went to Israel and eventually the United Kingdom. His passport was apprehended by US officers and he lost his American citizenship, thus becoming the most notable expatriate American scientist of the times. In the late 1950s he broke his ideological ties with Marxism and turned to Eastern thinkers, becoming an iconic figure in the New Age culture of the 1960s and 1970s (Freire Junior 2015).

In order to assess Bohm's legacy to physics this chapter asks what was Bohm's persona inside the expert circle of physicists? He was a long-standing critic of the standard interpretation of quantum mechanics, thus a quantum dissident, and devoted much of his professional time to an alternative interpretation of quantum theory, the causal interpretation. He suggested it at a time, the early 1950s, when research on the foundations of quantum mechanics was poorly viewed among physicists as most of them considered foundational issues already solved by the founding fathers of this physical theory. As time went by interest in foundational issues widened, indeed it became a regular field for physical research, and Bohm's works went on to be considered more positively, or at least with more tolerance.

This chapter goes beyond this framework, which considers Bohm's prestige as a kind of mirror of the prestige from the research in foundations, to look in a more detailed manner at Bohm's achievements and how these achievements were received by the public, in this case, other physicists. This subject uses scientometric data, mainly the number of citations of papers over time, adding to these data some qualitative information concerning how these papers were seen by other physicists.[3]

The second section of this chapter presents and analyzes the list of Bohm's most cited scientific papers. The following sections, third to fifth, successively analyze the three groups of Bohm's most cited papers, namely, Collective Variables, Reinterpretation of Quantum Mechanics, and Quantum Effects and Experiments. The sixth section is dedicated to considering Bohm's persona through some qualitative comments from his contemporary fellow physicists. The concluding remarks assess Bohm's legacy to physics.

---

[3] An extended version of this chapter will appear in my book "David Bohm - A life dedicated to understand the quantum world," to be published by Springer.

20.2 Bohm's Most Cited Scientific Papers

Today David Bohm has ten papers with more than 300 citations.[4] They are the
following: Aharonov and Bohm (1959), Bohm (1952a, b), Bohm and Gross (1949a,
b), Pines and Bohm (1952), Bohm and Pines (1951, 1953), Bohm and Vigier (1954)
and Bohm and Aharonov (1957). That is not bad for any scientist, in physics at least.
Figure 20.1 contains this list in the order of number of citations. The ensemble
of these papers reveals some patterns of Bohm's contributions to physics which
will be analyzed later. All these ten papers were published between 1949 and
1959.

For the moment, however, I want to comment that even if we amplify our search
for papers above 100 citations, there will be no changes in the patterns we are going
to analyze. Indeed, with more than 100 citations and fewer than 300 there are the
following papers: Aharonov and Bohm (1961a, b), Bohm and Bub (1966), Bohm
et al. (1987), Bohm (1953) and Bohm and Weinstein (1948). Except for Bohm and
Weinstein (1948), all these papers fall in the field of interpretation of quantum theory

Bohm's papers with over 300 citations

Paper	Authors	Year	Citations
1. Significance of Electromagnetic Potentials	Aharonov, Bohm	1959	3991
2. A Suggested Interpretation [...] Hidden Variables .1	Bohm	1952	2659
3. Collective Description .3	Bohm, Pines	1953	1270
4. A Suggested Interpretation .2	Bohm	1952	1260
5. Collective Description .2	Pines, Bohm	1952	727
6. Theory of Plasma .A	Bohm, Gross	1949	595
7. Collective Description .1	Bohm, Pines	1951	377
8. Model of the Causal Interpretation	Bohm, Vigier	1954	364
9. Theory of Plasma .B	Bohm, Gross	1949	338
10. Discussion of Experimental EPR	Bohm, Aharonov	1957	311

Fig. 20.1 Bohm's papers with over 300 citations

---

[4]The data on citations were collected from Web of Science on June 1, 2017.

and quantum effects and experiments. Except for the 1987 paper by Bohm, Hiley & Kaloyerou, they roughly cover the period between the late 1940s and the early 1960s.

Let us now come back to the 10 papers above 300 citations and published between 1949 and 1959. As Bohm got his PhD during World War II and was active in physics till his death in 1992, our first conclusion is that Bohm's most influential stage in physics came a little after his PhD, lasted for 10 years or so and then receded to less influential works.

These data allow us to deal with two thorny issues which appear when one reflects on Bohm's public image. First, Bohm is widely acknowledged for his dialogue with Eastern thinkers, particularly Jiddu Khrisnamurti, and these dialogues have influenced some cultural circles. However, hints about the influence of these exchanges on Bohm's more influential papers are scant. The reason for this is related to the fact that these dialogues began in the early 1960s and there is no work from that time which appears in the list of Bohm's ten most cited papers. Thus, if these dialogues had any influence on Bohm's physics, this would not be detected through the approach used in this work.

Second, Bohm worked for years on a philosophical program he called implicate order (Bohm 1980).[5] In scientific terms this program led to attempts to obtain quantum mechanics and geometry deriving from some primary algebraic structures. In this program Bohm worked collaboratively with Basil Hiley, his long-standing assistant at Birkbeck College in London. Interesting as this program might be, and it *is* in my opinion, for the future development of physics, it has yet to bear fruit in terms of its influence when citations are concerned. Indeed, none of Bohm's papers resulting from the implicate order program appears among his most cited papers. Furthermore, other evidence of influence or fecundity, such as the number of people working in this direction gives a poor result. Thus, neither the dialogue with Eastern thinkers nor the implicate order program are good candidates to gauge the influence of Bohm's works among his fellow physicists. We arrive at this conclusion constrained by our decision to look for Bohm's legacy to physics through the data from the number of citations of his papers, that is scientometrics, a traditional source for historians of science.

Now let us rearrange these data and group these ten papers according to the subject covered. In the first group we have the five papers relating to Plasma and Methods for Condensed Matter, which we call Collective Variables – Plasma and Metals. They were produced by Bohm and his doctoral students Eugene Gross and David Pines and were published from 1949 to 1953. They are the papers number 3, 5, 6, 7, and 9. The second group concerns the Reinterpretation of Quantum Mechanics; they are the three papers 2, 4 and 8. The eighth was co-authored with

---

[5]In his *Wholeness and the Implicate Order*, David Bohm (1980, 188) states, "a *total order* is contained, in some *implicit* sense, in each region of space and time. Now, the word 'implicit' is based on the verb 'to implicate'. This means 'to fold inward' (as multiplication means 'folding many times'). So we may be led to explore the notion that in some sense each region contains a total structure 'enfolded' within it."

Jean-Pierre Vigier. They were published between 1952 and 1954. The last group concerns Quantum Effects and Experiments. The papers resulted from Bohm's work with his doctoral student Yakir Aharonov. They were published from 1957 to 1959 and they are the two papers number 1 and 10.

20.3   Collective Variables: Plasma and Metals

Let us now take the five papers related to Collective Variables. Bohm worked on plasma with Eugene Gross, resuming work he had begun during World War II, and went on to extend this approach to metals with David Pines. In addition to these five papers, two with Gross and three with Pines, there was another paper, by Pines alone (1953). These were the main papers where the collective variables approach was developed. The historian of physics Kojevnikov (2002) has studied Bohm's early approach to plasma and his evolution to the use of "collective variables" to deal with the phenomena of plasma and electrons in metals as well as the philosophical assumptions behind Bohm's approach. Kojevnikov also made an analysis of the long-range influence of this approach among condensed matter physicists and on the particular forms this approach received through its later circulation. These papers have been steadily cited since their publication. In Fig. 20.2, we take the evolution of number of citations over the years for one of these papers just to illustrate its steady flow of citations.

Collective description, initially intended to deal with plasma and electrons in metals, was further extended to other systems including superconductivity and nuclear physics. It proved rich and directly contributed to the wide problems in physics related to many-body systems, for instance in superconductivity and nuclear physics. Evidence of the circulation of these ideas is reported in Ben Mottelson's Nobel speech, when he addressed the evolution of the models of nuclear physics

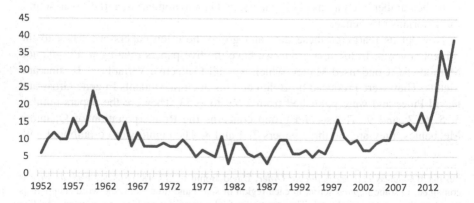

Fig. 20.2 A collective description of electron interactions – 2 collective vs individual particle aspects of the interactions (Pines and Bohm 1952) – 727 citations

and recalled: "it was a fortunate circumstance for us [Mottelson and Aage Bohr] that David Pines spent a period of several months in Copenhagen in the summer of 1957, during which he introduced us to the exciting new developments in the theory of superconductivity."[6] This interaction had been earlier recorded in the joint paper by Bohr et al. (1958). The wide impact of the collective description was also acknowledged by the Japanese physicist H. Nakano, who nominated both Bohm and Pines for the Physics Nobel Prize in 1958.[7] These works using collective variables approach are therefore part of Bohm's lasting contribution to physics.

20.4   Reinterpretation of Quantum Mechanics

Now we examine the three papers on the reinterpretation of quantum physics, also in chronological order of publication. The first two were published together and the number of citations of the first one is the best index of their influence. This is because many people may have cited it thinking of the causal interpretation in general. The third paper is a development of this interpretation. Thus, let us consider the evolution of citations of the first paper, plotted in Fig. 20.3. After an initial number of citations, most criticizing the suggested interpretation, the paper went ignored for almost 10 years. It was revived from the 1970s on thanks to the publication of Bell's theorem, in 1965, and the appearance of its early experimental tests, which contrasted quantum mechanics with local realism.[8]

In the 1980s interest in Bohm's proposal continued to increase due to other factors. On the one hand, it may have been a side effect of interest in Alain Aspect's 1981–1982 experiments. These experiments reinforced physicists' trust in quantum mechanics predictions, against the possibility of local realist theories.[9] However, the debates triggered by Aspect's experiments recalled Bohm's interpretation. Indeed, it had survived these tests as it was as nonlocal as quantum mechanics was. On the other hand, this interest was reinforced when Philippidis et al. published the first

---

[6]Ben Mottelson, The Nobel Lecture, available at: https://www.nobelprize.org/nobel_prizes/physics/laureates/1975/mottelson-lecture.html

[7]See: https://www.nobelprize.org/nomination/archive/show_people.php?id=10862

[8]On the reception of Bohm's ideas on the reinterpretation of quantum mechanics, see Freire Junior (2015, Chap. 2).

[9]Incidentally, I should remark that Mario Bunge no longer followed these developments in physics and did not realize that Bohm's early ideas had survived these experimental tests. Thus, in 1998, he still thought experiments on Bell's theorem had discarded the full family of hidden variable theories. "However, I am not convinced that the experiments that refute the Bell inequalities concern only non-locality (or non-separability). So far as I can remember, the result is more general than that: it refutes the entire family of hidden variables theories." Mario Bunge to the author, 24 August 1998.

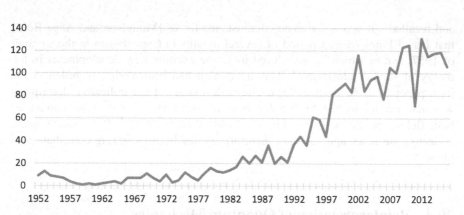

Fig. 20.3 A suggested interpretation of the quantum theory in terms of hidden variables (Bohm 1952a, b) – 2659 citations

diagrams of the causal interpretation paths as well as of Bohm's quantum potential (Philippidis et al. 1979). These diagrams were produced by computers.

Pictures are always powerful tools for communication, in particular as a way to gain insights on weird or unknown physical phenomena, as emphasized by historian Peter Galison in his book on the material culture of twentieth century microphysics (Galison 1997), and by historian David Kaiser in his study of the dispersion of Feynman's diagrams (Kaiser 2005). These diagrams are now an iconic representation of Bohm's proposal and many argue that they brought more intelligibility to quantum physics. Finally, from the 1990s to nowadays, citations of this paper have continued to increase, influenced by the appearance of Bohmian mechanics, a slightly modified presentation of Bohm's early ideas advanced by Sheldon Goldstein and by the wide interest in foundations of quantum mechanics following the blossoming of quantum information in the mid-1990s.

20.5 Quantum Effects and Experiments

We finally move to the papers which emerged from the collaboration between Bohm and his doctoral student Yakir Aharonov. Their first paper addressed the question: is there an experiment whose results may be used to portray the EPR argument? They analyzed an experiment conducted by the Sino-American physicist Madame Chien Shiung Wu and concluded that the EPR argument ran counter to quantum mechanics. In the first years after its publication, this paper did not attract much attention but later, after interest arose in Bell's theorem experiments, the paper received more citations. It is mostly used as an example of the first attempt to check the EPR argument against quantum mechanics predictions.

Two years later, in 1959, Bohm's most cited paper was published. It presented what is known as the Aharonov-Bohm effect. Aharonov was its first author reflecting

Fig. 20.4 Significance of electromagnetic potentials in the quantum theory (Aharonov and Bohm 1959) – 3991 citations

his share of work in the solution of the problem, the subject of his doctoral thesis. Aharonov and Bohm challenged the idea, current in classical electromagnetism, according to which the vector potential has no physical significance. They showed that an electron beam travelling in a region free from any electrical field but near a solenoid where a magnetic field is confined may still undergo influence due to the variation in the confined magnetic field.

This paper started an industry devoted to discussing its theoretical and experimental implications. Theoretical issues concerned whether this effect was strictly quantum or not, if it was topological in nature, and if it was an illustration of the quantum non-locality. Experimenters produced a string of experiments, mainly led by Murray Peshkin and Akira Tonomura, whose results convinced physicists of the real existence of such an effect. It is this whole industry which is responsible for the huge and ever-increasing number of citations of this paper (see Fig. 20.4). The paper has been the basis for the nominations both of Bohm and Aharonov for several scientific accolades, which included the Wolf Prize in Physics for Aharonov in 1998.

20.6   Qualitative Comments

All this scientometric data may now be tempered with evidence from the views of some of Bohm's contemporaries on his work. Evidence from these data corroborates the views on Bohm's achievement from some of his greatest collaborators. In 1989, when Basil Hiley, who was Bohm's long-standing assistant at Birkbeck College, was asked by his head of department to send some background information in order to support Bohm's nomination for the Nobel Prize for quantum non-locality, he chose precisely the following three fields to single out Bohm's deeds. According to Hiley's letter, "the first major contribution made by Bohm was in the area of plasma" and "the second area in which Bohm has contributed significantly is our understanding of quantum mechanics." In this latter area Hiley cited the paper on the Aharonov-Bohm effect and the papers on the reinterpretation of quantum mechanics.[10]

Let me concentrate now on the case of his papers concerning the reinterpretation of quantum mechanics. Well known are John Bell's reminiscences about the influences shaping his deep interest in the foundations of quantum mechanics. In 1982, he wrote "Bohm's 1952 papers on quantum mechanics were for me a revelation," and "In 1952 I saw the impossible done." Thus, Bohm's causal interpretation is part of the history of Bell's theorem.

Earlier, Bohm had been compared to Kepler. Indeed, in 1958, in a fierce exchange of correspondence with Léon Rosenfeld about Bohm's works and their worth, Lancelot Whyte wrote: "Naturally you are fully aware ... that valuable results may spring from mistaken motives and reasoning. Kepler is a good example. But this awareness is not evident in your review." It is not bad for a physicist to be compared to Kepler. Jumping in time, after Bohm's passing, evidence of the influence of his works may be seen in the very fact that there is an intellectual diaspora of physicists who were influenced by Bohm's causal interpretation. Just to illustrate, physicists such as Sheldon Goldstein, Basil Hiley, and Antony Valentini disagree about the meaning and reach of Bohm's original works on the reinterpretation of quantum mechanics. Only major thinkers may have this kind of diaspora as their offspring.

However, recognition of Bohm's contribution to the enlightenment of the foundations of quantum mechanics closely follows the line of citations of his 1952 papers. Both reflect the changing views of physicists about the role and value of research on these topics. It is against the backdrop of these changing views that we may understand Melba Phillips' comment: "It is too bad, very sad indeed, that he did not live to see how his reputation has shot up recently. His interpretation of quantum mechanics is becoming respected not only by philosophers of science but also by 'straight' physicists." Phillips, who was his close friend for most of his life, wrote this in 1994, and Bohm had passed away in 1992.[11] This was the time when the field of quantum information began to take off, drawing from the resources exploited in the research on the foundations of quantum mechanics.

While Bohm's work on the reinterpretation of quantum mechanics was responsible for most of his influence on physics, one should not underestimate the influence of the other two camps where Bohm excelled, Collective Variables and Quantum Effects. As for the work on Collective Variables, we have seen the acknowledgment of Ben Mottelson at the occasion of the Nobel Prize, and Nakano's nomination of Bohm and Pines for the Nobel Prize. On the work on Quantum Effects, the number of citations of the paper by Aharonov and Bohm is enough to tell about its influence among fellow physicists.

---

[10]Basil Hiley to Sessler, 9 January 1989. Folder A172, David Bohm Papers, Birkbeck College, London.

[11]For all the previous quotations, see Freire Junior (2015). On the relation between Bohm and Phillips, see their letters in Talbot (2017).

20.7   Conclusion

The above analysis justifies the following claims:

- Bohm left lasting and influential contributions in all three domains we considered here: Collective Variables, Quantum Effects, and Interpretations of Quantum Mechanics. These contributions rank him as one of the most important theoretical physicists in the twentieth century.
- His most influential contributions were published from 1949 to 1959 and there was a decline in the production of influential papers from 1959 on.
- The wholeness program (1960s –), responsible for Bohm's wide intellectual influence, has not borne similar fruits – yet – among scientists.

Bohm was a physicist who made many and lasting contributions. However, as for his legacy for quantum physics, rather than by one specific and lasting contribution, I think he should be acknowledged for his pointing out the relevance of the research on the foundations of this theory, which includes both his works on Interpretation of Quantum Mechanics and Quantum Effects. For 60-plus years, the relevance of this foundational research has also been championed by Mario Bunge.

Acknowledgement I am thankful to the Brazilian CNPq for the funding (443335/2015-0) and Italo Carvalho for helping me with the figures.

References

Aharonov, Y., & Bohm, D. (1959). Significance of electromagnetic potentials in the quantum theory. *Physical Review, 115*(3), 485–491.
Aharonov, Y., & Bohm, D. (1961a). Time in quantum theory and uncertainty relation for time and energy. *Physical Review, 122*(5), 1649–1658.
Aharonov, Y., & Bohm, D. (1961b). Further considerations on electromagnetic potentials in quantum theory. *Physical Review, 123*(4), 1511–1524.
Bohm, D. (1952a). A suggested interpretation of the quantum theory in terms of hidden variables – I. *Physical Review, 85*(2), 166–179.
Bohm, D. (1952b). A suggested interpretation of the quantum theory in terms of hidden variables – II. *Physical Review, 85*(2), 180–193.
Bohm, D. (1953). Proof that probability density approaches (Psi)2 in causal interpretation of the quantum theory. *Physical Review, 89*(2), 458–466.
Bohm, D. (1980). *Wholeness and the implicate order*. London: Routledge.
Bohm, D., & Aharonov, Y. (1957). Discussion of experimental proof for the paradox of Einstein, Rosen, and Podolsky. *Physical Review, 108*(4), 1070–1076.
Bohm, D., & Bub, J. (1966). A proposed solution of measurement problem in quantum mechanics by a hidden variable theory. *Reviews of Modern Physical, 38*(3), 453–469.
Bohm, D., & Gross, E. (1949a). Theory of plasma oscillations. A. Origin of medium-like behavior. *Physical Review, 75*(12), 1851–1864.
Bohm, D., & Gross, E. (1949b). Theory of plasma oscillations. B. Excitation and damping of oscillations. *Physical Review, 75*(12), 1864–1876.

Bohm, D., & Hiley, B. J. (1993). *The undivided universe: An ontological interpretation of quantum theory*. London: Routledge.

Bohm, D., & Pines, D. (1951). A collective description of electron interactions. 1. Magnetic interactions. *Physical Review, 82*(5), 625–634.

Bohm, D., & Pines, D. (1953). A collective description of electron interactions. 3. Coulomb interactions in a degenerate electron gas. *Physical Review, 92*(3), 609–625.

Bohm, D., & Vigier, J. P. (1954). Model of the causal interpretation of quantum theory in terms of a fluid with irregular fluctuations. *Physical Review, 96*(1), 208–216.

Bohm, D., & Weinstein, M. (1948). The self-oscillations of a charged particle. *Physical Review, 74*(12), 1789–1798.

Bohm, D., Hiley, B. J., & Kaloyerou, P. (1987). An ontological basis for the quantum-theory. *Physical Reports, 144*(6), 321–375.

Bohr, A., Mottelson, B. R., & Pines, D. (1958). Possible analogy between the excitation spectra of nuclei and those of the superconducting metallic state. *Physical Review, 110*(4), 936–938.

Bunge, M. (1955). Strife about complementarity. *British Journal for the Philosophy of Science, 6*, 1–12. & 141–154.

Bunge, M. (1956a). La interpretación causal de la mecánica ondulatoria. *Ciencia e Investigación, 12*, 448–457.

Bunge, M. (1956b). Beitrag zur Diskussion über philosophische Fragen der modernen Physik. *Deutsche Zeitrschrift für Philosophie, 4*, 467–496.

Bunge, M. (1967). *Foundations of physics*. Berlin: Springer.

Freire Junior, O. (2015). *The quantum dissidents – Rebuilding the foundations of quantum mechanics 1950–1990*. Berlin: Springer.

Galison, P. (1997). *Image and logic – A material culture of microphysics*. Chicago: Chicago University Press.

Kaiser, D. (2005). *Drawing theories apart: The dispersion of Feynman diagrams in postwar physics*. Chicago: Chicago University Press.

Kojevnikov, A. (2002). David Bohm and collective movement. *Historical Studies in the Physical and Biological Sciences, 33*, 161–192.

Philippidis, C., Dewdney, C., & Hiley, B. J. (1979). Quantum interference and the quantum potential. *Nuovo Cimento della Societa Italiana di Fisica B – General Physics, Relativity, Astronomy and Mathematical Physics and Methods, 52*(1), 15–28.

Pines, D. (1953). A collective description of electron interactions: IV. Electron interaction in metals. *Physical Review, 92*, 625–636.

Pines, D., & Bohm, D. (1952). A collective description of electron interactions. 2. Collective vs individual particle aspects of the interactions. *Physical Review, 85*(2), 338–353.

Talbot, C. (Ed.). (2017). *David Bohm: Causality and chance, letters to three women*. Berlin: Springer.

Part IV
Cognitive Science and Philosophy of Mind

Chapter 21
Mario Bunge and Contemporary Cognitive Science

Peter Slezak

Abstract Bunge's writings on the mind-body problem (Bunge 1980, 1991, 2010) provide a rigorous, analytical antidote to the persistent anti-materialist tendency that has characterized the history of philosophy and science. Bunge suggests that dualism can be neutralized "with a bit of philosophical analysis" (Bunge 1991) but this is clearly too optimistic in view of the recent revival of dualism as a respectable doctrine despite a vast industry of philosophical analysis. The conceivability of zombies (Chalmers 1996) leads to the possibility of dualism and thereby to the falsity of materialism. Bunge relies on his general case that "arguably all the factual ("empirical") sciences only study concrete (or material) entities, from photons to rocks to organisms to societies" (Bunge 2010). Bunge's immunity to philosophical extravagance is to be commended, but he is perhaps like someone who rejects Zeno's paradoxes as physical absurdities and thereby leaves the puzzle itself untouched. While philosophers need to be cured of their paradoxes, perhaps Bunge's strategy of just getting on with real scientific inquiry is, after all, the best approach.

Bunge has mastery of an improbably vast range of disciplines. His writings provide a manifesto for the materialist conception of mind and body, placing ideas in their rich historical context. Few philosophers have written in such depth and breadth as we see in Bunge's work and there is much to cheer for someone who shares Bunge's views and prejudices. However, the very dazzling comprehensiveness and deftness in Bunge's analyses often means a correlative lack of sufficiently detailed argumentation for its many insights, bold claims and criticisms. Accordingly, I have focused my following comments selectively on Bunge's more controversial claims in order to engage with them seriously as his eminent status deserves.

P. Slezak (✉)
School of Humanities and Languages, University of New South Wales, Sydney, Australia
e-mail: p.slezak@unsw.edu.au

© Springer Nature Switzerland AG 2019

363

M. R. Matthews (ed.), *Mario Bunge: A Centenary Festschrift*,
https://doi.org/10.1007/978-3-030-16673-1_21

21.1 Functionalism, Behaviourism, Dualism

Following an erudite discourse on the various kinds of matter (physical, chemical, living, thinking, social and artificial), it is difficult to dispute Bunge's jaundiced observation: "The reader is invited to compare this rich crop to the contributions made by metaphysicians to both their own discipline and to science during the same period" (Bunge 2010, p. 83–4). Indeed, despite recent heroic efforts (Stoljar 2017) to demonstrate progress in philosophy, a wide consensus among professional philosophers since Russell (1912) concurs with Bunge's scepticism (Slezak 2018).

Nevertheless, a difficulty arising from Bunge's provocative critiques is that not all the targets of his admitted "bashings" (2010, p. xi) are equally deserving. For example, Bunge suggests "most contemporary philosophers of mind are indifferent to psychology, or are remarkably uninformed about it" (2010, p. ix). This charge cannot be sustained today in light of the work of such eminent philosophers as Stich, Fodor, Cummins, Dennett, Churchlands, Thagard, Nersessian, Bechtel, Egan, Bickhard, Elster and dozens of others. Bunge's diagnosis is quite misplaced as, for example, when he suggests that in the philosophy of mind "few of its practitioners bother to keep up to date with the science of mind" (Bunge 2010, p. x).

Such sweeping *ad hominem* pronouncements lead Bunge to hand-waving dismissals of important, subtle doctrines. Thus, for example, he says "the functionalist view of the mind, favoured by most contemporary philosophers of mind, should be dropped as being both scientifically shallow and medically hazardous" (Bunge 2010, p. 155).

Functionalism in philosophy of mind is the doctrine that a mental state is identified, not on the basis of its material composition but on the basis of its functional role in the system of which it is a part. The doctrine may be seen in Aristotle who begins his *De Anima* asking:

> A … problem presented by the affections of soul is this: are they all affections of the complex of body and soul, or is there any one among them peculiar to the soul by itself? To determine this is indispensable but difficult. (Aristotle 1941, p. 536)

Bunge asks the same question "Are mind and body two separate entities?" (Bunge 1980, xiii). As Bunge himself notes, Aristotle recognized that the question whether body and soul are one is as misleading as the question of whether the wax and its shape are one (Bunge 1980).

The modern version of functionalism derives from insights into computational or abstractly specifiable states that are not intrinsically dependent on their composition or realization. An early statement of the view was Fodor's *Psychological Explanation* (Fodor 1968) in which he acknowledged particular inspiration from Chomsky's (1965) highly idealized, mathematical models of linguistic competence. In particular, functionalism is, therefore, neutral between materialism and dualism because mental states are identified by their abstractly specified role rather than their substance or multiple possible realizations. The case for materialism then becomes the general scientific case for the only possible physical realizations of the relevant functional roles. Dualism is ruled out on this conception because there are no scientific grounds for believing that an immaterial substance is a possible realization for the relevant functional states.

Although the conception of the physical realization has changed, the mind-body problem has remained essentially the same for centuries. The functionalist solution is captured loosely in the familiar distinction between hardware and software in a computer. That is, the mind is to be understood in a literal sense as the software rather than the hardware of the brain. Pylyshyn's (1984) landmark *Computation and Cognition* is perhaps the *locus classicus* of this doctrine that takes computation to be more than merely a metaphor and, rather, when appropriately spelled out, a literal account of information processing in the brain.

To be sure, functionalism remains open to a variety of philosophical criticisms but cannot be dismissed in Bunge's manner (Levin 2010). Characterizing modern information processing psychology as "brainless cognitive science," Bunge suggests that, seen in historical perspective, "computationalism is a sophisticated version of behaviourism" (Bunge 2010, p. 227). On the face of it, this is a surprising claim because the 'Cognitive Revolution' and its computational paradigm has been seen as emerging from the downfall of behaviourism (Gardner 1987) once internal, mental representations and processes became legitimate theoretical constructs again. However, Bunge (2010) explicitly identifies computationalism with the behaviourism it has displaced. He offers the following schematic diagram to illustrate the parallel:

(a) Classical behaviourism

Stimulus ➔ Black box ➔ Response

(b) Computationalism

Stimulus ➔ Program ➔ Readout

However, the "program" in computational models of cognition cannot be compared with the "Black Box" of behaviourism precisely because the "program" constitutes the theoretical postulation of internal representations that were eschewed by behaviourism. The 'box-and-arrow' models and formalisms of computational theories are precisely hypotheses about mental events and processes that were taboo for Skinner on the grounds that they must be question-begging.

Daniel Dennett has been foremost in elucidating the significance of AI and computation for the philosophy of mind. For example, he has shown how modern computational approaches can avoid the difficulty that Skinner posed – that is, how to avoid the paradox that psychology *with* homunculi seems circular, whereas psychology *without* homunculi seems empty. The successive decomposition of tasks "discharging the loan on intelligence" with progressively stupider homunculi shows how psychological theories may posit internal states without falling afoul of Skinner's worries. Dennett explains that AI is a form of psychological theorising adopting the "Intentional Stance" and differing from traditional philosophy in that "the AI worker pulls his armchair up to a console" (Dennett 1978a, p. 58).

AI shares with philosophy (in particular, with epistemology and philosophy of mind) the status of most abstract investigation of the principles of psychology. But it shares with psychology in distinction from philosophy a typical tactic in answering its questions. (Dennett 1978a, p. 60)

In a seeming inconsistency, on the one hand, as we have just noted, Bunge takes computationalism to be behaviourist but, on the other hand, he also takes high-level computational theories to be dualist. For example, Bunge suggests that "advocates of this view [linguistic naturalism or biolinguistics] have adopted Cartesian mind-body dualism" (Bunge 2010, p. 112), but this is, at best, a caricature of Chomsky's program (see Jenkins 2001; Di Sciullo and Boeckx 2011) for which Bunge gives no argument or justification. Bunge thinks that inquiry that stays at the higher level of mental phenomena rather than neuroscience and "objective brain facts" is evidence that "psycho-neural dualism prevails" (Bunge 2010, p. 154).

This might be charitably understood as simply characterising the distinction between different levels of analysis rather than the usual Cartesian connotation of the term "dualism," but Bunge appears to intend the latter pejorative reading. This reading is further suggested in his warning "that the dualist philosophies of mind are hazardous to mental health because they divert the researcher's and the therapist's attention from the brain to an immaterial and therefore inaccessible item." (Bunge 2010, p. 155). On the contrary, the biolinguistics program has been explicitly conceived as integrating high level theories with their biological, neurological substrate.

Fodor (1968) and Pylyshyn (1984) articulated the rationale for a high-level inquiry with its proprietary vocabulary and explanatory principles, just as we see elsewhere in science. One can fully acknowledge the reality of many phenomena without falling into dualism or ascribing them to the basic, elementary physical constituents of the world. As Fodor (1989) has argued in countering the ailment of "epiphobia"; rivers, sails and mountains are no less real for not figuring in basic physics – Bunge's own reason for being a materialist rather than a physicalist, as he uses these terms. Indigestion, inflation and other kinds of "being" in good standing are surely real without belonging to properties of elementary particles.

Thus, for example, whatever may be its other failings, economics is not committed to dualism or occult entities by virtue of seeking generalizations above the level of the individuals who make up the economic system. This was, of course, Durkheim's (1898) famous conception of social facts as "things" widely, but unjustly, seen as some kind of mysticism but, in fact, an anticipation of Fodor's (1989) criticism of "epiphobia" – the fear of postulating theoretical entities.

In this spirit, Chomsky has referred to his abstract idealizations as adopting a 'Galilean' approach to science (see Pylyshyn 1972, 1973). Chomsky writes:

> ... we are keeping to abstract conditions that unknown mechanisms must meet. We might go on to suggest actual mechanisms, but we know that it would be pointless to do so in the present stage of our ignorance concerning the functioning of the brain. ... If we were able to investigate humans as we study other, defenceless organisms, we might well proceed to inquire into the operative mechanisms ... (Chomsky 1980, p. 197)

Chomsky's functionalist view was unmistakable in his *Aspects*, where he wrote:

> The mentalist ... need make no assumptions about the possible physiological basis for the mental reality he studies. ... One would guess ... that it is the mentalistic studies that will ultimately be of greatest value for the investigation of neurophysiological mechanisms, since they alone are concerned with determining abstractly the properties that such mechanisms must exhibit and the functions they must perform. (Chomsky 1965, p. 193, fn. 1)

Of course, philosophers are not the only ones tempted by the barren doctrine. Bunge notes, the foremost neuroscientists have been avowed dualists including Sherrington, Penfield, Sperry and Eccles. Thus, Bunge's advice that researchers should stick to the brain is questionable if the alternative is not an avowed Cartesian ontological dualism but only high-level, top-down theory. The virtues of such an 'Intentional Stance' (Dennett 1989) or semantic, knowledge-level (Newell 1990) analysis above the level of cognitive architecture have been seen as providing the rationale for the enterprise of cognitive psychology. The approach is seen paradigmatically in Marr's (1982) distinction between the levels of computation, algorithm and implementation, corresponding roughly to Chomsky's competence-performance distinction. Only the implementation or "realization" level is concerned directly with the physical, neural substrate that Bunge appears to insist on as the only respectable level of analysis.

21.2   Materialism, Physicalism and Dualism

Bunge declares "I am an unabashed monist" and "I am a materialist but not a physicalist" (2010, p. vii). These latter terms have been often used interchangeably and so it is important to understand Bunge's specific meaning. By "physicalist" Bunge means someone who holds that the laws of physics are explanatory for all phenomena. Bunge explains that his own expertise as a physicist led him to appreciate "that physics can explain neither life nor mind nor society" nor "chemical reactions, metabolism, color, mentality, sociality, or artifact" (Bunge 2010, p. vii). Bunge's mission is "to reunite matter and mind" at a time when its materialist message is more timely than it would have been a decade or two earlier. Bunge's writings on the mind-body problem (Bunge 1980, 1991, 2010) are intended to provide an antidote to a persistent anti-materialist tendency that has characterized the history of philosophy and science. He concludes:

> ... psychoneural dualism is worse than barren: it is an obstacle to the advancement of science and medicine. Fortunately, this obstacle can easily be removed with a bit of philosophical analysis. (Bunge 1991, p. 520)

This remark is clearly too optimistic in view of the revival of dualism as a respectable doctrine despite a vast industry of philosophical analysis. The materialist orthodoxy of the mid twentieth century has been eroded and dualism has, indeed, regained a certain respectability (Chalmers 1996). However, besides taking a passing swipe (Bunge 2010, p. 177), Bunge does not address the principal arguments based on the Method of Conceivability' and Zombies (see also Kirk 2005).

Furthermore, there is a certain irony in the fact that Bunge's criticism of functionalist theories places him in the same camp as leading critics of materialist theories such as Strawson (2006, 2018) who charge functionalism with leaving out the essentially subjective, first person, phenomenal, qualitative features of experience. For example, Strawson (2006, 2008) holds materialist philosophers to be guilty of "the silliest view ever held by any human being" (Strawson 2008, p. 8). He construes the Lucretius world-view of Dennett and others as a grievous error exceeding the implausibility of "every known religious belief." He says,

> For this particular denial is the strangest thing that has ever happened in the whole history of human thought, not just the whole history of philosophy. It falls, unfortunately, to philosophy, not religion, to reveal the deepest woo-woo of the human mind. (Strawson 2006, pp. 5–6)

Albeit for different reasons, Bunge joins Strawson (2006) and Searle (1997) in heaping scorn on philosophical adversaries such as Dennett who are said to deny the most obvious reality of their own experience. Searle parodies the title of Dennett's book as *Consciousness Denied* instead of *Consciousness Explained*. If Searle is right, Dennett has managed an intellectual achievement that Descartes showed to be impossible. This suggests that these rhetorical features of the debate about consciousness are not irrelevant matters of polemical style but rather symptoms of the peculiarity of the views at stake.

Thus, Searle charges materialists with making "stunning mistakes" (Searle 1992, p. 246) and "saying things that are obviously false" (Searle 1992, p. 247). Block (1990, p. 129), too, suggests that Dennett's (1991) book would be more aptly titled *Consciousness Ignored*. "Such authors pretend to think that consciousness exists, but in fact they end up denying its existence" (Searle 1992, p. 7). Searle (1997) includes Armstrong (1968, 1980) among such deniers. Searle writes acidly "I regard Dennett's denial of the very existence of consciousness not as a new discovery or even as a serious possibility but rather as a form of intellectual pathology" (Searle 1997, p. 112). Since these accusations are directed at our foremost philosophers, we are confronted with a peculiar situation that deserves attention as something more than mere *ad hominem* rhetorical excess.

21.3 Fantasy Worlds

Despite the difficulties of articulating a version of materialism that is immune from philosophical objections, from the 1950s there had been a consensus on materialism and the progress from early 'Identity' versions to the more recent 'functionalist' accounts (Fodor 1968). Thus, for example, originally Thomas Nagel (1965) avowed an intellectual commitment to materialism as the only scientifically respectable account, while confessing a psychological discomfort because of its deep intuitive, introspective implausibility. Before his apostasy, Nagel (1965) took it as a datum of subjective experience that materialism has a deep, intuitive implausibility that is independent of its overwhelming systematic merits as a true scientific, philosophical thesis. Bunge makes a typically acerbic remark: "responsible people do not mistake conceptual possibility, or conceivability, for factual possibility or lawfulness; and they do not regard the ability to invent fantasy worlds as evidence for their real existence" (Bunge 2010, p. 177).

It is perhaps understandable that philosophers will elevate their only research tool to a pre-eminent status as a guide to metaphysical possibilities, but the tendency gives grounds for Bunge's jaundiced view of their discipline. Indeed, the balance has become reversed with intuitions coming to dominate systematic scientific considerations. Thus, Nagel has been among those who have shifted their allegiance to various forms of 'Mysterianism' (McGinn 1989), outright dualism or even panpsychism (Strawson 2006). Of this latter doctrine, in a related context, Bunge remarks that it "illustrates the cynical principle that, given an arbitrary extravagance, there is at least one philosopher capable of inventing an even more outrageous one" (Bunge 2010, p. 167).

This is the sentiment that Descartes had expressed in his *Discourse* remarking that "nothing can be imagined which is too strange or incredible to have been said by some philosopher" (1637/1986, p. 118). The modern shift in the consensus has been due, not to any new scientific revelations that would provide grounds for doubting the broad materialist picture but rather to the increased weight placed on philosophical intuitions. Despite having been elevated to the status of an official "Conceivability Argument" (Stoljar 2001) or "Method of Conceivability" (Chalmers 2002), it is difficult to find anything other than a bare description of the faculty itself as "a kind of rational intuition or intellectual presentation of a possibility: a clear and distinct idea" (Stoljar 2001, p. 393). Indeed, as Levine puts it, "The conceivability of zombies is ... the principal manifestation of the explanatory gap" (Levine 2001, p. 79). Yablo too, notes that for Chalmers (1996), "Almost everything in *The Conscious Mind* turns on a single claim ... that there can be zombie worlds" (Yablo 1999, p. 455). The conceivability of zombies leads to the possibility of dualism and thereby to the falsity of materialism.

Bunge suggests that this argument "does not even distinguish between conceptual and physical possibility" (Bunge 2010, p. 23) and, although this is not accurate, it captures something important about the extravagance of such contemporary philosophy which Bunge characterizes as "just *jeux d'esprit*" (Bunge 2010, p. 23). Current arguments concerning the conceivability of zombies and the "explanatory gap" are little more than unwitting, often verbatim, rehearsal of Descartes' own reasoning in his *Meditations*. However, John Cottingham has bluntly remarked that Descartes' argument from conceivability to dualism "is, or ought to be, regarded as one of the most notorious nonsequiturs in the history of philosophy" (Cottingham 1992, p. 242). The alleged metaphysical, ontological implications of conceivability intuitions have become a central topic of philosophical debate (Gendler and Hawthorne 2002) but the question to be considered is, in Loar's words echoing Bunge, whether "we have managed to break out" (Loar 1999) of purely conceptual premises to metaphysical conclusions.

In fairness to Descartes, it is worth noting that he had good, essentially scientific reasons for his dualism and even his cogito meditations have an important logical structure that has not been properly recognized (See Slezak 1988, 2010). The criticism of Cottingham and Loar applies more to modern dualists than to Descartes himself.

21.4   Exorcising the Ghost or the Machine?

Regarding the thesis of materialism, Bunge makes the important and perhaps surprising point that "there is no generally accepted concept of matter" and, therefore, "We do not have a generally accepted materialist theory of mind" (Bunge 2010, p.xvii). Although Bunge is harshly critical of Chomsky's views on various issues concerning language, on this question Chomsky has made a similar point. Chomsky (2000) appears to undermine the entire philosophical mind-body enterprise as it has been traditionally conceived. Chomsky's makes the surprising suggestion that the problem cannot even be formulated coherently. Whereas the problem is universally seen as the mystery of the *mind* and how it might be explained in material terms, Chomsky reverses the puzzle as one about the *body*. He suggests that since Isaac Newton "the theory of body was demonstrated to be untenable" (Chomsky 2000, p. 84). Ironically, he notes "Newton eliminated the problem of "the ghost in the machine" by exorcising the machine; the ghost was unaffected" (Chomsky 2000, p. 84).

With this development, "the mind-body problem disappeared, and can be resurrected, if at all, only by producing a new notion of body (material, physical etc.) to replace the one that was abandoned" (Chomsky 2000, p. 84). For this reason, there is generally no reduction of one science to another but rather the reducing science changes to permit unification of the previously recalcitrant theory. This is a remarkable analysis of the puzzle of consciousness that turns everything on its head. If Chomsky is right, as Bunge would appear to concur, there is no more a mind-body problem than there was a valence-atom problem or electricity-matter problem. Chomsky writes: "the traditional mind-body problem became unformulable with the disappearance of the only coherent notion of the body (physical, material, and so on)" (Chomsky 2009, p. 189).

While recognizing that commitment to computationalism is the central dogma of modern cognitive science, Bunge's critique seems to miss its mark here. He writes:

> ... computers are not exactly natural. Worse, unlike live human brains, they are limited to performing algorithmic operations. They lack spontaneity, creativity, insight (intuition), the ability to feel emotions, and sociality. Indeed, computers have to be programmed; there can be no programs for coming up with original ideas ... (Bunge 2010, p. 110)

Elsewhere Bunge explains further, "And, of course, by definition of "original," an original design is one that has never been described before – that is, one that is so far unknown" (Bunge 2010, p. 228). Ironically, given his own harsh criticism of social constructivists, here Bunge falls into the error seen notoriously in Brannigan (1981) who sees creativity and originality in science as a matter of social achievement and priority as if this precluded rational, cognitive, intellectual processes. Bunge cites exactly the same social notion of originality and thereby entirely side-steps the key question of whether computers can do what we do when we make original inventions or discoveries regardless of whether they happen to have been anticipated in an uninteresting sociological sense.

21.5   Programming Original Creativity

On the more fundamental issue, Bunge's critique of computer creativity and originality is the so-called "Lady Lovelace" objection addressed in Alan Turing's classic article 'Computing Machinery and Intelligence' (Turing 1950). Aside from issues of principle, the empirical facts refute the charge that computer algorithms are incapable of originality, spontaneity and creativity as humans are. The AI programs of Newell and Simon (1972) developed further by Langley et al. (1987; see Slezak 1989) are an existence proof of original, creative scientific discovery by computer. For example, the BACON program has discovered Boyle's Law and Kepler's Law from the observational, numerical data which deserves to be regarded as original in the relevant sense that the result was not programmed but found by heuristic problem-solving methods that are essentially the methods that underlie human creative thought.

Bunge's suggestion that computers "are limited to performing algorithmic operations" (Bunge 2010, p. 110) fails to recognize the crucial distinction between algorithmic and heuristic problem solving as developed by Newell and Simon (1972). Although even the latter are strictly algorithmic by virtue of being fundamentally computer programs, they apply techniques of problem solving that are not guaranteed to find a solution since many interesting problems are not susceptible to algorithmic solution.

Apart from such actual developments in AI, Bunge's argument overlooks the fact that human beings are also strictly subject to programming in the sense that we are deterministic machines whose brains are a complex combination of inheritance and learning – all forms of programming, albeit not by a human or other independent intelligence. Unless human originality is ascribed to some mysterious, inexplicable indeterministic source, our own creative discoveries must also be due to describable cognitive processes that are ultimately products of the information processing physically embodied in the brain. On Bunge's own materialist view of human minds, he must be committed to just such a view of originality and creativity as describable, that is, programmable.

21.6   Plato's Problem: Nature or Nurture?

Perhaps the most sophisticated computational theory of mental phenomena is Chomsky's generative grammar of language, that Bunge acknowledges to be a "naturalization project" (Bunge 2010, p. 112). However, far from being a kind of Cartesian dualism (Bunge 2010, p. 122), the generative program is a vindication of materialism by showing how a physical, biological system might embody the special properties of language such as compositionality and recursiveness. Moreover, Bunge's response to Chomsky's claims for the innateness of Universal Grammar (UG) does not address the fundamental grounds for the claim – the idea that the human brain has an initial state that includes the principles underlying all humanly learnable languages and, therefore, a species-specific aspect of the human genetic endowment. Purporting to answer Chomsky, Bunge asserts "all knowledge is learned" (Bunge 2010, p. 166). Bunge complains:

> Unfortunately no one has bothered to explicitly state the rules of UG, and geneticists have not found the presumptive UG gene(s). Nor is there any reason to expect such findings, for languages are highly conventional .... (Bunge 2010, p. 112)

First, Bunge overlooks the fact that Chomsky's claim at this level of generality should be uncontroversial because even the most extreme empiricist or behaviourist must agree that something is innate to permit language "learning" at all. The only question at issue is how much. Moreover, the "innateness" claim for language is essentially the same as for other cognitive systems such as vision in mammals which are not fully determined by genetically determined structures in the brain but partially fixed by innate factors and partly determined by experience.

More specifically, the "reason to expect such findings" includes the "poverty of the stimulus" argument or what Chomsky refers to as "Plato's Problem" – namely, that of explaining how we can know so much on the basis of so little data. Bunge asserts "there is no evidence whatsoever that anything learnable is encoded in the genome" (Bunge 2010, p. 183), but he doesn't address the persuasive evidence from acquisition of the complex structures of language in all children by the age of three without effort, without instruction, without adequate evidence. This phenomenon is quite different from "learning" and rather typical of biologically determined maturation along a pre-determined course of development, triggered by experience but not learned from it.

And, of course, despite Bunge's assertion that "no one has bothered to explicitly state the rules of UG," modern generative linguistics in the 'Minimalist' program is precisely stating the rules of UG, of course, as always, conjectured provisionally as in any other branch of empirical science. Bunge makes the surprising remark that "idealists like Chomsky and his followers ignore empirical linguistics" (Bunge 2010, p. 136) by which he means "real speakers and linguistic communities" rather than "abstract systems" (Bunge 2010, p. 136).

In a certain sense Bunge is right, but this is merely the "Galilean" approach to science that Chomsky has championed on the basis of his famous "competence – performance" distinction (see Pylyshyn 1973; Slezak 2014). Galileo ignored the empirical evidence of real pendulums and real projectiles in favour of mathematically idealized "abstract systems" for the same reason. Bunge complains that "brainless psychology" can only describe mental phenomena but not explain them "because genuine explanation involves revealing mechanisms" (Bunge 2010, p. 159). However, this would rule out Newton's law of gravitation and Kepler's law of elliptical planetary orbits, *inter alia*, which famously did not reveal underlying mechanisms. Like Marr's (1982) computational model of vision, Chomsky's competence model of tacit knowledge abstracts and idealizes from underlying mechanisms and, thereby provides the most fruitful approach to ultimately discovering them.

21.7 Conclusion

Bunge's work is both exhilarating and exasperating at the same time. Its characteristic scope, insight and erudition is joined with a refreshing impatience for the many varieties of nonsense in the academy. I confess to considerable sympathy for Bunge's jaundiced view of those "professors who play parlour games instead of tackling serious problems" (Bunge 2010, p. 11) and I share his conviction that much modern philosophy is guilty of this kind of lapse. However, as stated at the outset, there are points where his insights, bold claims and criticisms need much finer detail and attention to contemporary literature in philosophy of mind and cognitive science.

References

Aristotle (1941). *De Anima* (R. McKeon, Trans.). *The basic works of aristotle* (pp. 535–561). New York: Random House.

Armstrong, D. M. (1968). *A materialist theory of mind*. London: Routledge & Kegan Paul.

Armstrong, D. M. (1980). *The nature of mind*. Sydney: University of Queensland Press.

Block, N. (1990). Consciousness ignored. Review of Daniel Dennett consciousness explained. *The Journal of Philosophy*, 181–193.

Brannigan, A. (1981). *The social basis of scientific discoveries*. Cambridge: Cambridge University Press.

Bunge, M. (1980). *The mind-body problem: A psychobiological approach*. New York: Pergamon Press.

Bunge, M. (1991). A philosophical perspective on the mind-body problem or, why neuroscientists and psychologists should care about philosophy. *Proceedings of the American Philosophical Society, 135*(4), 513–523.

Bunge, M. (2010). *Matter and mind: A philosophical inquiry. Boston studies in the philosophy of science volume 287*. Dordrecht: Springer.

Chalmers, D. (1996). *The conscious mind: In search of a fundamental theory*. Oxford: Oxford University Press.

Chalmers, D. (2002). Does conceivability entail possibility? In T. Gendler & J. Hawthorne (Eds.), *Conceivability and possibility* (pp. 145–200). Oxford: Oxford University Press.

Chomsky, N. (1965). *Aspects of the theory of syntax*. Cambridge, MA: MIT Press.

Chomsky, N. (1980). *Rules and representations*. New York: Columbia University Press.
Chomsky, N. (2000). *New horizons in the study of language and mind*. Cambridge: Cambridge University Press.
Chomsky, N. (2009). The mysteries of nature: How deeply hidden? *The Journal of Philosophy, 106*(4), 167–200.
Cottingham, J. (1992). *"Introduction" The Cambridge companion to Descartes*. Cambridge: Cambridge University Press.
Dennett, D. C. (1978a). *Brainstorms*. Vermont: Bradford Books.
Dennett, D. C. (1978b). Current issues in the philosophy of mind. *American Philosophical Quarterly, 15*, 249–261.
Dennett, D. C. (1989). *The intentional stance*. Cambridge, MA: MIT Press.
Dennett, D. C. (1991). *Consciousness explained*. London: Penguin.
Descartes, R. (1637/1985). Discourse on the method In *The philosophical writings of Descartes, volume 1* (J. Cottingham, R. Stoothoff & D. Murdoch, Trans.). Cambridge: Cambridge University Press.
Di Sciullo, A. M., & Boeckx, C. (Eds.). (2011). *The biolinguistic enterprise: New perspectives on the evolution and nature of the human language faculty*. Oxford: Oxford University Press.
Durkheim, E. (1898). Individual and collective representations. *Revue de Métaphysique et de Morale*, vi, May. Reprinted in D.F. Pocock, 1974, translator, *Sociology and Philosophy by Emile Durkheim*. New York: The Free Press.
Fodor, J. A. (1968). *Psychological explanation*. New York: Random House.
Fodor, J. A. (1989). Making mind matter more. *Philosophical topics, 17*, 59–79. Reprinted in Fodor, *A theory of content and other essays* ([DATE] pp. 137–160). Cambridge, MA: MIT Press.
Gardner, H. (1987). *The mind's new science*. New York: Basic Books.
Gendler, T., & Hawthorne, J. (Eds.). (2002). *Conceivability and possibility*. Oxford: Oxford University Press.
Jenkins, L. (2001). *Biolinguistics: Exploring the biology of language*. Cambridge: Cambridge University Press.
Kirk, R. (2005). *Zombies and consciousness*. Oxford: Oxford University Press.
Langley, P., Simon, H., Bradshaw, G., & Zytkow, J. (1987). *Scientific discovery: Computational explorations of the creative processes*. Cambridge, MA: MIT Press.
Levin, J. (2010). Functionalism. In E. N. Zalta (Ed.), *The Stanford Encyclopedia of Philosophy (Summer 2010 Edition)*. http://plato.stanford.edu/archives/sum2010/entries/functionalism/. Accessed 10 June 2018.
Levine, J. (2001). *Purple haze: The puzzle of consciousness*. Oxford: Oxford University Press.
Loar, B. (1999). David Chalmers's the conscious mind. *Philosophy and Phenomenological Research, 59*(2), 465–472.
Marr, D. (1982). *Vision a computational investigation into the human representation and processing of visual information*. Cambridge, MA: MIT Press.
McGinn, C. (1989). Can we solve the mind-body problem. *Mind, 98*, 891.
Nagel, T. (1965). Physicalism. *The Philosophical Review*, 74, 336–356. Reprinted in J. O'Connor ed., *Modern Materialism*. New York: Harcourt Brace & World.
Newell, A. (1990). *Unified theories of cognition*. Cambridge, MA: Harvard University Press.
Newell, A., & Simon, H. A. (1972). *Human problem solving*. New Jersey: Prentice-Hall.
Pylyshyn, Z. (1972). Competence and psychological reality. *American Psychologist, 27*, 546–552.
Pylyshyn, Z. (1973). The role of competence theories in cognitive psychology. *Journal of Psycholinguistic Research, 2*(1), 21–50.
Pylyshyn, Z. (1984). *Computation and cognition*. Cambridge, MA: MIT Press.
Russell, B. (1912/1967). *The problems of philosophy*. London: Oxford University Press.
Searle, J. R. (1992). *The rediscovery of the mind*. Cambridge, MA: MIT Press.
Searle, J. R. (1997). *The mystery of consciousness*. London: Granta Books.
Slezak, P. (1988). Was Descartes a liar? Diagonal doubt defended. *British Journal for the Philosophy of Science, 39*, 379–388.

Slezak, P. (1989). Scientific discovery by computer as refutation of the strong programme. *Social Studies of Science, 19*(4), 563–600.

Slezak, P. (2010). Doubts about Descartes' indubitability: The cogito as intuition and inference. *Philosophical Forum, 41*(4), 389–412.

Slezak, P. (2014). Intuitions in the study of language: Syntax and semantics. In L. M. Osbeck & B. S. Held (Eds.), *Rational intuition: Philosophical roots, scientific investigations*. Cambridge: Cambridge University Press.

Slezak, P. (2018). Is there progress in philosophy? The case for taking history seriously. *Philosophy, 93*(4), 529–555. https://doi.org/10.1017/S0031819118000232

Stoljar, D. (2001). The conceivability argument and two conceptions of the physical. *Philosophical Perspectives, 15*: *Metaphysics*, 393–413.

Stoljar, D. (2017). *Philosophical progress: In defence of a reasonable optimism*. Oxford: Oxford University Press.

Strawson, G. (2006). *Consciousness and its place in nature*. Exeter: Imprint Academic.

Strawson, G. (2008). *Real materialism and other essays*. Oxford: Clarendon Press.

Strawson, G. (2018). *Things that bother me*. New York: NYRB Press.

Turing, A. (1950). Computing machinery and intelligence. *Mind, 59*, 433–460.

Yablo, S. (1999). Concepts and consciousness. *Philosophy and Phenomenological Research, 59*(2), 455–463.

Chapter 22
Is Consciousness an Epiphenomenon?

Ignacio Morgado-Bernal

Abstract The nature of consciousness remains one of the main unsolved questions in neurobiology. Although recent advances suggest that sooner or later we will be able to understand the neural mechanisms underlying awareness, it seems very difficult to understand how neural activity becomes a subjective experience, the so-called hard-problem of consciousness. The apparent intractable nature of this problem causes some scientists to avoid it altogether and deal only with the neural correlates of consciousness. However, for others, consciousness is an epiphenomenon, that is, something without a direct function, like the redness of blood – a characteristic which was not selected for, but was a consequence of the mechanism selected to deliver oxygen. In that view, qualia, the phenomenological experiences, correspond to internal discriminations that are reliable correlates of underlying neural mechanisms. Consciousness itself is not causal. It is the neural structures underlying conscious experience that are causal. In contrast, a hypothesis is proposed here for which the functional integration of cortical circuits could generate the conscious experience as a feedback mechanism that allows the brain to continuously alter its ongoing operation in order to get a very precise adjustment of the organism to its internal and external environment. This means that without consciousness the brain function would lose versatility and effectiveness.

Contrary to what most of us commonly believe, scientists such as the late Nobel laureate, Gerald Edelman, and the respected Spanish neuroscientist, Joaquin Fuster from the University of California in Los Angeles, affirm that consciousness is an epiphenomenon (Edelman et al. 2011). That is, consciousness is a side effect of the physiological activity of our brain, with little or no practical value, something equivalent to the noise of the engine of a car or a machine.

We humans know that much of our behaviour takes place automatically and unconsciously, but we feel that our voluntary behaviour is ruled by conscious

I. Morgado-Bernal (✉)
Institut de Neurociències, Universitat Autònoma de Barcelona, Bellaterra (Barcelona), Spain
e-mail: Ignacio.Morgado@uab.cat

© Springer Nature Switzerland AG 2019
M. R. Matthews (ed.), *Mario Bunge: A Centenary Festschrift*,
https://doi.org/10.1007/978-3-030-16673-1_22

thoughts. With our conscious thoughts, we analyse situations and their backgrounds, value things, set options, take decisions and make plans. Then, based on this reasoning, we act. Therefore, we have the feeling that consciousness is a determinant of our conduct, that is, a causative agent, something that is denied by Edelman and Fuster with their epiphenomenonalist stance (Edelman et al. 2011).

Their denial was in part justified by a surprising observation by the researcher Benjamin Libet and colleagues from the Physiology Department of the University of California in San Francisco. Their experiment (Libet et al. 1983) was meant to demonstrate that the unconscious electrical processes in the brain, called the readiness potential (or *Bereitschafts potential*) and discovered by Lüder Deecke and Hans Helmut Kornhuber in 1964, precede conscious decisions to perform volitional, spontaneous acts. That is, the brain activity related to the preparation of a movement takes place before the subject reveals an awareness of their intention to make that move. This could mean that unconscious neuronal processes precede and potentially cause volitional acts that are retrospectively felt to be consciously motivated by the subject.

The experiment was controversial because it challenged the belief in free will. However, today that conclusion is questioned on the basis of some criticisms of its implicit assumptions and also as a result of new experiments using more precise techniques. In this context, recent studies indicate that when the subjects claim to be aware of their intention to move, it might be only the end of a preceding and not instant process of conscious deliberation. That is, "unconscious neural decision processes build up until they cross a threshold which then enables the instantaneous appearance of a full-blown conscious intention" (Guggisberg and Motazz 2013, p. 8).

22.1 What Is Consciousness?

For Mario Bunge, the word "consciousness" denotes a large variety of mental processes, such as reactivity or sensitivity to some physical or chemical agents, being able to identify or discriminate some internal or external stimuli (awareness of phenomenal consciousness), or being self-aware, that is, being aware of one's own feelings, perceptions or actions (Bunge and Ardila 1987, Bunge 2010). To distinguish between consciousness and awareness, Bunge considers primarily the capacity of thinking. Animals can be *aware* of certain stimuli and pay attention to them without thinking. However, thinking about these stimuli should be a *conscious* activity (Bunge 2010, p. 211).

Bunge's view is that consciousness is the collection of brain states that allows us to be aware of, and think about, the things that happen (Bunge 2010). As water may present in gaseous, liquid and solid states, without losing its nature, the mind and its processes can present themselves consciously or unconsciously, without losing its nature either. Implicit memory, for example knowing how to ride a bicycle, is an unconscious memory; and explicit memory, such as knowing who Martin Luther

King was, is a conscious memory. Both represent a different state of memory, a mental process.

However, consciousness is not the perception of mental processing itself, which is totally unconscious and based on physiological mechanisms of the brain. It is the means by which the brain presents the result of that mental processing. In the words of Christof Koch, "You can never directly know the outer world. Instead, you are conscious of the results of some of the computations performed by your nervous system on one or more representation of the world" (Koch 2004, p. 302). Consciousness is, moreover, a particular mental state, personal and subjective. Subjectivity is the principal feature of consciousness, to the point that there is no way to prove that the other people with whom we live are conscious beings like us; they might be zombies so perfect that their behaviour is indistinguishable that of a conscious being. Nor is it possible to know if other people, in the event that they are also conscious beings, feel the conscious experience as we do. There is no way of knowing, for example, if, when one person sees the colour red, another perceives it just like the first. Even an identical physiological recording, say EEG or MRI, of both brains when they perceive the same colour, wouldn't be a definitive test of perception identity, because it is still not possible to know if every brain generates subjectivity and imagination from physiology in the same manner.

Nevertheless, it seems difficult to imagine a robotic individual having the behavioural flexibility and adaptive capacity comparable to those of a conscious being. These capacities could infer that qualia, that is, the contents of consciousness or the phenomenological consciousness, contain a large amount of information able to be implicitly and immediately captured in every conscious experience (Koch 2004).

22.2   How the Brain Creates Consciousness?

Some experimental discoveries and clinical observations indicate that the conscious state is generated by the functional integration of different neural circuits of the cerebral cortex involved in the processing of information (Tononi 2005; Tononi et al. 2016). It could be thought of as a kind of teamwork of the cortical neural networks. Recently, it has also been shown (Koch et al. 2016) that the anatomical bases of the neural correlates of consciousness are primarily localized in a restricted temporo-parietal-occipital hot zone with additional contributions from some anterior regions. Nevertheless, for the integration of information to take place, a background condition is necessary. To reduce the thresholds of excitation of the neurons that make up these networks, there needs to be a process that is in charge of different subcortical and brain stem structures, such as the brain stem reticular formation and the paramedial thalamus and its intralaminar nuclei, reciprocally connected to wide regions of the cerebral cortex (Koch 2004; Koch et al. 2016).

The conscious state can be compared to the light reflected on a screen by a projector. If the projector, comparable to the thalamus and other subcortical structures, is not turned on, there is no light on the screen, that is, there is no cortical activity which generates consciousness. However, during surgery, it has been shown that anaesthesia can deactivate the neurons of the cerebral cortex up to 10 min before the neurons of the thalamus, the patient being already unconscious prior to the deactivation of the latter (Velly et al. 2007). This suggests that, despite the fact that under normal conditions the activation of the thalamus is necessary to cause consciousness, the latter is associated with the functional state of the cerebral cortex.

22.3   The Hard Problem

The explanation above refers to brain structures and mechanisms that must be activated to create the conscious state, but it is a different matter to describe how this activation works, that is, how the integrated activity of the cerebral cortex produces conscious imagination and what constitutes the latter. In short, how objective matter becomes subjective imagination, the so-called *hard problem* of consciousness, which no one until now has been able to solve or clarify. For Mario Bunge (Bunge 1980, 2010), the more important alternative to classical dualism is the psychoneural identity theory, the reductive conjecture that all mental processes are neural processes. In that sense, consciousness is identical to the brain activity that makes it possible, or, more specifically, the feeling of pain is identical to the firing of the C-fibers of the nervous system.

Not everyone resolves the issue in the way that Bunge does. In discussing the hard problem, the Nobel laureate Francis Crick questioned whether qualia can be explained by modern science (Foreword in Koch 2004). Other scientists, recognizing the apparent intractable nature of this problem, avoid it altogether and deal only with the neural correlates of consciousness.

However, in a paper entitled "Biology of consciousness", Edelman and two of his colleagues attempted to refute the idea that the phenomenal "feel" of conscious experience cannot be explained in scientific terms (Edelman et al. 2011). In their opinion, the hard problem of consciousness does not require a solution, but a cure. Qualia, they explain, correspond to internal discriminations that are reliable correlates of underlying neural mechanisms. Consciousness itself is not causal. It is the neural structures underlying conscious experience that are causal. The conscious individual can therefore be described as responding to a causal illusion. Importantly, Edelman and his colleagues talk about correlates, and not identity, of consciousness, as Bunge does, with its neural underlying mechanisms. They, therefore, believe that consciousness is an epiphenomenon; it is something like the redness of blood, a characteristic which was not selected for but was a consequence of the mechanism selected to deliver oxygen.

Bunge (2010) describes epiphenomenalism as the dualist doctrine which asserts that mental events are the passive by-product of brain activity and he considers this mistaken because a product is generally thought of as stuff separable from its source and incapable of influencing it. Qualia, he says (Bunge 2010, p. 174), are an emergent property of the nervous system.

Regarding consciousness as an epiphenomenon, in personal correspondence to Edelman, his colleague Gally replied, confirming the view of qualia (that is, the conscious or phenomenical experience) as being non-causal illusions, resulting from the same activity of the nervous system that generates the internal discrimination correlating with them. Qualia, he supposes, are epiphenomena, like the redness of blood or the noise of an engine. Conscious feeling of qualia was not selected for, but resulted from the neural mechanisms selected to make internal discriminations, the ones that distinguish, say, the experience of blue from the experience of warmth. Gally also said that consciousness has evolved as an inevitable consequence of the device selected in evolution to process thoughts, in the same way that the red colour of the blood is a consequence of the device selected by evolution to transport the oxygen to body organs. More or less, he said that consciousness has evolved as by chance, a conception which, without doubt, diminishes the idea and appreciation that we humans have of phenomenological consciousness.

Gally also emphasized that qualia are non-causal illusions saying that he cannot see how those illusions (qualia or consciousness) per se can cause even a single neuron to fire, causing it to be activated and generate behaviour. Feeling the conscious experience as causal, he also said, is something like succumbing to an illusion. Certainly, knowing how consciousness could affect neurons is something difficult to imagine, but no more than the other way around; that is, how neuron firing can give rise to qualia, which is the essence of the hard problem. Surprisingly after all those considerations, Gally added that knowing the neural mechanisms that make the illusion of consciousness possible is the greatest challenge for science in the twenty-first century.

All in all, it is hard to believe that a phenomenon so important in our lives as conscious perception might be a fluke of nature. If qualia are non-causal and innocuous illusions, we would have to admit also that a completely unconscious being, such a perfect zombie, might have the same capacity of behavioural flexibility and adaptation to its environment as a conscious being, in the same way as a noiseless engine might be as efficient as a noisy one. In evolutionary terms, we would say that the zombie could have the same fitness or ability to pass on its genes to offspring that a conscious human has. Would this be possible? Would somatic pain, as an example, have the same adaptive value if it was not felt consciously? However, there is currently no way to test this hypothesis, since we cannot eliminate qualia, that is the conscious experience, without jointly deactivating the same neural mechanisms that produces the internal discriminations.

Can we conceive an efficient discriminative nervous system without illusions of consciousness? Of course, we can, but even so, there would be other related issues to resolve. Certainly, it is possible that just the mental processes of the highest level, such as those that involve solving difficult problems, creativity or innovation, have the ability to activate the neural mechanism of consciousness. But it is also possible that only these high-level processes need consciousness and that specificity would then be an evolved improvement, and not a collateral or irrelevant consequence in the process of natural selection. Moreover, all kind of mental processes involved in giving flexibility to behaviour tend to be conscious. Even if we admit this kind of flexibility in a zombie and if all experiences took place unconsciously as in the life of an unconscious being, would its life have the same sense of human life as we understand it, or would it be something different? Perhaps it is not the conscious experience that gives meaning to our lives? If that is the case, then consciousness does not have a causal value. Finally, if we accept that consciousness is an epiphenomenon, what do we gain in the explanation of brain and behaviour? This approach does not solve the hard problem, it just removes it. In addressing these questions an alternative hypothesis is proposed to explain the nature of consciousness.

22.4 An Alternative Hypothesis

The functional integration of cortical circuits could generate conscious experience as a feedback mechanism that allows the brain to alter its operation in order to get a very precise adjustment of the organism to its internal and external environment. That is, the conscious mind might be the device that natural selection has promoted so that the brain can effectively exercise its main adaptive and survival functions. This means that without consciousness these brain functions would lose versatility and effectiveness. Consequently, conscious perception *is not* an independent causal agent or something independent of brain activity, or imposed on it, but a biological complement which is a part of the functional system that makes intelligence and adaptive behaviour possible for evolved beings.

A metaphor for a better understanding of this proposal is the mirror. When a person looks at it and corrects her hairstyle, it is not the mirror, as an independent agent, who orders her to do it, but the mirror provides the relevant information to do it correctly. Importantly, the person asks the mirror h*ow am I?* and not *how do you think I am?* That is why we can say that the mirror is only a tool of the true causal agent, which is the brain of the person who is looking at it. Therefore, there is not dualism in looking in a mirror and acting accordingly.

In the same way that we can feel and admit that the image in the mirror determines or influences our behaviour, we can also feel and admit that our consciousness does so. But it does so in response to the functional work of the cerebral cortex, the true causal agent, which used the mirror image as an intrinsic feedback, as an auxiliary element to perform its role effectively. This assistance increases the learning and adaptive flexibility of the brain, some of which is lacking in even the more sophisticated robots. At the same time, consciousness gives a special meaning to our life.

There is, however, a problem in supporting this hypothesis, as Gally (personal correspondence) pointed out: How could the conscious feedback exercise its effect on neurons? How could consciousness, being nothing more than an illusion, influence the brain? Or is it more than just an illusion? Although the process may be unexplained, the situation is similar to the related case in which it is known that the brain is able to create qualia and awareness, even when we do not know how it does it. In addition, the "illusion" of consciousness exists as a common experience. Then, unless that we were to assume a dualistic explanation, consciousness must be, somehow, generated by the brain. As explained earlier, a solution to the hard problem of consciousness can be the psycho-neural identity hypothesis of Bunge, wherein qualia are identical to the brain activity that generates them.

However, we can also question whether our brain has enough capacity to understand what is subjectivity and how matter becomes imagination? If so, the answer to the question could be useful not only in explaining the hard problem, but also in addressing how conscious experience could affect neurons and produce behaviour, being therefore causal. If we knew how the brain makes awareness and qualia possible, we might also know how conscious experience affects the brain. Similarly, if we know how water becomes steam we have keys to understanding how the steam converts into water.

Unfortunately, these hypotheses about consciousness are not verifiable in a scientific manner. Firstly, in checking how effective a totally unconscious mind would be, it is not possible to selectively disable consciousness without jointly turning off the causal agent that generates it, namely the cerebral cortex. In addition, it is also the case that when we sleep we are unaware although our brain and mind continue working, but they have less adaptive power than they have when we wake up and become conscious.

A life without consciousness certainly would have a very different kind of sense, if any, that a conscious one. But it could also be the case that the human brain has not evolved enough to understand how the subjective matter becomes imagination, that is, to solve the hard problem. There are things we do not know, but we can guess how they are. For example, we may think that there are intelligent beings in other planets and imagine how they could be, even conceding that we may be wrong. But with respect to awareness and subjectivity, we are not even able to figure out an explanation in advance of its nature. When we ponder the neural mechanisms underlying consciousness, we seek to determine the spatial and temporal dynamic of the brain that makes consciousness possible. However, when we wonder how the brain turns neural activity into subjectivity, what kind of answer are we expecting? Would an informatics algorithm, a mathematic formula or something related to a new kind of energy satisfy our scientific requirement? We may not actually know what we are looking for.

Crick questioned whether qualia can be explained by modern science. In the same way, we cannot deny a priori the value of awareness because if we don't know what something is, we can hardly say whether this thing is useful or not. Just as the brain of a chimpanzee does not have capacity to understand and solve square roots, the human brain might not have enough capacity to understand the nature of consciousness.

The ability to understand consciousness might have not evolved because it may not have adaptive value, as the following metaphor illustrates. In order to produce a tasty meal, we need a recipe, the right ingredients and cooking. Of course, the sequence and temporality of the cooking process is critical, but what then could add to the final result knowing why or how the mixture of ingredients and cooking gives rise to such a good taste? Would that knowledge make the meal taste better? Would it provide the meal with any other kind of advantage or property? Probably it wouldn't. Therefore, in the same way that knowing how the combination of ingredients of a meal originates does not improve its flavour, knowledge of the nature of consciousness may not make it more powerful; it would not improve its adaptive capacity, its fitness. There does not yet appear to be a full convincing solution to the hard problem of consciousness.

Natural selection tends to promote useful things. Figuring out the neural mechanisms of consciousness will have useful consequences in clinical or educational domains, but knowing about the intimate nature of subjectivity, apart from satisfying our scientific curiosity, could be of little or no practical use. Consciousness evolved in response to the challenges of the environment. That is, in order to survive, the animal had to develop flexibility in thinking and behaviour. We do not know whether, rather than a knowledge of the nature of consciousness, it may actually be its ignorance that has adaptive value. In a fearful and hostile world, few things help survival more than a large dose of supernatural beliefs that induce the mysteries of consciousness.

22.5   Conclusion

Common arguments that deny the role of phenomenological consciousness as a causative agent of human behaviour are plausible, but they are not conclusive. They do not rule out the possibility that, as well as neurons are capable of causing imagination, imagination, in turn, could retroactively influence the brain and the behaviour of the person. Our capacity to understand consciousness will evolve when new environmental conditions, possibly after millions of years, make that understanding necessary. Coping with new emergent difficult questions could then be the price that future society will need to pay for such a relevant promotion.

References

Bunge, M. (1980). *The mind-body problem. A psychobiological approach*. Oxford/New York: Pergamon Press Ltd.
Bunge, M. (2010). *Matter and mind. A philosophical inquiry*. New York: Springer.
Bunge, M., & Ardila, R. (1987). *Philosophy and psychology*. New York: Springer.
Edelman, G., Gally, J. A., & Baars, B. J. (2011). Biology of consciousness. *Frontiers in Psychology, 2*(4), 1–7.
Guggisberg, A. G., & Motazz, A. (2013). Timing and awareness of movement decisions: Does consciousness really come too late? *Frontiers in Human Neuroscience, 7*(385), 1–11.

Koch, C. (2004). *The quest for consciousness. A neurobiological approach.* Denver: Roberts and Company Publishers.

Koch, C., Massimini, M., Boly, M., & Tononi, J. (2016). Neural correlates of consciousness: Progress and problems. *Nature Review Neuroscience, 17,* 307–321.

Libet, B., Gleason, C. A., Wright, E. W., & Pearl, D. K. (1983). Time of conscious intention to act in relation to onset of cerebral activity (readiness-potential). The unconscious initiation of a free voluntary act. *Brain, 106,* 623–642.

Tononi, J. (2005). Consciousness, information integration and the brain. *Progress in Brain Research, 150,* 109–126.

Tononi, J., Boly, M., Massimini, M., & Koch, C. (2016). Integrated information theory: from consciousness to its physical substrate. *Nature Review Neuroscience (advanced on line publication), 17*(7), 450–461.

Velly, L. J., Rey, M. F., Bruder, N. J., Gouvitsos, F. A., Witjas, T., Regis, J. M., Peragut, J. C., & Gouin, F. M. (2007). Differential dynamic of action on cortical and subcortical structures of anesthetic agents during induction of anaesthesia. *Anaesthesiology, 107,* 202–212.

Part V
Sociology and Social Theory

Chapter 23
Bunge and Scientific Anthropology

Marta Crivos

Abstract Mario Bunge has made a monumental contribution to many fields of science and philosophy over decades. He is an ongoing inspiration to critical thinkers in those disciplines. His work in social studies and his political commentary are equally significant. This paper is written as a personal recognition of his high standing in the area of my own work and career as a professor and researcher in ethnographic studies. That work has been conducted at the National University of La Plata, the only academic institution in Argentina which provides for the many specializations in the field of anthropology, and which shares Bunge's philosophy in its development of a systemic and evolutionary approach to studying human beings as a natural species.

23.1 Introduction

In anthropology at the National University of La Plata (UNLP) to believe in the possibility of a science of the human being has been and still is a stigma that has hindered the career of many 'politically incorrect' anthropologists. There have been obstacles and criticisms systematically endured by those who supported this naturalist programme in the last decades. This enlarges the breach between naturalistic and humanistic anthropologists, and accounts for the high degree of specialisation and the absence of disciplinary integration in anthropology. It is, therefore, at this point that Mario Bunge and his work gain value and prominence on both personal and institutional levels at the Faculty of Natural Sciences of UNLP.

M. Crivos (✉)
Laboratory for Research in Applied Ethnography (LINEA), Faculty of Natural Sciences and Museum, National University of La Plata, National Scientific and Technical Research Council (CONICET), Buenos Aires, Argentina
e-mail: crivos@fcnym.unlp.edu.ar

© Springer Nature Switzerland AG 2019
M. R. Matthews (ed.), *Mario Bunge: A Centenary Festschrift*,
https://doi.org/10.1007/978-3-030-16673-1_23

23.2 Bunge and Didactics of Human Sciences

Regarding didactics, Bunge's contributions turn out to be crucial to the scientific training of generations of Argentinians. Since his classic *La ciencia. Su método y su filosofía* [*Science: Its Method and its Philosophy*] (Bunge 1966), his work has been the required reference in introductory courses for a wide range of scientific disciplines. Even from viewpoints that criticise his ideas, the reference to Bunge has proved to be invaluable. Therefore, generations of students and professionals benefited from his work, making possible their access to a clear and persuasive presentation of the bases and the scope of the scientific endeavour, and the approach to the intricate relations that connect and differentiate the various branches of such endeavour to philosophy.

In the area of theory and anthropological methodology, versions of positivism and historicism alternate according to the tendencies and preferences of anthropologists, and the discussions around them have a privileged scenario in the classrooms of our university. However, the grounds in favour of and against one tendency or the other, and the rational debate of its bases and consequences are not always central to these discussions. Instead, no reference to these debates is commonly made or it becomes diluted to the point where it is even difficult to minimally define our professional activity. I can still remember the phrase 'anthropology is what anthropologists do' often repeated as a leitmotif during the student assemblies in the 1960s every time the rationale behind anthropological work was questioned. That *ad hoc* statement would seemingly make us immune to criticism, unsuspected of scientism, positivism and all the 'isms' restricting the semiotic rationality that would then allow us to shape and reshape our research into countless, rigorous interpretations.

Along with the Messianic political component that characterised much of the period of our vocational training, semiotic-structural alchemy converted the human phenomena approach into an extraordinary endeavour only based on a few conditions of adaptation to scientific standards of the time and striving to achieve a Batesonian balance between imagination and rigour (Bateson 1976). As time went by, the second component – rigour – was simply eliminated by these positions to the benefit of the first – imagination – encouraging the uncritical proliferation of multiple interpretations that a certain state of affairs can inspire. This makes the endeavour of scientific knowledge of human events less viable and increasingly unreachable. In addressing this situation, reading Bunge's writings has an almost therapeutic effect. It is about the inclusion of the objective knowledge of nature, including human nature, in a clear and unique strategy: the scientific method.

The 'Bunge Effect' is immediate: discussion is presented in a rational context. In our particular case, this approach, whereby Bunge's literature is presented as an alternative to dogmatic thought and to all types of irrationalism displayed along the intellectual history of the West, has the value of showing the rules of a game. The game is supposed to be played by those who belong to institutions promoting scientific activity: universities, the National Council of Scientific and Technical

Research, the Agency of Scientific and Technical Promotion, etc. Paradoxically, many anthropologists live off the support, prestige and professional development possibilities that come from our membership of those institutions. However, only a few consider the nature of our endeavour in reference to the conditions that qualify it as scientific.

Therefore, from a pragmatic point of view, the application of Bunge's perspectives in the different classrooms of our university has always led to setting the rules for 'the science game'. This game, proposed by Bunge in all his writings, does not try to impose its rules, but to clearly specify its scope, thus differentiating itself from other games oriented to the understanding of the multiple aspects in which human affairs are considered. Nevertheless, in the classrooms of any faculty of science, laboratories and scientific research centres, this is the game to be played: the one concerning the search for conceptual representations of the structure of the facts and the improvement of the human control over these facts (Bunge 1981).

In this sense, the criticism commonly levelled against the Bungean model of science may lead into a trap pointed out by philosopher Ludwig Wittgenstein as one of the biggest mistakes that can be made in this field: to confuse the language games and to try to apply the rules of one of the games to the other (Wittgenstein 2014). Each game has its own rules and limitations, and everything will work correctly if they are properly followed. For example, the religious experience of the miracle would not be a suitable subject of scientific consideration. Wittgenstein explains:

> For it is clear that when we look at it in this way everything miraculous has disappeared; unless what we mean by this term is merely that a fact has not yet been explained by science which again means that we have hitherto failed to group this fact with others in a scientific system. (. . .) The truth is that the scientific way of looking at a fact is not the way to look at it as a miracle. (Wittgenstein 2014, pp. 51–52)

This is a scientific view of the facts that prompts field anthropologists to move continuously from the observation of habits to the understanding of their structure. In other words, when the anthropologist (ethnographer) does fieldwork, he or she investigates the individual behaviour to discover the social structure, and this structure to understand the former. The field worker comes and goes between micro and macro levels, action and structure (Bunge 1999). Bunge's work shows, on the one hand, the belief in the possibility of scientific anthropology and, on the other, that the setting where the debates about such possibility can be recognised and eventually resolved is the factual sphere whose most genuine expression is exemplified in the ethnographic fieldwork (O'Meara 1989; Aunger 1995).

23.3 Bunge and Ethnography

Ethnographic fieldwork aimed at describing human lifestyles continues to be the endeavour that brings ethnography closer to the facts and to the data of reality. In this sense, ethnography constitutes an irreplaceable instance of elaboration and testing of hypothesis in all the branches of anthropology. Therefore, the ethnographic register represents the main source and evidence that help us prove the inextricable systemic articulation of different aspects of human behaviour. In addition, this register allows for the identification of the individual and group trajectories setting the human life strategies of different enclaves, going through and associating different actors and institutions whose interrelations are not fully understood by macroanalytical approaches of other social and human sciences.

Currently, there is some consensus about originality –which distinguishes anthropology from other sciences of man – being a particular methodological approach mainly characterised by ethnographic fieldwork. In addition, this approach is oriented to finding variables relevant to the description of lifestyles from a holistic perspective considering the interdependencies of all the aspects of human life. Therefore, the problems related to the anthropological research of sociocultural reality will emerge from that descriptive and heuristic stage (Crivos and Denegri 1997).

The theory of systems is a general perspective, a way of analysing the relation between variables, which is strongly related to the traditional anthropological holism. These systems constitute a group of covarying entities and, thus, Bunge finds reference to the work in the anthropological field – ergo, ethnography –as a manifestation of the most conscientious and coherent systemism of all social sciences. Since ethnographers describe and analyse particular local settlements so as to enrich the inventory of human variability, very specific and precise theories present the most widely accepted use of systemic models in current anthropology. The large-scale models that characterise general theories are frequently used by archaeologists and prehistorians (Rodin et al. 1978).

In our ethnographic investigations, we move towards the systemic approach about the lifestyles of populations long settled in specific environments.[1] In this sense, systemic supporters believe that every society is imbricated in a natural environment (Bunge 1999). From this perspective, developed throughout almost three decades of research, it was possible to consider human variability and adaptability in the context of routine practices in which conceptions and experiences are updated and reshaped around an environment that is, in turn, transformed.

In this context, the ethnographic research carried out in different regions of the rural area of Argentina reveals the relation between humans and their natural environment, taking daily activities in the domestic unit as a reference. Although the scope of activities, as interrelated systems of principles and behaviours, is an empirical problem that must be tackled in each case, in our work we privilege its functional meaning (Hill 1966; Howard 1963). In other words, the activity is mainly considered as a unit which is pertinent to the characterisation of the strategies for the resolution of different types of problems, especially those that

---

[1]"Caracterización antropológica del modo de vida. Implicancias teórico-empíricas de las estrategias de investigación etnográfica". Project *Incentivos* for teachers and researchers. Secyt-UNLP. 1995-present.

involve the subsistence of a settlement in a specific environment. Therefore, the relationship between humans and their environment can specifically be identified in those practices oriented towards the resolution of everyday problems. It is in the context of everyday experiences that the ideas about the environment are formed and are put to the test. In fact, daily domestic activities are defined by routines, generated by expectations that are developed over time, and performed in settlements that are, in turn, altered by these same activities. As such, they offer an adequate starting point from which to consider material, social, and symbolic aspects of human lifestyles in different contexts (Lave 1995).

It has been assumed in our study that the ideas about the environment that a community constructs can be inscribed, modified, and tested precisely in the context of these everyday practices. Far from being fixed and immutable –as defined by some inventories of folk categories and native taxonomies– these perspectives a community has of its environment are flexible and open to modification as specific chores and tasks are performed (Crivos and Martínez 1996).

Our focus on individual and group trajectories in relation to the subsistence activities of the domestic group allows us to identify and map physical and social spaces set or affected by them, and to trace relational networks that connect the micro level (individual decisions and actions) with the macro level (projections of those actions at a local, regional, and global scale). These facts have strengthened the conviction that anthropological research needs to be related with the 'naturalistic' version of ethnography, which is deemed as an essential empirical-descriptive instance. Having a heuristic role, naturalistic ethnography also tests theoretical concepts and statements in different specialisation areas of the discipline, such as palaeoanthropology, biological anthropology, archaeology, and folklore. In this context, the 'interpretive' versions of ethnography turn out to be no more than unverified assumptions or hypotheses (Bunge 1996).

In 1999 I was a visiting professor at The London School of Economics and Political Science. Sir Raymond Firth was part of the doctoral thesis seminars organized by the Anthropology Department. One day, a young anthropologist was presenting his 'ethnographic' work conducted in Australia while developing a fluent presentation easily associated to the community of postmodern 'continental' anthropologists. Showing great stoicism, Sir Raymond squinted and seemed to make efforts to deal with the far-fetched and sometimes incomprehensible turns of the lecturer's speech. Opening his eyes wide, he suddenly addressed the young anthropologist and said: 'Ethnography concerns the evidence, young man. What are you doing?' Despite the paralysing – and, for many, reassuring – effect that expression had on the audience and, especially, on the young lecturer, everyone was impressed by the value of synthesis of the statement: 'Ethnography concerns the evidence'. Ethnography as an anthropological discipline had abruptly lost its autonomy to play a central role in the construction of anthropological knowledge from and on the facts. These facts were conceived as stimulus and evidence for anthropological ideas, being prudent assumptions at best or consecrated tenets at worst. Mario Bunge would have enjoyed and probably agreed with Firth's statement.

Nevertheless, this role of ethnography is recognised neither inside nor outside the anthropological community. In general, research in this field is more committed to explore the interchanges with other disciplines of natural and human sciences – interdisciplinarity – than to study the interaction between anthropological disciplines – intradisciplinarity. Consequently, the attempts to exchange ideas among archaeologists, ethnographers, and anthropobiologists have been scarce in terms of determining and tackling the challenges inherent to the anthropological approach, issues that will surely benefit from the contribution of the empirical and systemic dimension of the ethnographic fieldwork. Instead, and through the setting of the new disciplinary scope of ethnoarchaeology, archaeologists try to integrate the work of ethnography to their own domain, thus avoiding the exchange with ethnographers who are allegedly skilled and competent in strategies that may favour the access to actual information that is relevant to the interpretation of the archaeological record.

Recent evidence of this state of affairs is characterised by the decision made by the National Scientific and Technical Research Council (CONICET) to split the Membership Advisory Committee for Anthropology – Archaeology and the Biological Anthropology on the one hand and Sociocultural Anthropology (including Ethnography) on the other.[2] This decision underlies the assumption that ethnography contributes little to the archaeological or anthropobiological research and, ultimately, that these anthropological specialisation areas have virtually nothing to do with each other. This situation pursues the loss of a common object approached from an intradisciplinary perspective and to the worrying return of the natural/social dichotomy as a device for the idealistic philosophy, whose main goal is to block the path towards the scientific study of reality. This dichotomy deviates from the methodological unit of sciences as well as from the existence of a set of disciplines that are both natural and social, as in the case of anthropology (Bunge 1999, p. 63).

At the same time, and beyond the field of anthropology, a poor version of ethnography is appropriated, one that is limited to the erratic use of different interview and survey techniques by the so-called ethnosciences –ethnobotany, ethnozoology, ethnoecology, etc. The latter have invaded the professional market from both the anthropological field and from disciplines that account for the '-emic' approach to their knowledge domains. In addition, these versions are characterised by the field use of some qualitative research techniques, and those who implement them are suddenly turned into 'experts' – even without having any ethnography training or professional qualifications – in the study of the local conceptions on a wide range of knowledge domains, which 'accidentally' correspond to the ones defined by western science.

The use of the prefix 'ethno' in a range of disciplines belonging to the natural sciences, especially to biological sciences, suggests much more than an inventory of the knowledge of specific human groups regarding the components of their environment and its correlation, more or less forced, with the scientific knowledge. However, the increasing appropriation of this research field by the community of

---

[2]http://evaluacion.conicet.gov.ar/comisiones-asesoras-disciplinarias/?c=1

biologists has shortened the reach and depth of the investigation in terms of the strategies different communities adopt when using and handling environmental resources. Research carried out on the interactions between humans and other species constitutes a complex interdisciplinary or transdisciplinary endeavour which, from an anthropological perspective, requires a set of conditions and aptitudes that make up for the centre and content of naturalistic-oriented training programmes for ethnographers.

23.4 Conclusion

When considering ethnosciences as a specialization area for biologists and not as an interdisciplinary field for biology and ethnography, all the benefits of the contribution of the ethnographic approach to this interaction are lost. Consequently, Bunge asserts that systematicity implies an increasing interdisciplinarity, since focusing only on the boundaries between scientific disciplines obstructs progress, blocking the flow of different perspectives on one specific subject. While a macroanalytical approach will overlook certain aspects and dimensions of the relationship between a community and its environment, an in-depth, careful study of the everyday activities of each community will allow scientists to gain better insight into such a relationship. As regards other disciplines in the field of natural sciences, ethnography constitutes a heuristic practice to study the contexts in which human groups select, characterise, process and use elements of their environment.

Without having Mario Bunge as an upright and coherent intellectual referent, the project of defending and extending scientific anthropology in the National University of La Plata and elsewhere in Argentina and Latin America would be much weakened.

References

Aunger, R. (1995, February). On Ethnography: Storytelling or science? *Current Anthropology, 36*(1, Special Issue: Ethnographic Authority and Cultural Explanation), 97–130.

Bateson, G. (1976). *Pasos hacia una ecología de la mente*. Buenos Aires: Ediciones Carlos Lohlé.

Bunge, M. (1966). *La ciencia. Su método y su filosofía*. Buenos Aires: Siglo XXI Editores.

Bunge, M. (1981). *La investigación científica. Su estrategia y su filosofía*. Barcelona: Editorial Ariel.

Bunge, M. (1996). *Finding philosophy in social science*. New Haven: Yale University Press.

Bunge, M. (1999). *Las ciencias sociales en discusión. Una perspectiva filosófica*. Buenos Aires: Editorial Sudamericana.

Crivos, M., & Denegri, G. (1997). Modelo pluriteórico e investigación empírica. *II Congreso de la Sociedad de Lógica, Metodología y Filosofía de la Ciencia en España*. Universidad Autónoma de Barcelona (UAB). Printed in reports, pp. 129–132.

Crivos, M., & Martínez, M. R. (1996). Las estrategias frente a la enfermedad en Molinos (Salta, Argentina). Una propuesta para el relevamiento de información empírica en el dominio de la etnobiología. *Contribuciones a la Antropología Física Latinoamericana (Memoria del IV Simposio de Antropología Física "Luis Montané")*, (pp. 99–104). Instituto de Investigaciones Antropológicas – UNAM/Museo Antropológico Montané, Universidad de La Habana, in referred journal. ISBN 968-36-5689-7.

Hill, J. N. (1966). A prehistoric community in Eastern Arizona. *Southwestern Journal of Anthropology, 22*(1), 9–30.

Howard, A. (1963). Land, activity systems and decision-making models in Rotuma. *Ethnology, II*(4), 407.

Lave, J. (1995). *Cognition in practice*. Cambridge: Cambridge University Press.

O'Meara, J. T. (1989). Anthropology as empirical science. *American Anthropologist, 91*, 354–369.

Rodin, M., Michaelson, K., & Britain, G. M. (1978, December). Theory of systems in anthropology. *Current Anthropology, 19*(4), 747–762.

Wittgenstein, L. (2014). *Lecture on ethics*. New York: Wiley.

Chapter 24
Social Mechanisms and the Logic of Possibility Trees

Leonardo Ivarola

Abstract Mechanisms are usually understood as sequences of events that allow us to explain regular observed behaviors. This mode of thought presupposes an ontology of stable causal factors. In the present paper a critique to this conception in the field of social sciences will be carried out. In particular, it will be argued that social phenomena should be understood under the logic of "possibility trees" or "open-ended results". Finally, on the basis of Bunge's distinction between conceptual and concrete systems, a distinction between two kinds of mechanisms will be made: *theoretical* and *material* mechanisms.

24.1 Introduction

Explaining a phenomenon involves answering the question "why does it happen?" or "how does it happen?". The covering-law model developed by Hempel and Oppenheim (1948) considers that, for an explanation to be considered "scientific", it must include at least a universal or statistical law. This is the way in which a coherent explanation differs from mere descriptivism. In addition, there is a symmetry between "explanation" and "prediction", in the sense that any law-based explanatory argument is also useful for predictive purposes. Thus, we can say that the fall of my pen to the floor is *explained* by the fact that there is a scientific law that states that in vacuum the acceleration of all bodies in free fall is constant, as well as we can *predict* that if we drop a steel ball from 84 m high it will reach the floor in approximately 4.14 s.

However, the covering-law model does not provide information about the linkage between the identified phenomena or the regular behaviors. As Bunge puts it (Bunge 1997, 2004), it offers "black box" explanations: we know the inputs and outputs, but

L. Ivarola (✉)
Economics Faculty, University of Buenos Aires, Buenos Aires, Argentina
e-mail: l.ivarola@conicet.gov.ar

© Springer Nature Switzerland AG 2019
M. R. Matthews (ed.), *Mario Bunge: A Centenary Festschrift*,
https://doi.org/10.1007/978-3-030-16673-1_24

we do not know the process that links them. We know *what* happens, but not *how* it happens. The appeal to *mechanisms* fills this explanatory gap.

During the last decades, the concern about the nature of mechanisms and their role in the scientific investigation has grown considerably. This has given rise to the so-called *New Mechanistic Philosophy* (NMP) (Skipper and Milstein 2005), which maintains as a general principle that a great variety of real world phenomena are the product of the operation of mechanisms.[1]

Several mechanistic approaches understand mechanisms in terms of *robustness*. Glennan (2002, 2008), for example, proposes a mechanistic approach removed from the idea of mechanisms as sequences of interconnected events (see also Railton 1978; Salmon 1984). However, the notion of invariance or robustness can be put into question as long as we speak of social mechanisms. More precisely, there is an ontological problem which involves the difficulty of discovering stable causal factors in the realm of the social sciences. In other words, the causal nexus between social variables are people's actions, which do not always respond in the same way, but according to how they interpret the signals from the world, what expectations they form, how they are influenced by institutional and cultural factors, etc. As an alternative, it will be argued that social phenomena are better suited to a logic of *possibility trees* or *open-ended results* rather than to the logic of stable or robust factors: given a certain event, there are different paths or alternatives. Any of them is in principle plausible. The final process will depend on how people form their expectations, how they are influenced by the cultural, institutional and socioeconomic framework, etc.

Based on this problem this paper argues that the way we understand social mechanisms strongly depends on the kind of system such mechanisms belong to. In this respect, Bunge's distinction between a "conceptual system" and a "concrete system" allows us to elucidate two possible categories of mechanisms: *theoretical mechanisms* (mechanisms that could occur or could have occurred given our available knowledge) and *material mechanisms* (mechanisms that take place in a concrete system, in other words, the materialization of a logical path in the possibility tree).

24.2 Mechanisms and the New Mechanistic Philosophy

Although there are discrepancies in the way in which various NMP authors conceive what a mechanism is (for a typology of mechanisms see for example Gerring (2008) and Mahoney (2002)), they all share certain basic ideas, such as the role mechanisms play in the scientific explanation, the importance of a causal approach in the

---

[1] see Bechtel and Abrahamsen (2005), Bunge (1997, 2004), Darden (2006), Elster (1989), Gerring (2008), Glennan, (1996, 2002, 2008), Hedström and Swedberg (1998a, b), Machamer et al. (2000).

comprehension of mechanisms, the characteristic of mechanisms being automatic processes, etc.

NMP is often seen as an alternative stance to logical empiricism. According to the latter, scientific theories must be grounded on statements that describe regularities at the level of events, i.e., empirical laws. These laws can be used for different scientific purposes. For example, in the *Covering-law model* developed by Hempel and Oppenheim (1948) it is asserted that a good explanatory argument should appeal to laws. To explain a phenomenon scientifically is to subsume it under a law (Suppe 1977). However, this mode of thinking has some drawbacks. For example, there is a substantial difficulty in the discovery of laws outside physics or astronomy. In this framework, the NMP has replaced the *nomothetic* approach by another based on a looser idea: that of "stable" or "invariant" regularities (Woodward 2003). The domain of laws is universal, that is, it is neither limited spatially nor temporally. In contrast, invariant regularities only work in bounded domains. Laws are also related to the idea of behaviors without exceptions: "whenever X occurs, Y will occur". In contrast, regularities emerging from the behavior of mechanisms are prone to exceptions. Specifically, mechanisms behave in a regular manner within a given structure. However, some random perturbations can significantly distort the conditions that provide stability to such mechanisms.

An additional problem that the nomothetic approach entails is its disconnection from scientific practice, concerning both the social and natural sciences, which usually put together their explanatory arguments on the basis of mechanisms, not laws. Science discovers patterns of behavior, but then goes on to try to find an explanatory argument that illuminates the patterns. This argument is precisely the mechanism that connects the correlated variables. Let us take for example the relationship between lung cancer and cigarette smoking. The correlation between these two variables is very strong: 95% of people with lung cancer are or have been smokers. Science not only intends to discover this covariational pattern, but also to explain it, and such explanation does not lie in looking for a law of nature, but in looking for one or several mechanisms. Following the example, in tobacco smoke several carcinogenic substances have been found. These substances produce specific alterations in cells, which end up proliferating abnormally and as a consequence they give rise to malignant tumors. Also, nicotine inhibits the functioning of certain receptors that suppress the growth of tumors.

Another reason why the NMP wishes to break with the nomothetic tradition is that its standard model of scientific explanation –the covering law model– provides *black box* explanations: we know the inputs and outputs, but we do not know the process by which when using these inputs we obtain such outputs. We know *what* works, but not *how* it works. In this regard, Bunge (1997, and also 1964, 1967, 1968, 1983) distinguishes between three types of theories or scientific hypotheses, according to their depth or explanatory power:

1. Black box, descriptive, or phenomenological, which answer only questions of the "What is it?" type.

2. Gray box, semiphenomenological, or semitranslucent, which give only sketchy or shallow answers to questions of the "How does it work?" kind.
3. Translucent box, mechanismic, or dynamical, which answer in detail questions of the "How does it work?" type.

The covering-law model provides *black box* explanations: it shows the correlation between two variables –inputs and outputs– but it does not show the way in which the output is produced from the set of inputs. The law of ideal gases belongs to the "black box" category since it shows the relationship between volume, temperature and gas pressure, but it does not specify the way in which the change in one variable contributes to the change in another. On the other hand, a *gray box* hypothesis represents the interiors of its referents in a schematic way. Network models of social systems fall into this category. They represent the composition and structure of social systems, and relate micro levels to macro levels. However, because they are static, such models cannot account for growth or decay, let alone any process of cohesion or decomposition. Finally, a *translucent box* makes explicit the mechanism by which, given some inputs, certain outputs take place. A translucent box theory is deeper than the rest because it does not just say what happens, but also what causes something to happen.

Now, what aspects characterize a mechanism? Leaving aside any definition or characteristic of mechanisms belonging to some specific discipline, there exist at least two fundamental notions of mechanisms implicit in almost all the approaches belonging to the NMP: *activity* and *automaticity*. In relation to the former, one of the most cited papers in the new mechanistic literature is Machamer et al. (2000), where they define mechanisms as "entities and activities organized such that they are productive of regular changes from start or set-up to finish or termination conditions" (Machamer et al. 2000, p. 3). According to Machamer and colleagues, a mechanism is composed of both *entities* and *activities*. Activities are the producers of change. They are not a mere description of the kind of changes that occur, but they are in fact responsible, in a causal sense, for the changes that take place in a mechanism. Entities, on the other hand, are the things that engage in activities. Entities have specific properties that make possible the exercise of specific activities.

Regarding the notion of *automaticity*, mechanisms are usually conceived as "automatic processes": once the triggering factor has been activated, a sequence of events will take place until it arrives at a certain result. Only one type of intervention is required: a manipulation at the start or set-up conditions. In this sense, the concept of automaticity means that there is no need to intervene permanently in order to achieve a desired result. The effect of aspirin's mechanism in organisms is a good example. Aspirin inhibits the production of prostaglandins (substances that "inform" the nervous system of the existence of pain or discomfort). Such inhibition brings about analgesic, anti-inflammatory and antipyretic effects. In this case, the intervention just consists in ingesting the aspirin. After this, a sequence will continue to hold until the final stage is reached. No other intervention is required *during* the sequence. *Automaticity* and *activity* are two concepts that go hand in hand. To the extent that activities are stable, the internal process of the mechanism will be able to operate without interruption until its final stage. In other words, the fact that activities are stable is a necessary condition of the mechanisms' automaticity.

24.3 Mechanisms' Invariant Behavior

Mechanisms are usually understood as systems whose behavior is stable, robust or invariant. Following Glennan (2002), this is what allows us to differentiate mechanisms from sequences of unique and rarely repeatable events. Specifically, Glennan (2002) proposes an approach to mechanisms that moves away from the ideas of Salmon (1984) and Railton (1978) to understand mechanisms as interconnected sequences of events. According to this idea, a mechanism is a chain of events related to a particular phenomenon, such as the chain of events that led to the breakage of a vase after a child bounced a tennis ball inside his house. Glennan calls this class of processes "fragile sequences": those that involve a succession of events that very rarely can happen again. In particular, the conditions that favored a certain result to take place (e.g., the breaking of the vase) are very restrictive, in the sense that any change in these conditions –no matter how small– can modify the final result. For example, a minimal change in the force driving the thrown ball can cause it to bounce in a different place than the initial one, thus making it follow a different path.

"Fragile" sequences are not the product of mechanisms. For example, the sequence of events that began with the visit of the archduke of Austria to Sarajevo, the murder of him and his wife, the Austrian demand to investigate the crime in Serbian territory, the refusal of Serbia (supported by the Russian Empire), the declaration of war of the Austria-Hungary empire to Serbia, and the consequent alliance of European countries on both sides that led to the First World War is not mechanistic, simply because it does not appeal to any mechanism in the explanatory argument. This sequence of events is "unique", in the sense that it is practically impossible to repeat it again. In contrast, the sequence of events that occur inside a car engine is a stable or robust process, and this is so since it is the result of the operation of mechanisms. The parts of a fragile sequence can be "robust", as were the Archduke and his wife, the city they attended, etc. Where there is no robustness is in the configuration of the sequence's parts, which, as said before, can be unique. Thus, mechanisms can be understood as interconnected sequences of events, although not every sequence of events is associated with mechanisms. For such a sequence to be mechanistic, a system must have a stable configuration of its parts (see Glennan 2002).

The notion of invariance and its relationship with mechanistic approaches has been conceptualized in different ways. One of these is Nancy Cartwright's "nomological machines" approach (Cartwright 1995, 1997, 1999). Since the world is both "messy" and "dappled", constant conjunction of events are observable only in rare circumstances, i.e., those where a particular system of components is properly "shielded" from external influences. Cartwright calls them *nomological machines*, which are defined as "a fixed (enough) arrangement of components, or factors, with stable (enough) capacities that in the right sort of stable (enough) environment will, with repeated operation, give rise to the kind of regular behavior that we represent in our scientific laws" (Cartwright 1999, p. 50).

A good example of what a nomological machine is are vending machines. They are machines whose repeated functioning gives rise to a law or a regular association between properties. This process begins after the customer inserts currency or a token into the machine and selects the wanted article. There are a series of mechanical processes inside the machine that end up with obtaining the selected product. The "law" emerges from the satisfactory and repeated functioning of the vending machine: *if X* (the coin inserted) *is to occur, Y* (the good obtained) *will take place.* However, for this to happen, the machine must be shielded or isolated from anything that might disturb the internal operation. This is precisely what happens with a vending machine: the mechanism is shielded from several (though not all) types of external influences.

Woodward (1996, 2002, 2003) has suggested a different way of conceptualizing this knowledge. According to the author, scientific knowledge should be based on generalizations that describe patterns of counterfactual dependence of a particular class, which Woodward calls "active" (Woodward 1996) or "interventionist" (Woodward 2002) counterfactuals. For this to occur, a generalization must be invariant under interventions in the independent variables.

If a generalization is *invariant*, it means that it is stable under changes in certain conditions or circumstances. However, invariance is not per se sufficient for using a generalization for interventionist, predictive or explanatory purposes. Instead, what is important is to obtain information that allows us to understand what would happen to the dependent variable once the independent variable was intervened physically (either by human agency or by a natural process).[2]

For instance, let's suppose an equation where rainfall is expressed in terms of the height of the mercury column in a barometer. Such equation will exhibit a regular behavior between the dependent and the independent variable. Nevertheless, it is not invariant under manipulations in the independent variable. In this juncture, Woodward asserts that the right way to recognize which generalization will be useful for scientific purposes is through the justification of active or interventionist counterfactuals. Such counterfactuals involve hypothetical interventions: "If an intervention on X were to occur (such that the value of X was modified), it would produce a change in Y". Thus, and contrary to the equation where rainfall is expressed in terms of the height of the mercury column in a barometer, a regularity that may properly justify interventionist counterfactuals should exhibit a relationship between atmospheric pressure (independent variable) and the height of the mercury column of a barometer (dependent variable). This interventionist

---

[2]For Woodward, such manipulations must be understood in terms of human agency only in a "heuristic" sense.

counterfactual would be expressed as follows: "If the atmospheric pressure was manipulated, the height of the mercury column in the barometer would change".

24.4 Concrete and Conceptual Systems

Within the NMP there is a considerable controversy about whether mechanisms are processes/entities of the real world or they are only analytical constructions. To understand this discussion, it is necessary to take into consideration the type of system where a mechanism takes place. To begin, Bunge (2004, p. 188) defines system as "a complex object whose parts or components are held together by bonds of some kind". Bunge (1997) offers a typology of systems, namely:

(a) *Natural*, such as a molecule or an organism;
(b) *Social*, such as a school or a firm;
(c) *Technical*, such as a machine or a TV network;
(d) *Conceptual*, such as a theory (hypothetical-deductive system) or a legal code; and
(e) *Semiotic*, such as a language.

Every kind of system is characterized by properties of its own, and one kind is not reducible to another, even though it may be composed of items of a different type (Bunge 1997, p. 415). The first three systems are associated with the material or real world. Bunge labels them "concrete systems". A *concrete system* is a bundle of real things held together by some bonds or forces, behaving as a unit in some respects and embedded in some environment. Atoms, molecules, crystals, stars, cells, multi-cellular organisms, ecosystems, cohesive social groups—such as families, firms, and entire societies—are concrete systems. So are all material artifacts.

By contrast, theories, classifications, and codes are *conceptual systems*, and systems of signs, such as languages, are semiotic ones. On the other hand, mere collections of items, even if they are of the same kind, are not systems, for they do not hang together. For example, cohorts, same-income groups, and social classes are not social systems but aggregates, best called "human groups" (Bunge 1997, p. 415). For the purposes of the present paper, two categories of mechanisms will be distinguished:

1. Mechanisms as entities or processes in conceptual systems.
2. Mechanisms as entities or processes in concrete systems.

In the first category there are advocates for the "analytical" conception (see, for example, Hedström and Swedberg 1998a), who consider that mechanisms are models that are useful for explaining concrete phenomena. Whereas authors such as Bunge (1997, 2004) and Glennan (1996, 2002, 2008) assert that mechanisms refer to causal contributions, where causal contributions are factors that do not operate in conceptual systems but in concrete ones.

Within the *analytical* approach it is argued that mechanisms are unobservable analytical constructions that provide hypothetical links between observable events (Hedström and Swedberg 1998b). Broadly speaking, given a regularity or observed phenomenon, what a scientist aims to discover is the mechanism that explains it. However, such a mechanism does not necessarily have to operate in the real world. On the contrary, it is a *plausible* mechanism, and its acceptance by the scientific community depends on its predictive and explanatory capacity.

Now, if mechanisms are understood as analytical constructions, then there may be infinite possible mechanisms to explain the same phenomenon, while different phenomena can be explained by the same mechanism. However, the choice among different mechanisms can never be guided by their true value: since they are models, they will always distort the reality they intend to describe. Therefore, such choice should be guided by how useful the various analytical models are for the purposes at hand. Since this class of mechanisms belongs to the field of conceptual systems, to speak about "mechanisms" is equivalent to speak about "mechanistic models". The construction of such mechanisms involves, in principle, only those elements that are believed to be essential for the current problem. Specifically, since the understanding of social phenomena is acquired little by little, we must abstract many details of reality and just focus on the relevant aspects that provide an adequate explanation (Hedström and Swedberg 1998b).

The "analytical" conception of mechanisms has been developed mainly in the field of the social sciences. In contrast, within the natural sciences the notion of mechanisms as entities or processes in *concrete* systems seems to predominate. Within the philosophy of science, one of the greatest critics of the "analytical" vision is Mario Bunge. For Bunge (1997, 2004), a mechanistic model is composed of (1) a mechanism (a process in a concrete system); (2) its composition (the set of parts of a concrete system); (3) the environment; and (4) the structure (the set of bonds or ties that hold the components of the concrete system together). Contrary to social mechanistic approaches like Stinchcombe (1991) and Hedstrom and Swedberg (1998b), for Bunge, mechanisms are not pieces of reasoning but pieces of the furniture of the real world (1997, p. 414).

The difference between conceptual and concrete systems can be understood through the notion of *causality*. In concrete systems, the relationship among parts of a mechanism is causal: we say that the growth of the monetary base brought about an increase in prices, or that fexofenadine inhibited the production of histamine. All these are phenomena in which parts are causally linked. However, in a conceptual system, relationships are not causal, but *logical*. Let's suppose the equation $Y = 2X$. If we assign different values to X, the value of Y will be modified at a rate of change of 2 to 1. However, there is no causal association; the relationship between X and Y is purely logical. In this regard Bunge says:

> Because change occurs only in concrete complex things, it makes no sense to talk about mechanisms in pure ideas or abstract objects, such as sets, functions, algorithms, or grammars, for nothing happens in them (when taken in and by themselves). In other words, the concept of a mechanism is alien to logic, mathematics, and general linguistics, none of which know of time. This is why logic, mathematics, and general linguistics explain nothing by themselves. (Bunge 1997, p. 418)

However, Bunge does not deny the possibility of modeling mechanisms. But unlike the analytical approach, which assumes that mechanisms are models, Bunge argues that models are just a representation of the real processes that takes place in concrete systems.

24.5 Possibility Trees

It has been argued above that several mechanistic approaches presuppose an ontological assumption that establishes that mechanisms are stable processes or systems, where such stability is observed both in the mechanism composition and in its behavior. However, this notion of stability or invariance is unlikely in the social realm. In order to argue this point, let us take the dualistic ontology proposed in Machamer et al. (2000) and in Machamer (2004). According to those authors, causal relationships are mediated by *activities*. In the social realm, *activities* are plainly individuals' decisions and actions. These activities are influenced by several factors like the socio-cultural sphere, the information agents receive from the world, the expectations they form about the evolution of certain variables, etc. Depending on what types of activities people carry out, different results will be obtained.

At this juncture, an interesting difference in relation to the approach adopted by Machamer and colleagues can be mentioned. In this approach "activity" refers to a *singular* notion: once a factor F is triggered, an activity A will start to operate. Such an activity is involved in the production of a result R. However, in the social sciences activities are not singular but "plural": once a factor F is triggered, a set of potential activities $(A_1, A_2, \ldots, A_n)$ may start to work. Each of these activities is associated with the production of a different result $(R_1, R_2, \ldots, R_n)$ The prevailing result will depend on several factors, namely how people form their expectations, how they interpret the information from the economic and political world, the socio-cultural institutions, the changes in the social structure, etc. In other words, the basic structure of social phenomena responds to the logic of *possibility trees* or *open-ended results*: when a causal factor is activated, there are multiple possible decisions. Depending on which actions people carry out, different results will be achieved (Ivarola 2017; Ivarola et al. 2013).

In economics, a good example that represents the logic of possibility trees is the so called "Keynes effect". According to this, an increase in money supply leads to a decrease in the interest rate, stimulating investment and consequently employment and production. In some interpretations it is assumed that an increase in the quantity of money leads automatically and in a stable way to an increase in employment and production. However, this is not true. On the contrary, depending on the contextual framework and people's interpretations and expectations, different causal paths are plausible. Aware of these limitations, Keynes says:

For whilst an increase in the quantity of money may be expected, *cet. par.*, to reduce the rate of interest, this will not happen if the liquidity preferences of the public are increasing more than the quantity of money; and whilst a decline in the rate of interest may be expected, *cet. par.*, to increase the volume of investment, this will not happen if the schedule of the marginal efficiency of capital is falling more rapidly than the rate of interest; and whilst an increase in the volume of investment may be expected, *cet. par.*, to increase employment, this may not happen if the propensity to consume is falling off. (Keynes 1936, p. 155)

In the Keynes effect, economic variables are linked through people's activities. In other words, these actions are responsible for enabling the causal link between economic variables. Nonetheless, people's activities are not linear, but they may change for lots of reasons. When this happens, a modification in the causal links will be observed. Since these changes are plausible, the feasibility of speaking about stable causal factors in the social realm can be questioned. Particularly, there are two notions that will be put into question: *invariance* and *automaticity*.

Regarding a mechanism's invariance or robustness, it has been shown that this is a consequence of the stable behavior of its constituent parts. In other words, a mechanism's behavior is invariant because the activities carried out within such mechanism are invariant. Let us think of any biological mechanism. The mechanism's internal activities are not unstable; on the contrary, they denote a high degree of robustness. In contrast, the activities that take place in social phenomena are people's actions, which tend to be very volatile. For example, any change in the interpretation of the signals from the world may result in a significant change in the process of expectations formation; any institutional change may re-direct the people's course of action towards other paths; etc. And since the activities carried out within a social mechanism are not stable or invariant, the overall mechanism's behavior will not be invariant either.

Here it is important to emphasize that social phenomena do not belong to either of the two kinds of processes mentioned by Glennan (2002). They are not *robust* sequences because in each stage of the sequence different future outcomes are plausible. Given a sequence A→B→C, we say that it is robust when the arrows denote invariant relations between parts: "most of the time in which A occurs, B will occur". The same happens between B and C. However, if this sequence refers to a social phenomenon, then once A occurs, B, B′ or B∕∕ may occur. In a robust sequence, the result is almost always the same. But this is not the case for social phenomena. The final result will depend on the specific circumstances that occur at a particular moment.

Nevertheless, the above argument does not imply that it is not possible to observe regular sequences at the level of social events. Not only are they plausible, but also history tells us about their feasibility. This is the reason why it is not correct to label social phenomena as *fragile* sequences. Remember that a fragile sequence is a unique and unrepeatable sequence, like the path that a tennis ball follows after being bounced into a room. There is the chance that social phenomena involve regular behaviors. However, this not because they are the consequence of transfactual mechanisms operating in concrete systems, but because there is low volatility in the process of expectations formation, stable institutional arrangements over time, etc.

Let us take as an example the case of the "Phillips curve". The economist William Phillips (1958) published an article entitled "The Relation between Unemployment and the Rate of Change of Money Wage Rates in the United Kingdom, 1861–1957", in which he asserted the existence of a negative correlation between the rate of unemployment and inflation. Two years later, Samuelson and Solow found the same statistical record for the United States between 1900 and 1960. The explanation for this inverse correlation between inflation and unemployment is straightforward: given the expected prices, a reduction in unemployment causes a rise in nominal wages, which in turn brings about a rise in prices. Therefore, a reduction in unemployment causes a rise in prices in comparison to last year, that is, an increase in inflation.

However, this regularity broke down between 1960 and 1970. The inflationary process was accentuated in those years, which modified the way in which people formed their expectations. During periods of low and non-systematic inflation both workers and companies tended to ignore past inflation, so they assumed that in the coming periods prices would not differ significantly from current prices. But in a framework of systematic inflation, agents began to assume that in the next periods prices would continue to rise, which caused the inverse relation between inflation and unemployment to disappear. The example shows that if certain conditions remain stable –in this case low inflation expectations– a regularity may emerge at the level of events. However, this is not the consequence of robust relations among mechanisms' elements but of a set of specific circumstances with limited temporal validity. When these circumstances or conditions are no longer met, the regularity will disappear.

Another condition that social processes do not meet is *automaticity*. There is a certain consensus in conceiving mechanisms as automatic processes: by activating a certain causal factor, a sequence of events will start to work until the expected result is achieved. For example, Machamer and colleagues define mechanisms as processes that go from set-up or start conditions to finish or termination conditions. Only one type of intervention is required: at the start conditions. Unless significant disturbing factors operate, this intervention guarantees that the mechanism's final stage will be reached *automatically*, that is, without any kind of interruption or deviation.

In this sense, the notion of automaticity means there is no need to intervene successively in order to achieve a desired result. Let us suppose some examples of mechanisms in the natural sciences: the ingestion of ecstasy produces an increase of serotonin, which causes euphoria. Aspirin inhibits the production of prostaglandins (substances that "inform" the nervous system the presence of discomfort), generating analgesic, anti-inflammatory and antipyretic effects. In both cases, interventions are understood as the ingestion of ecstasy or aspirin. They activate a sequence that will continue without interruptions until its final stage. Unless their effects are counteracted by other causal factors, additional interventions will not be necessary.

The idea of automaticity does not only concern the field of natural sciences; it is also present in many social mechanisms. Let us take for example the economic mechanism called "Pigou effect": a decrease in prices raises the real value of liquid assets, which increase the real value of consumers' wealth and as a consequence the consumption of goods and services. Only one –and not necessarily anthropomorphic– type of intervention is required: that prices go down. The activation of demand will come automatically.

Contrary to this position, it is easy to realize that social processes can be *interrupted* (e.g., they stop at some intermediate stage of the sequence), as well as they can deviate from the intended target. The Keynes effect mentioned above clearly illustrates this lack of automaticity in social processes. It is wrong to think that a positive change in the real amount of money will *automatically* lead to a decrease in the interest rate, this to an increase in investment, and therefore to an increase in the level of employment and national income. On the contrary, depending on the contextual framework and on the people's interpretations a social process will be able to take different ways. Think of this in terms of *interventions*. In an automatic process only one intervention is required: at the starting or set-up conditions. In a process like the Keynes effect, more than one intervention will have to be necessary in order to reach the expected result. Thus, it is highly probable that in order to achieve an increase in national income economic authorities should intervene not only in the amount of money, but also generating signals that positively influence expectations formation both of consumers and entrepreneurs.

Neither are social processes the product of nomological machines. In building a nomological machine it is necessary that its constituent elements be correctly assembled and protected from any perturbing factors, so that whenever X is to occur, Y will be the obtained result. Let's assume for a moment that it is possible to set up the machine in the social realm. The problems will arise when such machine is intended to be protected from perturbing factors. In a vending machine the internal mechanism is protected from many external factors. This allows it to exhibit a regular behavior, in the sense that (almost) every time coins are inserted into the slot the selected product is provided by the machine.

However, it is a bit unrealistic to think that nomological machines are plausible in the social realm. To begin with, social processes are the kind of phenomena that take place in open systems (see Lawson 1997). That means any unexpected exogenous factor can disrupt the functioning of the machine. In such systems uncertainty prevails, so we cannot know for sure what will happen in the near future. We can predict with some confidence the occurrence of certain events, but they are not guaranteed at all. Even worse, we do not even think about the factual possibility of a myriad of eventual phenomena. If we are not able to know what factors are to occur in the future, then it seems inappropriate to try to *shield* a machine from unknown factors. How we can nullify or counteract the effect of something we do not know?

Even if we assume the feasibility of shielding a nomological machine, there is the chance that, because of "endogenous" problems, such machine may yield unstable results. For instance, the Keynes effect may be understood as a nomological machine provided that certain antecedent conditions are met. Let's suppose that the policy maker has been doing everything he can to make it work in the real world. If so, then a high positive correlation between money supply and national income will be observed. However, let's suppose now that, at some point during the year, a little group of businessmen believes that they could sell much less than expected, and because of this the level of investment decreases. Suppose further that this strategy spreads towards other businessmen. The wider the scope, the greater the negative effect on employment and production. Clearly, such instability is not the result of failures in the shielding or in the assembling of parts. Rather, it has its origin in an "endogenous" problem.

24.6   *Theoretical* and *Material* Mechanisms

In *The Dappled World: A Study of the Boundaries of Science* (1999), Cartwright addresses her critique to the nomothetic approach to science. According to the latter, the real world is an ordered world where law-like regularities prevail. This is the legacy of those philosophers of science who took physics as a model to follow. Cartwright is opposed to this vision of science. For the author, a good part of what happens in nature happens not by the operation of laws but by chance. Because of this lack of laws, the world is both "messy" and "dappled". There is no guarantee that a regularity that prevails in one period will also prevail in the next one. The only way to account for regularities is through the fulfillment of very specific conditions. For this reason, Cartwright addressed her analysis to the study of capacities (or natures) and nomological machines.

There is some compatibility between Cartwright's "dappled world" approach and the approach developed in the present paper. In the first place, the real world is understood as a system where there are no per se regular connections or constant conjunctions of events. If these appear, it is because certain factors remained invariant at that moment of time. Also, the notion of "dappled world" emphasizes the importance of the domain or range of applicability in which the scientific models are understood. More precisely, according to this notion it is possible that several theories –apparently contradictory– are simultaneously "true", as long as they belong to completely different domains. It is interesting to highlight this aspect, since if we understand social processes as derivations of possibility trees, then the drive to develop a general theory becomes irrelevant. Contrary to this idea, for each situation there is a model that will provide information about the conditions that would be needed to meet in order to reach a specific result.

Let's illustrate this situation with a simple example. Figure 24.1 shows the possible alternatives that follow from the activation of a node. Both the alternatives and the nodes represent social phenomena, such as the increase in public spending, racial segregation in neighborhoods, etc. In order to reach any of these nodes it is necessary to meet certain conditions S_i. Let's now suppose that we are in node A and we want to arrive at node X. We have two ways:

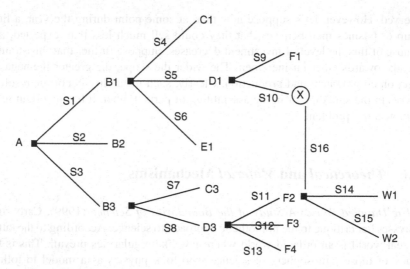

Fig. 24.1 Possibility tree

$$A \rightarrow B1 \rightarrow D1 \rightarrow X$$

$$A \rightarrow B3 \rightarrow D3 \rightarrow F2 \rightarrow X$$

Let's suppose that we choose the first option. To do this, we must first get to B1, and to get to it, we must meet conditions S1. This is important, since if we did not do it we could go to B2 or B3. Let us suppose that these conditions are satisfied and we arrive at B1. The next step is to go to D1. For this to happen, conditions S5 must be fulfilled. Otherwise, we could end up in C1 or E1. Finally, once in D1 we should meet conditions S10 to get to X. Nevertheless, if some of all these conditions cannot be established in the real world, we could opt for the second option. In this case, to reach B3 conditions S3 should be fulfilled; to reach D3, conditions S8 should be met; to arrive at F2, conditions S11 should be fulfilled; and finally, to culminate the journey at X, conditions S16 should be met.

In the standard mechanistic approach, it is assumed that there is a pre-established and stable connection between cause and effect, e.g., between aspirin and headache-relief, between public expenditure and national income. Contrary to this, the present paper argues that there is no such connection in the social realm. By increasing public spending, the result may not be economic growth but inflation. By increasing the amount of money, the result might not be the purchase of bonds but hoarding. Certain specific conditions must be fulfilled so that the connection between two variables is fruitful. This contrasts sharply with the notion of mechanisms as stable contributions. In this approach the connection is already given; all we must do is trigger the right factor and the mechanism will start to work. In contrast, in a social process there is no given or pre-established connection. It all depends on the characteristics or background conditions that prevail at the moment a causal factor is triggered.

Therefore, if we understand social phenomena under the logic of possibility trees several social sciences models will not be representations of mechanisms that work regularly in the social realm. The models obtained are, instead, *blueprints* to close the possibility tree in order to reach a certain result. In this regard, a model construction consists in designing a blueprint B. Based on a set of propositions P, an outcome R is inferred. The set of propositions P fulfills the role of closing the possibility tree towards a certain path that ends with the derivation of R.

Given the myriad of possible scenarios, a blueprint may be used both for knowing those conditions that lead to a certain result, and for restricting the feasibility of other possible results. Likewise, it is highly probable that more than one path or alternative to get the final result may exist. Thus, there will be more than one blueprint that leads to the desired goal. When this occurs, a choice among them will have to be made. By way of example, let's suppose that government authorities want to increase the national income through an increase in the aggregate demand. One possible way is the Keynes effect: to the extent that the money supply is increased, a decrease in the interest rate will take place. This change will stimulate investment and consequently employment and production. Nonetheless, for this to happen certain conditions have to be met: that the increase in the liquidity preference is lower than the increase in the quantity of money, that entrepreneurs form good expectations about their future sales, that the marginal propensity to consume does not decrease, etc. If some of the conditions specified in the Keynes effect are not met in the real world, then the model will not be a good blueprint for this particular policy. Fortunately, this is not the end of the story, but other ways can lead to an increase in national income. For example, if the marginal propensity to consume is high, then it is likely that government authorities design a blueprint where aggregate demand will increase through an increase in consumption. Finally, if for different reasons this path is not feasible either, the policy maker can find a way through an increase in the public expenditure, as this is another component of the aggregate demand.

What kind of knowledge do the blueprints provide? In what follows it is argued that such knowledge does not express a *causal* but a *logical* contribution. In doing so, it is important to recall Bunge's distinction between concrete and conceptual systems. Blueprints may be understood as thought experiments where one creates a model for isolating causal relationships in a theoretical way (see for instance Mäki 1992, 2011). More precisely, if conceived as an isolated causal contribution this could imply that, under very specific circumstances (those that neutralize the action of perturbing factors), C will *necessarily* produce R. And in a more general sense, C will contribute to bring about R, although there exists the possibility that R does not appear if a set of disturbing factors occur.

Contrary to this, in the present paper it is argued that the model or blueprint contribution is purely *potential*: if the set of conditions C is met in the real world, an outcome R could emerge. When saying that the contribution of C to R is potential, the argument here is that the relationship between C and R is logical.

However, there is the chance that such relationship may become causal. For this to occur, the passage from the conceptual to the concrete system must be effective. In relation to this, it must be noted that what is asserted in the blueprint must be (approximately) met in the real world.[3] For instance, if a policy designed on the Keynes effect is expected to be implemented by the policy makers, then certain conditions (e.g., there are constant returns to scale, the marginal propensity to consume remains constant throughout the changes in income, changes in primary employment in the investment industries bring about proportionally greater changes in total employment, etc.) must be met.

Therefore, we are not speaking about stable causal factors here, as it may be the case that C does not contribute to R but to some other (perhaps undesired) results. The relationship becomes causal once the mechanism successfully operates in the real world. The relationship between money supply and national income postulated in the Keynes effect is only a logical connection; it will turn to a causal relationship when, once the policy has been implemented, a connection between money supply and national income is observed in the real world.

Thus, social mechanisms should be understood as blueprints of processes that could take place in the real world. They do not refer to something that actually happens, but to something that *could* happen if the conditions stipulated in the blueprint were met in the real world. These models inform us not about stable causal contributions, but about logical relationships that *could become* causal. This is not a criticism of the NMP, but only of an ontological assumption that predominates in several approaches of the NMP. It should be noted that, according to this new perspective, the notion of mechanism is not put into question. In other words, the existence of social mechanisms is not denied; what is denied is that these mechanisms are stable contributions. Bunge's distinction between conceptual and concrete systems is useful for differentiating two kinds of mechanisms: *theoretical mechanisms* – those that are modeled or belong to the conceptual system, and that their existence in such a system is not a sufficient condition of their existence in the real world – and *material mechanisms* – those that take place in the concrete systems. According to Bunge, the difference between one and the other is due to the type of connection between their interacting parts: while in theoretical mechanisms their parts are related logically, in material mechanisms such relations are causal.

24.7 Conclusion

In the present article it has been argued that within the social sciences the logic of invariance or stable causal factors –a widespread notion in several approaches to the New Mechanistic Philosophy – should be replaced by the logic of possibility trees

---

[3]The term "approximately" is used because the blueprint restrictions are idealizations, and as such they cannot be fully reproduced in the real world.

or open-ended results. Since people have several possible courses of action and do not always act in the same way, it is naive to think that social mechanisms will be as stable as natural sciences' mechanisms. Also, and on the basis of Bunge's distinction between conceptual and concrete systems, it has been shown that there are two ways of understanding social mechanisms: as theoretical mechanisms –those whose parts are logically linked– and as material mechanisms –those whose parts are causally linked. Knowledge of theoretical mechanisms only tells us about something that could occur in the real world. However, there is no safe bridge from the logical links to the causal ones, that is to say, from the conceptual to the concrete system. For this to be effective, the conditions stipulated in the theoretical mechanisms must be met.

References

Bechtel, W., & Abrahamsen, A. (2005). Explanation: A mechanist alternative. *Studies in History and Philosophy of the Biological and Biomedical Sciences, 36*(2), 421–441.

Bunge, M. (1964). Phenomenological theories. In M. Bunge (Ed.), *The critical approach to science and philosophy: In honor of Karl R. Popper* (pp. 234–254). New York: Free Press.

Bunge, M. (1967). *Scientific research, 2 vols.* Berlin: Springer.

Bunge, M. (1968). The maturation of science. In I. Lakatos & A. Musgrave (Eds.), *Problems in the philosophy of science* (pp. 120–137). Amsterdam: North-Holland.

Bunge, M. (1983). *Treatise on basic philosophy: Vol. 6. Understanding the world.* Dordrecht: Reidel.

Bunge, M. (1997). Mechanism and explanation. *Philosophy of the Social Sciences, 27*(4), 410–465.

Bunge, M. (2004). How does it work? The search for explanatory mechanisms. *Philosophy of the Social Sciences, 34*(2), 182–210.

Cartwright, N. (1995). Ceteris Paribus Laws and Socio-economic machines. *The Monist, 78*(3), 276–294.

Cartwright, N. (1997). Models: The blueprints for Laws. *Philosophy of Science, 64,* Supplement. Proceedings of the 1996 Biennial Meetings of the Philosophy of Science Association. Part II: Symposia Papers, S292–S303.

Cartwright, N. (1999). *The dappled world.* Cambridge: Cambridge University Press.

Darden, L. (2006). *Reasoning in biological discoveries.* New York: Cambridge University Press.

Elster, J. (1989). *Nuts and bolts for the social sciences.* Cambridge: Cambridge University Press.

Gerring, J. (2008). The mechanismic worldview: Thinking inside the box. *British Journal of Political Science, 38*(1), 161–179.

Glennan, S. (1996). Mechanisms and the nature of causation. *Erkenntnis, 44*(1), 49–71.

Glennan, S. (2002). Rethinking mechanistic explanation. *Philosophy of Science, 69*(S3), S342–S353.

Glennan, S. (2008). Mechanisms. In S. Psillos & M. Curd (Eds.), *The Routledge companion to philosophy of science* (pp. 376–384). Abingdon: Routledge.

Hedström, P., & Swedberg, R. (Eds.). (1998a). *Social mechanisms. An analytical approach to social theory.* Cambridge: Cambridge University Press.

Hedström, P., & Swedberg, R. (1998b). Social mechanisms: An introductory essay. In P. Hedström & R. Swedberg (Eds.), *Social mechanisms: An analytical approach to social theory* (pp. 1–31). Cambridge: Cambridge University Press.

Hempel, C., & Oppenheim, P. (1948). Studies in the logic of explanation. *Philosophy of Science, 15*(2), 135–175.

Ivarola, L. (2017). Socioeconomic processes as open-ended results. Beyond invariance knowledge for interventionist purposes. *Theoria. An International Journal for Theory, History and Foundations of Science, 32*(2), 211–229.

Ivarola, L., Marqués, G., & Weisman, D. (2013). Expectation-based processes. An interventionist account of economic practice. *Economic Thought, 2*(2), 20–32.

Keynes, J. (1936). *The general theory of employment, interest and money*. New Delhi: Atlantic.

Lawson, T. (1997). *Economics and reality*. London: Routledge.

Machamer, P. (2004). Activities and causation: The metaphysics and epistemology of mechanisms. *International Studies in the Philosophy of Science, 18*(1), 27–39.

Machamer, P., Darden, L., & Craver, C. (2000). Thinking about mechanisms. *Philosophy of Science, 67*(1), 1–25.

Mahoney, J. (2002). *Causal mechanisms, correlations, and a power theory of society*. Paper presented in the American Political Science Association Meetings, Boston. http://www.allacademic.com/meta/p66368_index.html. Accessed 15 Oct 2017.

Mäki, U. (1992). On the method of isolation in economics. In: C. Dilworth (Ed.), *Idealization IV: Intelligibility in science, Poznan Studies in the Philosophy of the Sciences and the Humanities* (Vol. 26, pp. 319–354). [Reprinted in: Davis, John (ed.) (2006) Recent Developments in Economic Methodology, Volume 3. Edward Elgar, 3–37.]

Mäki, U. (2011). Models and the locus of their truth. *Synthese, 180*(1), 47–63.

Phillips, W. (1958). The relation between unemployment and the rate of change of money wages in the United Kingdom, 1861–1957. *Economica, 25*(100), 283–299.

Railton, P. (1978). A deductive-nomological model of probabilistic explanation. *Philosophy of Science, 45*(2), 206–226.

Salmon, W. (1984). Scientific explanation: Three basic conceptions. *PSA: Proceedings of the Biennial Meeting of the Philosophy of Science Association*, 1984(2). Symposia and Invited Papers, 293–305.

Skipper, R., & Milstein, R. (2005). Thinking about evolutionary mechanisms: Natural selection. *Studies in History and Philosophy of Biological and Biomedical Sciences, 36*(2), 327–347.

Stinchcombe, A. (1991). The conditions of fruitfulness of theorizing about mechanisms in social science. *Philosophy of the Social Sciences, 21*(3), 367–388.

Suppe, F. (Ed.). (1977). *The structure of scientific theories*. Urbana: University of Illinois Press.

Woodward, J. (1996). Explanation, invariance, and intervention. *Philosophy of Science, 64*(Proceedings), S26–S41.

Woodward, J. (2002). What is a mechanism? A counterfactual account. *Philosophy of Science, 69*(S3), S366–S377.

Woodward, J. (2003). *Making things happen: A theory of causal explanation*. Oxford: Oxford University Press.

Chapter 25
Cultures as Semiotic Systems: Reconceptualizing Culture in a Systemic Perspective

Andreas Pickel

Abstract This paper incorporates a broad understanding of culture into Mario Bunge's systemist philosophy. Cultures are viewed as semiotic systems. Semiotic systems are symbolic systems along with their users. This approach makes it possible to relate immaterial symbolic systems to real material social systems via semiotic systems, which are neither purely material nor purely ideal, but combine elements of both. The goal is to sketch an ontology of culture that is consistent with emergentist materialism, and useful for the integration of culture into social science analysis.

25.1 Why Semiotic Systems?

This paper reformulates the problem of culture and society in systemic terms. The long-standing debate in the social sciences on culture is complex, controversial and often conceptually confused. In some approaches, culture is everything, in others it is nothing. In between, there are many more or less interesting and fruitful approaches. How should culture be viewed in a systemic perspective? And why bother in the first place? Many basic questions in the debate on culture can be addressed in a conceptually different way in the systemic perspective proposed here. By integrating culture into a realist and materialist ontology, the study of culture becomes part of the scientific enterprise. How can any serious conceptualization of culture as ideas, narratives, scripts, schemas, repertoires, that have a real impact on social life be developed with an ontology that is anti-constructivist (i.e. realist) and anti-idealist (i.e. materialist)? Having tried to follow Mario Bunge's systemist philosophy for over two decades in various explanatory contexts, but above all in the sociology and political economy of national culture, I have grappled with this philosophical issue in practical terms (Pickel 2003, 2004b, 2013).

A. Pickel (✉)
Political Studies, Trent University, Peterborough, Canada
e-mail: apickel@trentu.ca

© Springer Nature Switzerland AG 2019

M. R. Matthews (ed.), *Mario Bunge: A Centenary Festschrift*,
https://doi.org/10.1007/978-3-030-16673-1_25

415

Bunge has a central place for culture in his systemism (1981, 1996, 1998), but I found his "sociological" conception, which he contrasts with a broad anthropological conception, too narrow. The scholarly study of culture has made significant strides over the past two decades. This has occurred without shared philosophical presuppositions or an explicit framework. Instead, a number of concepts have been proffered that facilitate the empirical study of key dimensions of "culture" – culture as practice (Bourdieu 1990; Lizardo 2011), culture as frames (DiMaggio 1997), culture as narratives (Franzosi 1998; Wierzbicka 2010), culture as cultural capital (Bourdieu 1990, 1991; Lizardo 2005), culture as schema (Nishida et al. 1998), cultural repertoires (Swidler 2001), cultural scripts (Goddard and Wierzbicka 2004), ideocultures (Fine 2012), organizational cultures (Morrill 2008); culture as discourse (Schmidt 2008), collective imaginaries (Taylor 2004), social representations (Jodelet 2008), and others. As Lamont and Small write:

> Readers will note that we have not defined "culture," at least not explicitly. We have taken this approach because the literature has produced multiple definitions, and consensus is unlikely to emerge soon. Given this state of affairs, the best approach is a pragmatic one. Today, many cultural sociologists have examined empirical conditions using specific and (often) well-defined concepts, such as frames or narratives, that in one way or another are recognizable as "cultural." (Lamont and Small 2008, pp. 76–77)

Building on previous studies and in pursuit of a systemist agenda (Pickel 2004a, 2006, 2007), this paper will eschew the pragmatic route, proposing instead basic elements of a philosophical foundation of culture.

Bunge's view of culture as one subsystem of society alongside economic and political subsystems does not capture what many in the social sciences with an interest in the role of culture believe to be of central importance. In contrast to Bunge, this paper will argue that cultures should not just be viewed as a particular kind of social system that creates and reproduces culture, such as scientific and educational institutions or the arts. A conception of culture should incorporate the many languages and codes, natural and artificial, that are used in all social systems, not just in so-called cultural systems.

Bunge argues that such a broad use of the concept, where we speak of economic cultures, political cultures, or organizational cultures – that is, where culture is everywhere – leads to conceptual problems. One of his arguments is that if we speak of culture in this broad sense, we will get entangled in paradoxes such as speaking about the culture of a culture and lose our conceptual bearings. Of course, cultural systems such as universities not only create and reproduce certain cultural products and practices, they also *have* cultures in the broader sense, some of which they share with other economic and political institutions. Conceptually, that does not seem to pose an insurmountable problem. It should therefore be possible to develop and use such a broad conception of culture while retaining Bunge's systemic philosophy based on a comprehensive and explicit ontology. This is the task of this paper.

Systemism, as propounded by Bunge, claims that the world is a system of systems. Systemism is materialist, but it defends an emergentist materialism. That is, while all real or concrete systems like atoms, cells, and societies are material or physical, many material systems are physical but not just physical – they are chemical systems, or biological systems, or social systems, each with an additional layer of particular and irreducible emergent properties. Human social systems have evolved languages, above all, natural languages – the most fundamental type of semiotic system. In addition to natural languages, there is a myriad of artificial languages – symbolic systems such as religions and writing, technical instructions and organizational cultures. Symbolic systems are conceptual and thus immaterial, but they are parts of real systems by way of their users. Symbolic systems along with their users are referred to as semiotic systems (Bunge 2003). In other words, semiotic systems are a combination of material and symbolic, or real and conceptual, systems. This basic definition is the core of my proposal for a systemist conceptualization of culture – culture as semiotic systems.

25.2 What Are Semiotic Systems?

Social systems are systems whose components are gregarious animals, in particular but not exclusively human beings (Bunge 2003). Semiotic systems are a combination of symbolic systems and material systems. *More precisely: semiotic systems are symbolic systems along with their users.*

In the systemist ontology, all systems, whether material or conceptual, natural or artificial, are defined in terms of their composition, environment, and structure (CES). Thus, symbolic systems are composed of symbols (like words, musical notes, graphics) (C), their symbolic context or environment (the symbolic reference frame in which they are interpreted) (E), and the relations between those symbols, i.e. their structure (the rules according to which they are ordered) (S). Material systems, in addition to composition, environment and structure, have mechanisms (M), i.e. real or material processes that make them work. In symbolic systems, no material processes occur. Symbolic systems are conceptual systems rather than natural systems: they are constructed, human-made. They originate in biopsychosocial systems, that is, they are constructed by humans, though not necessarily by design. Once they exist as artefacts, they can be used and further developed by anyone who learns how to use the symbolic system. In the absence of knowledgeable users, symbolic systems are defunct or "dead" even if their material carriers (stone, paper, digital file) survive. A semiotic system, to repeat, is composed of a symbolic system as well as its users.

25.2.1 A Few Examples for Initial Clarification

Are there any symbolic systems that are not part of semiotic systems? In general, since symbolic systems cannot function or exist without the activity of human brains, only dead or obsolete symbolic systems (former semiotic systems) should be considered "without semiotic systems". Such symbolic systems are thus lost symbolic systems. Archeologists, anthropologists, linguists and other specialists may however work on reviving dead symbolic systems. They become the "new users" by trying to reconstruct how "traditional users" applied such symbolic systems.

Even though symbolic systems cannot exist without semiotic systems (i.e. users), the distinction between symbolic systems and semiotic systems is of fundamental importance. From within a semiotic system, the corresponding symbolic system is treated as real by its users. Whether a natural language, musical score, or a legal code, they are "fused" with its various kinds of users when they are being used, practiced, enacted. A written natural language needs to be spoken, a musical score has to be composed and performed, a legal code has to be drawn up, socially established, and administered. "Fusion" here is simply a metaphor for the act of learning, practicing, and further developing a symbolic system. Invention is of course a prior step, though in the case of complex symbolic systems many individual actors over time contribute to what later can be seen as the invention of a symbolic system. Such a process may be spontaneous, unintended; it may be planned and by design; or some of both.

Most of us in the second decade of the twenty-first century are familiar with the use of apps and software in our digital devices. Every app is a designed symbolic mini-system, invented by specialists for users with a specific purpose. Before its release, the app is a semiotic mini-system with few users other than its inventors and testers. The advantage of apps over previous generations of software programs is their ease of use. Apps follow certain templates with which most potential users are familiar. Not all functions in the digital universe can be packaged in simple apps. A program like Adobe Photoshop, for example, which provides advanced editing functions for digital images, has a much steeper learning curve. In order to become a user, you have to invest time in acquiring the corresponding skills. The distinction between "knowing" a symbolic system and "knowing how to use it" will be addressed again below.

"Symbolic system", as the phrase is used here, refers to sets of symbols, symbolic fragments, symbolic practices and symbolic contexts.[1] Unlike social systems, symbolic systems are not real, their elements are held together by purely symbolic bonds. Symbols "become" real when they are part of brain processes (subjective reality) and when they are commonly held or known by a group of people (intersubjective or social reality). For the symbolic content of particular brains to be meaningful, it has to be embedded in symbolic contexts. Individuals may have a perfect, near-perfect, high, intermediate or low command of a language, e.g. English. But regardless of their level of linguistic competence, we assume that individuals are practicing the same language, i.e. we think they are all speaking

---

[1]Bunge's definition of the concept of a system is as follows: "Complex object every part or component of which is related to at least one other component. Examples: an atom is a physical system composed of protons, neutrons, and electrons; [. . .] a valid argument is a system of propositions held together by the relation of implication and the rules of inference; a language is system of signs held together by concatenation, meaning, and grammar" (Bunge 2003, p. 282).

English. We come to this conclusion because we have in mind a "perfect" model of the English language (a symbolic system), with respect to which we can assign degrees of competence. Such a formal linguistic classification is an abstraction from the practice of actual speakers in real social contexts, yet the idea that languages exist as such is widely held. We will return below to this common kind of "reality illusion" – a basic semiotic mechanism.

25.2.2 Natural Languages

A natural language is the most important and complex type of semiotic system in social life. In evolutionary terms, the emergence of natural languages in the species has been a precondition for the emergence of other semiotic systems. In terms of individual development, a natural language is the first symbolic system humans learn. It is a precondition for learning most other symbolic systems since a natural language functions as a master frame or meta-system. Language acquisition, a biological capacity, opens the door to learning a specific semiotic system (our "native culture"), with the help of which we are able to acquire other semiotic systems.

The power and significance of natural languages for all of social reality is documented and theorized in social linguistics, linguistic anthropology, discourse studies, and other empirical research areas concerned with the relations between language and society. The fundamental importance of natural language for social reality is therefore not at issue. Language is both carrier, facilitator, and component of "culture", though there is no consensus on how to conceptualize culture (Lamont and Small 2008). In the systemic perspective developed here, natural languages as well as other symbolic systems can be said to exist only as semiotic systems – as symbolic systems along with their users. Semiotic systems should be clearly distinguished from social systems as a precondition for their conceptual fruitfulness in the discussion of "culture and society", as will be further discussed below. The fact that the concept of semiotic system contains both real material components (users) as well as immaterial conceptual components (symbols) creates the ontological space for a concept of "culture" that is neither purely social nor purely symbolic. The two conceptual poles represent the basic ontological tension characteristic for any non-reductionist perspective on culture.

The members of any social system use a variety of semiotic systems. Some are more general, such as those related to status and gender, ethnicity, and class. Others are more specific, such as those related to place, family, and profession. The most fundamental semiotic system, as just noted, is a natural language. It is usually shared by most members of a state-society[2] where it is an official language, though not

---

[2]"State-society" refers to modern societies contained in a territorial state, the standard sociopolitical system globally today.

necessarily by all. While the semiotic mechanism of a shared natural language is crucial for any social order, social integration is possible without a shared natural language. Think of a tribe capturing slaves from another tribe with a different natural language. Or colonial societies importing populations from other continents. Or modern societies with linguistic minorities. The primary ties in these social systems (tribe, colonial society, modern society) are not semiotic – though of course they will have some ways of communicating with non-speakers of their language, if only with the use of a club or a rifle. The fundamental ties are political (e.g. domination), economic (e.g. exploitation), and biological (e.g. rape).

25.2.3   Semiotic Systems and Technology

Technological systems, whether for communication or other purposes, have built-in symbolic systems, from construction blueprints to user instructions. A cell phone network is a techno-social-symbolic system, though we usually take the social and symbolic elements in technological systems for granted. We purchase a piece of technology since we assume that it is a tool for social communication based on symbolic systems such as icons, texts, and language. We experience the cell phone as the key material reality because, like a book, it is the physical thing with which we work. We tend not to be aware of the technological infrastructure behind it, unless it is malfunctioning; or of the semiotic systems from social networks to codes for text and talk, unless we fail to understand them or don't know how to make them work. In short, while semiotic systems consist of symbolic systems and their users, they frequently are based on or require technological artefacts to work. In other words, in order to use a symbolic system, in addition to theoretical knowledge, the user needs practical skills to operate the tool through which it can be applied. The practice aspect is usually not distinguished from the theoretical dimension of symbolic systems – once again we take it for granted. While our theoretical knowledge of a language may be quite advanced, our practical ability to speak it may be very limited. For the native speaker, the reverse tends to be true: the practical ability to speak exists without explicit theoretical knowledge of the language, such as the ability to write or read it, which is learned at a later stage in life.

To the extent possible, semiotic systems should be described and modeled in the way they are ordered and employed by their users – since this is the way they actually work. That is, users subjectively experience symbolic systems as real and often even material things. We could call this a "reality illusion", but it appears as an illusion only from our particular philosophical vantage point, i.e. emergentist materialism. In this view, we make an ontological distinction between immaterial symbolic systems and their material equivalents in individual brains and social user groups. This philosophical distinction, which transcends the normal subjective and intersubjective experiences of symbolic systems as real, may give us explanatory leverage for analyzing the working of cultural processes as both material and immaterial phenomena. It is an ontological distinction that is rarely made in social science analysis. At the same time, it is a distinction that does not negate the importance of subjective and intersubjective – mental and social – realities qua social facts.

Take the analysis of cultural phenomena such as values, attitudes, and beliefs. The most relevant semiotic systems for this purpose, i.e. the crucial context, are natural languages, the semiotic metasystems in which cultural phenomena of this type are embedded. Yet values, attitudes and beliefs are widely treated as distinct things existing everywhere and as the same class of things independent of their ethnocultural context. As concepts thus reified, they are simply translated from the target language to the analyst's language (often English) as if the context tacitly assumed in the analyst's own language had universal applicability in other languages. The analyst's symbolic realism, the same reality illusion that is commonly held by users within particular semiotic systems, thus becomes part of the scientific analysis of cultural things (beliefs, values, and so on) and often leads to questionable results (Wierzbicka 1993). The ontology proposed here, on the other hand, makes it clear that we cannot simply ignore how symbolic systems in particular semiotic systems are interpreted and used, proceeding with analysis as if we all shared the same symbolic systems in different linguistic and cultural contexts.

25.2.4 Portability

Notwithstanding their linguistic substrate, most semiotic systems, unlike social systems, are portable in time and space. The users of symbolic systems can carry these symbolic systems to other social systems, other semiotic contexts, and into the future. But the effect of such transport depends on the semiotic systems to which a user moves a symbolic system. If we are talking about a natural language, as the symbolic mover you will depend on the existence of people in the new language context who are actually capable of using your language. In the case of most languages, this means your local expat community. In the case of widely spoken languages, above all English as a second language, you may be able to find a large number of interlocutors across the globe. Other languages, such as Chinese, Russian, or French have a large but regionally more circumscribed semiotic reach.

Natural languages are a special case of semiotic systems. Other types of semiotic systems can be moved across linguistic boundaries more easily. Professions such as doctors, lawyers, teachers, scientists, and artists can in principle plug into their semiotic systems in other cultures and languages. Often, this is not a question of semiotic transferability of existing knowledge and skills, but one of political and social barriers to entry.

There are, on the other hand, much simpler symbolic systems that are nevertheless of great social importance. Take traffic rules as a symbolic system that all participants in public traffic are expected to use, violation of which may result in serious injuries and fatalities. Many of the written rules of traffic are represented by symbols, some with text, others iconic – such as the stop sign, one-way street, speed limits, traffic lights, train and pedestrian crossings. Like most regulations and laws in this world of territorial states, traffic rules are legally codified at state level. While many traffic rules and symbols are shared internationally, others are specific to a particular country. Generally, the same traffic rules can be represented in different languages and symbols without much ambiguity or loss of content. Along with the important assumption of practical driving skills, this is why drivers licensed in one country are usually permitted to drive in other countries. They are expected to be able to master the differences in symbols that may exist between countries. The true challenge often lies in understanding the informal rules of a traffic system which local drivers share but which are not formally codified. They may not be part of the formal symbolic system of traffic rules, but they are part of actual traffic conventions of the semiotic system in practice. This example is also another illustration of the *two basic dimensions of semiotic systems: the knowledge dimension and the practice dimension.*

Semiotic systems differ, among other things, in complexity and scope. Learning symbolic systems such as traffic rules is considerably less challenging than learning a second language. While both natural languages and traffic rules can be studied at a theoretical level, the point is to be able to use them. The practical aspect that many semiotic systems have suggests a distinction between this type of propositional "knowing" of a system and being able to practice a semiotic system. "Knowing a semiotic system" can stay entirely at the theoretical level. Mastering music theory does not require the playing of a musical instrument. The former requires theoretical knowledge, the latter requires practical skill. Conversely, playing a musical instrument does not require any theoretical knowledge. The skill can be acquired entirely "by doing", by imitation and practice. Playing classical music, on the other hand, requires a high level of both knowledge and skill. Listening to music requires no theoretical knowledge, though appreciation of a musical piece or genre benefits from it.

25.2.5   Semiotic Systems Tend to Be Functional

Some semiotic systems are structured around the fulfillment of specific social roles and tasks; others are designed to facilitate the use of particular technologies (see above). Semiotic systems also have political effects and implications. Approaching semiotic systems in this fashion reminds us that they are rarely acquired for their own sake. Let us further keep in mind the fundamental fact that most semiotic systems are firmly embedded in and dependent for their use on a natural language. The same symbolic system may of course be used on the basis of different natural languages. This may not be a problem in the case of traffic rules or technical instructions. But in the case of translating a work of fiction from one language to another, the challenge is great. It is usually not self-evident which semiotic and social systems are necessary for a particular technology or social policy to be transferred successfully from one context to another.[3]

We acquire the semiotic systems relevant to the fulfillment of our biopsychosocial species needs while growing up in our native social system. Food, security, love, attracting mates, and maintaining or improving our social status are examples of such individual needs. While responding to the same basic needs, such semiotic systems vary to some extent between different societies and communities. It may be useful here to underline the general point that the existence and working of semiotic systems and mechanisms does not determine social processes or necessarily even play a significant role in them. Social systems and mechanisms are often more decisive than the independent effect of semiotic systems. Yet because they are distinct, omnipresent and in principle have great potential to "move" material reality, the role of semiotic systems can never be discounted. The moving, of course, is done by the material components of semiotic systems, i.e. their users, not by any independent symbolic dynamic.

The phrases "political culture" and "economic culture" are often invoked to explain the fact that the same type of political systems such as a liberal democracy and economic systems such as twenty-first century capitalism are embedded in different cultures, i.e. driven by different semiotic systems. There are semiotic systems that are functional for an economic system in question, such as expertise and skills required for production, sales, and organization of a firm. Other semiotic systems that may or may not facilitate the economic system's functioning will be simultaneously in use – above all one or more natural languages, possibly in addition English as a business language – as well as semiotic systems concerning management-worker relations, leadership style and ethics, and marketing and public relations. Such *semiotic systems may be global*, as is the case for many used in large multinational corporations, corresponding to key physical and social technologies employed, *while others may be used exclusively at national, sectoral, or local levels*.

25.2.6 Sciences and Arts

Unlike the area of basic human needs or the field of technology, the basic sciences are semiotic systems involved in the invention and application of symbolic systems without any direct material function. Of course, the sciences as social systems would hardly survive if at least indirectly they did not follow social rules and mental habits, and if they failed to have clear material consequences. Basic science, however, is concerned only with conceptual systems designed to explore how the world works. Peer recognition and social status are the main social mechanisms of material reward for scientists. Some of the basic scientific knowledge produced

---

[3]This problem of "institutional transfer" has been widely discussed in development and post-communist transformation debates.

becomes the foundation for the development of new and powerful technologies that are sought after by actors in political and economic systems because of their practical application.

Even more than basic science, the arts are occupied with producing symbolic systems and artefacts that do not have any obvious function or material consequences. Of course, artistic artefacts, especially if they are visible or audible, can serve a variety of psychic and social needs. The entertainment function of the arts in the democratic age has given rise to large economic systems specializing in the production and sale of cultural products. Semiotic systems, even those that appear to be practiced for their own sake, for their inherent value as a practice or for their aesthetic value, may serve a variety of purposes in the material world.

The multiplicity and overlap of different semiotic systems in the same social system presents a fundamental challenge for any explanation that attempts to take into account and understand how different semiotic systems relate to each other and to the social system in which they are being used. Let us approach this problem by focusing on how social systems and semiotic systems differ and how they relate.

25.3 Social Systems and Semiotic Systems: How They Differ and How They Relate

A semiotic system is unlike a social system. A semiotic system is fundamentally defined by or in terms of its symbolic system, i.e. its parts, context, structure and the ties between its symbols (its "logic") rather than by the social ties between its users, which may or may not exist and can be of many different kinds. This is of course also the way in which we as individual users approach a symbolic system – as something real and often even material (a book, a text, letters, notes, but also personalities, places, stories). The members of a semiotic system share symbolic bonds. Emergentist materialism claims that only material things are real. A symbolic system is conceptual, not material, and therefore not real. But a symbolic system in a human brain from where it originates and where it is used (i.e. conceived, known, believed, or felt), is real by virtue of being part of a material system – the brain processes of individuals sharing a symbolic system, on the one hand, and the practice of a group or the discourse of a society, on the other. The physical carriers of symbolic systems such as books, musical scores, and digital files are material artefacts, but they do not make the symbolic system represented on these carriers any more real. They are only technical components of a communication system. The symbolic system as recorded on a particular physical medium remains meaningless until it is processed by a human brain that understands the symbols and their symbolic context or environment.

Let us recap. A symbolic system qua immaterial system does not have any material existence. It exists only conceptually. The fact that we can think and talk about it as if it existed is the fundamental semiotic mechanism that turns something purely symbolic into a social reality, i.e. the social reality of a group sharing symbolic bonds. But is this not pure social constructivism of a radically idealist kind? Radical social constructivist theorists do in fact what social actors do when dealing with symbolic systems. They embue them with an independent reality. So, the social constructivist does what any social actor does, only at a philosophical or meta level. This is where the problem comes in. Just because this is what social actors do and believe does not mean that this is what is really going on.

If we viewed semiotic mechanisms as the only mechanisms creating social reality, the answer to the question, is this not pure social constructivism, would be yes. Then economic production and political power would exist only to the extent that they are semiotically produced. They would be "only" semiotic facts. Shared illusions. What can mislead us here is the fact that there are no social facts or social reality in which semiotic systems don't exist. But those semiotic systems do not necessarily shape, let alone create, this piece of social reality. The question is how and to what extent semiotic systems actually do play a role in creating social reality alongside economic, political, and biopsychic mechanism – and this is a theoretical and empirical question.

Let us consider some additional examples to illustrate the fundamental distinction between semiotic systems and social systems. Islam is defined first by the doctrines laid down in the Koran. The users or followers of this religious system are all Muslims, but Muslims as a whole do not form a real or concrete social system. There are many social systems, from entire states to local communities, where Islam as a semiotic system is practiced, but what distinguishes them as social systems such as states or families are primarily social ties, not the symbolic ties of the religious doctrine. Examples of social ties are sexual and familial love, economic cooperation and exploitation, and political domination and loyalty. The large degree of overlap between a social system like the state-society of Saudi Arabia and a semiotic system like Islam can make distinguishing between the two types of systems in analysis and explanation difficult. A semiotic system like a religion does not operate separately from the social systems in which it is practiced. In fact, it often plays a central role in reinforcing social ties, or indeed in weakening or dissolving them. The challenge for social science is to find out how these different types of systems work together. Thus when analyzing the power of the state of Saudi Arabia, it is necessary to identify political and economic mechanisms domestically and globally, and how Islam as a semiotic system is involved in these material processes.

Different fundamental interpretations of the Koran certainly have to do with the properties of this symbolic system, including its inconsistencies, ambiguities, omissions, and openness. But interpretation is not a property of the symbolic system but a material process in a human brain and in a community of interpreters. Symbolic systems like religions or ideologies are full of instructions for action, many of which are not controversial among its users. It can therefore appear like the symbolic system has a direct, unmediated effect on the social world. All religions require from the faithful prayer, and all ideologies require their followers to have conviction. But the fact that religious believers can lose faith and stop praying and ideological believers can lose conviction and abandon the cause shows that real mental and social mechanisms, not symbolic demands, mediate between a symbolic system and social reality.

We know that faith and conviction can have powerful effects on social reality, regardless of the content of the faith or ideology in question, because they enable or reinforce mental and social mechanisms. As the famous Placebo effect seems to suggest, the real mechanism at work may not be a particular medication that is being administered but the biopsychic resources mobilized by a believing patient and their effects on the physical state of the body. It is not the symbolic power of the faith that has a direct, immaterial or spiritual effect on successfully fighting the illness. Rather, a particular positive mental state may have a general effect on certain of the patient's biological processes that facilitate or strengthen the mobilization of the body's defense and self-repair mechanisms.

The Placebo effect, however, is a general side-effect of a biopsychic mechanism that may or may not be helpful in dealing with a particular somatic disorder. The existence of the Placebo effect does not refute the existence of the biochemical effects that most medications have, regardless of whether or not the patient has faith in their efficacy. The two are instances of different biochemical mechanisms. Yet the Placebo effect underscores the general fact that mental states – emergent properties of brain activity – can directly affect other biochemical systems. The various mechanisms operating in human biopsychosocial systems seem to work in all directions. Biological processes may have a causal effect on mental and social systems; mental processes may have a causal effect on biological and social systems; and social processes may have a causal effect on mental and biological systems. How such causal processes work requires an understanding of both the different systems involved and the different mechanisms at play. We will discuss the role of mechanisms in semiotic systems in a separate section below.

Semiotic systems are composed of symbolic systems and biopsychosocial systems, i.e. the users of symbolic systems. Symbolic systems have a composition, environment and structure, but unlike material systems they have no mechanisms. In other words, they have no dynamic element of their own. To play a role in real, material processes, symbolic systems require the brains of real people who live in social systems. Whatever dynamic may be at play – influence, causality, random effect – a symbolic system exists only by virtue of its use – conscious or not – by actual people in concrete circumstances. A symbolic system has implications for social reality because individuals use it as a guide to action or a background program for thinking, talking, feeling, and doing. Thus symbolic systems matter for social reality, but only as parts of semiotic systems.

The relationship works both ways: individuals use symbolic systems in standard, unreflected, habitual ways. This can have the appearance of individuals being used by symbolic systems. At the same time, however, individuals can also manipulate and modify symbolic systems consciously and creatively. So sociological reductionism in principle is not justified (Pickel 2012).

25.3.1   Hölderlin in the Amazon

Semiotic systems matter for social reality because their users also belong to social systems. Imagine an aficionado of German Romantic literature who lives in a remote Amazon village where he works as a medical doctor. His use of this particular semiotic system (or subsystem), that is, reading Hölderlin every night before going to sleep, will have no direct effect on the social reality in which he lives and works. No one else in the village is capable of using this particular semiotic system. The only effect it has on the village as a social system is through how the doctor's use of the literature affects his state of mind during times of social interaction with his fellow villagers. On the whole, this particular semiotic system used in this particular social context remains the private activity of its sole user. When the village becomes connected to the Internet, the doctor will have an opportunity to join virtual communities interested in German Romantic literature and interact with other users. This too will probably not have any direct effect on the social reality of the village, unless the doctor cuts down on his working hours in order to be able to spend more time on the Internet.

German Romantic literature is an example of a semiotic system (SS1) that has little effect on the social system of the village where it is being used by one individual. Let us consider another semiotic system, i.e. the Panoan language (SS2) spoken by the villagers. The doctor, who was raised and educated in Lima, Peru, has his place in the village as a result of his mastery of the semiotic system of modern Western medicine (SS3). His first language is Spanish (SS4), his command of the local language is still limited. Most of the villagers only speak their Panoan language, though a handful speak some Spanish. Both semiotic systems, i.e. Panoan and Spanish, have a powerful effect on the village social system. Panoan is the natural language of the villagers, and as such their semiotic metasystem. The same is true for Spanish, a semiotic metasystem that is used only by the doctor, and occasionally by a small number of other villagers, but only as a second language. The third semiotic system of importance for the local social system is that of Western medicine. While in principle this semiotic system is universal in its applicability, it is usually dependent on at least one natural language to allow for communication between doctor and patient. The language of the villagers through which any health problems they experience is conceptualized and communicated, is missing linguistic equivalents for some of the key concepts and terms of Western medicine. While the doctor is able to successfully treat many health problems, let us assume that communication problems emerge in the area of child birth. The doctor has tried to introduce some new techniques that the pregnant women and the villagers in general found offensive for reasons the doctor was unable to comprehend. Mutual incomprehension between the villagers and the doctor led to growing mutual distrust, frustration, and the doctor's eventual departure from the village. One thing seems certain: Hölderlin cannot be blamed. Symbolic systems matter for social reality, but only as parts of semiotic systems. Yet not all semiotic systems in use in a social system will matter for social reality.

In short, social systems are not semiotic systems, and semiotic systems are not social systems. Most social systems contain a number of semiotic systems, while semiotic systems not used in any social systems are bound to disappear. The coexistence and codependence of social systems and semiotic systems, however, does not eliminate the need to carefully distinguish them. Indeed, some explanatory power will derive precisely from a clearer understanding of the different systems and mechanisms involved in making some semiotic systems of eminent significance in a social system while making others virtually irrelevant.

Can there be and is there ever a direct correspondence between a social system and a semiotic system? Can the users of a particular semiotic system all be tied into one and only one social system? This question is difficult to answer in general terms. In one sense, each concrete social system will have a corresponding semiotic system. In other words, all components of the social system are members of a corresponding semiotic system through which they communicate with and recognize each other. But the system's symbolic content may be almost identical with the content of semiotic systems used in other social systems of the same type, for example families, schools, firms, or churches in the same civilization, culture, economy or religion.

Arguably each concrete social system has its own particular semiotic system. On the whole, social systems of a particular kind in a particular systemic context will show minor variations on a basic template. Social systems do not invent or reinvent entire symbolic systems but adapt them to some degree in their own practice. Take, for example, types of family systems, which differ fundamentally between but not within "civilizations" (Therborn 2011). Yet each individual family may vary to a small extent in the way it understands and practices the civilizational family template. From the subjective viewpoint of family members, such minor variations may well be of major importance. Between the civilizational layer and that of the individual family, other semiotic systems may come into play, based on national, geographical, regional, and a variety of other cultural differences and corresponding semiotic systems. As I have tried to show in my own work, national cultures are semiotic systems of overriding importance in the world today (Pickel 2013).

Significant or radical change in a semiotic system will begin in a small part of the user community. It will not be driven by the nature of the symbolic system but by changing conditions in individual lives and social systems where it is used, as well as by creativity and chance. The fundamental transformation of the Christian family system as a semiotic system in parts of the Western world from extended families to single-parent households, unmarried couples and gay married couples have occurred first in a small but subsequently increasing number of family systems, eventually growing into a social movement seeking recognition in a variety of semiotic systems from public opinion, national culture, the law, morality, and religion.

All material systems have one or more basic mechanisms that make them function in a characteristic way. Thus the mechanisms of production and distribution are fundamental in economic systems, while the mechanisms of control and allocation of power and resources are fundamental in political systems. By contrast, symbolic systems like religious doctrines, political ideologies, and scientific theories have no mechanisms of their own – faith, conviction, and commitment to truth are psychosocial mechanisms that work in semiotic systems because semiotic systems have material parts, viz. individual users of the symbolic system.

All human social systems incorporate a number of semiotic systems, and as a result the dynamic in which semiotic systems are caught up is determined by the mechanisms at work in a particular social system. This is not to say that the semiotic system becomes merely an instrument of that social system. The social system does not fully control the semiotic system, even though its symbolic system has no mechanisms of its own. But the symbolic system usually has many other individual users in addition to those in one particular social system. The value and effects of a semiotic system depend on its wider use beyond the adopting social system, i.e. on being a semiotic system that is not contained in a particular social system's boundaries. Natural languages, for example, can function as official languages of a political system and the society it governs. A natural language with a relatively small number of speakers such as Estonian (1.1 million) functions as an official language for a state just as well as does English for the United States. The usefulness and power of English, an official language in a number of states, goes far beyond the respective societies, i.e. as international language of science, business, and politics. The same is of course not true for a language like Estonian.

Semiotic systems that are enclosed by a particular social system give that social system a kind of monopoly over the symbolic system, a situation that can be a source of economic or political power for that social system. Language policies are part of political mechanisms that affect the living conditions for particular linguistic communities. Copyright laws are part of political and economic mechanisms that give owners a monopoly over the use of a particular semiotic system, whether a technology or an individual work of art, that can be exploited economically.

25.4 Semiotic Mechanisms

Using a common natural language is a fundamental semiotic mechanism. This semiotic mechanism is often coupled with two basic social mechanisms: inclusion and exclusion. Semiotic systems like natural languages can pose serious barriers to social interaction. Throughout human history, sex, commerce, and war – biological, economic, and political mechanisms, respectively – have demonstrated that state-societies can more easily be penetrated than a semiotic system such as a shared language. State-societies are threatened, however, by large numbers of users of foreign natural languages arriving and settling on their own territory, for example migrant workers. This is the case especially if they form ethnolinguistic communities that can pose a challenge to other social systems. Where assimilation or integration does not occur spontaneously or as a result of policies, minority linguistic groups may begin to pursue nationalist goals based on this powerful semiotic system, i.e. their own language and ethnoculture. Such semiotic mechanisms, above all in ethnolinguistic systems, but also in religious or ideological systems, are symbolic levers for the exertion of various forms of political power. A *semiotic mechanism* (language use) here is coupled with a number of *social mechanisms*: the primary use of a different natural language in a language community growing in numbers (*demographic mechanism*), the mobilization of this community by internal leaders (*psychosocial mechanism*); political claim-making on behalf of the community vis-a-vis the territorial state where they live; and the response of the state (confrontation, escalation, de-escalation, accommodation) that is being challenged (*political mechanisms*).

A semiotic system such as a natural language has as its major semiotic mechanism the communication between its users. If at least some of these users do not form their own social subsystem with its own social mechanisms (e.g. an expat community), additional mechanisms at work will be primarily psychosocial and occur at the level of individuals and small dispersed groups. Such *psychosocial mechanisms* can range from personal attempts at assimilation, living one's own ethnolanguage to the extent possible under given circumstances (accommodation), and/or feeling excluded and marginalized from the dominant culture (marginalization). If at least some users do form a social system, such as an organized ethnic group, their semiotic system and its mechanisms just described will have wider effects. The classic case is the combination of a dominant natural language with a state-society, which will make that semiotic system integral and instrumental for the social system's functioning. Another typical instance is the combination of a minority language with sub-state political institutions in the pursuit of greater autonomy (e.g. Quebec, Catalunia).

Even though there are no symbolic mechanisms working on their own, it would of course be wrong to suggest that this renders the content of a symbolic system irrelevant. A natural language comes with specific historical narratives, traditions, traumas, attitudes, and behaviours that are sure to play some role whenever such a semiotic system is coupled with social and political mechanisms. But the symbolic system, bereft of mechanisms of its own, is not an actor or agency of any kind in the social processes just discussed. Linguistic conflicts, language policies, marginalization and exclusion or assimilation can become powerful and disruptive social phenomena, especially in modern states that derive their legitimacy and integrity from the assumption of a unified nation. Of course, natural languages are not the only type of semiotic system that can give rise to powerful and conflictual consequences in political and economic systems. Neoliberalism today, or Keynesianism in a previous age, are important examples of highly influential ideologies on a global scale with enormous social and political consequences.

25.4.1 Summary

It has been argued up to this point that symbolic systems have no internal mechanisms driving them. This is an implication of our realist and materialist ontology: only material systems are real and have mechanisms that make them work. Symbolic systems, by contrast, are immaterial. Rather than discarding them as irrelevant for the purposes of social science explanation, which after all seeks to explain social reality, the reality of symbolic systems should be understood in their material context, i.e. the individuals using these symbolic systems. As the constitutive material components of social systems, individuals put those symbolic systems to use in particular systemic contexts. Symbolic systems considered as just that, viz. conceptual systems with their particular components, structure and context, can be studied, interpreted, evaluated, modified, applied and practiced only by their users. Qua symbolic systems, they are otherwise dead like a once spoken language that disappeared long ago along with the death of its last users. Our interest in the role of symbolic systems in social reality entails that we focus on semiotic systems – the intersection between symbolic systems and social systems.

25.4.2 "Reality Illusion": Symbolic Bonds as Social Bonds

Natural languages are fundamental for communication in any society. However, they also work as *key social bonds* for members of particular social systems, even though these bonds are *symbolic* rather than social. That is, the symbolic bond of shared nationality based on a common language maintains the belief in a real social bond – a belief that facilitates national communication and the formation and maintenance of social bonds in other social systems of state-society. States adopt one or more semiotic systems of this kind as their official language that, like the official monetary currency, is backed by real social power. To function in a state's legal system, for example, requires the use of the official language, i.e. access to the legal "currency". Conversely, state power is legitimated by the symbolic bonds of national culture, as all states claim to be the prime representative of the nation and to act in the national interest.

Widely held beliefs are of course major social facts. The content of a belief is part of a symbolic system. It may seem that the symbolic system itself is therefore transformed into a social fact. But there is no such ontological transformation. True, subjectively we may experience such a transformation – that which we believe in along with others becomes real for us. But in objective terms, a mental representation that is socially shared (i.e. a semiotic system) does not give autonomy and agency to the content of that mental representation. The "reality illusion" which we have identified as a central semiotic mechanism remains firmly rooted in its biopsychic and social agency.

Obsession with an idea is not a property of the symbolic system to which the idea belongs. Obsession is a mental state, one that may be collectively shared. It is a potential implication or effect of a symbolic system, but actual effects will become manifest (real, material) only under specific biopsychic and social conditions. It is only in this indirect sense that symbolic systems can be dangerous or can be said to have any other property that is really the property not of the symbolic system but of the – real, material – biopsychosocial systems. The difference between our subjective and intersubjective experience of the reality of ideas (the "reality illusion") and the objective situation our ontology postulates can be explained as follows. When we deal with ideas in personal or public life, they are in fact real to us (and a "reality illusion" only in objective terms) because they have been activated – in our own minds and collectively in a shared symbolic context. We could also say that ideas have no effects unless and until they are being practiced in one way or another. At the same time, it is important to note that symbolic systems do have their own properties, i.e. conceptual properties, in particular semantical and logical properties (on "conceptual existence", see Bunge and Mahner 2001, 95–102).

Racist ideas, for instance, become dangerous when they are held and acted upon by racists, i.e. believers. The same ideas are harmless, or will be undermined and rejected, when they are the subject of critical study or critical practice. It is important to note that, as critics, we share the symbolic system of racism with racists: we understand it, and we consider it to be real, psychologically and socially, but we do not endorse it and instead act against it. It is highly unlikely that in a society in which racism is widespread, anyone would be unaware of this symbolic system. But in addition to advocating for or opposing a symbolic system such as racism, instead of taking a position it is of course also possible to deny its relevance. These are semiotic – mental, social, and political – implications of a symbolic system, i.e. very different and mutually opposed implications of the same symbolic system. Here we encounter again the practice dimension of semiotic systems.

25.5 Semiotic Systems: The Practice Dimension

A symbolic system, even one as simple as racism, does not act by itself or force a particular type of action. If we reduce racism as a symbolic system to its core message that one particular race is superior to others, this does not entail anything in particular. True, it could be used as part of a justification for genocide. It could also be attacked as a pernicious doctrine that has to be opposed and eliminated. We don't usually refer to people holding the latter view as racists but rather as anti-racists. However, both are reactions to the same *symbolic* system which can be practiced along a continuum defined by two diametrically opposed *political* views. There is, in addition, the possibility to ignore or dismiss this symbolic system as irrelevant or outdated. Knowing yet ignoring a symbolic system that is still being practiced by a significant number of people will not help to end the practice, though it may make the practice invisible, and thus lend it strength. Only by way of addressing the practice of a symbolic system, such as by endorsing, ignoring, or rejecting it, can a symbolic system have a real material effect.

If you know and understand a symbolic system, you are a member of the corresponding semiotic system. Whether and how you practice this knowledge is not predetermined merely by the fact of having knowledge. It will depend on semiotic mechanisms that process and apply a particular symbolic system. Semiotic mechanisms have biological, psychological, and social dimensions. Practicing a symbolic system refers to all three dimensions. Racists claim that race is a biological fact, such as when native intelligence is taken to be a function of race. In contrast to early and mid-twentieth century scientists who shared that view, human genetics today remains mired in confusion over the concept (Fullwiley 2014). Racism's continuing social and political power is a result of psychological and social mechanisms. This is the relevant terrain and context for the study of semiotic mechanisms related to racism, not the pernicious implications of the doctrine as such.

Let us briefly explore the practice dimension of semiotic systems in the context of new social media. Facebook, an online social network with over one billion members, is a techno-social symbolic system. Knowing about Facebook's existence does not make you a member of the corresponding semiotic system. But you could know what it is and how it works without being a formal subscriber, and thus still be a member of the semiotic system. The overwhelming majority of members of the semiotic system Facebook are of course formal members and profiled users of the technology. In other words, these are people who practice the semiotic system by using it (*semiotic mechanism*). The owners and operators of this technology provide a clearly organized, gradually changing framework for its use by private users, businesses, political organizations, and a myriad of other social groups and networks (*technosocial mechanism*). The social network is privately owned and managed, financed by commercial advertising and the sale of user information to third parties (*economic mechanism*). Membership has become de facto mandatory for any individual and organization that seeks to be a recognized presence in the world today (*psychosocial mechanism*).

Facebook is therefore not just a semiotic system, but indeed a tightly organized and professionally run social system, composed of its owners, managers and staff, commercial customers, and account holders. More specifically, Facebook is one of the world's largest profit-generating economic systems. It is also a technosocial system offering a free communications platform to its members. Facebook as a semiotic system can be practiced in a variety of ways, depending on your status in Facebook as an economic system. As owner, operator, or commercial client, you use the semiotic system for economic purposes. Operators under the control of owners create and govern Facebook as a technosocial system. In principle, Facebook is regulated by a variety of national and regional political authorities, though none with Facebook's global reach and power. All of the above actors practice the semiotic system Facebook in line with conventional economic and political mechanisms, such as profit-making, political mobilization, and private communication. At the same time, Facebook as a semiotic system has opened up a world of new practices. True, users remain subject to existing psychological and social mechanisms, but some practices Facebook makes possible are new – with profound psychological and social consequences.

What used to be the privilege of large organizations has become possible for any Facebook user: create a public image of yourself using text, audio, photos, videos and links to others. The size of your public may in fact not go far beyond family and friends, but potentially your reach is global. In any event, the fact that we can construct and constantly modify our public image in many sophisticated ways has created some fundamentally new conditions of existence. This is true in social terms, that is, how and to whom we can reach out in direct communication, as well as in psychological terms, that is, how we see and create ourselves. The general and most widely used semiotic mechanisms and forms of symbolic representation that Facebook makes possible are of course not of the regular user's own making. They are either technologically predetermined by the corporation or socially preconstructed by trendsetters. But the practice of creating specific symbolic content within those semiotic, technical, and social structures is up to the individual user. Effects such as obsessive use of the technology result from psychological and social mechanisms that Facebook as a semiotic system may facilitate but does not cause. The same applies to other psychological consequences of Facebook use, which of course does not diminish their seriousness or deny the need for regulation.

Another practice with significant social and political consequences made possible by Facebook is the possibility of large-scale mobilization for political, economic, and cultural purposes. It is the network effect of Facebook through which smaller groups of friends are interlinked with others that makes possible the spread of specific information and calls to action "in real time". Of course, Facebook as a symbolic system provides only the technical preconditions for such actions; it is the semiotic practice of formulating a convincing message for a receptive audience that will matter. For this reason, it is mistaken to explain large-scale political mobilizations and even revolutions as the outcome of new technologies like Facebook. Semiotic mechanisms of this kind will not have significant effects in the real world unless social, political, and psychological mechanisms combine in a context in which there is receptiveness for particular political messages. Yet it is true that political rulers who feel threatened may try to shut down the technosocial systems that facilitate the mobilization of oppositional activity.

25.6 Methodological Implications

The same semiotic system is compatible with the full range of types and modes of social relations, from peaceful forms of co-operaton to violent conflict. Nationalism, for example, is often held responsible for violent actions committed in its name. But this is not a necessary implication of nationalism as a semiotic system. Many nationalists are non-violent in principle, and nationalist doctrines concerning violence are usually defensive rather than aggressive. The causes for nationalist violence lie in the combination of nationalism with other – political and psychological – mechanisms (Pickel 2015). These cases illustrate that semiotic systems can have a broad range of outcomes, in particular with respect to how its users relate to others. Does this suggest that those dismissive about the role of culture in social and political affairs are right? Is culture perhaps epiphenomenal after all?

Our systemic conception provides an answer to this question: symbolic systems outside of social systems are in fact epiphenomenal, but semiotic systems – i.e. symbolic systems with their users – are not. Semiotic systems link symbolic systems with real social systems. In other words, the users of a semiotic system, who will be simultaneously members of many different social systems, bring "their" symbolic systems into the various social systems to which they belong. This fact alone does not explain whether and how specific symbolic systems influence and affect those social systems. To take our standard example of a powerful semiotic system, a natural language plays a key role in most social systems from being a basic means of communication and a marker of identity to functioning as a store of national culture and as a semiotic metasystem. However, a natural language of which there is only one speaker in a social system will, qua symbolic system, have no effect on that social system.

This may seem obvious, yet even a major semiotic system such as a natural language does not receive systematic attention in most social science research. Even in explicitly culturally oriented studies, such as those on values, attitudes, and beliefs, the role of the natural language as central semiotic system is often ignored. True, where the natural language used by the social scientist is the same as that of the social systems under study, nothing significant may be gained by problematizing this semiotic system. However, since increasingly social science research globally is transacted in English, the question about the significance of using this semiotic system in describing and analyzing social realities with different natural languages should be raised more often. This is true *a fortiori* for research conducted by native English speakers in and on social systems with a different natural language. This problem has been recognized, but it continues to be ignored or marginalized in much comparative research (Wierzbicka 2014).

Where our primary interest is not "culture" but in how concrete social systems and mechanism (e.g. profit-making, military rule) work, is there any reason to be concerned about semiotic systems? In other words, do we need to take into account semiotic systems in order to produce good explanations of how social systems work? The general answer is that not all semiotic systems in use in a particular social system are important for the functioning of the social system. At the same time, there will be key semiotic systems without which the functioning of a particular social system cannot be explained adequately. In other words, some semiotic systems in use matter more than others for the working of a social system. Some may be irrelevant, but others may be crucial. Where, on the other hand, our primary research interest falls clearly in the area of "culture", semiotic systems will be automatically at the centre of analytical attention. The materialist conception of culture presented here may then facilitate the inclusion of non-cultural – political, economic, psychic – mechanisms into a cultural analysis.

25.7 Conclusion

This paper has proposed a conceptualization of culture as semiotic systems. It is rooted in Mario Bunge's philosophy of science, in particular his systemic perspective, or systemism. The question of what is culture is addressed in this ontological context. The paper has identified building blocks for a systemic perspective and framework – in particular the categories of symbolic system, semiotic system, and social system. Semiotic systems, it was argued, are the central link between real material social systems, on the one hand, and immaterial symbolic systems that encompass much of what is discussed under the heading of culture, on the other. Since semiotic systems are defined as symbolic systems along with their users, semiotic systems relate and combine material and symbolic dimensions, real systems and conceptual systems. The two poles reflect the basic ontological tension contained in any non-reductionist perspective on culture. The concept of "semiotic system" creates a clearly defined ontological space for a conception of culture that is neither purely social nor purely symbolic.

In subjective and intersubjective experience, symbolic systems appear as real – real to me or real to us. Moreover, as users we experience symbolic systems not only as real but often even as material things. This "reality illusion" appears as such only from our particular philosophical vantage point of emergentist materialism. In this view, we make an ontological distinction between immaterial symbolic systems and their material equivalents in individual brains and social user groups. But subjectively and intersubjectively, such a distinction is superfluous since symbolic *experiences* are psychic and/or social facts. As social scientists, we have to treat symbolic experiences as real – but not the symbolic systems underlying them. This philosophical distinction should give us explanatory leverage for analyzing the working of cultural processes as simultaneously material and immaterial phenomena. In order to understand how symbolic systems affect social systems, we assume that real psychic and social mechanisms, not symbolic demands or logics, mediate between a symbolic system and social reality. A symbolic system *qua* immaterial system does not exist as such. The fact that we can think and talk about it as if it existed is the fundamental semiotic mechanism that turns something purely symbolic into a social reality, i.e. the social reality of a group sharing symbolic bonds.

Most semiotic systems tend to be functional, i.e. they are involved in the satisfaction of needs and in the pursuit of goals. In other words, they play a role in material processes – biological, psychic, and social. This is the material context in which such semiotic systems have to be examined. At the same time, the ontology proposed here cautions us that we cannot simply assume that our understanding of a symbolic system – say a religious doctrine or a political ideology – as analysts is shared by actors in different linguistic and cultural contexts and therefore appropriate and relevant for analysis. Symbolic systems under study thus remain problematic – social science cannot take them for granted or ignore them.

There are no social facts or social realities without semiotic systems. However, the coexistence and codependence of social systems and semiotic systems does not eliminate the need to carefully distinguish them. Indeed, some explanatory power will derive precisely from a clearer understanding of the different systems and mechanisms involved in making some semiotic systems of eminent significance in a social system while making others virtually irrelevant. A key question here is how users of particular semiotic systems are positioned in particular social systems.

The use of a symbolic system is a basic semiotic mechanism. In the absence of semiotic mechanisms, symbolic systems have no dynamic elements of their own. The properties of a symbolic system, i.e. its content and internal logic, exist conceptually, but not materially. They have no reality on their own. Only individual and collective users can transform symbols into semiotic reality, which in turn can make them into active components of social processes. Semiotic mechanisms activate symbolic systems in social reality. The conception of culture proposed here encompasses all three types of systems and mechanisms – symbolic, semiotic, and social. A conception is not a concept, let alone a definition, of culture. The conception presented here offers an ontology in a systemic context that may be more or less consistent with existing approaches to culture. Whether and how this conception of culture as semiotic systems may be helpful remains to be explored.

References

Bourdieu, P. (1990). *The logic of practice*. Cambridge, MA: Polity Press.

Bourdieu, P. (1991). *Language and symbolic power*. Cambridge, MA: Harvard University Press.

Bunge, M. (1981). *Scientific materialism*. Dordrecht: Reidel.

Bunge, M. (1996). *Finding philosophy in social science*. New Haven: Yale University Press.

Bunge, M. (1998). *Social science under debate: A philosophical perspective*. Toronto: University of Toronto Press.

Bunge, M. (2003). *Philosophical dictionary* (Enlarged ed.). Amherst: Prometheus Books.

Bunge, M., & Mahner, M. (Eds.). (2001). *Scientific realism. Selected essays of Mario Bunge*. Amherst: Prometheus Books.

DiMaggio, P. (1997). Culture and cognition. *Annual Review of Sociology, 23*, 263–287.

Fine, G. A. (2012). *Tiny publics: A theory of group action and culture*. New York: Russell Sage Foundation.

Franzosi, R. (1998). Narrative analysis – Or why (and how) sociologists should be interested in narrative. *Annual Review of Sociology, 24*, 517–554.

Fullwiley, D. (2014). The 'Contemporary Synthesis': When politically inclusive genomic science relies on biological notions of race. *Isis, 105*(4), 803–814.

Goddard, C., & Wierzbicka, A. (2004). Cultural scripts: What are they and what are they good for? *Intercultural Pragmatics, 1*(2), 153–166.

Jodelet, D. (2008). Social representations: The beautiful invention. *Journal for the Theory of Social Behaviour, 38*(4), 411–430.

Lamont, M., & Small, M. L. (2008). Culture matters. The role of culture in explaining poverty. In D. Harris & A. Lin (Eds.), *The colors of poverty: Why racial and ethnic disparities persist* (pp. 76–102). New York: Russell Sage Foundation.

Lizardo, O. (2005). Can cultural capital theory be reconsidered in the light of world polity institutionalism? Evidence from Spain. *Poetics, 33*(2), 81–110.

Lizardo, O. (2011). Bourdieu as a post-cultural theorist. *Cultural Sociology, 5*(1), 25–44.

Morrill, C. (2008). Culture and organization theory. *Annals of the American Academy of Political and Social Science, 619*(1), 15–40.

Nishida, H., Hammer, M. R., & Wiseman, R. L. (1998). Cognitive differences between Japanese and Americans in their perceptions of difficult social situations. *Journal of Cross-Cultural Psychology, 29*(4), 499.

Pickel, A. (2003). Explaining, and explaining with, economic nationalism. *Nations and Nationalism, 9*(1), 105–127.

Pickel, A. (2004a). Systems and mechanisms: A symposium on Mario Bunge's philosophy of social science. *Philosophy of the Social Sciences, 34*(2&3), 169–181.

Pickel, A. (2004b). *Homo nationis*: The psychosocial infrastructure of the nation-state order. *Global Society, 18*(4), 324–346.

Pickel, A. (2006). *The problem of order in the global age: Systems and mechanisms*. New York: Palgrave.

Pickel, A. (2007). Rethinking systems theory: A programmatic introduction. *Philosophy of the Social Sciences, 37*(4), 391–407.

Pickel, A. (2012). Between Homo Sociologicus and Homo Biologicus: The reflexive self in the age of social neuroscience. *Science and Education, 21*(10), 1507–1526.

Pickel, A. (2013). Nations, national cultures, and natural languages: A contribution to the sociology of nations. *Journal for the Theory of Social Behaviour, 43*(4), 425–445.

Pickel, A. (2015). Nacionalismo y violencia: una explicación mecanísmica. Con especial referencia a las teorías de Charles Tilly y Michael Mann. In *Cultura y representaciones sociales. Revista electrónica de ciencias sociales.* (Mexico D.F.) número 18 (marzo), 26–62.

Schmidt, V. (2008). Discursive institutionalism: The explanatory power of ideas and discourse. *Annual Review of Political Science, 11*, 303–326.

Swidler, A. (2001). *Talk of love: How culture matters*. Chicago: University of Chicago Press.

Taylor, C. (2004). *Modern social imaginaries*. Durham: Duke University Press.

Therborn, G. (2011). *The world. A Beginner's guide*. Cambridge, MA: Polity Press.

Wierzbicka, A. (1993). A conceptual basis for cultural psychology. *Ethos, 21*(2), 205–231.

Wierzbicka, A. (2010). Cross-cultural communication and miscommunication: The role of cultural keywords. *Intercultural Pragmatics, 7*(1), 1–23.

Wierzbicka, A. (2014). *Imprisoned in English: The hazards of English as a default language*. Oxford: Oxford University Press.

Chapter 26
Bunge on Science and Ideology: A Re-analysis

Russell Blackford

Abstract Mario Bunge has provided a useful analysis of the phenomenon of ideology, dividing ideologies into religions and sociopolitical ideologies and showing how both can be analyzed into very similar elements. This approach illuminates why sociopolitical ideologies so often bear the trappings of religion, and how they can play a similar role in their adherents' lives. Importantly, both contain cognitive content that includes one or another view of human nature. Science can threaten religions and sociopolitical ideologies by undermining their credibility and their specific claims, though science can also inform sociopolitical ideologies in ways that are potentially beneficial. Unfortunately, ideologues often insist on an arrow of causation that goes from ideology to science, rather than from science to ideology. That is, ideologues make judgments about science by using their own partisan beliefs, procedures, and epistemic standards, rather than allowing scientific findings to inform the emergence of an ideology grounded in reason. In this respect, ideologies of all kinds can become enemies of free scientific inquiry.

26.1 Ideologies: A First Look

In everyday English, an ideology is a system of ideas relating to politics, economics, and the functioning of society, and used to guide policy deliberations and political activism. The system may be comprehensive and systematic, though many of its adherents will not have studied its details. For them, it is more like a set of slogans and broad goals.

In *Political Philosophy: Fact, Fiction, and Vision* (Bunge 2009), Mario Bunge provides a useful analysis of the phenomenon of ideology, emphasizing the practical role of ideologies. He views them, in the first instance, as belief systems that integrate sweeping factual claims (such as 'Most citizens in our country are poor')

R. Blackford (✉)
School of Humanities and Social Science, University of Newcastle, Callaghan, Australia
e-mail: russell.blackford@newcastle.edu.au

439

with related value judgments (such as 'Poverty is abhorrent'), social goals ('Let us eradicate poverty'), and means to achieve them ('Let us adopt democratic means to eradicate poverty') (Bunge 2009, p. 142). As he explains, the claims may or may not be true, the value judgments may or may not be well-grounded, the goals may or may not be attainable, and the means may or may not be realistic.

As he develops the analysis, Bunge divides ideologies into two branches: religions and what he refers to as sociopolitical ideologies. This might seem surprising, since, in everyday English, we often distinguish between religions and ideologies, but it also has analytical advantages. It enables him to discuss the distinctive aspects of religious understandings of the world, while also illuminating the resemblances between ideologies, in the ordinary sense, and religions. Bunge's analysis of the elements of religions and sociopolitical ideologies, and how they both operate in social contexts, suggests why sociopolitical ideologies often display the hallmarks of religion and how they can play a similar role in their adherents' lives.

Although religions and sociopolitical ideologies are distinguishable, it is note-worthy that some religions have strong political elements, while some sociopolitical ideologies favor or disfavor particular religions. For example, a religion may enjoin its congregants or believers to mandate certain practices, or perhaps their entire system of belief and conduct, through secular political power. The medieval Church exercised authority and influence that pervaded all aspects of life in European Christendom, and something of the kind applies today with the dominant Islamic sects in the Middle East.

For their part, sociopolitical ideologies can favor or disfavor particular religions, as with American-style conservatism, institutionally represented by the Republican Party, which openly favors traditional Christian religiosity and morals. Sociopolitical ideologies start to *resemble* religions when they engage in prominent symbolism and ritual, offer esoteric knowledge and arcane vocabularies to their adherents (or at least those who are taught the details of the system), and instil deep psychological transformations in committed adherents. Political awakenings or radicalizations can show much commonality with religious conversions. The most radical and comprehensive sociopolitical ideologies resemble millenarian religions when they predict – and may attempt to bring about – massive social transformations.

26.2 Bunge on Religions and Sociopolitical Ideologies

Bunge analyzes religions and sociopolitical ideologies in a way that draws attention to their similarities. In each case, he identifies eight relevant elements, and he makes their equivalences obvious. First, he focuses on religion, and more narrowly on theistic religions, thus excluding Confucianism and original Buddhism, which he describes as 'secular ideologies' (Bunge 2009, p. 143).[1] By implication his analysis also excludes various kinds of animism and polytheism. He describes the nature of theistic religions, and their operation in social contexts, using the following eight elements:

- A group of congregants or believers.
- The society that hosts their religion, tolerantly or otherwise.
- The believers' general outlook or worldview.
- Their specific religious beliefs.
- The conceptual and practical issues addressed by the religion.
- The aims of the believers (for example, personal salvation).
- The policies that they advance.
- The means they employ to obtain their aims.

Bunge writes that religious believers accept authorities such as those of holy texts and priests, and 'are therefore expected to be both dogmatic and conformist, at least in religious matters' (Bunge 2009, p. 143). He goes on to provide a rich account of what is involved in a theistic religion. For instance, its general worldview will include a supernaturalist metaphysics, an objectivist and dogmatic epistemology, a theory of value that ranks beings according to their goodness and their closeness to God, and a specifically religious system of deontological ethics. As to the supernatural metaphysics, 'a religious metaphysics consists in a collection of doctrines about the supernatural and our relations to it, as well as about the immaterial soul. Such metaphysics revolves around two foci: God and man' (Bunge 2009, p. 144). The conceptual and practical issues addressed by a religion will involve how to know the deity, how to live as it wishes, and how to achieve spiritual salvation, as well as issues to do with church maintenance, interaction with co-religionists and with non believers, and attitudes to the surrounding social order.

On Bunge's account, religious means to achieve salvation are varied: they may include such practices as prayer, fasting, and charity, but they can also take less gentle forms:

> However, censoring publications, beating up heretics, or even burning them at the stake, placing bombs in markets, temples or abortion clinics, or even waging holy war, are not excluded. All means are allowed when the goal is either to earn a place in Heaven or to promote the interests of the faith. (Bunge 2009, p. 145)

Bunge's analysis could easily be extended, with minor modifications, to belief systems with more than one god, or with no gods at all, but including belief in other phenomena that are commonly regarded as supernatural, such as reincarnation, a spiritual afterlife, astral influences, and magic. A system of belief including some of these might have many of the characteristics of a monotheistic religion, though

---

[1] This is a questionable way to classify Buddhism, even its original form, with its doctrines of rebirth, *nirvana*, and spiritual enlightenment. However, little turns on this for current purposes. Likewise, for the supernatural elements accreted by, and often associated with, the Confucian tradition.

the Abrahamic monotheisms arguably stand out for their record of intolerance and persecutions (Blackford 2012, pp. 20–33). Bunge takes a dim view of theistic religion, which he portrays as dogmatic, politically reactionary, and potentially extreme and dangerous.

Bunge views sociopolitical ideologies as general conceptions of the social world, 'That is, the general propositions (principles), value judgments, and practical proposals of a sociopolitical ideology concern the social order and ways to conserve or transform it' (Bunge 2009, p. 148). In parallel with his analysis of religions, he analyzes sociopolitical ideologies and their operation in a society using the following elements:

- A group of like-minded partisans and sympathizers.
- The society that hosts their ideology, tolerantly or otherwise.
- The group's general outlook or worldview.
- Its specific sociopolitical beliefs.
- The issues addressed by the ideology (for example, poverty or war).
- The political aims of the group (for example, to shore up or to destabilize an existing regime).
- The policies that the group advances.
- The means it employs to implement its policies and obtain its aims.

Groups supporting various sociopolitical ideologies can range greatly in their degree of organization and cohesiveness. Bunge sees the general outlooks of sociopolitical ideologies as incorporating an ontology, epistemology, axiology, ethics, and praxiology. The ontology more or less explicitly 'states something about human nature (material or spiritual, biological, or biosocial) as well as about society (aggregate of individuals, indivisible whole, or system).' The inherent epistemology 'is a set of principles about social knowledge: that it is possible or impossible, objective or subjective, scientific or humanistic.' The axiology 'involves individual values (such as certain virtues), social values (such as peace), or both,' while the ethics of the ideology is a moral doctrine such as contractarianism, emotivism, utilitarianism, or Kantian deontology. Finally, the praxiology of the group members 'is a (usually tacit) set of principles about human action, both individual and social' (Bunge 2009, p. 149).

A political group's aims 'can be narrow or broad, short-term or long-term, realistic or utopian, altruistic or selfish, and so on' and their realization might involve 'social maintenance, improvement, reform, or revolution.' As with a religious group, a political group's methods might vary widely: they can include 'education, mass mobilization, or both; taxation or re-engineering; invoking God's help or resorting to armed struggle; organization or mass deception; parliamentarism or dictatorship, and so on' (Bunge 2009, p. 150).

As becomes evident, Bunge is less adamantly hostile to sociopolitical ideologies than to religions. He hopes for the emergence of a socially beneficial ideology grounded in reason. Nonetheless, he views all sociopolitical ideologies to date as unscientific and otherwise flawed.

26.3 Human Nature

Religious belief systems understand human nature in the context of a twofold cosmic order: the world known to us through our senses stands in a defined relationship to a supernatural or transcendent order of things, and there is likewise a transcendent dimension to human life. Religions typically identify some ultimate good above and beyond ordinary human flourishing, postulate deep, personal transformations to ensure that this ultimate good can be achieved, and view human life as extending in one way or another beyond the natural world (Taylor 2007, pp. 15–20). It follows that religions focus on supernatural transformative powers and our alignment with them. Bunge's analysis of religion is consistent with all this. Of Christianity, he remarks: 'Christians believe in Original Sin, the relentless decay of humankind since the expulsion from Paradise, work as a curse, and suffering as the best way to atone for sin' (Bunge 2009, p. 51).

Like religions, sociopolitical ideologies may emphasize personal transformations, but in the context of achieving social aims. These aims might take the form of desired social changes, though successful resistance to change might be the primary aim of a conservative ideology. At any rate, sociopolitical ideologies have their own views of human nature:

> All political ideologies involve some view of human nature: that we are either born or made good or bad, aggressive or cooperative, obedient or rebellious, lazy or industrious, educable or stubborn, generous or selfish, redeemable or incorrigible, and so on. (Bunge 2009, p. 51)

It is not, unfortunately, straightforward to settle disputes involving rival accounts of human nature. Even if we rule out viewpoints that postulate a supernatural dimension to human life, no strictly scientific account of the subject currently prevails. Furthermore, the search for a compelling theory is hampered by the observable complexity and diversity of human motivation and behavior. Bunge is well aware of this problem, though he welcomes some scientific inputs to our understanding of ourselves.

Importantly, Bunge rejects biological reductionism, by which he means the idea that human nature at the level of individuals and above is determined directly by biological evolution and genetic hardwiring. This leads him to regard projects such as human sociobiology and evolutionary psychology with evident scorn.[2] He rules out biological reductionism on the basis that *Homo sapiens* evolved to be able to create distinct cultural groups with their own symbols, institutions, practices, and so on. Thus he writes, 'Biological evolution has certainly equipped us to think and act as humans, but it does not tell us *what* to think and do: only experience can fill this gap' (Bunge 2009, p. 51; Bunge's emphasis). He explains this in terms of neural plasticity, which enables us to live in new ways, although when we do so our existing institutions might not be suitable to the new condition, and thus might need to be changed.

---

[2] See the sub-section below entitled *Bunge on the Sociobiology Controversy.*

Bunge sees the leading, and rival, scientific theories of human nature as the idea that we are automatons, programmed by our genomes, and an equally extreme idea that we are balls of putty shaped by our environments. He favors neither of these:

> I submit that neither [should be adopted] because both are incompatible with contemporary science. Indeed, biology and psychology show that we are the product of two intertwining factors: genes and experience. A baby with serious genetic disorders cannot become a normal adult, and one with the best genome will not make it if neglected or abused for a long time. (Bunge 2009, p. 53)[3]

For Bunge, we cannot conclude glibly that human development is the product of genes multiplied by environment, but he suggests that this sort of formula is not entirely misleading. Though its variables cannot be quantified, the formula conveys the important truths that, first, human nature is real, since our species has a distinctive genome and not every environment favors human life, and second, the environment is changeable, and human nature changes with it. Thus, our nature is plastic, though not infinitely malleable. Bunge concludes, 'There is neither genetic predestination nor environmental omnipotence' (Bunge 2009, p. 53). He proposes that we attempt to modify the environment to benefit everybody: 'let us fashion it so as to give good genes a chance, and to compensate for bad genes with extra family, medical, and educational support' (Bunge 2009, p. 53).

As this suggests, Bunge's own view of human nature is generally optimistic: his positive assessment of the chances of social progress is based on a lively sense of our ability to plan, change ourselves, and reconstruct our existing arrangements. He especially emphasizes our prosocial emotions and tendencies: we are inherently cooperative animals, primed to feel empathy, and have a sense of fairness. This understanding of human nature informs Bunge's caustically critical attitude to much work in the social sciences, especially economics, insofar as it models the participants in economic systems as fundamentally competitive and driven by self-interest.

26.4 Science, Ideology, and Epistemic Standards

In science and historical scholarship – and in everyday problem solving – we use procedures and standards that we take to be truth-tracking. In particular, we have a practical idea of what counts as good evidence, and we have standards for logical validity and argumentative cogency. Some of our epistemic standards, including at least some of the rules of inference identified in formal logic, cannot be given any

---

[3] We might, however, wonder who actually takes the extreme determinist view that Bunge describes, wherein we are automatons shaped by our genomes. Bunge appears to classify Steven Pinker's *The Blank Slate* (2002) as one example of such a view. However, any fair reading of this book will show that Pinker views individual character, capabilities, and behavior as deriving from an interaction of genes and environment. His quarrel is with social scientists and others who give the human genome almost no role in human behavior.

deeper justification, but that is the nature of any standard of reasoning that we take to be fundamental. Any attempt to give rational justification to our most fundamental standards would have to rely at some point on those same standards. As a further problem, it has proved difficult to identify and explain our fundamental standards in a comprehensive, ordered, and uncontroversial array. Attempts at this to date have generated a vast literature in fields such as philosophical logic and scientific epistemology.

Despite these difficulties and complications, we often make progress toward the truth. As we do so, we are well-advised to believe whatever might be the logical implications of our existing knowledge, to reject claims that are self-contradictory or broadly inconsistent with current science, to trust the evidence of our senses in conditions where experience tells us that our vision, hearing, and other senses are reliable (and to study the circumstances in which their reliability is doubtful), and to give little credence to idiosyncratic testimony about inexplicable, scientifically anomalous events. We do well to accept a proposed hypothesis as more likely to be true if it contains an economy of elements that explain a wide range of observed phenomena, and especially if it makes novel predictions of events that are subsequently observed. Again, we are wise to doubt any hypothesis that is saved from falsification only by an auxiliary hypothesis introduced, ad hoc, for that very purpose.[4] Yet again, we should not commit to hypotheses that have competitors that cannot be ruled out as false or obviously implausible.

None of this is meant to be an exhaustive or definitive account of the rules governing rational inquiry or scientific argument. It shows, however, that we are not helpless when it comes to identifying some truth-tracking procedures and standards that are largely agreed in practice. As science advances, it applies these and related standards as precisely as possible to the best data available. Scientific procedures may involve the use of observational instruments, sophisticated quantitative methods, and carefully controlled conditions for isolating variables and obtaining evidence, as with double-blind studies in medical research. All of this is continuous with the methods used in everyday inquiry, and with those of serious, but non-scientific, forms of investigation and scholarship. For Susan Haack, science is much like these, except 'more so' in its systematic efforts to overcome human frailties and epistemic disadvantages (Haack 2007, pp. 24–25, 99–109).

We are, at all times, hampered by cognitive and perceptual limitations. But as Georges Rey describes, we make adjustments during the process of rational inquiry. For example, beliefs based on our own fallible memories can be checked against the memories of others, their testimony can be checked against that of still others, and all of this can be cross-checked against current evidence from our senses, including written records. Rey observes that 'almost everyone knows all this' – in

---

[4]However, an originally ad hoc hypothesis sometimes turns out to be independently warranted. When this happens, it tends to vindicate, or at least provide additional support for, the hypothesis that it was introduced to save.

everyday life, the need for adjusting and cross-checking is taken to be obvious and uncontroversial (Rey 2007, p. 252).

Despite the difficulties, then, we have good reasons to trust that rational inquiry, particularly scientific inquiry, is progressively illuminating a reality independent of ourselves.[5] We can also witness the great success of science in allowing us to predict and control the world, a success that seems inexplicable unless science's findings are largely true. Furthermore, science has shown an ability to attract converts to its theories. Around the world – across human cultures that disagree on much else – scientific procedures, theories, and terminologies converge, driving out traditional explanations for natural phenomena.

Rational inquiry must proceed step by step, and resolving disagreement is possible only if the parties can appeal to mutually shared assumptions and to mutually intelligible examples and analogies. If, however, there is sufficient common basis for inquiry, striking observational data and cogent arguments can be used to overthrow even some assumptions that appear, locally, to be part of common sense. In the early seventeenth century, Galileo challenged the entrenched physics and astronomy supported by the Catholic Church, along with the commonsense assumption that the earth is stationary. He employed a mix of empirical observations and ingenious arguments that used examples and analogies which his contemporaries understood. In particular, he was able to draw upon knowledge and experience that was available within his culture in order to demonstrate the reliability of telescopic observations and the reality of the phenomena that they revealed, such as the moons of Jupiter (Chalmers 1999, pp. 22–24; Kitcher 1993, pp. 227–233).

Sometimes, in the process of inquiry, we make – and argue from – incorrect personal or cultural assumptions. This is one reason why it is good for science to involve numerous participants coming from many intellectual, cultural, geographic, and demographic backgrounds, provided they are acting in good faith by applying widely accepted epistemic standards to the best of their ability. This is assisted by some of contemporary science's institutional characteristics, such as anonymous peer review and post-publication scrutiny by experts from across the world (compare Levy 2017).

By contrast, religions and many sociopolitical ideologies come furnished with their own local and partisan procedures and standards. Consider an extreme example. Stanley Fish has contrasted the epistemic standards of reason with those of fundamentalist Christianity, all in the service of his argument that there is no such thing as liberalism. Fish denies that there are canons of reason that transcend local, partisan 'assumptions and agendas' (Fish 1994, p. 135). Accordingly, he maintains that a Christian fundamentalist who sees evolutionary theory as obviously false, since it conflicts with the literal Genesis record, is in no worse an epistemic position than a scientifically informed secular liberal who views the Genesis record, interpreted literally, as false because it conflicts with robust findings grounded in the methods of scientific inquiry.

---

[5]This paragraph draws on and summarizes the line of argument in Levy (2002, pp. 37–39).

For Fish, this is merely a clash between two rival faiths, 'or if you prefer (and it is my thesis that these two formulations are interchangeable) between two ways of thinking undergirded by incompatible first principles, empirical verification and biblical inerrancy' (Fish 1994, p. 136). He concludes that liberalism is not tolerant, since it does not tolerate – in a very strong sense meaning 'take seriously' or even 'cherish' – views such as those of Christian fundamentalists. It does not matter to Fish that secular liberals are tolerant in an obvious and familiar sense. They usually refrain from employing the traditional methods of suppressing disliked viewpoints: imprisonment, exile, torture, execution, state censorship, and other forms of persecution.

A more plausible conclusion than Fish's claim that liberalism does not exist is that some groups, such those who adopt biblical inerrancy as an axiomatic principle, may lie beyond the reach of persuasion by ordinary, cross-cultural standards of reason and evidence such as refined and put to work in the practice of science. This is one example of a more general idea, that it is always possible, with sufficient ingenuity, to intepret evidence so as to preserve a favored belief. As Jonathan Glover expresses the point, 'No matter how absurd, any belief *can* be preserved if you are prepared to make sufficient adjustments to the rest of the system' (Glover 1988, p. 155; Glover's emphasis).

If necessary, we can save our favorite beliefs from falsification by inventing conspiracy theories, radical new systems of physics, powerful and deceptive supernatural beings, or whatever else might come in handy. Why do this? As Glover explains, we might be sufficiently committed to a belief out of religious or political commitments or for any number of emotionally compelling personal reasons (for example, we might cherish an idea that we associate with our families, lovers, or dear friends, or with happy periods in our lives). If we go down this path, however, we will 'pay a price' (Glover 1988, p. 157) in the sense that others who lack our commitments will find our arguments untenable, or even absurd.

Accordingly, Christian fundamentalists will pay a price. By using their local, partisan epistemic standards, they might, to their own satisfaction, avoid falsification of their doctrines. But this comes at the risk of shutting themselves out of legitimate scientific inquiry and losing credibility in public debate within modern liberal democracies. Christian fundamentalists are an extreme case, but they are not the only religious and ideological dogmatists who use their own partisan standards for seeking truth.

26.5 Science, Religion, and NOMA

Science is a threat to religion. In that sense, the instincts of seventeenth-century leaders of the Church were correct when they attempted to stifle discussion of the Copernican heliocentric theory. From their viewpoint, the theory had shocking implications. Thus, Cardinal Bellarmine (later canonized as Saint Robert Bellarmine) acted decisively in 1615 to warn Father Paolo Antonio Foscarini, a supporter of

Galileo, not to teach heliocentrism as the literal truth. The Copernican theory that the earth revolved around the sun, while rotating on its axis, contradicted the literal words of numerous biblical texts and their interpretations by ancient and contemporary commentators. If it were true, the credibility of the scriptures and the Church would be seriously undermined.

In writing to Foscarini, Bellarmine intimated that the Church would need to move circumspectly to revise its scriptural interpretations if heliocentrism were ever decisively proved, but he observed that no such proof was yet available. In early 1616, following deliberations by the Roman Inquisition, and at the behest of Pope Paul V, Bellarmine directed Galileo himself not to hold, teach, or discuss heliocentrism. Galileo was later tried by the Inquisition in response to his *Dialogue Concerning the Two Chief World Systems*.[6] In 1633, he was summoned to Rome, tried for heresy, threatened with the Inquisition's torture instruments, forced to recant, and placed under house arrest until his death in 1642. Here, we see religious hierarchs exercising secular political power in an effort to suppress ideas that cast doubt on their system of ideas.

As long as religion makes claims about the empirical world, it is open to falsification by science. On any particular occasion, the falsification may relate to a relatively minor doctrinal point that is shown to be incorrect. Nonetheless, these occasions can accumulate to undermine the religion concerned – and perhaps religion more generally. In the twenty-first century, few Christians are concerned by the implications of heliocentrism, but some are very concerned by theoretical findings in geology and biology that relate to the age of the earth and the origin of living species.

In the USA in particular, this has prompted an ongoing struggle by fundamentalists and creationists against evolutionary science. One well-meaning response from a prominent scientist was the late Stephen Jay Gould's principle of Non-Overlapping Magisteria (or 'NOMA'). Bunge, who vigorously rejects this supposed principle, describes it as 'the opinion that there is no conflict between science and [religious] opinion because they deal with non-overlapping domains: science deals with the material world, whereas religion deals with intangibles such as values and souls' (Bunge 2009, p. 146). In his 1999 book, *Rocks of Ages*, Gould develops the idea in detail. He argues that science and religion possess separate, limited, and non-overlapping 'magisteria,' or domains of intellectual and teaching authority, and so can never come into legitimate conflict. Rather, any conflict between science and religion must involve at least one stepping beyond its area of authority.

If we accept Gould's analysis, the magisterium of science relates to 'the factual construction of nature' (Gould 1999, p. 54), which means that science has the authority to document natural phenomena and theorize about explanatory mechanisms. By contrast, religion has authority in respect of 'ultimate meaning and moral value' (Gould 1999, p. 3) or 'moral issues about the value and meaning of life' (Gould

---

[6]This was completed and approved for publication by the Vatican in 1630. It was printed and distributed in 1632.

1999, p. 55). But this creates problems, some of which Bunge identifies. Perhaps most fundamentally, Gould renders what he considers legitimate or authorized religious claims unfalsifiable by science – but only by severely limiting the role that religion has played in the past (and into the present). As analyses of religion such as Bunge's make clear, it has never confined itself to providing unfalsifiable answers to questions of purpose, meaning, and conduct. When he wrote *Rocks of Ages*, Gould was well aware that religion's historical role was far wider:

> At earlier periods of most Western cultures, when science did not exist as an explicit enterprise, and when a more unified sense of the nature of things gathered all 'why' questions under the rubric of religion, issues with factual resolutions now placed under the magisterium of science fell under the aegis of an enlarged concept of religion. (Gould 1999, p. 104)

This shows Gould rejecting the historical and popular concept of religion (Blackford 2000, p. 10). It is not that an original, narrow conception of religion, such as offered by Gould, was later enlarged. Rather, the 'enlarged' conception of religion that Gould mentions was the original and continuing one. Gould's not-so-enlarged conception of religion is, therefore, blatantly revisionist. By contrast, Bunge correctly understands religions as, roughly, systems of belief that interpret human experience and the natural world in terms of their relationship to a supernatural or transcendent realm. In practice, religions have much to say about the workings of the supernatural order upon the natural world and human history. As a result, there is ample scope for religious teachings to conflict with scientific findings about ourselves and the world we live in. For Bunge, then, the idea of NOMA is well-intentioned but intellectually untenable. It overlooks a core of Big Questions in which religion and science both take an interest:

> Did the universe have an origin and, if so, how did it begin? What was the origin of life? How did the many biospecies originate and evolve? What is the mind and how does it originate both in the course of evolution and in that of individual development? How do religions emerge and decline, and what functions do they perform? (Bunge 2009, pp. 146–147)

These questions are, as Bunge argues, handled by science and religion with radically different procedures and standards. Religion relies on faith and grace, the words of sacred scriptures, compatibility with existing dogma, and ultimately claims of divine revelation to founding religious leaders. It changes through textual exegesis and the resolution of conflicts between rival sects. By contrast, science makes discoveries by offering, testing, and revising hypotheses.[7]

Jerry Coyne labels this contrast of procedures and standards as a methodological incompatibility between science and religion (Coyne 2012, p. 2656). Although this makes intuitive sense, it also merits some words of qualification. In theory, such a

---

[7]This may not adequately capture the nature and the full range of scientific arguments, as discussed in the previous section. But on any plausible analysis, when compared to religions, with their holy books, priesthoods, revelations, textual exegeses, and so on, the sciences follow very different procedures and employ very different epistemic standards.

methodological difference might not have been fatal to the credibility of whatever was the true religion. It might, indeed, not have amounted to an incompatibility. There is a sense in which the methods of religion and science *could* have turned out to be compatible and complementary. Prior to the impressive explanatory successes of modern science, it was readily conceivable that scientific inquiry would confirm the general worldview of Christianity, while adding detail about the multiplicity of natural phenomena and the laws governing God's creation. Instead, science did not confirm religious claims about the universe or the human situation. Quite the opposite.

Bunge states that there are further Big Questions that science does not tackle, such as whether the Creator left traces from which we can infer his intentions, and why an omnicompetent deity tolerates evils such as earthquakes and human threats to the environment. He says that these are legitimate questions for skeptics to put to believers, but they are not scientific questions 'because they presuppose the existence of the Creator' (Bunge 2009, p. 147). We need, however, to be careful here. Taken in isolation, some questions related to religion might defy any scientific investigation. That could be so of questions about a hands-off deity that created the universe and then retired without meddling with its handiwork. But at least some questions in the vicinity of those mentioned by Bunge actually are open to scientific investigation.

For example, we can investigate the actions of an alleged deity who would have left specific traces that ought to be detectable. That is, we can check to see whether any such traces are found when we go and look. Thus, whether or not the existence or activity of a divine Creator is scientifically testable depends on exactly what conception of this Creator has been put forward by theologians. More specifically, it depends on what claims have been made about how this Creator regularly interacts, or has interacted on specific occasions in the past, with the observable world.

Bunge also makes the important point that NOMA, if true, 'blocks any fruitful interactions between science, on the one hand, and value theory and ethics on the other' (Bunge 2009, p. 148). The problem here is that Gould's theory hands value theory and ethics solely to religion, cutting them off from any relevant considerations from science or even from secular philosophy. This is an impoverished and unhelpful way to think about ethics and everything that goes with it: not just value theory in the abstract, but questions relating to political philosophy and the nature of a good or just society. It tends to deny that empirical knowledge contributes usefully to moral and political discussion. Bunge offers a good counterexample: he reports that psychology, sociology, and epidemiology have identified ill effects for individual health and social cohesion from large social inequalities. That being so, and assuming that health and social cohesion are worth pursuing,[8] our moral standards and political decisions should be adjusted accordingly.

---

[8]On this occasion, there is no need to unpack different senses – objective or subjective – of 'worth pursuing.'

This is not to deny that at least *some* of our ultimate ethical assumptions must derive from a source other than the domain of impersonal empirical facts. But that source is a matter for philosophical conjecture and debate. Religion, in its usual, non-revisionist meaning, is an unlikely source of our ultimate ethical assumptions.[9]

In the upshot, much of the content of Christian scriptures and Church tradition has been discarded, even by most theologians, because it is inconsistent with newer scientific explanations or because, in any event, it has been rendered irrelevant by the advance of science. This process of theological accommodation to science can go on indefinitely. Even without the NOMA principle, religions can adapt and avoid scientific falsification. That is, they can remove from their teachings any direct contradictions of the best-supported scientific theories of the time. For example, a religion that once taught that our planet is several thousand years old can come to teach that it is really a few billion years old. The superseded doctrine need not even be completely repudiated. It can be reinterpreted as an allegory, or as a story that is worth retaining, not as literal or even allegorical truth, but at least as a venerable record of what was believed in past times by people striving sincerely after knowledge of the divine order. In the limit, almost all of a religion's narratives and founding doctrines might come to be rationalized as metaphors, moral exhortations, venerable stories, and the like.

However, this is a difficult process. Many systems of religious belief are too closely integrated to tamper with easily. A theological system might have an inner logic that prevents its doctrines – at least the more important ones – being jettisoned without reimagining the system as a whole. This can provoke disillusionment, intolerance, and zealous resistance. When theological accommodation to the advance of science does take place, it can also provoke cynicism from doubters and outsiders. We can reasonably question whether the system had any good warrant in the first place.

All of this has a cumulative effect: to an objective viewer, religious explanations of the world and its phenomena now appear speculative, implausible, and premature (in the sense that they were fashioned and promulgated prior to any relevant evidence). Theological systems and traditional mythologies can be elegant and brilliant, and some may contain elements of psychological insight, but they are not credible sources of knowledge in a scientific age.

26.6 Science and Ideology: The Arrow of Causation

For Bunge, sociopolitical ideology cannot legitimately change the content of science: 'The investigators who bow to ideological pressure cease to do scientific work, because they replace the search for truth with the acceptance of dogma' (Bunge

---

[9]There is an extensive body of literature, pro and con, discussing theological approaches to ethics. For one critical account of such approaches, see Blackford (2016, pp. 79–94).

2009, p. 173). This approach is clearly correct. Science has its own procedures and epistemic standards that have incrementally revealed facts about the natural world. By contrast, science is distorted when conformity to the doctrines of one or another contestable ideology becomes, in itself, an epistemic standard. Thus, the arrow of causation should always run from science to ideology: worthwhile ideology should be informed by science, not vice versa.

When ideology is used to override science, the result can be, as Bunge reminds us, travesties such as the Nazi attacks on 'Jewish physics' and Stalinist attacks on 'bourgeois science' (Bunge 2009, p. 173). Current ideological threats include right-wing and fundamentalist attacks on evolutionary biology and stem cell research. More surprising have been dubiously (as Bunge sees them) left-wing characterizations of science as just one more ideology to be unmasked and contested, and feminist attacks on the allegedly phallocentric nature of science and logic (Bunge 2009, p. 173).[10] Bunge also provides examples of what he considers ideology in scientific garb, including monetarist economics and certain work in the field of political science, notably that of Samuel Huntington (Bunge 2009, pp. 175–179). He sees all this as empirically ungrounded, conceptually and methodologically weak, and based on false assumptions about human psychology. In addition, he argues, it is used to support policies that benefit giant corporations while exacerbating global inequalities, causing political and economic dislocation, and cutting social expenditures.

This raises large and murky issues about the value of contemporary economics in particular. It is clear, however, that free market ideology can mold its adherents' acceptance or otherwise of scientific findings. In their recent book on the American political scene, *Asymmetric Politics*, Matt Grossmann and David A. Hopkins comment that Republicans do not show an underlying hostility to science. Rather, they distrust the scientific consensus on specific issues such as evolutionary biology and anthropogenic global warming. In these cases, they reason backwards to reject science where its findings are ideologically inconvenient. In the case of global warming, Republicans reject proposed solutions that call for government intervention in free markets, and so they reject the underlying science (Grossmann and Hopkins 2016, p. 194). Here, the arrow of causation from science to well-informed ideology is reversed.

Like many scientifically oriented philosophers, Levy (2017) is concerned at the very large number of people in the USA who reject scientifically uncontroversial theories. In the case of climate change, as he reports, it is not poor reasoning skills (independently assessed) or ignorance that predicts the rejection of scientific consensus among political conservatives – rather, it is adherence to free market ideology. Simon Keller likewise argues that individuals who reject important findings from climate science may do so, at least in part, out of a prior distrust of environmentalism and a support for unregulated markets. As Keller explains, the

---

[10]Bunge views feminism as a political movement whose goals and achievements are worthy of respect. He is, however, conspicuously unimpressed by academic feminist theory.

revelation of dangerous warming of the atmosphere was not seen as surprising by environmentalists:

> It is a fact that fits quite smoothly with a picture of the world to which we were already committed. It is even tempting to say that the evidence about climate change stands as a vindication of that picture. (Keller 2015, p. 221)

Thus, the facts of anthropogenic global warming tended to vindicate a pre-existing ideology critical of unfettered economic growth. By contrast, it would be psychologically difficult for someone to accept the relevant science if that person began by supporting industrial activity and consumption, and technological innovation, by feeling no special love of nature, and with a distrust of government activity and international institutions. People will find climate science counterintuitive and alien if does not fit with their prior understanding of the social and natural world. They might then be persuaded by attacks on climate science funded by organizations with a material interest in opposing government regulation. However, Levy makes a further point. He provides evidence that American conservatives tend to reject climate science even when their ideology is more a traditional social conservatism than a faith in unregulated markets (Levy 2017). It appears that the intense politicization of climate change, stirred up by organizations with a vested interest in debunking the whole idea, has made acceptance of consensus science in this area into one of the markers of an ideological opponent and (hence) an ill-intentioned, untrustworthy person.[11]

If this is correct, the point may generalize beyond arguments about climate science. Perhaps any debate that becomes sufficiently politicized or moralized in the minds of some participants thereby becomes toxic. That is, those participants will even listen to their opponents, whom they view as inherently malevolent. In such an environment, ideology hinders rational public discussion and impedes scientifically informed policy deliberation.

26.7 Left-Wing Ideologues and the Sociobiology Debate

26.7.1 Ideological Zealotry on the Left

As Bunge's examples make clear, ideologically grounded hostility to inconvenient science does not show itself on only one side of politics. On the contrary, it can be found across the political spectrum, from Nazis who speak of Jewish physics to Stalinists who decry what they call bourgeois science. In that respect, Neil Hamilton's *Zealotry and Academic Freedom* (Hamilton 1995) provides an invaluable history of the successive waves of religious and ideological zealotry that

---

[11] Levy (2017) refers to previous findings that suggest we rely on cues relating to competence and benevolence in deciding whose testimony is reliable.

have threatened the work of American academics since the late nineteenth century. Hamilton begins with religious fundamentalism in the 1870s and concludes with left-wing political zealotry in the late 1980s and early 1990s. Various kinds of ideological zealotry continue to distort public discussion, public policy, and even the practice of science and scholarship.

Alice Dreger's recent book, *Galileo's Middle Finger* (Dreger 2015), recounts the experiences of present-day scientists who have come under ideological attack from the political and academic left. Her examples include: J. Michael Bailey, a psychology professor at Northwestern University whose research relating to transgender people led him to reject an account frequently relied upon by transgender advocates; biologists Randy Thornhill and Craig Palmer who set out their case that rape is, at least usually and predominantly, about sexual gratification, rather than power, anger, misogyny, and a male wish to dominate women; and Napoleon Chagnon, a distinguished anthropologist whose extensive field work with the Yanomamö people of South America led him to an unsentimental, almost Hobbesian vision of our evolved human nature – this was at least part of what got him into trouble with colleagues.

The journalist Patrick Tierney accused Chagnon (and the late James V. Neel, Sr.) of gravely unethical and harmful medical practices in their dealings with the Yanomamö. Chagnon was ultimately cleared of all charges of misconduct, but the impact on his personal life and his family was so severe that he was hospitalized on one occasion after collapsing with stress.

In all of these cases, candid, intellectually honest, and seemingly decent people were subjected to abuse ranging from misrepresentation of their viewpoints, to death threats, to bogus charges of misconduct. This was the price they paid for their honesty and a certain fearlessness that, as Dreger elaborates throughout her book, they shared with Galileo. For left-wing ideologues, such as those who harmed Bailey, Thornhill and Palmer, and Chagnon, the causal arrow ought to go from ideology to science. Scientists who don't defer to ideological dogma are viewed by such ideologues as fair targets for punishment.

Writing of the controversy that followed Edward O. Wilson's book *Sociobiology: The New Synthesis* (Wilson 1975b/2000), Ullica Segerstråle refers to the mentality of a witch hunt that can infect ideological attacks on science:

> Controversies involving sensitive political issues exhibit something of the social psychology of witch hunts. Once they have started, it does not help much that the targets themselves protest and try to demonstrate their innocence. The original interpretation tends to stick, and those who criticize it as incorrect or unfair – or worse, try to defend the target – run the risk of being identified themselves as supporters of the same unpopular cause that got the target in trouble in the first place. (Defend someone as not being racist, and you automatically come under suspicion for racism yourself.) This was exactly what happened in the sociobiology controversy. (Segerstråle 2000, p. 15)

Wilson's huge book synthesized much of what was known about social animals and the working of animal societies. The controversy arose from its final chapter, which treated *Homo sapiens* as one more social animal among others, suggesting that an evolved human nature lies beneath the diversity of human cultures. For many left-wing ideologues, this was a heresy that could not be tolerated.

26.7.2 The Sociobiology Dispute

Since the early 1990s, human sociobiology has morphed into the vigorous, though controversial, field of evolutionary psychology, announced most prominently in the flagship collection of papers *The Adapted Mind: Evolutionary Psychology and the Generation of Culture* (Barkow et al. 1992). Irrespective of terminology, however, and any nuances of approach, human sociobiology and evolutionary psychology study human nature in the light of evolutionary biology. Arguments about human behavioral tendencies are made by drawing analogies with the behavior of other species (especially other primates), comparing the social behavior of nonhuman primates with that of humans in present-day hunter-gather societies, pointing to consistencies between the identified human behavior and findings from elsewhere in the fields of biology, psychology, anthropology and history, and referring to experimental outcomes that appear to corroborate hypotheses based on evolutionary reasoning.

Scientists working in this field often frame hypotheses, and develop arguments, that employ the concept of inclusive fitness. In brief, this refers to the expected number of offspring that will be produced by a particular organism and genetically related organisms, weighted by the closeness of the relevant genetic relationships. Thus, an increase in inclusive fitness implies an increase in the extent to which the organism's genes[12] will be replicated. Genes that predispose an organism to behave in a particular way will tend to be passed down to later organisms so long as the behavior promotes inclusive fitness in the first organism's environment. The hypotheses, theories, and arguments of sociobiologists[13] can be tested using the ordinary procedures and standards of science, such as whether they are broadly consistent with existing, well-established scientific theories; whether they yield predictions that lead to further observations and experiments, and whether these prove to be corroborative; and how they compare with rival explanations.

In 1975, however, opponents of (human) sociobiology were not content to rely on this methodology. Their underlying fear was that sociobiology might identify race- and sex-related differences in cognition, emotional tendencies, and behavior that could then be used to rationalize racist or sexist beliefs and to justify social inequalities. For such opponents of sociobiology, it is not necessarily reassuring to

---

[12]More accurately, the component of its genome that distinguishes it from other organisms in which it is in evolutionary competition.

[13]This terminology is used for the sake of simplicity, and to reflect the time of the original dispute. Absent those considerations, it would now be more appropriate to write of evolutionary psychologists rather than sociobiologists.

be told that its hypotheses relate to psychological and behavioral differences only at the level of populations, and hence give no support for stereotyping individuals or denying them equal opportunity. That might be so, but it depends on the details, including the characteristics involved and the size of any differences identified.

On 13 November 1975, the *New York Review of Books* published a letter with 16 signatories, under the heading 'Against "Sociobiology"' (Allen et al. 1975). The signatories included the distinguished biological scientists Stephen Jay Gould and Richard Lewontin, both of whom were among Edward O. Wilson's colleagues at Harvard University. Much the same attack on Wilson and sociobiology was then elaborated in 'Sociobiology – Another Biological Determinism,' authored by Lewontin and others on behalf of a group calling itself the Sociobiology Study Group of Science for the People (1976). In both cases, the authors attempted to smear Wilson's views with guilt by association, alleging a similarity to, and continuity with, earlier theories supposedly endorsing biological determinism. The *New York Review of Books* letter alleged that such theories survive, in one form or another, because 'they consistently tend to provide a genetic justification of the *status quo* and of existing privileges for certain groups according to class, race or sex' (Allen et al. 1975).[14]

Whatever the scientific merits of the rival positions, the undoubted (and unhidden) intention was to strangle sociobiology at birth. Those involved viewed Wilson's project as a threat to their ideological (in this case Marxist) beliefs and aims. With considerable justification, Wilson protested:

> The issue at hand, I submit, is vigilantism: the judgment of a work of science according to whether it conforms to the political convictions of the judges, who are self-appointed. The sentence for scientists found guilty is to be given a label and to be associated with past deeds that all decent persons will find repellent. (Wilson 1976, p. 183)

This episode, followed by the continuing bitter resistance to sociobiology/evolutionary psychology in many quarters, brings to mind the efforts by Cardinal Bellarmine and other leaders of the seventeenth-century Catholic Church to block heliocentrism. This is not to predict that evolutionary approaches to the study of human cognition, emotional tendencies, and behavior will eventually prevail as heliocentrism did. Some controversial evolutionary hypotheses in this area may ultimately be rejected. Nor is there anything wrong with subjecting hypotheses about human cognition, emotions, and behavior – developed by sociobiologists or others – to careful scrutiny.

Neil Levy has expressed dismay at the sociopolitical implications of some hypotheses about human nature based on evolutionary reasoning, but he insists that he does not therefore regard these hypotheses as *ipso facto* false. Rather, he urges, it is all the more important to assess their truth 'by examining them as sets of empirical

---

[14]Wilson responded in both the *New York Review of Books* (Wilson 1975a) and *BioScience* (Wilson 1976). In each case, he objected vigorously to the criticism he'd received, pointed to various distortions of his actual position, and reproached what he saw as a form of self-righteous academic vigilantism.

claims; that is, by assessing the degree to which they are coherent, supported by evidence, have considered and eliminated alternative explanations, and so on' (Levy 2004, p. 139). This approach is correct in principle. No one can complain if scientific hypotheses that seem iconoclastic from one perspective or another are subjected to close scrutiny using science's ordinary procedures and standards.

This applies all round: it was legitimate in the seventeenth century for the Church to test the observations and arguments in favor of heliocentric theory (but not to punish its proponents as heretics); it was legitimate to test the observations and arguments supporting biological evolution; and it was, and remains, legitimate to test those supporting anthropogenic global warming. A time does come, however, when the scientific issues are settled, with no reasonable prospect that the consensus will ever be overturned. That point has been reached with heliocentrism and evolution, and probably also with global warming. It is too early to make such a claim about the most controversial ideas proposed by sociobiologists.

In itself, scientific scrutiny is healthy. As Segerstråle concludes in her study of the sociobiology controversy, science is improved by criticism. Reviewing the course taken by the controversy, she suggests that the field was eventually better off for critiques driven by moral and political concerns, though these should never have included the highly personal and inflammatory attacks that were endured by Wilson and others (Segerstråle 2000, pp. 404–408). The dispute led to many scientists expanding their skills in order to take part, to newcomers with relevant expertise joining in, to proponents of sociobiology addressing concerns and developing more sophisticated theories, and to Richard Dawkins' lucid explanations to the public of the concepts involved. None of this, however, was the intent of ideologically motivated opponents who collaborated in an effort to block the research program implicit in the final chapter of Wilson's *Sociobiology*.

If they'd had their way – perhaps if they'd used smarter tactics – they'd have succeeded in preempting, rather than improving, scientific debates about genes, culture, and human nature.

26.7.3 Bunge on the Sociobiology Controversy

To Mario Bunge's credit, he defends the proponents of sociobiology from the early accusations that it was 'a piece of reactionary ideology on a par with Social Darwinism.'[15] He correctly points out that 'sociobiology started as an earnest attempt to explain sociality in terms of modern biology' (Bunge 1998, p. 35). Nonetheless, he is scathing about the scientific merits of the enterprise to such an

---

[15]There is a separate question, beyond the scope of this chapter, as to whether Social Darwinism and its leading nineteenth-century proponent, Herbert Spencer, are treated fairly by their present-day critics. For a searching examination of Spencer's views, suggesting that they were more intellectually defensible, and perhaps more humane, than is usually acknowledged, see Richards (1987, pp. 243–330).

extent that he misrepresents its arguments. For example, he describes sociobiology as a discipline 'which attempts to reduce the entire field of social behaviour to biology and, in particular, to neo-Darwinian evolutionary theory' (Bunge 1998, p. 35). But this is tendentious at best.

In his 1975 foray, Wilson did make the provocative suggestion that the humanities and social sciences are branches of biology. In a sense, however, this is true: that is, we could set out an idealized typology of academic disciplines in which study of the social and cultural behavior of a particular species, *Homo sapiens*, is depicted as a sub-set of the biology of that species, and hence as branching from biology itself. It does not follow that attempts to understand human society and culture require any less time and effort than we currently see. *Homo sapiens* is, after all, a planet-dominating species with *very* complex behavior. Nor does it follow that all human social behavior can be reduced directly to biology, with no mediation from culture and consciousness. That was not Wilson's argument in 1975, and nor is it argued by his intellectual successors, contemporary evolutionary psychologists.

Among Bunge's other complaints, he alleges that sociobiology has failed as science because sociobiologists are 'like psychoanalysts and religious fundamentalists [. . .] obsessed with sex and reproduction' (Bunge 1998, p. 35). He immediately acknowledges that sex and reproduction are important, but he states that higher animals, particularly human beings, are more concerned with their own survival than with spreading their genes. In support, he cites research to the effect that most men are monogamous, that most families are planned, and that 'in modern human societies dominance and status are inversely correlated with fertility' (Bunge 1998, p. 35).

But sociobiology was never based on an assumption that human beings are consciously motivated to reproduce as much as possible (even if some dynasts and patriarchs have, indeed, thought and acted along those lines). To reproduce their genes, the ancestors of modern human beings needed to survive until they were fertile, have a liking for sex, and attract mates – but they did not need a conscious wish to reproduce. The point of evolutionary reasoning is not to argue that contemporary humans consciously seek to maximize their reproductive success. The issue, rather, is what phenotypical traits, including any genetically linked psychological and behavioral tendencies, might have enhanced inclusive fitness in our ancestral environments and thus persisted. Evolutionary reasoning about present-day human and animal behavior thus distinguishes between its proximate and ultimate causation.

In *Sociobiology*, Wilson discussed sex and reproduction only briefly, emphasizing the prevalence in human societies of permanent, but not necessarily monogamous, bonds. He also drew attention to the development in human beings of a dissociation between sex and reproduction. That is, he anticipated Bunge's complaint, right from the beginning:

> Sexual behavior [in *Homo sapiens*] has been largely dissociated from the act of fertilization. It is ironic that religionists who forbid sexual activity except for purposes of procreation should do so on the basis of 'natural law.' Theirs is a misguided effort in comparative

ethology, based on the incorrect assumption that in reproduction man is essentially like other animals. (Wilson 1975a, b/2000, p. 554)

At one point, Bunge dismisses human sociobiology as 'a failed science' that 'failed in skipping levels, just as atomic physics would fail if it attempted to explain earthquakes' (Bunge 1998, p. 36). Perhaps the enterprise will eventually fail, but that has not yet happened. In this instance, it is Bunge who has failed to appreciate the different levels of explanation employed by serious contributors to evolutionary reasoning about human behavior. Thus, he misunderstands and underestimates the resources available to practitioners in what has become the field of evolutionary psychology.

However, little of present interest depends on the outcome of this skirmishing. Whether or not Bunge, or any other particular critic, has given Wilson and those who have followed him a fair hearing, it is acceptable to subject any scientific research program to rational scrutiny. If the direction taken by particular research has iconoclastic implications – again, viewed from one or another perspective – the need for scrutiny becomes more urgent. The science must then stand or fall based on relevant evidence and argument.

26.8   Can We Live Without Ideology?

Perhaps all religions are false. That is, there is no supernatural or transcendent realm; we simply do not live in the kind of twofold cosmos described in religious teachings. Even so, there might be true sociopolitical ideologies: ideologies whose factual claims correspond to sociopolitical reality. Bunge's lifetime work can, indeed, be interpreted as an attempt to devise just such a factually true, as well as socially beneficial, ideology, one informed by science and employing no special, partisan procedures and standards to establish its truth-claims. Bunge envisages an ideology based on established social science. It would, for example, take account of human beings' inbuilt wish for fairness, the negative impacts on the individuals of unjustified discrimination, and the social damage caused by severe inequality (Bunge 2009, pp. 135–136).

To develop such an ideology, Bunge elaborates, it will be necessary to bring together a voluntary association of people of good will; a society with a modicum of scientific and technological communities; a scientific and humanistic outlook; a fund of up-to-date scientific and technological (including legal) ideas; and a range of social, environmental, etc., issues – but note that Bunge excludes what he considers 'phony and diversionary issues such as gay marriage, assisted suicide, and abortion' (Bunge 2009, p. 180).[16] Meanwhile, Bunge regards all extant ideologies as

---

[16]Here, however, there is a problem. Why dismiss these as phony issues, and who decides which issues are phony? Perhaps there are *bigger* issues, those with a greater impact on the sustainability of societies and on international cooperation upon fair terms. Nonetheless, the issues that Bunge

unscientific. As he remarks, traditional ideologies, both religious and secular, over-simplify reality. In addition, they often attract fanatics – intolerant extremists 'who divide humankind into friends and foes.' Though extremists usually fail, sometimes 'they triumph because they attain great military power or get hold of great wealth, and then cause grief all round' (Bunge 2009, p. 181).

Indeed, ideologies can be dangerous – the more so when their truth-claims amount to comprehensive understandings of the natural and social worlds, and especially when their programs for social change become apocalyptic, as with Nazism, Stalinism, and Pol Pot's agrarian socialism. Each of these attempted to force social and economic transformations on an immense scale, with no toleration for enemies or for any dissent. Comprehensive, apocalyptic ideologies offer their adherents a chance to participate in grand historical struggles that are represented as just and necessary. For many of their partisans, this may be inspiring, idealistic, and exhilarating, but it also leads to totalitarianism and atrocities.

Bunge denies that we can do away with all ideologies, since – he claims – all individuals need them to orient themselves in society and the world more generally, while political regimes need them to design policies and to 'mobilize or immobilize the citizenry' (Bunge 2009, p. 179). Hence, he sees the challenge as developing a humane and scientifically grounded ideology, rather than attempting to sweep all ideology aside. His observation about political regimes is plausible, and he is probably correct to state that we cannot do away with all ideologies. However, the claim that each individual needs an ideology is unconvincing. As Bunge himself notes, 'most activities in daily life are ideologically neutral' (Bunge 2009, p. 180). Indeed, as he also acknowledges, most people have become cynical about formal ideologies and place more value on various personal interests and bonds with others.

In a trivial sense, everyone has some kind of *worldview*. This will include some general understanding of human nature, how the individual's own society works, and that person's situation within it. Someone's worldview might contain an extensive stock of beliefs derived from everyday experience and from science, without being a *comprehensive* worldview: the person might have no opinion on a wide range of topics, perhaps because of a lack of interested in them or perhaps because the relevant evidence available seems incomplete or ambiguous. That worldview might not include the necessary supernaturalist content to count as a religion, and it might include very little in the way of sociopolitical beliefs or commitments to specific political aims, policies, and methods.

It follows that even a well-informed individual's worldview might not contain enough content with the right structure to match Bunge's description of a sociopolitical ideology. That worldview certainly need not resemble any of the well-known ideologies whose partisans wield or strive for power in current societies. It is, then, quite possible to live without an ideology, and it is definitely possible for an

---

dismisses are genuinely, reasonably, and deeply important to many people. For many severely and terminally ill people, for example, proposals for physician assisted suicide are far from being bogus or trivial.

individual to get by with a cynical attitude to whichever off-the-shelf sociopolitical ideologies happen to be prominent locally. More generally, we can adopt a mindset that accepts much incompleteness in our understanding of the world – in which case, we will be unlikely to commit to any extant religion or well-known sociopolitical ideology.

While embracing an extant ideology is not needed to orient ourselves to living socially, or to forming a viable worldview, some minimal ideological content might be needed to motivate sincere and active involvement in politics: involvement, that is, impelled by something more than our material interests. Whether we press for social change or attempt to defend the status quo, we are normally moved by the prospect of some political good to be obtained, or some evil to be overcome (perhaps a current injustice, or perhaps a threat to what we perceive as a benign social order). To make any progress, we will need to clarify our aims and policies, and we'll need to work with like-minded others. Taken together, this is at least the beginning of an ideology. What is *not* required, however, even for collective political action, is allegiance to any of the well-developed sociopolitical ideologies. In many cases, it is possible for adherents to a variety of ideological systems, along with individuals who do not subscribe to any identifiable ideology at all, to form coalitions for limited purposes. In particular, whenever there is an urgent need to oppose great evils, people with many ideologies and worldviews are likely to make common cause.

26.9 Conclusion

Though we can, as individuals, live without adopting an ideology of any kind, sociopolitical ideologies are required for the purposes of political regimes. More generally, they help to bind large numbers of people to pursue shared aims and policies. In addition, religions continue to attract adherents for many reasons, among them the psychological comfort and sense of transcendent purpose that religions offer, and perhaps the allure of learning cosmic truths that are otherwise inaccessible. For such reasons, there is no prospect of doing away with all ideologies. We can, however, study them, and we can point to their weaknesses and dangers.

As Bunge illuminates in his analysis, religions and sociopolitical ideologies have much in common, and accordingly they can play similar roles in their adherents' lives. One point of commonality is that, to say the least, religions and sociopolitical ideologies have a poor record in justifying their claims. Moreover, comprehensive, apocalyptic ideologies of all kinds should be handled with caution. Even comparatively moderate ideologues can inspire zealotry and dogmatism in pursuit of their aims and policies, and in attempts to preserve their cherished truth-claims. When confronted by inconvenient scientific ideas, ideologically motivated people frequently respond by reasoning backwards, rejecting whichever hypotheses, theories, or research projects have implications that they find troubling.

Ideologues may, therefore, judge science against their own partisan beliefs, procedures, and epistemic standards, rather than allowing scientific findings to inform the emergence – anticipated and encouraged by Bunge – of an ideology grounded in reason. In this respect, ideologies, whether religious or secular, can become enemies of free scientific inquiry.

References

Allen, E., Beckwith, B., Beckwith, J., Chorover, S., Culver, D., Duncan, M., Gould, S., Hubbard, R., Inouye, H., Leeds, A., Lewontin, R., Madansky, C., Miller, L., Pyeritz, R., Rosenthal, M., & Schreier, H. (1975, November 13). Against 'sociobiology'. *New York Review of Books*. http://www.nybooks.com/articles/1975/11/13/against-sociobiology/. Accessed 27 Feb 2018.
Barkow, J. H., Cosmides, L., & Tooby, J. (Eds.). (1992). *The adapted mind: Evolutionary psychology and the generation of culture*. Oxford: Oxford University Press.
Blackford, R. (2000). Stephen Jay Gould on science and religion. *Quadrant, 365*, 8–14.
Blackford, R. (2012). *Freedom of religion and the secular state*. Chichester/West Sussex: Wiley-Blackwell.
Blackford, R. (2016). *The mystery of moral authority*. Basingstoke: Palgrave Macmillan.
Bunge, M. (1998). *Social science under debate: A philosophical perspective*. Toronto: University of Toronto Press.
Bunge, M. (2009). *Political philosophy: Fact, fiction and vision*. New Brunswick: Transaction Publishers.
Chalmers, A. F. (1999). *What is this thing called science?* (3rd ed.). St Lucia: University of Queensland Press.
Coyne, J. A. (2012). Science, religion, and society: The problem of evolution in America. *Evolution, 66*, 2654–2663.
Dreger, A. (2015). *Galileo's middle finger: Heretics, activists, and the search for justice in science*. New York: Penguin.
Fish, S. (1994). *There's no such thing as free speech and it's a good thing, too*. Oxford/New York: Oxford University Press.
Glover, J. (1988). *I: The philosophy and psychology of personal identity*. London: Penguin.
Gould, S. J. (1999). *Rocks of ages: Science and religion in the fullness of life*. New York: Ballantine.
Grossmann, M., & Hopkins, D. A. (2016). *Asymmetric politics: Ideological republicans and group interest democrats*. New York: Oxford University Press.
Haack, S. (2007). *Defending science – Within reason: Between scientism and cynicism*. Amherst: Prometheus.
Hamilton, N. (1995). *Zealotry and academic freedom: A legal and historical perspective*. New Brunswick/London: Transaction Publishers.
Keller, S. (2015). Empathising with scepticism about climate change. In J. Moss (Ed.), *Climate change and justice* (pp. 219–235). Cambridge: Cambridge University Press.
Kitcher, P. (1993). *The advancement of science: Science without legend, objectivity without illusions*. Oxford: Oxford University Press.
Levy, N. (2002). *Moral relativism: A short introduction*. Oxford: Oneworld.
Levy, N. (2004). *What makes us moral? Crossing the boundaries of biology*. Oxford: Oneworld.
Levy, N. (2017). Due deference to denialism: Explaining ordinary people's rejection of established scientific findings. *Synthese*. https://doi.org/10.1007/s11229-017-1477-x.
Pinker, S. (2002). *The blank slate: The modern denial of human nature*. New York: Viking Penguin.
Rey, G. (2007). Meta-atheism: Religious avowal as self-deception. In L. M. Antony (Ed.), *Philosophers without gods: Meditations on atheism and the secular life* (pp. 243–265). Oxford: Oxford University Press.
Richards, R. J. (1987). *Darwin and the emergence of evolutionary theories of mind and behavior*. Chicago/London: University of Chicago Press.

Segerstråle, U. (2000). *Defenders of the truth: The sociobiology debate*. New York: Oxford University Press.

Sociobiology Study Group of Science for the People. (1976). Sociobiology – Another biological determinism. *BioScience, 26*(182), 184–186.

Taylor, C. (2007). *A secular age*. Cambridge, MA: Harvard University Press.

Wilson, E. O. (1975a, December 1). For sociobiology. *New York Review of Books*. http://www.nybooks.com/articles/1975/12/11/for-sociobiology/. Accessed 27 Feb 2018.

Wilson, E. O. (1975b). *Sociobiology: The new synthesis* (25th anniversary ed.). Cambridge, MA/London: Harvard University Press.

Wilson, E. O. (1976). Academic vigilantism and the political significance of sociobiology. *BioScience, 26*(183), 187–190.

Part VI
Ethics and Political Philosophy

Chapter 27
Ethical Politics and Political Ethics I: Agathonism

Michael Kary

Abstract Mario Bunge calls his ethical system *agathonism*, to signify "seeking the good". For Bunge the good is above all else the satisfaction of human needs. In this way, human needs act as the foundation stone of Bunge's ethical system. The aim of this chapter is to reset the foundation stone of agathonism, while allowing the valuable portions of the edifice to remain. Space limitations leave fulfilment of the latter as an exercise for the reader. Consequently, this chapter is mostly of a critical nature. New characterizations of rights, moral duties, and the relation between rights and duties are proposed.

27.1 Introduction

The great project of Mario Bunge's *Ethics: The Good and the Right* (volume 8 of his *Treatise on Basic Philosophy*; Bunge 1989) is to build an objective system of morality based on a scientific understanding of the human condition. The great project of his *Political Philosophy: Fact, Fiction, and Vision* (Bunge 2009) is to advance a political system and a society based on that system of morality. Bunge sees an urgent purpose to these two projects: to rescue mankind. This principally from two great evils, social injustice and extinction. The latter, the extinction of mankind specifically, not any other species, is for him the greatest possible evil of all. For Bunge, social injustice is inherent to capitalism and liberal democracy, and to a lesser extent even to social democracy; the extinction of our species, an imminent consequence of at least the first, if not of all three.

Another project, implicit in these works but no less evident, is another rescue operation: to rescue the socialism of his father and of his youth—from irrelevance, from obsolescence, from its senescence; from its temptress, social democracy; and from the bad company it has kept, for more than a century now.

M. Kary (✉)
Montreal, Canada
e-mail: mkary111@gmail.com

© Springer Nature Switzerland AG 2019
M. R. Matthews (ed.), *Mario Bunge: A Centenary Festschrift*,
https://doi.org/10.1007/978-3-030-16673-1_27

Along the way of these endeavours Bunge undertakes his usual job of sketching out the lay of the land, clarifying and defining various key concepts of the field, and systematizing his own postulates. Also as usual, or rather more so, and especially when arguing for his societal project and against others, Bunge writes as if bursting at the seams, overflowing with ideas and opinions that he is unable to pour out fast and fully enough before moving on to his next ambitious projects, all in the endeavour to find container enough for that overflowing volume.

As one might expect from such a cataract, the flow is uneven. Just within the rush of either offhand or more or less considered opinions, Bunge seeks to, if not rescue us from, at least open our eyes to a litany of lesser evils drawn from his own personal list of *bêtes noires*. These include electric guitars, rock music more generally (or rather as he hears it, rock noise), professional sports, tycoons, vulgar TV comedians, the unplanned life, solitude, economic parasites rich and poor, playing ghetto blasters (pronounced a cultural crime), tachiste painting (pronounced toxic, his example of pseudo-, bogus, and junk art, three terms for the same thing all in the same paragraph), charity, harmless nobodies, and of course religion. Alongside these digressions Bunge also explains, at least in outline, how to solve the environmental crisis, ensure a just society, and put an end to war. Some of his pronouncements can be read as comic relief (perhaps unintended, perhaps not), but many of the opinions and ideas on and off these lists have a source in Bunge's practical politics, his scientism or his socialism, his optimism or his pessimism, his skepticism or his cynicism or his credulity, with all of these being interlinked. Every sentence is some combination of controversial, insightful, objectionable, enlightening, exasperating, and sometimes even absurd (such as his proposal to mine the moon for metals and minerals)—in total interesting throughout.

This analysis of Bunge's ethics and political philosophy is split into two companion chapters. The present one deals predominantly with his ethics. No chapter and not even two chapters can do justice to all their diversity, nor to all their merits and demerits. Instead, the emphasis is on a selection of fundamentals, to include the foundation stone.

A question raised by all of this is what role Bunge's prescriptions are expected to fulfil, and so how to evaluate them. For example, if taken as friendly advice, it is hard to find fault with much of Bunge's ethics, including his supreme norm, "Enjoy life and help others live" (to which more recently he has added a parenthetical "an enjoyable life"). But although Bunge surely sees his *Ethics* and *Political Philosophy* as admirable self-help books, for the individual and society, he also sees them as much more: he adopts his father's view that "politics ought to be the arm of morality" (Bunge 1989, p. xv; 2009, p. 8), and even "the strong arm of morality" (Bunge 1990a, p. 657). He flags "moral errors, that is, crimes of various sizes" (Bunge 2009, p. 204); asserts that failure to do one's duty is a moral crime (Bunge 1989, p. 274); contends that the law should conform to morality; and says that helping others to live, be healthy, and perhaps even to be reasonably happy, is a requirable moral duty (Bunge 1989, p. 97).[1]

"Enjoy life" is benign as a (suitably qualified) suggestion, "help others live (an enjoyable life)" only as mixed a bag of potential harms and benefits as any other well-intentioned call to action, but—especially considering Bunge's systemic biological-psychological-social concept of health, his advice to "face it: there are few if any strictly private vices" (Bunge 1990a, p. 656), and his view that "a sick saint is as useless to society as a sated pig" (Bunge 1989, p. 104)—they take on a whole other significance if not doing them is considered a crime of any sort. This implication is related to Bunge's political philosophy, and so is revisited in the conclusion to the companion chapter.

True, contrary to Bunge's assertion that "there is no law ordering us to help one another" (Bunge 1989, pp. 96–97), Good Samaritan laws already do exist, including in Bunge's own Canadian home province; but their application is far more limited than might be the case following Bunge's system. Good Samaritan laws only require us to offer assistance to those in imminent physical danger. But Bunge commands us to enjoy life, and requires us to help others do the same. Is being sour (let alone depressed and unmedicated) and isolated—thus flouting the supreme norm—really a supreme crime? Not that Bunge explicitly draws this conclusion himself. But it is hard to see how he could reject it and remain completely consistent.

Since, if taken as friendly advice, there is not all that much to complain about Bunge's ethics, and much more to like, here they are considered more seriously.

27.2 Some Fundamentals of Bunge's Ethical Doctrine

27.2.1 Some Deciphering

In keeping with the rest of the *Treatise*, Bunge presents his ethics in an axiomatic format. One of the merits of the axiomatic method is that it should clearly distinguish between assumptions, proposals, conventions, and deductions. Bunge admits though that postulates may be disguised as definitions, something he does explicitly in one instance (Bunge 1989, p. 187). Further, unlike the rest of the *Treatise*, Bunge relies on many terms from ordinary life used idiosyncratically, and sometimes inconsistently. All this combined with many layers of definitions upon definitions often obscures or confuses what he is really claiming or proposing. In order to analyze Bunge's ethics, his system is in need of some deciphering.

The most fundamental concepts of agathonism are survival, well-being, need, and evaluation and thus value, which Bunge first interrelates as follows:

---

[1] Bunge leaves the force of the requirement unspecified, but in context it appears he intends that requirable moral duties are those that morality demands be enforced by law.

> We take it that all animals evaluate some things and some processes... (Bunge 1989, p. 1)

> We also take the commonsensical view that all normal animals strive to attain or retain a state of well-being – which, however, is not the same for all. Consequently normal animals value positively, i.e. they find good, anything they need for their well-being and, in the first place, for their survival. (Bunge 1989, p. 4)

> ...there are no values in themselves but there are valuable items wherever there are organisms. According to this view values emerged on our planet about four billion years ago together with the first organisms capable of discriminating what was good for them, i.e. what was favorable to their survival, from what was not. (Bunge 1989, p. 13)

> Whatever favors the survival of an organism, hence is biovaluable to it, contributes to its health; conversely, whatever is detrimental to its health is biologically disvaluable to it even though it may be valuable in other respects.[2] (Bunge 1989, p. 21)

> Only a few basic values, first of all survival, "are in our genes". (Bunge 1989, p. 23)

Early on Bunge specifies the basic needs to be both survival and everything needed for it (Bunge 1989, p. 5). Together these form what Bunge calls the primary values. What he specifies as "legitimate" wants are well-being and everything needed for it, giving as examples of the latter love, security, activity, and advancement. These items, to include well-being itself, being only wants, are classified as secondary values (Bunge 1989, p. 5). But further on (p. 11) Bunge instead says that well-being is a survival condition, and on the same page appears to contradict his earlier want-implies-value linkage by asserting "No organisms, no needs, hence no values." Likewise, later (p. 35) Bunge instead defines the basic needs to be composed of not just what he calls the primary needs, survival and its necessities, but also of what he now calls the secondary needs, health and its necessities.

In any case, only much further on does Bunge explain what he takes to constitute well-being, in the form of a postulate disguised as a definition:

> DEFINITION 2.1 An animal (in particular a human being) is in a state of well-being (or welfare) if and only if it has met all of its basic needs (Definition 1.12). (Bunge 1989, p. 44)

Bunge's Definition 1.12 is:

> DEFINITION 1.12 Let x be a biological, psychological or social deficit of a human being b in circumstance c. We call x
>
> (i) a *primary need* of b in c if and only if meeting x is necessary for b to stay alive under c in any society;

---

[2] In fact health and survival, or longevity, can conflict. Examples: The physical exercise required for longevity frequently results in overuse injury. Occasional sickness in infancy and childhood, in the form of immunological challenge, is necessary for development of the immune system and thus longevity. Escaping predation may require getting injured, or hiding for long periods without food or water, or sheltering in pathogenic circumstances. Relatively severe caloric restriction, which in mice increases or reduces longevity depending on the strain (Liao et al. 2010), has a variety of effects (such as reduction in core body temperature) whose full consequences for health are still to be determined, but which in any event have to be investigated independently of the effect on longevity (Redman and Ravussin 2011).

(ii) a *secondary need* of *b* under *c* if and only if meeting *x* is necessary for *b* to keep or regain health under *c* in *b*'s particular society;

(iii) a *basic need* if and only if *x* is a primary or a secondary need. (Bunge 1989, pp. 34–35)

In short then, Bunge's formal definition of well-being is the state of being alive and healthy. Since presumably one can only be healthy while alive, this can be further reduced to: well-being is the state of being healthy.[3]

Bunge uses these ideas to form in several steps his most important derived concept, the aforementioned and startling idea of a "legitimate" want. In telegraphic form: an item is *psychologically valuable* if someone wants it; if psychologically valuable, it is a *psychologically legitimate want* if it contributes to their long-term health, otherwise it is *psychologically illegitimate*; if psychologically legitimate it is a *legitimate want* (or *desire* or *aspiration*) if it can be met without hindering the satisfaction of any of anyone else's basic needs, and without endangering the integrity of society or any of its valuable subsystems (sequentially definitions 1.6, 1.7, and 1.13, Bunge 1989, pp. 27, 35).

Reduced to lowest terms, in agathonism "legitimate" wants (desires, aspirations) are those whose satisfaction contributes to the long-term health of the individual, without hindering the survival or health of anyone else, and without endangering the "integrity" of the individual's society.[4] Put another way, for Bunge the only legitimate wants are those which innocuously satisfy basic needs.[5]

This idiosyncratic usage, combined with the layers of definitions upon definitions, is confusing not just for the reader. Further on, instead Bunge repeatedly says something contrary. For only a few examples:

> ... whereas the satisfaction of primary needs is a matter of life and death, and that of *secondary needs a matter of health or sickness, the satisfaction of legitimate wants is one of happiness or unhappiness.* (Bunge 1989, p. 36, emphasis added)

> We group needs and wants into four clusters: primary needs (to be met for sheer survival), *secondary needs (to be satisfied in order to enjoy good health), legitimate desires (to be met for the sake of reasonable happiness),* and fancies. (Bunge 1989, p. 40, emphasis added)

> NORM 2.1 Long-term well-being and, a fortiori, reasonable happiness, calls for the following ranking: Meeting primary needs (survival) ought to precede meeting *secondary needs (health), which in turn should precede meeting legitimate wants,* which ought to dominate the satisfaction of fancies. (Bunge 1989, p. 48, emphasis added)

> DEFINITION 1.14 Let *x* be a thing, a property of a thing, or a process in a thing (in particular a mental process in someone's brain). We attribute *x* (i) a primary value for human

---

[3]Bunge spontaneously re-defines well-being to be something different, the state of being comfortable—or rather, the state of unwell-being as the state of being uncomfortable (Bunge 1989, p. 45). This variant definition does not appear to be the one he is using anywhere else.

[4]What constitutes the integrity of a society is not explained in either Bunge's *Ethics* or his *Political Philosophy*. In Volume 4 of his *Treatise*, Bunge briefly describes system integrity in general (Bunge 1979, pp. 35–38). How Bunge would define it in such a way as to allow social and political reform—to include unification with other societies or the splitting or reconstituting of a single society or any of its subsystems—is an open question.

[5]When basic needs are as in Bunge's Definition 1.12, not as given on his p. 5.

beings in circumstance $c =_{df} X$ contributes to satisfying at least one primary need of any humans, in any society, when in circumstance c; (ii) a secondary value for human beings in circumstance $c =_{df} X$ contributes to meeting at least one of the secondary needs of humans under c in their particular society; (iii) a tertiary value for human beings in circumstance $c =_{df} X$ contributes to meeting at least one of the legitimate wants (or desires or aspirations) of humans in circumstance c; (iv) a quaternary value for human beings in circumstance $c =_{df} X$ contributes to meeting a fancy, i.e. a desire that is not a legitimate want; (v) a basic value $=_{df} X$ has either a primary or a secondary value. (Bunge 1989, p. 36)

NORM 11.1 The one and only morally legitimate function of the economy is to help people meet their basic needs and fulfil their legitimate aspirations. (Bunge 1989, p. 367)

In this contrary formulation, Bunge takes legitimate wants to be not matters of innocuously obtained basic needs (survival and health), but of something new and as yet undefined, "(reasonable) happiness". Bunge is famously down on happiness (Bunge 2016, p. 411), so it should come as no surprise that he also has an idiosyncratic definition of reasonable happiness (likewise appearing only several pages after its first use). Even though Bunge recognizes that happiness is a psychological state, instead the definition is a disguised norm specifying what ought to suffice us for it:

DEFINITION 2.3 An individual is reasonably happy if and only if she is (i) in a state of well-being, and (ii) free to pursue her legitimate wants. (Bunge 1989, p. 48)

Reduced to lowest terms this gives: to be reasonably happy is to be healthy and free to innocuously pursue one's long-term health. Or again: to be reasonably happy is to have met and be free to innocuously pursue one's basic needs (survival and health).

Obviously this was not what Bunge had in mind when he spontaneously re-defined legitimate wants. In many later uses of the terms, Bunge clearly uses both concepts to mean something else undefined. Bunge repeatedly uses what would otherwise be the redundant phrase "basic needs and legitimate wants", legitimate wants being as defined thus far only the fulfilment of basic needs. Equally obviously, these other, implicit concepts of legitimate want and reasonable happiness are what he has in mind in his political philosophy. The project of a world order where only health is legitimate is a non-starter, not just in general, but also for Bunge himself:

... the exclusive pursuit of health will turn a man into a health freak, that of love into a Don Juan, that of knowledge into a physical cripple, and so on ... (Bunge 1989, p. 58)

And indeed it is only in a mention in the chapter summary, and then two decades later in Bunge's *Political Philosophy*, that he finally does give the definition he is missing, albeit inconsistently:

1.5 Summary
Animals are said to be in a state of well-being only if they have satisfied all of their primary and secondary needs. And humans may attain – within bounds and over certain periods – reasonable happiness if, in addition to enjoying well-being, they can work to meet their legitimate wants. The latter are the desires that can be satisfied without seriously harming anyone else. (Bunge 1989, pp. 52–53)

... legitimate desires—those whose satisfaction do not prevent anyone else from meeting their primary needs (Bunge 1989). (Bunge 2009, p. 93)

A justifiable desire is one whose satisfaction does not prevent anyone from meeting his or her basic needs (Bunge 1989). (Bunge 2009, p. 330)

A consistent choice between the, shall we say two, different concepts of "legitimate" wants—on the one hand, as those which innocuously contribute to the long-term health of the individual, and on the other, those which are merely innocuous—can be no easy choice for agathonists. The concept of a legitimate want is woven into the system. As Bunge says,

Every cluster of needs or wants is the root of a group of values: primary, secondary, tertiary, and quaternary. Only the first three are genuine or legitimate. See Figure 1.3. This categorization will have momentous consequences for our views on rights and duties, i.e. our ethical doctrine. (Bunge 1989, p. 40)

If Bunge chooses the limited concept (pursuit of health), he will be advocating in his own terms a freakish world order. Earlier in his *Ethics*, Bunge had emphasized this even more:

Only very sick people, like Nietzsche, would claim that health alone counts … Certainly, *primum vivere, deinde philosophari,* but human life is not only eating and drinking: it is also caring for others, learning, playing, and much more. (Bunge 1989, pp. 22–23)

But if Bunge takes these considerations into account and chooses the freer concept (any innocuous pursuit), he will undermine his system. Consider:

On the other hand in our axiology all values are relative to the evaluator, but while some are objective others are not. Equivalently: There are objective standards or canons for determining whether a given item is really good or bad for someone's well-being. On the other hand there are no standards or canons in the case of likes and dislikes that fail to originate in basic needs or legitimate wants. (Bunge 1989, p. 68)

… we need something positive and empirically accessible to justify our valuations. Basic needs and legitimate wants supply just such positive grounds. Moreover, since such deficits can be the subject of scientific investigation, the corresponding values can be discovered rather than invented. (Bunge 1989, p. 75)

In which sense is Bunge using the term here? On the one hand he would seem to be using the freer concept, because otherwise the combination of basic needs and legitimate wants is redundant. On the other hand he seems to be using the limited concept, because otherwise there is no difference between legitimate wants and what he derisively calls "fancies" or whims, for which he says there are no standards or canons, and which elsewhere he says are beyond the ken of rights and duties (Bunge 1989, p. 97). (These statements should be qualified with the caveat that whims might sometimes be noxious, in that circumstance making them more amenable to objective evaluation, not less.)

When faced with the task of untangling a knotty problem there is always another more radical solution, to be examined after the following section.

27.2.2   Some Clarifications and Corrections

Bunge started his *Ethics* by claiming all animals "evaluate", when really what we know is that they (and plants[6]) all "discriminate" (or "identify"), and act according to those discriminations. For example, plants typically grow their leaves toward the light and their roots toward the soil. Evaluation is not just a process of distinguishing between items, or locking on to some single one, but also of assigning them a value: "values are not things, states of things, or processes in things: these can only be value-bearers of objects of valuation. Values are relational (or mutual) properties *attributed* to objects of certain kinds *by* organisms of certain types and in certain states" (Bunge 1989, p. 15, emphasis changed). Attribution is an intermediate conceptual stage, one that is in our brains not our genes.[7] As Bunge says elsewhere: "in order to be able to attribute a property to an object we must form some concept of such property... attribution is a psychological process" (Bunge 1983, pp. 165–166).

Considering values this way, as conceptual objects, that is as *ideas* about the suitability or desirability of items,[8] has at least three distinct advantages over Bunge's other view that they emerged on our planet some four billion years ago with the first organisms capable of discriminating what was good for them. First, it is completely standard. Second, it coheres with Bunge's philosophy of psychology and his epistemology. Third, it is not clear at all that *any* organism is truly capable of discriminating what is good for it—see below.

The value-free, objective concepts to use in all these circumstances have already been supplied: they are those of *function* and *role* (Mahner and Bunge 2001; Kary and Mahner 2002). In particular, the role of an item in a larger process—such as the role of food in the survival of an organism—consists of all its effects on any of the other things involved in that process. With these concepts, there is no need to put an ethical twist on the relationship between fruit flies and fruit: unless their psychology is substantially deeper than we know (Gibson et al. 2015), fruit flies do not find fruit good, instead fruit has a specific role in their lives. Roles emerged with the first systems, but values only emerged with the first organisms capable of conceiving them.

Even in animals capable of conceiving values, discriminatory behaviour need only be automatic, such as when reflexively spitting out something foul tasting. An item famously favourable for their survival and well-being, but nevertheless equally famously reflexively rejected by children, is cod liver oil. This brings up some questions about Bunge's claims: are animals really capable of discriminating (at least some of) what is good for them, and do all normal animals naturally strive

---

[6]Bunge also regarded his restriction to animals as too strong, noting parenthetically that at least some bacteria can discriminate between some things, as in chemotaxis (Bunge 1989, p. 24).

[7]For a similar assessment see Obiedat (2019) in this volume.

[8]In Bunge's system and here, an "item" can be anything: thing, process, idea, or even absence of something, such as peace and quiet as the absence of conflict and sound.

to get that? Is valuing survival really "in our genes"? Put another way, do all normal animals really have an instinct for survival?

An instinct for survival is an epistemological impossibility, something that would require instinctual omniscience. The idea that all organisms struggle for survival is a myth. Only sentient organisms can struggle, and only sentient organisms that can conceive of death or life and envision either as an outcome of their actions or inactions can struggle for survival. What most sentient organisms struggle for, including ourselves much of the time, is not survival, but to feel—not good, nor even comfortable, as in Bunge's variant definition of well-being—but *to feel the way we want to feel*, a complex state with many aspects, and which may vary substantially even when external circumstances are constant. In fact Bunge agrees with this: "ultimately *what is valued is being in certain states*" (Bunge 1989, p. 25, emphasis in the original).

What we and other sentient organisms want or like to feel is to greater or lesser extents based on aversions to pain and analogous sensations (e.g. anxiety, itch, nausea), attractions to pleasure and analogous sensations (e.g. satisfaction, relief, exhilaration), aversions to being on the receiving ends of violence, and many other aversions and attractions inherited or acquired, to include for example an apparently inborn aversion to snakes (Hoehl et al. 2017). There is also plenty of normal cross-mixing of aversions and attractions: in many circumstances there are attractions to some painful or related sensations, and aversions to some pleasurable or related sensations. A seemingly universal example is scratching, which is both comforting and uncomfortable, or even painful and damaging. All together, these have worked somewhat well, and often enough well enough, to allow *Homo sapiens* to survive *pro tempore*.

Yes, we can taste salt, find a certain amount of it palatable, and a certain amount more inedible; but the only way we know that it is good or bad for us is through inference, not perception. And even with all the inferential might of modern science behind us, even for this simplest of food items, we still do not really know salt's ideal pattern of consumption, in general or for us as individuals, or if there is either such thing; only that the amount many people seem to find good seems far too much.

In other words, we and other organisms have *some* ability to discriminate or identify, not on the basis of *what is good for us*, but on the basis of *what worked well enough for our ancestors*—or else by pure coincidence, *what might work for us today*.

27.3   Some Analysis and Criticism

It was not for nothing that Bunge wanted to find values and goodness in the adventures of organisms appearing four billion years ago. It was for the same reason that he defined legitimate wants as the (innocuous) pursuit of health, which he in turn presumed to be a requirement for survival: by making the ultimate basis of morality the need to survive, basic morality would become objective and universal. By becoming objective and universal, it would then become the basis for an objectively right world order—the Utopian socialist ideal. But the attempt to get moral objectivity and universality from basic needs and "legitimate" wants, however conceived, leads to many conundrums, of which there is space here to consider briefly only two.

27.3.1 Is It Legitimate to Want to Father or Bear a Child?

Clearly in many circumstances, for Bunge the answer to the latter half of that question is NO. With good reason, Bunge considers the world overpopulated, and overpopulation a threat to the survival of mankind. Bunge never makes clear exactly how far he would go to solve the problem, and his views of the matter are not completely consistent (Bunge 1989, p. 312), but he certainly has approved of going very far indeed. That though concerns only Bunge's restrictions on the right of women to bear (not on men to father) children under certain circumstances. Thus, for the time being, only for poor societies does Bunge prescribe reductions in the number of births per woman through "Demographic dictatorship" and coercion (Bunge 1989, p. 364, 2009, p. 108). The question here is instead: under agathonism, is it *ever* legitimate to want to father or bear children?

Recall that Bunge has defined legitimate wants in at least two, really four, different ways: as those whose satisfaction contributes to the agent's long-term health without hindering anyone else from getting what they need to be healthy, and without endangering society; as those that can be satisfied without seriously harming anyone else; as those whose satisfaction does not prevent anyone else from getting what they need to survive; and as those whose satisfaction does not prevent anyone from getting what they need to be healthy. In largely sequential order:

Does bearing children contribute to a woman's long-term health? It radically endangers the woman's immediate health, to the extent of endangering her immediate survival—and with immediate death there is by definition neither immediate nor long-term health. Childbirth always has been and always will be a mortal danger to both mother and child. As for the effects of childbearing on maternal health and longevity once past the immediate crisis, a trade-off between parental longevity and the production of offspring is integral to the disposable soma theory of the evolution of ageing, and such a trade-off has been demonstrated experimentally in fruit flies (Westendorp and Kirkwood 1998). Reduced female longevity as a function of "reproductive success" was found epidemiologically in the British aristocracy (Westendorp and Kirkwood 1998) and in rural Poland (Jasienska et al. 2006). Using data from 153 countries, and in consideration of confounding factors such as ethnicity, disease occurrence, population geography, and religion, the relationship between female longevity and fecundity was found to be negative (Thomas et al. 2000).[9]

Can women bear children without hindering anyone else from getting what they need to be healthy; or at a minimum, can it be done without seriously harming anyone else, or preventing anyone else from getting what they need to survive? For a start, everyone needs adequate sleep to be healthy, and newborns certainly hinder everyone in the immediate vicinity from getting adequate sleep. But even minimally, every child is bound to eventually catch a communicable disease, and transmit it to someone other than the mother. Influenza is a prime example: children spread it like wildfire, it seriously harms just about everyone who catches it, and seriously kills a good fraction of them.

Can women bear children without endangering society? On the contrary, every new generation of young people endangers society: youth are rebellious, disproportionally the perpetrators of violent crime, and of reckless and generally anti-social behaviour.

What about Bunge's Norm 4.7, "Seek the survival of humankind"? Here Bunge's ontology must be brought in to help: humankind, as a kind, is a concept, not an organism that can survive or not. What Norm 4.7 must really be commanding is: "Continue the human lineage eternally—like it or not." But the only justification for this norm is Bunge's claim that it is equivalent to his supreme norm (Norm 4.6), "Enjoy life and help others live". Bunge never defines or describes or explains enjoyment anywhere in his *Ethics*; it just pops up out of the blue. In context though the only reasonable interpretation of it is "reasonable happiness". In turn this was only defined as being healthy and free to innocuously pursue one's long-term health; or minimally, being healthy and free to pursue one's innocuous wants. But even only as already described, childbearing is detrimental to both health and freedom, and far from innocuous. Consequently Norm 4.6 proscribes childbearing, and on top of that is proved inconsistent with Norm 4.7 (see also Zecha 1990 for other conflicts between the two norms, and Bunge 1990b in the same volume for Bunge's rebuttal).

Unmentioned in all of this so far is that there is an individual who is *always* eventually deprived of health and life as a result of the endeavour: the child. Procreation eventually and inevitably leads to the periodic sickness and eventual death of the offspring, who would not have faced either otherwise. Is it legitimate to have a desire whose satisfaction will put someone through this? One might think that agathonism has an escape clause, namely that in Bunge's view the fetus gains personhood and thus moral standing no earlier than upon birth (Bunge 1989, p. 164). But no: Bunge holds that

abortion should be recommended if a child is not wanted, or tests show that it would be severely handicapped. The reason for this recommendation is that it is morally wrong to

---

[9]There are some conflicting results and the conflicts have not been resolved. Nevertheless under agathonism potential parents must resolve them in order to know if the children they desire can legitimately be considered valuable, and so whether any desire for them is legitimate or not.

make three people – the unwanted child and his parents – unnecessarily miserable. (Bunge 1989, p. 165)

As for fatherhood, if only by the damaging effects of childbearing on mother, child, and society, under agathonism the desire for it is straightforwardly illegitimate too.

It might be argued that some trade-offs are involved, between immediate danger and long-term help, for parents or society. Or between the inevitable sickness, suffering, and eventual death of any offspring, and the highlights of their life. Never mind the objection that there is no guarantee for any of it—that sometimes offspring also abandon their parents, leave them destitute, drive them crazy, or kill them; or do something similar for society; or endure thoroughly miserable lives of ill-health themselves. The matter is simplified because agathonism is primarily a system of thresholds, not calculi, and wanting to bear children fails to clear any of them.

In short and obviously, the production of offspring is literally dangerous, sickening, and draining, and this is why evolution has to trick us and every other animal species into it—so to speak.

It may surprise some to learn that Bunge long ago recognized the basics of all this. It is precisely why in subsequent work he accepted a redefinition of biological value in terms of favouring an organism's ability to undergo its species-specific life history, rather than in terms of its health (Mahner and Bunge 1997, p. 159). However, this move is of no help with the problem of human procreation, which like almost everything else is optional in human life history. Nor is it of use more generally in human ethics, if only because we spend so much of our resources fighting our species-specific life history—or at least the back half of it.

Clearly the problem with all of this is the concept of a "legitimate" want itself. Considering the earlier conclusion, that it either undermines agathonism or renders it freakish, why not call off the thought police, and just dispense with it? Why not leave matters of legitimacy and crime for the law, and have morality suffice with those that are proper to it, namely right and wrong and good and bad? Bunge astutely got it right that crime is to be addressed by legislation not defined by it, but if there are extralegal crimes there will be extrajudicial punishments, and these will come with no institutional restraint, and no judiciousness.

Three decades ago, in a *Festschrift* marking the completion of Bunge's *Treatise*, Garzón-Valdés (1990) suggested that Bunge give up the concept of a psychologically legitimate want, although for different reasons. Bunge's reply in the same volume was curious:

> ...Garzón-Valdés suggests that I give up the notion of a psychologically legitimate want. How could I, if I want to keep the distinction between wants and needs, and between socially harmless and socially harmful actions? (Bunge 1990a, p. 656)

His reply is curious first of all because the distinction between wants and needs has nothing to do with the former's legitimacy. Nor does the distinction between socially harmful and harmless actions need to depend on what motivated them. Instead that could be added as a potentially extenuating or incriminating circumstance, just as is done in law.

What this shows is just how attached Bunge is to the idea of judging a person's desires to be legitimate or not. Why? An answer may be found in Bunge's political philosophy, examined in the companion chapter.

27.3.2   The Need to Survive

Although Bunge analyses the concepts of primary, secondary, higher order and basic needs, rights and duties, he does not do so for the concepts of needs, rights or duties themselves. Rights and duties will be dealt with briefly here, before moving on to the matter of needs.

In his *Ethics*, Bunge equates legal rights with freedoms, and moral rights with abilities (Bunge 1989, pp. 96–97). He goes so far as to say that a polar bear who conquers a certain territory has gained the natural right to control it (Bunge 1989, p. 95). He also describes the polar bear as discharging a natural duty when caring for her offspring, although without explaining how that counts as a duty or what her moral status is if she rejects it, as often happens in various species. He equates legal duties with the absence of the right to refrain from performing them, but gives no characterization of moral duties, except to say that he conceives of both moral rights and duties as norms or rules. In his *Political Philosophy*, Bunge has rights as licences to do something (Bunge 2009, p. 105), which is perhaps defining the obscure with the just as obscure.

Yet it is duty that is the foundational concept. A duty is nothing more than an obligation to do or refrain from doing something. Moral duties are felt obligations. The feeling is either actual, expected, or exhorted. It is generally hoped that legal duties will also be felt as obligations, but if not in the end they are enforced obligations, to mean sanctioned when unfulfilled.

As for rights: rights *are* duties. From a logical point of view the concept of a right is entirely redundant, since for one person to have a right is exactly for others to have a duty, or in other words to be constrained or obligated. That a person has a duty does not imply they have the right to perform it: Catch-22.[10] That one person gets a right does not automatically imply that others get it too, nor that the right-holder has any duty to others. These have to come instead from some proposal that rights or duties should be universalized. The point of this analysis is not that it would have saved Bunge a long and complicated set-up to a problematic proof that rights imply

---

[10]The doctor in the novel of the same name had the duty to ground any flyer who was crazy, but did not have the right to, because of a chain of regulations that end in Catch-22.

duties.[11] Instead it alters Bunge's previously somewhat *ad hoc* or idealistic pairings of duties with rights, and so the analysis of which rights are viable or moral.

As described earlier, the foundation of agathonism is the need to survive, and everything needed for it. This leads to the concept of basic human needs: first, survival and its necessities, one of them assumed to be health; and so second, health and its necessities. The pursuit of these necessities is made the root of values, which in turn are the sources of rights and duties. In agathonism, it is the objective reality of basic human needs that makes the basic values, norms, rights, and duties of agathonism objective. But what is the basic nature of needs?

Any analysis shows that needs are always instrumental. Therefore they cannot serve as the foundation of anything, without circularity or infinite regress. In particular, survival is not a need in itself. That survival is no necessity, or in other words that nobody needs to survive, is proved simply: nobody does. Bunge dispenses with this problem in half a page, accepting circularity in the guise of sociality as a fitting solution:

> However, there is one *prima facie* counterexample to the identification of value with need or want satisfaction, namely no less than life itself. Indeed, one may well ask what need or want does life satisfy. Put negatively: Why should we cherish life, let alone regard it as inviolable or even sacred, if living satisfies no obvious need but, on the contrary, is the source of all needs and wants? [. . .] The puzzle evaporates in a systemic perspective, where every human being is regarded as (actually or potentially) playing some role in a society, and thus being able to satisfy some of the needs and wants of other people. (Bunge 1989, pp. 37–38)

One would hope that is not all there is to it. Making everyone's life only valuable insofar as it is or might be to the benefit of others puts us all on short notice. It would be remiss not to note that this was the working philosophy of the Gulag (Applebaum 2003, p. xxxix). Even outside dystopia, in the world of ordinary practical politics, where Bunge hopes his ethics will eventually be applied, short notice leads to short shrift. The chain of reasoning behind the second half of the supreme norm, "help others live"—rather than kill or exploit them—is thin. Bunge says eventually that his supreme norm "suffers no ifs or buts and offers no reasons or causes: it is the yardstick or standard we use to evaluate morally every action and every other norm" (Bunge 1989, p. 300), but of course it was not spontaneously created out of nothing. Instead the duty to help others live is grounded in the idea that (at least during crucial parts of our life history) we need them in order to live ourselves. But those faced

---

[11] There are several problems with the proof. It is based in part on Norm 4.2, which is or at least is supposed to be of the following form. If one person has a right, then: another person has the duty to help them exercise it, if and only if they are the only other person who can do so. (In fact the norm has, instead of if and only if, equality by definition.) Bunge's proof that rights imply duties incorrectly translates this as follows. If one person has a right, then: if one person has a right, then another person has the corresponding duty . . . ; and likewise for the other person. The deduction that rights imply duties, and the corollary that everyone has some duties, then follow from the pre-postulated existence of rights, in turn derived from the pre-postulated existence of needs.

with a conflict between the right to life and the duty to help others live are free to choose either (Bunge 1989, p. 101).

Bunge's argument begs the question: it assumes people have needs, period. Because all needs are instrumental, those purported needs were instead proposed in the service of survival. The whole question though is whether or not that survival is needed in the first place. If survival is not needed then any items needed for survival—including other persons—are not needed either.

Let us face it: if all mankind were to vanish, only the dogs would truly miss us— and even their goodwill rests only on their not knowing the whole story. The rest of the biosphere would either rejoice or be indifferent, and the rest of the universe would not notice. There is no need to survive, for us or any other organism. We live not out of need, but only out of either habit, external compulsion, inability to attain or fear of the alternative, or in the last analysis, *want*—either our own, or that of whomever is keeping us from dying. If living required no effort then even the completely blasé could just continue on; but living requires *someone's* exertion and in the last analysis exertion requires the will. The chain of reasoning is not no organisms, no needs, no values (Bunge 1989, p. 11); but *no wants, no needs*—as the Stoics, with whom Bunge sympathizes, understood. Wants, not needs, are the foundation of morality, if only because wants are the *only* wellspring of needs. Our lives are valuable not just insofar as they serve others, but also and independently, insofar as we want to live them. Bunge concludes the previous quotation by almost saying as much himself:

> ...we normally enjoy satisfying our own needs and wants as well as some of those of others – and this is what life is all about. (Bunge 1989, p. 38)

Among our wants the desire to live is *usually* first and foremost, leading to the needs for survival, and thus much of Bunge's system; but usually is far from always. In particular, the desire to live is secondary whenever someone puts their life on the line—something people do perhaps surprisingly often, for any number of noble, trivial, or hate-filled reasons.

27.4 The Good and the Right

Bunge began his *Ethics* with a consideration of needs, from which he derived values and hence the good. He then declared right actions to be those that promote the good. From this he concluded that "the good is in every regard – conceptually, biologically and technically – prior to the right" (Bunge 1989, p. 6). The later discussion in his *Political Philosophy* was both more nuanced and more ambiguous, to the point of suggesting the matter was a pseudo-problem (Bunge 2009, pp. 105–111). Does resetting the foundation stone of morality from needs to wants change any of this? We come with both moral sentiments and appreciations of the good, each on their own. Often though what we find good we therefore find to be right, and what we find right we therefore find to be good. Derived from wants, neither the good nor the right comes first, rather both have parts that are freestanding, parts that go hand in hand, and parts that conflict.

At the beginning of his chapter on morality changes, Bunge (1989, p. 133) explains that morals change because needs and wants change. But he holds that only our wants change quickly, not our basic needs, so only our tertiary and quaternary values, and so our secondary norms, change quickly. But is that really the case, or does even basic morality, even for large social systems and not only selected individuals or organizations, also sometimes turn on a dime? Historically, unambiguously the answer is yes: consider the Gulag (Solzhenitzyn 1974; Applebaum 2003), the Holodomor (Klid and Motyl 2012; Applebaum 2017), China's one-child policy (Jian 2013), Mao's Cultural Revolution or Mao's Great Famine or Mao's China more generally (Chang and Halliday 2005; Dikötter 2010a, b; Zhou 2013), Nazi Germany, the killing fields of Cambodia, the Inquisition, or uncounted other such turnings.

It is important to compare these great shifts with those that did *not* occur in other catastrophes. For example, although nowhere near as catastrophic as the contemporaneous period in the Soviet Union, the Great Depression was a time of extreme duress in the United States, Canada, and elsewhere in the west. Yet, basic morality hardly changed if at all in response. If anything, moral sentiments for the unfortunate were refined. Why the difference? An obvious consideration: nobody *wanted* the Great Depression. It did not come about because of any change in ideology or belief. But every one of the ultra-catastrophes mentioned in the previous paragraph came about precisely because of that. Ultimately, one or more people *wanted* them to happen[12]—and with the aid of ideology, were able to convince enough others likewise.

27.5 Conclusion

The objectivity of Bunge's ethical system was intentionally limited by construction. It includes wants of some form, and though founded on basic needs, Bunge's conception of them is wider than most, and makes them contextual. He also allows

---

[12]Some might consider Mao's Great Famine to be the one exception on this list. Mao's personal physician, Li Zhisui, recalled otherwise:

> I did not immediately understand, because it was so hard to accept, how willing Mao was to sacrifice his own citizens in order to achieve his goals . . . It was not until the Great Leap Forward, when millions of Chinese began dying during the famine, that I became fully aware of how much Mao resembled the ruthless emperors he so admired. Mao knew that people were dying by the millions. He did not care. (Zhisui 1994, p. 125)

China during Mao's Great Famine, like the Soviet Union before it during Stalin's greatest famine (the Holodomor and associated famines in the North Caucasus, Volga region, and elsewhere in the Soviet Union), was a net exporter of grain until it ended the famine in 1961 by means of imports (Krawchenko 1984; Wemheuer 2013; Meng et al. 2015; Applebaum 2017).

an escape clause that makes his ethics even less objective: "a moral norm that were to ignore moral sentiments, or were at variance with the well known sociological finding that moral rules are human creations that affect (positively or negatively) human relations, could be discarded out of hand" (Bunge 1989, p. 312). Putting needs in the service of wants limits this objectivity even more.

Curiously, in addition to the presumed foundational objectivity of basic needs, Bunge advanced an argument for ethical objectivity based on intersubjectivity:

> Thus, the Universal Declaration of Human Rights (1948) is the graveyard of moral relativism. As Bobbio (1985) wrote, that was the first time in history that universal agreement was reached on any value system. From then on people have been condemning certain practices... as violating the basic human rights enshrined in a document formally subscribed by all nations. What holds for moral philosophy also holds for legal philosophy... because it absolutizes the law of the land. From 1948 on we can evaluate every single piece of legislation as being objectively just or unjust, according as it matches or violates human rights or international law. (Bunge 2009, p. 118)

There is one small problem with this argument: the Universal Declaration of Human Rights did not reach universal agreement. Ten of the 58 UN member states did not sign on. Two did not show up for the vote, but eight had obvious reasons: Byelorussia, Czechoslovakia, Poland, the Russian Federation, Saudi Arabia, South Africa, Ukraine, and Yugoslavia.

But socialist, racist, Islamist, and sexist regimes hardly make for the most notable exceptions. Two more of interest: the American Anthropological Association (The Executive Board, AAA 1947); and one Mario Bunge.

Provision 1 of Article 16 of the Universal Declaration reads as follows:

> Men and women of full age, without any limitation due to race, nationality or religion, have the right to marry and to found a family. (United Nations 1948)

Bunge's view:

> In short, the world population has become so huge, and the resources so strained, that parenthood has ceased to be a human right. (Bunge 1989, p. 164)

Only once the wants are agreed upon, or assumed, can ordinary objective analysis be applied to the means for achieving them, to include moral norms. Wants are not objective, but the most important of them are *largely* intersubjective, especially within cultures. In principle this should leave much of Bunge's ethical system intact.

Acknowledgement I thank Joseph Kary for reading and commenting on the manuscript, and Michael Matthews and John Forge for their respective contributions in shepherding the manuscript through its various stages.

References

Applebaum, A. (2003). *Gulag: A history*. New York: Doubleday.
Applebaum, A. (2017). *Red famine: Stalin's war on Ukraine*. New York: Doubleday.
Bobbio, N. (1985). *L'età dei diritti*. Torino: Einaudi

Bunge, M. (1979). *Ontology II: A world of systems (Treatise on basic philosophy, vol. 4)*. Dordrecht: Reidel.

Bunge, M. (1983). *Epistemology & methodology I: Exploring the world (Treatise on basic philosophy, vol. 5)*. Dordrecht: Reidel.

Bunge, M. (1989). *Ethics: The good and the right (Treatise on basic philosophy, vol. 8)*. Dordrecht: Reidel.

Bunge, M. (1990a). Garzón Valdés on needs, wants, and legitimacy. In P. Weingartner & G. Dorn (Eds.), *Studies on Mario Bunge's treatise* (pp. 655–657). Amsterdam: Rodopi.

Bunge, M. (1990b). Zecha's acid test of agathonism. In P. Weingartner & G. Dorn (Eds.), *Studies on Mario Bunge's treatise* (pp. 660–663). Amsterdam: Rodopi.

Bunge, M. (2009). *Political philosophy: Fact, fiction, and vision*. New Brunswick: Transaction Publishers.

Bunge, M. (2016). *Between two worlds: Memoirs of a philosopher-scientist, with an appendix by Marta Bunge: My life with Mario*. Switzerland: Springer.

Chang, J., & Halliday, J. (2005). *Mao: The unknown story*. London: Jonathan Cape.

Dikötter, F. (2010a). *Mao's great famine: The history of China's most devastating catastrophe, 1958–62*. London: Bloomsbury.

Dikötter, F. (2010b, December 16). Mao's great leap to famine. *The New York Times*. http://www.nytimes.com/2010/12/16/opinion/16iht-eddikotter16.html. Accessed 11 Feb 2018.

Garzón Valdés, E. (1990). Basic needs, legitimate wants and political legitimacy in Mario Bunge's conception of ethics. In P. Weingartner & G. Dorn (Eds.), *Studies on Mario Bunge's treatise* (pp. 471–488). Amsterdam: Rodopi.

Gibson, W. T., Gonzalez, C. R., Fernandez, C., Ramasamy, L., Tabachnik, T., Du, R. R., Felsen, P. D., Maire, M. M., Perona, P., & Anderson, D. J. (2015). Behavioral responses to a repetitive visual threat stimulus express a persistent state of defensive arousal in *Drosophila*. *Current Biology, 25*(11), 1401–1415.

Hoehl, S., Hellmer, K., Johansson, M., & Gredebäck, G. (2017). Itsy bitsy spider…: Infants react with increased arousal to spiders and snakes. *Frontiers in Psychology*. https://doi.org/10.3389/fpsyg.2017.01710.

Jasienska, G., Nenko, I., & Jasienski, M. (2006). Daughters increase longevity of fathers, but daughters and sons equally reduce longevity of mothers. *American Journal of Human Biology, 18*(3), 422–425.

Jian, M. (2013). *China's barbaric one-child policy*. https://www.theguardian.com/books/2013/may/06/chinas-barbaric-one-child-policy. Accessed 11 Feb 2018.

Kary, M., & Mahner, M. (2002). How would you know if you synthesized a thinking thing? *Minds and Machines, 12*(1), 61–86.

Klid, B., & Motyl, A. J. (2012). *The Holodomor reader: A sourcebook on the famine of 1932–1933 in Ukraine*. Edmonton/Toronto: Canadian Institute of Ukrainian Studies Press. https://holodomor.ca/the-holodomor-reader-a-sourcebook-on-the-famine-of-1932-1933-in-ukraine/. Accessed 5 Jun 2018.

Krawchenko, B. (1984). The man-made famine of 1932–1933 in Soviet Ukraine. *Conflict Quarterly, 4*(2), 29–39. https://journals.lib.unb.ca/index.php/JCS/article/viewFile/14623/15692. Accessed 11 Feb 2018.

Liao, C.-Y., Rikke, B. A., Johnson, T. E., Diaz, V., & Nelson, J. F. (2010). Genetic variation in the murine lifespan response to dietary restriction: From life extension to life shortening. *Aging Cell, 9*(1), 92–95.

Mahner, M., & Bunge, M. (1997). *Foundations of biophilosophy*. Berlin/Heidelberg: Springer.

Mahner, M., & Bunge, M. (2001). Function and functionalism: A synthetic perspective. *Philosophy of Science, 68*(1), 75–94.

Meng, X., Qian, N., & Yared, P. (2015). The institutional causes of China's great famine, 1959–1961. *Review of Economic Studies, 82*, 1568–1611. https://doi.org/10.1093/restud/rdv016.

Obiedat, A. Z. (2019). How can Mario Bunge's scientific-humanistic ethics engage Islamic moral law? In M. R. Matthews (Ed.), *Mario Bunge: A Centenary Festschrift*. Cham: Springer.

Redman, L. M., & Ravussin, E. (2011). Caloric restriction in humans: Impact on physiological, psychological, and behavioral outcomes. *Antioxidants & Redox Signaling, 14*(2), 275–287.

Solzhenitzyn, A. I. (1974). *The Gulag archipelago, 1918–1956: An experiment in literary investigation.* (T. P. Whitney, Trans.). New York: Harper & Row.

The Executive Board, American Anthropological Association. (1947). Statement on human rights. *American Anthropologist, 49*(4), 539–543.

Thomas, F., Teriokhin, A. T., Renaud, T. F., de Meeûs, T., & Guégan, F. (2000). Human longevity at the cost of reproductive success: Evidence from global data. *Journal of Evolutionary Biology, 13,* 409–414.

United Nations. (1948). *Universal declaration of human rights.* http://www.un.org/en/universal-declaration-human-rights/. Accessed 11 Feb 2018.

Wemheuer, F. (2013). Collectivization and famine. In S. A. Smith (Ed.), *The Oxford handbook of the history of communism.* Oxford: Oxford University. https://doi.org/10.1093/oxfordhb/9780199602056.013.023.

Westendorp, R. G. J., & Kirkwood, T. B. L. (1998). Human longevity at the cost of reproductive success. *Nature, 396*(6713), 743–746.

Zecha, G. (1990). Which values are conducive to human survival? A Bunge-test of Bunge ethics. In P. Weingartner & G. Dorn (Eds.), *Studies on Mario Bunge's Treatise* (pp. 511–528). Amsterdam: Rodopi.

Zhisui, L. (1994). *The private life of chairman Mao: The memoirs of Mao's personal physician, Dr. Li Zhisui* (T. Hung-chao, Trans.). New York: Random House.

Zhou, X. (2013). *Forgotten voices of Mao's great famine, 1958–1962: An oral history.* New Haven: Yale University Press.

Chapter 28
How Can Bunge's Scientific-Humanistic Ethics Engage Islamic Moral Law?

A. Z. Obiedat

Abstract In the eighth volume of his *Treatise on Basic Philosophy* (Bunge M, Treaties on Basic Philosophy. Vol. 8, *Ethics: the Good and the Right*. Reidel, Dordrecht, 1989) Mario Bunge articulates a detailed theory establishing ethics on the basis of a scientific philosophy. Analogously, the Islamic legal theorist of Andalusia, Shāṭibī (1320–1388 C.E.), sought to resolve massive Islamic doctrinal disagreements by establishing a common denominator that could unify the Islamic legal theories of his time. This paper argues that Shāṭibī's ethical theory when compared with Bunge's scientific-humanistic ethics, will (1) transform our understanding of the secular-religious divide in ethics and (2) will assist contemporary Islamic ethicists in finding common grounds with scientific-humanistic ethics.

Not all religious ethics are based on theological mythology or dominated by superstitious rituals, as Shāṭibī's theory demonstrates. Equally, not all secular ethics lead to subjective relativism, anarchy, or loss of meaning as Bunge's theory establishes. The symmetry between these two theories is seen in Shāṭibī's concept of "moral hierarchy." According to Shāṭibī, norms are not perpetually fixed. Rather, "necessities" take precedence over "needs" and these over "desirables." Necessities are "pentagonal", to mean the preservation of "social existence" takes precedence over the protection of the "self"; that over "mind"; that over "lineage"; and that over "property."

In contrast, in Bunge's ethics the satisfaction of "life" takes precedence over "health" and that over "happiness." Doing the right according to Bunge is a social harmonization project by which everyone achieves their welfare without jeopardizing anyone else's. The path from Shāṭibī's Islamic ethics to Bunge's modern one is a matter of updating the naturalistic and sociological grounds while abstaining from outdated theological language. This comparison should be of interest to scholars of cultural studies and global ethics that often operate under an assumption of inherent irreconcilability between premodern and non-western cultures on the one hand and western and modern cultures on the other.

A. Z. Obiedat (✉)
Middle East & South Asia Studies Program, Wake Forest University, Winston-Salem, NC, USA
e-mail: obiedaaz@wfu.edu

© Springer Nature Switzerland AG 2019
M. R. Matthews (ed.), *Mario Bunge: A Centenary Festschrift*,
https://doi.org/10.1007/978-3-030-16673-1_28

487

28.1   Clash of Ethical Worldviews and the Need for Dialogue

Not all pre-modern ethics are based on theological mythology or dominated by superstitious rituals, as demonstrated by the medieval legal theorist of Andalusian Muslim Spain, Abū Isḥāq Ibrāhīm al-Shāṭibī (1320–1388 C.E.). Shāṭibī articulated this view in his work, *The Reconciliation of the Fundamentals of Islamic Law* (Shāṭibī 2011). Equally, not all secular ethics lead to subjective relativism, anarchy, or loss of meaning, as Bunge's ethical theory establishes (Bunge 1989). This comparison should be of interest to scholars of cultural studies and global ethics, who often operate under an assumption of inherent irreconcilability between premodern and non-western cultures on the one hand and western and modern cultures on the other (Halliday 2002, p. 107).

Modern scientific philosophies have achieved a great deal of thorough logical theorizing and empirical investigation. Given the complexity and the needed level of education to access such philosophies, most of the world's population is still uninformed about the fruits of these scientific philosophies. Bunge's ethics is one example of these highly articulate philosophies. My aim here is not only to describe his theory, but also to demonstrate the capacity of Bunge's philosophy to lend its ideas to non-western and pre-modern ethical systems. The reader will observe a change of tone and of quality of arguments while moving between the modern and premodern ethical systems, but this is the current intellectual scene we face today and we have to make the best of it. There is a strong effort to popularize science to the public and this paper analogously aims to popularize scientific philosophy to current Muslim communities who base their thoughts on medieval scholastic literatures. Yet, there is more to engaging Bunge's ethics with Islamic moral law than popularization. In line with the "global ethics project" (Küng 1991), creating a bridge between medieval systems of thought and modern ones is a fruitful intercultural task in its own right. One of Bunge's writings indicates that epistemologically he sees no such compatibility (Mahner and Bunge 1996), but I shall demonstrate a different possibility in the ethical realm.

The clash between the modern western powers and Muslim peoples from eighteenth century Colonialism to the modern-day "War on Terror" is mainly a factor of economic struggle and geopolitical domination (Rogan 2011, pp. 8–10). This is why a surface understanding might deem this clash to be rooted in a conflict of values and worldviews rather than anything else. However, this paper will reveal a deeper understanding that challenges the view of an intrinsic cultural clash between western and Islamic cultures. One of the arguments adduced to support this clash is that the west is generally irreligious and progressive while the Islamic east is religious and intellectually impoverished. From a technological, economic, administrative, and governmental point of view, this claim of course has strong weight. Yet, an important distinction should be drawn, considering that the west has produced secular movements such as Colonialism, Communism, and neo-liberalism that are also objectionable by today's intellectual and ethical standards. This criticism of the historical experience of secular modernism cannot lead to the total condemnation of secular worldviews, since not all secular movements are totalitarian, aggressive, or dogmatic movements. The western historical experience rather shows the inherent seeds of change and self-correction, the very seeds that grew to reject Nazism and Stalinism. Therefore, humanist, rationalist, and naturalist secular movements should be in stark opposition to the irrational and unscientific secular examples.

In the same analytic spirit, neither all aspects of Islamic culture are demagogically populist nor politically totalitarian, as is the case with many aspects of the Iranian or Saudi cases. The reason for this is that Islamic culture also bears the seeds of change, the same ones that contributed, to some extent, to the advancement of the West itself at the verge of the Renaissance: Arabic science and mathematics (Saliba 2011), Islamic capitalism (Heck 2006), and educational institutions (Makdisi 1990). This advancement is particularly seen in a strong Islamic interest in Aristotelian logic and natural science as proposed by Averroes (Fakhry 1997), who differentiates between the levels of religious and philosophical truths. Therefore, Islamic ideas that abide by logic, the science of the day, and universal human spirit should be in harmony with the humanist, rationalist, and naturalist secular movements.

There is little effort needed to justify focusing on Shāṭibī, because the fifteenth century is already the peak of accumulation of Islamic legal theory, known as *uṣūl al-fiqh*. Contemporary Islamic theorists still consider him an authoritative master who has not been surpassed (Hallaq 1997, p. 132). In contrast, the reason for selecting Bunge's theory for this comparison is found in the Enlightenment promise of a comprehensive and systematic worldview (Matthews 2012, p. 1401). Few contemporary philosophers can be ideal representatives of secular modernity, either because they deny the possibility of building philosophical systems or, if they accept such a possibility, they did not succeed in building one. The variety of contemporary philosophy is hardly capable of performing its classical task as the "venture in rational inquiry whose mission is to provide tenable answers to our big questions regarding human being, the world, and our place within its scheme of things" (Rescher 2001, p. 3).

This sought-after comprehensive worldview is particularly evident in Bunge's synthesis of the *quadrivium* of human knowledge, i.e., rationalistic, natural, social, and humanistic fields. This synthesis is articulated in the semantic, ontological, epistemic, and ethical volumes of his *Treatise on Basic Philosophy*. Let us see how Bunge articulates ethics in this light.

28.2 On the Symmetry Between Scientific Humanist Ethics and Islamic Ethics

An attempt to construct an ethical theory is a very problematic endeavor not only because it gives rise to severe political and cultural disagreements, but also because the presumed premises of any ethical theory – such as the nature of the world (ontology), the nature of human beings (anthropology), and the nature of human knowledge of that world (epistemology) – are also controversial. Historically, many ethical proposals have been formed on a religious basis; an example is the Ten Commandments. In contrast, scientific humanism is a philosophy that considers human knowledge, as founded on reason and experience, the justification for ethics – in contrast to supernatural divine knowledge, superhuman prophecy, and cultural ideology or racial supremacy. Scientific humanism is a version of humanism, the one most integrated with contemporary scientific findings and technology. This choice is in contrast with non-scientific humanistic and secular philosophies such as Existentialism.

According to Bunge, ethical theory is composed of the triad of axiology, morality, and praxeology (Bunge 1989, p. 2). "Axiology (or value theory) is centrally concerned with the good, morality with the right, and praxeology [or action theory] with actions that are both efficient and right." (Bunge 1989, p. 5) Accordingly, the anatomy of every single ethical action reveals three dimensions: a goal, a social code, and an application for the merger of the goal and the social code. In this context, the *goal* is a psychological intention or cognitive objective to be fulfilled, i.e., it relates to value(s). Secondly, the *social code* is a compilation of implicit or explicit instruction(s) that, based on some value, determines the validity or invalidity of human actions, i.e., norms. A proper *application* of the goal or value in the relevant social code or norm in a given situation is generally the combination of values and norms in a particular situation taking efficiency into consideration.

The three dimensions of ethical action can be illustrated as follows (Fig. 28.1).

In other words, to be ethical, in the scientific humanist context, is to (1) uphold good values, (2) construct and abide by the right norms, and (3) satisfy the goals with efficient actions. To do otherwise implies some combination of evil values, wrong norms, or inefficient actions, any of which are unethical.

For the purposes of comparison, good values are the natural goals that human actions seek to reach. This is symmetrical with the primordial nature of things (lit. *fiṭrah*) that should be followed in an Islamic ethics. Similarly, moral norms that are designed to balance rights and duties in social reality, and that are supposed to be standards for the validity or invalidity of human actions, are comparable to the five-level model known as *al-aḥkām al-sharʿiyyah*, (i.e., prohibited, repugnant, permitted, recommended, and obligatory, in short form known as *aḥkām*) (Hallaq 1997, p. 40). Although the study of efficient actions is associated with contemporary economic and management strategies, contextual judicial rulings (lit. *aqḍiyah*) and legal-moral verdicts (lit. *fatāwā*) are, to a great degree, traditional Islamic strategies

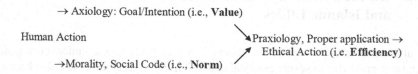

Fig. 28.1 Ethical action

Table 28.1 Two ethical systems

Scientific humanist ethics	Ethics in islamic legal theory
Good values	Primordial nature of things (*fiṭrah*)
Right norms	Scriptural or interpretive (*aḥkām*)
Efficient actions	Judicial rulings, *aqḍiyah*, and moral verdicts, *fatāwā*

for assuring the applicability and efficiency of *aḥkām* in the proper situation (Hallaq 1997, p. 123).

Therefore, to be ethical in the Islamic moral law is to (1) seek the primordial nature of things, *fiṭrah*, (2) abide by one of the five norms, *aḥkām*, and (3) choose the proper contextual rulings and moral verdicts, *aqḍiyah* and *fatāwā*, in space and time. To do otherwise would be following misguided values, corrupted norms, or improper *aqḍiyah* and *fatāwā*, i.e. counter-Islamic ethics.

The following table outlines the symmetry between the two ethical systems (Table 28.1).

Needless to say, the scientific humanist ethical triad of axiology, morality, and praxeology is divergent in content and mechanism from the Islamic one. Nonetheless, there is a degree of convergence. This essay attempts to reveal the elements of convergence and divergence between these two ethical systems (Sect. 28.6). Notably, Bunge's ethics is conceptually crystal clear (Sect. 28.3). However, we will borrow the findings of other research (Obiedat 2012, pp. 110–20) that extracts and conceptualizes the basis of value theory in Islam (Sect. 28.4).

28.3 An Analysis of Scientific Humanist Ethics

The domain of value theory according to Bunge is centrally concerned with the notions of 'good' and 'evil' prior to the investigation of moral norms, as the latter are concerned with 'right' and 'wrong.' "Good is [. . .] conceptually prior to the right" (Bunge 1989, p. 6) as he puts it, since a right action is one promoting the good, whereas a wrong one promotes evil. In other words, we need the guidance of the *good* to do the *right* and the misguidance of the *evil* to do the *wrong*. This priority of good to right, however, is not enough to sort out right from wrong automatically. Sorting out right from wrong is the specific task of morality based on value awareness. Being as it may, the questions are, "what is the nature, content, and structure of 'value' that value theory studies?" and "what is the 'norm' that morality studies?"

"What is the nature of 'value'?" The answer equally applies to morals, as follows. Bunge notes, values "are relational or mutual properties" (Bunge 1989, p. 13). In other words, a value or moral is a *joint property of some material thing or process and an organism, in some context,* rather than an independent self-existing object. Thus, "values and morals do not exist by themselves" (Bunge 1989, p. 11), but exist only in relation to something else according to a living being. Interestingly, we can say that values only existed when life, i.e., biology, did. Hence, before the emergence of living organisms, there were neither values nor morals and after the disappearance of living organisms there will be none. To be precise, Bunge states that "values emerged on our planet about four billion years ago together with the first organisms capable of discriminating what was good for them" (Bunge 1989, p. 13).

Although the biological foundation of value is reasonable, attributing the process of valuation to all biological beings, as Bunge does, may not hold. Living beings practice valuation in an instinctive or already programmed (prewired) behavior rather than a cognitively calculating manner; thus, they can be tricked quite easily according to their level of cognitive evolution. So, we may modify Bunge's statement by saying *values emerged on our planet millions of years ago together with the first organism capable of conceiving of what is good for itself.*

The analysis of value offered by Bunge is the following: "*a* is valuable, in *respect b* for *organism c* in *circumstance d* with *goal e* in the light of the body of *knowledge f*" (Bunge 1989, p. 14). According to this analysis, the value *a* can never exist in itself; rather, it goes along with the other *relata b, d, e, f* in its relation to organism *c*. In other words, a value necessarily needs a living organism in some circumstance capable of forming some goal based on some knowledge, in order to emerge. Without the aforementioned material carriers that form such relational properties, a 'value' can neither emerge nor maintain its existence by itself.

The difficulty in understanding the nature of value, which Bunge resolves, resides in perceiving it as an *intersection of temporal elements* rather than a self-existing *independent* or *permanent* element. To be sure, Bunge says "there are only valuable or disvaluable objects [...] for some organisms in certain states" (Bunge 1989, p. 11). Hence, "'good' and 'bad,' 'right' and 'wrong' [are treated] as *adjectives* not *nouns*," since these words denote the properties of entities (Bunge 1989, p. 3, emphasis added) not entities as such. This understanding of the nature of value is the opposite of the Platonist or Idealist position that perceives good or value in itself. "In the real world, there are no values in themselves, anymore than there are shapes or motions in themselves" (Bunge 1989, p. 13), as they are a shape of something and a motion of some moving object. "Likewise, there are no morals in themselves. Instead, there are animals which, when behaving according to certain patterns [e.g., according to certain values and norms], contribute to the welfare of other animals" (Bunge 1989, p. 11).

Bunge notes that he has learned this analysis from Aristotle's *Nicomachean Ethics*, in which the latter "demolished ontological value absolutism, arguing that man does not know any goods in themselves, let alone *the* ultimate good which Plato rambled around: he can only know good actions and good things" (Bunge 1989, p. 64). The following example will better explain the nature of value as a relational property.

> When saying that well-being is a biological and psychological value, we mean that, being a survival condition, we evaluate positively some states of physical health and psychological contentment: i.e. we say that well-being is good for us. Likewise, when saying that honesty is a moral and social value we mean that we assign honest behavior, nay, honest people, a

positive role in social life as well as in keeping our own peace of mind. No organisms, no needs, hence no values. No society, no social behavior, hence no social values, whence no need for moral norms. (Bunge 1989, p. 11)

Having presented Bunge's understanding of the ontological status of value as a relational or relative property and its opposition to the idealist ontology, we note that Bunge also opposes the subjectivist tradition (Blanshard 1962, p. 104). Interestingly, for Bunge value is objective, though only indirectly, since it "is relativistic but not subjective" (Bunge 1989, p. 16). Relativism and subjectivism should not be confused since a relative property is still an objective one. This point is misunderstood by many current fashionable postmodern philosophies that equate Einstein's Theory of Relativity with absolute subjectivity, i.e., loss of objectivity. The Theory of Relativity is just the opposite. It is objective, but in a complex form rather than a simple one. Similarly, changes in the various variables that relativize value a, i.e., in respect b for organism c in circumstance d with goal e in the light of the body knowledge f could deem the same object or process as valuable or disvaluable (Bunge 1989, p. 17). Hence, "although we can class *values* (or value functions), we cannot class the value-bearers" (Bunge 1989, p. 17). Therefore, we cannot say that an apple objectively possesses the property of good in itself, but rather an apple is only good in relation to the variables b, c, d, e, f. In this vein, an apple for a healthy and hungry person that likes apples is good for him. This understanding is epistemologically precise when it admits that "All values are subject-rooted even though not all of them are subjective. Likewise, motility and vision occur only in animals yet they are perfectly objective" (Bunge 1989, p. 67).

In this context, Bunge does not adopt Hume's is-ought gap or Rudolf Lotze's famous value/fact barrier, as if the two were totally different and incompatible domains. In contrast, "value [is] an abstraction from [the act] of valuation, and the latter [is] a special kind of fact"; thus, value is accessible to rational as well as empirical study (Bunge 1989, p. 71). Indeed, value is a fact, but not all facts are values. Values or "value judgments" as in Lotze's wording can be justified or criticized, rather than accepted or rejected dogmatically, intuitively, or emotively, when they are rooted in basic needs or legitimate wants (Bunge 1989, p. 74). Bunge states "we indulge deliberately and cheerfully in what idealists and intuitionists call the *naturalist fallacy*, for we naturalize some values and socialize others. In this regard, our axiology is in the tradition of such diverse thinkers as the Greek atomists and the Stoics, Aristotle, Spinoza, Hobbs, Hume, Bentham, Marx, Durkheim, [and] ... Russell" (Bunge 1989, p. 72).

The objectively complex or relative nature of value, however, does not prevent us from classifying values. The key point is that values are relativized according to biological and social contexts. On the one hand, biological values can be classified according to: (1) environmental values, e.g., clean air, (2) visceral values, e.g., adequate food, and (3) mental values, e.g., needed feelings and emotions. On the other hand, social values can be classified according to: (1) economic values, e.g., productivity, (2) political values, e.g., self-government, and (3) cultural values, e.g., advancement of knowledge. In turn, every genus may be split into several species.

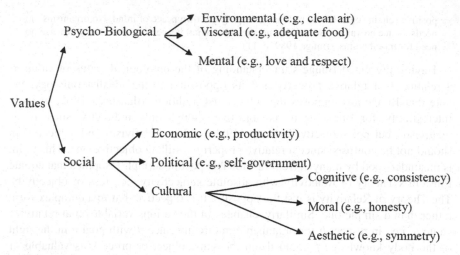

Fig. 28.2 Classification of values

For instance, cultural values may be grouped into cognitive, moral, and aesthetic values (Bunge 1989, p. 16). The following diagram shows a possible classification for general kinds of values (Fig. 28.2).

The merit of this classification is that there is no such thing as a single value ranking, let alone a single value function (Bunge 1989, p. 18), since many values could be equally crucial. For example, a bouquet of flowers and a meal may bear the same price tag; yet, their aesthetic and biological worth or value are not the same: they are hardly comparable according to a single value ranking (Bunge 1989, p. 18). Having analyzed Bunge's stance on the ontological nature and kinds of value, we are still faced with the question, 'what is *valuable*?', but this is an ill-posed question. The correct question should be, "what are the circumstances *b*, *c*, *d*, *e*, *f* that make an item valuable?"

Valuable or non-valuable items are the ones that may help or hinder the organism in two ways: internally (psychologically) and externally (environmentally) (Bunge 1989, p. 20). In this approach, Bunge deals with value as a natural or factual phenomenon. Values are, exclusively, seen among members of the animal kingdom, whose psychological faculties enables them to evaluate in contrast to e.g. fungi or plants. For the social animals, the roots or sources that establish the valuable are: (1) biological, (2) psychological, and (3) social. The latter is exclusively composed of individuals having the first two sources, but socializing adds additional considerations such as economics, politics, and culture. In this vein, the general anchor that makes item *x* valuable to organism *b* is *the ability to meet someone's need or want*.

So, we can define three kinds of value, that is, bio-value, psycho-value, and socio-value. Of *bio-value* it can be said: let *x* be an item (thing, process, or lack of thereof) in organism *b* or in the environment of *b*.[1] Then *x* is *biologically good* for *b* if and only if *x* contributes to keeping *b* in *good health* (Bunge 1989, p. 22). As for *psycho-value*: "let *x* be an item (thing or process) internal or external to organism *b* endowed with mental abilities. [Then] *x* is *psychologically valuable* for *b* in circumstance c if and only if *b* desires or wants *x* when c is the case" (Bunge 1989, p. 27). Moreover, in consideration of the prior condition of bio-value, a psycho-value is defined to be *legitimate* or genuine if and only if it contributes to the long-term health of *b*, otherwise it is illegitimate. Finally, *socio-value* combines the definitions of the previous bio-value and psycho-value: a social group is *socially valuable* only if it helps its members attain or retain their *good health* and meet their *psychologically legitimate wants* (Bunge 1989, p. 32). A corollary of this definition is that the most valuable of all the actual or possible social systems, in any given society, is the one that best serves all members of its society. In other words, the most valuable social system is the world system, i.e., the one encompassing all human kind.

Let us briefly remind ourselves of the results of Bunge's classification of values compared to other theories. According to a theologically-oriented value theory, God or his scriptures are the sources of all values (Bunge 1989, p. 32). Needless to say, Bunge's tripartite grounds of value, i.e., bio-psycho-social, are richer and much more objectively complex than traditional axiological theism. Bunge's theory also surpasses secular ethical views such as nihilism and idealism. In axiological nihilism, as in the case of Nietzsche's thought, the denial of values leaves the door open for individually false or power-oriented sources to generate values; hence, the destruction of the social domain. As for axiological idealism, e.g., Kant's criticism or modification of values seems impossible as they are ideals predetermined by intuition or *a priori* claims. The same holds, *mutatis mutandis*, for theologically-oriented axiology.

Let us now see how the sources of value characterize the valuable in the human context. In the human domain, valuable properties originate in human needs and wants, which form all types of values, i.e., the bio-psycho-social values. Needs emerge instinctively, by internal and external causes, within the human body and society in reaction to the environment. The same holds for wants where they emerge consciously or subconsciously within the human neuro-endocrinal system, i.e., the psyche, and its environment. Axiologically speaking, the phenomenon of bio-value is primarily the result of *needs*, psycho-value of *wants*, and socio-value a compatible mixture of supra-individual *needs* and *wants*. The following table exemplifies such needs and wants according to their roots (Bunge 1989, p. 35) (Table 28.2).

This table exemplifies the bio-psycho-social sources of value anchored in human needs and wants. This leads us to deal with some further ramifications of the value sources. Bunge divides *needs* into primary and secondary and *wants* into also tertiary and quaternary. Primary and secondary needs form our *basic needs* while the tertiary and quaternary wants, despite being important, are what could be called

---

[1] An "item" in this view could be a thing, like water, a process, such as exercise, or the absence of dangerous processes such as natural catastrophes.

Table 28.2 Different values

Bio-value	Psycho-value	Socio-value
Clean air and water	Being loved	Peace
Adequate food	Loving	Company
Shelter and clothing	Stimulation	Mutual help
Health care	Recreation	Social security

extra basic. What makes a need primary is its being necessary to keep a person *alive*, while what makes a need secondary is its being necessary to keep or regain *health* (Bunge 1989, p. 35). Needless to say, a person may stay alive without having good health; yet, being alive while being permanently sick definitely impedes human functionality, besides being unpleasant. Thus, the fulfillment of both primary and secondary needs is necessary and therefore they are considered *basic* human needs.

Still, humankind is not all about staying alive and healthy. In fact, we have individual and cultural wants that are not derived instinctively. They are the outcome of our personal psychological development and social psychology that make us want, desire, and aspire. The person-made reality of our psychological *wants* is totally unrestricted either in relation to oneself or society, contrary to the natural restriction of our biological *needs*. One may want (1) to be a swimmer or a gangster, (2) to love and marry or embrace celibacy, or (3) to be rationally consistent or obscurely inconsistent. All the former are examples of wants; however, Bunge investigates the *want* in its axiological relevance, i.e., legitimacy. Here he says, *x* is:

> a *legitimate* want *b* in circumstance *c* in a society *d*, if and only if, *x* can be met in *d* without (i) hindering the satisfaction if any *basic needs* of any other member of *d*, and (ii) without endangering the integrity of any *valuable subsystem* of *d*, much less that of *d* as a whole". (Bunge 1989, p. 35 emphasis AZO)

In other words, one's psychological *wants* are legitimate once they respect *basic needs* of one's self and others and the social integrity that contributes to these basic needs. In short, whereas satisfaction of primary needs is a matter of *life and death*, and that of secondary needs is a matter of *health and sickness*, the satisfaction of legitimate want is one of *happiness and unhappiness* (Bunge 1989, p. 36).

The previous assertion of basic human needs and extra-basic wants as the roots of bio-psycho-social values leads us symmetrically to the hierarchy of values. So, a value *x* as a mental process in someone's mind in circumstance *c* can be primary, secondary, tertiary, or quaternary as follows. The value *x* is a primary one if it contributes to satisfying at least one *primary need* of any human in any society. The same holds for *x*, *mutatis mutandis*, as (1) a *secondary* value if it contributes to satisfying at least one of the *secondary needs*, (2) a *tertiary* value if it contributes to satisfying at least one of the *legitimate wants*, and (3) a *quaternary* value if it contributes to meeting a *fancy*, that is, a desire or whim (Bunge 1989, p. 36). Bearing in mind, as Bunge points out, that a legitimate want is one that preserves and does not hinder any *basic need* or *valuable subsystem*, a *fancy* does not contribute to this preservation but neither does it necessarily lead to any illegitimate want.

To illustrate, the participation in the sport of football and competing for the World Cup can be a legitimate want insofar as (1) it leads to the satisfaction of one's basic needs in terms of physical training and income and, (2) does not hinder the social value system, e.g., competing for a nationalist or racial supremacy cause. However, wearing a million-dollar shoe is a fancy insofar it does not contribute to the satisfaction of any basic need and does not hinder the integrity of any valuable social subsystem, otherwise it is illegitimate fancy.

We have to note that the full realization of any given value is incompatible with that of some other; hence, *nobody can be completely happy* (Bunge 1989, p. 48). For example, the full attainment of knowledge-seeking could make the learner physically weak; or oppositely, the full attainment of athletic values could leave no time for knowledge-seeking. As a result, the individual ought to be *reasonably happy*, i.e., to be in a state of well-being and free to pursue legitimate wants. A ramification of this analysis, the definition of *"good"*, which Plato mystically posed, is, according to Bunge, that of *"a predicate of some object* b *in circumstance* c *for human beings having [legitimate] primary, secondary, or tertiary values"* (Bunge 1989, p. 36). The following figure schematizes the hierarchy of human values and their relational objective roots (Fig. 28.3).

Based on this interpretation of Bunge's thoughts, I shall note that whereas *survival* is the meeting of all basic needs (Bunge 1989, p. 44), *happiness* is the meeting of all needs and the reasonable meeting of legitimate wants (Bunge 1989, p. 44). In this line, primary and secondary values, i.e., survival, are universal and cross-

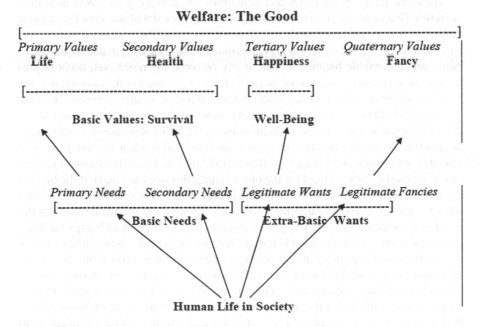

Fig. 28.3 Human values

cultural based on the universality of human nature, while tertiary and quaternary values, i.e., happiness and fancies, are circumstantial due to the particularities of human socio-psychology. An example of a universal necessity for survival is clean air and water and adequate food, which are therefore all universally good to all humans, whereas the particularities of *happiness* encompass wants, such as poetic or aesthetic aspirations. Clearly, world literatures, visual arts, and music have different aspects if not contradictory values in many instances. In other words, extra-basic values are more circumstantial than basic ones. However, all basic and extra-basic values are relational properties.

A consequence of this hierarchy is a Bungean norm that "meeting primary needs, *survival*, ought to precede meeting secondary needs, *health*, which in turn should precede meeting legitimate wants, [happiness] which ought to dominate the satisfaction of *fancies*" (Bunge 1989, p. 48). Additionally, every bio-psycho-social value has its own primary, secondary, tertiary, and quaternary hierarchy. In other words, every primary, secondary, tertiary, and quaternary value has within itself a bio-psycho-social triangle; thus, there is no hierarchy among the bio-psycho-social sources of values themselves, for they are inter-independent (Bunge 1989, p. 39). Thus, there is no preference for a healthy *biological* body compared with overcoming continuous *psychological* terror or *social* insecurity. All are necessary at once for bio-psycho-social beings, i.e., humans, since social insecurity can cause biological death or lead psychologically to a state of terror, therefore physical deterioration. As a result, the maximal or optimal value is not just biological survival but rather full biological and psychological satisfaction in society.

There are many possibilities and actualities for achieving the good in human societies. One may infer that the highest good would consist in attaining the greatest natural richness combined with the most efficient distributive social justice, i.e. *collective welfare*. Such a supreme good would enable everyone to attain their well-being and reasonable happiness. This is my personal inference; yet, according to Bunge the survival of humankind ought to be the supreme good, *summum bonum*, of all human beings. "Everything else, even social justice, comes thereafter" (Bunge 1989, p. 59). Bunge's rationale is simple: none of the sought-for values can be achieved without the survival of humankind. This being the case, the immediate necessities to secure the supreme good are universal nuclear disarmament and planetary environmental protection (Bunge 1989, p. 59). Granted the conditionality of the survival of human kind for the other values, this does not make it the highest good. I think that ensuring the survival of humankind is a necessary step for reaching a higher good, but not the supreme good itself. As we have shown, Bunge sees the good as a predicate of some object *b* in circumstance *c* for human beings fulfilling legitimate primary, secondary, and tertiary bio-psycho-social values. Therefore, it is inconsistent to give primary and secondary values, i.e., the survival of human beings, supremacy in identifying the highest good. An integral stance that encompasses all primary, secondary, tertiary, and quaternary values would be more consistent with Bunge's own philosophical systemism. Therefore, *integral social welfare* ought to be the supreme good for all human beings. I suggested this critical modification to Bunge in 2006, and he accepted it.

This search for achieving the good in human life is what determines Bunge's ethical doctrine; so, he calls it Agathonism, from *agathon*, Greek for "good" (Bunge 2001, p. 201). Bunge consistently builds over his *summum bonum*, i.e., the survival of mankind, or rather, integral welfare as just described. His moral maxim is *"enjoy life and help live"* (Bunge 1989, p. 104). This maxim is a prescription made of excellent logico-ethical engineering. It combines the satisfaction of *wants* with that of *needs,* whence satisfaction of wants, *enjoy*, is directed at needs, *life*. It harmonizes *needs* with *wants* by making wants a means to the end, needs. Moreover, the compatibly between the *self* and the *other* is the second part of the maxim, *help live*. It is the duty of the self to help others, whence achieving one's wellbeing is based on total collective wellbeing; therefore, stipulating the social aspect. The norm, *enjoy life and help live*, synthesizes egoism and altruism, self-interest and other-interest, egocentrism and socio-centrism, autonomy and heteronomy. It may therefore, be called *selftuist* (Bunge 1989, p. 104). Perhaps we may improve the ethical maxim, according to our modification of Bunge's highest value, by saying *enjoy welfare and help improve the welfare of others*. Having presented the general elements of Bunge's value theory, we turn now to the medieval Islamic theory.

28.4 Foundations of Islamic Ethics

Islam has a worldview guiding actions and legislation, but is there a value theory in Islam? And if there is, what would the 'good' or 'valuable' be according to it?

Shāṭibī's (1388/2011) theory, known as "higher objectives of Islamic moral law", *Maqāṣid al-Sharī'ah*, serves as one of the best theorizations of Islamic moral theory, as we will see in Sect. 28.6. Yet, this theory does not rationalize its foundation beyond inducing the patterns. This process according to Shaṭibī draws from the particulars of Qur'anic and prophetic legislation the most general norms or the "spirit of laws". This is why I shall rely on another investigation that does construct an Islamic value theory (Obiedat 2012, pp. 330–340). Unlike the modern scientific and logical language of Bunge, the following will utilize premodern discourse. Yet, we shall find amazing convergence.

There are three major value types in the Qur'an: authoritarian, utilitarian, and naturalist (Obiedat 2012, p. 110). The authoritarian value system justifies its norms through reference to the status of the divine, the utilitarian by the assured worldly or after-worldly reward, and the naturalist by rationally referring to the nature of things, i.e. *fiṭra* (Obiedat 2012, pp. 113–115). The authoritarian or submissive definition of good is corroborated by several Qur'anic verses. Their gist can be formulated in the statement 'whatever God commands or finds valuable is good' (Obiedat 2012, p. 111). In contrast, the second, i.e., utilitarian justification, is also substantiated by many Qur'anic verses. It could be captured by the claim 'what generally brings about utility, *mṣlahah,* to the largest number of people and most times is indeed good.' Clearly this is a utilitarian value orientation (Opwis 2010). The third, naturalist notion, is based on another group of Qur'anic verses. They lead to the claim 'whatever agrees with the essential nature of things conforms to God's purpose of creation, *fiṭra*; therefore, it is de facto 'good' (Obiedat 2012, p. 114). In this way, the reader is stuck with three justifications for the good, i.e. authoritarian, utilitarian, and naturalist. These justifications are by no means similar in their implications and need to have their relations clarified and systemized, in accordance with the general stance of Muslim jurists and theologians that interpretations of the Qur'an have to be made consistent.

There is a particular degree of interrelatedness between the utilitarian and naturalist justifications of good, for the naturalist justification can be shown to include the utilitarian one. For example, a person finds utility in quenching their thirst because it is the nature of human beings to do so. Likewise, a mother breastfeeds her child because it is the nature of mothering to do so. This way of designating values by referring to primordial behaviors that humans must conform to is very common in Islamic discourse. Here, we see how utility is derived from a naturalist stance and both are considered good. Similarly, a seller abides by the terms of the contract he signs because it is a social requirement or the nature of contracts and stable transactions to do so, otherwise trade will no longer be possible and social trust and cooperativeness will eventually perish. The analogy can be expanded to cover everything else in Qur'anic legislations. In short, what meets the nature of things is utility.

The converse situation is highly problematic, for nature is inherent but utility is contextual. We cannot reason the other way around, to derive the nature of things from utility, for one's person utility is not always natural as it could be rooted in fancies or misunderstanding of priorities. Also, a thief's utility is in conflict with that of the property owner. In short, the nature of things is not always derived from utility since the latter is changing in a contradictory and conflicting manner, personally and socially.

The utilitarian and naturalist Qur'anic justifications of good are reconciled, I suggest, in the principle 'good is what is generally utilizable, granted it is derived from the nature of things'. There is however a problem reconciling this assertion with the authoritarian justification. According to the authoritarian definition, following what God dictates is good in itself, even if it lacks any utility or opposes the nature of things! Yet, this cannot be held to be true here since the Qur'an cannot be contradictory, as Muslim theologians and jurists keep reminding us. Still, if the utilitarian definition could indeed be reduced, in principle, to partially agree with the naturalist stance, the pressing question would be, how can we reconcile both of these with the authoritarian notion?

A possible reconciliation for the three Qur'anic justifications of good is the following synthesis: 'what is generally utilitarian, granted it is derived from the nature of things, is good; therefore, authoritatively demanding submission.' So, in this sense, the authoritarian is not a justification but rather a motivation to follow the utilitarian-naturalist justification. In other words, the authoritarian verses have to be understood as a rhetorical or pedagogical shortcut, not an essential definition for good. The following verse exemplifies the way that submission to divine authority, or responding to it, is interpreted in terms of utilitarian notions, i.e.

reviving humanity: "O you who believe! Respond to God and His Messenger when he calls you to that which gives you life" (Q. 8:24).[2]

The aforementioned synthesis for the three justifications of good is not agreed upon by the majority of Muslim scholars, but it is a tenet of the Muʿtazila theological school, which holds that "God is bound to do what is the best (*al-aṣlaḥ*) for humans" (Watt 2008, p. 51). It is also agreeable to a great number of later legal theorists from most camps that "the *sharīʿah* has been formulated for the interest of the humans" (Shāṭibī 2011, p. Vol. 2, 3). The Qurʾan further illuminates that the authoritarian notion could be reduced to the utilitarian one: "whoever acts righteously benefits himself; and whoever acts evilly, it is against himself: and your Lord is not at all a tyrant to His servants" (Q. 41:46).

In other words, the authoritative notion, tyranny, or absolutism of the divine, in that verse is not the goal, but rather the utilitarian-naturalist one is the goal. The authoritative notion is rather a rhetorical way of discourse that assumes the utilitarian-naturalist basis. The same verse is repeated in identical or similar wordings at Q. 3:182, Q. 8:51 and Q. 21:10, all confirming that values in the Qurʾan are rooted or founded in the synthesis above.

The Qurʾanic clarification of the purpose of creating mankind is "your Lord said to the agents of nature [angels]: I am going to place in the earth a *successor* [to Myself]" (Q. 2:30–1). Humans are divine successors on earth, bestowed with the capacity of naming, that is, the use of language or utilization of consciousness. This clarification is in sharp contrast to the purpose of unconscious agents of the universe, such as stars or mountains, and could be central to the interpretation of other verses which discuss values. In this verse, a human being receives a godlike status capable of succeeding the creator of nature by understanding that nature, although it is expected to do otherwise sometimes. This succession to the creator of nature implies that humans are capable of sensing the nature of things and acting generally in a utilitarian manner in order to do good. This clearly negates the notion of blind obedience and confirms the maxim: 'what is generally utilizable, granted it is derived from the nature of things, is good; therefore, authoritatively demanding submission'. We may therefore conclude with the maxim of Qurʾanic ethics as seen in Q. 8:24: "Respond [. . .] when He calls you to that which gives you life." Here, the response is to the values of nature, *fiṭra*, which brings good in this life. Obviously, the contemporary study of nature is the sole function of modern science and technology. In this case, science becomes the central arbitrator of the utility and the nature of things and actions whenever controversial ethical questions arise.

Although the appeal to supernatural deity and metaphysical scenarios in the classical Islamic discourse can justifiability be accused of being inconsistent and empirically falsified superstition, the ethical maxim *act by what gives you life* can still lead to the same axiological results starting from the atheistic stance. Let us see how the former maxim has strong moral applications converging between ethics of scientific humanism and classical Islamic legal theory.

---

[2]For alternative scholarly English translation of these Qurʾanic verses, see Nasr (2015).

28.5   Morality According to Bunge's Scientific Humanism

It is a fact of human behavior that when one values something essential to his or her life – either by nature or nurture – one normally tries to obtain it (Bunge 1989, p. 93). The normality of humans' valued objects is conditioned in the social domain, since human groups exist solely because "every human being has needs and wants that can only be met with the help of others" (Bunge 1989, p. 95). This explains the emergence of moral codes from valued objects as an abstraction of material and institutional processes known as moral culture. Such moral culture facilitates spotting moral problems and communicating and guiding rational thinking about them. The necessity of morality is simple: if every individual were self-reliant from birth there would be no point in sociality, let alone in reciprocity or mutual help, and in the associated system of the rights and duties made explicit in moral and legal codes (Bunge 1989, p. 95). In other words, everyone needs the help of someone else to meet all of her needs and some of her wants (and thus realize all of her primary and secondary values as well as some of her ternary and quaternary values) (Bunge 1989, p. 102).

This is why the value of basic needs, i.e., life and health, does not always need socially-based knowledge to make the individual aware of them. They are instinctive in every individual, unlike norms which are a social reaction to basic and extra-basic values, the latter being legitimate wants. In short, morality is about how to live properly in society (Bunge 1989, p. 129). Conversely, humans that fail to comply with certain norms succumb early in life and leave no descendants (Bunge 1989, p. 121), which confirms that some duties are in the genes (Bunge 1989, p. 119).

Like values, moral norms have bio-psycho-social roots, simply because the purpose of moral norms is *to help realize (or inhibit) human values* (Bunge 1989, p. 94). Since *needs* and *wants* are key concepts in value theory, *rights* and *duties* are the key concepts in morality. The relationship between rights and duties, in principle, is strictly mutual and dynamic since "every right implies a duty" (Bunge 1989, p. 102). The following chart lists the types of moral rights and duties and an example for each forming a system of morality (Bunge 1989, p. 100) (Table 28.3).

Bunge views norms as the social actualization of values (Bunge 1989, p. 98), in which rights imply duties and vice versa (Bunge 1989, p. 101). Primary and

Table 28.3 Rights and duties

Moral Roots	Examples of Rights	Examples of Duties
1- Environmental	Clean environment	Environmental protection
2- Bio-psychological	Survival	Help others survive
3- ┌→ Cultural	Learning	Teaching
4- Social├→ Economic	Work	Faithfulness in workmanship
5- └→ Political	Liberty	Popular participation

secondary values guide corresponding primary and secondary rights, where the latter imply corresponding primary and secondary duties. Hence, a human being in a society has *basic moral rights* to something if it contributes to his *well-being* and he has a *secondary moral right* in something if it contributes to his reasonable *happiness*, granted in both cases that he neither jeopardizes his primary rights or anyone else's primary or secondary rights. Bunge's formal definition of a *moral right*: If x is a human being in a society y, and z is a thing or process in or out of z, then: (1) x has a basic moral right to z in y, if, and only if, z contributes to the *well-being* of x without hindering anyone else in y from attaining or keeping things of the same kind as z; (2) x has a secondary moral right to z in y, if, and only if, z contributes to the *reasonable happiness* of x without interfering with the exercise of the primary rights of anyone else in y (Bunge 1989, p. 97–98).

For example, an individual has the *primary bio-psycho-social rights* to shelter, the opportunity to find love, and social security respectively. She also has the *secondary bio-psycho-social rights* to participate in sports, express opinions, and seek leadership positions, granted that she neither jeopardizes her primary rights or anyone else's primary or secondary rights. Conversely, if a human being has already achieved her primary rights, that is, survival, then she has a *moral duty* to help another human do the same if no one else can perform such duty. Equally, a human being has a *secondary moral duty* to help another human exercise their secondary rights if no other can perform such duty.

In other words, if any individual has the *primary right* to shelter, recreation, and social security and if collective taxation alone is capable of providing these rights, then it is a *primary duty* of citizens to pay their taxes. In addition, if some individuals have *secondary rights* to participate in sports, express opinions, and seek leadership positions, and if civil society alone is capable of facilitating them, then, inferring from Bunge, it is the *secondary duty* of that civil society to do so. Although basic needs and legitimate wants generate rights and these duties, the latter in turn restrict rights (Bunge 1989, p. 103). In other words, the limitation of rights is caused by the symmetrically generated duties. Hence, rights and duties are in a state of mutual restriction or equilibrium. The more rights one has, the more duties they would have. The only way to minimize the burden of duties is to decrease the gained rights. For example, a graduate student may have a right to a socially subsidized educational and vocational loan so long as she returns to her community to provide service. The greater the loan, the greater the expectation of the service. If that graduate student is incapable of fulfilling such commitment, she should not increase the size of her loan, or not take it initially. This explains the mechanism of mutual equilibration of rights and duties in a socially interrelated manner. This is shown below (Fig. 28.4).

The aforementioned structure is the scientific humanist foundation of morality based on the previously presented value theory. At this point of analysis, the road between the morality of scientific humanism and the Islamic one has been intellectually paved.

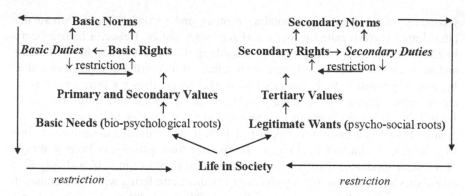

Fig. 28.4 Rights and duties

28.6 Morality Between Scientific Humanism and Shāṭibī's Islamic Ethics

The ethical maxim Bunge's scientific humanism is *enjoy welfare and help improve the welfare of others* and the maxim for the Islamic one is *act by that which gives you life*. The *good* in scientific humanism is a predicate of some object *b* in circumstance *c* for human beings seeking the satisfaction of *basic needs*, i.e., life and health, and *legitimate wants*, i.e., happiness. The good in the *sharī'ah*, as philosophized by jurists in the many major Islamic schools,[3] is the social satisfaction of necessities (*ḍarūriyyāt*), needs (*ḥājiyyāt*), and desirables (*taḥsīniyyāt*) (Shāṭibī 2011, Vol. 2, pp. 13–14). *Necessities* in the Islamic maxim are things or processes that if they cease to exist then society or the individual will perish accordingly. In contrast, *needs* are extra-qualities that could be added to the necessities that make life easier, but their inexistence would not make life impossible. Finally, *desirables* are additional qualities that could be added to the needs that make life more enjoyable and beautiful to live. Obviously, absence of desirables neither makes life uneasy nor impossible. Unlike Bunge's dual dichotomy of needs and wants, the Islamic one has a tripartite dichotomy of necessities, needs, and desirables, which adds flexibility and more detail to the dichotomy.

Medieval Muslim legal theorists presented a detailed view on the nature of human necessities. In particular, necessities are hierarchically the preservation of religion, self, mind, lineage, and property (Shāṭibī 2011, p. Vol.2, 10.). Preservation of religion refers to the maintenance of the society against aggression. Dispensing with

---

[3] Although these three hierarchic values originate in the Shāfi'ī school, they are widely accepted by the majority of the *sunnī* Islamic jurisprudence schools, i.e., the Mālikī, Ḥanafī, and Ḥanbalī. However, since this theory was a latter development in Islamic jurisprudence as envisioned first by the Persian Muslim Jurist, al-Juwaynī (1028–1085 C.E.) (Ṣaghīr 1994, pp. 399 and 431), the proper application of these three values was not taken seriously in the body of the law. Therefore, one may not claim that these values are mainstream in the strict sense.

this social system was seen as leading to the collapse of security, productive, and peaceful interactions. Such attention to society corresponds to Bunge's preservation of "the integrity of any *valuable subsystem* of [that society] as a whole" (Bunge 1989, p. 35). The preservation of (1) *religion* is aimed at protection of society from obliteration; (2) of *self* is aimed at protection of life from murder, torture, or injuries; (3) of *mind* is aimed at preserving the intellect from the hallucinations of drugs, alcohol, or the like; (4) of *lineage* is aimed at sustaining one's offspring and performing parental responsibilities; and (5) of *property* is aimed at preventing theft, confiscation, negligence, or damage.

These pentagonal values of social existence, life, consciousness, family, and wealth are the *necessities* that lead to other supplementary values, which are the *needs* which in turn may lead to a third category, *desirables*. For example, breathing is a *necessity* for preserving the self, but having proper ventilation in one's house is a *need* that adds to that satisfaction of that necessity, while having a nice aroma is a *desire* that complements the *need* of ventilation. Similarly, the necessity of maintaining the ethical-legal system requires courts and judges. So, if there were no violations at hand and conflicts or controversies were not arising, appointing full-time judges would not be a necessity. Villages are an example of where judges are unnecessary, contrary to mega-cities. Maintaining beautiful architecture for courts is desirable in this case but not needed for judicial procedures. Symmetrically, wellness is necessary for preserving the *self* but fitness is a need, consciousness is necessary for the *mind* but primary education is a need, parental support is necessary for *lineage* but compassion in marital relationship is a need, and security is necessary for *property* but registration of property is a need. These exemplifications of needs that are subsidiaries of the five basics necessities confirm that they vary and proliferate based on social and contextual circumstances, and also demonstrates that they are less important than the necessities they seek to sustain.

The Islamic ethical maxim, *act by which gives you life*, properly sets a clear basis for the triad of Islamic values, necessities, needs, and desirables. It orients moral action towards the various aspects of life whether necessary, needed, or desired. In Bunge's terminology, these value-classes are aimed at life in the integral sense, socially and individually, primarily, secondarily, and tertiarily. The synthesis of the maxim 'act by which gives you life' is concerned with the balance between the individual and the social, where 'you' addresses both. Although 'act by which gives you life' is directed at action, it presumes mental activities in which seeking the means to life includes knowledge, planning, science, technology, etc. These dynamics situate Islamic ethical thinking on a similar factual and technical horizon as the scientific humanistic ethics. Here, Bunge provides a good modern heir to the medieval Islamic proposals. Shāṭibī's axiological value-class is seen in the following chart (Fig. 28.5).

This triad system of values, i.e., necessities, needs, and desirables, was proposed by many scholars of the *sunnī* schools of Islamic moral law and was a result of inducing the totality of the moral norms and legal codes commanded or suggested in the Qur'an and the prophetic traditions (Shāṭibī 2011, p. Vol. 2, 44). The symmetry between the pentagon of necessities of Islamic axiology, i.e., social existence,

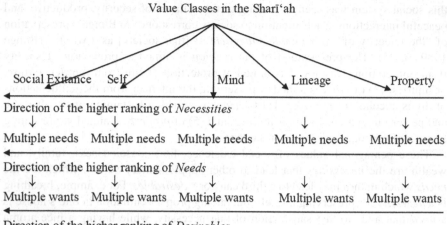

Fig. 28.5 Shāṭibī's value classes

self, mind, lineage, and property, and that of the bio-psycho-socio roots of values in scientific humanism is strong. Attaining social order and property are mainly *social* values, protecting the mind is a *psychological* value, and preserving lineage and self (in the physical sense) are *biological* values. The recognition of the bio-psycho-socio sources of value in Islam is due to the Qur'anic notion *fiṭrah* that calls for observing natural processes and things. The bio-psycho-social sources of values in scientific humanism are more comprehensive than Shāṭibī's and offer a precise dichotomy as they may include items that are not considered necessities in the traditional Islamic axiology, such as environmental preservation.[4] This current concern falls under biological values. However, the difference in era and knowledge-update do not disqualify the Islamic medievalist value theory altogether but call for a scientific update and more rigorous conceptualization in the Islamic value theory.

Some scholars consider this classification as a purely legal one. However, it is manifest that *desirables* are not a purely legal category and this applies to many needs also (Coulson 1964, pp. 83–84). Therefore, an Islamic judge would not persecute a person for not seeking their desirables. Furthermore, it is prohibited to envy or wish someone evil according to Islamic values as it hinders compassion, social cohesion, and cooperation. Still, these acts do not have any civil punishment confirming that prohibiting envy is merely a moral value, not a legal one. Just like scientific humanist ethics, this Islamic value system corresponds to a wide moral conception in which law is only part, while the rest of the values are left for the self or the society to uphold in their relevant contexts.

---

[4]For contemporary alternative classification for the "higher objectives of Islamic moral-law *sharīʿah*", see: ʿAbd al-Raḥmān (1994, pp. 93–110), ʿAtiyyah (2003, pp. 91–105), and al-Najjar (2008, pp. 34–38).

Rights and duties in this Islamic axiology have a somewhat unique conception. This may in part be due to the fact that the Arabic word for right, i.e., *ḥaqq*, can also be the same for duty (Ibn Manẓūr 1984, p. Vol. 9, 941.). The difference goes in the auxiliary preposition. So, it said right to me (*ḥaqq lī*) and right upon me (*ḥaqq 'alay*) to express consecutively 'my right' and 'my duty'. Consequently, every norm in Arabic, linguistically and conceptually, already inherits the right-duty mutual balance that Bunge seeks to affirm in western axiological thought. In this vein, Islamic morality goes under the five-prong model of *al-aḥkām al-shar'iyyah*, moral norms. These are classified in the continuum of prohibited, repugnant, permitted, recommended, and obligatory.

This morality model explains why rights and duties are already fused and inseparable. As a result, one's *duty*, such as testimony in court, is obligatory to do and prohibited to dismiss. Equally, one's *right*, such as preserving one's health, is *obligatory* to obtain and *prohibited* to decline. On a different level, one's desirable or legitimate want, such as dressing elegantly in social activities, is *recommended* to achieve and *repugnant* not to achieve. Finally, if something is out of the realms of needs and wants such as choosing between a yellow apple and a red one it would be neutral to choose, i.e., permitted or *ḥalāl*.

Deriving from the triad of Islamic values, all the rights and duties associated with the five *necessities* and their multiple needs are obligatory to do and prohibited to dismiss. For example, in preserving the second necessary category, i.e., the self, the norm dictates that it is obligatory to feed oneself and prohibited to kill others. However, it is recommended to eat delicious food and repugnant not to prepare one's guest the best of available feasts (al-Nawawī 1997, p. 60). Thus, rights and duties already appear fused in order to perform one's obligation or avoid one's prohibition. This is the meeting between Islamic axiology and morality. In this line, values in the triad of necessities, needs, and desirables are equally penetrated by the five moral norms of prohibited, repugnant, permitted, recommended, and obligatory. Muslim legal theorists long before Shāṭibī are quite clear that the application of these five norms is not fixed and perpetual, but changing and dynamic. This is why they say "a single action can have any of these [five norms] or some based on its nature" (Khallāf 1978, p. 116). This means that any event or processes is potentially subject to be prohibited, repugnant, permitted, recommended, and obligatory based on the context. So, the prohibition of alcohol is lifted if one needs to quench one's thirst in the absence of water or for medical purposes. Similarly, obligation to penalize theft is forgiven if there is a famine, as evidenced by the ruling of the second caliph, 'Umar (al-Duraynī 2013, p. 14). Therefore, nothing is perpetually prohibited or obligatory but based on the dynamics of the situation of fulfilling the five necessities.[5]

---

[5] Unfortunately, this knowledge of Islamic jurisprudence is not currently clear in Muslims' public practice since it is presumed that prohibitions or obligations are perpetually unchanging. However, this public unawareness does not affect the clarity of the medieval legal theory on this issue.

Fig. 28.6 Duties and rights
in scientific humanism

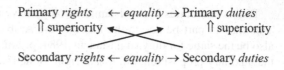

Primary *rights* ← *equality* → Primary *duties*

⇑ superiority   ⇑ superiority

Secondary *rights* ← *equality* → Secondary *duties*

As for the hierarchy of moral rights, in scientific humanism, they are *rights to realize primary, secondary, and tertiary values.* Quaternary values, which are products of fancy, remain beyond the ken of rights and duties, particularly in a poverty-stricken society (Bunge 1989, p. 97). Thus, the primary category includes rights and duties in primary values, i.e., *basic needs* and the secondary category includes rights and duties in secondary values, i.e., *legitimate wants.* The hierarchy of morals is explained in the following figure, where the vertical or diagonal arrows show superiority and horizontal ones show equality (Fig. 28.6).

In other words; (1) primary *rights* take precedence over secondary *rights*, so the primary right to have a shelter supersedes the secondary right to have a means of transportation. (2) Primary *duties* take precedence over secondary *duties*, hence the primary duty to feed one's children overtakes the secondary duty to pursue one's legitimate aspirations. (3) Primary *duties* take precedence over secondary *rights*; thus, the primary duty of popular political participation supersedes the secondary right of recreational gathering with friends. (4) An individual faced with a conflict between primary right and duty or secondary right and duty is morally free to choose either, subject only not to jeopardize anyone else's right (Bunge 1989, p. 101).

For example, when a person is faced with the conflict between the primary duty of forest protection and the primary right of consuming wood for home construction and furniture, he is free to choose either. Needless to say, a creative and sustainable reconciliation between conflicting cases would be necessary if feasible, such as consuming an equivalent amount of wood to that planted every year. Similarly, when she faces the predicament of choosing between the secondary duty of pursuing advanced knowledge or the secondary right of enjoying sports, she is free to choose either. A creative reconciliation between conflicting cases would be required if feasible in this case. The previous four rules, which encompass all particularities of moral life, are the mechanism for determining the priorities and resolving the conflicts between moral norms in scientific humanism. This mechanism of applying general norms to the particularities of human action is known in the Islamic moral-law as *fatwā.*

A learned philosophy reader may ask, along with Plato, of any religion that espouses what it considers to be the moral good: *Is it good because the gods approve or do the gods approve because it is good?* In the case gods approve of good because it is good, a modern person would ask; *does that make scripture superfluous?* The classical Mu'tazilah theological camp is peculiar in holding that God approves of the good because it is intrinsically so. Consequently, when religious scripture violates this objectively recognizable good, then the scripture has to be interpreted in a way to align with the good, as a reference point, not otherwise. Theologically, Shāṭibī is not as clear as the Mu'tazilah on this, but the final legal outcome is identical. In

Shāṭibī's view, the particulars of the scriptures have to be interpreted in a way to align with the five necessities outlined above.

This understanding from the Muʿtazilah and Shāṭibī implies that the scripture is merely a reminder or an indicator of the good not a comprehensive source for the good. In the historical case of predominantly oral Semitic cultures in antiquity, segments of the Bible were the book (or scroll), because it seems to have been the first book there was in the relevant language. Equally, for the predominantly illiterate Semitic Arab communities, the Qur'an was also *al-kitāb*, which literally means the book or ironically the bible. In both Jewish and Islamic cases, the historical significance of the scripture is that it was not only one reminder out of many, but also the first and possibly the only written reminder or indicator of good. However, with the evolution and current flood of humanistic and scientific knowledge, the proportional domination of the scripture either as reminder or indicator becomes much smaller. A simple comparison between the centrality of scriptures in medieval universities around the Mediterranean and the marginalization of these very scriptures in its modern counterpart suffices to explain the cognitive and moral weight of scriptures.

28.7 Conclusion

To recapitulate, ethics is about good values and right morals. The *good* in scientific humanism is about recognizing *needs*, that is, life and health, and *wants*, i.e., happiness. Symmetrically, the good in the *sharīʿah* is the social satisfaction of necessities (*ḍarūriyyāt*), needs (*ḥājiyyāt*), and desirables (*taḥsīniyyāt*). Morality in the first system is mutual restriction of rights and duties to realize primary, secondary, and tertiary values. Morality in the *sharīʿah* is the double sided rights and duties including the dynamic application of the five norms, prohibited, repugnant, permitted, recommended, and obligatory. The ethical maxim for the first ethics is *enjoy welfare and help improve the welfare of others* and for the second is *act by which gives you life*.

Bunge's secular ethics are neither absolutely permissive nor individualistic, as is instead the case in some currently fashionable liberal and postmodern tendencies. It is not licentious because whatever rights one gains are automatically restricted by equal duties, in addition to the hierarchic restriction of primary over secondary norms. Scientific humanism is not a purely individualistic path also because both primary and secondary norms are equally derived from bio-psycho-social sources of values. Thus, rights bestowed upon the individual by his society entail their social duties also.

Symmetrically, a great portion of Islamic morality is issuing *fatwā* for the public and assuring moral self-awareness in particular situations, i.e., deciding the applicable moral norms and determining their priorities and resolving possible conflicts between them. The hierarchy of morals in Islamic ethics, as suggested by Shāṭibī and others, is as follows: preservation of *social existence* takes precedence over the protection of the *self*; that over *mind*; that over *lineage*; that over *property* (al-Ghazzālī et al. 1990, p. Vol. 2, 562).[6] Within each of the former five items, preserving their *necessities* takes precedence over their *needs* and that over their *desirables*. Consequently, preserving one's source of *wealth* comes lower than protecting *familial relations* from breakdown, which comes below preserving one's *mind* (or psyche) from mental disorder. Still, escaping a situation causing mental disorder is less important compared to caring for one's potentially fatal injury or malnutrition. Finally, defending one's family or society is worth sacrificing life. This norm application mechanism, *fatwā*, is quite similar to that in scientific humanism. The difference is that in the secular ethical system, the bio-psycho-social sources of values always come together, which puts the individual's life on par with the social life. In the Islamic system a particular superiority is given to the society over the individual. By contrast, Bunge does not see anything worth sacrificing life that would make it a primary duty since nothing supersedes the value of life, but it can be voluntarily done.

Obviously, Shāṭibī's Islamic ethics is neither "religious dictatorship" (Marinoff 2007, p. 453) nor spiritual asceticism (anti-material or unworldly). It is not totalitarian because four of the five values, the self, mind, lineage, and property compose 80% of their moral categorization, which are initially devised to serve the individual. Shāṭibī's Islamic morality is not unworldly or ascetic also, as one might think of religious ethics, because the unworldly cannot give rights to be requested or duties to be fulfilled, and because nothing is immaterial about the five values.

Knowing the good, in sound value theory, is neither a magical endeavor nor a supernatural revelation. In scientific humanism, knowing the good is reliant on understanding needs and wants while depending on the human rational capability to devise strategies and tools achieving these values. In a nutshell, knowing the good involves gaining knowledge of nature and the constructive intelligence of reason; in transcendental terms, constructing the sacred from within the worldly. Symmetrically, knowing the good in the Qur'anic hermeneutics articulated in this chapter is based on a particular Islamic legal tradition not far away from the former naturalist-humanist stance. Knowing the good in this Islamic ethics is neither blind submission to literal understanding of holy texts, as the great majority of literalist Muslims would prefer, nor absolute ascetic seeking of the hereafter as some religious mystics would favor. Valuation is clearly set towards preservation of the major elements of human socio-individual life, i.e., social existence, self, mind, lineage, and property, all of which are worldly subjects *par excellence*.

Doing the right is neither an authoritative and obscurely superstitious endeavor nor anarchist chaos. In scientific humanism, doing the right is a fine balance between the rights of attaining basic needs and legitimate wants and the duties derived from obtaining these rights. Therefore, it is a social harmonization project by which everyone would achieve his/her rights without jeopardizing anyone else's rights. The same holds, *mutatis mutandis*, for Shāṭibī's Islamic ethics, where norms are

---

[6]There is no consensus amongst medieval jurists on this hierarchy for some think that the "mind" should be lower than "lineage" in priority.

not perpetually fixed rulings. Rather, norms are circumstantial variables aiming to balance social realization of the three value-classes of necessities, wants, and desirables.

In the end, one might wonder why secular ethics and religious ethics are deemed unbridgeable. The moral theory of scientific humanism is neither absolutely permissive nor individualistic, which makes interaction between Shāṭibī's ethics and Bunge's quite reasonable. Equally, Islamic morality is neither absolutely totalitarian nor unworldly, which makes its interaction with secular ethics fruitful. Could it be said that Bunge was making scientific ethics more religious and al-Shāṭibī was making Islamic ethics more secular? This is probably a sloganeering misconception because once naturalism and rationality are combined to ground any system of thought, the results should be more objective, consistent, and novel. Such a system of thought, thus, would be universally more acceptable regardless of religiosity or irreligiosity. The difference here between the scientific humanist ethics and the Islamic one is rather a difference of knowledge update and logico-conceptual articulation.

Would all Muslim ethicists agree on the synthesis proposed here? Definitely not. The same holds true for scientific humanist ethics, which is by no means a point of consensus among individualists, irrationalists, and anti-science secularists. The conclusion is that merging naturalism (as a way to explore and experiment with the nature of things) with rationality (as a method to define, construct, and devise strategies and tools) is the way to know the good and do the right. Such a path should be a golden rule for international cooperation and intercultural dialogue.

Acknowledgement I am grateful to the Canadian mathematician, Dr. Michael Kary, for kindly reading this chapter and offering several corrections.

References

'Abd-al-Raḥmān, Ṭ. (1994). *Tajdīd al-Manhaj fī Taqwīm al-Turāth*. Beirut: al-Markaz al-Thaqāfī al-'Arabī.

al-Duraynī, M. F. (2013). *al-Manāhij al-Uṣūliyyah fī al-Ijtihād bil-Ra'y fī al-Tashrī' al-Islāmī*. Mu'assasat al-Risālah.

al-Ghazzālī, A. Ḥāmid M., Bihārī, M.A. ibn 'Abd al-Shakūr, al-Ansārī, 'Abd al-'Alī Muḥammad. (d. 1111/1990). *al-Mustaṣfā fī 'Ilm al-Uṣūl wa ma'ah Fawātiḥ al-Raḥamūt bi Sharḥ Musallam al-Thubūt*. Beirut: Dār Ṣādir.

al-Najjār, 'Abd. a.-M. (2008). *Maqāṣid al-Sharī'ah bi-Ab'ād Jadīdah*. Beirut: Dār al-Gharb al-Islāmī.

al-Nawawī, A.Z.Y. ibn S. (d. 1277/1997). *Forty Hadith Qudsī*. Cambridge: Islamic Texts Society.

'Atiyyah, J. a.-D. (2003). *Naḥwa Taf'īl Maqāṣd al-Sharī'ah*. Amman: al-Ma'had al-'Ālamī lil-Fikr al-Islāmī.

Blanshard, B. (1962). *Reason and goodness*. New York: G. Allen & Unwin.

Bunge, M. (1989). *Treaties on basic philosophy. Vol. 8, ethics: The good and the right*. Dordrecht: Reidel.

Bunge, M. (2001). *Philosophy in crisis: The need for reconstruction*. Amherst: Prometheus Books.

Coulson, N. J. (1964). *A history of Islamic Law*. Edinburgh: Edinburgh University Press.

Fakhry, M. (1997). *Averroes, Aquinas and the rediscovery of Aristotle in Western Europe*. Washington, DC: Georgetown University.

Hallaq, W. B. (1997). *A history of Islamic legal theories: An introduction to Sunnī Uṣū al-Fiqh*. Cambridge: Cambridge University Press.

Halliday, F. (2002). *Islam and the myth of confrontation: Religion and politics in the Middle East*. London: I.B. Tauris.

Heck, G. W. (2006). *Charlemagne, Muhammad and the Arab roots of capitalism*. Berlin: W. de Gruyter.

Ibn Manẓūr (d. 1312/1984). *Lisān al-'Arab*. Cairo: Dār al-Ma'ārif.

Khallāf, 'Abd. a.-W. (1978). *'Ilm Uṣūl al-Fiqh*. Kuwait: Dār al-Qalam.

Küng, H. (1991). *Global responsibility: In search of a new world ethic*. New York: Crossroad Pub. Co.

Mahner, M., & Bunge, M. (1996). Is religious education compatible with science? *Science & Education, 5*, 101–123.

Makdisi, G. (1990). *The rise of humanism in classical Islam and the Christian West: With special reference to scholasticism*. Edinburgh: Edinburgh University Press.

Marinoff, L. (2007). *The middle way: Finding happiness in a world of extremes*. New York: Sterling Pub.

Matthews, M. R. (2012). Mario Bunge, systematic philosophy and science education: An introduction. *Science & Education, 21*, 1393–1403.

Nasr, S. H. (2015). *The study Quran: A new translation with notes and commentary*. New York: Harper Collins.

Obiedat, A. Z. (2012). Defining the good in the Qur'an: A conceptual systemisation. *Journal of Quranic Studies., 14*, 110–120.

Opwis, F. M. M. (2010). *Maṣlaḥa and the purpose of the law: Islamic discourse on legal change from the 4th/10th to 8th/14th century*. Leiden: Brill.

Rescher, N. (2001). *Philosophical reasoning: A study in the methodology of philosophizing*. Malden/Oxford: Blackwell.

Rogan, E. (2011). *The Arabs: A history*. London: Penguin Books.

Ṣaghīr, 'Abd. a.-M. (1994). *al-Fikr al-Uṣūlī wa-Ishkāliyyat al-Sulṭah al-'Ilmiyyah fī al-Islām: Qirā'ah fī Nash'at 'Ilm al-Uṣūl wa Maqāṣid al-Sharī'ah*. Beirut: Dār al-Muntakhab al-'Arabī.

Saliba, G. (2011). *Islamic science and the making of the European renaissance*. Cambridge, MA: The MIT Press.

Shāṭibī, A.I.I. ibn M. (d. 1388/2011). *The reconciliation of the fundamentals of Islamic law*. Reading: Garnet Pub and Centre for Muslim Contribution to Civilization.

Watt, W. M. (2008). *Islamic philosophy and theology*. New Brunswick: Routledge.

Chapter 29
Ethical Politics and Political Ethics II: On Socialism Through Integral Democracy

Michael Kary

Abstract The world order envisioned in Mario Bunge's political philosophy is one that combines a flowering and sustainable industrial society, a clean environment, full social justice, full realization of individual potential, and perpetuation of the human lineage. The problem is that these objectives are not jointly satisfiable, and more than one is individually impossible. Socialism is the hero and the villain of the piece. Along the way to its conclusion new conceptions of politics, socialism, capitalism, and today's social democracy are proposed.

29.1 Introduction

A conventional definition of politics is as the pursuit of power. This, essentially what Bunge calls "puny politicking" (Bunge 2009, p. 312), may be politics as we often get it, made to order for megalomaniacs and egomaniacs, but it is not the politics the rest of us want. We want politics to be a certain part of the attempt to solve problems of certain types. A view of which types of problems are to be addressed, what types of attempts should be made to solve them, and which part politics of what kind should play in the endeavour, is in essence a political philosophy. In practice political philosophies are usually sketchy, largely tacit, sometimes concealed, and riddled with inconsistencies. What better candidate for Bunge's philosophical treatment?

Bunge's political philosophy is an outgrowth of his ethics. Some key features of Bunge's ethical doctrine, agathonism, are examined in a companion chapter in this *Festschrift*. It is also fair to say that his ethics are an outgrowth of his political philosophy, and that both are outgrowths of his politics—and again the other way around. Bunge's father was a socialist from socialism's early modern days, and a socialist senator in the Argentine legislature. The young Bunge (though not his father) was a member of the communist party from 1935 to 1947, when, predictably,

M. Kary (✉)
Montreal, Canada
e-mail: mkary111@gmail.com

© Springer Nature Switzerland AG 2019
M. R. Matthews (ed.), *Mario Bunge: A Centenary Festschrift*,
https://doi.org/10.1007/978-3-030-16673-1_29

they expelled him. He was also a Marxist, a believer in dialectics, and even receptive to Lysenkoism.[1] In his memoirs, he recounts how the "fascist's heinous crimes favored the spread of communism and made many people, my father and myself included, discount the news about Stalin's crimes as sheer bourgeois propaganda" (Bunge 2016, p. 39).

Bunge's socialism has come a long way since his youth. He has long since turned on Marxism (but not entirely against Marx), Lysenkoism, dialectics, and communism. His gradual rejection of communism—what Bunge also calls state socialism—came about from his development as a scientist and philosopher, and from an eventful life well lived. Still though the problem set of socialism remains near the top of his political philosophy, only down one notch thanks to the rise of factors that threaten the human lineage, versus only making it miserable.

29.2   A Sketch of Politics

Politics, or the specific activity of political systems, can be thought of as a means for achieving adequate consent for action, despite perennial conflicts over both political philosophy and details of implementation. In turn a political system is any group of organisms capable of having their own thoughts guide their actions, having a diversity of opinions about what to do, yet having some of those actions deliberately coordinated for some purpose. The difference between politics and management is a matter of which is the central problem: the diversity of opinion or the coordination of action. Thus, there are management problems in politics and political problems in management.

In authoritarian politics "consent" and thus coordination is theoretically achieved through loyalty, capitulation, and fear, but in practice there is always behind the scenes manoeuvring, if not more open forms of accommodation. In democratic politics consent, though for those who did not win often grudging right from the beginning of a mandate, is achieved through electoral contests, horse trading, habituation, winning over, and policy compromises of various sorts. Thus at its heart politics is or should be about conflict resolution, in the service of beneficial action. Much as with authoritarian regimes, in democratic societies one conflict—over who is allowed to govern—can paradoxically be resolved through conflict aggravation or exaggeration, and thus dividing and democratically conquering the electorate.

An election itself serves to temporarily resolve some conflicts—by virtue of convention, and through exhaustion—and also sometimes to aggravate them.

---

[1]In 1951 he wrote: "In modern times, it is in classical genetics that the metaphysical categories of chance and necessity fight each other inconclusively... on the one hand the laws of heredity are immutable and we can do nothing to modify them save in a negative way, by eugenics. If Lysenko claims to have modified this fatal destiny in a constructive way, this is decried as Soviet propaganda. On the other hand, the result of crossing sexual cells is as chance-like as dice-throwing... Bourgeois science does not know of a way out." (Bunge 1951, pp. 216–218)

Electoral campaigns can rely on conflict exaggeration, so what the overall balance is between resolution and aggravation depends on the particular circumstances. This is why smooth-running democracies value civility in campaigning, magnanimity in victory, graciousness in defeat, and compromise and consultation in governance.

As befits a book-length treatment as well as many earlier writings on the subject, Bunge brings more variety to his understanding of politics. A sample:

> Politics is the struggle for, or management of, political systems on all scales. (Bunge 2009, p. 214)

> Politics is the struggle for power as well as the exercise of power in social systems of all kinds and sizes. It is also the art of conflict resolution in both contention and administration. (Bunge 2009, p. 11)

> ...politics, though fought largely with words, is about material interests, not ideas in themselves, let alone words. (Bunge 2009, p. 27)

> A consequence of the above results and suggestions for political science is that material interests explain only political actions of certain kinds. (Bunge 2009, p. 57)

> Politics is about managing heterogeneity: about either securing or controlling social divisions. (Bunge 2009, p. 67)

> Politics is a moral minefield because it is about protecting rights and enforcing duties. (Bunge 2009, p. 251)

> ...politics is all about social problems. A social problem is a persisting difficulty that affects a whole sector of society and calls for new policies or bigger resources or deafer moral ears. (Bunge 2009, p. 249)

Some of these, especially the second and the last, are more or less compatible with the preceding sketch of politics. As for the role of material interests, they are important, but in practice any idiotic interest or disinterest can be the object of political contention. Of course which issues count as idiotic depends on one's political philosophy. For example, in Bunge's, gay marriage, assisted suicide, and abortion are all classified as "phony and diversionary" (Bunge 2009, p. 163).[2]

The problems Bunge describes as being of the first magnitude, for affecting everyone and so potentially terminating the human lineage, are the arms race (especially the existence of nuclear weapons), overpopulation, the depletion of non-renewable resources, and environmental degradation (Bunge 1989, p. 114). Bunge says that these, although moral problems for concerning the survival of mankind, can only be solved with the right political measures. Bunge has many ideas about what those are, but there is no space here to do justice to either their virtues or their faults. As with the companion chapter on his ethics, instead the attention is on only a few fundamental features of his political philosophy.

---

[2] This is a change from Bunge's earlier view that abortion is a "genuine" and certainly "real serious problem", but of the third order of magnitude, for affecting only a small fraction of humankind, and so paling in comparison with those of the first and second magnitudes (Bunge 1989, pp. 114, 158, 362).

29.3   Capitalism Versus Socialism

The purely political part of Bunge's political philosophy is not so unusual at the national level, and effectively absent at the municipal level. Bunge favours democracy but leaves the matter of which electoral form to political science. The latter absence is regrettable, because it could use Bunge's philosophical consideration: every electoral form has serious, major problems, and even tiny perfectly reasonable variations would often make for complete reversals of outcome. This takes "the will of the people", "government by the people and for the people", "national self-determination" and so on from being not just obvious ontological nonsense but seemingly also political nonsense as well. Be that as it may, the most striking parts of Bunge's political philosophy, nationally and internationally, are instead his views on political economy.

Bunge argues that capitalism and (liberal) democracy are suicidal institutions, and moreover mutually incompatible (Bunge 2009, p. 318). By contrast Bunge considers socialism to be "one of the great inventions of the early nineteenth century" (Bunge 2009, p. 154). Perhaps ironically, Bunge has the former Soviet Union as suicidal only in its last years; but perhaps correctly, for it being only democidal until then. But more on that later.

Before considering any of Bunge's views on capitalism or socialism, a fundamental question needs to be answered: what are capitalism and socialism?

Socialism is associated with manifestos, and various other explicit proposals for what it is or should be, of which Bunge's own political philosophy is but one. Socialism is also Utopian, in the sense of being a project for an ideal or right society. Consequently, socialism is diverse. Despite, or is it because of, its emphasis on fraternity, the varieties of socialism are *frères ennemies*: bloody fratricidal conflict has been characteristic of socialism. Indeed, socialists have imprisoned, enslaved, starved, tortured and killed vastly more socialists than have capitalists, whose efforts have been puny in comparison.[3]

The question "what is capitalism" has first a large historical answer, with still ongoing evolution and diversity, rather than better a philosophically encyclopedic one as for socialism. Underlying all its diversity is the fundamental idea of capitalism: *freedom of association of persons and capital, for the purpose of profitable enterprise*. This formulation takes for granted that people have the right to get and keep capital, something not generally the case under socialism. Profit and enterprise are both integral to capitalism, but in any particular firm or individual, one or the other may be the dominant consideration, with extreme skewness in either direction generally detrimental to the endeavour.

No one should underestimate how important the evolution of freedom of business association was for the evolution of freedom of association more generally. For example, in Spain, home of the Mondragon cooperative whose example is so

---

[3]See Rummel (1992, 1994, 1997), Becker (1996, 2005), Applebaum (2003, 2017), Chang and Halliday (2005), and Dikötter (2010a, b).

important to Bunge's vision of the economy, the Penal Code of 1822 forbade all unauthorized meetings of four or more people. Freedom of association and assembly were not enshrined in the Spanish constitution until the liberal reforms of 1869. The requirements of business enterprise, including the ability to form corporations (and cooperatives as a sub-species), were important contributors to this evolution (Guinnane and Martínez-Rodríguez 2011).

The evolution of capitalism divides it into two forms, raw and enlightened. Raw capitalists were most notorious for not being fussy about how capital was acquired: enclosure, slavery, colonization—all at one time or another got the job done with presumably little loss of sleep. Enlightened capitalism moved on to the still evolving variety of fundamental protections and endowments of social democracy, which can be described as *the right to capital.*

Capital in a limited sense is a sum of money used to make money. Of course money only makes money when put to work, so to speak, so capital in a looser sense is any asset that can be used for productive purposes. Thus in a large sense an individual's capital can be thought of as also including the intangible assets of their education, their health, and maybe even their opportunities. The enlightened capitalism of the social democracies implements the right to capital by ensuring open access to health care, education, and other fundamental opportunities; by cooperating with or supporting the educational, health, and business sectors to ensure there are opportunities; and working to help figure out and solve the problems when any of it fails. Contrary to Bunge (1989, pp. 33, 51), the welfare state since Bismarck is not well-described as a relief state: it does not operate soup kitchens, or throw coins at beggars.

In his *Political Philosophy* Bunge has much to say about the evils of capitalism, but little as to what it is. He identifies capitalism with the free market (Bunge 2009, p. 322), but also says that the free market is the child of the marriage of capitalism with government (Bunge 2009, p. 320). One has to turn to his earlier *Social Science Under Debate* for his full characterization:

> *Capitalism* is the socioeconomic order characterized by the private ownership of the means of production, free (individual or collective) contracting of labour, and markets that operate competitively at set prices. (Bunge 2000, p. 420)

The characterization of capitalism previously offered here obviously implies free contracting of labour and (at least the possibility of) private ownership of the means of production. It also implies free markets, because enterprise can only be profitable if the products of enterprise can be sold, and buying and selling involves an association of persons. It does not though require either competition or set prices. Considering the reverse direction of inference, Bunge's characterization does not imply either freedom of association or the purpose of capitalism, profitable enterprise. The two characterizations overlap but are not equivalent.

As for socialism, Bunge both boils it down to an essential idea, and characterizes its essential features. For Bunge the essence of socialism is equality (Bunge 2009, p. 141). Yet hardly any or maybe no socialism, to include Bunge's, has ever advocated total true equality. Nor has any ever implemented it, for good reason: different people have different requirements whose costs are not the same. Equal reward therefore gives unequal benefit, and equal benefit requires unequal reward. Perfect equality is not just a naive ideal; upon heterogeneity it is a logical impossibility.

This means equality better not be the essential idea of socialism. In fact Bunge does give the essential idea himself, although seemingly inadvertently (Bunge 2009, p. 327). The highest value, or the principal ideal of socialism, is not equality, but *equity*, or *social justice*: a distribution of benefits and burdens according to some judgement of what is right. That judgement comes from the particular socialist ideology, but always from some variation of the fundamental socialist slogan, *From each according to their abilities, to each according to their needs.*

The essential features of socialism, or its proposed mechanisms for achieving its ideal, vary from socialism to socialism. In Bunge's they are:

(a) the socialization (not nationalization) of all the means of production, exchange, and credit; (b) the administration of the economy by the workers; and (c) government programs aiming at decreasing social (in particular economic) exclusions and inequalities. (Bunge 2009, p. 154)

Or almost equivalently, in an earlier formulation:

Authentic socialism is a classless social order, that is, one wherein all the goods and burdens are equitably distributed. It is characterized by (a) the cooperative ownership of the means of production, trade, and credit and (b) the self-management of all business firms – that is, democracy in the workplace. Shorter: Authentic socialism = Cooperative property together with self-management. (Bunge 2000, pp. 431–432)

In turn Bunge wants to put this nationally within an overall economic order most similar to that described in Vanek (1975), that is with the support of some sort of national planning agency and other such economic infrastructure. Bunge also wants to extend this order internationally, to include international resource management by world government agencies.

These characterizations are in terms of generalities. In what follows three specifics of Bunge's socialism are examined.

29.3.1 Cooperatives

In Bunge's socialism, cooperatives are the only permitted business form. To this rule Bunge allows only three exceptions: family-operated concerns (Bunge 2009, p. 331), because Bunge believes they have no salaried employees (Bunge 2009, p. 190); and state-owned firms that either dispense unprofitable social services and strategic public goods (such as infrastructure, communications, and energy); or manufacture weapons.

A short analysis: there is nothing wrong with cooperatives—when they work.

Now a longer analysis. Bunge gives examples where "cooperatives beat capitalism without firing a single shot", and extolls the success of cooperatives despite being "usually in hostile economic and political environments" (Bunge 2009, pp.

323, 332). The problem with these views is that cooperatives are part of capitalism, and have remained so since their invention. Indeed, cooperatives are more a part of capitalism than they are of socialism.

First, cooperatives long predate socialism. Bunge himself gives an example of one founded in Spain in 960, thus under Islamic capitalism (for which see e.g. Banaji 2007). Hundreds of thousands have prospered uneventfully not just under western capitalism but even under fascism. In France, Italy, Portugal and Spain, cooperative law was just a sub-species of corporate law, and the Mondragon cooperative was founded under the legal regime of Franco's own *Ley de Cooperación* of 1942. Although it put cooperatives under the same political control as every other economic institution in Spain (Romero 2004), Franco's 1942 law allowed cooperatives a special advantage, since under it cooperatives were classified as not-for-profits and cooperants paid no income tax (Ministerio de Trabajo 1944, p. 3; Mondragon, undated). The *Estatuto Fiscal de las Empresas Cooperativas* of 1969, also under Franco, specified a number of tax advantages for cooperatives (Romero 2004).

Second, cooperatives are not a part of every socialism, they were rejected by many socialists, and kept or kicked out of the major socialist countries. On the other hand they are perfectly acceptable and unremarkable in every capitalist economy, to the point where even today's United States Department of Agriculture still makes available a helpful booklet on how to start one (USDA 2011). In short, cooperatives were born and have prospered in capitalist and even fascist societies, but have faced by far the greatest hostility amongst socialists and within socialist societies. No wonder that while Mill said, through cooperatives workers become their own masters, Marx instead said they become their own capitalists, and Walras that the essence of cooperatives was being a means of enabling workers to acquire capital through saving (Jossa 2005).

If cooperatives are fine when they work, when do they not work? Consider only one crucial problem, as illustrated next (see Bright 2014; Intel 2017).

In 2011 the Intel corporation announced it would build a new microprocessor fabrication plant. The cost was projected to be US$5 billion, and it was to have 1000 employees. Set up as a worker's cooperative, it would have required membership fees of $5 million per employee. In Bunge's socialist economy, supposedly nobody is poor, but surely nobody is going to have that much money available for investment either. It is not even clear whether Bunge's system permits the accumulation of personal savings for capital formation. In theory there might be another source of start-up capital, namely debt. But any enterprise supposed to float for years before any sales in $5 billion of debt would drown in it instead.

Even all that is only the beginning. By 2014, the building to house the manufacturing equipment had been completed, at a cost till then of approximately $1.67 billion. But by then the personal computer market had changed, and the completion of the plant, still years in the future, could not be justified. As a cooperative, each member-worker would have been stuck with either $1.67 million of sunk savings, or a treadmill of $1.67 million of interest-bearing debt, without even the prospect of any offsetting income.

By 2017 market conditions changed again, and Intel announced it would complete the plant in still more advanced form, to employ now 3000 people. This required $7 billion more investment for a total of approximately $8.67 billion, and three to four more years construction, for completion in 2020 or 2021. In the form of a cooperative, this would have meant now membership fees totalling $2.89 million per worker, the fruits only beginning to be realized a decade after the initial owner-worker's investments—perhaps, as any such fruits are still all in the future. It is no coincidence that capitalism, not the cooperative variant of market socialism, excels at such ventures, for they require exactly what capitalism was specifically designed to provide: *capital*.

According to Bunge, a necessary condition for an ideal to be worthy is for it to be feasible (Bunge 1989, p. 58). In turn according to Bunge's Definition 2.7, an ideal is "feasible (or attainable, or realistic) if and only if it can be reached with the available means or with means that can be made available within the foreseeable future" (Bunge 1989, p. 58). This means that Bunge's proposal to base modern industrial society on the cooperative model is not a worthy ideal.

The above was based on an example from a capital-intensive industry. What about for light industry, such as in the Mondragon cooperative?

The current membership fee for joining Mondragon is approximately €15,000 (Mondragon undated). Although Mondragon can arrange for this to be paid over a period of 3 years, using deductions from salary, it is still a staggering commitment for any worker. It is of a size and type not suitable for everyone, and certainly not at every stage of everyone's life. When Mondragon hits tough times the value of this investment diminishes. If it were to go bankrupt, it would vanish. Mondragon seems set to endure but Bunge envisions a world of all and only diverse and innovating cooperatives—or in other words a world of many failed cooperatives, for as Bunge well knows, such is the nature of innovation.

29.3.2 Rewards

Bunge's general idea of socialism includes a principle of "the administration of the economy by the workers". Comparing with his earlier formulation, this seems to mean at least workplace democracy, or the "self"-management of all business firms. As with all democracies this self is no self at all, rather just some form of majority rule. Bunge further says that full-fledged social justice, that is to say the equitable distribution of burdens and benefits, cannot come from above, but should and could be constructed from below (Bunge 2009, p. 331). Contrary to any of these ideas, Bunge instead specifies an interlocking top-down list of criteria. This list can only conceivably be administered by the state:

NORM 6.1 (i) The exercise of basic rights, and the performance of basic duties [includes doing one's job well; Bunge 1989, p. 4], deserve no special rewards or punishments. (ii) Supererogatory virtuous actions deserve reward, and undererogatory sinful actions deserve punishment. (iii) A just reward of the virtuous meets one of his legitimate aspirations. (Bunge 1989, p. 188)

NORM 6.2 [. . .] (iii) The sole inequalities justified in the distribution of goods and services are those which are to the benefit of all – namely the rewards of merit and punishments of misdeeds in accordance with Norm 6.1. (Bunge 1989, p. 182)

. . . the problem of designing a morally just society boils down to a sociotechnical problem, namely that of adjusting the distribution of benefits and burdens over a population in agreement with the available resources as well as [being treated solely on the strength of one's rights and merits, and getting the rewards of merit and the burdens of punishment due to supererogatory virtuous and undererogatory sinful actions]. (Bunge 1989, p. 188)

In a just society everyone gets the living minimum required to meet his basic needs, whereas persons with above average ability earn more but may have more burdens than benefits . . . (Bunge 1989, p. 188)

Those who render distinguished service to society, e.g. by performing dangerous or dirty jobs, deserve rewards beyond their rights. (Bunge 1989, p. 181–182)

. . . the miner should be compensated for performing hazardous and insalubrious tasks, whereas the mathematician finds his greatest reward in doing his job well—whence he should be offered a bonus only when teaching dull students who just want passing grades. (Bunge 2009, p. 329)

If you comply with this social contract, you should enjoy life. But if you shirk your responsibilities, you will only get what you need to survive . . . (Bunge 2009, p. 330)

By the first and fourth citations, Bunge takes doing one's job well to be the normal duty expected of people with average abilities. What all this boils down to is that everyone who does their job well—mathematician, carpenter, artist—gets exactly the same remuneration, which is the amount required to meet basic needs, or in other words that required for survival and health. Those who do less get only what they need for survival, not health. In this classless society, only two classes of people get more, and so enough to satisfy in addition some "legitimate" wants: those who do their job better than well, and those who take on dangerous or dirty jobs.

Bunge defines a poor society as one having "the (economic, cultural and political) resources required to meet the basic needs but not the legitimate wants of all of its members—whence it can secure the well-being but not the reasonable happiness of all of them" (Bunge 1989, p. 50). Wealthier societies can afford more, but under the ideal social justice of authentic socialism and integral democracy, any surpluses must be apportioned according to the socially just rewards for merit. If we assume a normal distribution of merit, or any where the average is also the median, then no matter how wealthy the society, authentic socialism and integral democracy demand that half the population—that is to say, everyone of average merit or less—should be poor.

29.3.3 Private Property

Socialists distinguish personal property, or property used for personal consumption, from private property in general. Presumably in all socialisms, a toothbrush is personal property, but the distinction is not always so clear. Thus the communes imposed by Mao had communal kitchens, communal food, and expropriated for public use and later for smelting, everyone's kitchen utensils. A car could be an item of personal consumption, and the same car at the same time could also be used as a tool for work (e.g. by offering taxi service on the way to one's own destination). The same could be said of work clothes, such as a ballerina's tutu. Should it belong to the ballerina or the public? Bunge's overall view of the matter:

> NORM 11.3 In a just society everyone may own whatever she requires to meet her basic needs and legitimate aspirations, as well of those of her dependents, and nobody may own more than this. (Bunge 1989, p. 374)

> ...in order to satisfy our basic needs and legitimate aspirations, we only have to have access to certain things without necessarily owning them. (Bunge 1989, p. 374)

> Everyone should be able to... buy or rent adequate living quarters in a clean and secure environment. (Bunge 2009, p. 350)

> The ownership of the necessaries of life is so important to physical and mental well-being, that everyone should enjoy it. Given that necessaries are limited or even scarce, their universal access calls for limiting their private appropriation. [...] Wherever private property fails, collective management should be tried, because it combines property with equality. (Bunge 2009, pp. 107–108)

> ...my owning A prevents you from using A, which is unfair if you happen to need A to survive. The solution would of course be to share A if A happens to be sharable, as is the case with the means of production, trade, finance, transportation, and communication, as well as with culture. In other words, where property is shared rather than sequestered, no property conflicts flare up. (Bunge 2009, p. 109)

Some of his rationale for restricting property rights includes:

> private property worship has all but eliminated any restrictions on the protection of the environment—our most cherished public property—thus leading to global warming, aquifer depletion, deforestation, pollution, a dramatic decrease in biodiversity, and the consequent disinheritance of our progeny. (Bunge 2009, p. 231)

> Wherever there is private property there is bound to be exploitation of some kind, from slavery to the self-exploitation of the self-employed farmer, craftsman, or trader. There is also likely to be violence in defense of one's property or in the attempt to expand it at the expense of others. (Bunge 1989, p. 183)

> Firstly, as long as land is privately owned, the great majority of people are deprived from the space they need to live in good health. [...] Land, particularly in an overcrowded world, should become the common property of humankind, only to be leased to firms, preferably cooperatives, as long as they make rational and socially beneficial use of it. (Bunge 1989, p. 183)

There are obviously many problems with this collection, and it is not easy to draw out from them a consistent view of property rights. Clearly Bunge is conflicted over them, for on the one hand he has a humanist's appreciation of the need for private property, while on the other he has a socialist's visceral distaste for it.

Bunge is absolutely correct that property is valuable, and that owning it confers advantages not shared by the propertyless. Much the same can be said for anything limited, and this includes for example higher education. Faced with this problem, two solutions present themselves: take away rights and make everyone worse off;

or extend them and make everyone better off. Land is finite, but renting instead of owning does not change the amount of it. For shortages of any kind, sharing at best only meliorates them up to saturation; at worst it generates rather than assuages conflicts. Among other ways of meliorating shortages is something Bunge says we need to do anyway: radically reduce the size of the human population.

In general, every problem listed by Bunge that has something serious to do with property is either not solved by collective ownership, or can be ameliorated by regulation. But not every system of desiderata and constraints yields a solution, of which more later.

29.4   A Glimpse at the Track Record

Before taking any promissory notes, the skeptical citizen will try to ascertain whether the signatory has a good track record, whether he has not made too many lavish promises to too many people, and whether the nostrum he offers is supported by any evidence. (Bunge 2009, p. 206)

Although every murder is reprehensible, it is the more serious, the greater the number of victims. (Bunge 1989, p. 177)

29.4.1   Socialism

Socialism or something like it has been tried non-violently in numerous individual Utopian communities, their variety spanning from New Lanark to the kibbutzim. The trend has been to eventually turn capitalist or evaporate, socialist principles and all. All were small-scale attempts, often amongst people who were already mostly friends, family, or otherwise closely associated.

Authentic socialism, that is to say Bunge's version, is supposed to give a classless society, one where all the benefits and burdens are equitably distributed, through cooperative ownership and democratic management. In other words, it is supposed to achieve full social justice. In Bunge's view authentic socialism has no track record: it has never been implemented in any society, certainly not on a large scale.

Though correct, that assessment is not the whole story. In many attempts over the past century, people have *tried* to implement socialism on a large scale, in particular something like Bunge's authentic socialism. Every attempt began with similar calls for social justice, a classless society, self-management, and everything attractive that goes with them. What was the result? Table 29.1 gives only a hint of what no chapter can do justice to.[4]

---

[4]For book-length histories of only some individual components, see Solzhenitzyn (1974), Becker (1996, 2005), Applebaum (2003, 2017), Chang and Halliday (2005), Dikötter (2010a,b), and Zhou (2013).

Table 29.1 Numbers murdered by various political regimes, on the basis of group identity including political affiliation or status, indiscriminately, or from reckless and wanton disregard for life, excluding battle deaths; twentieth century battle deaths through 1987; and battle and all-cause deaths during World War II (Rummel 1992; 1997, Table 16A.1; Undated; for World War II specifically, US National World War II Museum, Undated)

Regime type	Regime	Number murdered by regime, best estimate	Estimate range (if available)
State socialist or market socialist (Yugoslavia) or variant socialist	Maoist China, 1928–1987	76,702,000	45,837,000–152,363,000[a]
	USSR, 1917–1987	61,911,000	28,326,000–126,891,000
	Khmer Rouge Cambodia, 1975–1979	2,035,000	635,000–3035,00
	North Korea, 1948–1987	1,663,000	710,000–3,549,000
	Titoist Yugoslavia, 1944–1987	1,072,000	595,000–2,130,000
	Romania, 1948–1987	435,000	245,000–920,000
	Albania, 1944–1987	100,000	25,000–150,000
	Castro Cuba, 1959–1987	73,000	35,000–141,000
Fascist or other non-communist authoritarian, or chaotic (Mexico)	Batista Cuba, 1952–1959	1000	1000–20,000
	Stroessner Paraguay, 1954–1987	2300	1800–4300
	Marcos Philippines, 1972–1986	15,000	10,000–25,000
	Argentina, 1976–1982	20,000	15,000–40,000
	Franco Spain, 1939–1975	275,000	210,000–350,000
	Portugal, 1926–1982	741,000	331,000–1,851,000
	Mexico, 1900–1920	1,417,000	618,000–3,290,000
	Nazi Germany, 1933–1945	20,946,000	15,003,000–31,595,000
Total, all socialist regimes world-wide	1917–1987[b]	148,286,000	78,000,000–297,432,000[a]
Battle deaths, all wars	1900–1987	35,700,000	
World War II battle deaths	1939–1945	15,000,000	
World War II civilian and military deaths, all causes	1939–1945	60,000,000	

[a]Does not include possible modification in error estimate due to recent upward revision of Chinese famine death estimate to 38,000,000, 1958–1962

[b]Rummel (1997, Table 16A-1) gives the applicable years for the line total as 1958–1980, but this appears to be an error

Not every socialist regime has been a bloodbath: for example, Venezuela has so far only been sliding into dictatorship, brutality, total poverty, and chaos (Human Rights Watch 2017). Others were either forcibly aborted early on and so of unknown eventual trajectory (e.g. Chile), or variegated and, like Israel, social democratic overall.

Bunge gives particular credit to Soviet communism for the following achievements: rapid industrialization, thanks to the 5-year plans; for being the only country in the world to escape the Great Depression, and having full employment; having the smallest income inequality in the world, as measured by the Gini index; having one of the highest education levels; and for defeating the Nazis for us at enormous cost. The purpose of this chapter is to consider Bunge's view of political philosophy, not of political history, but the two inform each other, and it would be remiss not to examine these claims.

If the Soviet Union did defeat the Nazis for us, did it first enable them? Certainly the Soviets and Nazis first colluded to divide and devastate Eastern Europe between them: this was the Hitler-Stalin pact of 1939. Snyder has argued that the mutual aiding and abetting of the two regimes went beyond that, even as foes (Snyder 2010; Applebaum 2010). Consider just the economic aspect: trade between the Soviet Union and Nazi Germany continued all the way through the interwar period, and Soviet supplies were important or critical to Germany's rearmament. In particular, they broke the Allied blockade, *and were essential to the German invasion of the USSR:* "without Soviet deliveries of these four major items (oil, grain, manganese, and rubber), however, Germany barely could have attacked the Soviet Union, let alone come close to victory" (Ericson 1999, p. 182). Once Germany did invade, why was the cost so enormous? Besides Stalin's overall military incompetence, in 1937 he once again purged the Soviet military: of the high command, 60% of the marshals, 88% of the army commanders, 100% of the admirals, 90% of the corps commanders, 68% of the divisional commanders, and 56% of the brigade commanders were executed. In total some 30,000 officers were purged. "This action constituted nothing short of a catastrophe for the Red Army, because when the Germans invaded in June 1941, they found the Red Army weakly led by incapable but politically stable Soviet officers" (Doyle 2010, p. 248). And had it not been for the American lend-lease programme, what was left of the Soviet military after Stalin's predations would have had multiple billions of dollars less worth of materiel to fight with (Ericson 1999, p. 182).

How much should educational attainment count as a Soviet achievement? Prior to the reforms of 1861, the bulk of the population was illiterate and hardly any primary education was available for peasants or labourers. But following these and the later Witte reforms, education and literacy spread rapidly. Russian universities of the Czars left a legacy of intellectual achievement and Nobel prizes, and the literacy rate amongst military inductees from the general population rose from 22% in 1874–1883 to 63% in 1904–1913. The revolution brought all this progress to a several-years long halt (Sorokin 1944, pp. 145–146). Afterwards schooling at all levels did undergo a frenzied revival and expansion, but had to suffer the ideological prison of Marxism-Leninism.

Next, consider income inequality. The easiest way to eliminate it is to impoverish everyone. Nor is income worth anything if there is nothing to buy. Thus, income inequality in any form is a meaningless statistic without some idea of absolute levels and the availability of goods. Nor does it take into account the distributions of burdens, which in all communist countries were beyond measurement. Because of the shortages and closed opportunities that were typical of the communist systems, their success in achieving even the little component of equity that is normally indicated by income equality cannot be reckoned in those terms. As Khrushchev recounted in his memoirs:

> It was a hungry time back then, even for people like me who held fairly high positions. We were living more than modestly, and even in your own house you couldn't always eat your fill. And so when we went to the Kremlin, we filled ourselves to the brim. When the break was announced, we all headed to the glutton's den, as we jokingly called it among ourselves. (Khrushchev 2004, p. 54)[5]

This was during the multimillion-death famine of the Holodomor, the North Caucasus, the Volga region, and elsewhere in the Soviet Union.

The idea that the Soviet Union had no mass unemployment is a myth (Jones 1930–1935; Muggeridge 1933), and not just because being a slave in the Gulag does not count as employment. To address in only the most understated terms the claim that the Soviet Union escaped the Great Depression—again the time of the Holodomor: "It seems to me the year was 1932. Hunger was stalking Moscow... A fierce battle was underway then—for bread, for food products in general..." (Khrushchev 2004, pp. 55–56). The Great Depression was bad, but it never got to cannibalism. Instead cannibalism was resorted to in at least Lenin's Great Famine, the Gulag, Stalin's Greatest Famine, Mao's Great Famine, and North Korea's Most Recent Known Great Famine.[6] The iconic image of the Great Depression is Dorothea Lange's photograph *Migrant Mother* (Lange 1936); those of the contemporaneous ultra-catastrophes of state socialism, the American Relief Administration's and Fritjof Nansen's photographs from Lenin's Great Famine, of piles of emaciated and naked corpses, desperately stripped to clothe the still barely alive; of the barely clothed or naked walking dead; and of not-yet fully eaten human flesh (Nansen 1921–1923; Patenaude 2002, pp. 324–332). State socialists learned from that experience: almost no photographs of the subsequent Soviet famines escaped confiscation, none from Mao's Great Famine, and similarly from North Korea's (Becker 2005; Boriak 2008; Dikötter 2010a). For these the only images we have are those evoked by eyewitness accounts (e.g. Becker 2005; Klid and Motyl 2012; Zhou 2013).

Did the Soviet Union industrialize rapidly under Stalin? No one has written more intelligently about measurement than Bunge, so no one knows better than Bunge that

---

[5]The Soviet Union's system of non-income, class-based privileges extended far beyond the Kremlin (see e.g. Jones 1930–1935; Wells 1933/2012; Utley 1940/2012; Kravchenko 1946/2012).
[6]See Williams (1934), Applebaum (2003, 2017), Becker (1996, 2005), Patenaude (2002), and Dikötter 2010a, b).

such statements require a yardstick. If the yardstick is the rate at which industrial society was invented and originated, then Soviet industrialization was rapid. If it is the rate of other latecomers like Singapore, Taiwan, and South Korea, who started with less and got to more, their industrialization paying for prosperity rather than the people being methodically starved and worked to death to pay for it—then the Soviet achievement is impressive more for its catastrophic than constructive aspect.

In any case, as Bunge knows, it is not plans that accomplish anything, but people. Who were the people responsible for the Soviet industrialization of that era? Largely American managers and engineers, typically working for American corporations; and the slave labour first of the Gulag, and second of the regular Soviet worker:

> ...American experts played the decisive role in the reindustrialization and economic reorganization of Russia. The greatest construction feats of that period, like the Dnieprostroi and Magnitogorsk projects, as well as the development of great new industrial centers, were made under the direct planning, advice, and guidance of American experts. An American Colonel, Hugh L. Cooper, builder of the Dnieper Dam, was the first foreigner to be decorated by the Supreme Soviet of the U. S. S. R. Practically all the main branches of industry in the Soviet Union have been remodeled and shaped under the guidance of American experts. (Sorokin 1944, p. 166)

> ...we were taken down to Kharkiv for the opening of the Dnieper dam. There was an American colonel who was running it, building the dam in effect. "How do you like it here?" I asked him, thinking that I'd get a wonderful blast of him saying how he absolutely hated it. "I think it's wonderful," he said. "You never get any labor trouble." (Muggeridge, interviewed by Carynnyk 1983)

29.4.2 Capitalism

Bunge's criticisms of capitalism are of two types: for what it does do, and for what it doesn't. As he effectively acknowledges (Bunge 2000, p. 425), criticism of what it does not do is more criticism of its proselytisers than of capitalism itself. In complete contrast with socialism, the examples of e.g. today's China, today's Vietnam, Dickensian England, and over the course of their history Scandinavia, the United States, Chile, Japan, Mexico, South Africa, Switzerland and the Philippines have shown that capitalism is compatible with just about any social order or disorder, as long as business enterprise and capital accumulation and aggregation are allowed. Who and where it prospers is another question.

Like any other freedom, freedom of business association is always restricted. The question is not whether, but how do we want business to be regulated. Since "we" are never of one mind let alone an unchanging one, the result is variety and on-going evolution.

Despite this variety, Bunge considers all capitalisms to have suffered or rather inflicted the same faults. Of his serious criticisms, most relate not to capitalism itself, but to an option within capitalism: the choice between the corporate or cooperative forms of business enterprise. Even within the corporate form, problems can be addressed by regulation. For example, a constitutional amendment requiring

Table 29.2 The distinctions between fascism and communism, summarized from Bunge (2009, pp. 179–180)

	Fascism	Communism
Marketing strategy	Empty rhetoric about this	Empty rhetoric about that
In the interest of	Wealthiest members of society, leaders and their sponsors	Ruthless political elite
Politics	Total submission of the bulk of a people to a strong state	Total submission of the people to the dictates of the nomenklatura
Modus operandi	Military or paramilitary organizations	Relentless purging and bloodletting

a maximum 12:1 pay ratio was put to a referendum in Switzerland in 2013. It failed to pass, but if it had, Switzerland would have remained capitalist. Earlier, other related measures did pass (Garofalo 2013). However, and importantly, inequality is also deliberately inherent to Bunge's system, and more on this in the conclusion.

Finally, Bunge is justly alarmed over environmental degradation and resource depletion. He identifies capitalism as the problem, but state socialism was far worse (see e.g. Zhou 2013, pp. 121–132; Shapiro 2001), and Bunge does acknowledge that Soviet-style communism was environmentally unsustainable. But there is a problem beyond either: *industrial society*. This though is another matter left for the conclusion.

Several authors have noted similarities between communism and fascism, and Bunge has acknowledged these but emphasized differences between the systems. This section ends with a summary of those distinctions (Bunge 2009, pp. 179–180). They form a fitting justification for the last sentence of Orwell's *Animal Farm* (Table 29.2).

29.5 Conclusion

At the time when Bunge's father became a socialist, socialism made sense. It had only been tried, apparently with some success, in a few small Utopian communities, amongst people who were already close. Socialism was also the only political philosophy to attach itself to science and atheism, as Bunge's father had already boldly done himself. What else claimed to offer anything to the underclasses, except perhaps the opportunities and remaining free land of the New World? The social democratic compromise was only nascent, and at best the New World could not be for everybody. In any case the robber barons, the gilded age, and imperialism were already in full swing, the United States had been through a bloody civil war, and Latin America was poor and politically problematic. Socialism might really have seemed like a shining New Civilization.

Although the outside world didn't know it, that all ended when Lenin came to power in 1917, with Trotsky and Stalin nearby. The Soviets were diligent at covering up their crimes, with secrecy and a sophisticated paid network of international deception. But eventually the truth came out, starting with the reporting of Gareth Jones (1930–1935), Rhea Clyman (1933), and Malcolm Muggeridge (1933) (see also Thevenin 2005; Klid and Motyl 2012; Applebaum 2017). From then on, driven home again and again, still to this day looking at North Korea and Venezuela, the task of the progressive political philosopher switched from expounding socialism to figuring out its tragic flaw, and to proposing something different.

Solzhenitsyn's view was that the problem was ideology, which gave evildoing justification and the evildoer steadfastness and determination. The particular ideology of the communists was equality, fraternity, and the future, but other evildoers had their own and it didn't seem to matter which. Jones and Muggeridge, reporting from the ground as it happened, ascribed it to something similar: Jones, to idealism; Muggeridge, adherence to the General Idea. Home again in England, Muggeridge also recalled Dzerzhinsky's obituary notice in *Pravda*: it praised him for rising above petty bourgeois emotions like pity, respect for justice, or for human life. Thus Muggeridge also ascribed the problem to a hatred of happiness and civilization, something recognizable in other brutalities. He and Jones also noted a special feature of communism, class warfare, but with the class enemy being in effect everyone.

Of all these aspects, only the war on everyone seems unique to socialism, and thus too the sheer volume of destruction. All of history's other holocausts, archetypically as the Holocaust, were directed at a circumscribed *other*. Only the socialist holocausts also turned in on themselves, to discover an endless supply of compromisers, shirkers, parasites, bloodsuckers, traitors.

Bunge's analysis is that the faults of Soviet-style communism were nationalization instead of socialization of the economy, total submission of the people to a political elite, and "relentless purging and bloodletting—Stalin's original contribution to statecraft" (Bunge 2009, p. 180). But Stalin only perfected what was started by Lenin, Trotsky, their lieutenants, and every socialist who enthusiastically followed their orders. The Red Terror Decree was issued by Lenin in 1918, the Gulag was born under Lenin's order in 1919 (Applebaum 2003; Remnick 2003), and it was Trotsky who enthusiastically cried "Long Live Civil War!" (Trotsky 1918). In that same speech transcribed, is found this exchange:

> ...the Russian revolution has not yet known terror, in the French sense of the word. The Soviet power will now act more resolutely and radically. It issues this warning... this whole game of yours can end in a way that will be tragic in the highest degree! [Martov, from his seat: "We did not fear the Tsarist regime and we're not afraid of you, either." (Trotsky 1918)

Bunge has no analysis of what was common to all the diverse forms and circumstances of socialism that produced similar fates (including Tito's market socialism); nor any view of how the total submission came to be. The common tragedies of different socialisms across diverse cultures raise the suspicion that the problem is at its core. Why else has it been that upon the death or retirement of its singular leader, upon foreign conquest or mere collapse, hard-core fascism transformed with few stumbles into functioning and prosperous democracy, while communism lost every ideal except hypocrisy, kleptocracy, and authoritarianism?

The society envisioned in Bunge's political philosophy combines full social justice, full realization of individual potential, a flowering and sustainable industrial society, a clean environment, and perpetuation of the human lineage. Is any of it feasible, under socialism or capitalism or anything else? Is there anything about it that is not desirable?

Industrial society is not sustainable, because the planet is finite and industry requires non-renewable natural resources. Of course the pace to the finish line can be slowed, but industry, no matter how supposedly green, is inevitably polluting and not sustainable.[7] Surely Bunge realizes it, and this is why the socialist Bunge resorts to ideas the physicist Bunge knows make no sense, such as extra-terrestrial mining.

Similarly, the socialist Bunge upholds the Marxist rhetoric of the full realization of human potential. But the humanist and scientist Bunge knows that life is short, art is long, and individuals are pluripotent, so no matter how favourable the environment, they are unable to realize all of their abilities. As for perpetuation of the human lineage, it too is out of reach: less than five billion years from now, our sun will erase whatever remains of it.

The remaining item on Bunge's list is full social justice. Apart from any problems with the specific mechanisms Bunge proposes to achieve it, is full social justice really possible, and is the pursuit of it really desirable?

For socialism, social justice is matching each person's needs, and abilities or merits (in Bunge's version also their "legitimate" aspirations), with their benefits and burdens. Especially in Bunge's version, it assumes merit varies, so inequality of outcome is built into the system and itself considered one of its virtues.

Today's social democracy works for an utterly different conception, better described not as social justice but as social good: assured basic outcome, equality of fundamental opportunity, and widespread prosperity. To assess burdens, instead of judging merit, or the ability to work, it evaluates the ability to pay. Instead of metering reward, it sets ground rules for economic transactions, mediates major conflicts, and does whatever else works its way onto the list, whether from bureaucracy or democracy. Instead of judging the legitimacy of individual desires, it regulates goods and services, and from there lets people live their own lives in their own way. Judgement of individuals is left for criminal actions, while in daily life it is the products, processes, and services that are judged, not individual merits.

By contrast the motto of socialism could just as well be *judge and be judged*. It is a logical consequence and required companion of the socialist's conception of social justice. Surpluses out of line with individual virtues and vices must be

---

[7]There are some things Art does better than Science. One of them is to convey the enormity of a problem. The film *Manufactured Landscapes* (Baichwal 2006) does this almost wordlessly, by exhibiting the manufacture of the most mundane of household products, and the horror not just of resource extraction, but of both not recycling *and* recycling.

"surrendered". Everyone has a quota, and not meeting it must be denounced and punished, in Bunge's version by being reduced to mere subsistence (2009, p. 330). In sum: the total submission of the people that Bunge decries is socialism's very core.

In his review of Bunge's *Political Philosophy*, Agassi (2011), following Hayek, came to a similar conclusion: the cult of excellence brings relentless conformism, meritocracy requires an authority to adjudicate merit, and "the rest is history" (Agassi 2011, p. 559).

Socialism has though been very valuable, most of all to those who have not lived under it. Perhaps something equivalent to social democracy would have been invented without socialism, but it was the Swedish socialists who were on the scene and ready to forge the compromise. A century before that, it was the wealthy Robert Owen and his New Lanark community, a precursor of socialism and enlightened capitalism, that amazed the world, showing that industry could be run productively, profitably, and happily. The value of socialism as an ideology has been to draw attention to its problem focus, and insist those problems can and should be solved. Through the fear that it might take power, starting in Bismarck's Germany, and through compromise starting in Sweden, socialism has contributed greatly to their solution. But the time is long past when socialism's own track record in power can be blamed on a few bad men, capitalist encirclement, floods or droughts.

It would be wrong though to say that socialism's only value has been in opposition. It has done much to form and inform the lives of many individuals, helping to make the meaning and purpose of their lives. In particular, we, especially those of us contributing to this volume, should appreciate it for having helped to make the one and only Mario Bunge.

Acknowledgement I thank Joseph Kary for reading and commenting on the manuscript, and Michael Matthews and John Forge for their respective contributions in shepherding the manuscript through its various stages.

References

Agassi, J. (2011). Bunge nevertheless. *Philosophy of the Social Sciences, 43*(4), 542–562.
Applebaum, A. (2003). *Gulag: A history.* New York: Doubleday.
Applebaum, A. (2010). The worst of the madness. New York Review of Books, 11 November. http://www.nybooks.com/articles/2010/11/11/worst-madness/. Accessed 21 Mar 2018.
Applebaum, A. (2016, April 7). The victory in Ukraine. *New York Review of Books.* http://www.nybooks.com/articles/2016/04/07/the-victory-of-ukraine/?sub_key=56ecc95f3747c. Accessed 21 Mar 2018.
Applebaum, A. (2017). *Red famine: Stalin's war on Ukraine.* New York: Doubleday.
Baichwal, J. (2006). *Manufactured Landscapes (film).* Toronto: Mercury Films.
Banaji, J. (2007). Islam, the Mediterranean and the rise of capitalism. *Historical Materialism, 15*, 47–74.
Becker, J. (1996). *Hungry ghosts: China's secret famine.* London: John Murray.

Becker, J. (2005). *Rogue regime: Kim Jong Il and the looming threat of North Korea*. Oxford: Oxford University Press.

Bright, P. (2014). *Intel closes AZ chip factory before it even opens—Upgrading existing plants is deemed a better use of capital*. https://arstechnica.com/information-technology/2014/01/intel-closes-az-chip-factory-before-it-even-opens/. Accessed 11 Feb 2018.

Bunge, M. (1951). What is chance? *Science and Society, 15*(3), 209–231.

Bunge, M. (1989). *Ethics: The good and the right* (Treatise on basic philosophy, Vol. 8). Dordrecht: Reidel.

Bunge, M. (2000). *Social science under debate: A philosophical perspective*. Toronto: University of Toronto Press.

Bunge, M. (2009). *Political philosophy: Fact, fiction, and vision*. New Brunswick: Transaction Publishers.

Bunge, M. (2016). *Between two worlds: Memoirs of a philosopher-scientist, with an appendix by Marta Bunge: My life with Mario*. Switzerland: Springer.

Carynnyk, M. (1983). Malcolm Muggeridge on Stalin's famine—Conclusion. *The Ukranian Weekly*, 51(23). http://www.ukrweekly.com/old/archive/1983/238322.shtml. Accessed 11 Feb 2018.

Chang, J., & Halliday, J. (2005). *Mao: The unknown story*. London: Jonathan Cape.

Clyman, R. (1933). [Various newspaper articles] http://uamoderna.com/shafka-dok/balan-rhea-clyman-holodomor. Accessed 5 June 2018.

Dikötter, F. (2010a). *Mao's great famine: The history of China's Most devastating catastrophe, 1958–62*. London: Bloomsbury.

Dikötter, F. (2010b, December 16). Mao's Great Leap to famine. *The New York Times*. http://www.nytimes.com/2010/12/16/opinion/16iht-eddikotter16.html. Accessed 11 Feb 2018.

Doyle, R. C. (2010). *The enemy in our hands: America's treatment of enemy prisoners of war from the revolution to the war on terror*. Lexington: University Press of Kentucky.

Ericson, E. E., III. (1999). *Feeding the German eagle: Soviet economic aid to Nazi Germany, 1933–1941*. Westport: Praeger.

Garofalo, P. (2013). What we can learn from Switzerland's CEO pay cap vote. *US News*. https://www.usnews.com/opinion/blogs/pat-garofalo/2013/11/25/the-importance-of-switzerlands-112-ceo-pay-cap-vote. Accessed 11 Feb 2018.

Guinnane, T. W., & Martínez-Rodríguez, S. (2011). Cooperatives before cooperative law: Business law and cooperatives in Spain, 1869-1931. *Revista de Historia Económica / Journal of Iberian and Latin American Economic History (Second Series), 29*(1), 67–93.

Human Rights Watch. (2017). *Crackdown on dissent: Brutality, torture, and political persecution in Venezuela*. https://www.hrw.org/sites/default/files/report_pdf/venezuela1117web_0.pdf. Accessed 21 Mar 2018.

Intel. (2017). *Intel supports American innovation with $7 Billion Investment in Next-Generation Semiconductor Factory in Arizona*. https://newsroom.intel.com/news-releases/intel-supports-american-innovation-7-billion-investment-next-generation-semiconductor-factory-arizona/. Accessed 11 Feb 2018.

Jones, G. (1930–35). [Various newspaper articles] In: *Gareth Jones' Newspaper Articles from 1930–35*. http://www.garethjones.org/published_articles/published_articles.htm. Accessed 11 Feb 2018.

Jossa, B. (2005). Marx, Marxism and the cooperative movement. *Cambridge Journal of Economics, 29*, 3–18. https://doi.org/10.1093/cje/bei012.

Khrushchev, N. (2004). In S. Khrushchev (Ed.), *Memoirs of Nikita Khrushchev, Vol. 1, Commissar [1918–1945]* (G. Shriver, Trans.). University Park: Pennsylvania State University Press.

Klid, B., & Motyl, A. J. (2012). *The Holodomor reader: A sourcebook on the famine of 1932–1933 in Ukraine*. Edmonton/Toronto: Canadian Institute of Ukrainian Studies Press. https://holodomor.ca/the-holodomor-reader-a-sourcebook-on-the-famine-of-1932-1933-in-ukraine/. Accessed 5 Jun 2018.

Kravchenko, V. (1946/2012). I chose freedom: The personal and political life of a soviet official. Excerpted in B. Klid & A. J. Motyl (Eds.), *The Holodomor reader: A sourcebook on the famine of 1932–1933 in Ukraine* (pp. 91–92, 111– 13, 118–19, 130). Edmonton/Toronto: Canadian Institute of Ukrainian Studies Press. https://holodomor.ca/the-holodomor-reader-a-sourcebook-on-the-famine-of-1932-1933-in-ukraine/. Accessed 5 Jun 2018.

Lange, D. (1936). *Migrant mother* [photograph]. In: *Dorothea lange's "migrant mother" photographs in the farm security administration collection: An overview.* United States Library of Congress, Prints and Photographs Division, Prints and Photographs Reading Room. http://www.loc.gov/rr/print/list/128_migm.html. Accessed 21 Mar 2018.

Ministerio de Trabajo, España. (1944). *Reglamento de la Ley de cooperación (2 enero 1942).* https://bibliotecadigital.jcyl.es/es/catalogo_imagenes/grupo.cmd?path=10071992. Accessed 11 Feb 2018.

Mondragon. (undated). *Mondragon corporation/co-operative experience/FAQs.* https://www.mondragon-corporation.com/en/co-operative-experience/faqs/. Accessed 11 Feb 2018.

Muggeridge, M. (1933). [Various newspaper articles] Linked from: *Gareth Jones' Newspaper Articles from 1930–1935.* http://www.garethjones.org/published_articles/published_articles.htm. Accessed 11 Feb 2018.

Patenaude, B. M. (2002). *The big show in Bololand: The American relief expedition to soviet Russia in the famine of 1921.* Stanford: Stanford University Press.

Remnick, D. (2003, April 14). Seasons in hell—How the Gulag grew. *The New Yorker.* https://www.newyorker.com/magazine/2003/04/14/seasons-in-hell-4. Accessed 11 Feb 2018.

Romero, C. (2004). *De la Ley de Cooperativas de 1942 al Reglamento de Sociedades Cooperativas de 1978: Un análisis crítico.* http://www.mapama.gob.es/ministerio/pags/Biblioteca/Revistas/pdf_ays%2Fa018_02.pdf. Accessed 11 Feb 2018.

Rummel, R. J. (1992). *Lethal politics: Soviet genocide and mass murder since 1917.* New Brunswick: Transaction Publishers.

Rummel, R. J. (1994). *Death by government.* New Brunswick: Transaction Publishers.

Rummel, R. J. (1997). *Statistics of democide: Genocide and mass murder since 1900.* New Brunswick: Transaction Publishers.

Rummel, R. J. (Undated). *China's bloody century.* https://www.hawaii.edu/powerkills/NOTE2.HTM. Accessed 11 Feb 2018.

Shapiro, J. (2001). *Mao's war against nature: Politics and the environment in revolutionary China.* Cambridge/New York: Cambridge University Press.

Snyder, T. (2010). *Bloodlands: Europe between Hitler and Stalin.* New York: Basic Books.

Solzhenitsyn, A. I. (1974). *The Gulag Archipelago, 1918–1956: An experiment in literary investigation* (T. P. Whitney, Trans.). New York: Harper & Row.

Sorokin, P. (1944). *Russia and the United States.* New York: E. P. Dutton.

Thevenin, E. (2005). France, Allemagne et Autriche face à la famine de 1932–1933 en Ukraine. *James Mace Memorial Panel, International Association of Ukranian Studies Congress, Donetsk, Ukraine, June 29.* http://www.garethjones.org/ukraine2005/Etienne%20Thevenin.pdf. Accessed 5 June 2018.

United States Department of Agriculture, Rural Development. (2011). Understanding cooperatives: how to start a cooperative. *Cooperative Information Report, 45,* Section 14. https://www.rd.usda.gov/files/CIR45-14.pdf. Accessed 11 Feb 2018.

United States National World War II Museum. (Undated). *Research starters: Worldwide deaths in World War II.* https://www.nationalww2museum.org/students-teachers/student-resources/research-starters/research-starters-worldwide-deaths-world-war. Accessed 11 Feb 2018.

Utley, F. (1940/2012). The dream we lost: Soviet Russia, then and now. Excerpted in B. Klid & A. J. Motyl (Eds.), *The Holodomor reader: A sourcebook on the famine of 1932–1933 in Ukraine* (pp. 50–57, 86–87). Edmonton/Toronto: Canadian Institute of Ukrainian Studies Press. https://holodomor.ca/the-holodomor-reader-a-sourcebook-on-the-famine-of-1932-1933-in-ukraine/. Accessed 5 Jun 2018.

Vanek, J. (1975). *Self-management: Economic liberation of man—Selected readings.* Harmondsworth: Penguin.

Wells, C. (1933/2012). Kapoot: The narrative of a journey from Leningrad to Mount Ararat in search of Noah's ark. Excerpted in B. Klid & A. J. Motyl (Eds.), *The Holodomor reader: A sourcebook on the famine of 1932–1933 in Ukraine* (pp. 113–16, 120–22). Edmonton/Toronto: Canadian Institute of Ukrainian Studies Press. https://holodomor.ca/the-holodomor-reader-a-sourcebook-on-the-famine-of-1932-1933-in-ukraine/. Accessed 5 Jun 2018.

Williams, W. (1934, February 24). My journey through famine-stricken Russia. *Answers*, 16–17. http://www.garethjones.org/soviet_articles/whiting_williams_1934.htm. Accessed 21 Mar 2018.

Zhou, X. (2013). *Forgotten voices of Mao's great famine, 1958–1962: An Oral history*. New Haven: Yale University Press.

Part VII
Biology and Philosophy of Biology

Chapter 30
A Reconstruction of the Theory of Ecology Based on Mario Bunge's Mechanistic Epistemology and Systemic Ontology

Carolina I. García Curilaf and Guillermo M. Denegri

Abstract This work proposes a structural reconstruction of the main laws, hypotheses and models sustaining ecological theory, by taking into account Mario Bunge's mechanistic epistemology and systemic ontology. In order to do this, in the first sections, we will develop Mario Bunge's philosophical principles, and in the following sections, we will use these principles to epistemologically analyse ecology. Ecology has taken the theory of evolution by natural selection as a theoretical framework. Through this theory, ecology explains the relations between organisms of the same and different species, and the environment. Ecology has constructed a theory that allows coordination of all the levels involved in the theory; and to link and relate all the levels of the reality it studies, thereby constituting a practical, applicable and improvable theory.

30.1 Introduction

The main objective of this paper is to analyze the results of the scientific activity of ecology, exposing its philosophical and epistemological presuppositions, such as concepts, hypotheses and laws, which organize the theory of ecology. In order to reconstruct the main laws behind the current theory of ecology, Mario Bunge's principles of realist epistemology and systemic emergent-materialist ontology will be applied.[1] Ecology has moderated the dispute between rationalism and

---

[1] See Bunge (1959, 1968, 1974, 1979, 1983, 1985a, b, 1997a, b, c, 1999a, b, c, 2004) and Mahner and Bunge (1997, 1998).

C. I. García Curilaf (✉) · G. M. Denegri
Institute of Research in Production, Health and Environment (IIPROSAM), Faculty of Exact and Natural Sciences, National University of Mar del Plata, Mar del Plata, Argentina

National Council of Scientific and Technological Research (CONICET), Mar del Plata, Argentina

M. R. Matthews (ed.), *Mario Bunge: A Centenary Festschrift*,
https://doi.org/10.1007/978-3-030-16673-1_30

empiricism through the gradual inclusion of law-like hypotheses, that is, through the construction and systematization of a theory with a ratio-empiricist epistemological and systemic ontological position, such as Bunge proposes. For realist epistemology, every scientific theory is a hypothetical-deductive system made of general and abstract theoretical principles, and laws, which are objective guidelines that describe, explain and predict the regularities of nature through terms referring to unobservable entities (Bunge 1959, 2004). As high-level laws have unobservable properties, it is necessary to deduce from them observable subsidiary models and hypotheses, some of which may still have some unobservable properties. Subsidiary hypotheses are used to indirectly perform an empirical test on the most theoretical levels and to describe specific systems. Scientific progress implies an increase in the systematization or coordination of all the levels that make up a theory (Bunge 1959, 2004).

This paper aims at proving that philosophy and science are not hermetic compartments that are applied separately; on the contrary, based on Mario Bunge's (2004) theory, it will be shown how ecology makes epistemological and ontological assumptions through which it gets to know the object of its study. The philosophical principles lay the foundations for ecology. Establishing, for example, that a theory is the best way of getting to know reality, which is part of an epistemological decision that logically excludes other possible ways of getting to know reality. Another epistemological assumption adopted by every factual science, especially ecology, is the renowned gnoseologic dualism based on the definitive division between a subject and an object, in which the object is seen as an Other that can only be known through hypotheses, concepts and theories; that is to say, indirectly through the creation and invention of a priori structures that are empirically tested. For scientists, the world is not known by untested intuition or spiritual apprehension.

Among the ontological theses presupposed by ecology is the view that the structure of reality is formed by several levels. This ontological hypothesis implies that reality is divided into several levels or sectors, each characterized by its own set of properties and laws (Bunge 2004). The main levels recognized and studied by ecology are organisms, populations, communities and ecosystems (Begon et al. 2006; Curtis et al. 2008).

Mario Bunge's ontological approach (Bunge 1979, 1995, 1999a, b, c; Mahner and Bunge 1997) states that every concrete object is connected with another in different ways. This means that every component of the system interacts with other components of the system either directly or indirectly. Besides, it affirms that every object is a system or part of a system with emerging properties that its components lack (Bunge 1999a). From a systemic ontological point of view, ecology is defined as the study of the processes that regulate the distribution and abundance of organisms and the interactions among them, and also, the study of the way in which such organisms act as a means to transport and transform energy and matter through the biosphere; that is to say, the study of the design of the structure and the function of the ecosystem (Begon et al. 2006; Curtis et al. 2008).

The law-like proposition that supports ecological theory and is integrated with systemic ontology is the following: *Organisms that form populations and/or com-*

munities reciprocally interact with one another and with their physical environment,
transforming and transporting energy and matter through a system that feeds from
them. Based on this law-like proposition, ecologists study the structure and functioning of communities and ecosystems, as well as of populations and organisms.
Ecology is a science that studies the connections and emerging properties that result
from such relationships. In order to achieve this aim, ecology deduces models and
hypothesis with observable terms from the law-like proposition. This enables the
description, explanation, and prediction of the behavior of certain part of reality, for
example, a lake.

Interrelations produce emerging properties in an ecosystem, such as: diversity,
structure (vertical), trophic structure, biomass, productivity, regulation, and ecological succession (Begon et al. 2006; Curtis et al. 2008). Within a community,
interspecific relations produce emerging properties, such as: rivalry, predation,
parasitism and mutualism. Different interactions imply the generation of natural
selection processes and co-evolution (Begon et al. 2006; Curtis et al. 2008).

At a population level, intraspecific relationships produce emerging properties,
such as: density, age and gender structure, spatial distribution, biomass, coverage,
birth and death rates, growth pattern, migration patterns, life strategies (Begon et al.
2006; Curtis et al. 2008).

At the individual level, ecology studies the connections between the environment and the organism. There is a correspondence between organisms and the
environment they live in; the former adapts to a certain environment through
natural selection. However, the organism responds to the environment generating
emerging properties, such as: irritability, acclimatization, physiological adaptation,
phenotypic plasticity. There are several reasons why organisms live in certain
environments, for example: the environment offers suitable conditions and resources
for their growth, development and reproduction (Begon et al. 2006; Curtis et al.
2008).

30.2 Scientific Theories from Mario Bunge's Perspective

Factual science seeks to create maps of the structure of reality, that is, laws. The
conceptual reconstruction of reality's objective structure is a scientific law; a system
of such law-like statements is a scientific theory. Factual science is a conceptual
reconstruction of the objective structures of events (Bunge 2004).

Factual sciences learn about reality through theories, which are methodically
based on ideas connected to each other in an organic relationship, and theories
and hypotheses are so coupled that a change in a hypothesis can cause changes
in theories. Science tries to explain facts, but experience is not the only—and not
even the main—object of investigation. If experience is scientific, it constitutes an
indispensable means of testing theories, but it does not provide all the content, nor
the meaning of all theories. In addition, to explain human experience, we need some
knowledge of the natural world we inhabit; and this world, generally not seen, nor

touched, is reproduced gradually through testable theories that go beyond what can be the object of experience. Science tends to construct conceptual reproductions of facts and structures, that is, factual theories.

Scientific knowledge does not seek final or incorrigible truth. What science asserts is that:

1. It is truer than any non-scientific model of the world;
2. That it is capable of proving such a truth claim by subjecting it to empirical test;
3. That it is capable of discovering its own deficiencies, and,
4. That it is capable of correcting its own deficiencies, that is, of constructing increasingly adequate partial representations of the structure of the world.

What allows science to achieve these objectives (the construction of partial and increasingly true reconstructions of reality) is its method (Bunge 2004). Scientific method poses a problem, constructs hypotheses deduced from available theories in the corresponding science, makes predictions on the basis of hypotheses, and empirically tests them in order to formulate conclusions. In case a hypothesis is refuted, or predictions or experiments are wrong, these steps are repeated, and where results are otherwise, conclusions are added to theories.

Science also accepts a series of philosophical ontology-related hypotheses; for instance, it presupposes the existence of real objects, that is, it considers that things exist independently of the knowing subject. When a factual hypothesis is constructed to cover a set of facts, it is assumed that the facts are real. Tests aiming at the factual truth of a presupposed hypothesis assume that there is something outside the internal world of the subject and that it will agree to some extent with the proposition in question or disagree with it. Factual science does not prove the existence of the external world, but presupposes philosophical hypotheses. There would be no way to experiment or to theorize about the world if it did not exist by itself; a factual theory refers to something other than the subject and the verification of the theory presupposes the manipulation, and sometimes even the modification, of the theory's referent or object. No corrections to factual theories would be necessary after testing if they were mere conventional constructions that did not attempt to reflect reality in some symbolic way.

In fact, far from asserting relations among phenomenal predicates, scientific theories contain non-phenomenal ones; moreover, science explains appearance in terms of (hypothesized) objective facts. Scientific knowledge needs to relate through theory those elements given separately in experience. Without any theoretical guide, all we can observe are individuals and a moving crowd, or William James' 'blooming, buzzing, confusion'. Science requires ideas and theories consisting of a system of mutually interconnected hypotheses of different levels; the more representational the theories used for description, the deeper the description will be.

30.3   Ecology Adopts a Hierarchy of Ontological Theses

One of the ontological theses adopted by ecology is that reality has a structure of levels. This is an ontological hypothesis supported by modern science and holds that reality as we know it today is not a solid homogenous block but rather is divided into several levels, each characterized by its own set of emergent properties and laws.

Another ontological assumption adopted by ecology is that the higher levels imply the lower ones, historically and contemporarily, that is, the higher levels are not autonomous, but depend for their existence on the lower levels, and have been formed from these consequent to a number of evolutionary processes (Bunge 2004).

Ecology adopts the thesis of ontological determinism, which holds that science can only explain things through theories and laws. A perfect explanation can never be obtained, that is why there are no definitive explanations. Every explanation can be improved, and in general, the explanatory potential of a theory can increase without limits, precisely because its defects can be detected. The possibility of progress can only be suggested by the history of science and by the way in which theories are critically constructed and examined, but there is no law of scientific progress and nothing guarantees its persistence.

According to the systemic proposal, all objects are systems or system components, and this applies to all material things, as well as ideas. Bunge's systemic ontological approach (Bunge 1979, 1995, 1999a; Mahner and Bunge 1997) argues that all concrete objects are connected to each other in different ways. This means that every component of the system interacts with other components of the system either directly or indirectly. This alternative surpasses atomism and ontological holism, since the former sees the trees but not the forest and the latter sees the forest but not the trees. The systemic approach, however, facilitates the perception of the trees and their components and environment, as well as the forest (Bunge 1999c). It implies a concrete system as a composite unit where each of its components is capable of changing and acting on other components of the unit; or being modified by them.

A system is composed of multiple things that interact with each other influencing their behaviour and composition. Since the components of the system interact with one another, the system has an endo-structure. This involves all relations (particularly, links or connections) among those components. Since the system has an environment, it also has an exo-structure, or collection of links to things in its environment. (Obviously, the system's inputs and outputs are included in its exo-structure.) The union and logical sum of the endo-structure and the exo-structure will be called the total structure of the system.

According to the systemic ontological approach, a system presents emergent properties that its components do not have. A property of a system is said to be emergent if and only if it is not exhibited by any part of the system. For example, the ability to form ideas is a property of certain neuron systems, not individual neurons or the entire nervous system. When a system is formed by two individuals, atoms or people, it exhibits emergent properties. In addition, each of its components acquires a property that it did not have before.

A systemic point of view aims to distinguish the organism level from the ecosystem level; and establish relations among them. Moreover, it shows how organisms combine with each other and how, in turn, individual behavior is influenced (inhibited or stimulated) by the organism's environment. Based on this, it can be said that every organism belongs to a biosystem and behaves, at least partly, as a member of it. The task of ecology is to study the way in which organisms combine to form the biosystem, how they in turn stimulate or inhibit individual action, and how diverse biosystems interact, either directly or indirectly.

30.4   Ecology and Its Field of Action

Ecology is the scientific study of the processes regulating organisms' distribution and abundance and their interactions, as well as the study of how, in turn, these organisms serve as means for transferring and transforming energy and matter through the biosphere. In other words, ecology is concerned with the design of the structure and the function of the ecosystem (Begon et al. 2006).

Ecology studies interactive systems, and the effects of these interactions on organisms and the environment. A system is defined as a set of interacting elements with specific function(s). Any change in any of its elements implies a change in all the others, and therefore, a modification of the whole set. Systems have the following main characteristics: they fulfil a specific function, they are influenced by the environment in which they are located, they influence the environment that surrounds them, and the products the system sends to the environment provoke a response (feedback) of the environment to the system. By environment we understand the set of physical, chemical, biological and social components capable of causing direct or indirect effects, in a short or long term, on living beings and human activities.

Ecology works with different levels of reality. Although the levels of reality in ecology vary among the various authors' viewpoints[2] the most generalized opinion on this issue[3] points out that ecology presents four levels of organization: (a) the individual organism; (b) the population formed by individuals of the same species; (c) the community formed by a more or less large number of populations; and (d) the ecosystem formed by the species that make up a community and the biotic environment.

---

[2] See McMahon and Mein (1978) and Niles (1985).
[3] See Mahner and Bunge (1997), Begon et al. (2006), Curtis et al. (2008), and Krebs (2008).

30.5   The Theory of Natural Selection and Its Relation to Ecology

The theory of evolution by natural selection guides ecology in the course of its investigations, and constitutes its theoretical framework—cementing it and providing it with a general idea of what nature is like and what emergent properties arise from the interrelationships between organisms and the environment.

The theory of evolution by natural selection (Klimovsky 2001; Begon et al. 2006) presents a hypothetical deductive system of hypotheses of different levels, ranging from the most theoretical to the most observable. Below we will develop the hypotheses that constitute this theory and we will relate them to models and hypotheses used by ecology.

The first hypothesis, called "hypothesis of exponential growth", holds that all populations have the potential to populate the entire earth and would do so if all individuals survived and if each individual produced the maximum number of offspring. But this does not occur, many individuals die before they reproduce and most of them, if not all, reproduce at a lower than maximum rate. Population ecology considers this type of growth and uses "exponential growth models".

The second hypothesis called "population growth" holds that the number of members of a population in a closed habitat is limited, it cannot be exceeded due to resources limitations; therefore, population growth cannot be exponential; based on this hypothesis, ecology develops a logistic model of population growth. The logistic model indicates that the number of individuals in a population is not determined by the reproductive potential, as described by the exponential model of population growth, but by the environment. A given environment can only support a limited number of individuals in a given population, in any specific set of circumstances. The size of the population will be around this number, which is called the environmental load capacity. It is an average number of individuals that the environment can sustain under a certain set of conditions. For animal species, the carrying capacity may be determined by the availability of food or access to shelter sites. For example, in plants, the determining factor can be access to sunlight or water availability when the population begins to increase. Unlike exponential growth, the logistic growth slows down as the population approaches the carrying capacity, and finally, the population stabilizes at or near the carrying capacity, although there may be fluctuations around the carrying capacity.

The third hypothesis is called "variations" and indicates that offspring presents variations, that is, characteristics that the parents do not exhibit and that are in most cases inheritable. These are casual and come from an accidental modification of the genetic configuration; variations can occur in different ways, some are favourable for survival and others not, and they are also heritable.

From the first and second hypothesis, the fourth hypothesis can be deduced. This hypothesis, called "competition hypothesis", argues that if resources are limited and species tend to increase exponentially, there will come a time when there are more candidates than available resources; this derives in a struggle for resources where some will defeat their competitors; individuals try to improve their abilities to outdo others, and thus, a competition for resources is established. Ecology uses this hypothesis for explaining competition relations among individuals of the same or different species, called intra and interspecific competition respectively. The competition principle is also used by the theory of the ecological niche and the factors that affect population growth which are dependent and independent of density.

The fifth hypothesis called "the survival of the fittest" is deduced from the fourth, the "competence hypothesis", and it states that, among competitors, individuals having more favorable characteristics will survive. The process of disappearance can be fast or slow, according to the characteristics that define each individual in a given habitat when they are in favorable or unfavorable conditions. Ecology considers this hypothesis, since many of its explanations maintain that in the competition process or, for instance, in a predator-prey relation, the individuals that survive are the fittest and the ones leaving more offspring.

From the fifth hypothesis (survival of the fittest) and the third hypothesis (existence of favorable and unfavorable variations), the sixth hypothesis, called "principle of natural selection", is deduced: the emergence of a favourable and in-heritable characteristic ends up changing the species through different generations. The sixth hypothesis explains the appearance of new species from the old through a process that does not involve teleology; favourable characteristics appear purely coincidentally, and constantly in order to allow adaptation and natural selection. What is favorable or unfavorable depends on the environment. Some individuals will reproduce and leave more offspring in certain environments, but not in another. In this sense it can be said that nature is selective. If some individuals leave more offspring than others, proving to be more apt to their environment, and the hereditary characteristics of a population change from generation to generation, it is said that evolution by natural selection occurs. Ecology considers this hypothesis, for example, for dealing with mutualism, competition, niche hypothesis, etc., which involve processes of coevolution and natural selection, the fittest individuals being the ones that survive and leave more offspring.

It would be reasonable to say that the organisms of a certain generation are adapted to the environments of previous generations. Previous environments act as filters, but it seems that organisms are adapted (adjusted) to their current environment only because this environment tends to be similar to the previous environment. Organisms were not foreseen for the present and the future, nor are they adapted to these, they are consequences of the past and therefore they have been adapted by past conditions.

By definition, the fittest organisms are those leaving a greater number of off-spring, compared to the offspring left by other less fit individuals in the population. Those individuals leaving a greater number of offspring in a population exert a greater influence on the hereditary characteristics of this population. No population of organisms can contain all the possible genetic variations that could exist and influence on efficacy. As a consequence, it is unlikely that natural selection results in the evolution of perfect individuals, or maximum fitness.

Ecology takes the theory of natural selection as a fundamental pillar for understanding interactions among organisms and the environment, and as the essential explanation for organisms' distribution and abundance, as well as the basis on which models of population and communities' growth were built. For example, the theory of the ecological niche is deduced from the competition principle, the same as intra- and inter-specific competitions, which formulate models for exponential and logistic growth, or MacArthur and Wilson's *Theory of Island Biogeography*. The process of ecological succession implies the competition principle, since species in the early stages of a succession will be r-selected, prevailing in environments with little or no competition.

It is important that the species in the early stages of the succession have high reproductive rates. The species in the late stages are K-selected, they are good competitors and tend to prevail in their environment. That is why their reproductive rates are lower and invest more energy in offspring. Because of evolutionary processes, ecosystems have trophic structures, and the flow of matter and energy among different links of the trophic chain is the result of interactions developed among organisms of the same species and of different species, and with the environment.

30.6 The Reconstruction of Ecological Theory

As we have pointed out, the theory of evolution by natural selection is the theoretical framework of ecology. This theory explains, through a series of high-level hypotheses, the ways in which organisms and the environment interrelate, taking competition as a constitutive principle of these relations, which derives in natural selection of those which have favorable characteristics in relation to the environment.

By considering the theory of natural selection, ecology studies emergent properties arising from systemic relations between individuals of the same species or different species, and their environment. The law-like high-level hypothesis underpinning ecology is the following:

> The organisms in a populations and/or communities interact with each other and with their surrounding physical environment, transforming and transporting energy and matter through a system that feeds back on itself.

From this hypothesis, which includes the hypothesis of natural selection since organisms adapt to the surrounding environment, ecology deduces other hypotheses of intermediate level and through which constructs models and explains how different ecosystems work. The ecological theory constitutes a hypothetico-deductive system formed by hypotheses that are deduced from the theoretical framework of the theory of evolution by natural selection. The ecologist formulates hypotheses, collects data, tests hypotheses by carrying out experiments or sampling, analyzes data, makes comparisons with other similar investigations, and draws conclusions and formulates new questions.

Based on the theory of evolution by natural selection, ecology deduces hypotheses and models for each ontological level of organization of reality, which are ecology's objects of study. The different interactions occurring at the organism, population, community and ecosystem levels will develop processes of natural selection and co-evolution.

30.7   Main Hypotheses of the Ecological Theory at the Organism Level

In this section, we will develop the main hypotheses supporting the ecological theory at the organism level. The theoretical principles explaining how individuals are affected by the biotic and abiotic environment, and by the way in which individuals affect the environment, are taken directly from the theory of evolution by natural selection.

Individuals are each of the living beings that make up the different spaces of the planet Earth and that can vary enormously in form, characteristics and primordial elements. All organisms assume the presence of matter as well as a constant interaction between inside and outside through different types of biological relations. There is a correspondence between the organisms and the environments in which they live; this does not mean that organisms are "designed/created" for a certain environment, but that they are adapted to it. This adaptation is a response to selection by nature (hypothesis of Darwinian evolutionary adaptation). Individual selection is the most common and is based on the fact that each individual tries to maximize their fitness, which is the value of survival and reproduction of a given genotype in relation to others. This does not mean that the organism does not respond to the environment (irritability, acclimatization, physiological adaptation, and phenotypic plasticity), but that its characteristics allow adaptation.

Organisms live in a certain environment if it offers conditions and resources appropriate for their growth, development, and reproduction. Organisms adapt to the environmental conditions that allow them to develop a population at optimum levels; this fact determines and explains the distribution and abundance of the species in different biomes, according to different temperatures.

Ecology states that organisms have different ways of obtaining the resources that allow them to grow and reproduce; hence they are classified as primary producers, decomposers, parasites and predators; and, in turn each type of organism fulfils a specific function in the trophic chain enabling the flow of energy and matter in each of its links. Some organisms will develop physical, chemical, morphological and/or behavioral defences against possible attacks and predation. These defences are aimed at reducing the likelihood of a meeting with a consumer and/or increasing the chances of surviving the encounter. But interaction does not necessarily end there. A well-defended food resource exerts selection pressure on consumers. Those consumers that are best fitted to combat such defences will leave more offspring and their characteristics will be more likely to spread in the consumer population. It is said continuous evolution occurs. In these cases, where evolutionary pressures reciprocally act making evolution of each *taxon* partially dependent on the evolution of the other, the interaction receives the name of co-evolution. Any feature that increases the energy a consumer invests in discovering or capturing prey constitutes a defence if, as a consequences of this, the consumer makes less use of this resource as food.

At the organism level, ecologists study organisms' life history, constituted by those aspects of their life cycle that affect survival and reproduction. Some components of the life history of an organism are the age and size of reproductive maturation, the attempts at reproduction, the way it distributes available energy among reproduction, growth and body maintenance, the way it distributes resources among its progeny, their number, etc. (Begon et al. 2006). The study of life histories pays special attention to the way in which adaptation processes and diverse constraints interact, determining a certain kind of adaptation or Darwinian fitness that enables the survival and proliferation of those organisms that employ successful strategies.

Demographic aspects are the basis for a comparative study of life histories; they help to make inferences about their evolution, and to determine populations' viability and the design of strategies for their conservation, management, and sustainable use of natural resources without damaging the ecosystems.

30.8   Hypotheses Underlying Ecology at the Population Level

A population is a group of inter-fertile organisms of the same species that interbreed and coexist in space and time. Knowledge of population dynamics is essential for studying the diverse interactions among organism groups; and can be practically used for formulating policies regarding the use of adequate and sustainable resources. Some of the emergent properties arising from relations among individual population individuals are: the patterns of growth, mortality, age structure, density and spatial distribution.

Population ecology analyzes a population growth rate, that is, the increase in the number of individuals in a given unit of time per each individual. Exponential growth is the simplest model of population growth, which draws from the hypothesis of the theory of evolution by natural selection and indicates that the number of individuals increases at a constant rate. A key aspect of exponential growth is that, although the rate of increase per capita remains constant, the rate of population growth increases rapidly as the number of breeding individuals increases. While exponential growth is typical of small populations with abundant resources available, exponential growth cannot exist without having a decrease in the size of the population.

The logistic model, which takes into account the carrying capacity, describes one of the simplest population growth patterns observed in nature. In many populations, the number of individuals is not determined by reproductive potential alone, but by the environment. A given environment can only support a limited number of individuals from a given population in any specific set of circumstances. The population size will be around this number, which is known as environmental carrying capacity. It is an average number of individuals in a population that the environment can sustain under a certain set of conditions. Regarding animal species, the carrying capacity of an environment is determined by the availability of resources. Logistic population growth considers intra-specific competition as a factor implying evolutionary processes of the individuals in a population, since environmental carrying capacity constitutes a struggle for resources and it is the fittest individuals that will survive.

Populations also have characteristic patterns of mortality with a variable risk of death at different ages. Age structure is an important factor in predicting population growth. In every population there are two other interrelated properties: density and pattern of spatial distribution. Density is the number of individuals per unit area or volume, while the spatial distribution pattern describes the spatial location of organisms. The factors affecting a population's size and density are, among others: light, temperature, available water, salinity, space for nesting, and scarcity (or excess) of necessary nutrients. If any essential requirement is scarce, or any environmental characteristic is too extreme, it is not possible for the population to grow, even if all other needs are met.

Those factors causing changes in birth or death rates, as population density changes, are called *denso-dependent* and imply the hypothesis of inter-specific competition, since individuals compete for available resources.

Many factors operate on populations in a density-dependent manner. As population increases, food reserves are reduced and competition among population members becomes greater. Eventually, this leads to a higher mortality rate or a lower birth rate. Predators are attracted to high prey density areas to capture a greater proportion of the population. Similarly, diseases can spread more easily when population density is high.

Environmental disturbances often act as density-independent factors. Within the so-called population life histories, the most interesting and variable properties are the reproduction strategies, consisting of groups of co-adapted characteristics affecting survival and reproduction. After a number of studies, biologists have noticed that reproduction strategies vary from one individual to another within a population and also from one population to another among related organisms. In other words, patterns include genetically determined variations which are subject to natural selection.

Throughout their life, organisms have to carefully administrate the amount of time and energy allocated to different activities, i.e., a great deal of time and energy spent on an activity (e.g. search for food) implies a reduction in the time and energy available for other activities (for example, offspring care). A certain balance in the distribution of energy among different functions results in the particular adaptive strategy of an organism; and determines the environmental conditions in which it will be competitively successful.

At the population level, intra-specific relations develop emergent properties such as: density, age and sex structures, space available, biomass, coverage, birth rate, mortality, pattern of growth, migration patterns, life strategies (Begon et al. 2006; Curtis et al. 2008). Populations live as part of a community (a group of organisms inhabiting a common environment and having reciprocal interactions). There are three main types of specific interaction in communities arising from natural evolutionary processes: competition, predation and symbiosis.

The more similar the organisms are as regards their requirements and lifestyles, the more intense competition between them will be. As a result of competition, the global fitness of the interacting individuals can be reduced.

Predation means feeding on living organisms; predator-prey interactions affect populations' evolution, they also influence population dynamics, and they can increase the diversity of species by reducing competition among prey. Predator tactics are under intense selective pressure and it is likely that the individuals that obtain food more efficiently will leave the largest number of offspring. From the potential prey side, it is likely that the individuals that are most successful in avoiding predation leave the most offspring. Thus, predation affects the evolution of both the predator and the prey. It also affects the number of organisms in a population and the diversity of species within a community.

Symbiosis is an intimate relation among organisms of different species. It can be beneficial for a species and indistinct for the other (commensalism), beneficial to one and damaging to the other (parasitism), or beneficial for both species (mutualism). Long-term symbiotic relationships can produce deep evolutionary changes in the organisms involved.

30.9   The Competition Hypothesis in Population and Community Ecology

Competition is the interaction among individuals of the same species (intra-specific competition) or different species (inter-specific competition) using the same resource, which tends to be in limited quantity. As a result of competition, the reproduction of the interacting individuals may be impaired. The resources for which organisms may compete include food, water, light, living space, nesting sites or burrows. Competition may be of two types: interference or exploitation.

It is common to find ecologically similar species living together in the same community. This observation led us to consider the degree of similarity needed between two or more species for them to coexist in the same place and at the same time. This took us, in turn, to the concept of ecological niche. The analysis of situations in which similar species coexist have shown that resources are often subdivided, or they are distributed by coexisting species, thereby exhibiting each species' fundamental niche. In communities where there is niche overlap, and as a result of selective pressures exerted by inter-specific competition, natural selection can result in an increase in the differences between the competing species, a phenomenon known as *character displacement*. According to this interpretation, competition among organisms whose ecological niches overlap implies a selection of individuals whose characteristics overlap, leading to a divergence observed among species.

30.10   Community Composition and the Stability Problem

Seen from a global perspective, ecological communities often seem to be in equilibrium. Many species remain in large areas for many generations. However, when communities are examined at local level, it is evident that they, just as the populations integrating them, are not often in a state of equilibrium. There are two questions concerning community composition. First of all, what is it that determines the number of species in a community? And, secondly, which factors explain the changes occurring in community composition over time?

Small islands tend to be excellent natural laboratories for the study of evolutionary and ecological processes due to their size and relative isolation. American researchers R. MacArthur and E.O. Wilson formulated the hypothesis that the number of species in a given island remains relatively constant over time, though these species are constantly changing. According to their proposal, known as the equilibrium theory of island biogeography, there is equilibrium between the rate at which species immigrate to a new island and the rate at which a locally existing species extinguishes. Although the number of species is in equilibrium, composition is not because when a species is extinguished, it is usually replaced by a different species.

Immigration rate decreases as more species come to the island, because existing species are already settled and, thus, more equipped to compete with newly arrived species. Extinction rate increases faster when there is a large number of species due to greater inter-specific competition. The equilibrium number among species is determined by the intersection of immigration and extinction. According to the island biogeography model, the two most important variables influencing specific diversity are the island size and distance from a source, usually the continent that can provide the colonizers.

In accordance with the intermediate disturbance hypothesis, the diversity of species in a community is determined by the frequency of environmental disturbances. When disturbances are very common or very rare, species diversity is low. On the contrary, when the frequency of disturbances is intermediate, species diversity is high. It is thought that the primary factor in this decline is inter-specific competition; yet, although all species were competitively equal, the most resistant to adverse effects of physical extreme conditions, predation or disease, would ultimately dominate the community.

Ecological succession comprises those changes occurring in community composition after some disturbance is interrupted. Many observations have shown that recolonization starts with short-lived plant species of rapid growth which are then replaced by other species of longer cycles. As photosynthetic components of the ecosystem change, animal life associated to them also changes. Finally, the community would reach a mature stable state, which is called climax. This model of species replacement is called facilitation because the first colonizing species create favourable conditions for other species to settle. Two other alternative mechanisms that could determine successional process have been proposed: tolerance and inhibition. According to the inhibition hypothesis, the first species avoids colonization by another species. Yet, eventually, the first colonizing species will be replaced by the last to arrive, and these species, in turn, may avoid colonization by others, until they are replaced, or until a subsequent disturbance reduces their number.

Another model, the tolerance hypothesis, suggests that pioneer species neither facilitate nor inhibit colonization by the last-arrived species. The dominant species in each stage are those which can better tolerate existing physical conditions and availability of resources. Currently, it is suggested that these three models are not mutually exclusive; they can operate simultaneously on different pairs of species within the community, gaining more or less importance at different stages of the succession. In other cases, cyclic succession schemes have been observed (Begon et al. 2006; Curtis et al. 2008).

30.11 Hypotheses Underlying Ecology at the Population Level

An ecosystem is a unit of biological organization consisting of all the organisms in a given area and the environment in which they live. It is characterized by interactions among living (biotic) and not living (abiotic) components, connected by a unidirectional flow of solar energy through autotrophs, heterotrophs, and the recycling of minerals and other inorganic materials. The ultimate source of energy for most of the ecosystems is the Sun.

The flow of energy through ecosystems is the most important factor concerning organization. The flow of energy from one organism to another takes place along the trophic or food chain, that is, a succession of organisms related to each other as prey and predator. Within an ecosystem there are trophic levels (producers, primary consumers, detrivores and decomposers).

Inorganic substance movements are known as biogeochemical cycles, because they involve geological as well as biological components of the ecosystem. The geological environment components are: lithosphere; atmosphere; heterosphere. Biological components include producers, consumers and decomposers. As a result of decomposers' metabolic activity, inorganic substances from organic components are released to soil or water. These substances are reincorporated from soil or water into primary producers' systems, are used by consumers and detrivores and then by decomposers, and the cycle is repeated with the substances being absorbed by plants.

Interrelations develop emergent properties in an ecosystem, such as: diversity, trophic structure, biomass, productivity, regulation, ecological succession (Begon et al. 2006; Curtis et al. 2008).

30.12   Conclusion

As it is evidenced throughout this paper, ecology is theoretically grounded on the theory of evolution by natural selection, and uses the assumptions of this theory to formulate explanations about interactions occurring between organisms and the environment, or among organisms of the same species or different species and the environment, and the evolutionary processes occurring from the individual level to the ecosystem level; by means of this theory, ecology establishes the theoretical-conceptual bases through which patterns and processes of each level can be described and explained. Ecology studies interactions occurring within a system and the emergent properties that arise from these relations, among organisms of the same or different species, and the environment in which these organisms live. To understand these interactions, it applies the theory of evolution by natural selection.

Ecology draws from mechanistic epistemology since its theory comprises a hypothetico-deductive system (Bunge 1959, 2004) formed by general theoretical principles, which are taken from the theory of evolution by natural selection, and through which it explains many things: co-evolutionary processes between populations and communities, diversity, population density, organisms distribution and abundance in certain environments, trophic webs, niches, similar species co-existence with different niches, ecological succession, intermediate disturbance hypothesis, ecosystems dynamic equilibrium, interrelations such as mutualism and competence, among others. Ecology assumes a unified and systemic theory, the theory of evolution by natural selection, and continues elaborating on it. This theory provides ecology with explanations on the relations among the different levels of the organised reality it studies—individual(s), population, community and ecosystem levels.

References

Begon, M., Harper, J. L., & Townsend, C. R. (2006). *Ecology: Individuals, populations and communities* (3rd ed.). Brookline: Blackwell Science Ltd.

Bunge, M. (1959). Causality. In *The place of the causal principle in moderns science*. Cambridge: Harvard University Press.

Bunge, M. (1968). The maturation of science. In I. Lakatos & A. Musgrave (Eds.), *Problems in the philosophy of science* (pp. 120–137). Amsterdam: North-Holland.

Bunge, M. (1974). *Semantics I. Sense and reference*. Dordrecht: Reidel.

Bunge, M. (1979). *Ontology II. A world of systems*. Dordrecht: Reidel.

Bunge, M. (1983). *Epistemology and methodology II. Understanding the world*. Dordrecht: Reidel.

Bunge, M. (1985a). *Seudociencia e ideología*. Madrid: Alianza Universidad.

Bunge, M. (1985b). *Teoría y realidad*. Barcelona: Ariel.

Bunge, M. (1995). *Sistemas sociales y filosofía*. Buenos Aires: Sudamericana.

Bunge, M. (1997a). *Ciencia, técnica y desarrollo*. Buenos Aires: Sudamérica.

Bunge, M. (1997b). Mechanism and explication. *Philosophy and social sciences, 27*, 410–465.

Bunge, M. (1997c). *La ciencia, su método y su filosofía*. Buenos Aires: Sudamérica.

Bunge, M. (1999a). *Las ciencias sociales en discusión*. Buenos Aires: Sudamérica.

Bunge, M. (1999b). ¿Qué es filosofar científicamente? *Revista Latinoamericana, 24*, 159–169.

Bunge, M. (1999c/1964). Phenomenological theories. In M. Bunge (Ed.), *Critical approaches to science & philosophy* (pp. 234–254). New Brunswick: Transaction Publishers.

Bunge, M. (2004). *La investigación científica, dos volúmenes*, México: Siglo veintiuno.

Curtis, H., Barnes, S., Schnek, A., & Massarini, A. (2008). *Curtis biología*. Buenos aires: Médica Panamericano.

Klimovsky, G. (2001). *La desventura del conocimiento científico. Una introducción a la epistemología*. Buenos Aires: A-Z.

Krebs, C. (2008). *The ecological world view*. Collingwood: CSRO publishing.

Mahner, M., & Bunge, M. (1997). *Foundations of biophilosophy*. Berlin: Springer.

Marone, L., & Bunge, M. (1998). La explicación en ecología. *Boletín Asociación Argentina de Ecología, 7*, 35–37.

McMahon, T. A., & Mein, R. G. (1978). *Reservoir capacity and yield*. Amsterdam: Elsevier.

Niles, E. (1985). *Unfinished synthesis. Biological hierarchies and modern evolutionary thought*. New York: Oxford University Press.

Chapter 31
Mechanismic Approaches to Explanation in Ecology

Rafael González del Solar, Luis Marone, and Javier Lopez de Casenave

Abstract The search for mechanisms has been a common practice in scientific research. However, since the empiricist critique of causality, and especially during the second third of the twentieth century, other non-mechanistic perspectives—especially deductivism—gained predominance. But the sustained effort of authors such as Michael Scriven, Mario Bunge and especially Wesley Salmon contributed to restoring the respectability of causality and mechanisms in philosophy of science. Some members of the causal family, usually lumped under the name of "new mechanistic philosophy", emphasize the description of mechanisms, especially causal ones, as a central aspect of explanation and other research practices in several areas of science. This approach offers viable solutions to the various ontological and methodological objections that are opposed to the two traditional approaches (the purely deductive and the purely causal). In this work the basic characteristics of three philosophies that highlight the description of mechanisms as a central element to explanation and their suitability for the science of ecology are discussed.

R. González del Solar (✉)
Philosophy Department, Autonomous University of Barcelona, Barcelona, Spain
e-mail: rafael.gonzalezd@e-campus.uab.cat

L. Marone
Facultad de Ciencias Exactas y Naturales, Universidad Nacional de Cuyo y CONICET, Mendoza, Argentina
e-mail: lmarone@mendoza-conicet.gob.ar

J. Lopez de Casenave
Departamento de Ecología, Genética & Evolución, Facultad de Ciencias Exactas y Naturales, Universidad de Buenos Aires, Buenos Aires, Argentina
e-mail: casenave@ege.fcen.uba.ar

© Springer Nature Switzerland AG 2019
M. R. Matthews (ed.), *Mario Bunge: A Centenary Festschrift*,
https://doi.org/10.1007/978-3-030-16673-1_31

31.1   Mechanismic Explanations

For most ecologists, mechanismic[1] explanations are conceptual devices—models or theories—that explain facts by describing the mechanisms that produce those facts. But the task to elucidate what a mechanism is and how it does the explaining belongs in the realm of metascience (Marone and González del Solar 2000). There are currently several philosophical projects centered in elucidating the precise nature of mechanisms and their role in science. Such attempts are collectively known as the "new mechanical philosophy" (NMP; Craver and Tabery 2017), which is about two decades old and considers that mechanisms are complex things. However, there are previous attempts by authors such as Wesley Salmon, Peter Railton, and especially Mario Bunge who have fiercely defended the centrality of mechanisms in scientific inquiry since the middle of the twentieth-century and conceive of mechanisms as processes. This larger group of philosophical projects, the one comprising work by NMP authors, as well as by Salmon, Railton, Bunge and a number of other philosophers, may be called "contemporary mechanismic philosophy" (CMP, González del Solar 2016) and in this work we will explore how some of those projects fare with respect to the practice of Ecology using the analysis of ecological facilitation as a case study.

31.1.1   Explaining with System-Mechanisms

The system-mechanism view identifies mechanisms with complex objects of sorts, and mechanical (or mechanistic) explanations with the description of the component objects and interactions that bring about the behavior of the mechanism.

Glennan, an NMP author who has developed this philosophical perspective, distinguishes two varieties of system-mechanisms: robust ones and ephemeral ones. In both cases, "a mechanism for a behavior is a complex system that produces that behavior by the interaction of a number of parts, where the interactions between parts can be characterized by direct, invariant, change-relating generalizations" (Glennan 2005, p. 445).

A mechanistic explanation consists of a "mechanistic model" made up of two elements: a *behavioral description* (explanandum) and a *mechanical description* (explanans). In the case of robust mechanisms, the former is a generalization describing the mechanism's overall *repeatable* behavior, while the latter shows how the parts of the mechanism and their interactions *regularly* bring about such behavior in virtue of their *stable* spatial and temporal configuration. A model of a robust mechanism describes a *type* that may comprehend any number of *tokens*.

---

[1] We use the adjective 'mechanistic' when referring strictly to new mechanical philosophy's projects; 'mechanismic' for those in the more comprehensive contemporary mechanismic philosophy.

Type identification relies on some sort of physical similarity as in the following example:

> For instance, the human central nervous system contains around a trillion neurons. There are lots of human beings, as well as lots of other organisms, that have neurons whose structure is similar to that of human neurons. Consequently, one can develop a general model of neurons that subsumes countless neural events. (Glennan 2002, p. S345)

Regularity confers mechanistic explanations properties somewhat similar to those of covering-law explanations (Glennan 2002, p. S348), but the former require the description of a mechanism to be considered and explanation, while the latter require scientific laws in its more traditional sense, i.e., as universal, unbounded generalizations. Direct, invariant, change-relating generalizations are robust, but not exceptionless.[2]

In the case of ephemeral mechanisms—which the author also calls "ephemeral processes"—, the explanandum describes a *unique* fact that results from a *temporally and spatially contingent* array of component parts. Yet, the interactions of the parts are robust—i.e., describable by invariant, change relating generalizations— and this feature justifies considering them as mechanisms (Glennan 2010).

Since mechanisms are not their models, one behavioral description may be explained by different mechanical descriptions. And while most generalizations are mechanically explicable, there are some laws that are not. This is the case of Maxwell's equations and other laws in fundamental physics, for which no mechanisms are known (Glennan 2002, p. S348). This fact constrains Glennan's project to define causation as the operation of a mechanism to the realm of the so-called special sciences.

31.1.2 Explaining with Entities and Activities

In their much-quoted paper, Machamer et al. (2000, p. 3) conceived of mechanisms as "entities and activities organized such that they are productive of regular changes from start or set-up to finish or termination conditions" and of explanations as descriptions of mechanisms. Such descriptions may be provided verbally or graphically, among other ways, and may consist in schemata or sketches. A *schema* is an abstract description of a *type* of mechanism, while a sketch of a mechanism is an abstraction that lacks the descriptions of "bottom out" entities and activities— i.e., entities and activities from the lowest level relevant for a given research—or has important gaps in its stages.

Descriptions of mechanisms include *set-up conditions* (descriptions of entities, their properties, and enabling conditions), *intermediate stages* (intervening entities

---

[2]Originally, Glennan (1996, p. 55) recurred to a Goodmanian notion of a causal law, i.e., a generalization that provides counterfactual support. Later, he borrowed the idea of direct, invariant generalizations from J. Woodward's counterfactual, manipulative account of causation.

and activities that produce the end from the beginning), and *termination conditions* (i.e., "privileged" states—rest, equilibrium, emergence of a new product—of interest, identified by practical considerations). All such descriptions are general and idealized, i.e., they refer to types of mechanisms and to normal conditions and/or situations simplified by (often implicit) *ceteris paribus* assumptions.

For Machamer and collaborators (MDC) mechanisms are regular, thus, mechanistic descriptions support counterfactuals. Yet, in contrast to covering law models, intelligibility is not achieved by invoking a regularity, but in virtue of the (description of the) *productive continuity* of the connections between stages, from set up to termination conditions, because descriptions of intermediate activities show "how the actions of one stage affect and effect those of successive stages" (*ibid.*, p. 12) and "are more accurately viewed as continuous processes" (*ibid.*, p. 13). Intelligibility, for MDC, is independent from the correction of the explanation, because it arises "from an elucidative relation between the explanans (the set-up conditions and intermediate entities and activities) and the explanandum (the termination condition or the phenomenon to be explained)" (*ibid.*, p. 21).

Since mechanisms occur in nested hierarchies—i.e., they are composed of entities and activities that belong in different levels of organization—descriptions of mechanisms are often multilevel and bottom out in the lowest-level mechanism considered relatively fundamental or unproblematic for a particular research. Explaining with mechanisms is not necessarily a reductive operation, and it may consist in exhibiting how a certain phenomenon results from the activities of entities that belong *either in lower or in higher levels of organization* and the integration of both levels may be essential for rendering certain phenomena intelligible.

Machamer and collaborators find that mechanism schemata are used in many ways similar to theories, since they are used to describe, explain, and predict phenomena, as well as to serve as blueprints to design experiments and interpret experimental results. The fate of a sketch may be to become a schema, or to be substantially modified or replaced. Like schemata, sketches are useful for designing observations and experiments, since they point to tasks that are still to be done.

31.1.3 *Explaining with Specific Processes in Systems*

Bunge (1979) states that the world is made of systems, and that a system is a complex object with properties that its components do not possess. Systems, then, have a composition (the collection of entities that make up the system), but also an environment (the collection of entities outside the system, though related to it), a structure (the relations among the components of the system and those among such components and the entities in the environment), and mechanisms (the specific processes that make the system emerge, persist, change—or remain unchanged—, and eventually disintegrate). Thus, scientific research is about systems, and to understand them researchers need to describe the composition, the environment, the structure, and especially the mechanisms of the system. Some of these descriptions are explanations if they satisfy a number of conditions.

For Bunge, explaining facts and their patterns is the main rationale for inventing and testing hypotheses, laws, and theories (Bunge 1998b). In order to explicate what explanations are, Bunge begins by defining a general, subsumptive form of scientific explanation (SE):

> A *scientific explanation* of a formula q is an answer to a well-stated scientific problem of the why-kind, consisting in an argument showing that q follows logically from a scientific theory (or a fragment of scientific theory, or a set of scientific theories), auxiliary hypotheses, and scientific data, not containing q. (Bunge 1998b, p. 19)

Note the following theses embedded in this claim:

(i)   *SEs are deductive arguments.* In fact, for Bunge all *rational* explanations are deductions, though different types of explanation include different kinds of generalizations as premises: laws in formal and factual science, and rules in technology.

(ii) *SEs are not about facts directly*, but about scientific propositions. A fact is explained indirectly through explaining the explanandum (or problem generator), which is a (usually quantitative) description of selected aspects of such fact.

(iii) *SEs are answers to well stated why-questions in the context of some scientific conceptual system.* This requirement reduces the ambiguity of the question and prevents the use of scientific resources for explaining largely non-confirmed or pseudoscientific explananda.

One subtype of rational explanation is mechanismic explanation—or SE *proper*—, a type of deductive explanation that answers the specific question "how does it work?", by including at least one mechanismic law in the explanans (Bunge 1997, 2004). Mechanismic laws$_2$[3] may be causal or probabilistic, since mechanisms—or portions of them—can be causal, stochastic or mixed. Mechanismic explanations can only be supplied by "translucid" (or representational) models as opposed to black-box (or phenomenological) ones, two extremes of a translucidity gradient (Bunge 1964, 1979, 1998a, b).

For Bunge, subsumptive and mechanismic explanations are not the horns of a dilemma, but stages of a research process, and mechanismic explanation is superior to merely subsumptive one on the following accounts:

(i)   *Heuristic power*: Inversely to deduction, the search for mechanismic explanations goes from the problem generator (explanandum) to a set of relevant propositions—including mechanismic generalizations—with explanatory potential "and this is why the demand to stop explaining concentrating on

---

[3]Bunge (1998a) distinguishes between laws$_1$—objective patterns of becoming—and laws$_2$, scientific descriptions of those patterns.

description or remain content with what has been explained leads to killing science" (Bunge 1998b, p. 8).

(ii) *Empirical content and depth*: Mechanistic models need descriptions of the relevant components, environment, structure, and specific processes of the systems (subsystems and supersystems) under study. Besides, they typically involve more levels of analysis than black-box models. All this makes mechanistic models empirically richer, deeper (Bunge 1998a, pp. 575–585), and more specific—thus more exposed to empirical tests—than phenomenological ones.

(iii) *Pragmatic usefulness*: Being symmetrical with prediction, subsumptive explanation provides a tool for manipulating the system of interest. Yet, its usefulness diminishes when the system behaves unexpectedly because of some factor hidden in the black box. Non-mechanistic laws will not suggest where or how to intervene for effectively altering the system. By contrast, mechanistic models offer a preliminary blueprint for understanding what might have happened, as well as some courses of intervention, since they describe the inner workings of the system. Consequently, mechanistic explanation provides a useful starting point for technological forecast and control.

31.2   Explaining Ecological Facilitation: A Case Study

Facilitation processes are positive ecological interactions[4] that are especially important in natural communities strongly shaped by environmental conditions (Connell and Slatyer 1977). Although ubiquitous in nature, and essential for succession theory and the early development of ecological science, theoretical ecologists have paid much less attention to positive than to negative interactions, especially competition (Bruno et al. 2003).

A common form of facilitation is habitat modification, where an organism changes its environment making it less stressful for other organisms. A classic example is that of taller plants providing shade (from sunlight) and shelter (from wind) to shorter, less tolerant plants. Typically, shade and shelter reduce UV radiation, respiration costs, transpirational demands, and maintenance of tissues below lethal temperatures, while increasing soil moisture through lower evaporative demand (Callaway 2007).

Baumeister and Callaway (2006) attempted to explain the pattern of spatial association of the stress-tolerant limber pine (*Pinus flexilis*) with Douglas firs (*Pseudotsuga menziesii*) and wax currants (*Ribes cereum*) in a harsh grassland-forest ecotone[5] of the Rocky Mountain Front. Climate conditions in the area included "extraordinarily high" warming winds, predominantly from the west-southwest,

---

[4]That is, ecological interactions that benefit all (mutualism) or one type (commensalism) of the organisms involved, while not damaging any of them.

[5]A transitional zone between two adjacent plant communities (Ricklefs 2008).

and high annual thermal amplitude (-40 to $37\ °C$). Annual precipitations averaged 70 cm.

Some preliminary observations confirmed the pattern of spatial association and suggested that facilitation was involved. The authors attempted to separate the effects of the likely aboveground *mechanisms*—snow accumulation (drift), wind amelioration, and shading—responsible for the facilitation function. They designed a three-way, fully factorial, blocked experiment at a level plateau dominated by *Festuca* grasses with scattered *P. flexilis*. Inside a 35×15 m rectangle encircled by a 2.5 m tall wire fence, 1×1 m plots were established with different treatments. Shade treatments ("shade", "shade + drift", "shade + no wind", and "shade + no wind + drift") were plots covered by 1.5×1.5 m propylene shade cloth (causing 48% reduction in photosynthetically active radiation, PAR). Drift treatments ("drift", "shade + drift", "drift + no wind", and "shade + no wind + drift") were plastic snow fences installed (yearly, October-April) directly windward of the plots. Wind was blocked (up to 80%), without increasing drift, with U-shaped polycarbonate fences. Five *R. cereus* and three *P. menziesii* seedlings were planted in each replicate quadrat for each of the nine treatment combinations.

Analysis of data on survival and growth of seedlings showed that only shade treatments produced significant differences either for *P. menziesii* or for *R. cereus*, suggesting a hierarchical operation of mechanisms, with shading on top. Other results further stressed the hierarchical effect of treatments.

Within the shaded plots, *P. menziesii* survival was lowest without wind protection and without enhanced snow accumulation, and highest with drift fences. With shade, wind reduction showed a significantly positive effect on seedling survival during the first winter, but not later. The highest positive effect of shade on seedling survival occurred in the first winter. For *R. cereus*, the positive effect of shade occurred in the first summer. The positive effect of shade on both species was corroborated by seedlings being taller, and root mass and root-to-shoot ratios higher, in shade than in no-shade treatments.

Summing up, *P. flexilis* trees protect *P. menziesii* individuals mainly by sheltering them from the wind, while the seasonal patterns of mortality suggest that shading is the main facilitation mechanism for *R. cereus* and sheltering has a secondary role. Results also showed that facilitation mechanisms interact so that the positive effects of some of them depend on the occurrence of another, mainly shading. Moreover, the operation of the facilitation mechanisms is hierarchical (Baumeister and Callaway 2006, p. 1828) and facilitative strength depends not only on the species involved, but also on circumstances such as season of the year.

Shading helps seedlings by moderating under crown temperatures and PAR. In summer, this reduces soil moisture evaporation and leaf evapotranspiration for *R*.

cereus. In winter, shade keeps temperatures higher than in the open, preventing damage from cold and high levels of photoinhibition for *P. menziesii*.[6]

Wind amelioration protects mechanically the seedlings located on the leeward side of *P. flexilis*: (1) Reduced wind velocities mitigate soil desiccation and evapotranspiration (through diminishing mechanical abrasion by soil particles and ice on the cuticular wax of leaves); (2) in winter, *P. flexilis* partially block warm winds that may increase above ground temperatures stimulating photosynthesis in seedlings growing on frozen ground, causing higher leaf temperature and evapotranspiration without the possibility of water replenishment by roots, favoring leaf desiccation. *P. flexilis* also increases drift, moderating temperatures, protecting seedlings against excessive irradiation and wind, and adding nutrients to the soil.

Schematically, the explanation provided by the authors is this:

Explanandum: Less tolerant plants tend to grow beneath and at the leeward side of stress-tolerant trees.

Explanans: Trees enhance survival and development of seedlings by acting like barriers that enhance snow accumulation, filter potentially harmful luminous radiation and mitigate wind speed reducing its kinetic energy and thus its ability to cause mechanical damages by collision and dehydration to seedlings (facilitation).

31.3 Contemporary Mechanismic Philosophy in the Light of the Ecological Facilitation Case

The mechanismic approach to explanation implies a strong relation between the ontology of explanation and its epistemology, so much so that—following Salmon (1984)—one of the champions of neomechanism insists in considering the approach as an *"ontic* account of scientific explanation"* (Craver 2014, our emphasis).

Indeed, according to Craver, not descriptions of mechanisms but mechanisms *themselves* are the objects that hold explanatory power. However, since explanation is an epistemic operation performed by a subject who manipulates symbols in order to represent real things and/or processes, it is difficult to accept Craver's contention. The idea that mechanism *description* is the main source of explanatory power is still ontologically committed and much less controversial.

---

[6] A light-induced depression of photosynthetic rate produced by an absorption of PAR higher than that the plant can effectively use, whose effect is increased by low temperatures (Germino and Smith 2000).

31.3.1 On the Nature of Mechanisms

Contemporary mechanismic philosophy (CMP) captures many if not all the features of the mechanisms of ecological facilitation. Mechanismic philosophers consider that mechanisms are systems—i.e., complex objects with a composition and interactions (Glennan) or activities (MDC)—or processes in systems (Bunge) (Table 31.1).

The ecological facilitation case study is consistent with the view that systems and mechanisms go together, a thesis explicitly defended by Bunge and common to all CMP authors (González del Solar 2017). In the case study, the composition of the system involves both biotic and abiotic entities. Biotic entities include individual organisms and collections thereof (populations, communities), all of them systems. Abiotic components include soil and ice *particles* (contributing to the harming potential of moving air molecules), and several *systems*, such as soil and water in the substrate, and electromagnetic radiation.

Table 31.1 Ontological aspects of mechanisms according to different authors in contemporary mechanismic philosophy

Ontology	Glennan (2002)	Machamer et al. (2000)	Glennan (2010)	Bunge (2000)
Nature	Complex systems	Entities and activities	Ephemeral "processes"	Specific processes in systems
Composition	Objects	Entities and activities	Objects?	Network of processes
Relation to phenomena	Produce phenomena	Produce phenomena	Produce phenomena	Produce phenomena
Modes of causation	Productivity	Productive continuity	Productivity	Productivity by transferring
				(a) Energy
				(b) Signal
Productivity	From causal interactions	From causal activities	From causal interactions	From networks of processes, causal or random
Regularity	Repeatable event	Regular "always or for the most part"	Unique	Lawful, repeatable or not
Organization	Determinant and robust configuration	Determinant, robust temporal and spatial	Determinant and not-robust configuration	Determinant, robust or not
Relation w/causation	Causes are mechanistic outside fundamental physics	Mechanisms are causal	Causes are mechanistic outside fundamental physics	Mechanisms may be causal, random or mixed

Adapted from González del Solar (2017)

However, a more detailed analysis suggests that, as ecologists view them, facilitation mechanisms are better understood as processes, not as complex objects or systems. To begin with, facilitation occurred *in* a system or *to* a system (an ecological community), but it was not a system. More precisely, facilitation is an equivalence class of processes that somehow change the system or prevent it to change. The equivalence criterion, i.e., the criterion for classing the processes denoted by the facilitation concept is functional: what makes a given process an instance of ecological facilitation is that it benefits at least one of the organisms involved in the process without harming the others.

That facilitation is a type of mechanisms means that it is a class of functionally equivalent concrete processes. Therefore, the claim that some mechanism is a type-mechanism should be understood as an ellipsis stating that it is a type of mechanisms. This supports Bunge's mechanismic perspective that mechanisms are not things themselves, but specific processes in things (systems).

Although facilitation is a process, it is not an ephemeral processes *sensu* Glennan (2010), because the behavior it brings about is not unique. Indeed, one may claim that, in a way, each thing and each process in the world is unique and that strict real identity does not exist, but this does not seem to be Glennan's approach.

Glennan offers the example of the ephemeral-mechanistic explanation of the demise of R. Barthes, the French literary critic, after being struck by a laundry truck. Barthes died while crossing a street in Paris, returning from a luncheon with President Mitterrand, and Glennan considers it a unique event. However, this view ignores the fact that while scientists always study particular facts, they also attempt to generalize such facts by means of abstractions and generalizations that blur or erase the differences among particular facts that researchers consider not to be relevant. Which differences are erased depends then on a variety of ontological, methodological, and even pragmatic criteria. Consequently, Barthes's death seems unique only if one focuses on Barthes's singularity—and ignores that he was just one in several million human beings walking the streets of Paris—, in which case the explanation of that fact would resemble a chain of unlucky causal events (González del Solar 2016). However, scientists do not usually remain content with such an approach. For example, a scientific researcher might view Barthes's death as an instance of a whole class of facts if she approaches that particular fact as an instance of the class of traffic casualties in big cities. The class could be further constrained to pedestrians older than a certain age, when attention and reflexes tend to diminish, without going all the way to the individual R. Barthes.

Another example. The extinction of dinosaurs may be seen as unique in that the complex processes leading to the death of the last member of the Superorder Dinosauria was somehow triggered by a meteorite impact. Yet, this is not the only possible approach to the problem of dinosaur extinction. In the first place, a fact such as a meteorite impacting the Earth may seem unique when one focus on the explanation of dinosaur extinction, but meteorites impact other astronomical objects on a more or less regular basis, so there is nothing intrinsically unique in the fact that one or more of them collided with Earth and were responsible for complex climate changes that eventually led to dinosaur extinction (González del Solar et al. 2014).

However, what is important, is that a very complex network of astronomical and ecological processes, not very different to those at work in our times—e.g., meteorite impacts, climate change, resource reduction and competition, predation, disease—can explain dinosaur extinction. The upshot is that the uniqueness of a fact heavily depends on the approach the researcher takes to study such fact and scientific researchers usually search for generality. To paraphrase a *cliché*, uniqueness is in the eye of the beholder.

Again, Bunge's mechanismic approach seems to fit better the nuances of generality of ecological mechanisms such as facilitation. The facts ecological mechanismic models intend to explain may not be universal, but neither are they unique. Indeed, they are repeatable enough to allow ecologists to build models and even theories about them (Resetarits and Bernardo 1998).

In a recent work, Pâslaru (2009) analyzed the mechanism of niche complementarity in the light of neomechanism and concluded that ecological mechanisms are not systems, or entities and activities, but insensitive networks of causal processes. This ontological thesis is in agreement with previous findings that ecological mechanisms are processes or networks of processes, not things (González del Solar 2016; Mahner and Bunge 1997). As for insensitivity, Pâslaru follows Woodward (2003) and takes insensitivity to mean that causal relationships "are not affected by modifications in the background conditions of variables X and Y or by changes in the actual circumstances of the relationship" (Pâslaru 2009, p. 834), where X and Y are the *relata* of a causal relationship. It would seem that insensitivity is not a property of the processes but a property of a collection thereof. Thus, a class of processes is said to be insensitive when different instances of the type maintain the functional equivalence in spite of changes in the background conditions. As a consequence, insensitivity is relative to those changes in background conditions and, in addition, should be a matter of degree (but Pâslaru does not develop this point); that is, some ecological (types of) mechanisms might be more insensitive than others. For example, in the facilitation case study, photoinhibition is heavily increased by the usually low temperatures in the area. Facilitation by wind amelioration would drastically change or even not take place in a region where winds are not strong.

The analysis of facilitation also provides support for the contemporary mechanismic thesis that mechanisms occur in nested hierarchies. Still, the hierarchical arrange of mechanisms can be interpreted in at least two ways. One of them is the view, common to all contemporary mechanismic philosophers, that mechanisms at one level comprise different (sub)mechanisms that take place at a lower level of organization. This is the *nested* aspect of mechanisms. For example, facilitation by shading, an ecological-level mechanism, comprises physical-level mechanisms responsible for the behavior of light in different circumstances and biochemical-level mechanisms that produce the phenomenon of photosynthesis. This type of nested organization is a consequence of the systemic nature of things (at least to the atomic level of organization). Strictly speaking, it may be not necessarily a hierarchy, for it is not clear that lower levels dominate in any sense over higher levels of organization (although it is clear that things in lower levels *precede* higher levels from a genetic point of view: things in lower levels *constitute* things in higher levels).

A second interpretation of the hierarchical arrangement of ecological mechanisms, the *hierarchy* aspect, is related to an altogether different phenomenon, namely that of the interactions among mechanisms. This aspect of mechanisms is nicely illustrated by Baumeister and Callaway's conclusion that, in the area, facilitation by shading is a necessary condition for other types of facilitation to be of significance. This kind of arrangement among mechanisms is hierarchical strictly speaking. More importantly, as the authors note, it suggests that, contrary to the usual practice, ecological mechanisms should not be studied isolated from other mechanisms, but paying attention to their interactions. It also shows that a purely functional account of an ecological fact invoking "facilitation" would miss the possible interactions among the relevant mechanisms.

Contemporary mechanismic philosophy states that components are connected through causal interactions, activities or causal/stochastic processes that *produce* the result of interest. Facilitation by *P. flexilis* includes causal connections: litter accumulation *produces* alterations in the environment of seedlings by adding nutrients to the soil. However, facilitation mechanisms that rely on blocking, such as shading and sheltering (from the wind), might challenge such productive connection, since they consist in *preventing* potentially harmful processes (Fig. 31.1). It is clear that *P. flexilis*' presence is relevant for explaining the survival of seedlings in virtue of what *would occur* in *P. flexilis*' absence. Ecologists frequently use counterfactuals to explain facts and to design their experiments, but this is a jump from ontology to epistemology that needs to be taken into account.

Machamer and colleagues (2000) state that activities can be types of causes, and that the general term 'cause' must be specified to become meaningful. How-

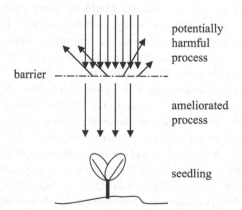

Fig. 31.1 Graphical representation of the type-mechanism of facilitation by blocking. The arrows represent any potentially harmful process (e.g., electromagnetic radiation, moving gases, vapor, soil and ice particles). The horizontal dotted line stands for a barrier, i.e., any object solid enough to interfere with the trajectories of the foregoing potentially harmful processes so as to reduce their potentially harmful capacity. (From González del Solar 2016)

ever, in facilitation by blocking there is really no "non-breaking" of seedlings' tissues produced by airborne soil or ice particles, nor any "non-burning" of the photosynthetic pigments in the seedlings by excessive PAR, i.e., there is a lack of productive *continuity*. Machamer and collaborators admit that even though counterfactuals occur in mechanistic explanation "they get at nothing ontological", because nonexistent activities cannot cause anything:

> Failures and absences can be used to explain why another mechanism, if it had been in operation, would have disrupted the mechanism that actually was operating. Maybe we should draw a distinction and say they are *causally relevant* rather than *causally efficacious*. (Machamer 2004, p. 35–36; emphasis in the original)

Glennan's solution is similar to MDC's: he takes counterfactuals to have explanatory import, but not any ontological substance. Bunge (1981) also deals with negative "causation" by distinguishing (genuine) *causation* from causally relevant *conditions*, but he deems counterfactuals "rhetorical tricks" and rejects them in both ontological and epistemic matters, because they "are not propositions, and consequently they cannot be assigned truth values" (Bunge 2006, p. 93). In sum, counterfactuals may have a heuristic role in research, but should not be taken to represent anything real.

These epistemological clarifications do not solve the ontological gap in facilitation by blocking. Is it possible to *cause* an event by impeding another event to occur? The answer is relevant for MDC's view that the source of intelligibility in a mechanistic explanation is the productive continuity between stages of the mechanism.

A discussion of "negative causation" exceeds the goals of this work; suffice it to say that mechanismic philosophers do not admit of negative causes and provide similar solutions to the problem: (a) they distinguish between causally *relevant* conditions and causally *efficacious activities*, and (b) they suggest the operation of more than one mechanism in cases of apparent causation by preemption (González del Solar 2016).

Thus, facilitation by blocking would involve at least two kinds of mechanisms: (a) those involved in the mitigation of harmful processes (i.e., shading and sheltering), and (b) all the normal metabolic processes of seedlings. Ameliorated environmental conditions brought about by mechanisms of the first kind are not causally *efficacious* of seedling survival, but causally *relevant* for it. Thus, some ecological regularities—such as the spatial association between *P. flexilis*, *P. menziesii* and *R. cereus*—are the product of not one but several mechanisms. These ontological considerations have epistemological consequences that challenge the views by Glennan and MDC: since mechanisms can interact, some ecological generalizations need to be explained by more than one mechanistic description. This is not unique of ecology; biologists explain homeostasis as a result of a multiplicity of mechanisms operating in an organism. On the other hand, Bunge's mechanismic perspective effortlessly accommodates these features of ecological research.

Contemporary mechanismic philosophers attribute a certain degree of regularity to the global behavior of mechanisms, except for Glennan's ephemeral systems, where regularity is restricted to interactions among components. Glennan characterizes regularities methodologically, as describable by invariant, change-relating generalizations—which are general but not exceptionless. Machamer and colleagues vaguely state that mechanisms are regular "always or for the most part" and Bunge defends that mechanisms are lawful.

Yet, the term 'regularity' may refer to different concepts involving repeatability:

(i) type/token repeatability: all instances of a type-mechanism behave approximately in the same manner in similar circumstances and this allows scientists to generalize and construct classes of mechanisms. For example, cellular respiration in foxes, tigers, chimpanzees, and humans are all instances (tokens) of a type-mechanism, mammalian cellular respiration.

(ii) repeatability along time: one particular mechanism behaves similarly in similar circumstances, and this allows scientists to generalize about its behavior in similar circumstances. For example, the mechanism of cellular respiration works regularly in individual foxes as long as glucose is available to their cells.

Baumeister and Callaway (2006) describe the spatial association of *P. flexilis*, *P. menziesii*, and *R. cereus* as a fact restricted to a region, but they also state that association by facilitation may be generalized for similar kinds of plants in similar kinds of environments. Their facilitation case study reveals a methodological difference between CMP authors. All of them would consider facilitation a mechanism in virtue of the interactions that bring about the association between *P. flexilis*, *P. menziesii* and *R. cereus* easily described by laws or invariant change-relating generalizations. However, the Glennanian view would consider it an ephemeral mechanism, while in the Bungian view facilitation is an instance of a mechanism type. In virtue of their attempts to generalize, Baumeister and Callaway do not seem to view facilitation as an ephemeral mechanism, but as a case of a type of restricted generality.

Glennan and, especially, MDC choose a non-inferential conception of explanation, while Bunge sticks to the deductive-nomological (D-N) format (Table 31.2). However, invariant, change-relating generalizations play a role similar to that of laws and thus may be part of explanatory devices similar to D-N explanation.

Although frequently ecologists describe mechanisms by means of generalizations, as in our case of facilitation, their explanations do not have an explicit D-N form, unless they come from the realm of theoretical ecology, whose models do not usually describe mechanisms (Cooper 2003). Consequently, it would appear that Bunge's requirement for explanations to be deductive arguments is too strong. Yet, Bunge allows for more or less informal explanations that do not have explicit inferential form. Thus, writing about narrative explanation in biology, he uncovers the bridge between this and full-fledged mechanismic explanation:

> Viewed at a closer range, many narrative explanations are not purely descriptive, for they either tacitly imply or explicitly invoke laws, causes, and mechanisms, even though they are not stated in the form of proper deductive-nomological arguments. [...] For example, narrative explanations often conjecture some adaptive scenario. The use of the notion of

Table 31.2 Epistemological aspects of mechanismic explanation according to different authors in contemporary mechanismic philosophy

Epistemology	Glennan (2005)	Machamer et al. (2000)	Glennan (2010)	Bunge (2004)
Logical form of explanation	Non-inferential	Non-inferential	Non-inferential	Deductive
Explanandum	Mechanically explicable invariant change-relating generalization	Termination conditions	Singular datum	Scientific empirical generalization
Explanans	Invariant change-relating generalizations	Start-up and intermediate conditions	Invariant change-relating generalizations	Mechanismic laws and boundary conditions
Generality	Type-mechanism	Type-mechanism	Unique event	Type of mechanism
Multiple realizability	Yes	–	No	Only functional accounts, not mechanismic explanations
Reductive	Token reducibility, not necessarily	Not necessarily	–	Not necessarily
Intelligibility provided by	Invariance	Productive continuity	Invariance	Deductive form (superficial) and mechanism description (deep)
Counterfactual support	Yes	Yes	No	Not relevant
Research strategy	Mostly analysis	Analysis and synthesis	Mostly analysis	Analysis and synthesis

Adapted from González del Solar (2016)

> adaptation, however, implies the application of the theory of natural selection, which, in turn, refers to a (general) mechanism of evolution. Moreover, reference to mechanisms and causes, in our strict as well as in the broad sense, presupposes the existence of laws, although they may not be explicitly referred to in the narrative. (Mahner and Bunge 1997, p. 111)

This argument is similar to that of Hempel's explanation sketch (Hempel 1965), but in a mechanismic explanation the source of intelligibility is the description of the mechanism. Glennan's minimal requirement to call an explanation mechanistic is that the interactions that produce the phenomenon described in the explanandum are couched in terms of invariant, change-relating generalizations. Bunge, in turn, requires those interactions to be lawful. However, his requirement should not be interpreted according to the traditional concept of a law but to a less strict notion of a law as in the following definition:

> DEFINITION 3.9. A factual statement is a *law statement* if, and only if,
>
> (i) it is general in some respect (e.g., it holds for a certain taxon);
> (ii) it is part of a (factual) theory (hypothetico-deductive system); and
> (iii) it has been satisfactorily confirmed (for the time being). (Mahner and Bunge 1997, p. 111, italics in the original)

This definition may still seem too strict for a science like ecology where scientific laws do not abound, unless we understand the explanatory practice as a progressive process of building different kinds of conceptual devices for understanding, in particular increasingly more powerful explanatory models.

In sum, CMP still needs to refine its views on the role of generalizations and deductive inference in mechanismic explanation.

31.3.2   On the Epistemology of Mechanisms

Descriptions of ecological mechanisms are part of the description of ecological systems. The usual motivation for such description of mechanisms is the search for an explanation of the behavior of the system of interest.

Mechanismic philosophers agree that a mechanismic explanation must include the description of the fact to be explained (explanandum) and a description of the processes (interactions/activities) and the entities involved in the production of the fact described in the explanandum. The analysis of facilitation suggests that ecologists advance their understanding of ecological facts by means of conceptual devices of different explanatory power. A first understanding of the spatial distribution of vegetation in the area studied by Baumeister and Callaway (2006) was approached through a functional account: the type of mechanism that explains the spatial relations between *P. flexilis*, on the one hand, and *P. menziesii* and *R. cereus* on the other is facilitation. Yet, the concept of facilitation designates a variety of subtypes of concrete processes, among them shading, wind amelioration, litter accumulation and snow accumulation, the description of which enhances explanatory power. The more detailed the description, the more explanatorily powerful it will be—and the less general. Ecologists have recognized this trade-off between detail and generality and they usually put it in terms of a gradient with "realism" in one extreme and "generality" in the other. Cooper (2003, p. 263) provides an adequate representation of the tension between these two epistemic values (representational realism vs. generality) relating three gradients: (a) "fidelity" (from high to low), (b) abstractness (from concrete to abstract), and (c) representational power (from causal/mechanical to phenomenological models) (Fig. 31.2).

An important feature of ecological research is that while ecologists usually deal with individuals and concrete situations, their explanations are given in more or less general terms. The step from individuals to types is made by abstraction and idealization. Because of their generality, mechanismic generalizations may support counterfactuals. Counterfactual support is important for forecast and may help in experiment design.

As already mentioned, Glennan requires that the mechanical description central to a mechanistic explanation be an invariant, change-relating generalization à la Woodward (2003). This kind of generalization—a causal invariant—is characterized by being invariant under ideal interventions. This is also a counterfactual notion. Pâslaru (2009) claims that counterfactual support (an epistemic category) is a result

Fig. 31.2 The tension between realism and generality represented in a conceptual space for models. (From Cooper 2003, p. 263)

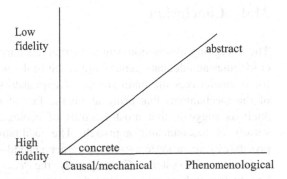

of the operation of mechanisms (an ontological category), which is mostly correct. However, counterfactual conditionals cannot be assigned truth values empirically, because they are not propositions, and this is a problem for a generalization intended to be a part of a scientific explanation (Bunge 2006). This is one of the reasons to prefer an approach to explanation couched in terms of ecological laws.

Yet, ecology does not seem to have general laws in the traditional sense of the word. Consequently, one has to choose between two possible strategies: either reject the possibility of general laws in ecology or redefine the concept of a law of nature. It is important to note that the first alternative does not amount to deem ecology unlawful. Ecology might lack ecological-level laws, but still be lawful because of all the laws at the physiological, biophysical and the biomolecular levels, i.e., there are no ecological miracles. Since ecological mechanisms are more often than not describable in terms of lower-level laws, ecological mechanismic explanation would still be using laws.

An alternative strategy would be to redefine the concept of a law of nature that, while offering some of the goods traditional laws provide, it is not as demanding as the latter. This is the avenue preferred by Woodward, Glennan, and Pâslaru, who rely on invariant, change-relating generalizations. This is also Bunge's choice, though he keeps the term 'law' of nature and defines it as an objective pattern that relates properties such that the scope of one of them is included in the scope of the other (Bunge 1998a). This would be the case of ecology if non-controversial ecological laws were finally found (for an attempt in that direction, see Dodds 2009).

What is clear is that generality is important for scientific explanation. Without generality (and repeatability) explanatory models become difficult to test and apply. Hence the interest of building models that seek generality, systematization, heuristic power, and predictive power, among other epistemic values, in addition to explanatory power.

For the time being, though, it should suffice to admit that ecological mechanismic explanation implies regularities of a certain kind and that the problem of how to precisely characterize them is still open.

31.4 Conclusion

The foregoing discussion suggests that contemporary mechanismic philosophy constitutes an adequate general approach to the problem of explanation in ecology for it emphasizes the main source of explanatory power, namely the description of the mechanisms that bring about the fact of interest. In addition, the present analysis suggests that most features of ecological facilitation support Bunge's variety of mechanismic approach. The facilitation mechanism is a network of processes—not a system—that is rather general—not universal nor unique—and that belongs in systems. Consequently, the "new mechanistic philosophy" would benefit from taking into account the contributions made by Bunge's systemism to the epistemology, but especially to the ontology, of scientific explanation.

References

Baumeister, D., & Callaway, R. (2006). Facilitation by *Pinus flexilis* during succession: A hierarchy of mechanisms benefits other plant species. *Ecology, 87*(7), 1816–1830.

Bruno, J. F., Stachowicz, J. J., & Bertness, M. D. (2003). Inclusion of facilitation into ecological theory. *TREE, 18*, 119–125.

Bunge, M. (1964/1999). Phenomenological theories. In M. Bunge (Ed.), *Critical approaches to science and philosophy* (pp. 234–254). New Brunswick: Transaction.

Bunge, M. (1979). *Ontology II. A world of systems*. Dordrecht: Reidel Publishing Co.

Bunge, M. (1981). *Scientific materialism*. Dordrecht/Boston: Reidel Publishing Co.

Bunge, M. (1997). Mechanism and explanation. *Philosophy of the Social Sciences, 27*(4), 410–465.

Bunge, M. (1998a). *Philosophy of science. Vol. 1, from problem to theory*. New Brunswick: Transaction.

Bunge, M. (1998b). *Philosophy of science. Vol. 2, from explanation to justification*. New Brunswick: Transaction.

Bunge, M. (2004). How does it work? The search for explanatory mechanisms. *Philosophy of the Social Sciences, 34*, 182–210.

Bunge, M. (2006). *Chasing reality. Strife over realism*. Toronto: Toronto University Press.

Callaway, R. (2007). *Positive interactions and interdependence in plant communities*. Dordrecht: Springer.

Connell, J. H., & Slatyer, R. O. (1977). Mechanisms of succession in natural communities and their role in community stability and organization. *American Naturalist, 111*, 1119–1144.

Cooper, G. (2003). *The science of the struggle for existence. On the foundations of ecology*. Cambridge: Cambridge University Press.

Craver, C. F. (2014). The ontic account of scientific explanation. In M. I. Kaiser, O. R. Scholz, D. Plenge, & A. Hüttemann (Eds.), *Explanation in the special sciences. The case of biology and history* (pp. 27–52). Dordrecht: Springer.

Craver, C. F., & Tabery, J. (2017). Mechanisms in science. In E. N. Zalta (Ed.), *The Stanford encyclopedia of philosophy*. https://plato.stanford.edu/archives/spr2017/entries/science-mechanisms/. Accessed 19 Dec 2017.

Dodds, W. K. (2009). *Laws, theories, and patterns in ecology*. Berkeley: University of California Press.

Germino, M. J., & Smith, W. K. (2000). High resistance to low temperature photoinhibition in two alpine, snowbank species. *Physiologia Plantarum, 110*, 89–95.

Glennan, S. (1996). Mechanisms and the nature of causation. *Erkenntnis, 44*(1), 49–71.

Glennan, S. (2002). Rethinking mechanistic explanation. *Philosophy of Science, 69*(S3), S342–S353.

Glennan, S. (2005). Modeling mechanisms. *Studies in History and Philosophy of Biological and Biomedical Sciences, 36,* 443–464.

Glennan, S. (2010). Ephemeral mechanisms and historical explanation. *Erkenntnis, 72*(2), 251–266.

González del Solar, R. (2016). Mechanismic explanation in ecology. Doctoral dissertation. Universitat Autònoma de Barcelona. https://www.educacion.gob.es/teseo/imprimirFicheroTesis.do?idFichero=66346. Accessed 28 Dec 2017.

González del Solar, R. (2017). Apuntes sobre a ontoloxia dos mecanismos. *Ágora de Orcellón, 32,* 191–212.

González del Solar, R., Marone, L., & Lopez de Casenave, J. (2014). Dos enfoques mecanísmicos de la explicación en ecología. In G. M. Denegri (Ed.), *Elogio de la sabiduría. Ensayos en homenaje a Mario Bunge en su 95° aniversario* (pp. 185–207). Buenos Aires: Eudeba.

Hempel, C. G. (1965). *Aspects of scientific explanation and other essays in the philosophy of science.* New York: The Free Press.

Machamer, P. (2004). Activities and causation: The metaphysics and epistemology of mechanisms. *International Studies in the Philosophy of Science, 18*(1), 27–39.

Machamer, P., Darden, L., & Craver, C. (2000). Thinking about mechanisms. *Philosophy of Science, 67*(1), 1–25.

Mahner, M., & Bunge, M. (1997). *Foundations of biophilosophy.* Berlin/Heidelberg/New York: Springer

Marone, L., & González del Solar, R. (2000). Homenaje a Mario Bunge o por qué las preguntas en ecología deberían comenzar con 'por qué'. In M. Denegri & G. Martínez (Eds.), *Tópicos actuales en Filosofía de la ciencia: Homenaje a Mario Bunge en su 80 aniversario* (pp. 153–178). Mar del Plata: Ed. Martín.

Pâslaru, V. (2009). Ecological explanation between manipulation and mechanism description. *Philosophy of Science, 76,* 821–837.

Resetarits, W. J., & Bernardo, J. (Eds.). (1998). *Experimental ecology. Issues and perspectives.* New York: Oxford University Press.

Ricklefs, R. E. (2008). *The economy of nature* (6th ed.). New York: W. H. Freeman & Co.

Salmon, W. C. (1984). *Scientific explanation and the causal structure of the world.* Princeton: Princeton University Press.

Woodward, J. (2003). *Making things happen. A theory of causal explanation.* New York: Oxford University Press.

Chapter 32
Bungean Systemic Ontology and Its Application to Research in Animal Parasitism

Francisco Yannarella, Mauro A. E. Chaparro,
José Geiser Villavicencio-Pulido, Martín Orensanz, and Guillermo M. Denegri

Abstract A mathematical model for understanding parasitism is proposed and developed. This model allows us to trace a distinction between the phenomena of parasitism and parasitic disease, respectively. The philosophical and theoretical framework for this model is Bunge's systemic ontology. Parasitism is understood as a biological subsystem, which can result in parasitosis or parasitic disease when destabilizing factors are present.

32.1 Introduction

Mario Bunge's systemic ontology (Bunge 1979) informs and strengthens our research on parasitism understood as a biological subsystem which is characterized by asymmetrical dynamic stability. The term "stability" refers to the reciprocal tolerance between parasite and host. "Dynamic" refers to a kind of interactive activity, defined as the set of metabolic processes that allow both the parasite and the host to exchange matter and energy with their environment. "Asymmetry" refers to the conditional aspect of the benefit; meaning that only the parasites benefits

F. Yannarella
Faculty of Veterinary Sciences, National University of La Plata, La Plata, Argentina

M. A. E. Chaparro
CEMIM, Faculty of Exact and Natural Sciences, National University of Mar del Plata, Mar del Plata, Argentina

J. G. Villavicencio-Pulido
Department of Environmental Sciences, Lerma Unit, Universidad Autónoma Metropolitana, Mexico City, Mexico

M. Orensanz · G. M. Denegri (✉)
Institute of Research in Production, Health and Environment (IIPROSAM), Faculty of Exact and Natural Sciences, National University of Mar del Plata, Mar del Plata, Argentina

National Council of Scientific and Technological Research (CONICET), Mar del Plata, Argentina
e-mail: gdenegri@mdp.edu.ar

© Springer Nature Switzerland AG 2019
M. R. Matthews (ed.), *Mario Bunge: A Centenary Festschrift*,
https://doi.org/10.1007/978-3-030-16673-1_32

575

from the relation, since it obtains resources from the host that enable it to develop its life cycle; but the parasite provides no benefit to the host. On the other hand, parasitosis is the alteration of the asymmetrical dynamic stability, which occurs when destabilizing factors intervene. This implies that not all parasitic relations generate parasitosis, to wit, parasitic diseases.

In wild, non-domesticated animals it is frequently observed that they harbor parasites without having any sort of illness as a consequence. In a previous essay (Yannarella 2014) titled *"Breves consideraciones sobre el parasitismo animal"* ('Brief Considerations on Animal Parasitism') this phenomenon is characterized as a dependent subsystem. From this line of reasoning we have arrived at the idea of attempting to construct a simulation model for the phenomenon of parasitism and, to postulate as a working hypothesis, that the most relevant variables are to be found in the interactive flow of energy between the components of the subsystemic processes. The complexity in this kind of biological systems resides in the intimacy of metabolic interactions, regulating the indispensable energetic resources for the sustainment of related life. Animal parasitism, understood as a dynamic and open subsystem, is sensitive to variations of states, which are generally compensated by the homeostatic mechanisms set in motion by the host. Parasitism-parasitosis are the phenomena that occur when there is a break in the sensitivity limit of the initial (natural) conditions, implying that minute and almost insignificant causes can produce considerable effects.

In this work we describe the construction of a mathematical model which was used to simulate animal parasitism understood as a subsystem. The construction of the model was oriented towards two clear objectives. The first objective is to determine the existence of a sustainable energetic level for the parasite-host relation by considering the energetic requirements for the maintenance of the relation itself. From the initial conditions, this energetic level or balance that results from the energetic flow between the two protagonists of the subsystemic process is used as a parameter. The second objective is to simulate the variations of state that reproduce different levels of relational commitment in agreement with the energetic demands of the parasite. The variables that were considered according to our objectives, and in relation to our theoretical framework, are the sustainable energetic level of the parasite-host relation and the evolution of relational compromise in according to the host's energetic cost.

These two variables, we believe, represent significant values for subsystemic dynamics: the dynamics of the energetic flow between the components of the parasite-host relation and their intersystemic activity, Biotope-Host-Parasite; meaning that the role of the energetic flow has a dimensional character. The changes of state that were observed in a same parasitism-parasitosis phenomenon occur in accordance with the effects of destabilizing factors that influence the energetic demand and supply; meaning that they act on the initial conditions that we have defined as asymmetrical dynamic stability.

The population of parasites presents a population dynamics that has been modeled according to a density-dependent model of a logistic type. The modification that we propose is that the rate of growth of the population is regulated by the availability of energy that the host supplies.

32.2 Philosophical Aspects

Parasitology as a biological discipline would seem, in principle, to be exempt from philosophical problems. However, when one examines with greater detail its theoretical, experimental and practical aspects, a conceptual richness is discovered, which turns it into a source of intellectual inspiration for parasitologists as well as for philosophers and epistemologists who want to study, deepen and take as an example for their research in philosophy of science. More than 50 years ago, Prof. Juan José Boero, perhaps the most conspicuous of the Argentinean parasitologists of the twentieth century, foreshadowed in his book *Parasitosis Animales* a statement that probably went unnoticed at that time and that today takes on an extraordinary relevance

> the world of parasites constitutes, perhaps one of the most exciting chapters of Biology, because it introduces the reader into the vast field of philosophical reflections. Boero (1967, p. 63)

To be sure, there are several works in the specialized literature that have dealt with the relations between parasitology and philosophy. For example, Schomaker and Been (1998) used Lakatos' (1983) philosophy of science in order to outline a Scientific Research Program for the study of plant-parasitic nematodes. Denegri (2008) used Lakatos' (1983) philosophy of science but with some modifications, to offer a philosophical foundation for parasitology in general. Likewise, Caponi (2003) claimed that the emergence of tropical medicine, based on the contributions of microbiology and parasitology, could be better understood as a Scientific Research Program.

Other authors have chosen instead, Kuhn's (1970) philosophy of science in order to dialogue with the theoretical aspects of parasitology. Ewald (1995) studied the concept of virulence within the framework of a Kuhnian paradigm in parasitology. Poulin (2000) identified a different paradigm in parasitology and he studied the way in which the concept of host manipulation is articulated within it. Dreyer et al. (2009), by recognizing a different paradigm in parasitology, examined how the case of filarial nematodes are studied with in the paradigm in question.[1]

These are just some examples of works that have articulated the two disciplines of parasitology and philosophy. Many more can be found in the specialized literature, but our point here is not to make an exhaustive review of them. Instead, we have suggested throughout this work that parasitology could greatly benefit from Bunge's (1979) systemic ontology.

---

[1] The preceding works have been reviewed in Orensanz (2017) and in Orensanz and Denegri (2017).

Fashion is also a part of philosophical activity, but we should follow the wise advice of our honored Mr. Mario Bunge who with straightforward words once wrote to us and pointed out that: *"I think you are following fashion. Remember that in philosophy, fashion seldom matches truth."* And this is to say that in general, philosophical and epistemological reflections in biology always or almost always refer to two or three disciplines that have been constituted and consolidated over time as the referents of meta-theoretical reflections, to wit: evolutionary biology, taxonomy, and population genetics, among others. That is why many believe that philosophy and parasitology have nothing to say to each other. The work that we have carried out throughout the years contradicts this statement and we share Mario Bunge's thoughts on this issue: *"... an epistemology that is not parasitic but strives to be useful to science ..."* Bunge (2009, p. 95).

The conceptual analysis allows us to elucidate a subtle but extremely important difference between the phenomenon of parasitism and parasitic disease; and this difference is important not only from a theoretical point of view, but from a practical one as well. The simultaneous presence of the parasite, the host and the environment is necessary and sufficient for the biological relationship known as parasitism to occur. On the other hand, the presence of these three entities are necessary but not sufficient for parasitic diseases to occur. For this last point, a fourth element is necessary, and this fourth element are the destabilizing factors (Denegri 2001). Therefore, what we are saying is that not all states of parasitism generate parasitic diseases, or parasitosis. In wildlife animals, it is often observed that they harbor parasites, but these do not damage the host in any way. Boero (1967) observed that many parasitic species do not cause any discomfort to the host and Johnstone (1971), studying the parasitic worms of the Patagonian animals, reached a similar conclusion, which he summed up in a famous phrase: *"happy worms in happy hosts."*

When parasites are considered to be health problems, the difference between parasitism and parasitosis becomes blurred. This not only implies underestimating theoretical problems, but also practical ones, since to treat a parasitosis it is not always sufficient to eradicate the parasites. Stated differently, one may eliminate the parasitic disease for the time being, but the parasites may still remain; if not in the hosts, at least in the external environment. This is why attention must be paid to the destabilizing factors associated with the disease. The observations of Canguilhem regarding the diseases and their treatment are in accordance with our line of reasoning: "Sometimes, even, therapeutics collaborates with the evil reinforcing what it had to weaken, multiplying what it had to reduce", (Canguilhem 2004, p. 21). And he continues: "Sometimes saying that the remedy is worse than the disease is little: the remedy is the disease itself" (p. 21).

Why is this so? Because if a massive deworming, for example, is carried out, this only eliminates the adult worms, but the environment is still contaminated with the parasitic worm's eggs. Even more so, a massive deworming may even increase the number of eggs in the environment. Far from permanently eliminating the worms from the scenario altogether, in this case the remedy multiplies and even reinforces the destabilizing factors.

According to the Bungean systemic approach, not only new levels of organization and their corresponding properties emerge, but new systems emerge. Instead of talking about emerging properties, we are talking about emerging systems instead. Each system has a composition, a structure, several mechanisms and an environment. What we call "parasitism" is an emerging biosystem. Its composition has two elements, the parasite and the host. Its structure is characterized by the relationship between the parasite and the host, a relationship whose fundamental emergent property is asymmetric dynamic stability. The environment of the parasite-host structure is the external environment.

Systemic ontology, as opposed to reductionism and applied to the interpretation of the phenomenon, explains the parasite-host relationship as a subsystem with properties that are not reducible to its components. Parasitism is defined as the interaction between the parasite, the host and the appropriate environment or biotope. Parasitism emerges as an emergent relational phenomenon. The passage from parasitism to parasitosis is determined by the presence of destabilizing factors (Denegri 2001, 2008; Yannarella 2014).

32.3 Building the Mathematical Model

The modeling of the interaction between host-parasite will be understood from a systemic point of view. The relation between parasites and hosts, in this model, is based on the consumption of energy, which for the host represents harboring a population of parasites that consume the energy in question, incorporated through feeding. Without considering the host's type of feeding habits, it is assumed that there exists a daily acquisition of energy. This is used for basal maintenance, growth, reproduction, and/or reserve.

The growth of the population of parasites is a function of the quantity of the host's energy, which is incorporated through the ingestion of food, and it self-regulates in relation to the available energy. In this work we propose a general model in which the maximum amount of energy which the host is able to accumulate will be modified as a response to the number of parasites that it harbors.

The proposed model for describing the interaction between parasite and host is represented by a system of differential equations. The system describes the population dynamic of the parasites ($P(t)$), and the energy that is present in the host, ($H(t)$). The amount of energy that is registered in the instant "t" is established in this model in a direct way in relation to the host's state, in which low values of $H(t)$ are associated with the host's impoverished conditions, and null values are associated with the host's death.

With these hypotheses it is stated that the dynamics of the two populations is modeled by:

$$\frac{dH}{dt} = r_H H \left(1 - \frac{H}{K_H(P)}\right) - \beta P H,$$
$$\frac{dP}{dt} = r_P(H) P \left(1 - \frac{P}{K_P(H)}\right) - dP.$$

Where r_H and $r_p(H)$ are the intrinsic rates of growth for $H(t)$ and $P(t)$ respectively. $K_H(P)$ and $K_P(H)$ are the carrying capacities regulated by the population of parasites and the available energy, respectively. The value β represents the rate of consumption of energy on behalf of the parasites. The value d is the rate of "natural" mortality of the population of parasites.

According to this model, the population dynamics of the parasites are associated with the amount of available energy that the host has, which in turn regulates the maximum amount of parasites as well as the intrinsic rate of growth. In addition, in this model we also take into consideration an additional rate of mortality of the parasites, which is independent of the amount of available energy.

In an analogous fashion to the preceding scenario, it can be stated that the amount of energy $H(t)$, is a function of the number of parasites in the host. This is described by the burden or carrying capacity in relation to the parasites. In addition, a decrease in energy is due to the fact that part of the energy is consumed by the parasites.

With this model, it is possible to propose different descriptions for the behaviors of the host in relation to its parasites. We seek to study the host's response in relation to its capacity for accumulating energy, and the presence of a given amount of parasites. This being so, we seek to study this reaction in a linear way, which means:

$$K_H(P) = K_H + \delta P$$

It is worthwhile to emphasize that with this function, there are three situations that can be derived in relation to the value of δ:

1. If $\delta > 0$ then the hypothesis is that there will be a decrease in the accumulation of energy as a response to the growth of the population of parasites.
2. If $\delta = 0$ then the hypothesis is that the presence of parasites does not produce any change in the host's habits.
3. If $\delta < 0$ then the hypothesis is that there will be a decrease in the accumulation of energy as a response which attempts to reduce the population of parasites.

In each of these models we will study the conditions for the existence of a state of balance, or equilibrium and, if possible, for their stability (Fig. 32.1).

32.4   Results

In each of the cases that will be analyzed, we consider that $K_P(H)=1+kH$ models the burden capacity of the host, which is a lineal response of energy. We also establish that $r_p(H) = r_p H$ will be the intrinsic rate of growth in the population of parasites. The objective of that formulation is that, in the case of the host's death,

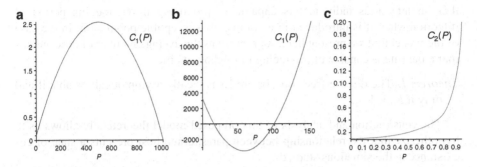

Fig. 32.1 Section (**a**) shows the behavior of $C_1(P)$ in the scenario in which $\delta > 0$, while Section (**b**) shows the case in which $\delta < 0$. Section (**c**) shows the behavior of $C_2(P)$

$H(t)=0$, the population of parasites will start to extinguish at a rate "d". With these specifications, the model is defined by:

$$\frac{dH}{dt} = r_H H \left(1 - \frac{H}{K_H + \delta P}\right) - \beta H P,$$
$$\frac{dP}{dt} = r_P H\, P \left(1 - \frac{P}{1+kH}\right) - d\, P.$$

The equilibrium points in the model will be those in which the system becomes simultaneously null. By an analytic resolution, the equilibrium points will be those that are simultaneously in accordance with:

$$r_H H \left(1 - \frac{H}{K_H + \delta P}\right) - \beta H P = 0$$
$$r_P H P \left(1 - \frac{P}{1+kH}\right) - d P = 0$$
$$\Rightarrow \left\{\{P = 0\}, \left\{P = \frac{H^2 k r_P + (r_P - dk)H - d}{r_P H}\right\}\right\}$$

$$\left\{\{H = 0\}, \left\{H = \frac{-P^2 \beta \delta + (\delta r_H - K_H \beta)P + K_H r_H}{r_H}\right\}\right\}$$

From here, it is easy to determine the following:

- Trivial equilibrium (extinction scenario): $E_0 = (0, 0)$; and
- Equilibrium in the absence of parasites: $E_1 = (K_H, 0)$.

These can be identified as "disease-free equilibriums".

The analysis of stability demonstrates that the equilibrium E_0 is always unstable for all the values of the parameters, while the equilibrium E_1 is locally asymptotically stable if and only if $r_P K_H - d < 0$.

At this stage we define $R_0 = \frac{r_P K_H}{d}$. This gives the following result. In mathematical models for immunology and epidemiology the parameter R_0 is called basic reproductive number, which represents the number of secondary infections that an infectious individual is capable of producing throughout its period of infection when it is introduced in a completely susceptible population. In the case of the model that we present here, R_0 represents the average number of descendants that a parasite is capable of producing throughout its life.

Theorem 1: The disease-free equilibrium E_1 is locally asymptotically stable if and only if $R_0 < 1$.

The construction and analysis of this proposed model theoretically allows us to hypothesize that the relationship between parasite and host presents a state of co-existence in the situations studied.

32.5 Conclusion

Synthesizing our view on animal parasitism, we can affirm that systemism allowed us to approach the phenomenon without the prejudices that are typical of reductionism. The reductionism to which we are referring to hinders the interpretation of the context as a functional whole. The concept of biosystem allows us to explore the phenomenon from an integrating perspective. This conceptualization shows a reality conformed by apparent disconnections, that in reality are linked by the underlying structured interaction. The mathematical modeling that we presented above allows us to clearly define both parasitism and parasitosis. This in turn allows us to reflect on the strategies to be defined when we are in the presence of a practical problem that we must solve in the field of medical and/or veterinary parasitology. If the development of these mathematical models would allow us in the future to further distinguish with greater clarity the subtle difference between parasitism as a biological phenomenon and parasitosis as a disease, then we could elaborate much more effective strategies for controlling and/or eradicating those parasitic species that cause pathologies in humans and in animals.

Bunge's systemic ontology provides conceptual structure in the elaboration of current knowledge of parasitism, and guidance from a sub-systemic and biological-relational perspective for future research in the domain.

References

Boero, J. J. (1967). *Parasitosis animales*. Buenos Aires: EUDEBA.
Bunge, M. (1979). *Treatise on basic philosophy volume 4. Ontology II: A world of systems*. Holanda: D. Reidel Publishing Company.
Bunge, M. (2009). *La Ciencia, su Método y su Filosofía*. Buenos Aires: Sudamericana.
Canguilhem, G. (2004). *Escritos sobre la medicina*. Buenos Aires: Amorrortu Editores.
Caponi, S. (2003). Coordenadas epistemológicas de la medicina tropical. *História, Ciencias, Saúde – Manguinhos, 10*(1), 113–149.

Denegri, G. (2001). *Cestodosis de herbívoros domésticos de la República Argentina de importancia en medicina veterinaria*. Mar del Plata: Editorial Martín.

Denegri, G. (2008). *Fundamentación epistemológica de la parasitología- Epistemologic Foundation of Parasitology* (Bilingüe ed.). Mar del Plata: Editorial de la Universidad de Mar del Plata (EUDEM).

Dreyer, G., Mattos, D., Figueredo-Silva, J., & Norões, J. (2009). Mudanças de paradigmas na filariose bancroftiana. *Revista da Associação Médica Brasileira, 55*(3), 355–362.

Ewald, P. W. (1995). The evolution of virulence: A unifying link between parasitology and ecology. *The Journal of Parasitology, 81*(5), 659–669.

Johnstone, I. L. (1971). *Enfoque ecológico para el control de la parasitosis ovina*. Buenos Aires: INTA.

Kuhn, T. S. (1970). *The structure of scientific revolutions* (2nd ed.). Chicago: University of Chicago Press.

Lakatos, I. (1983). *La metodología de los programas de investigación científica*. Madrid: Alianza Editorial.

Orensanz, M. (2017). Thomas Kuhn y la helmintología. *Análisis Filosófico, 37*(1), 55–77.

Orensanz, M., & Denegri, G. (2017). La helmintología según la filosofía de la ciencia de Imre Lakatos. *Salud Colectiva, 13*, 139–148.

Poulin, R. (2000). Manipulation of host behaviour by parasites: A weakening paradigm? *Proceedings of the Royal Society of London B: Biological Sciences, 267*(1445), 787–792.

Schomaker, C. H., & Been, T. H. (1998). The Seinhorst research program. *Fundamental and Applied Nematology, 21*(5), 437–458.

Yannarella, F. G. (2014). Breves consideraciones sobre parasitismo animal. In G. Denegri (Ed.), *Elogio de la sabiduría. Ensayos en homenaje a Mario Bunge en su 95° aniversario* (pp. 323–335). Buenos Aires: EUDEBA.

Part VIII
Mathematics

Chapter 33
Bunge's Mathematical Structuralism Is Not a Fiction

Jean-Pierre Marquis

Abstract In this paper, I explore Bunge's fictionism in philosophy of mathematics. After an overview of Bunge's views, in particular his mathematical structuralism, I argue that the comparison between mathematical objects and fictions ultimately fails. I then sketch a different ontology for mathematics, based on Thomasson's metaphysical work. I conclude that mathematics deserves its own ontology, and that, in the end, much work remains to be done to clarify the various forms of dependence that are involved in mathematical knowledge, in particular its dependence on mental/brain states and material objects.

33.1 Introduction

Bunge's position on the ontological status of mathematical objects has been clear from very early on and has not essentially changed since his first publication on the subject: mathematical objects do not *really* exist.[1] This negative claim is constant throughout his work and follows directly from his materialist outlook. But as Bunge himself has noticed many times, negation comes cheap. And one and the same negative claim can be consistent with many different, incompatible positive theses. Bunge's positive claims, for there is more than one, are somewhat more difficult to put together into one coherent position. Sometimes, Bunge says that mathematical objects *formally* exist. At other times, he emphasizes that they are creations of the human brain and thus, exist only in these brains. Therefore, their mode of existence

---

[1] It seems that the first implicit expression of this claim came in the first volume of Bunge's *Treatise* (See Bunge 1974). The detailed exposition appeared in the first part of the seventh volume of the *Treatise* (See Bunge 1985). The fundamentals of the position have not changed since then (See Bunge 1997, 2016).

J.-P. Marquis (✉)
Département de philosophie, Université de Montréal, Montréal, Canada
e-mail: Jean-Pierre.Marquis@umontreal.ca

© Springer Nature Switzerland AG 2019
M. R. Matthews (ed.), *Mario Bunge: A Centenary Festschrift*,
https://doi.org/10.1007/978-3-030-16673-1_33

is a mode of dependency: the existence of mathematical objects depends upon the existence of brains, human brains in communities. At other times, Bunge says that mathematical objects are *fictions*. The goal of this paper is to explore further the type of existence that characterizes mathematics according to Bunge and see whether and how these positive views can be put into a coherent whole.

33.2 Mathematics as a Science of Structures

Bunge's most explicit and complete presentation of his views on mathematics are in volume 7, part I, of his *Treatise* (Bunge 1985). I will not systematically go over the basic elements of his position (See Marquis 2011 for an overview and a critical presentation). I will focus on the relevant components for my presentation to be self-contained.

Bunge claims that contemporary mathematics is a *formal research field*. This means, basically, that its products are the result of a community of specifically trained individuals who share common methods, techniques, theories and who aims at solving a well-identifiable class of problems with those methods, techniques and theories and, by doing so, produce new definitions, theories, proofs, examples, counter-examples and algorithms according to certain standards of rigor.[2] Thus, mathematics is a science, but a science whose objects are *constructs* or *concepts*. Since the latter never come isolated, but in systems, mathematics is the study of *conceptual systems* (Bunge 1985, p. 19).

More precisely, mathematics studies the *structure* of these conceptual systems. Thus, and as far as I know this point has never been underlined even by Bunge himself, Bunge is endorsing a form of *structuralism* in mathematics. Here is one passage where Bunge explicitly expresses this view, worth quoting in full:

> Historians of mathematics have noted that, until around mid-nineteenth century, the bulk of mathematical research was concerned with individual constructs, such as particular figures, equations, functions, or algorithms. From then on, and particularly since the mid-20th century, mathematics has been conceived as the study of *conceptual systems*, such as groups of transformations (or even the whole category of groups in general), families of functions (or even entire functional spaces), and topological spaces (such as metric spaces in general) (Caution: Bourbaki, Bernays and others call 'structure' what others, e.g. Hartnett 1963, call 'system'. We stick to our convention … that every structure is the structure *of* some object: that it is the set of all relations among its components – the internal structure – plus those among the latter and the environment or context of the system, which can be empty – the external structure). (Bunge 1985, p. 19)

These systems take a more precise form in mathematics:

> Every mathematical system ("structure") can be characterized in either of two ways: (a) as a set equipped with a structure consisting of one or more operations or functions defined on

---

[2]In the present paper, 'mathematics' always refer to *contemporary* mathematics, for this is what Bunge has in mind in his analysis.

that set; (b) as a collection of objects together with one or more morphisms relating those objects – i.e. a category. Actually the second concept subsumes and supersedes the first). (Bunge 1985, pp. 19–20)

At first sight, Bunge seems to be endorsing either a kind of set-theoretical structuralism or a category-theoretical structuralism, though the foregoing quote suggests that he favors a category-theoretic point of view. In fact, the choice between them does not seem to be an issue. Bunge falls back on the current languages of mathematics, his position being that mathematics is about conceptual systems, no matter the particular language used. Indeed, Bunge always comes back to conceptual systems and never develops the set-theoretical or the category-theoretical perspectives. Thus, on the very next pages of the *Treatise*, one reads:

> A moment ago we stated that mathematics studies conceptual systems ("structures"). However, this is only a necessary condition: philosophy and the history of ideas too study conceptual systems, such as cosmologies and mathematical theories. (Bunge 1985, p. 21)

This is an odd claim to make, considering that mathematics studies the structure of conceptual systems. Does philosophy and the history of ideas study the structure of conceptual systems *as such*?

> What makes the mathematical study of conceptual systems unique is that (a) it is *purely conceptual* (i.e. does not make essential use of any empirical data or procedures) and it involves, at some point or other, (b) *positing or conjecturing the laws* (general patterns) satisfied by the members of those conceptual systems, as well as (c) *proving or disproving conclusively* some such conjectures (. . .).

> We may then define *contemporary pure mathematics* as the investigation, by conceptual (a priori) means, of problems about conceptual systems, or members of such, with the aim of finding (. . .) the patterns satisfied by such objects – a finding justified only by rigorous proof. (Bunge 1985, p. 22)

There is no mention of sets or categories in this definition. There is no mention of mathematical objects. Notice the very last claim: "finding . . . the *patterns* satisfied by such objects". We are back at structures. Mathematics is the study of the structures of conceptual systems. It is not entirely clear how mathematics differs from logic according to this definition. Bunge would probably fall back on sets or categories at this point, since these particular concepts allows us to move away from pure logic.

Why has Bunge not developed more explicitly a form of structuralism? In particular, why does Bunge fall back so quickly on mathematical *objects*? On page 23 of the same volume, for instance, we read: "The last problem in our agenda in [sic] whether mathematical objects are discovered or invented." Shouldn't Bunge ask whether *structures* are discovered or invented? Clearly, Bunge does not want to reify structures. As we have seen already, they are always structures of systems and systems are fundamentally made up of objects. That is a basic axiom of his ontology. In the case of conceptual systems, concepts are the objects. Clearly, a conceptual system is itself a concept and, thus, we are not really reifying structures. We are still talking about concepts and constructs. Furthermore, could it be that mathematics *is* formal, precisely because it deals with structures or relations *as such*? Of course, in the end, we will still have to turn to issues of existence, namely the existence of constructs or concepts. We would have nonetheless avoided the whole discussion surrounding the status of mathematical *objects*.

33.3 Mathematics as an Art?

Indeed, bringing mathematical objects in the picture suddenly brings contemporary mathematics closer to art than to science! To wit:

> ...in our view, mathematics is closer to art than to science as regards its objects and its relation to the real world, as well as regards the role of truth. (Bunge 2006, p. 195)

Mathematics is closer to art than to science from the ontological, epistemological and semantical perspectives! Logic seems to be saving the day ... Mathematics is decidedly a singular research field. Although the epistemological and the semantical components would deserve a careful treatment, I will simply ignore them in the present paper. From now on, only ontology matters.

Ontologically, mathematical objects and "artistic objects" are on the same plane.

> Mathematical objects are then ontologically on a par with artistic and mythololological [sic] creations: they are all *fictions*. The real number system and the triangle inequality axiom do not exist really any more than Don Quijote or Donald Duck. (Bunge 1985, p. 38)

We moved from the structure of conceptual systems to mathematical objects and Donald Duck. Again:

> In short, mathematicians, like abstract painters, writers of fantastic literature, 'abstract' (or rather uniconic) painters, and creators of animated cartoons, deal in fictions. To put it into blasphemous terms: ontologically, Donald Duck is the equal of the most sophisticated nonlinear differential equation, for both exist exclusively in some minds. (Bunge 2006, p. 192)

Donald Duck is not the problem. And it is not a priori ridiculous to compare Donald Duck to mathematical objects with respect to their ontological status. It is, in fact, rather fashionable these days and has been for some time. It certainly goes in the right direction, but one has to travel carefully to avoid certain pitfalls.

Of course, mathematics is *not* a form of art, despite the foregoing ontological, epistemological and semantical closeness to art. Bunge gave a list of ten differences between the two in numerous publications (See Bunge 1985, pp. 39–40, 1997, pp. 63–64, 2006, pp. 204–205. In the last publication, the list contains a few more elements. These were mentioned in the main text of the previous versions. The core of the list has not changed at all between these publications). We will not go over the differences presented by Bunge. Suffice to say that one element, already mentioned, stands out: Bunge insists on the *necessary* role of reason through logic in mathematics. Thus, it is tempting to say that one of the main differences between mathematical objects and artistic fictions is that whereas both are human creations, products of the imagination, the former is strictly bound by reason in its creation, developments and justification and the latter is not. It is this central role of reason that brings mathematics closer to the scientific territory. Reason is at the core of mathematics. Mathematics *must be* rational. It is the only conceptual domain, together with logic, that *can be fully and autonomously* rational. Even mathematical existence is bound by reason. Rationality is built-in. This is not to say, of course, that mathematics *is* logic. In contrast, artistic fiction does not have to be and, perhaps, cannot be[3] fully and autonomously rational.[4]

33.4   Real Fictions and Mathematical Fictions

I want to focus on the claim that mathematical objects are fictions. As such, this is highly ambiguous and could mean many different things.[5] Do fictions differ from ideal objects? Abstract objects? Imaginary objects? Does Bunge use the term in a different sense than, say, Leibniz when the latter talks about the fact that infinitesimals are fictions? In light of the last paragraph, should Bunge develop a philosophical theory of *logical* fictions and say that mathematical objects are logical fictions? This would make mathematical objects a special *kind* of fiction. I submit that, in fact, the idea of fiction does *not* play a central role in Bunge's philosophy of mathematics and that it could very well be dispensed with.[6]

The use of 'fiction' in Bunge's philosophy has two main purposes and they are, in my mind, rhetorical. The first one, already mentioned many times, is to point out that mathematical objects and fictions might very well be *only* constructs, creations of human brains. The second one is to fall back on pretense, on our capacity to treat certain concepts *as if* they had an autonomous existence, thus explaining the prevalence of various forms of Platonism among mathematicians.

> And in a widely publicized interview, the Princeton professor William Thurston stated that "Theorems just kind of exist, you know, just like mountains do". In our view this is an intelligent mistake. It is a mistake because formal existence is radically different from

---

[3]It could certainly be said that art *should not* try to be. An interesting question is where philosophy stands in this framework. Philosophy does not have the same conceptual autonomy as mathematics.

[4]This is not to say, of course, that mathematics and art have nothing in common. Historically, mathematics has been associated to a *technè* and I, for one, have argued that a large part of contemporary mathematics should be thought of as a systematic technology (Marquis 1997, 2006). I am here concentrating on the idea that mathematical objects and certain artistic objects, mostly literary 'objects', should be subsumed under the ontological category of *fictions*.

[5]Already in 1981, before the publication of Bunge's volume 7 of the *Treatise*, Roberto Torretti had already identified three different kinds of mathematical fictionalism (Torretti 1981). I must confess that I do not understand his classification and will therefore refrain from using it. His claim that Bunge's position might, in the end, be a form of idealism, is, however, not ridiculous. See also Robert Thomas' excellent papers on fiction and mathematics (Thomas 2000, 2002).

[6]In this particular regard, Bunge's position is not very different from what is now called 'mathematical fictionalism' in the literature. More about this link or, to be more exact, its absence, in the next section.

material existence: . . . . But it is intelligent because, as a matter of fact, the mathematician behaves in many regards *as if* constructs existed by themselves. He can do so because mathematical constructs, though human creations, do not bear the stamp of their creators: they are impersonal or intersubjective (though not objective). (Bunge 1985, p. 111)

Thus, Bunge wants to be able to resolve what seems to him to be a 'tension' between two poles: (1) the fact that mathematics is a creation of human brains *and*; (2) the fact that so many mathematicians describe mathematics and its objects as being totally independent of these human brains. Since everybody is familiar with fiction, at least literary fiction like stories, novels and plays, and that fictional character seem to typically resolve that kind of tension – they are undoubtedly created by humans *and* have a certain kind of autonomy –, it appears to be a convincing solution. *Voilà!* End of the story.

Be that as it may, Bunge does not rely on a theory of fiction to clarify the nature of existence of mathematical constructs. He does say quite a few things about existence and existence of mathematical constructs. It *is* surprising that he does not attempt to *derive* the desired properties of mathematical constructs from his conception of formal existence. That would allow him to avoid all reference to fictions, which, in the end, he seems unable to avoid for the reasons just mentioned. There seems to be something wrong with his notion of conceptual existence. Let us see.

33.5 Modes of Existence

Bunge has always resisted attributing any ontological traction to the existential quantifier and, in particular, Quine's approach according to which 'to be is to be the value of a bound variable' in a well-establish scientific theory.[7] He has steadfastly defended the idea that existence had to be represented by a specific predicate and, in contrast with a large literature in contemporary metaphysics, he has also claimed that there are different modes of existence.

To be real, for Bunge, is to be material or, in other words, it is to be mutable or changeable. This is his definition of *real existence*. Bunge used to oppose the latter to *formal existence*. However, he has recently introduced five different modes of existence: real, phenomenal, conceptual, semiotic, and fantastic (See Bunge 2016).

---

[7]The latter criterion is at the source of the vast literature on mathematical fictionalism. (See, for instance, Field (1980, 1989), Balaguer (1998), Yablo (2002), and Leng (2009)). Indeed, this criterion together with the so-called Quine-Putnam indispensability argument, seemed to provide good reasons for a certain form of Platonism with respect to mathematical objects. In this context, the claim that mathematics is a fiction is taken to follow from the claim that mathematics, like fiction, is not literally true, precisely because in both cases, these discourses literally fail to refer. Bunge has always resisted these Quinian arguments and he also very quickly pushed aside these fictionalist strategies, which he considers to be forms of nominalism and finds inadequate. Lately, Quine's arguments have been criticized and therefore the motivation for this form of mathematical fictionalism has somewhat shifted. See, for instance, Thomasson (2014).

Although it might appear to be a modification of his views, the basic distinction remains between the first mode and the four remaining modes: only the first mode, real existence, is absolute and context-independent; the others are relative and context-dependent. I set aside the real, the phenomenal and the semiotic and focus on the conceptual and the fantastic.

It is worth recalling Bunge's definition of a construct, as presented in the third volume of the *Treatise* Bunge (1977).

Definition: *x* is a *construct* if, and only if,

1. There exists (really) an animal capable of conceiving *x*;
2. The animal conceives *x* as a conceptual system or a member of such.

In other words, a thing is a construct if it can be thought by an animal as a conceptual system or in a conceptual system. Bunge treats this as if it clarified the notion of formal existence. As far as I can tell, this is only a definition and says nothing about existence.

A series of remarks is in order. First, the definition is two-dimensional. The notion of construct *depends* upon two ontologically perpendicular realms: (1) The realm of really existing animals and (2) The realm of other constructs.

Second, the definition presupposes distinctions between conceiving, thinking, feeling, imagining, desiring, etc. It is not clear, at least to me, that what I imagine is not a construct. If it is not, what is it? Are internal visual images constructs in Bunge's sense? Internal musical melodies? The last two do not belong to conceptual systems. Or, do they? In what sense?

Third, if it is taken as a definition of formal existence, I am not sure I see what we can infer from it. Since it is plausible to imagine that as soon as a human being capable of thinking was alive, that human was *capable* of conceiving mathematics, can we conclude that all mathematical concepts came into existence from that moment on? It all depends what one means by 'capable of conceiving' and 'conceived'. And, of course, no one can verify that mathematical objects came into existence that way.

Fourth, should we add that there was at least one animal that actually or really conceived a particular concept, at least once? Is it enough that only one animal can conceive it? Does it have to be communicated? Or communicable? If so, what does that presuppose, cognitively, culturally and socially?

Fifth, the animal has to conceive the construct *x* as a system or part of a system. Is it possible *not* to? What would that mean? How is that possible? The notion of an isolated construct seems to be an oxymoron.

Sixth, what about certain concepts that are part of mathematics but that seem to go beyond our capacities of conceiving them at a certain time? The history of mathematics is filled with examples of such constructs: 0, negative numbers, imaginary numbers, non-Euclidean geometries, higher-dimensional geometries, infinities, both great and small. For long periods of time, our best minds thought they had good reasons to doubt the existence/legitimacy of these mathematical objects, even though they were *capable* of conceiving them. For most of these objects and constructs, we now have fully developed theories and very few mathematicians (none?) would nowadays contest the existence of negative or complex numbers, not to mention quaternions, octonions, p-adic numbers, etc.

Seventh, what happens when what seemed to be a perfectly good theory turns out to be inconsistent? Do the constructs of that theory suddenly fail to exist? How does the discovery of an inconsistency affect formal existence? Do we have to suppose consistency?

Eight, to be capable of conceiving x, very often one has to master various cognitive tools, e.g. a written language, certain preliminary concepts y, z, etc. Thus, the capacity to conceive something might in fact depend on a complex network of concrete and conceptual technologies. Whence, conceptual existence might depend on much more than the existence of an animal and its cognitive capacities.

Ninth, one could actually imagine that the (independent) existence of these concrete and conceptual technologies could play a role in the conviction that the constructs that one finally conceives have an independent existence, simply because their existence depends upon a complex network of preexisting concrete and formal entities whose existence is hard to comprehend by a single animal.

Bunge then gives a specific definition of *mathematical existence*: if x is a construct, then x *exists mathematically*$=_{\text{df}}$ For some C, C is a set, class or category, such that (i) x is in C, and (ii) C is specified by an exact and consistent theory (Bunge 1985, p. 30).

We now have our answer for objects belonging to an inconsistent theory: they do not exist mathematically.[8] Bunge now fixes an underlying mathematical ontology, to be (mathematically) is either to be a set, class or a category or be in such an object (in a coherent fashion). We have moved from the existence of constructs *as such* to the existence of *mathematical* constructs *in mathematical* theories. These are very different cases and, from the philosophical point of view, the latter is well understood by mathematicians. Mathematicians know what they mean when they claim that such and such exist in such and such theory. There are *some* debates as to the methods that are legitimate to establish this existence of *some* objects and the nature of these debates is well understood, e.g. infinitely large cardinals for intuitionists and infinitesimals for some classical mathematicians.[9] It is, so to speak, an internal affair. Recently, Bunge has proposed that to exist conceptually means to be a constituent of a conceptual system (See Bunge 2016, p. 228). What about conceptual systems themselves? I suspect that Bunge's answer is the same as the one given in the *Treatise*: conceptual systems are usually parts of larger conceptual systems thus they exist whenever the latter is the case. Surely, this chain must end somewhere: there must be a conceptual system that contains them all. Bunge says that the latter question "makes no more sense than the question "Where is the

---

[8]Bunge is of course well aware that we cannot *prove* the consistency of most of our mathematical theories, in particular set theory and for a foundational categorical theory. Should we conclude that we simply cannot *know* that, in the end, our mathematical constructs exist?

[9]It is well known that Cantor and Russell resisted the introduction of infinitesimals for purely ideological reasons, even when they were perfectly acceptable objects in algebra at the time (See Ehrlich 2006).

physical universe?"'" (Bunge 1985, p. 30). The question of an overall mathematical conceptual system is *the* question of the foundations of mathematics, which we will leave aside.

At the end of the day, Bunge falls back on his fictionist stance. Thus, in his most recent publication, he claims again that "from the fictionist viewpoint, the debate over constructivity is a storm in a teapot. Indeed, whether or not there is a constructive proof of a given mathematical object, this is just as fictitious as Zeus or a talking dog" (Bunge 2016, p. 230). Thus, provided that we keep in mind the constraint brought by consistency, mathematical existence boils down to fantastic existence. According to Bunge, something exists fantastically if there is a work of fiction that contains or suggests that thing (Bunge 2016, p. 231). Bunge is clear that this type of existence includes music, plastic art, artistic cinema, as well as mathematics. Indeed, Bunge claims that "mathematicians and theoretical physicists are professional fantasizers. But their fantasies, unlike those of Hyeronimus Bosch or Maurits Escher, are bound by reason" (Bunge 2016, p. 232). Is Bunge being merely provocative here or does he *really* believe that mathematicians and theoretical physicists are, to use an image, rational writers?

I claim that Bunge does not need to invoke fictions at all and that, by doing so, he brings in unnecessary difficulties. Bunge *is* right that fictions and mathematics are both *abstract artifacts* and, as such, have an ontological status that differs from, say, electrons and trees. Thus, we preserve the basic ontological claim that the existence of mathematical concepts is a *dependent* existence. This, of course, is true of all abstract artifacts. I claim, however, that the *type* of dependence of mathematical concepts is not the same as the type of dependence of fictions. Moreover, we do not need the pretense, the *as if*, to make sense of their existence and their properties. One only needs to understand mathematics, how it is learned and how it is done, without any pretense.

33.6 Doubts About Fictions

There is a huge literature on fictions and their properties which I cannot do justice to in such a short paper. There is, however, one objection brought forward recently by Amie Thomasson that hits a soft spot.

When we deal with fictions, we do indeed *pretend* that certain objects or events are such and such. As I have indicated, this is one of the reasons Bunge appeals to fictions and draws a parallel between mathematical objects and fictions. For instance, when we go to the opera to see a performance of Puccini's *La Bohème*, we pretend that what we see is taking place in Paris around 1830. Of course, we are not committed to the claim that we actually *see* Paris on the stage. We pretend that the singer who plays *Mimi* dies, etc. We are willing to say that Mimi does die *in the opera*, although we know that the singer impersonating her does not. We are not committed to the *actual* death of the singer. What this pretense amounts to and how it varies has been analyzed in the literature and we do not need to go into details for our purposes. If we can pretend that someone dies on a stage, then we can certainly pretend that the Monster Group exists on some stage, so to speak. So far, so good.

The objection then goes as follows: when dealing with a fictional discourse, we *know* when we pretend and we know when we do not. In other words, we know which parts of the discourse have to be taken *as if* they were true and which ones are to be taken literally, e.g. that Puccini composed *La Bohème* is literally true, that Mimi dies is not.[10] Not knowing how and when to make the difference is a sign of psychopathology. If mathematical objects are fictions, we should be able to make similar distinctions in this case too. The fact is, we do not. Here is how Thomasson herself puts the objection:

> In the case of works of fiction or children's games of make-believe, there is a clear contrast to be drawn between committing oneself to the real content (the truth about the props) and committing oneself to the literal content: a difference between being committed to stumps versus bears, words on pages versus deaths on train tracks. That difference, however, is not obvious for the fictionalist about disputed ontological entities such a social entities, numbers, events, and properties. Committing oneself to the vows and paperwork being undertaken does seem to commit oneself to being married. Similarly, to the extent that it sounds redundant in English to say "there are five stumps *and* the number of stumps is five", being committed to the first claim does seem to commit one to the second, and so to there being a number. ... But then we cannot (...) take the latter claims, explicitly about numbers or propositions, to be *merely* pretending while the former are committing. (Thomasson 2014, p. 190)

In other words, in the case of natural numbers, the Monster Group, the homotopy type of the 3-sphere, is there a difference between the pretense and some underlying, literally true discourse? When I read about the Monster Group, I don't say to myself "I know that there is no such thing as the Monster Group, there is only these marks on paper. But the story (theory) says that such and such is true and I am willing to pretend that it is true, although I know that it is not." In fact, Bunge is suggesting that most mathematicians, at least those like Thurston, are pretending, but are not aware of the pretense. In the foregoing quote, Bunge says that mathematicians *behave as if* mathematical objects existed by themselves. Thus, Bunge explains mathematicians' behavior by attributing them an attitude towards the objects of their thought. Is this some sort of anthropological explanation? We study how mathematicians behave and the best explanation we come up with is that they pretend that what they talk about really exist. When we question them, they are not aware of this pretense. Notice that Thurston did not say "I know that mathematical objects are not like mountains, but I pretend they are. It allows me to do beautiful mathematics." However, he *did* say "*kind* of exist *like* mountains", indicating that he is well aware that there *is* a difference between the two, an obvious difference.

There is an additional puzzle. There *are* mathematicians who claim that mathematical objects are constructed and mental entities. They might even accept the

---

[10]Of course, that is one of the reasons why art is so powerful: even though we *know* that it is all a pretense, we *feel* emotions just as strongly as when it *is* real. Some people simply *cannot* watch horror movies, although they *know* they are watching movies, i.e. fictions.

claim that mathematical objects are fictions. However, I suspect that they would resist the statement that we thus pretend that they exist autonomously and can therefore be treated like any ordinary or real objects, that is ordinary classical logic can be used without restrictions. Imagine asking an intuitionist: why can't you pretend that mathematical objects exist autonomously, like all other mathematicians do, and use classical logic? Imagine telling an intuitionist: you might have not noticed this, but your classical colleagues do not *really* believe that mathematical objects exist, they only *pretend* that they do. As far as I am aware, neither the intuitionist nor the classical mathematician *decide* to pretend or know that they *have to* pretend.

I don't know what it would mean, in the case of mathematics, to stop pretending and fall prey to the literal interpretation of mathematical discourse (which, for ordinary fiction, leads us towards psychopathology). Would someone start attributing *real* properties to mathematical objects? Would the Monster Group be *really* frightening? We never have to tell our children "well, you know, we are sorry, we never told you this, but in fact, numbers do not *really* exist", whereas those of us who have decided to do as if Santa Claus existed had to have a conversation or at least make a verification at some point that our children have picked up on reality. Some children are *really* disappointed to learn that Santa Claus does not exist. It is an interesting thought experiment to imagine a child crying after learning that numbers do not *really* exist. In fact, no one has ever feigned that numbers *really* exist! What would *that* mean[11]?

Thus, at the very least, Bunge has to tell us how pretense works in the case of mathematical objects and how it differs from how pretense works in the case of ordinary fictions. If we are to explain the behavior of mathematicians by saying that they pretend that mathematical constructs exist by themselves, we have to be able to say how this pretense comes about, how it works and why it is the best explanation for that behavior. As far as I know, Bunge never went further than to suggest that one *could* explain this behavior that way. Perhaps one can. Perhaps there is a simpler explanation, and still within a materialist framework. We have to go back to the notion of dependent existence.

---

[11] That seems to be an easy exercise in Bunge's framework. If we were to feign that mathematical objects *really* existed, then it means that mathematical objects could be in various states. What exactly these states would be, that would have to be determined. Would they be more like physical objects or living organisms? It is up to your imagination to decide. I suspect that in some cases, the 'reality' would be expressed more in terms of an independence from the mind, the will of the subject, in contrast with the objects that we create. But this shows, once again, that if we pretend that mathematical objects really exist, we do it in a very selective fashion without having learned anything about it.

33.7 Abstract Artifacts and Varieties of Dependent Existence

To understand the various types of dependent existence, let us stick with fictions and literary works a little further. Amie Thomasson has developed an interesting theory of dependent existence that she has applied to clarify the ontological status of fictional characters and fictional works, among other things[12] (See Thomasson 1999).

According to Thomasson, the existence of a fictional character depends on (1) the creative act of a (really) existing author or authors and (2) on the existence of a literary work. In turn, the existence of a literary work depends on the acts of its creator or creators, but it also depends on some copy or memory of it and a (really) existing competent reader. Thus, the structure of dependence of a fictional character is a complex network of real existents, acts and intentions. In Thomasson's words, fictional characters "should be entities that depend on the creative acts of authors to bring them into existence and on some concrete individuals such as copies of texts and a capable audience in order to remain in existence" (Thomasson 1999, p. 12).

There are subtle points that we need not go into for our purposes.[13] The following remarks will suffice. According to this analysis, fictional characters have a history: they are born in certain historical, cultural and social contexts and by the act of a real human or a group of humans. In the case of literary fictions, there are writings or oral traditions, more generally literary works, in which the character first appear. It is important to notice, however, that for the character to be, that particular object, the original literary work, does not have to survive. Any copy or any faithful memory of it will do and will allow for the fictional character to exist. There is thus a certain independence from *particular* and *specific* real objects for them to exist and continue to exist. We all recognize that they can survive their creator and even the original book or work by which they were introduced. It is obvious that, according to Thomasson, fictional existence does not amount to the capacity of a human to conceive a fictional character. There has to be a creator (or creators) who not only creates the character, but does so by doing, building something, namely what we call a literary work. For the latter to be possible, one has to have a language and, in most cases, a written language (although one can argue that the latter is not necessary, as the various oral traditions clearly indicate), together with specific cognitive capacities, for instance a powerful enough memory (where, in most cases, one will find mnemonic tricks to help remember the stories) or the capacity to read a certain language and the latter has to be possessed by other humans afterwards.

---

[12]Thomasson seems to have move away from the specifics of her earlier theory. I stick to it simply because it provides an ontological analysis of fictions as dependent entities, thus an analysis that is close to Bunge's claims. I am *not* claiming that it is the most adequate analysis. In fact, I would be inclined to address these issues more in the spirit of Thomasson's recent work. That is another matter.

[13]Thomasson offers an interesting and rich classification of artefacts based on certain properties of the dependence relation. We refer the reader to her book for more.

There is, clearly, an intrinsic *social* component at work in this picture, since the existence depends on more than one human and even some cultural elements, which goes hand in hand with (neuro)*biological* capacities.[14]

This takes care of most of our intuitions about the existence of fictional characters. Thomasson introduces a more general framework to treat the different kinds of ontological dependence. It is worth looking at the dependence relations that she uses.

33.7.1 Thomasson's Ontological Categories

In her theory, Thomasson offers an explication of types of ontological dependence by combining four notions of dependence: rigid dependence, generic dependence, constant dependence and historical dependence. Thomasson introduces the distinction between constant dependence and historical dependence as follows:

> We can begin by distinguishing constant dependence, a relation such that one entity requires that the other entity exist at every time at which it exists, from historical dependence or dependence for coming into existence, a relation such that one entity requires that the other entity exist at some time prior to or coincident with every time at which it exists. These are not all of the different possible cases of dependence but merely describe some of the most interesting and general cases of dependence. (Thomasson 1999, p. 29)

Clearly, historical dependence is weaker than constant dependence. In other words, if x is constantly dependent on y, then x is also historically dependent on y. As we have already indicated, fictional characters historically depend on a creator or a group of creators to be, but they do not constantly depend on that creator. Examples of constant dependence are numerous. If consciousness is an emergent property of brains, then my consciousness is constantly dependent on my brain.

When the relation of historical dependence rests on a particular individual or a particular group of individuals, Thomasson qualifies this relation as being *rigid*. This qualification can be applied both for the constant case and the historical case. For example, I am rigidly historically dependent on my parents and *La Bohème* is rigidly historically dependent on Puccini's existence. My consciousness is rigidly constantly dependent on my brain. There are relations of dependence that are not rigid, but rather *generic*. In this case, the relation does not depend on a particular, singular individual, the latter being understood in a broad sense. To use Thomasson's example, "a given sample of alcohol is rigidly historically dependent

---

[14]Otavio Bueno has sketched a form of mathematical fictionalism based on Thomasson's views that is strikingly close to Bunge's. Bueno defends the idea that mathematical entities are *like* fictional characters since, according to him, they are created in a particular context and in a particular time and their existence depends upon the existence of written papers and competent readers. He even adopts an existence predicate and distinguishes it from the existential quantifier. However, in the end, his position differs both from Bunge's position and from Thomasson's. We cannot do it justice in such a short paper. See Bueno 2009.

Table 33.1 Dependence
relations

$$RCD \rightarrow RHD \rightarrow RD$$
$$\downarrow \quad \square \quad \downarrow \quad \square \quad \downarrow$$
$$GCD \rightarrow GHD \rightarrow GD$$

on the sugar from which it is formed, it is merely *generically* historically dependent on some yeast (or other appropriate catalyst)" (Thomasson 1999, p. 33). An example of a generic constant dependence is provided by the existence of a University. At any moment that the Université de Montréal exists, there must be persons who work at this particular institution, people who teaches, do research and other people that are registered as students, who attend classes, go to the library, etc. Of course, there is no particular person whom the University's continued existence requires.

Notice that Thomasson is well aware that there may be other cases of dependence and she does not claim to cover all possible cases.[15] The strongest relation of dependence is the category of rigid constant dependence (*RCD*). It entails all the others. Thus, it entails the categories of rigid historical dependence (*RHD*) and rigid dependence (*RD*). This can be pictured thus: $RCD \rightarrow RHD \rightarrow RD$.In turn, rigid constant dependence entails generic constant dependence (*GCD*). There is an obvious line of entailments between the generic dependences: the generic constant dependence (*GCD*) entails the generic historic dependence (*GHD*) which, in turn, entails the generic dependence (*GD*). Hence, the complete picture looks like this (Table 33.1).

We are not done. We have introduced the relations of dependence in general. We now specify two types of dependence that are ontologically fundamental for our purposes: the dependence on material or spatiotemporal entities and the dependence on mental states.[16] Each type of dependence yields a two-dimensional grid of 10 categories each therefore the whole space of ontological categories is four-dimensional with 100 possible categories. There is no need to present the whole system here. We refer the reader to Thomasson's book (Thomasson 1999, chap. 8).

---

[15] Thomasson argues that the relations of dependence, constant dependence and historical dependence are all reflexive and transitive. This suggests that the resulting ontology could be formalized using the mathematical theory of categories, by representing the relation of dependence by a morphism between objects. The links between the kinds of dependence can be represented by functors. In fact, the distinction between rigid dependence and generic dependence can also be captured via a specific type of adjunction. This is not surprising given the fact that mathematical functions capture a form of dependence. We even talk about dependent and independent variables. We leave this project for another paper.

[16] We will stick to the terminology of mental states instead of brain states, despite the fact that we are in a materialist ontology. The reason for this choice is that the term 'mental states' already suggests a certain independence from particular brains but still indicates a clear and well-understood realm of discourse. Of course, the question as to how mental states depend on brain states is fundamental and, at some point, we might be ready to talk about brain states. See, for instance, Piazza and Izard 2009. By saying that constructs are *equivalence classes* of brain states, Bunge himself introduces a different identity criterion for mental states than for brain states, thus introducing the possibility of treating them as a genuine category.

In a materialist framework, material entities are *not* dependent on mental states, neither rigidly nor generically. Mathematical realists or Platonists would probably claim that mathematical objects, although not material, neither depend on material entities nor do they depend on mental states. Thus, both material and mathematical objects would belong to the category of objects that are neither rigidly dependent nor generically dependent on mental states.[17] We are interested in the case where mathematical objects depend on mental states *and* on material entities, for this seems to be consistent with Bunge's claims. Where should they be in this framework?

One of the fascinating aspects of Thomasson's categories is that it opens the door to a multiplicity of ontological categories between the concrete and the abstract, a distinction that is usually considered to be a dichotomy. Of course, the latter depends on how these two terms are interpreted. Thomasson puts the distinction squarely within the space of material dependence. It seems reasonable to claim that concrete objects rigidly constantly depend on themselves. What about abstract objects or entities more generally? An obvious possibility in the present framework is to interpret the property of being abstract as being *historically independent* from material entities, which leaves open the possibility of being *generically* and/or *rigidly* dependent on material entities.

Thomasson herself proposes to take the weakest definition allowed by her system, namely that an entity is *abstract* if it is *not* rigidly constantly dependent on material (spatiotemporal) entities. This still leaves three possibilities: abstract entities can be rigidly historically dependent on material entities, rigidly dependent on material entities or, finally, not rigidly dependent on material entities. In the latter case, being abstract is not one single category. There are now various kinds of abstract entities. Nonetheless, if we accept the foregoing characterizations of concrete and abstract objects, we get back the familiar 'abstract/concrete' dichotomy.

On the one hand, fictional characters depend rigidly historically on mental states, since they rigidly depend on historical actors for their creation. They *also* belong to the category of generic constantly dependent entities, since after their creation, their existence depends on the mental states of agents that are capable of understanding the works in which they appear. This takes care of the dependence on mental entities. On the other hand, fictional characters are part of literary works. As such, they constantly depend on some material entities, i.e. copies of the work, and also rigidly depend historically on certain material entities to bring them into existence, i.e. the particular author who wrote the work at a particular time in a particular sociocultural context. Thus, the literary work of art itself, and the fictional characters it talks

---

[17] Thomasson herself distinguishes two opposites: the mental-material and the real-ideal. The first one would be reflected in the space of mental dependence, the material being independent of anything mental, and the second one would be placed in the space of spatiotemporal dependence, the ideal – which, from a Platonic point of view, would include numbers and similar entities – being independent of anything real. Notice that the material and the ideal are thus characterized purely negatively (Thomasson 1999, p. 125).

about have no spatiotemporal properties, but the work was created at a specific moment by a specific individual (or individuals). It is therefore an abstract object that nonetheless has a dependent existence and the dependence is both mental and material.

Thomasson mentions another case of abstract entity worth comparing with fictions and that will be useful for our own purposes: technological artifacts, like the telephone, the computer, etc. In these cases, it is not one specific object that we refer to, but a *type* of object with a specific *function*. In many cases, the type has many different material instantiations. Their existence is generically historically dependent on mental states, for they did not exist before a certain time, they had to be *invented*, but their coming into being does not depend on one unique and specific individual. The telephone was invented independently by many different people and so was the computer. There has to be *someone* to do it, but it does not have to be that particular person. Similarly, one could argue, as Thomasson does, that they generically constantly depend on material entities. If we were to lose all known exemplars of eight-tracks cassettes and machines, together with their plans and designs, then this technological artifact would cease to exist.

33.8 Mathematical Constructs as Abstract Artifacts

What about mathematical objects, or to borrow Bunge's terminology, mathematical constructs? They are undoubtedly abstract in any sense of that word and, thus, in particular in the sense proposed by Thomasson: they lack a spatiotemporal location or, in her terminology, they are not rigidly constantly dependent on any material entity. We can be more precise.

Let me immediately state in Thomasson's terminology what seems to be one of the claims repeatedly made by Bunge: mathematics is at the very least generically constantly dependent on mental entities. Although it does not depend on one particular mathematician for its existence, Bunge claims that mathematics requires the existence of some person capable of understanding it to continue to exist. Note that it is, in principle, absolutely impossible to verify this claim.

Is it generically constantly dependent on material entities? I claim that it is. Mathematics needs to be told, written, drawn, etc. It has always been accompanied by physical embodiments, tools (stones, wooden marks, compass, ruler, abacus, calculators and, nowadays, computers), symbolic system and notational devices of all kinds. Of course, it does *not* depend on one *particular* such material entity, but it *does* depend socially and culturally on the presence of symbolic representation in one form or another. Thus, if humanity were to disappear and all writings, marks, concrete models and mathematical tools were to disappear with it, mathematics would cease to be. Note once again, that it is, in principle, absolutely impossible to verify this claim.

I claim that there is an important difference with the case of technological artifacts described in the preceding section and it has to do with the role or the kind of dependence at play between the material entities involved in mathematics. The material objects embodying technological artifacts *are* real, genuine exemplars of these artifacts. A real, concrete and functioning turntable *is* just a turntable: it is a *real* token of the type. What I hold in my hand *is* a screwdriver. It was invented, build just to do what it does. It is a real token of the type. A drawn triangle is *not* a triangle, a constructed wooden dodecahedron is *not* a dodecahedron. The symbol 'π' is not the number π. No symbol, no sequence of digits, even thought of as a type, can be the number π, since the latter is irrational and transcendental. A written proof of a theorem is *not* the proof. The latter is an abstract object. There are no *real* tokens of mathematical types.

It is of course because of these facts that mathematical objects are traditionally considered to be *ideal* objects. Their dependence on material entities is of a different nature. We move away from technological artifacts and we move back to fictions. Most of mathematics is *written* and presents itself as a text. It can also be told, usually next to a blackboard or a piece of paper, thus with some written marks. Some mathematical texts are more akin to musical partitions in the sense that the notation tells you how to *do* things: it has to be performed, in the case of music, on an instrument, in the case of mathematics, in one's head or on a piece of paper. But one does not read a mathematical text like one reads a short story or a musical partition. When reading a mathematical paper, one usually needs a pen and a piece of paper and writes as she reads. While someone can take notes while reading a short story for various reasons, it is rarely in order to *understand* the story that someone would do it.

This brings us to the question of the rigid historical dependence of mathematics on humans and here we touch upon an important difference with fictions. Many mathematical theorems, proofs, constructions, theories, algorithms, etc. are identified by the name of the mathematician or mathematicians that introduced them: Gauss fundamental theorem of algebra, Gödel's incompleteness theorems, Wiles proof of the Shimura-Weil-Taniyama conjecture and, as a consequence, of Fermat's last 'theorem', Hamilton's quaternions, the Hopf fibration, Grothendieck toposes, Buchberger's algorithm, etc. It would be very easy to extend the list almost indefinitely. Still, although for sociological reasons the correct attribution of these items to their creators is important, they are nonetheless often thought as being ultimately independent of their creators, in contrast with, say, Offred in Margaret Atwood's novel *The Handmaid's Tale*.

I submit that this behavior, to go back to Bunge's terminology, is attributable to fundamental aspects of mathematical knowledge. First, a concept is sometimes introduced independently and differently by different mathematicians. These various presentations of the concept are then shown to be 'the same', in an appropriate sense of the latter term. One standard example of that phenomenon is the notion of computable functions. This shows that, in some cases at least, the notion of dependence is generic instead of being rigid.

Second, as already observed by Robert Thomas, a mathematical paper does not describe a series of events that happen in some possible space. It is not a narrative in the usual sense of that word (See Thomas 2000, 2002). Writing a mathematical paper is not like writing a fiction. The language of mathematics is such that it is possible for a mathematician to write a paper and erase any trace of a narrator or any reference to mental or material entities. Reading a mathematical paper is not like reading a fiction. In a sense, reading a mathematical paper is comparable to reading a blueprint: it guides you through concepts, examples, constructions, statements, proofs, computations, etc. via certain conventional signs and notational systems. Any competent mathematician who masters the concepts and the language used in a paper or a conference *reconstructs* in her mind the conceptual system that is presented. The particular way of describing that conceptual system does not matter (although the particular language might). Once it is understood and mastered, the mathematical content is in some sense *entirely* assimilated by the reader. It is from then on his or hers. Completely. There is nothing that seems idiosyncratic to the author, there is nothing that escapes the reader. What is more, the content can be *completely* represented in a different manner, even rewritten in a different language or framework. It can be extended by following its necessary logical conclusions. It can also be enriched, transformed, generalized, abstracted, applied, etc. Although a different writer can very well extend a given work, as we have convincingly seen in the latest extensions of Larsson's *Millenium series* by Lagercrantz, the new writer still has to adopt a style, characters, a history, conventions, etc. to go on and, as the example shows clearly, Lagercrantz's remarkable achievement cannot be presented as the *logical* and *necessary* extension of Larsson's work, nor can it be said that he has generalized it, or abstracted from it, or applied it, or expressed it by using different concepts.

Third, the natural numbers and elementary geometry have a special epistemological status which contribute to the conviction that mathematical objects have a distinctive ontological status. According to recent research in cognitive science, innate, non-linguistic, and universal cognitive capacities underlie the development of the natural number concept and of elementary geometry (See, for instance, Dehaene 2011; Dehaene and Brannon 2011; Dillon et al. 2013). The fact that these capacities are innate, universal and non-linguistic and that they serve as the cognitive bedrock for numbers and geometry certainly fuels the belief that what we are referring to in these cases is *in*dependent of mental entities or capacities. They are just given. Of course, this reinforces the idea that mathematics is not rigidly and historically dependent on mental entities. However, these findings certainly do not allow us to conclude that number *theory* and geometry as a *theory* are innate, non-linguistic and universal. That is where the dependences kick in and that we start attributing the creation of concepts, conjectures, proofs, calculations, algorithms to particular mathematicians.

These reasons explain why, at the end of the day, mathematicians refrain from claiming that mathematics is rigidly historically dependent on mental entities and even generically constantly dependent on mental entities. The case against a constantly historic dependence on material entities seems too easy to mention.

If I am correct, it is wrong to say that mathematical objects are fictions. In fact, *even ontologically*, there are substantial differences between the two. Thus, *pace* Bunge, mathematical objects are *not* on an ontological par with fictional characters.

That could be received as the bad news. The good news is that we do not need to talk about fictions nor do we need to talk about mathematical objects in any deep ontological sense.

33.9 Structuralism and Mathematical Objects

As it is clear from the definition given in Sect. 33.5 above, that Bunge tries to combine two different relations of dependence when he deals with mathematics. On the one hand, he clearly believes that constructs in general depend on brain states and therefore mathematical constructs in particular depend on brain states. This relation of dependence leads him to the claim that mathematical objects are *ficta*. On the other hand, mathematical constructs depend on other concepts in a singular way. There is a *conceptual* dependence between mathematical concepts whose nature is unique to mathematics.[18] When Bunge moves to this type of dependence, he switch to mathematical concepts as being conceptual systems or part of conceptual systems. This dualism is in fact inevitable if one wants to develop a form of structuralism *within a materialist framework*. The challenges consist in identifying the correct relations of dependence in both cases and how they should be articulated together into a coherent whole. I claim that the correct *foundational* or *metamathematical stance* is indeed a form of structuralism. This takes care of the conceptual dependence.

As to the dependence on mental/brain states, if one looks carefully at the remarks I have made in the previous section, then it should be clear that (1) there are some existence claims made concerning the ontological status of mathematical objects that go beyond what we can, even in principle, verify and, therefore, I suggest that we simply discard the underlying questions as being pseudo questions and (2) what *can* be said, over and above the internal existential questions settled by mathematicians themselves, can be taken care of by investigating mathematical practice and the human and social sciences, in particular the cognitive sciences, but not only them. Needless to say, I will only make sketchy remarks in this section.

Mathematical structuralism is the claim that mathematics is about structures. This thesis can be spelled out completely, through a metamathematical analysis of the notion of structure. Such an analysis was provided by Bourbaki already in the 1950s in a set-theoretical framework, although it was dismissed by most logicians, philosophers and historians of mathematics for various reasons.[19] A more recent and improved version was presented by the logician Michael Makkai (1998). It is not

---

[18]The reader might want to include logic here. That is another issue. For a long time, this conviction was captured by the claim that mathematics is analytic.

[19]One interesting exception is Erhard Scheibe who, in his tribute to Bunge in 1981, presents an analysis of invariance and covariance of physical theories based on Bourbaki's analysis. See Scheibe 1981.

necessary to present the technical details to understand the basic ideas underlying this form of structuralism.

The fundamental idea is straightforward. For a mathematical theory to be a structuralist theory, it should be possible to prove that the following claim is a meta-theorem: given any property P in the given language $\mathcal{L}$ of the theory T, for all objects X, Y of the theory, if $P(X)$ and $X \cong Y$, then $P(Y)$. In words, a theory is a *structuralist theory* if the provable properties of the theory are only those that are invariant under the proper notion of isomorphism. This says precisely that mathematics is *about* the properties and relations expressed in the proper language and that the underlying objects merely fill in the places to be filled in the relations of the theory. The specific nature of the objects is totally irrelevant. It is in this sense that mathematical objects are not the central concern and that they are always part of a system.[20]

33.10   Conclusion

It is ironic, in the present context, that the appropriate language for structuralism is a logic with *dependent* sorts (or types). This dependence reproduces the fact that a structure depends, for its existence, on previously given objects or systems. We are here dealing with a form of *conceptual* dependence. The appropriate language reflects a clear ontological hierarchy.

Structures themselves are given by a formula φ in a language $\mathcal{L}$. They are then interpreted in a system in which the notion of isomorphism plays *the key* role. Makkai's system covers set-theoretical structures, categorical structures, bicategorical structures, . . . , n-categorical structures, up to ω-categorical structures. Whether one has to go on is an open problem. How the foundational framework has to be develop and what will exist within it has to be settled by mathematicians and logicians. A philosopher must take note and see what follows. But it is first and foremost an internal affair to logic and mathematics.

How about the dependences of mathematical structures on mental/brain states and material entities? As I have already indicated, claims that mathematical structures originate with humans and that they would cease to exist if the latter were to disappear with the material production that comes with it are unverifiable. These claims are intuitively plausible for fictions and technological artifacts. They might be true, they might be false for mathematical structures. There is no way to know and it seems that there are just as many people who believe that they are false than they are true. We submit that it does not matter and that we do not have and cannot have the philosophical tools to settle the issue. They are, of course, consequences of a materialist ontology and, as such, have to be acknowledged. Furthermore,

---

[20]Makkai has developed a structuralist set theory using his framework. We refer the reader to his paper for details. See Makkai 2013 for the technical presentation, and Marquis (2012, 2018), for more on the philosophical ideas involved in the project.

ontological issues pertaining to abstract objects have concrete consequences. What *can* be settled are the types of dependence of mathematics on mental/brain states and material objects in its history, as it is practiced, as it is taught. Questions pertaining to these dependences *can* and *should* be investigated. The answers make a difference to real issues. This should not be a surprise to Bunge's readers: he has been calling attention to the real impact of ontological decisions all along.

Acknowledgement The author acknowledges the support of the SSHRC of Canada given for completion of this work.

References

Balaguer, M. (1998). *Platonism and antip-platonism in mathematics*. New York: Oxford University Press.

Bueno, O. (2009). Mathematical fictionalism. O. Bueno, ∅. Linnebo, New waves in philosophy of mathematics (59–79). London: Palgrave Macmillan.

Bunge, M. (1974). *Treatise on basic philosophy. Vol. 1. Semantics 1: Sense and reference*. Dordrecht: Reidel.

Bunge, M. (1977). *Treatise on basic philosophy. Vol. 3. Ontology I: The furniture of the world*. Dordrecht: Reidel.

Bunge, M. (1985). *Treatise on basic philosophy. Vol. 7. Philosophy of science and technology part 1*. Dordrecht: Reidel.

Bunge, M. (1997). Moderate mathematical fictionism. In E. Agazzi & G. Darwas (Eds.), *Philosophy of mathematics today* (pp. 51–71). Boston: Kluwer Academic.

Bunge, M. (2006). *Chasing reality: Strife over realism*. Toronto: University of Toronto Press.

Bunge, M. (2016). Modes of existence. *Review of Metaphysics, 70*, 225–234.

Dehaene, S. (2011). *The number sense* (2nd ed.). New York: Oxford University Press.

Dehaene, S., & Brannon, E. M. (2011). *Space, time and number in the brain: Searching for the foundations of mathematical thought* (Attention and performance, Vol. 24). London: Academic.

Dillon, M. R., Hu ang, Y., & Spelke, E. S. (2013). Core foundations of abstract geometry. *Proceedings of the National Academy of Sciences, 110*, 14191–14195.

Ehrlich, P. (2006). The rise of non-Archimedean mathematics and the roots of a misconception I: The emergence of non-Archimedean systems of magnitudes. *Archives for the History of Exact Sciences, 60*, 1–121.

Field, H. (1980). *Science without numbers*. Princeton: Princeton University Press.

Field, H. (1989). *Realism, mathematics and modality*. Oxford: Blackwell.

Leng, M. (2009). *Mathematics and reality*. Oxford: Oxford University Press.

Makkai, M. (1998). Towards a categorical foundation of mathematics. In J. A. Makowsky & E. Ravve (Eds.), *Logic colloquium '95: Proceedings of the annual European summer meeting of the ASL, Haifa, Israel* (pp. 153–190). Berlin: Springer.

Makkai, M. (2013). *The theory of abstract sets based on first-order logic with dependent types*. http://www.math.mcgill.ca/makkai/Various/MateFest2013.pdf

Marquis, J.-P. (1997). Abstract mathematical tools and machines for mathematics. *Philosophia Mathematica, 5*(3), 250–272.

Marquis, J.-P. (2006). A path to the epistemology of mathematics: Homotopy theory. In J. Gray & J. Ferreiros (Eds.), *The architecture of modern mathematics* (pp. 239–260). Oxford: Oxford University Press.

Marquis, J.-P. (2011). Mario Bunge's philosophy of mathematics: An appraisal. *Science & Education, 21*(10), 1567–1594.

Marquis, J.-P. (2012). Categorical foundations of mathematics: Or how to provide foundations for *abstract* mathematics. *The Review of Symbolic Logic, 6*(1), 51–75.

Marquis, J.-P. (2018). Unfolding FOLDS: A foundational framework for abstract mathematical concepts. In E. Landry (Ed.), *Categories for the working philosopher* (pp. 136–162). New York: Oxford University Press.

Piazza, M., & Izard, V. (2009). How Humans Count: Numerosity and the Parietal Cortex, in *Neuroscientist, 15*(3), 261–273.

Scheibe, E. (1981). Invariance and covariance. In J. Agassi & R. S. Cohen (Eds.), *Scientific philosophy today* (pp. 311–332). Dordrecht: Reidel.

Thomas, R. (2000). Mathematics and fictions I: Identification. *Logique et Analyse, 43*(171–172), 301–340.

Thomas, R. (2002). Mathematics and fictions II: Analogy. *Logique et Analyse, 45*(177–178), 185–228.

Thomasson, A. (1999). *Fiction and metaphysics*. Cambridge: Cambridge University Press.

Thomasson, A. (2014). *Ontology made easy*. New York: Oxford University Press.

Torretti, R. (1981). Three kinds of mathematical fictionalism. In J. Agassi & R. Cohen (Eds.), *Scientific philosophy today: Essays in honor of Mario Bunge* (pp. 399–414). Boston: Reidel.

Yablo, S. (2002). Go figure: A path through Fictionalism. *Midwest Studies in Philosophy, 25*, 72–102.

Chapter 34
On Leaving as Little to Chance as Possible

Michael Kary

Abstract Randomness was one of Mario Bunge's earliest philosophical interests, and remains as one of his most persistent. Bunge's view of the nature of randomness has been largely consistent over many decades, despite some evolution. For a long time now, he has seen chance as a purely ontological matter of contingency, something that does not result from either psychological uncertainty or epistemological indeterminacy, and that disappears once the die is cast. He considers the Bayesian school of probability and statistics to be pseudoscientific. Bunge upholds a fairly conventional view that chance is not any part of the purely mathematical theory of probability, and a thoroughly unconventional view that ontologically contingent processes are deterministic, though not classically so. This chapter examines Bunge's views on probability by investigating what any of the following have to do with each other: chance or randomness, likelihood, the mathematical theory of probability, determinism, independence, belief, psychological uncertainty, and epistemological indeterminacy.

34.1 On Randomness

It is often said that the theory of probability is just measure theory restricted to the unit interval. In the classical treatment of the finite case, this is the same as saying that probability theory is just finite set theory joined to the arithmetic of proper fractions and the whole.

This breakthrough in understanding is ascribed to Kolmogorov, but the view that this is what Kolmogorov's *Grundbegriffe* tells us is an exaggeration of Fréchet's explanation. What Fréchet said was that Kolmogorov's contribution was to bring together Borel's countable additivity and the calculus of probability (the work of many, including latterly Fréchet and Kolmogorov), unequivocally into one coherent axiomatization:

M. Kary (✉)
Montreal, Canada

© Springer Nature Switzerland AG 2019
M. R. Matthews (ed.), *Mario Bunge: A Centenary Festschrift*,
https://doi.org/10.1007/978-3-030-16673-1_34

It was at the moment when Mr. Borel introduced this new kind of additivity into the calculus of probability—in 1909, that is to say—that all the elements needed to formulate explicitly the whole body of axioms of (modernized classical) probability theory came together.

It is not enough to have all the ideas in mind, to recall them now and then; one must make sure that their totality is sufficient, bring them together explicitly, and take responsibility for saying that nothing further is needed in order to construct the theory.

This is what Mr. Kolmogorov did. This is his achievement. (Fréchet 1938, quoted in Shafer and Vovk 2006, p. 70)

If this achievement were exactly the same as reducing the theory of probability to measure theory on the unit interval, one would have to ask, for a start: why does Kolmogorov's axiomatization include the word "random" (Kolmogorov 1933/1956, p. 2), which does not occur in measure theory?

That this inclusion, however seemingly incidental and however glossed over by Kolmogorov, is essential, not accidental,[1] is illustrated by the following famous problem of pure mathematics, one which is neither solvable through, nor posable in, measure theory alone:

Bertrand's Paradox of the Great Circle
Let M and M' be two random locations on the surface of a sphere. What is the probability that the two are within 10 minutes of arc (on the great circle connecting them) of each other? (See Shafer and Vovk 2006)

Since Kolmogorov's presentation of Borel's solution, this is now generally known as the Borel paradox or the Borel-Kolmogorov paradox.

It might seem that Kolmogorov did exactly both those things—pose and solve the problem with measure theory alone—in his presentation of Borel's solution. That is only if one starts the story in the middle, at the stage where it could be solved by the computational apparatus of measure (and integration) theory: "Borel set the following problem: Required to determine 'the conditional probability distribution' of latitude θ, $-\pi \leq \theta < +\pi$, for a given longitude" (Kolmogorov 1933/1956, p. 51). Doing so neglects Kolmogorov's (and Borel's) setup of the underlying concepts and the specific problem. These show not just that the problem deals with random locations on the surface of a sphere, but that Borel actualized the meaning of random to be that the probabilities of events are proportional to their surface areas on the sphere.

This was also Bertrand's first solution to the problem. Bertrand's second solution—a different actualization, whose difference was not obvious (and so gave rise to the paradox) until resolved by Borel, is to have the probability of two locations being within a certain arc length of each other be twice the length of that arc as a fraction of the circumference of the great circle to which it belongs. This since the two locations must be somewhere along that great circle, while one location can be on either side of the other. (Put another way: Borel considered the great circle as a meridian, and asked for the conditional probability of latitude given longitude; while Bertrand's alternative was to consider the great circle as an

---

[1] In his *Foundations of Physics*, Bunge says "KOLMOGOROFF himself did employ a misleading terminology but it is as supernumerary as the observer and the ideal experiment in [quantum mechanics]... random samplings will occur in informal inferences and in applications of [the calculus of probability] not in its foundations." (Bunge 1967, p. 90)

equator, and ask for the conditional probability of longitude given latitude. Only Borel's version gives rise to probabilities uniform over surface area, that is to say, to Bertrand's first solution.)

Borel argued and most but not all accept that only that first solution is "the" correct one. So consider one of Bertrand's other paradoxes, not discussed by Kolmogorov but still one of pure mathematics, one with still more randomization options, all viable in their own way: Let C be a random chord of a circle. What is the probability that its length is greater than that of a side of an inscribed equilateral triangle? As with the paradox of the great circle, this problem famously does not have a unique or else not obviously unique solution (hence the paradox), because the solution depends on the particular implementation of "random chord", the three given by Bertrand being inequivalent: the chord between two random locations on the circumference; the chord whose midpoint is a location random with respect to area within the circle; the chord whose midpoint is a random location along a random radius.

Note that the problems cannot be stated as just "Let M and M' be two locations on the surface of a sphere..." or "Let C be a chord of a circle..."; nor as "any two", "two arbitrary", et cetera. Nor is the concept of (full or uniformly distributed) randomness just replaced by equiprobability: it has to be the equiprobability of events that are equivalent in some underlying sense—such as, in the paradox of the great circle, for being of equal area on the surface of a sphere. What makes the events equivalent varies from problem to problem (in the case of the paradoxes, from implementation to implementation), but any natural equivalence (that is, inherent to the problem rather than assigned arbitrarily) comes from arguments based on symmetry—geometric or logical or any other kind—or more simply, interchangeability. This is the equal likelihood of classical probability.[2,3]

Thus contrary to almost universal opinion (e.g. Shafer and Vovk 2006, p. 71), Kolmogorov's axiomatization did not replace the circular or vague concept of equal likelihood with measure. Rather it hid likelihood in the background, as part of the way the key term "random" has to be implemented when solving any particular problem. Nor need there be anything vague or circular about equal likelihood: it simply means the assignment of equal probabilities to measurable sets that are deemed equivalent. The nature of and reasoning behind the equivalence of the sets is particular to the nature of and reasons for their interchangeability in each problem.

For example, to solve the problem of the great circle, one of the two locations is first fixed arbitrarily. In this case *arbitrary* actualizes *random* because there are no distinctions between any locations, and likewise there is no alternative actualization. But the second location cannot be fixed arbitrarily—at least, without giving an

---

[2]Those who insist that equiprobability is all that is needed have two options: explain why the probabilities should be equal; or offer no explanation. The former is the same as explaining why the respective events should be considered equivalent, while the latter is the same as assigning their equivalence arbitrarily. Thus, there is no escaping the concept of equal (or else proportional) likelihood, whether arising naturally or assigned arbitrarily.

[3]The concept of likelihood (of events) in classical probability is not the same as the technical concept of likelihood (of model parameters) introduced by Fisher and as used in modern statistics, both Bayesian and non-Bayesian (see e.g. Reid 2013).

arbitrary meaning to "random". Instead, in Borel's solution and in Bertrand's first solution, the probability that it is inside any particular measurable subset of the surface is made proportional to that subset's surface area. Bertrand's alternative actualizes random in a different way.

Which solution corresponds to a material process of randomly selecting two locations on the surface of some real ball depends on the particular mechanism of the process. A process that corresponds to Bertrand's second solution would be as follows: spin the ball about a random axis, and by touching a pen to the surface at the corresponding middle, mark the corresponding random equator (or equivalently in this case, spin the ball about an arbitrary axis and mark with a pen the corresponding arbitrary equator). Stop the ball at an arbitrary time (or equivalently in this case, a random time) and mark with an "x" the corresponding random (or in this case equivalently, arbitrary) location on that equator. Keeping the pen on the surface starting at the centre of the "x", spin the ball randomly (*not* arbitrarily) in either direction about the axis. (Randomness at this and the subsequent step requires a physical randomizer of some sort: the mathematical theory specifies that the events have certain, in particular equal, probabilities, but it does not tell us how to make this come about in the real world, or even whether doing so is possible.) Then lift the pen from the surface at a random (*not* arbitrary) time, now marking the exit. The two random locations are the origin and exit, and if the experiment is repeated many times, they will be within 10 min of arc of each other in approximately the relative frequency given by Bertrand's second solution, ($[2 \times 10] \div [60 \times 360]$), which evaluates to 9.26×10^{-4}.

A process corresponding to Bertrand's first or Borel's solution, abbreviated this time, would be as follows: mark with a pen an arbitrary location on the surface of the ball, and consider it as a pole of some coordinate system of latitudes and longitudes. Keeping the pen on the surface, spin the ball arbitrarily either way about an arbitrary axis passing through the equator, or equivalently, have the pen move in a single random direction along the surface, away from the pole. In terms of the coordinate system, this corresponds to moving the pen randomly either way along an arbitrary line of longitude.

This time though it will not work to simply stop the motion at a uniformly random time, because directions of travel along the surface of a sphere away from any location on the surface diverge and reconverge. Put another way, it won't work because lines of longitude diverge towards the equator. In order to get the probabilities right, instead of stopping the motion at a uniformly random time, it should be stopped at a time whose probability is proportional to that divergence. That is to say, the motion should be stopped and the location marked at a time whose probability is proportional to the radius of a circle of latitude[4]: vanishing at the poles, and maximal at the two intersections with the equator. With the resulting

---

[4]That is, stopped at a time drawn from a probability distribution whose values are proportional to the radii of the circles of latitude crossed at the corresponding times, and so to the surface areas between corresponding differentially (infinitesimally) separated lines of latitude.

two random locations, and with the experiment repeated many times, they will be within 10 min of arc of each other in approximately the relative frequency $(1 - \cos[\pi/1080])/2$, which evaluates to 2.12×10^{-6}. This is the proportion of the surface area of a sphere occupied by a surface region of radius 10 min of arc, and it is a small fraction of the other solution.

Within the details of either process the mathematical theory tells us in what proportion the relevant probabilities should be, but it cannot not tell us how or even whether it is possible to construct a timer that will stop at times random according to any distribution, uniform or otherwise. The problem is far easier to solve either way in pure mathematics than it is to stage in the real world—if that be possible at all—and moreover the solution is exact.

34.2 Probability, in Generalities and Specifics

Rather than chance or randomness, it is the concept of time that does not occur in either the basic theory of probability or specific purely mathematical problems. Consequently, even though they may involve conditional probabilities, neither do the concepts of *becoming, transition,*[5] *change,* nor even of *occur* or *event* as used in science and philosophy (see Table 34.1 in Sect. 34.4).[6] Nor do the concepts of *state, selection,* or *choice,* although the problems could be rephrased that way. The only concept that is really necessary is the standard one of *mathematical fiat.*

In *Matter and Mind,* Bunge says "It is generally admitted that the probability calculus elucidates the concept of chance or randomness. This opinion is false..." (Bunge 2010, p. 54). If "elucidate" is taken to mean in specifics rather than in generalities, or to mean providing a ready-made explanation of *how* to achieve randomness in the real world rather than of *what* randomness is, then Bunge's view accords well with the preceding discussion. His explanation though does not accord quite so well:

> In short, the concept of chance, unlike that of probability, is non-mathematical. But of course it belongs in all of the factual sciences and technologies. In these, talk of chance is legitimate only if accompanied by a model of a randomization mechanism – which of course is a concrete thing that can produce only finite sequences of random events. (Bunge 2010, p. 54)

Earlier, in his *Chasing Reality,* Bunge was more expansive:

> Paradoxically, the general theory of probability does not include the concept of chance, even though it is applicable only to chance events...
> According to Humphreys (1985)... the probability calculus is not "the correct theory of chance." This remark is correct but incomplete. The failure in question shows that, to

---

[5]Transition probabilities are instead modelled using sequences of random variables indexed by a set used to represent time values. This extends the basic theory to the theory of stochastic processes (see Sect. 34.6).

[6]In purely mathematical problems, there is no reason for the conditional probability $P(A \mid B)$ to be read as "the probability of A, given that B has occurred". Instead it should be read as "the probability of B, given that A is the case"; or in other words, the probability of A renormalized to B as the sample space.

Table 34.1 Glossary of some terms used in the mathematical theory of probability

Term	Meaning in probability theory	Meaning when applied to reality	Clarification
(Random) sample space, (random) event space, universe	Set with at least one element, upon which a sigma algebra of measurable subsets is defined, and whose total measure is unity	Everything that might happen or be selected or be the case	In a chancy world unlike a fully deterministic one, there is more than one possibility
(Possible or random) outcome[a]	Element of the sample space	One thing that might happen or be selected or be the case	It can also be useful to consider as a single "outcome" a conjunction of more
	Example: a random location on the surface of a sphere; the outcome H of a single Bernoulli trial on the sample space $\{H, T\}$	Example: spontaneous emission of a photon; the state of equilibrium; a string of N heads in M tosses of a coin	basic components (see Sect. 34.5)
Elementary event[a]	Subset of the sample space consisting of one outcome. Consequently, all elementary events or outcomes are mutually exclusive	None; or same as for outcome	Mathematics is necessarily fastidious about distinguishing elements from subsets of the sample space; science is not
	Example: the event $\{H\}$ in a single Bernoulli trial on the sample space $\{H, T\}$		
Event	Measurable subset of the sample space	Circumstance, instance; secondarily, outcome	"Events" as changes of state (happenings or occurrences) are never disjunctive or negative;
		Example: never getting an ace or a deuce in N hands of M cards; getting a head on a single toss of a coin	"events" as used in probability may be
The certain event	The sample space	What must be the case	Everything that might be $=$ what must be

[a]Kolmogorov (1933/1956, pp. 2–4) used "elementary event" for an element of the sample space and "outcome" for an occurrence in the real world

be applied, the probability calculus must be enriched with non-mathematical hypotheses. At least one of these must be the assumption that the process in question has a random component. For example, the gambler assumes that the roulette wheel is not biased, and that its successive turns are mutually independent; in the kinetic theory of gases, one assumes that the initial positions and velocities of the particles are distributed at random...

The axiomatic probability calculus does not involve the notion of chance because it is just a special case of measure theory, which belongs in pure mathematics. [...]

In other words, the reason pure mathematics does not define the concept of chance is that every instance of genuine chance is a real process, hence one that can be understood only with the help of some specific factual hypothesis... (Bunge 2006, pp. 100–102)

Bunge's explanations diverge from mine over the following: the concept of randomness (or chance) is essential to the purely mathematical theory of probability; for this reason Kolmogorov included it, but because it varies in its specifics, he glossed over it without explicit elucidation; the required purely mathematical elucidation of the general concept of randomness is simply that the measurable sets are assigned probability values—either according to some characterization of events of equal or proportional likelihood, or else to induce such a characterization; a specific elucidation must be left to each specific problem; these specific elucidations are made either according to some specific interchangeability or proportionality characterization, or else to induce such a characterization; these problems unequivocally belong to pure mathematics, by virtue of ordinary mathematical fiat; but only by ontological or epistemological assumption to the real world, of which more later.

In other words, while Kolmogorov did free the purely mathematical theory of probability from considerations of the meaning of probability—other than as given by its mathematical behaviour—he did not do so for chance or randomness or likelihood (nor did he free statistics from the meaning of probability; Reid and Cox 2015). Rather, he only freed the theory of probability from being tied to their *specific* meanings, and likewise to their implementations and interpretations in the real world.

34.3 Randomization Mechanisms

The idea of a randomization mechanism, as invoked by Bunge, raises the following question: how exactly can a *mechanism* randomize? Within pure mathematics, the corresponding process is clear, as was illustrated in the discussions of Bertrand's paradoxes of the great circle and the chord: the "mechanism" is the assignment of probabilities to events, i.e. it is the specification of one or more probability distributions, by mathematical fiat and according to whatever criteria.[7] Thus as far as pure mathematics goes, it makes no difference whatsoever whether a probability distribution is derived objectively or subjectively or arbitrarily, as long as it is consistent.

In the real world, mechanisms are typically considered to be either of the classical (Newtonian, Laplacian) type, whose successive states can in principle be calculated

---

[7]Bunge says something similar in e.g. volume 5 of his *Treatise*, where he held that "A sequence of events can be said to be *random* if every event in it has a definite probability; otherwise the sequence is either chaotic or causal. This is not the standard definition of randomness—but then there is no standard definition." (Bunge 1983, pp. 365–366)

via formulas; or of the quantum type, where no mechanism for the *production* of chance is proposed, rather the *presence* of chance is assumed from the outset. From an ontological point of view, mechanisms of the latter type may seem remarkable. But in terms of the epistemology of chance they are uninteresting, because they short-circuit the problem: they involve randomness, but do not explain how it comes about.

One might consider along with Epicurus that there is a third type of mechanism, namely human (animal) agency. Bunge's ontology posits that there are biological laws not reducible to physical laws, and psychological laws not reducible to biological laws, and so suggests that this third type is genuine, albeit so far unexplained. Mechanisms of this type seem also capable of generating random behaviour, somehow related to input from the locus coeruleus into the anterior cingulate cortex (Tervo et al. 2014).

Bunge gave as an example of a randomization mechanism a roulette wheel, which is a strictly deterministic, classical machine. So too are the other classical randomizers, such as coin tosses. Their successive states are thus in principle calculable by the formulas of classical mechanics (for such calculations see e.g. Diaconis et al. 2007), and if they do indeed do as advertised and as Bunge claims— that is, randomize—then it would seem that there are formulas or at least algorithms for generating or describing sequences of random numbers.

Using an example supplied by Truesdell, of a ball striking a (frictionless) acute wedge head-on, Bunge asserts that because some techniques for solving problems in classical mechanics produce more than one solution, probability is involved in classical mechanical systems. It is unclear whether he means randomness is generated by such systems (Bunge 1979, p. 175), or that probability is needed to model such systems (Bunge 1985, p. 147), or only, as he says on the next page in the context of another example, that a moral of the story is that "randomness can be simulated by perfectly 'deterministic' systems" (Bunge 1985, p. 148). In any case just because some method gives two solutions does not mean that further analysis will not find mechanical considerations that decide between them, or even that find a third solution preferential to either. Some examples: if the frictionless wedge is a perfectly rigid body with a perfect acute angle at the apex, then it is perfectly sharp at the apex and must either slice the ball in two, or at least infinitesimally stab it, and thus prevent motion in all but the reverse direction. If it is objected that the ball too is infinitely hard, this is an argument that classical mechanics provides not two solutions but none. Or: the contact forces between the ball and the apex come from electrostatic repulsion, and the direction of this force against the ball when hitting the apex head-on runs along the bisector of the wedge angle, and away from the wedge (in mechanical terms: contact forces at a frictionless apex hit head-on are always along the line of symmetry, because a net contact force in any other direction would include a frictional component). The moral is that reality, like the trajectories of billiard balls, may not be accurately reflected in the mathematics of singularities, at least without further ado.

Illustrating with random sequences of zeros and ones, Bunge holds that

> Such strings of equiprobable events are not well-defined mathematical objects, for no formula can define them. (In other words, such sequences have no precise general term, such as $x^n/n!$ in the case of the exponential function.)... In short, Randomness = Disorder.[8] (Bunge 2010, p. 54)

And earlier:

> If the concept of chance were mathematical, it would be possible to concoct formulas for generating random sequences. But no such formula is conceivable without contradiction, since randomness is the opposite of mathematical lawfulness... (Bunge 2006, p. 102)

Before returning at the close of this chapter to the question of whether formulas can produce randomness, consider here as an entryway to it the question of whether well-defined mathematical objects must have precise formulas.

Definitions tell us *what* something is, not necessarily *how* to construct it, nor even how to recognize it in any of its potentially many forms. Is it really the case that an infinite sequence of numbers that has no formula for its general term is therefore not a well-defined mathematical object?

Consider the real number π, defined as the ratio of a circle's circumference to its diameter. As a real number, π can be represented by an infinite string of digits, specified by for example its decimal expansion. π has been well-defined since antiquity, but not because the ancients knew its value, or a formula to calculate it. It took Archimedes to bound π between two potentially infinite convergent series,[9] from which a sequence of successive approximations could be calculated. In ancient times, which were long before infinite decimal expansions were invented, the only way to write down a numerical value as we understand it was as an integer or a ratio of them. The ancients knew of no way to do this for π, other than as approximations, such as either AAA, or ΠAA divided by EAA (these using the Ionian numerals for 3, 22, and 7 respectively). Instead, even in ancient times, what made π a perfectly well-defined number was that this ratio was proved to be a single (albeit unknown) constant for all circles.

As for the string (or sequence) of its decimal digits, neither for π nor for the exponential function e^x is there known a "precise general term" for an arbitrary element of the sequence.[10] In the case of the exponential function e^x, what Bunge refers to as the general term $x^n/n!$ is not a general term of its *decimal* (or any other base) expansion: it is not a general term of the sequence or string of its digits, rather it is the general term for its *Taylor* expansion, which is instead a series whose partial sums successively better approximate e^x. Interestingly, Kirschenmann (1973, pp. 130–131) already said much the same about π in a volume edited by Bunge, although the formulas mentioned here in footnote 10 had not yet been discovered at

---

[8]This is a change from the view Bunge expressed in volume 3 of his *Treatise*, where he said "Randomness, being a special case of stochasticity, is a type of order" (Bunge 1977, p. 209).

[9]A series is a sum of (finitely or infinitely many) terms, a sequence (or string) is a list of them.

[10]There is though a formula that allows computation of any particular digit of the binary or hexadecimal expansion of π. It has been proven that there is no formula of its type for expansions in any base that is not a power of 2 (Borwein et al. 2004; Bailey and Borwein 2014).

the time. The apparent randomness of the sequence of decimal digits of π holds just as well more recently as it did then (Bailey and Borwein 2014).

Similarly, a string of equiprobable events is indeed a perfectly well-defined mathematical object: for events x_i and their corresponding probability measure P, it is any sequence of them $\{x_i\}$, $i \in I$, such that $P(x_i) = P(x_j)$ for all $i, j \in I$, where I is some countable index set. How to obtain such a string in the real world, or how to recognize whether we do or do not have one, are other matters entirely.

34.4 Terminology

Before proceeding any further there are some terminological matters to be dealt with. The theory of probability uses many words from both ordinary life and science and philosophy, but with different meanings. Table 34.1 provides a glossary of some of the key terms.

In numerous works Bunge describes Kolmogorov's theory as having two primitive concepts,[11] those of event and probability measure, with none of the concepts being new, all imported as-is from measure theory, only renamed. In turn Bunge has it that the philosophical challenge is to interpret the concepts of event and probability in factual terms. The view here is that the philosophical primitives are either *possible outcome*, or if preferred, the underlying set of them (*sample space*, i.e. the set of possible outcomes); and *random*. This is despite their seeming to disappear from Kolmogorov's presentation as quickly as they appeared, the theory instead being developed on the derived σ-algebra and the probability measure as defined on it. Likewise the philosophical challenge is to interpret the underlying concepts of possible outcome, sample space, and random. The interpretation of "events" and the probability function defined on them is instead a secondary matter. The interpretation of probability in particular follows naturally from the laws of large numbers.

34.5 Laws of Large Numbers

To prove the laws of large numbers it is convenient to define a new concept, a random variable, and to consider a sequence of N independent and identically distributed random variables on a single sample space. Here the laws will be taken for granted without proof. For the philosophical purposes of the present context, and to keep the discussion in line with the preceding, the laws will be applied in simplified form to a general setup that does not need any new concepts.

Let Ω be a finite sample space, with m elements ω_i. Let randomness be actualized by assigning arbitrary (up to the requirements of the axioms) probabilities to the

---

[11] In *Chasing Reality*, Bunge says there are three: event, probability space, and probability measure (Bunge 2006, p. 100). By "probability space" Bunge refers to the σ-algebra (field) of measurable subsets.

corresponding elementary events. Now consider a new sample space Ω_N, whose elements are all possible strings of length N of the ω_i. Thus for example if $\Omega =$ {head, tail, edge}, then using H for head and so on, an arbitrary element of Ω_4 could be *ETTH*.

In these new spaces, let randomness be actualized to mean that the probability of any particular string is the product of the probabilities of the component ω_i; i.e. so that the component probabilities are independent. In the case where there are only two possible components ω_1 and ω_2, the elements of Ω_N can be considered as Bernoulli trials of length N, or a model of successive heads and tails obtained when tossing a coin. If there were six possible components, and the probabilities chosen accordingly, this setup could be used to model N rolls of a fair or unfair die.

In any string, the relative frequency of any particular ω_i is the total number of times it occurs in the string, divided by the length of the string. Thus in the example for Ω_4, the relative frequency of T was $2/4 = 1/2$. In simplified form, the laws of large numbers say in effect that as N gets arbitrarily large, then for any particular component ω_i, the strings where the relative frequency of that component is not close to its probability comprise a set of negligible or zero measure.

The remarkable thing about the laws of large numbers is that they relate probabilities to relative frequencies, *even when the concept of probability is given no factual interpretation at all*—and even without having to repeat an experiment. This leads to a natural question: why should the relative frequencies in a string mimic the probabilities of each corresponding elementary event? The natural interpretive conclusion is that the probabilities of single events represent a tendency to be the case (where the specific nature of the tendency is as defined by the probability measure). Thus said as much Laplace, in the context of a discussion of Bernoulli's law of large numbers:

> Mais en y réfléchissant, on reconnaît bientôt que cette régularité n'est que le développement des possibilités respectives des évènemens simples, qui doivent se présenter plus souvent, lorsqu'ils sont plus probables. (Laplace 1814, p. lii)

This all has the effect of upholding, at least in general form, a propensity or objective interpretation of probability, as advocated by Bunge. But Bunge has a strict view of the class of real events it may apply to. Is there anything more to the matter?

34.6 On Bayesianism

In the Bayesian school of statistics, or more precisely in its subjectivist or personalist subtype, probabilities are said to describe or prescribe an individual's beliefs that the corresponding events are the case. In Bayesian usage, an outcome or event can be an ontological state or event, or instance (e.g. that of two locations being on opposite sides of an arbitrary equator); or a hypothesis or proposition, typically one that proposes or asserts that a possible event or state or instance is the case. Examples are the null hypothesis (the proposition that two statistics, e.g. the means, of two putatively different populations are equal), or the proposition that an individual committed a certain crime. Sometimes beliefs are short-circuited and Bayesians refer to the probability of a hypothesis (being the case) given the evidence, or similar. This is the rationalist subtype of Bayesianism. However, even non-Bayesians routinely refer to states, events, and instances interchangeably with a proposition or hypothesis asserting they are the case. Bunge sometimes admits this philosophically untidy usage, but only grudgingly (e.g. Bunge 1981, p. 309; Bunge 1985, p. 90).

Bayesianism introduces the concepts of *prior* and *posterior* distributions. In subjective Bayesianism, a prior distribution represents an individual's *ab initio* beliefs, while a posterior distribution updates the prior following data collection and via procedures having to do with Bayes' Theorem. In empirical Bayesianism, the prior is derived using data. Empirical Bayesianism can be considered a different name for, or a subtype of, a more general, "objective Bayesianism", a large and diverse category which may be too broad for the label (Berger 2006). In any type of Bayesianism, the forms of the various distributions involved may be derived from assumptions and reasonings about the underlying mechanisms presumed to generate them, and tested against their consequences. For example, the classical assumption of equal likelihood of outcomes for coin tossings and the like can be considered Bayesian (Berger 2006), while Poisson or Gamma distributions may be used for accident data. For a practical, important, and established example of empirical Bayesianism, see its use in road safety, as in Hauer (1997), Carriquiry and Pawlovich (2004), and Persaud and Lyon (2007).

Bunge has always excoriated Bayesianism, for reasons good and bad. Examples of good reasons for rejecting subjective Bayesianism are that science is about what really is the case, not what we believe to be the case (except when beliefs themselves are objects of the investigation); that there is no necessary connection between ontological indeterminacy and psychological uncertainty[12]; that it is simply not true that probabilities *are* beliefs or measures of them; and that it is simply not true that the probability calculus makes for a valid or useful theory of rational belief.

To elaborate some on that last claim before moving on to bad reasons. It is impossible to know what might make for rational beliefs without knowing what function beliefs serve in our lives. For a start, why do we need beliefs at all, and why should we believe in anything we do not know to be true, as subjective Bayesianism would have us do? Why should we not instead take a scientific attitude, and have knowledge, combined with hypotheses or assumptions, suffice? Or replace subjective estimates with postulation and calculation (Bunge 2003, p. 227)? Hypotheses or assumptions more generally need no psychological commitment, but are grounds for action anyway, to explore their consequences and test them for truth. Sometimes we have reasons for thinking they are true (is that the same as reasons for believing them?—a real theory of belief would clarify that), but often we put them to the test because we have reasons for thinking they are false, and want to know it rather than think it or believe it.

---

[12]In *Evaluating Philosophies*, Bunge says "objective indeterminacy implies subjective uncertainty—though not conversely" (Bunge 2012, p. 108). Yet the former implication is also false. It is so easy and common to be certain about outcomes which are not fixed in advance that the corresponding vocabulary is part of everyday life, to include *overconfidence, hubris,* and *counting chickens before they are hatched.*

Consider two evolutionarily plausible reasons for the existence of beliefs: when faced with adversity, believing firmly in unlikely success allows us to persevere; while naively optimistic beliefs about an inevitable and unalterably bad event reduce anxiety and allow life to go on. In either case, the proper function of beliefs would be in opposition to Bayesianism.

Equally, what is wrong with holding incompatible beliefs, or at least alternating between them? Is it not useful, and not just for politicians? What better way is there to understand both sides of an argument, to empathize and to sympathize? Is it not also part of the creative process, as we seek to understand the conflict, to perhaps reframe the problem? A real theory of belief would illuminate that.

For example, every major scientist of the classical era held at least two incompatible beliefs: that the world is fully deterministic; and that scientific investigation is possible. Scientific investigation relies on creative hypothesizing, rational choice of methods, correct decision-making, due consideration of alternatives, and creative experimentation. Yet in a fully deterministic world, there are neither choices nor decisions nor considerations nor creativity: every event is inevitable. Even today, in a quantum world, there is still no satisfactory explanation of how there can be such a thing as free will, nor is there currently any prospect for its explanation. Yet it is a capacity that scientists and investigators of any kind must take for granted. The juxtaposition within each individual of these two incompatible beliefs, on the one hand in the conclusions of science, on the other in the possibility of science, did not mark the failure of rationality. On the contrary, it was a creative inspiration that gave us among other things the classical theory of probability, which combined ontological determinacy with epistemological indeterminacy, mislabelled as psychological uncertainty.

In short, Bunge has been completely correct that the nature and function of beliefs is a matter for psychology and related sciences, and the idea that there should be an a priori theory of them, invented by mathematicians and curated by statisticians, let alone economists, is ludicrous.

Now to some bad reasons for opposing Bayesianism. These are chiefly Bunge's accounts of what Bayesian statistical practice must actually be like. In particular, one example he has repeated several times concerns probabilities relating to Acquired Immunodeficiency Syndrome (AIDS) and the Human Immunodeficiency Virus (HIV) (e.g. Bunge 2006, pp. 116–117, 2008, p. 176).

Bunge considers that everyone with AIDS also has HIV, but not everyone with HIV has AIDS. Diagrammatically, the situation is as in Fig. 34.1.

The Venn diagram of Fig. 34.1 becomes a representation of a (probabilistic or random) sample space (statistical population), consisting of random outcomes where HIV or AIDS or neither are the case, if the Universe is taken to be a population of individuals, and an individual from the population is, was, or will be selected at random. The tense is of no matter because time does not occur in the theory of probability. The diagram could also represent a random sample space if the Universe were either a population or an individual, and the development or spread of AIDS and HIV were random, within the individual or the population. This alternative scenario is normally factually incorrect. It could be correct though: for example, in a panmictic population, were drugs that prevented the development

Fig. 34.1 Venn diagram of
the relationships between the
circumstances of having
AIDS or being infected with
HIV, or neither

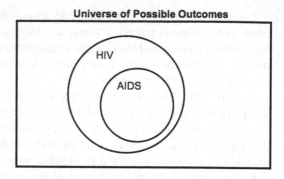

of AIDS also administered at random. The ontological differences between these and any other possible scenarios have no bearing on calculations made using the mathematical theory of probability, which is a part of pure mathematics and therefore ontologically carefree.

Consider then an individual with HIV who is, was, or will be selected at random from the general population.[13] With respect to this circumstance, one wants to know what the probability is of this random outcome also being one where AIDS is the case. According to Bayes' Theorem, $P(\text{AIDS} \mid \text{HIV}) = P(\text{HIV} \mid \text{AIDS})P(\text{AIDS})/P(\text{HIV})$.

For Bunge, who phrased the problem in terms of an individual who tests positive for HIV, the whole question and this answer are nonsense: the random selection has been actualized, contingency has been replaced by necessity, and *alea jacta est*. We are dealing with a specific individual, and this individual either has AIDS or not. Bayesianism would then lead to manifest error:

> Further, since the individual in question has tested HIV-positive, our Bayesian is likely to set $Pr(\text{HIV}) = 1$. And, since it is known that whoever has AIDS also has the HIV virus, $Pr(\text{HIV} \mid \text{AIDS}) = 1$. Thus, Bayes's formula simplifies to $Pr(\text{AIDS} \mid \text{HIV}) = Pr(\text{AIDS})$. However, this is known to be false: in fact, HIV is necessary but not sufficient to develop AIDS. (Bunge 2008, p. 176)

In fact that final clause is not the correct reason that the putative Bayesian conclusion is not generally true. For example, if everyone has HIV, then regardless of the fact that HIV is necessary but insufficient for AIDS, indeed $Pr(\text{AIDS} \mid \text{HIV}) = Pr(\text{AIDS})$: see Fig. 34.2 and the associated discussion. Instead the clause should read: (because) the prevalence of AIDS in the HIV population is not the same as the prevalence of AIDS in the overall population.

---

[13]Bunge has it instead as a random individual who tests positive for HIV, but this complicates the problem over his subsequent discussions, because positive tests have a probability of being false.

Fig. 34.2 Representation of a sample space consisting of random outcomes where HIV is the case and where AIDS or not-AIDS is the case. AIDS or not-AIDS is the case because individuals who have HIV may or may not have AIDS. HIV is the case because it was specified as the certain event. In this situation, $P(\text{AIDS} \mid \text{HIV}) = P(\text{AIDS})$

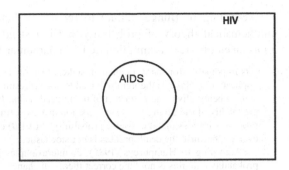

Yet disregarding that detail, Bunge's reasoning must still be wrong.[14] Bayes' theorem is a part of pure mathematics. It cannot fail, and especially not as a result of any ontological consideration about chance—even more so for Bunge, since for him chance does not occur in the mathematical theory.

In fact in solving this problem no probabilist and no statistician, Bayesian or otherwise, would reason as Bunge did. The probability of HIV is not 1, but instead corresponds to its prevalence in the population. Bayes' theorem does give the correct result, which in terms of the diagram is simply the relative size of the AIDS circle when compared to that of the HIV circle, rather than to the Universe oblong.

Nevertheless, if one were to reason as Bunge did—that is, as an actualist, rather than as a probabilist—Bayes' theorem would still give the correct result, as it is a theorem of pure mathematics. It cannot fail, regardless of any ontological consideration. Indeed following Bunge's reasoning, since he specified having HIV as being the certain event (for being actually the case), having HIV would by definition be the entire sample space. Diagrammatically the situation is as in Fig. 34.2, and the supposedly faulty inference is correct: having, through an actualist premise, disallowed any possibility other than having HIV, the probability of AIDS given HIV is no different from the probability of AIDS, regardless of any ontological consideration.

Of course the premise that having HIV is the certain event may be inconsistent with the rest of the set up—and as presented by Bunge, or at least with the caveats introduced here, it could be—but that is no matter for Bayesianism, rather a matter that probabilists of any kind can sort out. Not that actual Bayesianism would be totally innocent in this example either: many a Bayesian would be willing to replace a probability estimate derived from population prevalence with an estimate derived from opinion—albeit subject to *post hoc* revision, revision or updating or combining being central to Bayesian statistical methods.

---

[14]Bunge may have eventually realized this, as later he objected for different reasons:

> ...what is the use of this exercise? None, because what we need to know is the mechanisms of infection and immunity, and this is a matter for scientific research, not for logic. And if these mechanisms turn out to be causal rather than stochastic, we will not need the probability calculus either. We may call this objection *the argument from barrenness*. (Bunge 2010, p. 25)

According to Bunge, failure to recognize that chance does not occur in the mathematical theory of probability, but is instead something that must be added in as an ontological assumption tied to a particular factual model,

> ...is responsible for many popular mistakes. One of them is the belief that any mindless application of Bayes's theorem will lead to correct results... Here is one: Let A stand for the state of being alive, and B for that of being dead. Then, $Pr(B \mid A)$ may be interpreted as the probability of the Living $\rightarrow$ Dead transition in a given time interval. But, since the converse process is impossible, the inverse probability $Pr(A \mid B)$ does not exist in this case: it is not even nil, because the death process is not stochastic.
>
> According to Humphreys (1985), counterexamples like the preceding prove that the probability calculus is not "the correct theory of chance." (Bunge 2006, p. 101)

This reasoning must again be wrong: Bayes' theorem cannot fail, especially not as the result of any ontological consideration. Contrary to his explanation above and elsewhere, in revisiting the same example later, Bunge himself suggests the correct explanation: "this counterexample confutes the opinion that the transition probabilities calculated in quantum mechanics and other theories are conditional probabilities" (Bunge 2008, p. 172).

In fact Bunge's objection, suitably clarified, holds not just in quantum mechanics and other theories, but everywhere: transition probabilities from one outcome of a sample space to another outcome of the *same* sample space are *never* conditional probabilities on that sample space, *because such transitions never occur: all the outcomes that make up a sample space are mutually exclusive* (see Table 34.1). This is the reason for the theory of stochastic processes, where as mentioned in footnote 5, transition probabilities are instead modelled using a succession of random variables indexed by time. As in the earlier discussion here of the laws of large numbers (Sect. 34.5), a succession of random variables acts as a succession of events, or of sample spaces. In the theory of stochastic processes, conditional probabilities are used for transition probabilities, but now the conditions relate values of random variables further on in the succession with values of previous ones.

Yes, any individual outcome of a single sample space can itself be interpreted as an "occurrence", hence as a one-time transition—but from a previous state not specified by the theory, and having no role in it.

Consider the situation diagrammed in Fig. 34.3. The outcomes of this sample space can be random because they consist of individuals randomly selected from a

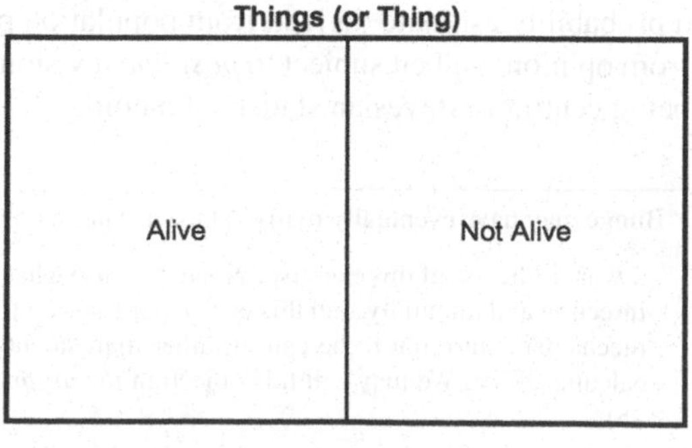

Fig. 34.3 Representation of a sample space consisting of random outcomes where being alive or not is the case

Things (or Thing)

Alive | Not Alive

population of alive and not-alive things; or because they consist of individuals, or an individual, randomly killed or left alive. The former set-up models transitions from *not-selected* to *selected*, and has nothing to do with transitions between alive and not-alive, because none occur. The latter two set-ups can be considered as modelling some stochastic transition or transitions from the state of being alive to being dead (because of a hidden assumption, that the only way to become either dead or alive is to have started out alive), but via absolute probabilities, not conditional ones. These ontological considerations of the set-up are irrelevant to correct probabilistic calculation: in any of the three set-ups, $P(\text{Alive} \mid \text{Not Alive})$ and $P(\text{Not Alive} \mid \text{Alive})$ *are both zero*, because the underlying intersection is null. That is to say, even in the case of a single individual randomly killed, the sample space consists of *both* of the two possible final outcomes, which are mutually exclusive—just as the sample space of a single coin toss consists of both of the two outcomes Heads and Tails, there being no transition between the two, and without reference to any previous state of the coin.

In other words, contrary to widespread interpretation (as in e.g. Humphreys 1985), conditional probabilities on a single sample space refer neither to transitions between the events they relate, nor to the generation of one of the events from the other, but—as the name suggests—to logical *conditions*, that is to say they are restrictions on the sample space, or in other words renormalizations.

34.7 Leaving as Little to Chance as Possible

The classical theory of probability was born and raised in a deterministic world. Or rather, within a deterministic worldview. This worldview reached its zenith with Laplace, master also of the classical theory of probability. The young Bunge explained this juxtaposition sarcastically, as a mark of mere inconsistency (Bunge 1951, p. 214). In this same work and many later ones though, Bunge held that "If everything is strictly predetermined, chance does not exist objectively and is but an aspect of our ignorance", "doomed to disappear with the progress of knowledge" (Bunge 1951, p. 213). By tying chance's doom in a fully deterministic world to the progress of knowledge, Bunge allows chance life in that same world as long as ignorance persists. If the theory of probability works when faced with ignorance in a deterministic world, why should it not work when faced with ignorance in a chancy world?

Consider another of Bertrand's famous paradoxes, the Paradox of the Drawers—but this time in its simplified modern form:

Monty Hall or Let's Make a Deal *Problem*
    Consider a game with two players, host and contestant. The host hides a valuable item behind one of three doors. The contestant guesses which door hides the item, and wins it if the guess is correct. Once the guess is made, but before revealing whether it was correct, the host does as follows: if the guess was correct, reveal that the item was not behind an arbitrary one of the other doors. If the guess was incorrect, reveal that the item was not behind the remaining incorrect door. Then ask the contestant if they want to remain with their original guess, or switch.

Problem: To determine which is the optimal strategy for the contestant: always stay with the original guess, always switch, switch or stay randomly, or use some other method.

The problem seems to have originally appeared in this form[15] in *Parade* magazine, as a question sent in by a reader, one Craig F. Whitaker of Columbia, Maryland, to *Parade's* columnist Marilyn vos Savant (1990). She affirmed that the best solution was to always switch. Thereupon thousands, many with PhDs, wrote in to correct her, usually to say it did not matter. Who was right?

Anyone who has read this far in this volume knows Bunge's answer: everyone is wrong. Rewards are a matter of social justice, and they should only be granted on the basis of merit and need, not luck. Do not gamble with justice, but go out and earn your living by doing something socially valuable.

Of course this answer will not satisfy everyone, or maybe anyone. But it does hold the truth: indeed, everyone is wrong. The best strategy is simply to stick with your original guess when you got it right, and switch when you got it wrong. Whether the guess started out right or not is another matter entirely.

This answer may seem facetious but it is not. The serious gambler leaves as little to chance as possible, and instead studies the game as it is played. The serious gambler looks for ticks or mannerisms that reveal the host's hand, or patterns in the choice of winning number that might reveal a failed randomization mechanism (Matthews 2003). Likewise the person seriously interested in not becoming "a statistic"—such as a victim of crime, or of a transportation accident—does not simply rely on platitudinous general statistics showing the result is rare: they study how the events, rare or not, come about, and look to avoid the generating circumstances.

34.8 Probabilistic Reasoning

Suppose in the Monty Hall problem the host's choice of doors is random, and the contestant's first guess also random. Or that only either one is random. Then the first guess will have probability 2/3 of being wrong, and the best strategy is to switch.

What though if both the choice of door and the first guess are always according to a fixed method—in particular, a constant choice? For example, the host might always put the prize behind door number 1, and the contestant might always guess door number 3. Is there anything probabilistic reasoning can bring to the situation?

---

[15]Bunge has mentioned this problem by name (Bunge 2012, p. 107), but without the second step that makes it interesting. He has also often commented on what sounds like it should be an equivalent problem, phrased in terms of three of the Apostles. It is inequivalent because the contestant does not make any selection, let alone two, only the "host" does. The paradox can be resolved similarly though. Bunge holds that problem to be effectively the same as the Monty Hall problem, and that in both cases no chance is involved and it makes no sense to talk about the probability of winning (e.g. Bunge 2003, pp. 229–230).

In Bunge's view, the answer must obviously be no: *alea jacta est*, or rather there never was any *alea* to begin with. Suppose though the game is not to be repeated, but played only once. The contestant may know the door is chosen according to some fixed, non-random method decided in advance, and the host may know the contestant does the same. But in any combination, neither host nor contestant knows what the other did. When playing the game, should the contestant stay with their original selection, or switch?

All that matters is that, however the non-random decisions were made, they were made *independently* of each other. In statistical terms, that their outcomes are not correlated. If so, then nothing has changed: the best strategy is to always switch. (Consider a hypothetical infinite ensemble of outcomes of the first guess, each one corresponding to one combination of choice by host and by contestant. By virtue of their choices being uncorrelated, the relative frequency of mismatched choices must be 2/3. Exactly as was done when considering the laws of large numbers in Sect. 34.5, the natural interpretive conclusion is that this is because in any single attempt, there is a 2/3 tendency for mismatch to be the case.) Thus ignorance, determinism *and independence* combine to make probabilistic reasoning relevant—just as in Laplace's time.

One may well ask by what physical processes events become independent. For events that start off somehow connected, the usual answer is by shaking, shuffling, exploding, or the like. Of course the situation is not so simple.[16] For example, probabilists know that in coin tossing, randomization may come not from the toss—which can bias the outcome, by anywhere from a small amount up to 100%—but from the initial and final conditions, something stage magicians are able to exploit (Diaconis et al. 2007). Likewise in card shuffling it is a famous result that to randomize a deck requires in general at least seven riffles—although stage magicians can defeat that too (Bayer and Diaconis 1992). Kolmogorov's answer:

> We thus see, in the concept of independence, at least the germ of the peculiar type of problem in probability theory. [...] In consequence one of the most important problems in the philosophy of the natural sciences is—in addition to the well-known one regarding the essence of the concept of probability itself—to make precise the premises which would make it possible to regard any given real events as independent. This question, however, is beyond the scope of this book. (Kolmogorov 1933/1956, p. 9)

34.9 Bayesian and Frequentist Statistics in Transition

Bayesianism today is in a state analogous to that of calculus in its early years. While calculus had its ghosts of departed quantities, Bayesianism has its personalist probabilities, and priors not derived from data or a scientific understanding of the

---

[16]At one time Bunge explicitly claimed it possible to go from non-probabilistic premises to probabilistic ones using the method of arbitrary functions (Bunge 1969, p. 445), but that was incorrect and in his writings since then he has consistently asserted the contrary.

underlying processes. In some ways Bayesianism is in better shape: the standard of mathematical rigour is higher, the understanding of the issues greater. In some ways it is in worse shape: the ghosts of departed quantities eventually materialized, but there are unresolvable technical difficulties with standard Bayesianism (Cox 2006; Cox and Mayo 2010), while personalist probabilities are at best no more valuable than any other guesses, with far from best being the typical case (Berger 2006). Nevertheless standard Bayesianism led to today's empirical Bayesianism (Berger 2006; Efron 2013), and even Fisher and Gini had presaged its formal development in their solutions to certain special problems (see Petrone et al. 2014). Moreover the Bayesian challenge has helped to clarify problems with standard frequentist methods, and vice versa, with both benefitting from the contest (Efron 2013).

Bunge has unwittingly given empirical Bayes methods his stamp of approval: in his *Medical Philosophy* (Bunge 2013), he cites approvingly a paper by Ioannidis (2005)—whose famous result relies on a Bayesian analysis, with priors that are reasonable enough factually. Under certain conditions Bayesian and frequentist inferences are the same (Reid and Cox 2015; Petrone et al. 2014).

Statisticians are famously divided between frequentists and Bayesians. Where are the propensity statisticians? The answer is a bit odd: nowhere; and everywhere. Nowhere, because if taken in the strictest terms (Bunge's[17]), it makes much of inferential statistics effectively impossible. This is chiefly because much of statistics is concerned not with future potentiality, but with estimating unknown parameters (Reid and Cox 2015), and usually without the benefit of randomization. For example, consider the problem of deducing the correct chronological sequence of archeological or historical artifacts (e.g. Cox and Brandwood 1959); or the problem, not of whether a finite table of numbers was *generated* randomly—impossible to solve by analysis of the table—but of whether it *is* random (Volchan 2002). The latter is randomness via patternlessness, because this route should lead to independence from whatever the table of putatively random numbers is to be used for—as long as the table is not used twice in the same way in the same endeavour. There seems to be no record of any statistician overtly advocating and practicing a propensity interpretation.[18]

On the other hand "propensity statisticians" are everywhere, because whether for past events or future, whether for estimating unknown parameters or prediction, whether in considerations of uncertain knowledge or hypothetical long-run frequencies, both practicing Bayesians and frequentists often end up talking about probabilities and chances as if they represented objective tendencies to be the case. Thinking from the start about probabilities in terms of tendencies might help frequentist statisticians get rid of some of even their unhelpful and misleading

---

[17]Bunge has also objected to the propensity label, preferring to call his own version the objectivist or realist interpretation (Bunge 2006, p. 103).

[18]"Propensity methods" is sometimes used as a description for propensity score matching methods as invented by Rubin, but they are not a reconceptualization of statistics in propensity terms, and include a Bayesian aspect. See Rubin (2007).

subjective orientation: for example, it might lead to their renaming *confidence* intervals as parameter *capture* or *coverage* intervals, statistical *significance* as statistical *resolution* or *discrimination*, *significant* as *resolvable* or *separable*, and *uncertainty* as *indeterminacy*, ontological or epistemological.

34.10 A Riddle

While Bunge dismisses probability from the Monty Hall problem, or when considering whether a given patient with HIV also has AIDS, he does allow that

> ... it is important to know the likelihood of having breast cancer when a mammogram is positive, as well as the likelihood that mammography reveals the presence of cancer. But it would be wrong to insert either in Bayes's formula, because neither cancer nor mammography tests are chance events. (Bunge 2003, p. 228)

There is a riddle here: how can there be likelihood without chance? Does not chance just mean opportunity? And if there is likelihood in the sense Bunge is using it, why is it not quantifiable, and why should it not behave mathematically exactly like probability?[19] Is it not the case that, as explained in Sect. 34.1, likelihood is the key to applying probability theory in any particular problem?

The solution to the riddle lies in *independence*. Whether or not the die is cast is only a matter of when independence gets introduced; and whether or not, through knowledge, it can get undone. Moreover, a physically random process need not be the only way to obtain independence. In this way, so far the only thing that stops the decimal digits of π from being a perfect sequence of random numbers is our familiarity with them (Kirschenmann 1973; Bailey and Borwein 2014). Similar considerations apply for any sequence generated by a formula, or for that matter any system of actualities: for uniformly, normally, or otherwise distributed randomness, the resulting outcomes need only follow the desired distribution, and not result in any unwanted correlations. It is not known whether any formulas or algorithms or deterministic natural processes for generating such arrangements exist—but so far, there is no reason to deem them impossible in principle.

34.11 Conclusion

All in all, and in interesting considerations too numerous and diverse to have captured completely here, Bunge's views on the nature of randomness mostly hold up, only he has been too strict in their application. Likewise he has been too ready to view a deformity as the plague, too unwilling to accept that disabilities in one

---

[19]Likelihoods as used in the technical statistical sense do not obey the laws of probability (see e.g. Cox and Mayo 2010).

area may heighten abilities in another, and too willing to abandon to an unanalyzed version of "likelihood" those problems that cry out for probabilistic reasoning, despite the underlying circumstances being fixed.

Acknowledgement I thank Michael Matthews and Paul McColl for their respective contributions in shepherding the manuscript through its various stages.

References

Bailey, D. H., & Borwein, J. (2014). Pi Day is upon us again and we still do not know if Pi is normal. *American Mathematical Monthly, 121*(3), 191–206. https://doi.org/10.4169/amer. math.monthly.121.03.191.

Bayer, D., & Diaconis, P. (1992). Trailing the dovetail shuffle to its lair. *Annals of Applied Probability, 2*(2), 294–313.

Berger, J. (2006). The case for objective Bayesian analysis. *Bayesian Analysis, 1*(3), 385–402.

Borwein, J. M., Galway, W. F., & Borwein, D. (2004). Finding and excluding *b*-ary Machin-type BBP formulae. *Canadian Journal of Mathematics, 56*, 1339–1342.

Bunge, M. (1951). What is chance? *Science & Society, 15*(3), 209–231.

Bunge, M. (1967). *Foundations of physics* (Springer tracts in natural philosophy, Vol. 10). New York: Springer.

Bunge, M. (1969). Corrections to foundations of physics: Correct and incorrect. *Synthese, 19*, 443–452.

Bunge, M. (1977). *Ontology I: The furniture of the world* (Treatise on basic philosophy, Vol. 3). Dordrecht: Reidel.

Bunge, M. (1979). *Ontology II: A world of systems* (Treatise on basic philosophy, Vol. 4). Dordrecht: Reidel.

Bunge, M. (1981). Four concepts of probability. *Applied Mathematical Modelling, 5*, 306–312.

Bunge, M. (1983). *Epistemology & methodology I: Exploring the world* (Treatise on basic philosophy, Vol. 5). Dordrecht: Reidel.

Bunge, M. (1985). *Epistemology & methodology III: Philosophy of science and technology—Part I—formal and physical sciences* (Treatise on basic philosophy, Vol. 7). Dordrecht: Reidel.

Bunge, M. (2003). *Emergence and convergence: Qualitative novelty and the Unity of knowledge.* Toronto: University of Toronto Press.

Bunge, M. (2006). *Chasing reality: Strife over realism.* Toronto: University of Toronto Press.

Bunge, M. (2008). Bayesianism: Science or pseudoscience? *International Review of Victimology, 15*, 165–178.

Bunge, M. (2010). *Matter and mind: A philosophical inquiry* (Boston studies in the philosophy of science, Vol. 287). Dordrecht/Heidelberg/London/New York: Springer.

Bunge, M. (2012). *Evaluating philosophies* (Boston studies in the philosophy of science, Vol. 295). Dordrecht/Heidelberg/New York/London: Springer.

Bunge, M. (2013). *Medical philosophy: Conceptual issues in medicine.* Singapore: World Scientific.

Carriquiry, A. L., & Pawlovich, M. (2004). From empirical Bayes to full Bayes: Methods for analyzing traffic safety data. Iowa Department of Transportation, Office of Traffic and Safety. https://www.iowadot.gov/crashanalysis/pdfs/eb_fb_comparison_whitepaper_october2004.pdf. Accessed 13 June 2018.

Cox, D. R. (2006). Frequentist and Bayesian statistics: A critique (keynote address). In *Statistical problems in particle physics, astrophysics and cosmology: Proceedings, PHY-STAT05, Oxford, UK, 12–15 September 2005*, World Scientific, Singapore, pp. 3–6. https://doi.org/10.1142/9781860948985_0001.

Cox, D. R., & Brandwood, L. (1959). On a discriminatory problem connected with the works of Plato. *Journal of the Royal Statistical Society, Series B (Methodological), 21*(1), 195–200.

Cox, D. R., & Mayo, D. (2010). Objectivity and conditionality in frequentist inference. In D. Mayo & A. Spanos (Eds.), *Error and inference: Recent exchanges on experimental reasoning, reliability and the objectivity and rationality of science* (pp. 276–304). Cambridge: Cambridge University Press. http://www.phil.vt.edu/dmayo/personal_website/ch%207%20cox%20%26%20mayo.pdf. Accessed 11 Feb 2018.

Diaconis, P., Holmes, S., & Montgomery, R. (2007). Dynamical bias in the coin toss. *SIAM Review, 49*(2), 211–235.

Efron, B. (2013). A 250-year argument: Belief, behavior, and the bootstrap. *Bulletin (New Series) of the American Mathematical Society, 50*(1), 129–146.

Hauer, E. (1997). *Observational before—after studies in road safety: Estimating the effect of highway and traffic engineering measures on road safety.* Amsterdam: Pergamon.

Humphreys, P. (1985). Why propensities cannot be probabilities. *The Philosophical Review, 94*(4), 557–570.

Ioannidis, J. P. A. (2005). Why most published research findings are false. *PLoS Medicine.* https://doi.org/10.1371/journal.pmed.0020124.

Kirschenmann, P. (1973). Concepts of randomness. In M. Bunge (Ed.), *Exact philosophy: Problems, tools, and goals* (pp. 129–148). Dordrecht: Reidel.

Kolmogorov, A. N. (1933/1956). *Foundations of the theory of probability (Second English Edition)* (N. Morrison, Trans.). New York: Chelsea Publishing Company.

Laplace, P. S. (1814). *Théorie Analytique des Probabilités* (2nd ed.). Paris: Courcier.

Matthews, R. (2003). QED: How to spot the pattern in a game of chance. *The Telegraph,* 17 July. https://www.telegraph.co.uk/technology/3310656/QED-How-to-spot-the-pattern-in-a-game-of-chance.html. Accessed 18 June 2018.

Persaud, B., & Lyon, C. (2007). Empirical Bayes before–after safety studies: Lessons learned from two decades of experience and future directions. *Accident Analysis and Prevention, 39,* 546–555. https://doi.org/10.1016/j.aap.2006.09.009.

Petrone, S., Rizzelli, S., Rousseau, J., & Scricciolo, C. (2014). Empirical Bayes methods in classical and Bayesian inference. *Metron, 72,* 201–215.

Reid, N. (2013). Aspects of likelihood inference. *Bernoulli, 19*(4), 1404–1418.

Reid, N., & Cox, D. R. (2015). On some principles of statistical inference. *International Statistical Review, 83*(2), 293–308.

Rubin, D. M. (2007). The design versus the analysis of observational studies for causal effects: Parallels with the design of randomized trials. *Statistics in Medicine, 26,* 20–36. https://doi.org/10.1002/sim.2739.

Shafer, G., & Vovk, V. (2006). The sources of Kolmogorov's *Grundbegriffe. Statistical Science, 21*(1), 70–98.

Tervo, D. G. R., Proskurin, M., Manakov, M., Kabra, M., Vollmer, A., Branson, K., & Karpova, A. Y. (2014). Behavioral variability through stochastic choice and its gating by anterior cingulate cortex. *Cell, 159*(1), 21–32. https://doi.org/10.1016/j.cell.2014.08.037.

Volchan, S. B. (2002). What is a random sequence? *American Mathematical Monthly, 109*(1), 46–63.

vos Savant, M. (1990). *Game show problem.* http://marilynvossavant.com/game-show-problem/. Accessed 11 Feb 2018.

Chapter 35
Dual Axiomatics

Reinhard Kahle

Abstract Mario Bunge forcefully argues for Dual Axiomatics, i.e., an axiomatic method applied to natural sciences which explicitly takes into account semantic aspects of the concepts involved in an axiomatization. In this paper we will discuss how dual axiomatics is equally important in mathematics; both historically in Hilbert and Bernays's conception as well as today in a set-theoretical environment.

35.1 Introduction

More than half a century ago, Mario Bunge (1967b) started out to axiomatize physical theories with a philosophical aim: "to rid them of the subjectivistic elements that had been smuggled into them by the logical positivists" (Bunge 2017b, p. 696). He opposed, in particular, "the structuralist [strategy] defended by Suppes, Sneed, Stegmüller, Moulines, and other philosophers [...] which ignores the semantic side of scientific theories" (Bunge 2017b, p. 701). As alternative he counterposed *Dual Axiomatics* which "concerns not just the formalism, but also the meaning (reference and sense) of the key concepts" (Bunge 2017b, Abstract). Bunge (1967a,b,c, 1976, 1980, 2009, 2017a,b) successfully applied and defended dual axiomatics for physical and other theories.

This work is partially supported by the Portuguese Science Foundation, FCT, through the projects UID/MAT/00297/2019 (Centro de Matemática e Aplicações) and PTDC/MHC-FIL/2583/2014 (Hilbert's 24th Problem), and by the Udo Keller Foundation.

R. Kahle (✉)
Theorie und Geschichte der Wissenschaften, Universität Tübingen, Tübingen, Germany
CMA, FCT, Universidade Nova de Lisboa, Caparica, Portugal
e-mail: kahle@mat.uc.pt

633

We recall here the first four of the seven main virtues of dual axiomatics as summarized by Bunge[1]:

DA1. *it preserves the form/content*, a priori/a posteriori, and rational/factual dualities;

DA2. *it unveils the tacit assumptions*, in particular the imprecise or false ones—that is, the place where the dog was buried, as a German would put it;

DA3. *it reminds one from start to finish which are the referents* of the theory, which prevents philosophical and ideological contraband [...];

DA4. *it helps disqualify extravagances* such as the many-worlds and branching universes fantasies [...]. (Bunge 2017b, §7)

In this paper, we show that dual axiomatics, understood as an axiomatics which takes semantics into account, is also the natural account in mathematics. It was, indeed, explicitly identified as such by Hilbert and Bernays in the Introduction of their magnum opus *Grundlagen der Mathematik* (Hilbert and Bernays 1934, 1939). But also modern set-theoretic structures serve dual axiomatics, if they are understood as "faithful representations" of "domains of actuality".

The paper finishes with three examples which show dual axiomatics at work.

35.2   Axiomatic Method

The axiomatic method is known since Euclid's time as a form to structure mathematical knowledge in a systematic way. Originally, it pretended to start with undisputed sentences as *evident truths* and to ensure that derivations by *correct rules* will, thus, produce only *true consequences*. Even outside mathematics this method was employed, most notable by Spinoza in his Ethics *more geometrico* (Spinoza 1677). It is Mario Bunge's merit to constantly underline the importance and fruitfulness of axiomatization in physics.

In mathematics, the understanding of axioms as evident truths was undermined with the discovery of non-euclidean geometry. However, the function of axiomatics as a tool to order and systematize an arbitrary area of knowledge gained ground. David Hilbert, who applied the axiomatic method masterly in his *Grundlagen der Geometrie* (Hilbert 1899), the precedent-setting reconstruction of euclidean geometry, fervently advocated the application of this method in- and outside mathematics (Hilbert 1900, 1918).

It is important to observe that Hilbert had a form of axiomatization in mind which was supposed to capture the properties of a given realm, like numbers, geometry, but also areas of physics: he explicitly mentions statics, mechanics, electrodynamics, and gases (Hilbert 1918, p. 405). We would like to call such a form of axiomatization, which refers to a predefined area of knowledge or facts,

---

[1]The labels are adapted for later references. The fifth ("*it exhibits the legitimate components of the theory as well as their deductive organization, and with it the logical status of each constituent*") and seventh ("*it eases the understanding and memorization of theories*") have to taken for granted in mathematics; the sixth one ("*it facilitates the empirical test of theories*") is, however, specific for empirical theories only.

characterizing axiomatics. This form has to be distinguished from the modern conception of abstract or *generalizing axiomatics*, where abstract properties of operations are formalized with an intentional neglect of the objects which are subject to such operations.[2] Generalizing axiomatics became dominant by modern algebra, with the notion of group as paradigmatic example, and it found some kind of finalization in Bourbaki's notion of structure.[3]

Hilbert also paved the way for reinterpretations of a given axiomatization, as expressed in his famous dictum about the axioms of geometry that "one must be able to say 'tables, chairs, beer-mugs' each time in place of 'points, lines, planes' ".[4] Bunge is open to such reinterpretations when he points to Hideki Yukawa's meson theory which "was long assumed to concern mu mesons, until it was found to describe pi mesons" (Bunge 2017b, p. 701). But these are still characterizing axiomatizations, where just the domain is changed. It is worth noting that Hilbert himself did not really studied abstract axiomatizations.[5]

He was, of course, aware of the development of abstract axiomatics, in particular in the hands of Emmy Noether in Göttingen. But Bourbaki's characterization of mathematics as "a storehouse of abstract forms" (Bourbaki 1950, p. 231) cannot be used as Hilbert's. Even in this case, when it comes to the application of these abstract forms in natural or other sciences, such an application is, however, founded in the addition of the previously neglected semantic part of dual axiomatics.

The examples in Hilbert's work concern, first of all, characterizing axiomatics.[6] Such a characterizing axiomatics comes itself in two forms, which Hilbert and Bernays discuss at the very beginning of the *Grundlagen der Mathematik*: *inhaltliche* and *formale Axiomatik* (contentual and formal axiomatics).

Contentual axiomatics can rely on a special epistemic relation to the specific subject area. But formal axiomatics—the result of the axiomatic method—diverges in two respects from contentual axiomatics: (1) it abstracts from the subject matter

---

[2]The distinction of characterizing and generalizing axiomatics is not new; one may find it in variations and under different names at several place. We like to mention Salt (1971) who uses, cum grano salis, *interpreted axiomatic theory* for our characterizing axiomatics and *abstract theory* for generalizing axiomatics; he also refers to Bernays's respective distinction of *material* or *pertinent axiomatics* and *descriptive axiomatics* (Bernays 1967). It is not by accident that Salt can use this distinction to defend Bunge against Freudenthal (1970) who apparently misread Bunge (1967b) when he took interpreted axiomatic theories as abstract ones.

[3]See, for instance, Corry (2002).

[4]Blumenthal (1935, p. 403), translated in Grattan-Guinness (2000, p. 208).

[5]This is very nicely illustrated by the following recollection of Saunders MacLane:

> In Hilbert's original work on integral equations, a point in a Hilbert space was an infinite sequence of complex numbers y_n with the sum of squares convergent. But von Neumann's lecture began with: 'Take a Hilbert space—an infinite-dimensional vector space over the complex numbers complete in a positive definite norm.' At the end of the lecture Hilbert asked Professor von Neumann, 'I would like to know, just what a Hilbert Space is.' Hilbert thought of his spaces concretely, not axiomatically. (MacLane 2005, p. 49f)

[6]For instance, a lecture of Hilbert of 1913 (Hilbert 1913) contains a chapter entitled "axioms of algebra"; but it does not axiomatizes Algebra in the modern sense, but just the real numbers.

and (2) involves simplifying idealizations of the actual state of affairs.[7] For this reason, formal axiomatics requires proofs of consistency, the subject of *Hilbert's Programme*. For the topic of this paper, however, we will focus on another, largely neglected, aspect of the axiomatic method, namely, that even formal axiomatics roots in contentual axiomatics:

> Die formale Axiomatik bedarf der inhaltlichen notwendig als ihrer Ergänzung, weil durch diese überhaupt erst die Anleitung zur Auswahl der Formalismen und ferner für eine vorhandene formale Theorie auch erst die Anweisung zu ihrer Anwendung auf ein Gebiet der Tatsächlichkeit gegeben wird. (Hilbert and Bernays 1934, p. 2)[8]

Thus, Hilbert and Bernays state that dual axiomatics is an integral part of the axiomatic method—and they vindicate DA1. Dual axiomatics is present in mathematics in the very same way as Bunge is advocating it in natural (and other) sciences.

35.3   Hilbert's *Inhaltlich*

At the time Hilbert started his investigation of axiomatics, an interpretation of a formal system required a "domain of actuality", as one can find it, for instance, in physics. In particular, geometry was traditionally seen as a discipline describing our physical space. Therefore, non-euclidean geometry was initially hardly acceptable as an alternative, as there is only one physical space. Only when mathematicians were providing models of non-euclidean geometries within euclidean geometry, the former ones became acceptable as now one was able to interpret them.

A key term in Hilbert's axiomatic conception is the one of "inhaltlich" (contentual). Whenever a "domain of actuality" is available, "contentual" will clearly refer to it. But for mathematics, in particular when it starts to deal with infinity, it is quite unclear what the proper "domain of actuality" should be; thus, with his axiomatic method, Hilbert set out to eliminate references to contentual assumptions

---

[7]See Hilbert and Bernays (1934, pp.1 and 2f.); the corresponding sentences read in German:

> Eine Verschärfung, welche der axiomatische Standpunkt in HILBERTs „Grundlagen der Geometrie" erhalten hat, besteht darin, daß man von dem sachlichen Vorstellungsmaterial, aus dem die Grundbegriffe einer Theorie gebildet sind, in dem axiomatischen Aufbau der Theorie nur dasjenige beibehält, was als Extrakt in den Axiomen formuliert ist, von allem sonstigen Inhalt aber abstrahiert.

and

> Andererseits können wir bei der inhaltlichen Axiomatik deshalb nicht stehenbleiben, weil wir es in der Wissenschaft, wenn nicht durchweg, so doch vorwiegend mit solchen Theorien zu tun haben, die gar nicht vollkommen den wirklichen Sachverhalt wiedergeben, sondern eine *vereinfachende Idealisierung* des Sachverhaltes darstellen und darin ihre Bedeutung haben.

[8]Formal axiomatics requires contentual axiomatics as a necessary supplement. It is only the latter that provides us with some guidance for the choosing the right formalism, and with some instructions on how to apply a given formal theory to a domain of actuality. (Hilbert and Bernays 2011, p. 2)

in mathematical axiom systems as much as possible and to replace them by purely formal considerations. His finitist mathematics was somehow the last "contentual kernel" he was wanting to accept.[9]

In this context, contentual mathematics is limited to symbolic manipulations of signs which can be overseen by simple inspections.[10] But one may note that Hilbert is promoting such a restriction only within his foundational programme as a tactical move against intuitionists: if higher mathematics could be justified on the base of finitist mathematics through consistency proofs, intuitionists—who would have to subscribe to finitist mathematics—would be committed to higher mathematics, too.

Here we will not discuss the fate of Hilbert's programme and its consequences for intuitionism,[11] but we like to point out that, outside his foundational programme, Hilbert is, of course, open to consider (higher) mathematics in contentual terms.[12] On the one hand, this is evident from his book "Anschauliche Geometrie" (translated as *Geometry and the Imagination*), written by Cohn-Vossen (1932; 1952). On the other hand, when criticizing Zermelo's set theory as containing "echte inhaltliche Annahmen" (Hilbert 1928, p. 2) (*real contentual assumptions*) he admits them as possible base; he just dismisses them as base for "absolute security" in a foundational perspective. The "contentual axiomatics" as mentioned by Bernays in the citation above is clearly meant to related to arbitrary domains, not just finitist mathematics.

---

[9]Weyl (1944, p. 640f.) comments on this:

> But how to make sure that the "game of deduction" never leads to a contradiction? Shall we prove this by the same mathematical method the validity of which stands in question, namely by deduction from axioms? This would clearly involve a regress *ad infinitum*. It must have been hard on Hilbert, the axiomatist, to acknowledge that the insight of consistency is rather to be attained by intuitive reasoning which is based on evidence and not on axioms.

And further on:

> Incidentally, in describing the indispensable intuitive basis for his *Beweistheorie* Hilbert shows himself an accomplished master of that, alas, so ambiguous medium of communication, language. With regard to what he accepts as evident in this "metamathematical" reasoning, Hilbert is more papal than the pope, more exacting than either Kronecker or Brouwer. [...] Elementary arithmetics can be founded on such intuitive reasoning as Hilbert himself describes, but we need the formal apparatus of variables and "quantifiers" to invest the infinite with the all important part that it plays in higher mathematics. Hence Hilbert prefers to make a clear cut: he becomes strict formalist in mathematics, strict intuitionist in metamathematics.

[10]Recall Hilbert's "*Am Anfang ist das Zeichen.*" (In the beginning is the sign.) and his discussion around this citation (Hilbert 1922, p. 163).

[11]See, for instance, Kahle (2015).

[12]Even less, he could be considered as a naive formalist, as already Bernays (1975, p. 2) stressed: "Yet it was certainly not Hilbert's intention that mathematics should consist only of proof theory (though some of his statements describing his attitude to metamathematics might suggest this view)." See also the discussion in Kahle (2019).

35.4   Set-Theoretical Structures

From a modern perspective, it is defensible to read Hilbert's "inhaltlich" as *semantically*.[13] When one has a clear semantic understanding of a mathematical domain, contentual axiomatics has simply to comply with this semantics. Here, semantics is understood in a broad sense, covering any "domain of actuality". But there is also a narrow reading of semantics, referring to set-theoretical structures which are used to interpret formal languages. Such structures were introduced in mathematical logic by Tarski at a time Hilbert had already left the stage.[14] But they can help to sharpen "contentual axiomatics" as it can be expressed now in set-theoretical terms. Set-theoretical structures fulfill already the two key requirements Bernays identified for formal axiomatics: (1) Abstraction from the subject matter and (2) simplifying idealizations. Thus, they can be placed as a layer between the formal axiomatics and the contentual axiomatics, the latter one understood as relating to a concrete "domain of actuality".

Using set-theoretic structures as semantics is a rather convenient way for mathematicians to acquit oneself of foundational questions: as essentially every mathematical theory can be recast in set-theoretical terms, consistency is reduced to consistency of ZFC (or the like). And as long as one has sufficient set-theoretic intuition, it serves perfectly as semantics for the mathematical realm under consideration.

Surely, this is not the original semantics but only a "faithful representation" of it. Moschovakis discusses this issue properly:

> A typical example of the method we will adopt is the "identification" of the (directed) geometric line Π with the set $\mathbb{R}$ of real numbers, via the correspondence which "identifies" each point P on the line with its coordinate $x(P)$ with respect to a fixed choice of an origin O. What is the precise meaning of this "identification"? *Certainly not that points are real numbers.* [...] What we mean by the "identification" of Π with $\mathbb{R}$ is that the correspondence $P \mapsto x(P)$ gives a **faithful representation** of Π in $\mathbb{R}$ which allows us to give arithmetic definitions for all the useful geometric notions and to study the mathematical properties of Π **as if points were real numbers.** [...] In the same way, we will discover within the universe of sets *faithful representations* of all the mathematical objects we need, and we will study set theory on the basis of the lean axiomatic system of Zermelo **as if all mathematical objects were sets.** The delicate problem in specific cases is to formulate precisely the correct definition of a "faithful representation" and to prove that one such exists. (Moschovakis 2006, Chapter 4: Are Sets All There Is?, p. 33f.)

Thus, set-theoretic structures serve as semantics, in the sense of dual axiomatics, only modulo Moschovakis's "identification".

---

[13] As semantics should provide *meaning*, this is in accordance with Georg Kreisel, who proposed to translate "inhaltlich" by "meaningful" (in a personal communication to Jan von Plato).

[14] Thus, it would be ahistorical to read Hilbert's "inhaltlich" just as semantic in terms of set-theoretical structures—still, such a reading will probably match to a large extent with Hilbert's understanding of the term.

35.5 Semantics Comes in

In the following, three examples are discussed where semantic preconceptions appear to interfere with an axiomatic approach. These examples will illustrate how the semantic side of dual axiomatics has to be taken serious, exemplifying the virtues DA2, DA3, and DA4.

35.5.1 The Parallel Axiom in Euclidean Geometry

The independence of the Euclid's parallel axiom from the others was a everlasting question in mathematics before the discovery of non-euclidean geometry resolved the question. While the independence was already discussed in antiquity, it was revived in Europe by Clavius's Euclid edition of 1574.[15] Clavius did not only report on the historic attempts, but provided a "proof" of the parallel axiom, based on the assumption that a curve in constant distance to a straight line is indeed also a straight line.

What is characteristic for this attempt—as well as many others—to prove the parallel axiom is that the author is convinced of the "absolute correctness" of a(nother) property which is used to prove the parallel axiom. Of course, this property is indeed correct in euclidean geometry, but not absolute in the sense that it can be derived from the other Euclidean axioms.

Taking the semantic part of dual axiomatics serious one should be, by virtue of DA2, judicious to identify such assumptions as the one of Clavius as "tacit".

35.5.2 Insane Models of Peano Arithmetic

Gödel's Incompleteness Theorems imply that first-order Peano Arithmetic admits non-standard models. Gaisi Takeuti reports on this:

> [Gödel's] way of teaching nonstandard models was an interesting one. It went as follows. Let T be a theory with a nonstandard model. By virtue of his Incompleteness Theorem, the consistency proof of T cannot be carried out within T. Consequently, T and the proposition "T is inconsistent" is consistent. There is, therefore, a natural number N which is the Gödel number of the proof leading to a contradiction from T. Such a number is obviously an infinite natural number. (Yasugi and Passell 2003, p. 3)

Such models, validating a formalized inconsistency statement of the form $\neg\mathsf{Con}_{\mathsf{PA}}$, are clearly not intended to be studied by an axiomatization of Arith-

---

[15]Clavius (1574); we follow here the general presentation of the problem in Scriba and Schreiber (2010, §6.4).

metic.[16] And the slightest semantic consideration will tell us that $\neg\text{Con}_{PA}$ is *false*.[17] In this way, we consider it as an instance of DA4. It is, in fact, just the shortcoming of first-order logic which doesn't allow us to fix the domain of the quantifiers to standard natural numbers.[18]

Bernays (1967) identified this problem—the existence of non-standard models for first-order theories—as the "limits of axiomatics". Dual axiomatics invites us to restrict the quantifiers in Peano Arithmetic to standard natural numbers, but one has to recognize that there are no tools to do so on the formal side. This, however, is just one more reason to value the semantic side, as Bernays (1978, p. 24) concluded:

> In allen diesen neueren Ergebnissen kommt eine gewisse Unvollkommenheit unserer heutigen Mathematik zum Ausdruck, und sie bewirken jedenfalls, daß neben den formalen Systemen der klassischen Theorien die inhaltliche, nicht-formalisierte Mengenlehre, wie sie in der Semantik angewandt wird, ihre Bedeutung behält.[19]

35.5.3 Tolerance Towards All Who Are Not Intolerant

The last example is no longer from Mathematics, but one which, we think, is in the spirit of Mario Bunge. This little toy example shows how an intended interpretation may interfere with a raw axiomatization. Recall one of Popper's "principle of humanitarian and equalitarin ethics":

> Tolerance towards all who are not intolerant. (Popper 1945, p. 205)

Raising this principle to a very definition of tolerance, one could try to give an implicit axiomatization of *tolerant* as a unary predicate expressing "x is tolerant" if x tolerates everything which is not intolerant; in formal terms:

$$\text{Tolerant}(x) :\Longleftrightarrow \forall y.(\text{Tolerant}(y) \rightarrow \text{tolerates}(x, y)) \wedge$$

$$(\neg\text{Tolerant}(y) \rightarrow \neg\text{tolerates}(x, y)).$$

Now, let us consider two parts of the world, one called West and another called East. West is proud of tolerating everything except East on the base that East does not tolerate West. Likewise, East claims to tolerate everything, except just West, as West does not tolerate East. Based on the specification of $\text{Tolerant}(x)$

---

[16]By Kikuchi and Kurahashi (2016) they are called, quite correctly, *insane* models of PA.

[17]Georg Kreisel liked to point to this formula as a false one, revealing in this way a quite strong semantic bias; see, for instance, Kreisel (1986, p. 143).

[18]Unfortunately, second-order logic is no way out, as it is no longer axiomatizable (Kahle 2019). For a discussion of the bias towards axiomatizable structures see also Kahle (2018).

[19]In all these newer results a certain imperfection of our contemporary mathematics is expressed; and these results cause at least that, next to the formal systems of the classical theories, the contentual, not formalized set theory, as it is applied in semantics, upholds its importance.

both, **West** and **East**, can be convinced to be tolerant and that just the other one is intolerant.

As a matter of fact, our axiomatization of tolerant appears to be rather symmetric with *intolerant* and, thus, not that what we would like to have. It is, indeed, the task of dual axiomatics to "unveil the tacit assumption" (DA2) that tolerance is supposed to be a *positive* opposite of intolerance, which still need to be made explicit; it also "reminds one from start to finish" of this reference (DA3).

In sum, *Dual Axiomatics* is the natural elaboration of Hilbert's axiomatic method, be it in mathematics, be it in other sciences. And we have to be thankful to Mario Bunge to call our attention to it.

References

Bernays, P. (1967). Scope and limits of axiomatics. In M. Bunge (Ed.), *Delaware seminar in the foundations of physics* (pp. 188–191). Berlin/New York: Springer.

Bernays, P. (1975). Mathematics as a domain of theoretical science and of mental experience. In H. Rose & J. Shepherdson (Eds.), *Logic colloquium '73* (Studies in logic and the foundations of mathematics, Vol. 80, pp. 1–4). Elsevier.

Bernays, P. (1978). Nachwort. *Jahresbericht der Deutschen Mathematiker-Vereinigung, 81*(1), 22–24.

Blumenthal, O. (1935). Lebensgeschichte. In: *David Hilbert: Gesammelte Abhandlungen* (Vol. III, pp. 388–429). Berlin: Springer.

Bourbaki, N. (1950). The architecture of mathematics. *The American Mathematical Monthly, 57*(4), 221–232.

Bunge, M. (1967a). Analogy in quantum mechanics: From insight to nonsense. *British Journal for the Philosophy of Science, 18*, 265–286.

Bunge, M. (1967b). *Foundations of physics* (Springer tracts in natural philosophy, Vol. 10). Berlin: Springer.

Bunge, M. (1967c). The structure and content of a physical theory. In M. Bunge (Ed.), *Delaware seminar in the foundations of physics* (pp. 15–27). Berlin/New York: Springer.

Bunge, M. (1976). Review of Wolfgang Stegmüller's the structure and dynamics of theories. *Mathematical Reviews, 55*, 33.

Bunge, M. (1980). *The mind–body problem.* Oxford: Pergamon.

Bunge, M. (2009). The failed theory behind the 2008 crisis. In M. Cherkaoui and P. Hamilton (Eds.), *Raymond Boudon: A life in sociology* (Vol. I, pp. 127–142). Oxford: Bardwell.

Bunge, M. (2017a). *Doing science.* New Jersey: World Scientific.

Bunge, M. (2017b). Why axiomatize? *Foundations of Science, 22*(4), 695–707.

Clavius, C. (1574). *Euclidis Elementorum Libri XV.* Romae.

Corry, L. (2002). *Modern algebra and the rise of mathematical structures* (2nd revised ed.). Basel: Birkhäuser.

Freudenthal, H. (1970). What about foundations of physics. *Synthese, 21*(1), 93–106.

Grattan-Guinness, I. (2000). *The search for mathematical roots, 1870–1940. Logic, set theories and the foundations of mathematics from Cantor through Russell to Gödel.* Princeton: Princeton University Press.

Hilbert, D. (1899). Grundlagen der Geometrie. In *Festschrift zur Feier der Enthüllung des Gauss-Weber-Denkmals in Göttingen, herausgegeben vom Fest-Comitee* (pp. 1–92). Teubner.

Hilbert, D. (1900). Über den Zahlbegriff. *Jahresbericht der Deutschen Mathematiker-Vereinigung, 8*, 180–184.

Hilbert, D. (1913). Elemente und Prinzipien der Mathematik. Vorlesung Sommersemester 1913, Universität Göttingen, available at the Max Planck Institute for the History of Science, Berlin.

Hilbert, D. (1918). Axiomatisches Denken. *Mathematische Annalen, 78*(3/4), 405–415. Lecture delivered on 11 September 1917 at the Swiss Mathematical Society in Zurich.

Hilbert, D. (1922). Neubegründung der Mathematik. *Abhandlungen aus dem Mathematischen Seminar der Hamburgischen Universität, 1*, 157–177.

Hilbert, D. (1928). Probleme der Grundlegung der Mathematik. In *Atti del Congresso Internazionale dei Matematici*, Bologna, settembre 3–19, 1928, Nicola Zanichelli, offprint.

Hilbert, D., & Bernays, P. (1934). *Grundlagen der Mathematik I*, (Die Grundlehren der mathematischen Wissenschaften in Einzeldarstellungen, Vol. 40). Springer; 2nd edition 1968.

Hilbert, D., & Bernays, P. (1939). *Grundlagen der Mathematik II* (Die Grundlehren der mathematischen Wissenschaften in Einzeldarstellungen, Vol. 50). Springer; 2nd edition 1970.

Hilbert, D., & Bernays, P. (2011). *Grundlagen der Mathematik I/Foundations of Mathematics I.* College Publications, bilingual edition of Prefaces and §§1–2 of Hilbert and Bernays (1934).

Hilbert, D., & Cohn-Vossen, S. (1932). *Anschauliche Geometrie* (Die Grundlehren der Mathematischen Wissenschaften in Einzeldarstellungen, Vol. XXXVII). Springer; English translation: Hilbert and Cohn-Vossen (1952).

Hilbert, D., & Cohn-Vossen, S. (1952). *Geometry and the imagination.* AMS Chelsea Publishing; English translation of Hilbert and Cohn-Vossen (1932).

Kahle, R. (2015). Gentzen's theorem in context. In R. Kahle & M. Rathjen (Eds.), *Gentzen's centenary: The quest for consistency* (pp. 3–24). Heidelberg: Springer.

Kahle, R. (2018). Structure and structures. In M. Piazza & G. Pulcini (Eds.), *Truth, existence and explanation* (Boston studies in the philosophy and history of science, Vol. 334, pp. 109–220). Springer.

Kahle, R. (2019). Is there a "Hilbert Thesis"? *Studia Logica, 107*(1), 145–165.

Kikuchi, M., & Kurahashi, T. (2016). Illusory models of Peano Arithmetic. *The Journal of Symbolic Logic, 81*(3), 1163–1175.

Kreisel, G. (1986). Proof theory and synthesis of programs: Potentials and limitations. In *Eurocal '85* (Lecture notes in computer science, Vol. 203, pp. 136–150). Springer.

MacLane, S. (2005). *A mathematical autobiography.* Wellesley: AK Peters.

Moschovakis, Y. (2006). *Notes on set theory* (Undergraduate texts in mathematics, 2nd ed.). New York: Springer.

Popper, K. R. (1945). *The open society and its enimies* (The spell of Plato, Vol. I). London: Routledge.

Salt, D. (1971). Physical axiomatics: Freudenthal vs. Bunge. *Foundations of Physics, 1*(4), 307–313.

Scriba, C. J., & Schreiber, P. (2010). *5000 Jahre Geometrie* (Vom Zählstein zum Computer, 3rd ed.). Berlin/Heidelberg: Springer.

Spinoza, B. (1677). *Ethica, ordine geometrico demonstrata.*

Weyl, H. (1944). David Hilbert and his mathematical work. *Bulletin of the American Mathematical Society, 50*(9), 612–654.

Yasugi, M., & Passell, N. (Eds.). (2003). *Memoirs of a proof theorist.* World Scientific; English translation of a collection of essays written by Gaisi Takeuti.

Part IX
Education

Chapter 36
Mario Bunge and the Enlightenment
Project in Science Education

Michael R. Matthews

Abstract This chapter begins by noting the importance of debates in science education that hinge upon support for or rejection of the Enlightenment project. It then distinguishes the historic eighteenth-century Enlightenment from its articulation and working out in the Enlightenment project; details Mario Bunge's and others' summation of the core principles of the Enlightenment; and fleshes out the educational project of the Enlightenment by reference to the works of John Locke, Joseph Priestley, Ernst Mach, Philipp Frank and Herbert Feigl. It indicates commonalities between the Enlightenment education project and that of the liberal education movement, and for both projects it points to the need to appreciate history and philosophy of science.

36.1 Introduction

The unifying theme of Bunge's life and research is the constant and vigorous advancement of the eighteenth-century Enlightenment project, and energetic criticism of cultural and academic movements that reject the principles of the project or devalue its historical and contemporary value.[1] Bunge is unashamedly a defender of the Enlightenment, while over the past half-century, many intellectuals, academics, educators, and social critics have either rejected it outright (postmodernists) or compromised its core to such an extent that it can barely give direction to the kinds of personal, philosophical, political or educational issues that historically it had so clearly and usefully addressed (multiculturalists). For many feminists,

---

[1] For his explicit endorsement of the Enlightenment project see Bunge (1994, reproduced in 1999, chap. 7).

M. R. Matthews (✉)
School of Education, University of New South Wales, Sydney, NSW, Australia
e-mail: m.matthews@unsw.edu.au

© Springer Nature Switzerland AG 2019
M. R. Matthews (ed.), *Mario Bunge: A Centenary Festschrift*,
https://doi.org/10.1007/978-3-030-16673-1_36

including educators, the very expression 'the Enlightenment' is derogatory and its advancement is thought misguided, misogynist and mistaken.

The practice of enlightened education has been a constant in Bunge's life since his founding in 1940 of a workers' school, (the *Universidad Obrera Argentina*) in Buenos Aires, and the writing of his first book, *Temas de Educación Popular* [Themes in Popular Education] (Bunge 1943).[2] As with the founding figures of the historical Enlightenment, Bunge saw that education had to be *critical, applied,* and promote *rationality* and a scientific habit of mind.

For much of the nineteenth and early-twentieth centuries support for the Enlightenment was the norm for most progressive thinkers; over the past half-century this has changed. With the rise of Romanticism, Feminist epistemology, Critical Theory, Postmodernism and Multiculturalism the repute of the Enlightenment waned. Its fundamental universalism in science and ethics is rejected. Many now regard it as the enabler of colonialism, racism, high-tech warfare, environmental despoliation and degradation of traditional cultures. Defence or rejection of the Enlightenment project has been a position-marker in major cultural and educational debates of the past half-century, such as: postmodernism, feminism, scientism, the Science Wars, multiculturalism, the appraisal of Western Civilization programmes in universities, constructivism in philosophy and education, promotion and state-support of alternative medicine, human rights across cultures, globalization, and so on down to state-sanctioned compulsory vaccination and water fluoridation.[3] In each case, commitment to the project leads in one direction, whilst rejection leads to another. Understandably these debates have spawned an enormous, and often vituperative, literature that has moved well beyond the academy and into popular culture.[4]

36.2   The Counter-Enlightenment in Education

The Enlightenment project is widely rejected in education. Consider some claims advanced by contemporary influential science educators.

For some, the task of teachers is to learn:

> ... how to deprivilege science in education and to free our children from the 'regime of truth' that prevents them from learning to apply the current cornucopia of simultaneous but different forms of human knowledge with the aim to solve the problems they encounter today and tomorrow. (van Eijck and Roth 2007, p 944)

---

[2] For the titles of the 15 chapters of this 99-page book, see Appendix.

[3] For accounts of the counter-Enlightenment tradition, see at least McMahon (2001) and Sternhell (2010).

[4] See for example Andersen (2017), Brown (2001), Gross et al. (1996), Koertge (1998), and Pinker (2018),

Others maintain that science is:

> mechanistic, materialist, reductionist, empirical, rational, decontextualized, mathematically idealized, communal, ideological, masculine, elitist, competitive, exploitive, impersonal, and violent. (Aikenhead 1997, p. 220)

Contributors to a major science education handbook hold that:

> ..one of the first places where critical inquirers might look for oppression is positivist (or modernist) science ...modernist science is committed to expansionism or growth ...modernist science is committed to the production of profit and measurement ...modernist science is committed to the preservation of bureaucratic structures ... Science is a force of domination not because of its intrinsic truthfulness, but because of the social authority (power) that it brings with it. (Steinberg and Kincheloe 2012, pp. 1487–88)

A proponent of constructivism, and former editor of a major research journal in science education, relates that:

> ...For constructivists, observations, objects, events, data, laws, and theory do not exist independently of observers. The lawful and certain nature of natural phenomena are properties of us, those who describe, not of nature, that is described. (Staver 1998 p. 503)

A hugely published researcher and former president of the National Association for Research in Science Teaching contends:

> In contrast to the mainstream of research in science education, I advocate a multilogical methodology that embraces incommensurability, polysemia, subjectivity, and polyphonia as a means of preserving the integrity and potential of knowledge systems to generate and maintain disparate perspectives, outcomes, and implications for practice. In such a - multilogical model, power discourses such as Western medicine carry no greater weight than complementary knowledge systems that may have been marginalized in a social world in which monosemia is dominant. (Tobin 2015, p. 1)

A much-awarded science educator maintains:

> ....we live forever in our own, self-constructed worlds; the world cannot ever be described apart from our frames of experience. This understanding is consistent with the view that there are as many worlds as there are knowers. ... Our universe consists of a plenitude of descriptions rather than of an ontological world *per se.*.... (Roth 1999, p. 7)

A feminist science educator asserts:

> Scientific knowledge, like other forms of knowledge, is gendered. Science cannot produce culture-free, gender-neutral knowledge because Enlightenment epistemology of science is imbued with cultural meanings of gender. This feminist critique of Enlightenment epistemology describes how the Enlightenment gave rise to dualisms (e.g., masculine/feminine, culture/nature, objectivity/subjectivity, reason/emotion, mind/body), which are related to the male/female dualism ... in which the former (e.g., masculine) is valued over the latter (e.g., feminine). (Brickhouse 2001, p. 283)

It is easy to multiply such examples. Opening any science education journal or book provides them in abundance. They give a sense of what is at stake; a sense of how philosophical positions have repercussions in educational policy, pedagogy and teacher training. The Enlightenment tradition simply rejects these claims and consequently curriculum development, pedagogy, and teacher education programmes predicated upon them. All of the foregoing claims are ill-informed and unsupportable,[5] but their rejection requires entering into philosophical argument and hopefully dialogue.[6] Bunge's corpus of work, with its richness, detail and clarity, make it a valuable resource for this engagement.

36.3   The Historical Enlightenment and the Enlightenment Project

It is important to make a distinction between the Enlightenment (noun) and the Enlightenment project (adjective). The historical Enlightenment occurred in Europe during the 'long' eighteenth century that stretched between the 'Glorious' English Revolution of 1688 and the French Revolution of 1789. By contrast the Enlightenment project, or the tradition to which it gave rise, has continued to the present day.[7] This chapter will adopt the convention of using the upper case 'Enlightenment' to refer to the eighteenth-century European events, debates, campaigns, and defining texts; and the 'Enlightenment project' to refer to the subsequent elucidation, elaboration, refinement, adjustment, and historically-informed defence of its basic commitments.

As well as the Enlightenment project there are also, lower case and plural, enlightenment projects. Europe did not have a monopoly on enlightenment principles; those principles so powerfully and clearly enunciated by different European writers, are also found, to one degree or another, in most other cultural traditions, perhaps most visibly in Islamic, Hindu and Chinese traditions. Divorced from, and independent of eighteenth-century Europe, such writers are contributing to an enlightenment project, but not to *the* Enlightenment project. Assuredly both upper and lower-case projects can learn from each other: Asia can learn from Europe, and Europe can learn from Asia and other cultures where enlightenment projects have flourished.

In China, for instance, both the 'Hundred Days Reform' of 1898 and the 'May Fourth Movement' of 1919 were animated both by European Enlightenment convictions, and by Chinese neo-Confucian convictions (Spence 1982; Tang 2015). In the Muslim world, the Arab Spring of 2011 was fuelled by natural anger and resentment, but it was also informed by both European Enlightenment arguments and arguments from the Islamic tradition. Ditto the incipient Saudi 'revolution' of 2018 begun by brave women demanding, and getting, the right to drive cars. Their appeals were part Western and part Islamic. In India Bhimrao Ramji Ambedkar's heroic mobilisation

---

[5] Arguments for this harsh appraisal can be found in Matthews (2015, chap. 8).

[6] For excellent critiques of the philosophical foundations of such constructivist and postmodernist writing in education, see: McCarthy (2018), Schulz (2007), and contributions to Matthews (1998).

[7] Among many high-quality books on the historical Enlightenment see: Anchor (1967), Ferrone (2015), Fitzpatrick et al. (2007), Gay (1966), Himmelfarb (2004), Israel (2001, 2006, 2011), Pagden (2013), and Porter (2000).

of the 'untouchables' and their campaign to end discrimination and oppression, was informed by Enlightenment sources and also by the secular Buddhist tradition (Matthews 2015, pp. 48–50, Mukherjee 2009). The Enlightenment project does not have a monopoly on enlightenment projects; the latter take place in all societies, but nowhere with the same comprehensive tool-kit of formulated arguments.

One of the first uses of the expression 'Enlightenment project' is in Alasdair McIntyre's *After Virtue* (McIntyre 1981); it occurs in the headings of three chapters. McIntyre decries the project, with Chapter 5 titled 'Why the Enlightenment Project of Justifying Morality Had to Fail'. He writes: 'The Enlightenment is consequently the period *par excellence* in which most intellectuals lack self-knowledge … in which the blind acclaim their own vision … ' (McIntyre 1981, p. 78). His basic claim is that the science-based, anti-teleological outlook of the Enlightenment cannot deal with the fundamentals of human living, namely intentional, ethical life and moral decision making. For McIntyre Aristotle could ground ethics rationally and ontologically because of his notion of 'embodied' natures that realize themselves towards their own 'perfection' in natural circumstances. But when Enlightenment philosophers rejected Aristotle, they lost an objective, rooted-in-nature moral compass. McIntyre's critique was continued in his *Whose Justice? Which Rationality?* (McIntyre 1988) where he rejects the Enlightenment's commitment to rationality in the singular as distinct from the plural rationalities.

Variants of McIntyre's arguments had been advanced by many in the nineteenth century Romantic reaction to the Enlightenment. And numerous epistemological and political arguments have been advanced through the twentieth century, with Horkheimer and Adorno's 1944 critique perhaps being the most influential (Horkheimer and Adorno 1944/1972). Kuhnian relativism, postmodernism and 'critical theory', have fuelled contemporary rejection of the project.[8]

It needs also to be appreciated that the singular term, *the* Enlightenment hides the reality that there were different national centres, each with their own religious, political and philosophical issues; and produced their own literatures, in their own time frames.[9] The most significant national centres being England,[10] Scotland,[11] France,[12] Italy,[13] Germany,[14] Holland,[15] and America.[16] The Enlightenment came

---

[8]See Berlin (1980), Fleischacker (2013, Pt.4), and Garrard (2006).

[9]Individual chapters in Porter and Teich (1981) are devoted to the Enlightenment in England, Scotland, France, Netherlands, Switzerland, Italy, Germany (Catholic and Protestant), Austria, Bohemia, Sweden, Russia and America. These chapters can be consulted in conjunction with the following national references.

[10]See Porter (2000).

[11]See Herman (2001).

[12]See Artz (1968) and Fitzpatrick (2007).

[13]See Venturi (1972).

[14]See Clark (1999).

[15]See Dunthorne (2007) and Schama (1981) and contributions to van Bunge (2003).

[16]See Cassara (1988), Commager (1977), Ferguson (1997), Koch (1961, 1965), and May (1976).

late and shone weakly in Catholic Spain and Austria, and did not shine at all in the Papal States. Climactic events such as the decade-long French Revolution (1789–1799) had different impacts on Enlightenment thought in different areas of Europe.[17]

The Enlightenment was not constituted by a static collection of authoritative texts providing timeless, non-contextual answers to political, social, religious, philosophical or scientific questions. In Kant's canonical 1784 essay 'What is Enlightenment?' he writes[18]:

> Enlightenment is man's release from his self-incurred tutelage. Tutelage is man's inability to make use of his understanding without direction from another. Self-incurred is this tutelage when its cause lies not in lack of reason but in lack of resolution and courage to use it without direction from another. *Sapere aude!* 'Have courage to use your own reason' – that is the motto of enlightenment. (Kant 1784/1995, p. 1)

A few years later in the Preface to the first edition of his *Critique of Pure Reason*, Kant enunciates the defining feature of eighteenth-century Enlightenment thinking, namely that it was critical and self-correcting:

> Our age is, in especial degree, the age of criticism, and to criticism everything must submit. Religion through its sanctity, and law-giving through its majesty, may seek to exempt themselves from it. But they then awaken just suspicion, and cannot claim the sincere respect which reason accords only to that which has been able to sustain the test of free and open examination. (Kant 1787/1933, p. 9)

The canonical texts were not exempt from criticism. Such criticism including adjustment in the light of social upheavals such as the French and American revolutions, the experience of European colonisation, and scientific accomplishments in fields such as electricity and chemistry – all fuelled the Enlightenment project.

The Enlightenment was at odds with ahistoric fundamentalisms of all kinds. There was, with some debate around the edges,[19] an identifiable canon of texts, but these did not constitute a scripture.[20] The texts did not become authoritative in virtue of being canonical; being in a recognised canon made them influential, but not authoritative. Unlike fundamentalism in religion, politics, and 'party' or institutionalised philosophy, a characteristic of Enlightenment analyses and debates is that they are not settled by quoting texts or reference to official interpreters. To the extent that scientific, political, religious or ethical debates are settled by authority,

---

[17]See contributions to Church (1974).

[18]Kant's 1784 essay, and the essay 'What is Enlightenment?' by Moses Mendelssohn to which Kant was responding, are contained in the Schmidt anthology (Schmidt 1996). Also included are 20+ eighteenth century contributions to the 'What is Enlightenment?' debate, and a dozen twentieth-century studies of the issue. Kant's essay, and its reception over the past two centuries, is well treated in Fleischacker (2013).

[19]Jonathan Israel (Israel 2006, p. 867) identifies and discusses 70 individual contributors to the formation of Enlightenment thought. Choosing who, 250 years later, might be included in the Enlightenment canon has its own problems.

[20]Among numerous anthologies of Enlightenment texts, see: Eliot and Stern (1979), Gay (1973), Hyland et al. (2003), and Kramnick (1995).

there is a corresponding departure from the ethos of the Enlightenment. All competent contributors to the Enlightenment project take the view that philosophical, political, or ethical positions are in the canon because they are correct, they are not correct because they are in the canon. This is the same attitude as taken by competent theologians about scriptural positions.

The seventeenth-century's new science – or 'natural philosophy' as it was then called, – of Galileo, Descartes, Huygens, Boyle and Newton caused a massive change not just in the science of the time, but in contemporary European philosophy.[21] This in turn had enduring repercussions for religion, ethics, politics, economics, law, literature, and culture.[22] The new science was instrumental in the birth of the modern world; it was the seed of which the Enlightenment was the fruit. It had a defining influence on eighteenth-century Enlightenment thinkers including: John Locke (1632–1704), Baruch Spinoza (1632–1677), Voltaire (1694–1778), Benjamin Franklin (1706–1790), Julien de la Mettrie (1709–1751), David Hume (1711–1776), Denis Diderot (1713–1784), Jean D'Alembert (1717–1783), Nicolas de Condorcet (1743–1794), Claude Adrien Helvétius (1715–1771), Immanuel Kant (1724–1804), Joseph Priestley (1733–1804) and Thomas Jefferson (1743–1826).

David Hume, in a much-quoted passage, captured the esteem and repute in which Newton was held – at least by the English – when he wrote in his *History of England*:

> In Newton this island may boast of having produced the greatest and rarest genius that ever rose for the ornament and instruction of the species. Cautious in admitting no principles but such as were founded on experiment, but resolute to adopt every such principle, however new or unusual. (Hume 1754-62/1879, vol. 6, p. 344)

Later in the same paragraph, Hume made a far more consequential claim, one that has echoed through philosophy in the subsequent centuries, and that bears upon one of the most contested of Enlightenment principles. He writes:

> While Newton seemed to draw off the veil from some of the mysteries of nature, he showed at the same time the imperfections of the mechanical philosophy, and thereby restored her ultimate secrets to that obscurity in which they ever did *and ever will* remain. (Hume 1754-62/1879, vol. 6, p. 344, emphasis added)

This is Hume the positivist and sceptic speaking. He enunciates the sceptical position concerning knowledge of nature's 'ultimate' constituency and composition. Hume's scepticism famously extended to all things not immediately perceivable or for which there are no sense impressions, including causal powers beyond constant conjunctions, angels, spirits, grace, miracles, Aristotelian forms, and so on. Enlightenment philosophers, of whom Hume was one of the greatest, divided on this epistemological principle of whether 'ultimate' or 'unseen' reality was knowable. There was a strong sceptical strain in the Enlightenment.[23] However nearly all

---

[21] For studies of the Scientific Revolution, see: Lindberg and Westman (1990), Osler (2000), and Wootton (2015).

[22] On the theme of science and the Enlightenment, see: Hankins (1985), Matthews (1989), and O'Hara (2010, chap. 7).

[23] See Garrett (2007) and contributions to Charles and Smith (2013).

believed that by adherence to the Newtonian method, as variously understood,[24] the empirical world, including importantly, the social world, could be known.

This conviction was led from the top. Newton in his *Opticks* said: 'If natural philosophy in all its Parts, by pursuing this Method, shall at length be perfected, the Bounds of Moral Philosophy will be also enlarged' (Newton 1730/1979, p. 405). David Hume echoed this expectation with the subtitle of his famous *Treatise on Human Nature* which reads, *Being an Attempt to Introduce the Experimental Method of Reasoning into Moral Subjects*.[25] In the preface he says he is following the philosophers of England who have 'began to put the science of man on a new footing' (Hume 1739/1888, p. xxi). The Marquis de Condorcet (1743–1794), a leading *philosophe* of the French Enlightenment said in his 1782 acceptance speech at the French Academy that: 'the moral [social] sciences' would eventually 'follow the same methods, acquire an equally exact and precise language, attain the same degree of certainty' as the natural sciences (Condorcet 1976, p. 6).

The epistemological division highlighted by Hume was about the knowability of the 'unseen' world of mechanisms and constituents; what today would be called the 'theoretical' or 'metaphysical' domain. This was the eighteenth-century debate about realism versus positivism/instrumentalism/empiricism; a debate which is on-going.[26] The debate continues within the Enlightenment project. Most adherents are realists, but no less a figure than Ernst Mach was a combative instrumentalist concerning theoretical terms in science. And there were 'in-house' disputes about the applicability of the Newtonian method to religious, scriptural and ethical questions. This prefigured contemporary debate about the pros and cons of 'Scientism'.

Nevertheless, despite geographic spread, local variations, and internal disputes over particular commitments, there are identifiable Enlightenment principles. Although there is some scatter, noise, localisation, and a few parochial outliers – philosophical lines of best fit can be ascertained for the Enlightenment package.

36.4   Bunge and Others on Enlightenment Principles

The Enlightenment party was a very mixed and heterogeneous group; it was a very Broad Church. Politically there were republicans, monarchists, constitutional monarchists and proponents of benevolent despotism; religiously there were atheists, Deists, Unitarians, Christians of varying kinds, and Jews; and among

---

[24]There was contemporary debate about just what was the method, what it allowed and did not allow, and what was its legitimate domain. For accounts of Newton's method, see Cohen (1980), Harper (2011) and contributions to Butts and Davis (1970).

[25]At the time, 'moral subjects and philosophy included present-day history, social sciences, politics, economics and ethics.

[26]See contributions to Agazzi (2017), Cohen et al. (1996), and Leplin (1984). The debate and literature is reviewed in Matthews (2015, chap. 9).

the religious, some were devout while others were cynical, supporting religion solely for its socially cohesive function; ontologically there were dualists, monists, materialists, and physicalists; epistemologically there were sceptics, fallibilists, and 'certainists'; culturally there were preservers and innovators, agitators and conformists. And as well as the well-known religious, politically conservative, and reactionary critics of the time, the Enlightenment had its own internal critics.[27] It was an enormously rich age; in Isaiah Berlin's estimation:

> The intellectual power, honesty, lucidity, courage and disinterested love of the truth of the most gifted thinkers of the eighteenth century remain to this day without parallel. Their age is one of the best and most hopeful episodes in the life of mankind. (Berlin 1956, p. 29)

Despite the heterogeneity, enough commonalities can be discerned to justify the gathering of so many individuals and texts into the Enlightenment family. Jonathan Israel, perhaps the foremost contemporary Enlightenment scholar, is prepared to tentatively assert that the European Enlightenment was a:

> single highly integrated intellectual and cultural movement, displaying differences in timing, no doubt, but for the most part preoccupied not only with the same intellectual problems but often even the very same books and insights everywhere from Portugal to Russia and from Ireland to Sicily. Arguably indeed, no major cultural transformation in Europe, since the fall of the Roman Empire, displayed anything comparable to the impressive cohesion of European intellectual culture in the late seventeenth and early eighteenth century. (Israel 2001, p. v)

The historical, philosophical, and educational task is to delineate just what were the commonalities, the guiding principles, that identify the Enlightenment family.

Bunge identifies the core principles of the historical Enlightenment as:

1. Trust in reason.
2. Rejection of myth, superstition, and generally groundless belief or dogma.
3. Free inquiry and secularism.
4. Naturalism, in particular materialism, as opposed to supernaturalism.
5. Scientism or the adoption of the scientific approach to the study of society as well as nature.
6. Utilitarianism in ethics, as opposed to both religious morality and secular deontologism.
7. Respect for praxis, especially craftmanship and industry.
8. Modernism, progressivism, and trust in the future.
9. Individualism together with libertarianism, egalitarianism (to some degree or other), and political democracy (though not yet for women or slaves).
10. Universalism or cosmopolitanism, for example, human rights and education for all 'free men'. (Bunge 1999, p. 131)

---

[27] For instance, the German Christian Erhard, wrote in 1789: 'Damned be the Enlightenment which exchanges blind trust in itself for blind trust in others' (Knudsen 1996, p. 270). This charge of blind trust, self-deception, if not arrogance, has been echoed in the following centuries by countless critics.

As a contribution to the Enlightenment project, each of these principles need to be connected to Enlightenment texts or sources, they need to be sufficiently elaborated for philosophical and policy purposes, and they need to be defended, and suitably adjusted or abandoned, if convincing contrary arguments are advanced.

Bunge's foregoing selection and distillation of the principles can profitably be compared to or triangulated with that of others. Carl Becker, a critic of the Enlightenment, in his 1932 *The Heavenly City of the Eighteenth-Century Philosophers* identified its four essential commitments as:

> (1) man is not natively depraved; (2) the end of life is life itself, the good life on earth instead of the beatific life after death; (3) man is capable, guided solely by the light of reason and experience, of perfecting the good life on earth; and (4) the first and essential condition of the good life on earth is the freeing of men's minds from the bonds of ignorance and superstition, and of their bodies from the arbitrary oppression of the constituted social authorities. (Becker 1932, pp. 102–3)

The physicist and philosopher Abner Shimony, in his 1996 Presidential Address – 'Some Historical and Philosophical Reflections on Science and Enlightenment' – to the US Philosophy of Science Association identified the core commitments of the historical Enlightenment as:

1. On matters of fact, whether particular or general, there is objective truth or falsity.
2. There is a universal human nature (except for abnormalities) in all places and times.
3. One aspect of this universal human nature is that the cognitive faculties of individual normal, human beings suffice in principle for determining the truth or falsity of propositions concerning matters of fact, though training and removal of superstitions, dogmas, etc. are needed for full realization of what is possible in principle.
4. The authority of socially-established experts and of social institutions, including those which claim divine sanction, is subordinate to judgments by natural human faculties.
5. As a corollary, no social institution has the right to control inquiry, or the communication of the results of inquiry, or the critical examination of claims to knowledge.
6. The basic natural sciences, particularly the physical sciences, provide exemplary instances of reliable methods of inquiry and reliable general results concerning matters of fact.
7. In particular, natural theology, essentially employing the methods of the natural sciences, is the primary mode of theological inquiry.
8. A corollary of the existence of a universal human nature is a universality of human goals.
9. Another corollary is that the basis for ethics is to be found in the constitution of every normal being, though there is disagreement concerning the exact character of this basis; among the prescriptions of the naturally-based ethics are

universal benevolence towards human beings and condemnation of punishment and constraint beyond what is needed for the common good.

10. Human cognitive faculties are capable in principle of devising good solutions to practical human social and political problems. (Shimony 1997, pp. S2–3)

It is useful to divide the personalities, ideas and movements into 'moderate' and 'radical' Enlightenments. The former include Locke, Rousseau, Voltaire and Kant; the latter Spinoza, Diderot, Holbach and Helvétius. The moderate Enlightenment favoured Reason, but not too much; it wanted clear evidence-based thinking, but mostly for science; it wanted social improvement but not political disruption; it denied the 'divine right' of kings, but not kingship; it wanted intelligent and non-superstitious religion, but not atheism.

Jonathan Israel has provided the most extensive and detailed study (800 pps) of the radical Enlightenment. He sees its fundamental principles as:

1. Adoption of philosophical (mathematical-historical) reason as the only and exclusive criterion of what is true.
2. Rejection of all supernatural agency, magic, disembodied spirits, and divine providence.
3. Equality of all mankind (racial and sexual).
4. Secular 'universalism' in ethics anchored in equality and chiefly stressing equity, justice, and charity.
5. Comprehensive toleration and freedom of thought based on independent critical thinking.
6. Personal liberty of lifestyle and sexual conduct between consenting adults, safeguarding the dignity and freedom of the unmarried and homosexuals.
7. Freedom of expression, political criticism, and the press, in the public sphere.
8. Democratic republicanism as the most legitimate form of politics. (Israel 2006, p. 866)

While there are other 'summations' of the Enlightenment,[28] the historian Philipp Blom writes: 'What makes the thinking of the radical Enlightenment so essential today is its power, its simplicity, and its moral courage' (Blom 2010, p. xvi). Bunge concurs with this estimation. In particular he affirms that the method of natural science needs be utilised in social science, that ethical principles cannot come from without, and there can be neither knowledge of the supernatural realm nor reason to believe there is such. But as with all serious advocates of the Enlightenment, he is not uncritical:

> Of course, the Enlightenment did not do everything for us: no single social movement can do everything for posterity – there is no end to history. For instance, the Enlightenment did not foresee the abuses of industrialization, it failed to stress the need for peace, it exaggerated individualism, it extolled competition at the expense of cooperation, it did not go far enough

---

[28]Kieran O'Hara lists six: 'new sources of authority, confidence and optimism, scepticism, universal reason, self-interest, elitism' (O'Hara 2010, chap. 1). The Appendix of Commager (1977) provides a good distillation of the thinking and commitments of the *Philosophes*.

in social reform, and it did not care much for women or for the underdeveloped peoples. However, the Enlightenment did perfect, praise, and diffuse the main conceptual and moral tools for advancing beyond itself. (Bunge 1999, p. 142)

These and other such corrections in the light of historical developments and philosophical critiques, are constitutive of the Enlightenment project; it is an intrinsically self-correcting enterprise. Corrections do not mean abandonment. Whilst recognising shortfalls in the Enlightenment project, Bunge correctly maintains that:

> we all ... are children of the Enlightenment: we all enjoy the benefits of secularism, free inquiry, rationality, objectivity, individual freedoms, and progress (in some respects). ... And this, the freedom to create, debate, and diffuse new ideas, is what the Enlightenment was all about. (Bunge 2000, p. 231)

Bunge is politically and intellectually pitted against all closed, authoritarian regimes, states and ideologies. He abhors, along with all liberals, censorship of scholarly work as is routinely practiced in China, most if not all Islamic states, Egypt, Turkey, and until recently all countries where the Catholic church exercised political power.

36.5   Education and the Enlightenment Project

It is unfortunate that education is not separately delineated in the foregoing 'fundamentals' lists. It deserves to be.

All eighteenth-century English, French and German Enlightenment figures saw education as essential for the reformation of their society and for the more radical thinkers, the creation of a new society. Locke,[29] Spinoza,[30] Priestley,[31] Rousseau,[32] Helvétius,[33] Kant,[34] all wrote works on education.[35] They all rejected religious and philosophical views that saw humans as essentially corrupted, Fallen, and incapable of learning; they were all committed to the improvement of life and society, to the possibility of progress; and to the efficacy of reason in ordering personal and national affairs. What enabled all of this was education. The Enlightenment education project has two major strands, they addressed two kinds of questions: philosophical and pedagogical.

The philosophical questions were to what extent and age should education be conducted? Who should control education – the State, churches, or parents?

---

[29]Locke (1693/1968) in Axtell (1968). See also Schouls (1992) and Tarcov (1989).
[30]Spinoza (1677/1910). See also Puolimatka (2001).
[31]Priestley (1765/1965; 1791).
[32]Rousseau (1762/1991). See also Trachtenberg (1993).
[33]Helvétius (1772/1810).
[34]Kant (1803/1899).
[35]See Parry (2008).

How should education be funded? Should education be classical or utilitarian? What role should the Church or churches play in education? Should religious teaching be allowed in state schools? What should be the content, curriculum, or programme of education? Should the state support private or religious schooling? In the pedagogical strand the questions were: What are the best methods of education? Are there natural or constitutional barriers to learning? Are learning difficulties remediable? What are the gains and losses of child-centered teaching?

The philosophical and pedagogical questions prompted lively debate. Answering the questions became part of the Enlightenment project. Some participants, for example Condorcet, contributed more to the philosophical issues, others, for example Rousseau, contributed more to the pedagogical issues, some, for example Dewey and Mach, contributed to both.

John Locke, Newton's self-described 'underlabourer' opens his hugely influential 1693 *Some Thoughts Concerning Education* with two central planks of the Enlightenment's educational programme.[36] First:

> A Sound Mind in a sound Body, is a short, but full Description of a Happy State in this World: He that has these Two, has little more to wish for; and he that wants either of them, will be but little the better for any thing else. (Locke 1693/1968, p. 114)

And second:

> I think I may say, that for all the Men we meet with, Nine Parts of Ten are what they are, Good or Evil, useful or not, by their Education. (Locke 1693/1968, p. 114)

Eighty years later, Helvétius captured the importance of education to Enlightenment thinkers, and to their policies for the remaking of society, when he wrote:

> If I can demonstrate that man is, in fact, nothing more than the product of his education, I shall doubtless reveal an important truth to the nations. They will learn that they have in their hands the instrument of their greatness and their felicity, and that to be happy and powerful, it is only a matter of perfecting the science of education. (Helvétius 1772/1810, chap. 1, 3; in Parry 2007, p. 230)

There was then, as now, dispute over what constituted the 'science of education'. John Locke (1632–1704), Jean-Jacques Rousseau (1712–1778), Joseph Priestley (1733–1804), and Heinrich Pestalozzi (1746–1827) were the most prominent contributors to the formation of a hoped-for educational science. Locke's empiricism, which linked concepts to experience and more specifically sensation, was developed as a guiding educational psychology or theory of learning. Such a psychology of learning provided an easy passage through to child-centred, experiential pedagogy.

In an essay 'Of Study' written in 1677 Locke expresses: 'The end of study is knowledge, and the end of knowledge practice or communication' (Axtell 1968,

---

[36] The book is of 200-odd pages, covering 215 sections. In English there were 40 printings of it as a separate book between 1693 and 1964. In French there were 23 translations and printings between 1695 and 1966. And there were American, German, Dutch, Spanish, Italian, Polish, Rumanian and Swedish printings. (Axtell 1968, pp. 98–104).

p. 406). To a degree he anticipates Kant's century-later, 1784 'dare to think for yourself' maxim when he writes:

> He that distrusts his own judgement in everything, and thinks his understanding not to be relied on in the search for truth, cuts off his own legs that he may be carried up and down by others, and makes himself a ridiculous dependence upon the knowledge of others, which can be possibly of no use to him; for I can no more know anything by another man's understanding than I can see by another man's eyes. (Axtell 1968, p. 419)

The Enlightenment commitment to education was manifest in the writings and practice of the two foremost scientist/statesmen of early America: Benjamin Franklin and Thomas Jefferson. As Governor of Virginia, Jefferson moved to establish a whole system of elementary and county-based secondary schooling. He reformed and reorganised the College of William and Mary, and when his reforms were frustrated, he moved to establish, and largely designed, the University of Virginia in Charlottesville. In a 1786 letter, 'A Crusade Against Ignorance', Jefferson writes of the constitution of the colony, that:

> I think by far the most important bill in our whole code is that for the diffusion of knowledge among the people. No other sure foundation can be devised for the preservation of freedom, and happiness. . . . Preach, my dear Sir, a crusade against ignorance; establish and improve the law for educating the common people. (Koch 1965, pp. 311-12)

He was animated to distance the American colonies from the European countries of their parentage where:

> ignorance, superstition, poverty and oppression of body and mind in every form, are so firmly settled on the mass of the people, that their redemption from them can never be hoped. . . . If all the sovereigns of Europe were to set themselves to work to emancipate the minds of their subjects from their present ignorance and prejudices, and that as zealously as they now endeavour the contrary, a thousand years would not place them on that high ground on which our common people are now setting out. (Koch 1965, pp. 311–12)

Benjamin Rush, a signatory to the Declaration of Independence, and conscious advocate of the Enlightenment project, wrote in his 1786 *Plan for the Establishment of Public Schools and the Diffusion of Knowledge in Pennsylvania* that:

> The golden age, so much celebrated by the poets, is already within reach; legislatures need only to establish proper modes and places of education in every part of the state. (Ferguson 1997, p. 153)

Although Enlightenment thinkers were dedicated to education, they differed over the reach and form of that education; in particular over the appropriate education of peasants in Europe and the working classes in Britain. There was an elitist and conservative strand in the Enlightenment; one that saw the best education as fitting a person to their 'station in society'. Locke's *Thoughts Concerning Education* could have been titled *Thoughts Concerning a Gentleman's Education* as he consciously acknowledged in the closing sentences of the book:

> I have touch'd little more than those Heads [topics], which I judged necessary for the Breeding of a young Gentleman of his Condition in general; and have now published these my occasional Thoughts with this Hope, That though this be far from being a compleat Treatise on this Subject, or such, as that everyone may find, what will just fit his Child in it,

yet it may give some small light to those … that dare venture to consult their own Reason, in the Education of their Children, rather than wholly to rely upon Old Custom. (Locke 1693/1968, p. 325)

In Germany, Adolf Freiherr von Knigge (1752–1796), a leader of the 'radical' Illuminati, wrote in 1788:

That one now gradually attempts to motivate the peasant to abandon many of his inherited prejudices in the methods of planting and indeed in the management of his household, that one hopes through purposeful schooling to destroy foolish fancies, stupid superstitions, and belief in ghosts, witches and similar matters, and that one now teaches the peasant to read, write, and calculate well – all this is indeed commendable and useful. But to give them all sorts of books, stories, and fables, to accustom them to transporting themselves into a world of ideas, to open their eyes to their own impoverished condition which cannot be improved, to make them discontented with their lot through too much enlightenment, to transform them into philosophers who blather about the uneven division of earthly goods – that is truly worthless. (Knudsen 1996, p. 276)

Two years later in Germany Johann Ludwig Ewald wrote in his *On Popular Enlightenment: Its Limits and Advantages*:

I would be very much misunderstood if one were to believe I intended to acquaint the peasant systematically with the full extent of these [new] sciences. That is neither possible nor useful. The slumbering mental capacities of these crude natural men could not comprehend such matters, and even if one were to do everything to awaken them, such learning would be neither intelligible nor useful to them. (Knudsen 1996, p. 276)

This politically conservative strand of educational thought occurred in all national Enlightenment traditions. But it was criticised internally for conflicting with the Enlightenment principle of equality which commonly translated into 'equality opportunity' and 'non-discrimination' policies in education. This was a slow process that worked itself out at different rates in different countries. This was all a part of the Enlightenment project.

In England, The Cavalier (Royalist) Parliament had passed the Corporation Act in 1661 and the Act of Uniformity in 1662. These Acts prohibited dissenters or 'nonconformists' (Presbyterians, Anabaptists, and later Methodists and Unitarians) from enrolling in Cambridge and Oxford which were then the only universities in England. The same strictures applied to Roman Catholics, Jews, Muslims and and of course Atheists.

In 1687, the year of completion of the *Principia*, the Catholic King James II asked Cambridge University to confer a degree upon a Benedictine monk and exempt him from taking the usual oath to uphold the Anglican faith. No less a figure than Isaac Newton was a leader of the successful fight against this proposal (Brooke 1991, p. 159).[37]

A similar history and struggles played out in all countries where Enlightenment writings and thought was found. In France, in 1687, the year of publication of

---

[37]Testifying to the slowness of educational reform, women were not granted full and equal rights at Cambridge until 1948.

Newton's *Principia*, the Edict of Nantes, which had given some measure of freedom and relief to Huguenots, was revoked by the Sun King, Louis XIV, and overnight one million French Protestants were made outlaws in their own country. Protestant services were banned, with those found taking part in them sent for life to the galleys as slaves; Protestants were banned from all government and educational employment; only Catholic marriages were recognised, so Protestant wives became concubines and Protestant children were made illegitimate and unable to inherit property; hundreds of protestant clergy were hanged. It is estimated that perhaps 200–500,000 Huguenots fled France for other lands (Goubert 1972, p. 160). Louis proudly boasted that he had rid France of heresy. He ruled till his death in 1715, being succeeded to the throne by his 5-year old grandson, Louis XV who ruled for a further 59 years till 1774, in turn being followed on the throne by Louis XVI who was guillotined in 1793 in the latter days of the French Revolution. There was no national system of education, all schools were private and under the control of local priests with natural philosophy (science) barely taught; *collèges* were under the control of either the Jesuit or Oratorian orders, and devoted almost entirely to theology, law and classics. One estimate is that through most of the eighteenth century, till the First Republic, fully 80% of the French population could neither read nor write, and no one saw this as a problem (Schapiro 1963, p. 197). This was the *ancien régime* against which Enlightenment thought struggled for a century, and then more through to modern day France.

Through the reign of Louis XV Jean D'Alembert (1717–1783) the encyclopedist, Louis René La Chalotais (1701–1785) the jurist, and Rolland d'Erceville (1730–1794) the educator, wrote articles, pamphlets and reports decrying the backward looking, classical-language obsessed, useless French education of the schools and *collèges*.[38] In a 1768 report to parliament, *Plan d'Education*, d'Erceville maintained:

> Each one ought to have the opportunity of receiving that education most suited to him; not every kind of soil responds to the same care and yields the same product; every mind does not require the same degree of culture nor do all men have the same needs or abilities; it is in relation to these abilities and needs that public education should be organised. (Kandel 1930, p. 184)

These were Enlightenment-inspired, pre-revolutionary interventions. With the French Revolution, many Enlightenment outsiders became policy-making insiders (Church 1974). Charles Maurice de Talleyrand-Périgord (1754–1838) in a 1791 Report to the National Assembly wrote:

> Instruction has in general the aim of perfecting man at all ages and to help ceaselessly to promote the advantage of each, and the benefit of society as a whole through enlightenment and experiment and to combat the errors of preceding generations. (Kandel 1930, p. 186)

---

[38]The educational writings and assembly reports of La Chalotais, Turgot, Diderot, and Condorcet are translated and published in English in de la Fontainerie (1932).

Nicolas de Condorcet (1743–1794) was the most thorough, consistent, and influential advocate of Enlightenment education in eighteenth-century France. His justly famous 400-page comprehensive *Report on Education* presented to the Legislative Assembly in 1792 is a landmark document in the history of education.[39] It has five philosophical papers on education and detailed curricula for all subjects in all schools from elementary (*petites écoles*) to university. Its opening sentence is: 'Public education is a duty that society owes to all citizens'. This was, and has remained, a rallying call of the Enlightenment education project; it assuredly is one of the 'hard core' defining principles of the project.

Condorcet elaborated an entire far-sighted Enlightenment education scheme, that though proposed 250 years ago, is strikingly modern and contemporary. His *Report* proposed, among other things: state-funding of all education and a four-tier system of schools culminating in university, compulsory elementary education for girls as well as boys, co-education, teaching mathematics and sciences from the elementary level, banning religious teaching in state schools, the teaching of non-religious based civics and ethics courses. Many other 'progressive' and 'liberal' reforms were proposed. Secondary schools were to teach many varied courses including pure and applied mathematics, experimental physics and chemistry, national and international history, logic, political constitutions, political economy, music and dancing. Indicative of the new way of thinking ushered in by the Enlightenment was the directive that the Constitution and the Declaration of Rights were to be taught as factual historical documents and so to be scrutinized, not adulated. Condorcet said the schools should avoid nationalism and patriotic excesses. The *Report* instituted a programme of competitive state scholarships (*élèves de la patrie*) to allow children of the poor to progress through boarding schools to the highest levels; it proposed appropriate adult education for farmers, workers and mothers; it made the teaching service independent of both the Church and the State having its own regulator and making its own regulations.[40]

By a decree of 1795 a system of central schools (*écoles centrales*) was established throughout France, one for each national department. Each school was required to have a public library, a garden, a natural history cabinet, and a laboratory for physics and chemistry.

Sadly, Condorcet suffered the same fate as Lavoisier. With a change of power in the revolutionary assembly, he was arrested and killed in prison. But his *Report* had been published and it was partly implemented in the brief years of the First Republic and Napoleon's rule. All its policies and programmes were shelved at the Bourbon Restoration. But progressively it was implemented through the later nineteenth century in France. Whilst the *Report* was a center-piece in the struggle between clericalism and secularism in nineteenth-century France, it was taken as a

---

[39] A 50-page portion of the lengthy text is in Fontainerie (1932). Reisner, an education historian, said of the Report that: 'Probably no finer ideal of education in a national state has ever been set forth' (Reisner 1930, p. 147).

[40] Condorcet's education writings are discussed in Schapiro (1963, chap. 11).

model for many other national and provincial systems right through the twentieth century.[41]

Condorcet in his influential *Sketch for a Historical Picture of the Progress of the Human Mind* (Condorcet 1795/1955) had given an early Enlightenment justification for social science, or the scientific study of society:

> The sole foundation for belief in the natural sciences is this idea, that the general laws directing the phenomena of the universe, known or unknown, are necessary and constant – why should this principle be any less true for the development of the intellectual and moral faculties of man than for the operations of nature. (Condorcet, 1795/1955, p. 173)

He thought that the impact of education could be charted, and appropriate 'experiments' or innovations, could be exported or generalised. If something worked in one department, it should work in another. In the following decades, especially during Napoleon's reign, there was intense debate and politicking about the curriculum, administration and control of the newly-established school system. The idea of a central state-controlled system was opposed by liberals and the Catholic Church. During the Restoration (1814–1830) the forces of Reaction rolled back most of the foregoing Enlightenment-motivated educational reforms. Edmond About (1828–1885) writing of his own Restoration-Era college education said:

> The serious studies in our day consisted in translating French into Greek and Latin and vice-versa, in handling a given subject in French or Latin and an elegant trifling in Latin verse. .... While the exact sciences it was good form to ignore unless one expected to enter St. Cyr and the *Ecole Polytechnique*. (Kandel 1930, p. 195)

In England, schooling was not much different. The philosopher C.E.M. Joad (1891–1953) typifies the circumstance some 50 years after About's lament for French education:

> I left my public school in 1910, an intelligent young barbarian. .... My acquaintance with the physical sciences was confined to their smells. I had never been in a laboratory; I did not know what an element was or a compound. Of biology I was no less ignorant. I knew vaguely that the first Chapter of Genesis was not quite true, but I did not know why. Evolution was only a name to me and I had never heard of Darwin. (Joad 1935, p. 9)

The foregoing pages give an indication of the centrality of education both for the Enlightenment and the Enlightenment project. They suffice to show that there were educational debates and disagreements among contributors to the project. Importantly, and obviously, the Enlightenment tradition had no monopoly on the promotion of enlightened education. Arguments for practicality, for modernity, for inclusion of local and national histories, for modern languages, for teaching mathematics and natural science, for curbing the educational power of State and Church, and so on – were made both inside and outside the Enlightenment tradition. Education was a natural sphere for 'popular front' campaigns. Proponents of Enlightenment education shared much with advocates of Liberal Education in the

---

[41] See Kandel (1930, chap. VI).

Anglo-American world[42] and with champions of *Bildung* in the Germanic and continental world.[43]

What distinguished Enlightenment-inspired educational proposals and programmes is their insistence on state responsibility for universal education (yet leaving open the question of State control of education),[44] the valuation of natural science, their efforts to have pupils appreciate the method of science, and to see its application to personal, social and cultural problems. Four examples of contributors to the educational strand of the Enlightenment project will be sketched so as to better appreciate these claims: Joseph Priestley, a Christian clergyman of the eighteenth century; Ernst Mach, a scientist, public figure and atheist of the nineteenth century; and Philipp Frank and Herbert Feigl, two scientist/philosophers of the twentieth century.

36.6 Joseph Priestley: An Eighteenth-Century Contributor to the Enlightenment Education Project

Joseph Priestley was born in Yorkshire in 1733 and died in Pennsylvania in 1804; his life spaned the core years of the European Enlightenment in which he played a significant role. He was an enormously gifted person, a polymath who made original and lasting contributions across a wide range of subjects. He wrote over 200 books, pamphlets, and articles in history of science (most importantly of electricity and optics), political theory, theology, biblical criticism, theory of language, philosophy of education, and rhetoric; as well as chemistry for which he is now best known (Priestley 1775).[45]

He was not just knowledgeable in many fields: there was an explicit interconnectedness to all his intellectual activity. For Priestley knowledge was not compartmentalised: his epistemology (sensationalism) related to his ontology (materialism), both related to his theology (Unitarianism) and to his psychology (Associationism); and these all bore upon his political and social theory (Liberalism). As with Mario

---

[42]Thomas Huxley's 'A Liberal Education; and Where to Find It', an address given at the 1868 opening of the South London Working Men's College, shows the overlap between nineteenth-century Enlightenment education and liberal education (Huxley 1868/1964). The alliance between Philipp Frank and James Conant in the 1950s and 1960s in the USA is an instructive twentieth-century example (Reisch 2017). Liberal education values the appreciation and transmission of knowledge; so also Enlightenment education.

[43]See Lövlie and Standish (2002).

[44]Joseph Priestley and fellow Dissenters wanted state support but absolutely opposed state control of education. The reconciliation of support with denial of control is a recurring question in the Enlightenment education tradition.

[45]Two definitive studies of Priestley are by Robert Schofield (1997, 2004). The latter contains a full bibliographic listing of his many books, pamphlets and articles. See also contributions to Anderson and Lawrence (1987), Birch and Lee (2007), Rivers and Wykes (2008), and Schwartz and McEvoy (1990).

Bunge two centuries later, Priestley was consciously a *synoptic* or *systemic* thinker: all components of knowledge (and political and personal life as a whole) had to relate together consistently.

Modern appreciation of Priestley has been blighted by the harsh and unfair judgement of Thomas Kuhn made in his best-selling *Structure of Scientific Revolutions* (Kuhn 1970). In a famous passage Kuhn writes of the irrationality of paradigm change in science and of old paradigms just dying out until 'at last only a few elderly hold-outs remain'. He then singularly names Priestley as an example 'of the man who continues to resist after his whole profession has been converted' and adds that such a man 'has *ipso facto* ceased to be a scientist' (Kuhn 1970, p. 159).

This outrageous charge 'blackened' Priestley's reputation in the academic world; Kuhn's has become the widely-accepted obituary for Priestley – the stubborn old man who held on to belief in a peculiar phlogiston substance and who resisted the dawning bright light of Lavoisierian chemistry. Pleasingly, some historians and philosophers have provided extensive studies that refute Kuhn's caricature of Priestley, but unfortunately their work is not translated into 20+ languages, nor set as class reading in countless thousands of courses and not read by millions.

A more generous and accurate assessment of Priestley was given by Frederic Harrison in his Introduction to a nineteenth-century edition of Priestley's *Scientific Correspondence*, as follows:

> If we choose one man as a type of the intellectual energy of the eighteenth century, we could hardly find a better than Joseph Priestley, though his was not the greatest mind of the century. His versatility, eagerness, activity, and humanity; the immense range of his curiosity in all things, physical, moral, or social; his place in science, in theology, in philosophy, and in politics; his peculiar relation to the Revolution, and the pathetic story of his unmerited sufferings, may make him the hero of the eighteenth century. (Bolton 1892, Introduction)

Priestley shared the Enlightenment conviction that a good education would benefit individuals and their societies. As he wrote in *The Proper Objects of Education*:

> All great improvements in the state of society ever have been, and ever must be … the result of the most peaceable but assiduous endeavours in pursuing the slowest of all processes – that of enlightening the minds of men. (Priestley 1791)

While many advocated and wrote about better and more widespread education Priestley was of the minority who practised what the Enlightenment preached: he had a life-long engagement in schooling, teaching and learning. Priestley's educational views were part of his overall systematic position: his theology, philosophy, epistemology, psychology, social theory and science were all parts of a coherent whole. He was under impressed with the state of English education, in particular education in natural philosophy, or science:

> I am sorry to have occasion to observe, that natural science is very little, if at all, the object of *education* in this country, in which many individuals have distinguished themselves so much by their application to it. And I would observe that, if we wish to lay a good foundation for a philosophical taste, and philosophical pursuits, persons should be accustomed to the sight of experiments, and processes, in *early life*. They should, more especially, be early

initiated in the theory and practice of *investigation*, by which many of the old discoveries may be made to be really *their own*; on which account they will be much more valued by them. (Priestley 1790, p. xxix)

This is one of the first endorsements of inquiry teaching, and more specifically of historical-investigative teaching – following in the experimental footsteps of those who have gone before.[46] This is in part why he wrote the first history of Optics (Priestley 1772)[47] and of Electricity (Priestley 1767/1775).[48] His assumption was that the habits and skills acquired in investigating nature – observing, hypothesising, seeking evidence for and against, experiments with controls – would flow on to the investigation of other matters: religion, revelation, politics, church history and so on. For Priestley, and a good many of the Enlightenment philosophers, science would be:

> the means, under God, of extirpating all error and prejudice, and of putting an end to all undue and usurped authority in the business of religion, as well as of science'. (Priestley 1775–1777, Vol. I, p. xiv)

Priestley had a good critical education at the Dissenting Academy at Daventry where he was exposed to lively debate and argument on all subjects. After ministry at Nantwich, he went on to teach at the famed Warrington Academy where he introduced physics and chemistry to the curriculum. The dissenting academies were a response by the non-conformist churches to the Anglican Church's monopoly on English school and university education; students of any faith, or no faith, could enrol. Robert Merton has been one of many to draw attention to the role of these Dissenting Academies in fostering and promoting science in England (Merton 1938/1970, p. 119). One commentator has said:

> It is in Non-conformist England, the England excluded from the national universities, in industrial England with its new centres of population and civilisation that we must seek the institutions which gave birth to the utilitarian and scientific culture of the new era. (Halevy, quoted in Brooke 1987, p. 11)

An historian of education has opined:

> Warrington Academy, was for 30 years arguably the finest educational establishment in the world, largely due to the input and influence of Joseph Priestley. (Rose 2007, p. 235)

This is a case where a significant part of the Enlightenment project, namely education was advanced by others, namely Christian believers. Newton at Cambridge inspired the Dissenters, but the Dissenters (and Catholics, Jews, Muslims and atheists) were forbidden to enrol there. In contrast, the Enlightenment's 'Free Inquiry' was the entrenched motto of the Dissenting Academies.[49]

---

[46]On the tradition of historical-investigative teaching of science, see Heering and Höttecke (2014).

[47]For the next 150 years this was the only English-language history of Optics.

[48]This authoritative work led to productive correspondence with Franklin, Volta and many others; it was instrumental in the birth of electrical science.

[49]On the contribution of the Dissenting Academies to English education and culture see Smith (1954) and Wykes (1996).

In 1758 at age 25 years Priestley took a pastor's position at Nantwich in Cheshire. While there he established a school with 30 boys and, in a separate room, six girls. He taught in the school for 3 years, 6 days a week, from 7 am to 4 pm, teaching Latin, Greek, English grammar and geography. In addition, he taught some Natural Philosophy and purchased an air pump and an electrical machine and instructed his pupils in their use. Priestley may well have been the first person to teach laboratory science to schoolchildren.

As well as some three decades of direct engagement in teaching, Priestley wrote a number of influential works on the theory and practice of education. His most famous work - *An Essay on a Course of Liberal Education for Civil and Active Life* (Priestley 1765/1965) – was written and published while teaching at Warrington Academy. It originally appeared as a pamphlet then it became a 25-page Preface to his *Lectures on History and General Policy* (Priestley 1788). In this incarnation it had 16 printings and was translated into Dutch (1793) and French (1798). In the American edition of 1803 Priestley adds a note to the above text:

> Since this was written, which is near forty years ago, few persons have had more to do in the business of education than myself; and what I then planned in theory has been carried into execution by myself and others, with, I believe, universal approbation. (Passmore 1965, p. 289)

This theme of connecting theory to practice runs through all Priestley's work, including his opposition to Lavoisier's new oxygen theory. Although he is neither a harbinger of Marxism nor a premature Positivist, Priestley was always suspicious of theory that ran too far in front of practice, or removed itself too far from the facts of the matter; for him, to use a later phrase, 'theory had to be proved in practice'. Priestley advocated a coordinated curriculum, saying that:

> When subjects which have a connection are explained in a regular system, every article is placed where most light is reflected upon it from the neighbouring subjects. (Passmore 1965, p. 293)

He advocated a structured and guided curriculum:

> The plainest things are discussed in the first place, and are made to serve as axioms, and the foundation of those which are treated of afterwards. Without this regular method of studying the elements of any science, it seems impossible ever to gain a clear and comprehensive view of it. (Passmore 1965, p. 293)

He stresses that liberal education for civil and active life needs to promote the understanding of the principles of subject matter, by saying:

> A man who has been used to go only in one beaten track and who has had no idea given him of any other ... Will be wholly at a loss when it happens that that track can no longer be used; while a person who has a general idea of the whole course of the country may be able to strike out another and perhaps a better road than the former. (Passmore 1965, p. 295)

As a teacher at the Dissenting Academy at Warrington Priestley insisted on students asking and answering questions, he promoted free engagement with all subjects including Divinity, he ensured that authorities on both sides of controversial issues be read and quoted. One of his Warrington students recalled that:

At the conclusion of his lecture, he always encouraged his students to express their sentiments relative to the subject of it, and to urge any objections to what he had delivered, without reserve. It pleased him when anyone commenced such a conversation. ... His object ... was to encourage the students to examine and decide for themselves, uninfluenced by the sentiments of any other persons. (Rutt 1817–1832, vol.1, p. 50. In Lindsay 1970, p. 15)

Priestley had some confidence that an educational regime such as he proposed and enacted, would result in the betterment of society. He said 'I cannot help flattering myself that were the studies I have here recommended generally introduced into places of liberal education, the consequences might be happy for this country in some future period' (Passmore 1965, p. 301). This was the *reformist* Priestley. But, with reason, he was also regarded as a *revolutionary*. His understanding of the flow-on effects of scientific investigation and of the flow-on effects of the acquisition of its associated mental and character dispositions led him in a sermon on 'The Importance and Extent of Free Inquiry', to proclaim from his Birmingham pulpit:

We are as it were, laying gunpowder, grain by grain, under the old building of error and superstition, which a single spark may hereafter inflame, so as to produce an instantaneous explosion; in consequence of which that edifice, the erection of which has been the work of ages, may be overturned in a moment and so effectually as that same foundation can never be built again. (Priestley 1785)

With Britain having just been defeated in the American Revolution (1775–1783) and with the first stirrings of the French Revolution (1787–1789) being felt in all European states and kingdoms, such words were not judicious. They led to his sobriquet 'Gunpowder Joe' and in 1791 to an enraged 'King and Church' mob ransacking his home, library and laboratory and his flight from Yorkshire to America.

Through Priestley's personal friendships with Benjamin Franklin, George Washington, John Adams and Thomas Jefferson, and the admiration they all had for him, there was a direct impact of Enlightenment ideas on late colonial and early independent US public life and education.[50]

Thomas Huxley (1825–1895), perhaps the most lucid and effective champion of Enlightenment causes and Enlightenment education in nineteenth-century England,[51] gave a speech in 1874 at the unveiling of Priestley's statue in Leeds. He said that Priestley was in large measure responsible for the intellectual/cultural advances that nineteenth century Britain had made over that of the eighteenth century:

Reason has asserted and exercised her primacy over all provinces of human activity; that ecclesiastical authority has been relegated to its proper place; that the good of the governed has been finally recognized as the end of government, and the complete responsibility of the governors to the people as its means; and that the dependence of natural phenomena in general on the laws of action of what we call matter has become an axiom.(Huxley 1874/1964, pp. 38–9)

---

[50]For Priestley's impact in early America, see Davenport (1990), D'Elia (1990), and Graham (2008),

[51]See Desmond (1994) and Jensen (1991).

36.7 Ernst Mach: A Nineteenth-Century Contributor to the Enlightenment Education Project

Ernst Mach (1838–1916), was one of the great philosopher-scientists of the late nineteenth and early twentieth centuries; and was major contributor to the Enlightenment project in education. Unfortunately, his contribution to education has been almost entirely ignored in the English-speaking world.[52] This is a pity, because current trends in the practice and theory of science education are in many respects repeating Mach's century-old arguments concerning the purposes and aims of science teaching, the nature of understanding, and the best ways to promote the learning of science. An obituary of a century ago, did draw attention to Mach the educator:

> It is Mach the *educationalist* whom we must here bring to the attention of our readers, particularly the younger ones, and not as someone who has passed on, but as a man whose seed is destined to put down ever further roots in physics teaching, and, with that, in all teaching about real things, and to fructify the whole spirit of this teaching. (Höfler 1916, W. A. Suchting trans.)

Mach was fluent in most European languages, an enthusiast of Greek and Latin classics, a physicist who made significant contributions to such diverse fields as electricity, gas dynamics, thermodynamics, optics, energy theory and mechanics; a historian and philosopher of science, a psychologist, Rector of the German University in Prague, a member of the Upper House of the Austrian Parliament and a writer of lucid prose. He was a person of strong character and convictions, a socialist and outspoken liberal-humanist in the centre of the archconservative Catholic Austro-Hungarian Empire. Einstein said of him that 'he peered into the world with the inquisitive eyes of a carefree child taking delight in the understanding of relationships' (Hiebert 1976, p. xxi). Mach made scientific and philosophical contributions across the whole temporal span from Darwin to Einstein. The first of Mach's 500 publications was in 1859, the year of Darwin's *The Origin of Species*; his last work was published 5 years after his death in 1921, the year of Einstein's *Relativity: The Special and General Theory*.[53]

Mach's understanding of science and philosophy bore upon his educational ideas. He was influenced by the ideas of the German philosopher-psychologist-educationalist Johann Friedrich Herbart. He applied Herbart's ideas in his first teaching assignment 'Physics for Medical Students', and in the text he wrote arising from this course (*Compendium of Physics for Medical Students* Mach 1863). Mach's concern here was with 'economy of thought', with getting across the general outline

---

[52]John Bradley, the English chemist and educator, organized his chemistry instruction on Machian principles (Bradley 1963–1968), and he wrote a useful book on Mach's philosophy of science (Bradley 1971). Mach the educator is discussed in Matthews (1990, 2015 pp. 33–37). The most comprehensive and best documented discussion of the subject is Siemsen (2014).

[53]An excellent documentary source of Mach's bountiful influence in science, philosophy and beyond is Blackmore et al. (2001).

of the conceptual modes of physics, and with overcoming the compartmentalism of physics.

Psychology was a long-standing interest of Mach's. At 15 years of age Mach had read Kant's *Prologomena* and signalled his subsequent positivist commitments – 'The superfluity of the role of the "thing-in-itself" suddenly dawned upon me' (Blackmore 1972, p. 11). His teaching was the occasion to unite pedagogical, psychological and scientific concerns. The first of his many science textbooks for school students, published in 1886, was widely used and went through several editions. Indeed, most of the major figures in European physics at the beginning of the twentieth century learnt science from Mach's school texts. These texts provided a logical and historical introduction to science, they sought to present students with the 'most naive, simple, and classical observations and thoughts from which great scientists have built physics' (Pyenson 1983, p. 34). Whilst at Prague he taught courses on 'School Physics Teaching'. In 1887 Mach founded and co-edited the world's second-published science education journal – *Zeitschrift für den Physikalischen und Chemischen Unterricht* (*Journal of Instruction in Physics and Chemistry*). He contributed regularly to this journal until a stroke forced his retirement in 1898.

Mach did not write any systematic work on educational theory or practice; his ideas are scattered throughout his texts and journal articles. However, there are three lectures where he addressed pedagogical issues. One of these is perhaps his most systematic treatment of education in general and science education in particular – 'On Instruction in the Classics and the Mathematico-Physical Sciences' (Mach 1886/1986), translated in his *Popular Scientific Lectures*. His other chief pedagogical papers are 'On Instruction in Heat Theory' (1887), and 'On the Psychological and Logical Moment in Scientific Instruction' (1890),[54] in volumes one and four respectively of his *Zeitschrift*.

As well as intellectual and practical interests in education, Mach had a notable Enlightenment-inspired political involvement in educational reform. The best of the Enlightenment thinkers connected thought to action. As Marx a century later would say, the point of philosophising was to change the world, not just to think about the world. Mach addressed teacher organizations, spoke in the Austrian Parliament on the need for school curricular change, and was active in the struggles to transform the entrenched German gymnasium pattern of separating schools for language and classics from those for science and mathematics. Mach championed the creation of the new *Einheitsschule* where integrated education in the humanities and the sciences could occur. There have been few scientists who have displayed such a wide-ranging interest in both formal (school) and informal (the reading public) education. Mach's relative neglect by English-speaking science educators is unfortunate.

---

[54]This last paper has recently, for the first time, been translated and published in English (Mach 1890/2018). Hayo Siemsen was translator and editor, who sadly died prematurely in 2018.

Well-founded curricular and pedagogical proposals in school science are based upon two foundations: views about the nature and scope of science, and views about the nature and practice of education. The Enlightenment project has contributed to both. There are of course other matters to be considered in drawing up curricula – political, social and psychological, to name just the obvious ones. But what one thinks, explicitly or implicitly, about the philosophy of science and about the philosophy of education will largely determine the form of the science curriculum. Mach's suggestions for the conduct of science education stem in part from his theory of science and his Herbartian theory of education. Some of the major themes of Mach's philosophy of science (his view of the nature of science) are the following:

- Scientific theory is an intellectual construction for economizing thought and thereby conjoining experiences.
- Science is fallible; it does not provide absolute truths.
- Science is a historically conditioned intellectual activity.
- Scientific theory can only be understood if its historical development is understood.

Mach's educational ideas are fairly simple and uncontroversial:

- Begin instruction with concrete materials and thoroughly familiarize students with the phenomena discussed.
- Aim for understanding and comprehension of the subject matter.
- Teach little, but teach it well.
- Follow the historical order of development of a subject.
- Tailor teaching to the intellectual level and capacity of students.
- Address the philosophical questions that science entails and which gave rise to science.
- Show that just as individual ideas can be improved, so also scientific ideas have constantly been, and will continue to be, overhauled and improved.
- Engage the mind of the learner.

Although a pre-eminent theorist, and concerned with economy of thought in education, Mach firmly believed that abstractions in the science classroom should, as Hegel said of philosophy, take flight only at dusk: 'Young students should not be spoiled by premature abstraction, but should be made acquainted with their material from living pictures of it before they are made to work with it by purely ratiocinative methods' (Mach 1886/1986, p. 4). This Enlightenment conviction goes all the way back to John Locke and beyond. A simple point, usually observed in its breach, as Arnold Arons has lamented:

> As physics teaching now stands, there is a serious imbalance in which there is an overabundance of numerical problems using formulae in canned and inflexible examples and a very great lack of phenomenological thinking and reasoning. (Arons 1988, p. 18)

Another of Mach's concerns was the tendency to overfill the curriculum. For him the principal aims of education were to develop understanding, strengthen reason and promote imagination. A bloated curriculum counteracted these aims:

> I know nothing more terrible than the poor creatures who have learned too much. What they have acquired is a spider's web of thoughts too weak to furnish sure supports, but complicated enough to produce confusion. (Mach 1886/1986, p. 367)

One hundred years later this lament is still being voiced about the USA's 'one mile wide and one inch deep' curricula.

Mach believed in presenting science historically, or as he put it, teaching should follow the genetic approach:

> every young student could come into living contact with and pursue to their ultimate logical consequences merely a *few* mathematical or scientific discoveries. Such selections would be mainly and naturally associated with selections from the great scientific classics. A few powerful and lucid ideas could thus be made to take root in the mind and receive thorough elaboration. (Mach 1886/1986, p. 368)

Mach's major textbooks on mechanics (1883), heat (1869) and optics (1922) all follow the genetic method of exposition. This was Priestley's method, and it is partly why Priestley wrote his histories. Mach realised that the logic of a subject was not necessarily the logic of its presentation – a point known to most school teachers, if not to administrators. The logic of a discipline and the logic of its pedagogy are not identical, as Mach's contemporary and fellow positivist Pierre Duhem also maintained:

> The legitimate, sure, and fruitful method of preparing a student to receive a physical hypothesis is the historical method ... that is the best way, surely even the only way, to give those studying physics a correct and clear view of the very complex and living organisation of this science. (Duhem 1906/1954, p. 268)

36.8 Philipp Frank and Herbert Feigl: Two Twentieth-Century Contributors to the Enlightenment Education Project

The two European émigré positivist philosophers Philipp Frank and Herbert Feigl flesh out more of the Enlightenment education project, specifically its implications for science education. Their writings and activities also show the large commonality between the Enlightenment project and the project of Liberal Education. Many things are shared, but the former has an intrinsic commitment to social and cultural change that is not central to the latter.

Philipp Frank (1884–1966) was born in Vienna in 1884 and died in Cambridge Massachusetts in 1966. In 1907 he received his doctorate in theoretical physics at the University of Vienna where he studied under Ludwig Boltzmann. Frank's first paper, published in 1907 at the age of 23 years – 'Experience and the Law of Causality' (Frank 1907/1949) – characterized his subsequent philosophical concern: namely prolonged and informed philosophical reflection on the structures, methodology and history of science. The meetings of the Vienna Circle that he instigated set the style of his subsequent intellectual career: there was a seriousness of purpose coupled with a genuine open-mindedness towards different opinions and traditions:

> This apparent internal discrepancy [in the group] provided us, however, with a certain breadth of approach by which we were able to have helpful discussions with followers of various philosophical opinions. Among the participants in our discussions were, for instance, several advocates of Catholic philosophy. Some of them were Thomists, some were rather adherents of a romantic mysticism. Discussions about the Old and New Testaments, the Jewish Talmud, St. Augustine, and the medieval schoolmen were frequent in our group. Otto Neurath even enrolled for one year in the Divinity School and won an award for the best paper on moral theology. This shows the high degree of our interest in the cultural background of philosophic theories and our belief in the necessity of an open mind which would enable us to discuss our problems with people of divergent opinions. (Frank 1949, pp. 1–2)

Frank published two explicitly educational papers: 'Science Teaching and the Humanities' (Frank 1950b) and 'The Place of Philosophy of Science in the Curriculum of the Physics Student' (Frank 1947/1950). He regretted that the 'result of conventional science teaching has not been a critically minded type of scientist, but just the opposite' (Frank 1947/1950, p. 230). In part this regret is because 'the science student who has received the traditional, purely technical instruction in his field is extremely gullible when he is faced with pseudophilosophic and pseudoreligious interpretations that fill somehow the gap left by his science courses' (Frank 1947/1950, p. 230). As a consequence, 'This failure prevents the science graduate playing in our cultural and public life the great part that is assigned to him by the ever-mounting technical importance of science to human society' (Frank 1947/1950, p. 231).

It is of course the history and philosophy of science that makes good these shortfalls; or rather, for Frank, just philosophy of science because this indeed consists of two inseparable components, 'logico-empirical analysis' and 'socio-psychologic' analysis (Frank 1947/1950, p. 248). The first is conceptual or semantic analysis, the second is careful historical analysis. He says that 'This analysis is the chief subject that we have to teach to science students in order to fill the gaps left by traditional science teaching' (Frank 1947/1950, p. 245).

Frank is an advocate of liberal education, affirming that a variety of subject matters should be mastered, and that as much as possible relations between the subjects should be brought out. He believes that humanities can be taught from *within* science, saying that:

> The student of science will get the habit of looking at social and religious problems from the interior of his own field and entering the domain of the humanities by a wide-open door … there is no better way to understand the philosophic basis of political and religious creeds than by their connection with science. (Frank 1950a, b, p. 281)

Herbert Feigl was born in 1902 in Reichenberg then in Austria-Hungary, a part of the Sudetenland which subsequently was incorporated in Czechoslovakia. He died in Minneapolis in 1988. At age 16 he read an article on the theory of special relativity and set about trying, without success, to refute it. He said that the attempt resulted in him learning a lot of mathematics and physics. At age 20 he went to the University of Vienna to study philosophy with Moritz Schlick (and additionally to study mathematics, physics and psychology). He was a founding member of the Vienna Circle established by Schlick in 1924 as a weekly evening discussion group, and he remained a member of the Circle until his emigration to the US in 1930. In 1927, Feigl presented his doctoral thesis on 'Chance and Law: An Epistemological Investigation of Induction and Probability in the Natural Sciences'. In the US he worked with Percy Bridgman at Harvard on the foundations of physics including the topic of operational definitions of theoretical terms. In 1940 he was appointed professor of philosophy at the University of Minnesota; in 1953 he established the Minnesota Center for the Philosophy of Science, a centre that would make a significant contribution to the articulation and spread of logical empiricist philosophy in the US and worldwide, especially through contributions to the many volumes of *Minnesota Studies in Philosophy of Science.*

Feigl published one explicitly educational paper: 'Aims of Education for Our Age of Science: Reflections of a Logical Empiricist' (Feigl 1955). Feigl regarded promotion of individual autonomy as the prime educational achievement:

> As long as education promotes the formation of intelligence and character in a manner that allows for free learning, rational choices, and critical reflection, human beings so educated will have an excellent opportunity for being masters of their own activities and achievements. (Feigl 1955, p. 322)

This is almost, and not accidently, a verbatim repetition of the opening sentences of Kant's 1784 'What is Enlightenment?' quoted earlier in this chapter. Not surprisingly, Feigl advocates teaching science in a historically and philosophically informed manner, saying:

> It is my impression that the teaching of science could be made ever so much more attractive, enjoyable, and generally profitable by the sort of approach that is more frequently practiced in the arts and the humanities. The dull and dry-as-dust science courses can be replaced by an exciting intellectual adventure if the students are permitted to see the scientific enterprise in broader perspective. Preoccupation with the purely practical values of applied science has overshadowed the intellectual and cultural values of the quest for knowledge. (Feigl 1955, p. 337)

And further, he embraces the orthodox liberal education position wherein: 'training in the sciences and in the scientific attitude should, of course, be combined with studies in history, literature, and the arts' (Feigl 1955, p. 338). As important as science is, it is not the only thing that Feigl treasures:

> I consider truly great music the supreme achievement of the human spirit...I am inclined to think that music expresses (even more than poetry) what is inexpressible in cognitive and especially in scientific language. (Cohen 1981, p. 5)

Feigl has a robust account of values and recognizes that they are an intrinsic part of education; that they mould and direct educational processes and are crucial to the establishment of educational aims. Feigl has an even more robust account of rationality and its place in education. He believes that the classical Aristotelean conception of man as rational animal 'may still be a good beginning' (Feigl 1955, p. 335), and then explicates the idea for education, stressing that rationality covers at least six virtues of thought and conduct:

- clarity of thought (the meaningful use of language and avoidance of gratuitous perplexities);
- consistency of reasoning (conformity with the principles of formal logic);
- reliability of knowledge claims (wherever the evidence is too weak, belief should be withheld);
- objectivity of knowledge claims (knowledge claims should be testable by anyone sufficiently equipped with intelligence and competence);
- rationality of purposive behaviour (maximum positive outcomes are to be gained at the cost of minimum negative outcomes);
- moral rationality (adherence to principles of justice, equity or impartiality, and abstention from coercion and violence in the settlement of conflicts of interest. (Feigl 1955, pp. 335–336ff)

Frank and Feigl made an indirect, but nevertheless significant, contribution to US and international physics education through their Harvard collaboration with physicist-philosopher-historian Gerald Holton who oversaw the much-used and influential *Harvard Project Physics Course* (Holton 1978). Holton's articulation of the philosophy of the course, and the course's structure, resonates with Enlightenment themes. Discrete topics in physics are linked to each other, to topics in other sciences, to mathematics, to philosophy, literature, and so on. Knowledge is a tapestry and should be presented as such to students. He distinguishes scientific training from scientific education, a distinction made by Mach and most other proponents of enlightened education. For Holton:

Training is achieved by imparting the most efficient skill for a scientific purpose. Education is achieved by imparting a point of view that allows generalization and application in a wide variety of circumstances in one's later life. (Holton 1978, p. 298)

36.9 Conclusion

Modern science is based on Enlightenment-grounded commitments: the importance of evidence; rejection of simple authority, especially non-scientific authority, as the arbiter of knowledge claims; a preparedness to change opinions and theories; a fundamental openness to participation in science regardless of gender, class, race or religion; recognizing the inter-dependence of disciplines; and pursuing knowledge for advancement of personal and social welfare. All of this needs to be manifest in science education, along with a willingness to resist the imposition of political, religious and ideological pressures on curriculum development, textbook choice and pedagogy.

These commitments are mostly made without awareness of their Enlightenment roots. It is important for educators to connect these contemporary commitments with their historical scientific-philosophical base; and to be aware of the trajectories and philosophical-political-religious buffeting that the commitments have experienced over time. If the past is known, it can be learnt from; and teachers can develop a sense of belonging to an open-minded, critical, scholarly tradition, and hopefully defend it. Some in this tradition take their inspiration from the Enlightenment, others from other sources. Defense of the tradition requires serious philosophical work. Questions of epistemology concerning the objective knowability of the world, questions of ontology concerning the constitution of the world, specifically regarding methodological and ontological naturalism, questions of methodology concerning theory appraisal and evaluation, and the limits, if any, of scientism, questions of ethics concerning the role of values in science – all need to be fleshed out, and Enlightenment answers defended against their many critics.

The Enlightenment education tradition has been advanced by numerous individuals. Just some – Priestley, Mach, Frank and Feigl – have been elaborated upon here. Other Anglo-Americans that warrant elaboration are Thomas Huxley, Frederick Westaway, John Dewey, and Gerald Holton. And there are numerous European, Latin American, and Asian contributors to the project. They all have a commitment to some constellation of the core Enlightenment principles that have been detailed above. The tradition is characterised by valuing the cultural importance of science[55]; by commitment to the growth of knowledge of the natural and social worlds; the diffusion of this knowledge by both formal and informal education; the utilisation of knowledge for the amelioration of social and cultural problems; and for the flourishing of personal life. For this to happen, the history and philosophy of science needs to be absorbed as science is taught, and more especially where science teachers are trained.[56]

That Enlightenment banner continues to be carried by Mario Bunge. He champions Enlightenment principles, adjusts them, and adds to them. In Latin America of the mid- and late twentieth century, he was one of the outstanding Enlightenment figures, and has been the same in the wider international academic community.

Appendix

Mario Bunge, *Temas de Educación Popular***, Buenos Aires: El Ateneo, 1943, 99 pages, Contents**

A Social problems

1. Technological education in Argentina
2. What kind of technologists should the popular universities train?

---

[55]Wallis Suchting provides a rewarding discussion of the cultural significance of science (Suchting 1994).
[56]The argument is developed, and literature canvassed, throughout Matthews (2015).

3. Women's technological education
4. Professional re-education
5. Patriotic action of the popular universities

B Didactic problems

6. Teaching the studying technique
7. Warning to the new technology teacher
8. Emulation and rivalry in the classroom
9. Commercial education in the popular universities
10. On the teaching of mathematics in technical schools

C Organization problems

11. Conditions the administration of a popular university ought to meet
12. Selection of the professoriat
13. Intervention of students and graduates in popular universities
14. Finances of the popular universities
15. Diplomas awarded by the popular universities

References

Agazzi, E. (Ed.). (2017). *Varieties of scientific realism: Objectivity and truth in science*. Dordrecht: Springer.

Aikenhead, G. S. (1997). Towards a first nations cross-cultural science and technology curriculum. *Science Education, 81*(2), 217–238.

Anchor, R. (1967). *The enlightenment tradition*. Berkeley: University of California Press.

Andersen, K. (2017). *Fantasyland: How America went haywire, a 500-year history*. London: Penguin.

Anderson, R. G. W., & Lawrence, C. (Eds.). (1987). *Science, medicine and dissent: Joseph Priestley (1733–1804)*. London: Wellcome Trust.

Arons, A. B. (1988). Historical and philosophical perspectives attainable in introductory physics courses. *Educational Philosophy and Theory, 20*(2), 13–23.

Artz, F. B. (1968). *The enlightenment in France*. Oberlin: Kent State University Press.

Axtell, J. L. (Ed.). (1968). *The educational writings of John Locke*. Cambridge: Cambridge University Press.

Becker, C. (1932/1960). *The Heavenly City of the eighteenth-century philosophers*. New Haven: Yale University Press.

Berlin, I. (Ed.). (1956). *The age of enlightenment: The eighteenth-century philosophers*. New York: Mentor Books.

Berlin, I. (1980). The counter-enlightenment. In H. Hardy (Ed.), *Against the current* (pp. 1–24). New York: Viking.

Birch, J. S., & Lee, J. (Eds.). (2007). *Joseph Priestley. A celebration of his life and legacy*. Lancaster: Scotforth Books.

Blackmore, J. T. (1972). *Ernst Mach: His work, life and influence*. Berkeley: University of California Press.

Blackmore, J. T., Itagaki, R., & Tanaka, S. (2001). Alois Höfler: Polymath. In *Ernst Mach's Vienna 1895–1930*. Dordrecht: Kluwer Academic Publishers.

Blom, P. (2010). *A wicked company: The forgotten radicalism of the European enlightenment*. New York: Basic Books.

Bolton, H.C. 1892, Scientific correspondence of Joseph Priestley. New York.

Bradley, J. (1963–1968). A scheme for the teaching of Chemistry by the historical method. *School Science Review* 44: 549–553; 45, 364–368; 46, 126–133; 47, 65–71, 702–710; 48, 467–474; 49, 142–150; 454–460.

Bradley, J. (1971). *Mach's philosophy of science*. London.

Brickhouse, N. W. (2001). Embodying science: A feminist perspective on learning. *Journal of Research in Science Teaching, 38*(3), 282–295.

Brooke, J. H. (1987). Joseph Priestley (1733–1804) and William Whewell (1794–1866): Apologists and historians of science. A tale of two stereotypes. In R. G. W. Anderson & C. Lawrence (Eds.), *Science, medicine, and dissent: Joseph Priestley (1733–1804)* (pp. 11–27). London: Wellcome Trust & Science Museum.

Brooke, J. H. (1991). *Science and religion: Some historical perspectives*. Cambridge: Cambridge University Press.

Brown, J. R. (2001). *Who rules in science: An opinionated guide to the science wars*. Cambridge, MA: Harvard University Press.

Bunge, M. (1943). *Temas de educación popular*. Buenos Aires: El Ateneo.

Bunge, M. (1994). Counter-enlightenment in contemporary social studies. In P. Kurtz & T. J. Madigan (Eds.), *Challenges to the enlightenment: In defense of reason and science* (pp. 25–42). Buffalo: Prometheus Books.

Bunge, M. (1999). *The sociology-philosophy connection*. New Brunswick: Transaction Publishers.

Bunge, M. (2000). *Social science under debate: A philosophical perspective*. Toronto: University of Toronto Press.

Butts, R. E., & Davis, J. W. (Eds.). (1970). *The methodological heritage of Newton*. Toronto: University of Toronto Press.

Cassara, E. (1988). *The enlightenment in America*. Lanham: University Press of America.

Charles, S., & Smith, P. J. (Eds.). (2013). *Scepticism in the eighteenth century: Enlightenment, Lumières, Aufklärung*. Dordrecht: Springer.

Church, W. F. (Ed.). (1974). *The influence of the enlightenment on the French revolution* (2nd ed.). Lexington: D.C. Heath & Co..

Clark, W. (1999). The death of metaphysics in enlightened Prussia. In W. Clark, J. Golinksi, & S. Schaffer (Eds.), *The sciences in enlightened Europe* (pp. 423–473). Chicago: University of Chicago Press.

Cohen, I. B. (1980). *The Newtonian revolution*. Cambridge: Cambridge University Press.

Cohen, R. S. (Ed.). (1981). *Inquiries and provocations: Selected writings of Herbert Feigl 1929–1974*. Dordrecht: Reidel.

Cohen, R. S., Hilpinen, R., & Renzong, Q. (Eds.). (1996). *Realism and anti-realism in the philosophy of science: Beijing international conference, 1992*. Dordrecht: Kluwer Academic Publishers.

Commager, H. S. (1977). *The empire of reason: How Europe imagined and America realized the enlightenment*. New York: Doubleday.

Condorcet, N. (1795/1955). *Sketch for a historical picture of the progress of the human mind* (J. Barraclough, Trans.), New York: Noonday Press.

Condorcet, N. (1976). In K. M. Baker (Ed.), *Selected writings*. Indianapolis: Bobbs-Merrill.

D'Elia, D. J. (1990). Joseph Priestley and his American contemporaries. In A. R. Schwartz & J. G. McEvoy (Eds.), *Motion towards perfection: The acheivements of Joseph Priestley* (pp. 237–250). Boston: Skinner House Books.

Davenport, D. A. (1990). Joseph Priestley in America: 1794–1804. In A. R. Schwartz & J. G. McEvoy (Eds.), *Motion towards perfection: The acheivements of Joseph Priestley* (pp. 219–236). Boston: Skinner House Books.

de la Fontainerie, F. (Ed.). (1932). *French liberalism and education in the eighteenth century*. New York: McGraw-Hill.

Desmond, A. (1994). *Huxley: From Devil's disciple to evolution's high priest*. New York: Basic Books.

Duhem, P. (1906/1954). *The aim and structure of physical theory*. (P. P. Wiener, Trans.). Princeton: Princeton University Press.

Dunthorne, H. (2007). The Dutch Republic: "That mother nation of liberty". In M. Fitzpatrick, P. Jones, C. Knellwolf, & I. McCalman (Eds.), *The enlightenment world* (pp. 87–103). London: Routledge.

Eliot, S., & Stern, B. (Eds.). (1979). *The age of enlightenment: An anthology of eighteenth century texts*. East Grinsted: Ward Lock Educational.

Feigl, H. (1955). Aims of education for our age of science: Reflections of a logical empiricist. In N. B. Henry (Ed.), *Modern philosophies and education: The fifty-fourth yearbook of the National Society for the Study of Education* (pp. 304–341). Chicago: University of Chicago Press. Reprinted in *Science & Education* 13(1–2), 2004.

Ferguson, R. A. (1997). *The American enlightenment, 1750–1820*. Cambridge, MA: Harvard University Press.

Ferrone, V. (2015). *The enlightenment. History of an idea*. Princeton: Princeton University Press (orig Italian, 2010).

Fitzpatrick, M. (2007). The age of Louis XIV and early enlightenment in France. In M. Fitzpatrick, P. Jones, C. Knellwolf, & I. McCalman (Eds.), *The enlightenment world* (pp. 134–155). London: Routledge.

Fitzpatrick, M., Jones, P., Knellwolf, C., & McCalman, I. (Eds.). (2007). *The enlightenment world*. London: Routledge.

Fleischacker, S. (2013). *What is enlightenment?* New York: Routledge.

Frank, P. (1907/1949). Experience and the law of causality. In his *Between physics and philosophy* (pp. 53–60). Cambridge, MA: Harvard University Press.

Frank, P. (1947/1950). The place of philosophy of science in the curriculum of the physics student. In his *Modern science and philosophy* (pp. 228–259). Harvard: Harvard University Press.

Frank, P. (1950a). *Modern science and its philosophy*. Cambridge, MA: Harvard University Press.

Frank, P. (1950b), Science teaching and the humanities. In his *Modern science and its philosophy* (pp. 260–285). Cambridge, MA: Harvard University Press.

Garrard, G. (2006). *Counter-enlightenment: From the eighteenth century to the present*. London: Routledge.

Garrett, A. (2007). Enquiry, scepticism and enlightenment. In M. Fitzpatrick, P. Jones, C. Knellwolf, & I. McCalman (Eds.), *The enlightenment world* (pp. 57–64). London: Routledge.

Gay, P. (1966). *The enlightenment: An interpretation* (2-Vols.). New York.

Gay, P. (Ed.). (1973). *The enlightenment: A comprehensive anthology*. New York: Simon & Schuster.

Goubert, P. (1972). *Louis XIV and the twenty million Frenchmen*. New York: Vintage Books.

Graham, J. (2008). Joseph Priestley in America. In I. Rivers & D. L. Wykes (Eds.), *Joseph Priestley: Scientist, philosopher, and theologian* (pp. 203–230). Oxford: Oxford University Press.

Gross, P. R., Levitt, N., & Lewis, M. W. (Eds.). (1996). *The flight from science and reason*. New York: New York Academy of Sciences (distributed by Johns Hopkins University Press, Baltimore).

Hankins, T. L. (1985). *Science and the enlightenment*. Cambridge: Cambridge University Press.

Harper, W. L. (2011). *Isaac Newton's scientific method: Turning data into evidence about gravity and cosmology*. Oxford: Oxford University Press.

Heering, P., & Höttecke, D. (2014). Historical-investigative approaches in science teaching. In M. R. Matthews (Ed.), *International handbook of research in history, Philosophy and Science Teaching* (pp. 1473–1502). Dordrecht: Springer.

Helvétius, C. A. (1772/1810). *A treatise on man; His intellectual faculties and his education* (trans: Hooper, W.). London: Venor, Hood & Sharpe. Original *De l'homme, de ses facultés intellectuelles et de son education* (1772).

Herman, A. (2001). *The Scottish enlightenment: The Scot's invention of the modern world*. London: Harper.

Hiebert, E. N. (1976). Introduction. In E. Mach (Ed.), *Knowledge and error*. Dordrecht: Reidel.

Himmelfarb, G. (2004). *The roads to modernity. The British, French, and American enlightenments*. New York: Alfred A. Knopf.

Höfler, A. (1916). Obituary for Mach. *Zeitschrift für den Physikalischen und Chemischen Unterricht, 29* (57, March).

Holton, G. (1978). On the educational philosophy of the project physics course. In his *The scientific imagination: Case studies* (pp. 284–298). Cambridge: Cambridge University Press.

Horkheimer, M., & Adorno, T.W. (1944/1972). *Dialectic of enlightenment*. New York: Herder and Herder.

Hume, D. (1739/1888). *A treatise of human nature: Being an attempt to introduce the experimental method of reasoning into moral subjects*. Oxford: Clarendon Press.

Hume, D. (1754–1762/1879). *The history of England: From the invasion of Julius Caesar to the revolution in 1688* (6 volumes). New York: Harper & Brothers.

Huxley, T. H. (1868/1964). A liberal education; and where to find it. In his *Science & education* (pp. 72–100). New York: Appleton 1897 (orig. 1885). Reprinted with Introduction by C. Winick, Citadel Press, New York, 1964.

Huxley, T. H. (1874/1964). 'Joseph Priestley'. In his *Science & education*. New York: Appleton, 1897 (orig. 1885). Reprinted with Introduction by C. Winick, Citadel Press, New York, 1964, pp. 9–39.

Hyland, P., Gomez, O., & Greensides, F. (Eds.). (2003). *The enlightenment: A source book and reader*. New York: Routledge.

Israel, J. I. (2001). *Radical enlightenment: Philosophy and the making of modernity 1650–1750*. Oxford: Oxford University Press.

Israel, J. I. (2006). *Enlightenment contested: Philosophy, modernity, and the emancipation of man 1670–1752*. Oxford: Oxford University Press.

Israel, J. I. (2011). *Democratic enlightenment: Philosophy, revolution, and human rights, 1750–1790*. Oxford: Oxford University Press.

Jensen, J. V. (1991). *Thomas Henry Huxley: Communicating for science*. Newark: University of Delaware Press.

Joad, C. E. M. (1935). *The book of joad: A belligerent autobiography*. London: Faber & Faber.

Kandel, I. L. (1930). *History of secondary education: A study in the development of Liberal education*. Boston: Houghton Mifflin Company.

Kant, I. (1784/1995). An answer to the question: What is enlightenment? In Kramnick, I. (ed.) *The portable enlightenment reader* (pp. 1–7). London: Penguin.

Kant, I. (1787/1933). *Critique of pure reason*, 2nd edit. (trans: Smith, N.K.). London: Macmillan (First edition, 1781).

Kant, I. (1803/1899). *Kant on education* (trans: Annette, C.). London: Kegan Paul.

Knudsen, J. B. (1996). On enlightenment for the common man. In J. Schmidt (Ed.), *What is enlightenment? Eighteenth-century answers and twentieth-century questions* (pp. 270–290). Berkeley: University of California Press.

Koch, A. (1961). *Power, morals, and the founding fathers: Essays in the interpretation of the American enlightenment*. Ithaca: Cornell University Press.

Koch, A. (Ed.). (1965). *The American enlightenment: The shaping of the American experiment and a free society*. New York: George Braziller.

Koertge, N. (Ed.). (1998). *A house built on sand: Exposing postmodern myths about science*. New York: Oxford University Press.

Kramnick, I. (Ed.). (1995). *The portable Enlightment reader*. New York: Penguin Books.

Leplin, J. (Ed.). (1984). *Scientific Realism*. Berkeley: University of California Press.

Lindberg, D. C., & Westman, R. S. (Eds.). (1990). *Reappraisals of the scientific revolution*. Cambridge: Cambridge University Press.

Lindsay, J. (1970). Introduction. In *Autobiography of Joseph Priestley* (pp. 11–66). Bath: Adams & Dart.

Locke, J. (1693/1968). *Some thoughts concerning education*. In Axtell, J.L. (ed.) *The educational writings of John Locke* (pp. 114–325). Cambridge: Cambridge University Press.

Lövlie, L., & Standish, P. (2002). *Bildung* and the idea of a liberal education. *Journal of Philosophy of Education, 36*, 317–340.

Mach, E. (1863). *Compendium de Physik für Mediciner*. Vienna: Braumüller.

Mach, E. (1886/1986). On instruction in the classics and the sciences. In his *Popular scientific lectures* (pp. 338–374). La Salle: Open Court Publishing Company.

Mach, E. (1890/2018). Über das psychologische und logische Moment im naturawissenschaftlichen unterricht. *Zeitschrift für den physikalischen und chemischen Unterricht 4*, 1–5. 'About the Psychological and Logical Moment in Natural Science Teaching' Hayo Siemsen (trans.). In Matthews, M.R. (ed) *History, philosophy and science teaching: New perspectives* (pp. 195–200). Dordrecht: Springer.

Matthews, M. R. (Ed.). (1989). *The scientific background to modern philosophy*. Indianapolis: Hackett Publishing Company.

Matthews, M. R. (1990). Ernst Mach and contemporary science education reforms. *International Journal of Science Education, 12*(3), 317–325.

Matthews, M. R. (Ed.). (1998). *Constructivism and science education: A philosophical examination*. Dordrecht: Kluwer Academic Publishers.

Matthews, M. R. (2015). *Science teaching: The contribution of history and philosophy of science: 20th anniversary revised and enlarged edition*. New York: Routledge.

May, H. F. (1976). *The enlightenment in America*. New York: Oxford University Press.

McCarthy, C. L. (2018). Cultural studies of science education: An appraisal. In M. R. Matthews (Ed.), *History, philosophy and science teaching: New perspectives* (pp. 99–136). Dordrecht: Springer.

McIntyre, A. (1981). *After virtue: A study in moral theory*. Notre Dame: University of Notre Dame Press.

McIntyre, A. (1988). *Whose justice? Which rationality?* London: Duckworth.

McMahon, D. M. (2001). *Enemies of the enlightenment: The French counter-enlightenment and the making of modernity*. New York: Oxford University Press.

Merton, R. K. (1938/1970). *Science, technology and society in seventeenth century England*. New York: Harper & Row.

Mukherjee, A. P. (2009). B.R. Ambedkar, John Dewey, and the meaning of democracy. *New Literary History, 40*(2), 345–370.

Newton, I. (1730/1979). *Opticks or a treatise of the reflections, refractions, inflections & colours of light*. New York: Dover Publications.

O'Hara, K. (2010). *The enlightenment: A Beginner's guide*. Oxford: Oneworldpublications.

Osler, M. J. (Ed.). (2000). *Rethinking the scientific revolution*. Cambridge: Cambridge University Press. S500/510.

Pagden, A. (2013). *The enlightenment and why it still matters*. Oxford: Oxford University Press.

Parry, G. (2007). Education and the reproduction of the enlightenment. In M. Fitzpatrick, P. Jones, C. Knellwolf, & I. McCalman (Eds.), *The enlightenment world* (pp. 217–233). London: Routledge.

Passmore, J. A. (Ed.). (1965). *Priestley's writings on philosophy, science and politics*. London: Collier Macmillan.

Pinker, S. (2018). *Enlightenment now: The case for reason, science, humanism, and progress*. New York: Viking.

Porter, R. (2000). *The enlightenment: Britain and the creation of the modern world*. London: Penguin Books.

Porter, R., & Teich, M. (Eds.). (1981). *The enlightenment in National Context*. Cambridge: Cambridge University Press.

Priestley, J. (1765/1965). *An essay on a course of liberal education for civil and active life*. In J.A. Passmore (ed.) *Priestley's writings on philosophy, science and politics* (pp. 285–304). London: Collier Macmillan.

Priestley, J. (1767/1775). *The history and present state of electricity, with original experiments,* second edition, J. Dodsley, J. Johnson & T. Cadell, London; 3rd edit., 1775 reprinted Johnson Reprint Corporation, New York, 1966, with Introduction by Robert E. Schofield.

Priestley, J. (1772). *The history and present state of the discoveries relating to vision, light, and colours.* 2 vols., London.

Priestley, J. (1775). *The discovery of oxygen. Part 1. Experiments by Joseph Priestley.* In Alembic Club Reprints No. 7. (1992). Chicago: University of Chicago Press.

Priestley, J. (1775–1777). *Experiments and observations on different kinds of air* (2nd ed., 3 Vols.) J. Johnson, London. Sections of the work have been published by the Alembic Club with the title *The Discovery of Oxygen* (Edinburgh, 1961).

Priestley, J. (1785). *The importance and extent of free inquiry in matters of religion,* to which is added *The Present State of Free Inquiry in this Country,* J. Johnson, Birmingham. In Rutt *Collected* Works, Vol. 18.

Priestley, J. (1788). *Lectures on history and general policy to which is prefixed, an Essay on the course of Liberal Educatioin for civil and active life.* Dublin: P. Byrne.

Priestley. (1790). *Experiments and observations on different kinds of air, and other branches of natural philosophy, connected with the Subject. Being the former six volumes abridged and methodized* (Vol. 3). Birmingham: J. Johnson.

Priestley, J. (1791). The proper objects of education. In J.T. Rutt (ed.) *The theological and miscellaneous works of Joseph Priestley* Vol.15, pp. 420–440.

Puolimatka, T. (2001). Spinoza's theory of teaching and indoctrination. *Educational Philosophy and Theory, 33*(3–4), 397–410.

Pyenson, L. (1983). *Neohumanism and the persistence of pure mathematics in Wilhelmian Germany.* Philadelphia: American Philosophical Society.

Reisch, G. A. (2017). Pragmatic engagements: Philipp Frank and James Bryant Conant on science, education and democracy. *Studies in East European Thought, 69*(3), 227–244.

Reisner, E. H. (1930). *The evolution of the common school.* New York: Macmillan.

Rivers, I., & Wykes, D. L. (Eds.). (2008). *Joseph Priestley: Scientist, philosopher, and theologian.* Oxford: Oxford University Press.

Rose, N. F. (2007). Science education in the 18th century. In J. S. Birch & J. Lee (Eds.), *Joseph Priestley: A celebration of his life and work* (pp. 234–238). Lancaster: Scotchforth Books.

Roth, M.-W. (1999). Authentic school science: Intellectual traditions. In R. McCormick & C. Paechter (Eds.), *Learning and knowledge* (pp. 6–20). London: Sage Publications.

Rousseau, J. J. (1762/1991). *Emile, or on education,* (B. Allan, Trans.). Harmondsworth: Penguin.

Schama, S. (1981). The enlightenment in the Netherlands. In R. S. Porter & M. Teich (Eds.), *The enlightenment in national context* (pp. 54–71). Cambridge: Cambridge University Press.

Schapiro, J. S. (1963). *Condorcet and the rise of liberalism.* New York: Octagon Books.

Schmidt, J. (Ed.). (1996). *What is enlightenment? Eighteenth-century answers and twentieth-century questions.* Berkeley: University of California Press.

Schofield, R. E. (1997). *The enlightenment of Joseph Priestley: A study of his life and work from 1733 to 1773.* University Park: Penn State Press.

Schofield, R. E. (2004). *The enlightened Joseph Priestley: A study of his life and work from 1773 to 1804.* University Park: Penn State Press.

Schouls, P. A. (1992). *Reasoned freedom: John Locke and enlightenment.* Ithaca: Cornell University Press.

Schulz, R. M. (2007). Lyotard, postmodernism and science education: A rejoinder to Zembylas. *Educational Philosophy and Theory, 39*(6), 633–656.

Schwartz, A. T., & McEvoy, J. G. (Eds.). (1990). *Motion toward perfection: The achievement of Joseph Priestley.* Boston: Skinner House Books.

Shimony, A. (1997). 'Presidential address: Some historical and philosophical reflections on science and enlightenment'. In L. Darden (ed.) *Proceedings of the 1996 PSA Meeting.* S1–14.

Siemsen, H. (2014). Ernst Mach: A genetic introduction to his educational theory and pedagogy. In M. R. Matthews (Ed.), *International handbook of research in history, philosophy and science teaching* (pp. 2329–2357). Dordrecht: Springer.

Smith, J. W. A. (1954). *The birth of modern education. The contribution of the dissenting academies 1660–1800*. Chicago: Independent Press Ltd. Alec, Allenson.

Spence, J. D. (1982). *The gate of heavenly peace: The Chinese and their revolution, 1895–1980*. London: Penguin.

Spinoza, B. (1677/1910). *Ethics & de intellectus emendatione*. London: J.M. Dent & Sons.

Steinberg, S. R., & Kincheloe, J. (2012). 'Employing the bricolage as critical research in science education'. In B. Fraser, K. Tobin & C. McRobbie (eds), *International handbook of science education* (pp. 1485–1500) 2nd Edition, Springer.

Sternhell, Z. (2010). *The anti-enlightenment tradition*. New Haven: Yale University Press.

Suchting, W. A. (1994). Notes on the cultural significance of the sciences. *Science & Education, 3*(1), 1–56.

Tang, Y. (2015). The enlightenment and its difficult journey in China. In Y. Tang (Ed.), *Confucianism, Buddhism, Daoism, Christianity and Chinese Culture* (pp. 279–284). Dordrecht: Springer.

Tarcov, N. (1989). *Locke's education for liberty*. Chicago: University of Chicago Press.

Tobin, K. (2015). Connecting science education to a world in crisis. *Asia-Pacific Science Education, 1*, 2. https://doi.org/10.1186/s41029-015-0003-z.

Trachtenberg, Z. (1993). *Making citizens: Rousseau's political theory of culture*. London: Routledge.

Van Bunge, W. (Ed.). (2003). *The early enlightenment in the Dutch Republic, 1650–1750*. Leiden: Brill.

van Eijck, M., & Roth, W.-M. (2007). Keeping the local local: Recalibrating the status of science and traditional ecological knowledge (TEK) in education. *Science Education, 91*, 926–947.

Venturi, F. (1972). *Italy and the enlightenment: Studies in a cosmopolitan century*. (trans: Susan, C.). New York: New York University Press.

Wootton, D. (2015). *The invention of science: A new history of the scientific revolution*. London: Penguin Random House.

Wykes, D. L. (1996). The contribution of the dissenting academy to the emergence of rational dissent. In K. Haakonssen (Ed.), *Enlightenment and religion: Rational dissent in eighteenth-century Britain* (pp. 99–139). Cambridge: Cambridge University Press.

Chapter 37
Cognition and Education: A Bungean Systemic Perspective

Ibrahim A. Halloun

Abstract Everything in the universe, according to Mario Bunge, is either a system or part of a system of particular structure and function. A systemic worldview helps us make sense of the world around us and flexibly systematize how we go about constructing, retaining, and deploying all sorts of knowledge. A four-dimensional system schema is proposed to conceive a system of any sort in meaningful and productive ways, especially in experiential learning. Experiential learning can be optimized when it involves systemic transactions, between a learner and objects of learning, consciously aimed at bringing about a systemic conceptual image of each object in accordance with the proposed schema. Formal education is then supposed to provide for systemic cross-disciplinary curricula designed and deployed under systemic pedagogical frameworks, like the Systemic Cognition and Education framework, to empower students with systemic profiles for lifelong learning and success.

Making sense of the world around us and developing our knowledge about this world in meaningful and productive ways are of prime importance to humankind. Formal education is supposed to help us systematize how we go about achieving these goals efficiently while preparing us for success in various aspects of life. The success of such endeavors depends mostly on how well we infuse order in both the physical world and our mental realm. Mario Bunge's systemism offers us the best framework, from both cognitive and pedagogical perspectives, to systematize our orderly quest for meaningful and productive knowledge, sustain such knowledge in memory, and optimize the efficiency with which we may retrieve it and creatively deploy it in any situation.

Bunge's systemism is a generic worldview that cuts across cultures and academic disciplines (Bunge 1967, 1979, 1983, 2000). Accordingly, everything in this world, whether natural or human-made, physical or conceptual, is a system or part of

I. A. Halloun (✉)
H Institute & Lebanese University, Beirut, Lebanon
e-mail: halloun@halloun.net

683

a system. Systems infuse order in the world, and help us reveal patterns in the universe, from the astronomical scale down to the subatomic scale, as well as in human body and mind. Systemic processes help us, as best as possible, systematize knowledge construction, retention, and deployment, and take full advantage of brain and mind patterns in any mental or sensory-motor process or product.

This chapter offers, in four sections, a particular perspective on Mario Bunge's systemism and how it helps us systematize both cognition and education. Our perspective emerges primarily from Bunge's philosophy and latest advances in cognitive science and especially the emerging field known as Mind, Brain, and Education (MBE) in neuroscience (Fischer et al. 2007; Knox 2016; Schwartz 2015). Section 37.1 outlines a systemic worldview whereby, in accordance with Bunge's systemism, the world within and around us is conceived as a world of dynamic systems that may interact with each other and affect the way each system is, operates, and evolves. A four-dimensional schema is proposed in this section for the construction of any system at the center of a middle-out organization of both the physical world and our mental realm, and a discussion of how to systematize our multifaceted, but especially realist-rationalist transaction with physical objects in experiential learning is presented. Section 37.2 discusses in the context of systemic cognition how such transaction involves primarily the perceptual deconstruction of any object of learning and its conceptual re-construction in two distinct but complementary and overlapping types of memory. These are working memory and short-term memory, both of which engage cerebral/mental systems of distinct functions so as to bring about learning outcomes that can be readily integrated with patterns of long-term memory. Section 37.3 subsequently argues for systemic education that explicitly and systematically: (a) aims at bringing students up with systemic profiles, (b) treats them as systemic beings with systemic minds, and (c) proceeds systemically to realize that aim in dynamic learning ecologies and in the context of cross-disciplinary curricula. The chapter concludes with a concise introduction of Systemic Cognition and Education (SCE), a generic pedagogical framework for student and teacher education of all levels that emerges from Bunge's systemism and MBE as discussed in this chapter.

37.1 A Middle-Out Systemic Worldview

We are constantly engaged in cognitive development, i.e., in the development of our content and process knowledge, and thus in changing the state of our mind (and brain) in certain respects, whether consciously or not and purposefully or not, and whether or not we are interacting with the outside world. Perhaps the most structured and most systematic cognitive development occurs under formal education, and the most meaningful development at all schooling levels occurs through experiential learning, i.e., through "transaction" – a term we borrow from John Dewey (Archambault 1964) and Mario Bunge (1967) – with real entities (objects and events included) in an appropriate learning ecology.

Fig. 37.1 Learning ecology
in experiential learning

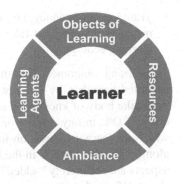

Experiential learning is about a learner's conscious and purposeful experience with one or more real world entities in a favorable learning ecology that includes, in addition to these entities (referred to hereafter as "objects of learning"), other elements that contribute to the course and outcome of the experience. In particular, experiential learning involves (Fig. 37.1):

- A *learner*, an individual student in formal education, who is engaged in the learning experience to fulfill specific purposes. In formal education, these purposes are typically set in a given curriculum in some form of content and process knowledge that students are expected to develop about, and in the context of, particular objects of learning.
- *Objects of learning*, various physical and/or conceptual entities about which, and in the context of which, the learner is expected to develop the expected knowledge (e.g., the human body or parts of it, a poem, a particular scientific concept or model).
- *Learning agents*, peers, teachers, and other people with whom the student may significantly interact during the learning experience.
- *Resources*, various physical tools, facilities, and/or information sources (textbook included) that are at the student disposal.
- *Ambiance*, classroom and school settings, other than resources, that set the overall perceptual and emotional atmosphere, and that might have direct or indirect effect on the course and outcome of the learning experience (e.g., light, temperature, student feelings).

Cognitive development takes place throughout the experiential learning experience, and the significance and meaningfulness of *learning outcomes* depend on all elements mentioned above. In particular, these outcomes depend on:

- The purpose(s) set for the learning experience.
- The sensory-motor and cognitive (mental and affective) state of the learner.
- The state of the objects of learning and all other elements in the learning ecology (Fig. 37.1).

- Transaction efficiency, i.e., the efficiency of all rational, affective, and sensory-motor exchanges that take place between the learner and all elements in the learning ecology.

Learning outcomes that the learner develops about any object of learning (denoted hereafter by O/L) at the outset of the learning experience make up a particular body of knowledge about this O/L. This body of knowledge is not a true copy of O/L in any respect or to any extent. It brings together perceived aspects which the learner focuses on in O/L and various elements in the learning ecology, along with other aspects in the learner's cognitive and sensory-motor state. Various aspects are not merely "added" together. They bring about an *emergent knowledge* about O/L with new conceptual elements in the learner's mind that cannot be attributed directly to either learner or ecology. Such emergent knowledge, as well as the entire learning experience, can best be explained by assuming the learner, objects of learning, and all other elements in the learning ecology as a set of interacting *systems* or parts of one complex system that may consist of all elements in Fig. 37.1.

37.1.1 Systemism and Systemic Transaction

This position resonates well with Bunge's "whole systemic worldview" whereby all:

"artifacts, whether physical like television networks, biological like cows, or social like corporations, are systems. Hence... they should be examined and handled as wholes, though not as blocs but as systems". In fact, "everything, whether concrete or abstract, is a system or an actual or potential component of a system". A system, Bunge argues, is "a complex thing whose components are bound together, as a consequence of which the whole has peculiar properties and behaves as a unit in some respects... systems have systemic (emergent) features that their components lack." (Bunge 2000, pp. 148, 149)

Emergent properties, according to Bunge, are either properties that a system possesses as a whole and that are not possessed by any of its constituents, or new properties that constituents may possess because of their interaction with each other and that they would not have possessed without such interaction. A person's experience with any object, Bunge also notes, "is not a self-subsistent object but a certain *transaction* between two or more concrete *systems*, at least one of which is the experient organism. Experience is always of *somebody and* of *something*" (Bunge 1967, p. 162, italics added). The resulting knowledge is experiential in the sense that it "is attained jointly by experience ... and by reason" (Bunge 1973, p. 170).

In addition to Dewey, the notion of transaction as held by Bunge is shared by numerous philosophers and cognitive scientists. For instance, Bachelard (1949) argues that what we know about the world results from a marriage between realism and rationalism. According to Johnson-Laird, "our view of the world, is causally dependent both on the way the world is and on the way we are" (1983, p. 402).

Lakoff and Johnson further argue that properties we attribute to physical objects "are not properties of objects *in themselves* but are, rather, interactional properties, based on the human perceptual apparatus, human conceptions of function, etc." (1980, p. 163).

Experiential learning, then, is primarily characterized by a purposeful, dynamic, multi-faceted transaction between a learner and object(s) of learning (O/Ls), as well as with various other elements that may exist in the learning ecology (Fig. 37.1). The transaction is multi-faceted in the sense that it involves a variety of cognitive and sensory-motor exchanges between the learner and various ecological elements, although it may involve primarily realist-rationalist exchanges with O/Ls. Cognitive exchanges engage elements and processes from the learner's content knowledge, reasoning skills (rational), and affects (motivation, emotions, interests, etc.). The transaction is realist because it is determined by the ontological state of both learner and O/Ls, and it results in "knowing" O/Ls in certain respects, and to a certain extent, the way these objects may exist in the real world. It is dynamic because it results in a temporary or permanent change in the cognitive and sensory-motor state of the learner, and possibly in the state of O/Ls and other ecological elements. Finally, the transaction is purposeful because the learner consciously aims at putting meaningful structure in what s/he perceives in O/Ls in order to make sense of these objects and develop that learner's knowledge state.

According to Poincaré (1902), things gain their significance from realist and rational perspectives only when related to each other. An entity, whether physical or conceptual, has little significance, if existing in isolation from other entities. The entity gains significance when it interacts with other entities, or when it is related, or connected to such entities in a well-defined structure. For instance, a heap of stones has little importance and utility. To borrow Bunge's words (1979, p. 3), the heap is an "*aggregate* or *assemblage* ... a collection of items not held together by bonds, and [it] therefore lacks integrity or unity". Stones become far more important when used in constructing a wall, and even better, a house. It is how the stones are stacked, how they are connected to each other, that turns the heap into a significant and useful structure, say a dwelling system.

A systemic perspective of the world allows us to bring cohesion and coherence to this world, as well as to our own thinking, and to understand certain aspects of this world that may not be easily conceived – and perhaps that may not be conceived at all – without such perspective. Such is the case, for example, with wholeness and holism. *Wholeness* is, for us, about the impact of any given entity or interaction (or connection of any sort) in a system on the entire system. Every constituent of the system interacts with (or is related/connected to) other constituents and somehow affects the state of the entire system. A change in any given constituent or in any given interaction or relationship between two constituents may result in a change of the entire system. Such a global impact can best be conceived and explained in the context of a system as a whole and not in terms of any clusters of relationships.

Holism is, for us, about the added value that a system as a whole brings to its constituents and the surrounding environment. A system is *holistic* in the sense that, as a *whole*, it is more than the sum of its parts. It has *emergent properties* (e.g., the shape of a house) and *synergetic functions* (e.g., dwelling) that no constituent (e.g., a stone) possessed individually before. The two holistic features may not be attributed to its individual parts and may not be fully understood and appreciated by simply breaking the system into such parts (by analysis or following a reductionist approach).

A caveat is quickly due at this point. Wholeness and holism, as we see them, do not deny the importance of individual constituents of a system within and outside the context of the system. The two systemic features do not necessarily imply a certain determinism or irreversibility in the state of system and constituents, especially not when of human or social nature. Under propitious conditions, a system may change its evolutionary course, and may as well recover from certain induced changes and return to its original state.

A systemic worldview is optimal for meaningful understanding of ourselves and the world around us, from the tiniest details to the big picture in any experience we might be engaged in. Research in cognitive psychology reveals that accomplished people, especially professional experts, are distinguished from other people more in the way they organize knowledge than in the type and amount of knowledge they hold in mind, and more in how they systematically deploy generic skills that cut across various professions than in how they follow idiosyncratic or profession-exclusive heuristics. System-based organization is in this respect among the most effective and efficient for structuring content knowledge, if not the optimal one. Similarly, systemic thinking, i.e., exploring the world purposely as a world of systems, and consciously constructing, retaining in memory, and deploying conceptual systems (e.g., scientific models), holds a superior standing when it comes to process knowledge (Halloun 2001, 2007, 2011, 2017a, and references therein).

37.1.2 System Schema

A system has been defined in a variety of ways in the literature, but definitions converge on the point that a system may consist of one entity (if simple) or many interacting or connected entities (if compound) confined within well-defined boundaries to serve particular purposes. The constituent entities, and thus the system, may be either physical, if consisting of material objects, or conceptual, if consisting of abstract elements (e.g., scientific models). A system, according to Bunge (1979), whether physical or conceptual, "may be said to have a definite composition, a definite environment, and a definite structure. The composition of a system is the set of its components; the environment, the set of items with which it is connected; and the structure, the relations among its components as well as among these and the environment" (p. 4). Later, Bunge (2000) added a fourth dimension, "mechanism".

The boundaries of a system are usually delineated by convenience, especially in terms of the purposes or function it is meant to serve, and so are chosen various elements and connections of interest within and outside the system boundaries

Fig. 37.2 System delineation

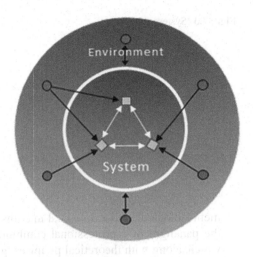

(Fig. 37.2). As shown in Fig. 37.2, the boundaries of a system and its environment (surroundings or settings in which it is embedded) are primarily determined by the purpose(s) the system is supposed to serve. The boundaries may then be conveniently delineated to account for certain entities and/or interactions (or connections) among entities of primary interest, and not others, in order to optimize what we are trying to achieve with the system (Halloun 2004/2006, 2007). Sometimes, these boundaries are delineated so that the system consists of a single entity with no internal interaction/connections, and at other times, to embody all entities of interest inside the system and end up with an isolated system with no environment to interact with. At all times, we are interested in specific interactions but not others within the system or with its environment. The arrows in the figure depict three such instances.

The two-sided arrows between system *constituents* (entities inside the system depicted with squares) indicate an interest in *mutual inter*actions or relationships between connected entities. The one-sided arrows between certain *agents* in the environment (entities outside the system depicted with disks) and constituents of the system indicate an interest only in the *action of* those agents on designated constituents, but not in the reciprocal action of constituents on agents (sometimes called reaction). The two-sided arrows between the system boundaries and agents in the environment indicate an interest in certain *mutual inter*actions between connected agents and the system as a *whole,* thus in the synergetic impact on the environment of all elements in the system acting together, and not the impact of individual system constituents.

We define a system of any sort, in both the physical world and the conceptual realm of human knowledge, in accordance with a four-dimensional schema (Fig. 37.3) that specifies the system's scope, constitution, and performance in the context of an appropriate framework (Halloun 2011, 2017a, b).

1. The *framework* of a system consists of all: (a) theoretical premises, like assumptions, principles, value system, and other ontological, epistemological,

Fig. 37.3 System schema

methodological, and axiological maxims and provisions typically spelled out in the paradigm of a professional community, and (b) ensuing strategic choices, which, along with theoretical premises, guide the specification and reification of the following three practical dimensions of a system.

2. The *scope* of the system specifies:

   (a) The system *domain*, or the field or area in which it exists and is of importance;

   (b) The system *function*, or the specific purposes it is meant to serve in that domain.

3. The *constitution* of the system specifies:

   (a) The system *composition*, its primary constituents which may be physical or conceptual entities (objects and their primary individual properties) inside the system that are relevant to its function, as opposed to secondary entities that may be part of the system but that may be ignored because we deem them irrelevant to the system function;

   (b) The system *structure*, primary connections (interactions or relationships) among primary constituents that significantly affect how the system serves its function;

   (c) The system *environment*, its primary agents or primary physical or conceptual entities outside the system, other systems included, along with their primary individual properties, that may significantly affect the system structure and function;

   (d) The system *ecology*, primary connections (interactions or relationships) between individual primary agents and constituents, and/or between the system as a whole and its environment, that significantly affect how the system serves its function (and affects the environment, if we are interested in the mutual system-environment impact).

4. The *performance* of the system specifies:

   (a) The system *processes*, dynamical actions (mechanisms or events) which constituents, and/or the system as a whole, might be engaged in, on their own (isolated system) and/or under external influence (of the environment), in order to serve the function of the system following specific rules of engagement;

(b) The system *output*, products, events, or any other effect (services included, when the system is, say, of social or industrial nature) that the system actually brings about, on its own or in concert with other systems as a consequence of its ecological interactions and processes, and that may fall within or beyond the scope originally set for the system.

37.1.3 Middle-Out, System Centered Epistemology

A systemic worldview allows us to systematize how we go about exploring and interacting with the world around us, meaningfully understanding this world, from the big picture in any situation we are interested in, down to the minute details and shuttling between the two structural levels efficiently and productively. It especially allows us to readily recognize morphological or phenomenological *patterns* (common regularities in space and time in the structure or behavior of different entities) that predominate in the universe at all levels, from the subatomic scale to the galactic scale, including the human mind, brain, and body, and that make our world interesting and comprehensible.

For instance, day and night recurrence and season cycles on Earth are examples of patterns, and so are the morphology and life cycles of humans and other species. The former earthly patterns are best understood in the context of our solar system (or the Earth-Sun subsystem), and the latter life patterns in the context of the species' ecological systems. Patterns also predominate in our thoughts and memories as we shall see later, and we have a natural tendency to look for patterns in the world around us, and even to rationally impose patterns on what we perceive in this world or conceive about it. A systemic perspective on the real world allows us to efficiently identify patterns of interest in this world, to rationally conceive of those patterns and to readily integrate corresponding knowledge in our memories in the manner discussed in the following sections.

As indicated in Fig. 37.4, systems occupy the middle of the rational hierarchy between a big picture (that may be a universal pattern) and specific details in a given situation. According to Lakoff (1987), humans organize their knowledge in middle-out structures whereby basic and most fundamental structures occupy the middle of the rational hierarchy between individual entities and the corpus of knowledge pertaining to those and similar entities. Systems, as we see it, are such basic structures. For example, a typical and crucial corpus of knowledge consists in science of a given theory or set of theories, and, in languages, of the various types and genres of discourse (or written text). A conceptual system in science, and more specifically a scientific model like Bohr's model of the atom, is to theory (the big picture) and concept (detail) what an atom is to matter and elementary particles. Each elementary particle at the bottom of the structural hierarchy is essential in

Fig. 37.4 Systems in the
middle-out hierarchy between
the big picture and individual
details in a given situation,
and between universal
patterns and local or specific
entities and/or connections
among entities

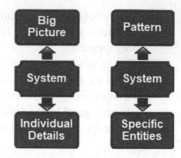

the structure of matter at the top of the hierarchy. However, the importance of an individual particle cannot be realized independently of that particle's interaction with other particles inside an atom. It is the atom in the middle of the hierarchy and not elementary particles that gives us a coherent and meaningful picture of matter, and it is the atom that displays at best the role of each elementary particle in matter structure.

The same goes for language. A sentence is a conceptual system that stands in the middle between discourse (or text) at the top of the hierarchy and phoneme (or even word) at the bottom. The sentence gives us a coherent and meaningful picture of any type of discourse, while, through corresponding semantics and syntax, it displays at best the meaning and role of each word in discourse structure (Halloun 2001, 2004/2006, 2007, 2011).

37.2 Systemic Cognition

A systemic worldview may serve us well to understand cognition and cognitive development meaningfully, both at the cerebral and conceptual levels, and more specifically to understand what learning outcomes (LOs) are about and how they come about. Our brain is the store of our memories (LOs) and the central processor of our perceptions, thoughts, affects, feelings (emotionally generated physical sensations), and actions, and it thus governs all our learning experiences. It is the major part of our nervous system – which has been traditionally and duly called a "system" –, an intricate system that is usually divided into two complex systems, the peripheral nervous system and the central nervous system. The latter includes the spinal cord and the brain. In the following, we concentrate our discussion on conscious cognitive processes that take place in the brain during experiential learning in typical settings of formal education.

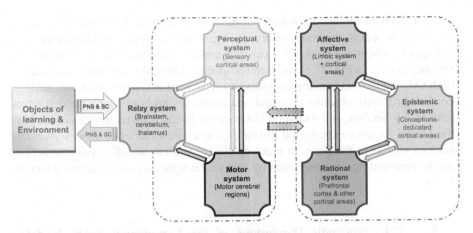

Fig. 37.5 An experiential learning model of the human brain

37.2.1 The Experiential Cognitive System

The brain is a complex system composed of billions of neurons grouped and interconnected in neural *networks*, each of specific structure and function. These networks form neural *patterns*, which allows us to conceive of them as finite sets of interconnected brain subsystems (or systems for simplicity) from either a morphological (structural) perspective or a phenomenological perspective (cognitive processes and learning outcomes). In this chapter, we concentrate our discussion on cognitive processes that are crucial for experiential learning and that highlight the importance of a systemic perspective on cognition.

From a phenomenological or functional perspective, the brain can be divided into a number of systems that may be delineated by convenience as indicated in Fig. 37.2, depending on the mental functions and processes we are interested in. Figure 37.5 shows a *model* of the brain, i.e., a *conceptual system* that provides a partial representation of how the brain is structured and operates during experiential learning. Box 37.1 outlines what we will discuss in this section about the dimensions of this system distinguished in the schema of Fig. 37.3, without rigidly structuring our discussion under the titles of these dimensions or in the order of their appearance in the box.

The experiential learning model of Fig. 37.5, like any other model, is a partial representation of the vast cerebral networks and flow of signals across the six brain systems involved in experiential learning. The model also partially represents the information exchange, via the peripheral nervous system and the spinal cord (PNS and SC), between these six systems, on the one hand, and objects of learning and the environment on the other.

The model represents a learning experience that involves input from concrete objects of learning (O/L) and the surrounding environment (Fig. 37.1). A concrete O/L may be an actual physical object or a physical representation of such object (e.g., a physical model, a computer simulation, a poster). The input detected by the senses is sent to the brain for processing, which subsequently induces memory changes in concerned areas of the cerebral cortex (mentioned between parentheses for each system in Fig. 37.5) and implies specific actions by the learner on either or both O/L and the surroundings if necessary. The input in typical classroom settings, consists primarily of a mix of visual and auditory signals to which we will return later in this section, and which do not necessarily instigate reciprocal action from the learner (passive experiential learning). The entire experience, from forwarding sensory signals to the brain, to memory change and commanding possible actions on the surroundings, is undertaken by vast neural networks spread across the entire brain.

Box 37.1: Schematic Dimensions of the Experiential Brain Model of Fig. 37.5

1. *Framework*. Premises drawn primarily from neuroscience, cognitive psychology, and the philosophy of science.
2. *Scope*. The model represents experiential learning with concrete objects of learning, and serves to describe and explain certain phenomenological aspects of neural networks in the brain regions distinguished in Fig. 37.5 so as to make sense of these objects and bring about specific learning outcomes in memory.
3. *Constitution*. The model consists of six interacting cerebral systems as shown in Fig. 37.5. The environment includes objects of learning, and possibly some other elements of Fig. 37.1, with which a learner interacts actively or passively in order to bring experiential learning to its desired ends.
4. *Performance*. Neural signals flow in specific ways across various constituents of the model so as to bring about learning outcomes in memory specified in accordance with the taxonomy of Box 37.2.

For convenience, we group these networks into six cerebral systems of six distinct broad functions that serve the purposes of our discussion of experiential learning in this section from mostly a phenomenological perspective. The systems that make up our brain model of Fig. 37.5 are the relay system, the perceptual system, the motor system, the affective system, the rational system, and the epistemic system. These constituent systems are delineated so as to reconcile the actual morphology of the brain and the taxonomy of learning outcomes (LOs) that have been developed (Halloun 2017b). This taxonomy classifies these outcomes along four multifaceted dimensions outlined in Box 37.2.

A learning experience may be triggered externally by an input from surrounding objects of learning as indicated in Fig. 37.5, or intrinsically in the absence of any external input (e.g., by a process of self-regulation). The model shown in Fig. 37.5

holds in the latter event, with intrinsic triggers substituting O/L in the leftmost box and some tweaking to the relay system if no physical interaction takes place with the environment. As indicated in this figure, once an external input is detected by the senses, appropriate neural signals are sent through the concerned parts of the peripheral nervous system (PNS) to the *relay system* of the brain, directly or through the spinal cord. This system relays somatosensory and other perceptual information, as well as motor information, to concerned parts of the brain, especially in the *perceptual* and *motor* systems respectively. It does so not in a passive manner, and not entirely on its own. It actually processes afferent sensory information (neural signals) that goes through it, and induces the *affective* and *rational* systems to get involved in focusing attention beforehand on particular sensory data emanating from O/L and surroundings, and thus in filtering perceived information. The latter two systems, as well as the *epistemic* system, get subsequently involved in processing filtered information.

The following is limited to a concise discussion of four processes that are most critical for determining how meaningful a transaction can be between a learner and O/L, i.e., how well the learner can make sense of and "understand" O/L, how readily s/he can bring about significant LOs, how long s/he can retain LOs in memory, and how productively s/he can deploy them in novel situations. Ample details about these and other processes can be found elsewhere (Halloun 2017a).

Box 37.2: Taxonomy and Systemic Assembly of Learning Outcomes (Halloun 2017b, in preparation)

Epistemic learning outcomes (LOs) pertain to various types of *conceptions* (concepts, laws, theorems, and other abstract constructs conceived to describe or explain morphological or phenomenological aspects in the physical world or mental realm), each of which may be classified in a number of categories (e.g., in science, laws comprise state, composition, interaction, causal, and quantification laws).

Rational LOs pertain to various types of *reasoning skills* (e.g., analytical reasoning, criterial reasoning, relational reasoning, critical reasoning, logical reasoning), each of which may be classified in a number of categories (e.g., analytical reasoning skills comprise surveying, differentiating, identifying regularities, describing, explaining, predicting).

Sensory-motor LOs pertain to various types of perceptual and motor skills, or *dexterities* (e.g., communication dexterities, digital dexterities, manipulative dexterities, artistic dexterities, eco-engagement dexterities), each of which may be classified in a number of categories (e.g., communication dexterities comprise listening, reading, speaking, writing, coordination of multiple representations).

(continued)

Box 37.2 (continued)

Affective LOs pertain to various types of *affects* (e.g., emotions, motives, interests, dispositions, values), each of which may be classified in a number of categories (e.g., dispositions comprise open-mindedness, risk taking, autonomy, curiosity, creativity).

Learning outcomes along some or all four dimensions may come together in systemic clusters of specific functions like metacognitive controls and competencies.

Metacognitive controls include reasoning skills and affects that monitor and regulate our thoughts and actions, and especially memory formation and retrieval.

A *competency* is a specific or generic cluster of all four types of LOs. A *specific* competency helps achieving a specific task like solving a specific problem about a particular system or situation. A *generic* competency allows the deployment of attained LOs in novel situations and in the development of new LOs (and subsequently new competencies).

37.2.2 Multi-stage Filtering of Afferent Sensory Information

When a conscious learning experience is triggered externally by sensory information from O/L, the peripheral nervous system sends detected information, directly or through the spinal cord, to the *relay system* for transfer to concerned cerebral systems (Fig. 37.5). However, not all input available to our senses is sent to our brain for conscious processing. The afferent sensory information (neural signals reaching the relay system) is filtered at different levels in the brain, before and while it is being processed there.

Filtering of afferent information begins in the relay system, first in a completely automatic and involuntary manner driven by our survival needs and our instinctual emotions. A small fraction of the signals forwarded to the relay system proceed through for conscious processing in applicable systems of Fig. 37.5 (only thousands out of the millions of information bits received in any given second are ushered in to the other cerebral systems). This is the fraction we "concentrate on" or "pay attention to", like the word or part of the word you "read" at any particular instant as you go through this text. The remaining signals (background information, including the part of the text that is not being read at a given instant) are blocked out.

Filtered information is sent to the perceptual system for processing. Meanwhile, the *affective* and *rational* systems are activated to provide *metacognitive controls* (Box 37.2) that govern the entire learning experience. Among others, metacognitive controls: (a) contribute to the filtering process, (b) sustain attention on specific perceptual information out of the already filtered information and/or redirect our senses to focus beforehand on specific perceptual data in the objects of learning, (c) determine which other systems of Fig. 37.5 will get involved in processing the information that makes it through, and to what end, and (d) regulate the formation of learning outcomes. The most critical metacognitive controls are carried by *innate* neural networks of the *affective system* that *instinctually* govern what Ekman (1992) and other psychologists call "basic emotions", or what Panksepp (1998, 2006) and other neuroscientists call "core emotions" (Gregory and Kaufeldt 2015; Panksepp and Biven 2012). Those controls may or may not allow affluent information to be processed in concerned cerebral systems, and determine, to a significant extent, the nature and quality of the cognitive and behavioral outcomes of our thoughts and actions. Some of these controls play a constructive role and allow meaningful learning to proceed, while others take over in a destructive way and may prevent any learning from taking place altogether.

For meaningful and sustained learning, metacognitive controls need to focus attention on engaging experiences that bring about emotionally significant outcomes for the individual's development. This requires the development of optimal, intuitive and almost automated means for directing attention to input that is significant to the situation we are in, and that positively and coherently engages our perceptual, motor, epistemic, and especially rational systems. Such means need to be induced and sustained by constructive metacognitive controls that originate typically in the *rational system*, and that are part of so-called executive functions handled by the prefrontal cortex (PFC) in this system, the highest level cognitive functions of the rational system.

37.2.3 Deconstruction and Reconstruction of Objects of Learning

Afferent sensory signals filtered in the relay system make up what we call a preliminary, partial *perceptual image* of the object(s) of learning (O/L). As shown in Fig. 37.6, the image is immediately analyzed in specialized areas of the perceptual system and broken down into an array of discrete and unimodal packages of perceptual information (neural signals). O/L is thereby cognitively deconstructed and a limited number of its features is retained for subsequent processing in the brain. Those features are then gradually synthesized to conceptually reconstruct O/L in specific respects that serve the transaction purposes. O/L reconstruction begins within specialized association areas (called unimodal association areas), and continues within more complex association areas (called multimodal association areas) located at the interface of various brain systems of Fig. 37.5 (Saper et al. 2000).

For illustration purposes, consider O/L deconstruction and reconstruction in a typical passive classroom experience involving, say, a teacher's audio-visual demonstration of a given phenomenon. Imagine yourself watching such a demonstration that Fig. 37.6 represents. Your eyes detect visual information that is processed in the relay system to tease out the visual constituents of the perceptual image (PI). Those constituents are then channeled to occipital cortical areas of the

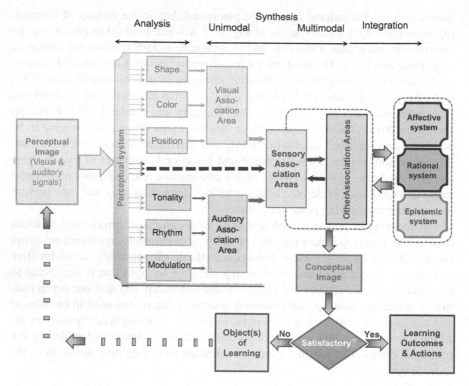

Fig. 37.6 Transaction with O/L that involves deconstruction and reconstruction of O/L

perceptual system where they are analyzed to form, say, three discrete packages of information pertaining to shape, color, and position or movement of any O/L in the demonstration.

Similarly, your ears detect auditory information (a mix of acoustic and linguistic data) that is filtered to sort out the auditory constituents of the perceptual image. Those constituents are then channeled to temporal cortical areas of the perceptual system where they are analyzed to form, say from acoustic and non-linguistic perspectives, three other discrete packages of information related to tonality, rhythm, and modulation. Similar analysis into unimodal packages takes place for the linguistic and other components of PI. Such additional analysis is represented by the thick dashed arrow emanating from the perceptual system in Fig. 37.6, and is not further discussed here to make our point with the minimum possible detail.

Once the analysis of PI ends, a gradual synthesis of the unimodal information packages begins that culminates in the formation of a conceptual image of the object(s) of learning. The three visual packages in our example are first sent in parallel along three separate neural pathways to a *unimodal* visual association area where they get integrated in a single visual package of neural signals. Similarly, the three auditory packages are sent in parallel along three separate pathways to a *unimodal* auditory association area where they get integrated in a single auditory package of neural signals. The two integrated unimodal packages (visual and auditory) are now sent, along with other unimodally integrated packages, to be integrated together in a single package within a *multimodal* sensory association area located in the posterior part of the cerebral cortex, at the interface of the parietal, temporal and occipital lobes. The newly integrated information is projected into other association areas for further processing and integration. The latter synthesis involves adduction of knowledge already in memory (i.e., neural input from various cerebral systems), and is controlled by the three rightmost systems of Figs. 37.5 and 37.6 (affective, rational, epistemic).

Analysis of the perceptual image into discrete visual, auditory, and other sensory packages in the perceptual system is automatic, unconscious and involuntary, and so is the unimodal synthesis that takes place in the corresponding association areas. Conscious and voluntary cognitive intervention (the actual transaction) begins with multimodal synthesis as indicated by the opposite arrows between different multimodal association areas in Fig. 37.6.

Multimodal synthesis takes place under the conscious control of the rational and affective systems and results in what we call a *conceptual image* of O/L (Fig. 37.6). The conceptual image is gradually constructed to fulfill the purposes originally set for the learning experience, and allow learning outcomes to be formed in memory and proper actions to be generated by the motor and perceptual systems to this end. The process is often a reiterative process that involves successive generation, analysis and synthesis of perceptual images, and gradual formation of a *panoramic conceptual image* of O/L, i.e., an image formed by successive refinement and superposition of many "shots" taken of O/L (Fig. 37.6).

The conceptual image (CI) of a given object of learning (O/L) formed by the end of a given multimodal synthesis course is evaluated by certain PFC areas in terms of its suitability to serve the purposes of the learning experience. When the evaluation outcome is not satisfactory – and usually it is not after the first PI-CI cycle – concerned PFC areas command our senses to refocus on specific aspects of O/L that help serving those purposes, and the PI-CI cycle is reiterated to refine the original CI or even replace it altogether. Reiteration continues until a panoramic CI is formed (by successive refinement and superposition) that satisfactorily helps fulfilling those purposes and that may eventually be sustained in memory.

PI faithfully, but partially mirrors the ontology of O/L in specific respects, hence the "realist" aspect of the learner's transaction with O/L. During PI analysis into unimodal bits of information, some PI details are lost. During unimodal, and especially multimodal synthesis, information is adducted from the learner's prior knowledge stored in the various cerebral systems of Fig. 37.5 and pertaining to various dimensions of the taxonomy of Box 37.2, hence the cognitive and especially "rationalist" aspect of the transaction. The emerging CI includes a blend of certain PI details (and hence of selective O/L details) and conceptual and affective details from the learner's own knowledge. In contrast to PI, CI is thus a holistic, non-positivist, non-gestalt brain-constructed image of O/L formed via the cognitive lenses of the learner. It serves to make sense of O/L and the entire learning experience.

CI *emerges* from a critical multi-faceted, but especially realist-rationalist trans-action between O/L and the brain/mind, and serves *synergetic functions* that O/L alone could not serve especially from a cognitive perspective. Most importantly, CI emerges from insightful dialectics between PI and adducted knowledge, i.e., from critical evaluation and regulatory negotiations that take place externally with O/L and internally within one's own knowledge in order to continuously refine PI, and insightfully regulate CI, until the purposes of the learning experience are met with satisfaction (Halloun 2004/2006, 2017a).

A "conceptual" image (CI) of a given O/L consists primarily of epistemic components. However, CI formation, from PI unimodal analysis to multimodal synthesis, involves practically all cerebral systems of Fig. 37.5, but especially the affective, rational, and epistemic systems. During a learner's transaction with O/L, various cerebral systems are engaged and developed to different degrees; and learning outcomes may come about from all four dimensions of the taxonomy of Box 37.2.

A learning outcome (LO) is a bit of content or process knowledge that narrowly pertains to a single conception corresponding to a specific aspect of O/L (epistemic LO in Box 37.2), or to a single reasoning skill (rational LO), dexterity (sensory-motor LO), or affect (affective LO) involved in the transaction with O/L. A CI may thus consist of any number of LOs of one type or another (especially epistemic) corresponding to O/L, and may induce the formation of other LOs not necessarily corresponding exclusively to O/L (e.g., the development of differential analytical thinking). Any mention of CI in this chapter thus implies any LO developed during the transaction with a given O/L. The LO may or may not correspond exclusively to the O/L in question and may be retained temporarily or permanently in memory.

37.2.4   Memory Formation

In each PI-CI cycle (Fig. 37.6), unimodal synthesis of discrete perceptual informa-tion brought about by unimodal analysis of the perceptual image results in data that need to be retained in memory for subsequent multimodal synthesis. The lifetime of synthesized unimodal data in memory is very short in a given cycle. It may last from a fraction of a second up to a very few seconds. Such a process, and perhaps an entire PI-CI cycle, can be represented by any of the "working memory" (WM) models proposed in the literature (e.g., Baddeley 2012; Baddeley and Hitch 1974; Cowan 2014; Pickering 2006). However, WM is thought to last for a maximum of 30 s. The PI-CI cycles may be reiterated for longer than that, and a series of panoramic CIs may be needed to have a comprehensive representation of O/L in any particular experience and to form meaningful information about objects of interest. For instance, watching a classroom demonstration for some minutes, and perhaps for a good portion of a class period, requires multiple panoramic CIs, and thus multiple WM episodes to form new knowledge (LOs) about the concerned O/L, and/or refine prior knowledge about that O/L and the entire experience at hand. This of course

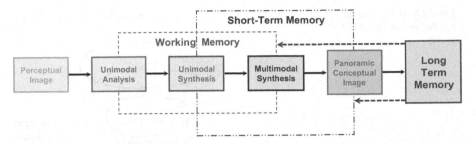

Fig. 37.7 Sequential memory formation with distinctive, but overlapping, working memory (WM) and short-term memory (STM)

assumes that the WM models are valid for any type of knowledge distinguished in Box 37.2, and not just for epistemic, and more specifically factual knowledge with which WM is traditionally associated.

New information (CI) emerging at the end of each PI-CI cycle (or WM episode) thus needs to be temporarily retained for an entire learning experience, which may last for many minutes, and perhaps hours. Given the limited WM lifespan, emerging information has to be retained *temporarily* in a different type of memory, namely for us what we refer to as short-term memory (STM). This position is being lately echoed in the neuroscience community where some have been arguing for what they call "prioritized long term memory" to handle such temporary, but relatively long storage of information that cannot be handled by WM (Rose et al. 2016).

Short-term memory (STM) consists, for us, of information temporarily maintained in mind beyond the lifespan of WM, and processed under the control of PFC in the rational system for the sole purposes for which the information has been retained, and only for as long as those purposes need to be served. STM pertains to the formation of a panoramic CI in reiterative PI-CI cycles, and may extend well beyond the conclusion of the reiterative process depending on the need for the image in question and emerging learning outcomes (Fig. 37.7).

Reiteration of PI-CI cycles takes place under the control of particular PFC areas in the rational system. With appropriate executive functions, PFC: (a) keeps all transaction processes and outcomes focused on the specified task, (b) retains primary or relevant information emerging at the end of each cycle and deletes secondary or irrelevant information, and (c) ensures that all cycles complement each other in order to form cohesive conceptual images that bring the learning experience to the desired ends. In each cycle, metacognitive controls to which the rational system is a major contributor govern all sorts of dialectics mentioned above in order to regulate the conceptual image constructed in the preceding cycle along all dimensions of the system schema of Fig. 37.3. The regulatory reiteration in question is helicoidal in the sense that appropriate dialectics continue to enhance CI from one cycle to the next until the desired panoramic image is satisfactorily constructed (Fig. 37.8).

Fig. 37.8 Helicoidal and
regulatory reiteration of PI-CI
cycles

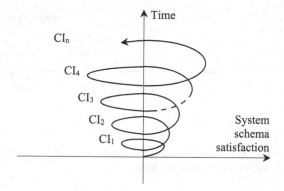

When the conceptual image (CI) is of a system object of learning that is an instance of a particular pattern, the image CI_j constructed by the end of a cycle j is developed and/or refined in the following cycle j+1 so as to better meet the system schema of Fig. 37.3, especially the system function and structure, and better reflect the pattern in question. The partially dashed line between CI_3 and CI_4 indicates that a given cycle (or WM episode) may be interrupted for any reason, but PFC would ensure that the information constructed by then is retained in STM, and would eventually bring the task back on track and ensure that it continues from where it got interrupted.

Information emerging at the end of each PI-CI cycle is temporarily retained – and processed – in memory under PFC control for as long as it is needed, and is subsequently dropped out of memory unless it makes its way to long-term memory. The lifetime of CI about a given O/L or learning experience can span from a few seconds (in WM) to a few minutes or even weeks (in STM). In formal education, the latter is the case when students retain some information (LOs) only because it is required in a given course, and not because they are convinced of its personal value, and when constructive affective and rational controls are not strong enough to induce such knowledge to be retained for good in long-term memory. Students retain required knowledge only long enough to pass a certain quiz or exam. Once the purpose is served, and the information is no longer needed, it is dropped out of memory altogether (or perhaps inhibited from being consciously remembered and retrieved). That is why, students who are able to do well on a certain quiz or exam are unable to do as well a short while afterwards on the same task. The learner is then said to have accomplished *transient learning* with LOs temporarily retained in STM. Alternatively, the learner accomplishes *sustained learning* when LOs are never dropped out, and are retained in long-term memory (LTM) instead of STM.

LTM is the resource a learner relies upon for meaningful transaction with objects of learning, whether these objects are physical or conceptual, and whether learning is experiential or not. LTM provides the cognitive lenses through which we see the world, no matter how passive and "objective" our transaction with this world is intentionally meant to be. This memory always determines what we see and what we hear, even when we encounter something or someone for the very first time in our lives. This is how, for example, we may determine that something is a plant or an animal, even when we have never seen anything like it before. Such categorization is carried out by systemic mapping of entities, properties, and especially relationships in the new experience to existing *patterns* in memory.

37.2.5 *Patterning*

Patterning is crucial to LTM, both at the anatomical/biological level of the brain and the conceptual level of the mind (Hernández et al. 2002; Kandel et al. 2013). LTM neural networks make up cerebral patterns (biological patterns in the brain), and connection among these networks and their activation involve patterns of neural signals flow and processing across various brain systems of Fig. 37.5. Transfer of the conceptual image (CI) of an object of learning (O/L) from STM to LTM depends to a large extent on how well CI is encoded in STM so that it can be readily integrated with LTM patterns.

Encoding of a conceptual image in neural networks begins with the unimodal analysis of the perceptual image and continues through multimodal associations (Fig. 37.6). Unimodal encoding is localized in specialized or dedicated areas of the perceptual system, and multimodal associations are spread across different cerebral regions and cortical areas, including specialized cortical areas and association areas. CI encoding begins in WM and continues in STM (Fig. 37.7). Transfer of CI into LTM, or rather its transformation from STM to LTM, its sustainability there, and the efficiency of its retrieval from LTM when needed depend primarily on how well encoded STM networks form a pattern, or part of a pattern, that can be incorporated with, or added alongside existing LTM patterns through appropriate rehearsal and consolidation processes. This all depends on: (a) the richness and depth of encoding in unimodal areas, and (b) the span of encoded multimodal networks across a variety of dedicated and association areas of the brain so as to bring about multiple conceptual representations of the corresponding O/L(s) that can be consciously negotiated with existing conceptual patterns.

Patterns predominate our conceptual realm as well as our biological cerebral networks. This realm includes content and process knowledge which we are consciously aware of and that we can communicate to, and negotiate with others. A correspondence appears to exist between cerebral patterns and conceptual patterns at the developmental level. Cerebral patterns favor conceptual patterns. The more we consciously focus on conceptual patterns in developing new knowledge, the better the chance for sustained learning, i.e., for the new knowledge to be encoded in LTM cerebral patterns.

Systems infuse order in the natural, social and conceptual worlds, and allow patterns to emerge meaningfully to and within the human mind.[1] As noted in the first section, systems are in the middle of the conceptual hierarchy between individual details and the big picture (especially a pattern) in any given situation (Fig. 37.4). As such, they are the best structured entities to ensure the coherence and efficiency of our corpus of knowledge at large, and to reveal syntactic and functional similarities and correspondence between entities of the same or different nature. When the physical world and the mental realm are conceived as interacting systems, patterns readily appear in the structure and behavior of various physical realities, from galaxies down to living organisms and microscopic matter, as well as in the structure of human knowledge, especially academic knowledge that is the object of formal education. As a consequence, it becomes easier for us to conceive the physical world and integrate conceptual learning outcomes with existing memory patterns (Halloun 2001, 2004/2006, 2017a).

37.3   Systemic Education

Bunge's argument for a systemic worldview on all aspects of life applies particularly in education (Matthews 2012). Many reformists in education have been calling, for decades now, for "systems thinking" and "systems-level understanding" of various topics taught at different levels of education.[2] Such calls have been heeded only recently with a systemic vision captured by Goleman and Senge (2014) in their account of Social and Emotional Learning (SEL):

> we feel SEL offers only part of what students need to be well prepared for life. In today's world of work and global citizenship, young people also need to comprehend the complexity of the problems they will face. Parallel to the development of SEL, for the past 20 years, innovative teachers have been working to introduce systems thinking into pre-K-12 schools to build a third intelligence—systems intelligence. (Goleman & Senge 2014, p. 1)

Some educators have begun integrating "systems thinking" successfully in their teaching.[3]

The implication is to call for a systemic education that explicitly and systematically adopts a generic systemic framework under which all actors (teachers included), organizations (schools and governing authorities included), mechanisms and products (curricula and learning outcomes included) are conceived as systems

---

[1] See Bunge (1979), Gee (1978), Gentner and Stevens (1983), Giere (1992), Glas (2002), Harré (1970), Hesse (1970), Johanessen et al. (1999), Johnson-Laird (2006), Lakoff (1987), and Wartofsky (1968).

[2] See Garcia et al. (2014), Goleman and Senge (2014), Johanessen et al. (1999), Laszlo (2015), and Liu et al. (2015).

[3] See Assaraf and Orion (2005), Hmelo-Silver et al. (2007), Mehren et al. (2018), Rodriguez (2013), and Waters Foundation (2010).

Fig. 37.9 4P profile of a
systemic well-rounded citizen

or parts of interacting systems that work in tandem to serve the major purpose of bringing up systemic citizens. In particular, such a framework that cuts across all educational levels and disciplines would: (a) aim at bringing students up with systemic profiles, (b) treat them as systemic cognizant beings with systemic minds and (c) proceed systemically to realize that aim in dynamic learning ecologies and in the context of cross-disciplinary curricula.

37.3.1 Systemic Student Profiles

Systemic education brings about learners with systemic profiles that embody professionals' patterns of success in modern life and that have at least four major general traits in common that would qualify them as 4P profiles. A 4P profile is the dynamic, constantly evolving profile of a systemic, well-rounded citizen empowered for lifelong learning and success in life, and characterized with progressive mind, productive habits, profound knowledge, and principled conduct (Fig. 37.9). The four P's are not absolute traits of a "one-size fits all" profile. They are universal "qualifiers" for distinct individual profiles which reliable research in cognitive science has constantly proven to be necessary for success – and excellence – in any aspect of life and in any era, especially our modern era (Halloun 2017a, b).

Progressive refers to an overall systemic and dynamic mindset with clear vision and determination to empower oneself and others for continuous growth and enhancement of various aspects of life. In this respect, and among other faculties, systemic education empowers every student to:

- Go after new ideas, and seek new means and methods to achieve what they are after.

- Engage in challenging tasks and take calculated risks, pursue what they are after with courage and perseverance, bounce back from any failure, and emerge with fruitful ends.
- Never follow blindly any authority, evaluate ideas critically and never accept them at face value, and appreciate and tolerate divergent points of view.
- Care about the welfare of others, whether at home, school, work, or community, and help empower them for success, even excellence in life.

Productive habits refer to practical and efficient cognitive and behavioral habits that are prone to systematic improvement and creative and advantageous deployment in various aspects of life. In this respect, and among other faculties, systemic education empowers every student to:

- Ask appropriate questions about any situation, devise flexible plans to deal with it, ascertain a plan's efficiency before carrying it out, and systematically carry out plans, evaluate them and refine them in the process.
- Identify or put together systems (in accordance with the schema of Fig. 37.3) to deal efficiently with physical or conceptual situations and identify structural and behavioral patterns within and across situations.
- Ascertain their own knowledge, consolidate their strengths, regulate their weaknesses, and resolve incoherence and inconsistency among their own ideas.
- Develop sound criteria and processes for selecting, using, and sharing appropriate resources, communicating ideas and cooperating with others.

Profound knowledge refers to a sound, essential, and coherent corpus of knowledge that readily lends itself to continuous development and efficacious and efficient deployment in various aspects of life. In this respect, and among other faculties, systemic education empowers every student to:

- Focus their content knowledge on a limited number of generic conceptions (concepts and conceptual connections) that are most meaningful for what they need to accomplish in any aspect life, and develop such conceptions coherently and efficiently.
- Maintain due balance between breadth and depth of sought after knowledge, avoid spreading thin across a wide corpus of knowledge, and revisit any acquired knowledge in novel contexts to help deepen it and broaden its scope without undue redundancy.
- Use acquired knowledge in creative ways not tried before within and outside its original scope, and go for new conceptions and processes that bring about innovative answers to certain questions or solutions to certain problems in everyday life.
- Develop generic and especially systemic tools for meaningful inception of new knowledge and comparison to prior knowledge, and for checking for internal coherence within one's own knowledge and for external consistency with what new knowledge is about in the real world and the conceptual realm of experts.

Principled conduct refers to productive and constructive conduct in all aspects of life, while intuitively driven for excellence and guided by a widely and duly acclaimed value system. In this respect, and among other faculties, systemic education empowers every student to:

- Value and convincingly implement high standards of achievement, efficiently control their negative emotions like fear and anxiety, and foster their positive emotions like motivation and interest.
- Appreciate and sustain elements of a universally acclaimed work ethic like integrity, honesty, responsibility, and accountability, and human values like honor, empathy, equity, and peace.
- Realize how individual humans' activities may have constructive or destructive ecological, cultural, and/or social impact, and contribute to constructive and sustainable solutions to related problems in their community.
- Appreciate and emulate distinguished figures behind constructive turning points in the history of humankind, objectively weigh the merits and risks on humanity of scientific findings and technological inventions, and decide whether or not such novelties should be sustained.

In systemic education, a 4P profile is translated in the program of study of any discipline or subject matter into learning outcomes (LOs) spreading across all four dimensions of our taxonomy (Box 37.2), and clustered in the form of systemic competencies required for achieving specific and generic systemic tasks, i.e., tasks pertaining to particular systems that relate to everyday life. Any such competency is then defined along all four dimensions of the taxonomy by correspondence to systemic objects of learning (O/Ls). Appropriate experiential learning tasks are explicitly designed and carried out around such O/Ls to help students consciously develop systemic conceptual images (CIs) that allow for the emergence of particular 4P traits that can be meaningfully sustained in LTM (Halloun 2017b).

37.3.2 Systemic Cognizant Students

Humans are dynamic complex systems whose minds and bodies are always involved in some form of learning. A learner's mental and physical parts and properties, and the way they relate to and affect each other internally and in relation to the learner's environment, determine the outcomes of any learning experience. In experiential learning, the nervous system is engaged in the manner shown in Fig. 37.5, along with concerned senses and other body parts. Latest research in neuroscience shows that movement of the entire body, including walking and exercising during or between class periods, facilitates sustained learning, and promotes cognitive development especially in elementary grades.[4]

---

[4]See Jensen (2005), Kiefer and Trumpp (2013), Kubesch et al. (2009), Osgood-Campbell (2015), and Sousa (2010).

Embodied cognition theory "assumes that cognition is essentially carried out in the sensory and motor brain systems" in concert with related senses and body parts, and that even all sorts of abstract concepts from numbers to affects, feelings, and social concepts like desire, pity, justice, and freedom, are "embodied in perception and action" (Kiefer and Trumpp 2013). Education has thus to explicitly attend to the needs of the entire body, but especially to cognitive and developmental needs that explicitly help optimize the brain constitution and performance and the overall systemic mindset.

Sustained and meaningful learning in LTM – as opposed to transient learning in STM – favors a systemic mindset, i.e., a mind that consciously thrives in exploring the world with systemic cognitive lenses, and that subsequently sets systemic purposes for any learning experience and brings about systemic LOs. As such, and as Bunge (1967) put it, experiential learning is a transaction between two systems, the learner and the object of learning. The outcomes of that transaction are for the purpose of constructing or consolidating a conceptual system (the conceptual image of the object of learning) that may readily reveal, and be integrated with, specific conceptual patterns in the mind of the learner.

Such a systemic mindset has the advantage of bringing about the desired LOs in any learning experience efficiently and with the least cognitive demands possible, i.e., with the least processing effort possible in various brain systems (Fig. 37.5), and especially during multimodal syntheses (Fig. 37.6). It also has the advantage of optimizing the cognitive efficiency of constructed knowledge, i.e., the efficiency with which it is retrieved from LTM, and the success with which it is deployed in novel situations (Halloun 2017a).

Systemic education explicitly attends to the cognitive needs of every student and scaffolds an entire learning experience to make it efficiently meaningful and productive. In the case of experiential learning (Fig. 37.6), this extends from the formation of the perceptual image (PI) of an object of learning (O/L) to the construction, consolidation, and subsequent deployment of the corresponding panoramic conceptual image (CI). CI formation, as mentioned at the end of Sect. 37.2.3, implies sustainable learning outcomes (LOs) in LTM that pertain not only to perceptual aspects of O/Ls and are thus not solely of epistemic nature, but that may cover all four dimensions of our taxonomy (Box 37.2) and belong to all cerebral/mental systems of Fig. 37.5. All systems are duly attended to in the process, and the learner is guided to be consciously aware of what every step of the way entails, from PI to CI and from WM to LTM (Fig. 37.7), and to optimize the entire process for a systemically rich and lean CI. Beginning with the formation of PI, attention is focused on aspects that are essential for systemic learning (e.g., deep and multiple encoding/representations of primary O/L constituents and properties), and care is taken to free PI and subsequently CI from unnecessary secondary features, noise or redundancy (Halloun 2017a, b).

37.3.3 Systemic Learning Ecologies

Systemic education provides systemic learning ecologies for systemic learners to come out with systemic learning outcomes (Fig. 37.3). In experiential learning, individual students interact (transact) systemically with O/Ls, as well with other people and entities that may be in their environment (Fig. 37.1). They consciously conceive of themselves and every other person and entity they interact with as dynamic systems, and take full advantage of such systems in their details and their wholeness, and of holistic transactions among them all. Most importantly, any transaction between a learner and any entity is open to change and regulation in every respect, from framework to performance in the four dimensions of the system schema (Fig. 37.3). This, of course, requires that the learner have a voice in the process, and that this voice be heard. As such, learner and environment make up a dynamic learning ecology in which they constantly interact and cause temporary or permanent changes in each other (STM or LTM respectively for learners).

For permanent changes, especially for learning outcomes (LOs) to be sustained in a student LTM, a learning ecology needs to be sustained in certain systemic form beyond the original experiential learning experience so that the CI originally *encoded* in STM can be *consolidated* and subsequently *stored* in LTM. Consolidation takes place through sufficient deployment/rehearsal of the original CI in familiar and novel contexts, with different O/Ls, so that this image gets strengthened and enriched through multiple encoding in different multimodal association areas (Fig. 37.6). For consolidation to efficiently take place and succeed in bringing about sustainable LOs, a number of conditions need to be satisfied among the most important of which are focused attention and critical and insightful regulation.

In a systemic learning ecology, experiential learning is focused on systemic aspects of O/Ls that bring about systemic CIs with a variety of LOs (Box 37.2) sustainable in different cerebral systems (Fig. 37.5). This is made possible by the affective and rational systems that consciously control any learning experience when the two systems work constructively in tandem to inhibit any distraction and keep the entire experience systemically focused, from purpose to outcomes, and especially from perceptual image (PI) to conceptual image (CI), including the adduction of necessary information from the epistemic and other systems.

Experiential learning is characterized especially by the fact that the multifaceted transaction it involves is consciously critical and insightful. The learner's rational system, and primarily PFC with some of its executive functions, critically monitors and evaluates all PI analysis and synthesis processes and outcomes (Fig. 37.6) in WM and STM (Fig. 37.7) to ensure that they satisfy the purposes originally set for the learning experience. The rational system, and particularly PFC, also determines and monitors the adduction of appropriate knowledge from STM and/or LTM, and evaluates validity of adducted knowledge for the situation at hand. It does the same for the emergent CI and insightfully regulates that CI so that appropriate LOs may be eventually stored in LTM.

Critical and insightful regulation is made in three respects. First, a realist or empirical *correspondence* is ensured between any CI and corresponding O/L(s) so that CI properly "represents" the corresponding O/Ls for the purposes set for the learning experience, and that it is free from any noise or unnecessary superfluous or redundant information. Second, a cognitive but especially rational *coherence* is internally established between CI and LTM patterns so that subsequent LOs can be integrated with/in such patterns. Third, a rational *consistency* is externally established with professional/academic knowledge so that LOs can significantly contribute to specific 4P traits (Halloun 2004/2006).

37.3.4 Systemic Cross-Disciplinary Curricula

For decades now, research has shown that students often complete their education at any level with fragmented, compartmentalized, and transient knowledge, and educators have called in vain for focused curricula that bring about meaningful and sustained learning. The latter unheeded call, often summed up in the nineteenth century proverbial phrase "less is more", is nicely expressed in a *Science* editorial by Bruce Alberts (2012):

> Research shows that the most meaningful learning takes place when students are challenged to address an issue in depth, which can only be done for a relatively small number of topics in any school year. But the traditional process of setting standards tends to promote a superficial "comprehensive coverage" of a field, whether it be biology or history, leaving little room for in-depth learning. The curricula and textbooks that result are skin-deep and severely flawed... At all levels of schooling, we need to replace the current "comprehensive" overviews of subjects with a series of in-depth explorations. (Alberts 2012, p. 1263)

Knowledge fragmentation can best be resolved in systemic curricula, and knowledge compartmentalization can best be avoided when such curricula are cross-disciplinary.

Systemic curricula are designed under systemic pedagogical frameworks to empower students with systemic 4P profiles, mandate to this end systemic programs of study, and provide for meaningful and insightful coverage of these programs in dynamic learning ecologies that rely on experiential learning and systemic assessment. Systemic curricula do not come in one-size fits all. They are flexible enough to account for cognitive and behavioral differences among learners of the same age group, and cater for the same learner to the distinctive cognitive needs imposed by different dedicated cerebral parts where learning outcomes of different types are encoded.

Systemic programs of study are explicitly conceived around well-defined systems that allow students to efficiently grasp the big picture in a given discipline, along with pertinent details, in a middle-out approach (Fig. 37.4). To this end, every item in a program of study is explicitly conceived for the purpose of constructing and/or deploying a specific system that reflects a particular physical or conceptual pattern. Furthermore, every system is conceived with only enough essential details and connections to bring learners efficiently to the target, and ready to lend itself to the integration with some other systems within the same and other disciplines.

Systemic programs of study rely on systemic assessment for deployment in the classroom and beyond. Systemic assessment does not consider assessment as an end by itself (assessment "of" learning), but as means of learning (assessment "as" learning) and instruction (assessment "for" learning). Every assessment task is considered a learning task whereby learners do not simply retrieve ascertained knowledge from memory and deploy it exactly as it used to be stored there, but actually regulate and change retrieved knowledge in the process of adapting it to the task at hand. Teachers use the outcome to track and regulate the evolution of individual students' profiles, and to evaluate and efficiently regulate instructional means and practices, and the entire learning ecology. Most importantly, all assessments are carried out as parts of an "assessment system" whereby "different types of information are collected throughout the year using a variety of assessment tools, [and] each type of information contributes to a bigger picture of student learning" (NASEM 2017, p. 22). Each of the system components "is designed with the same set of goals in mind, even if they are used for different purposes" (*ibid*, p. 91).

Cross-disciplinarity is about building bridges among different disciplines to serve practical, systemic purposes while recognizing that no "discipline or scholarly field is an island unto itself; it is created, evolves, takes shape and responds in certain cultural, social and intellectual circumstances" (Matthews 2012). Bunge's work is an ideal case of cross-disciplinarity. He coherently synthesizes a diversity of apparently distinct and remote disciplines like physics, biology, cognitive sciences, psychology, philosophy, sociology, economy, law, and politics. Philosophy, according to Bunge, "is the study of the most fundamental and cross-disciplinary concepts and principles". His "systemics" is about a "set of theories ... unified by a philosophical framework ... that focus on the structural characteristics of systems and can therefore cross the largely artificial barriers between disciplines ... [and thus] calls for a cross-disciplinary approach" (Bunge 1979, p. 1).

With systemic education, cross-disciplinarity is achieved when at least some systems in a given program of study are constructed with constituents coming from traditionally different disciplines, and when these systems are deployable in a variety of such disciplines, so as to bring coherence and consistency within and among disciplines, and facilitate transfer across disciplines and to everyday life.

Many professional organizations have recognized the importance of cross-disciplinarity and are working to make it a reality in practice. To this end, some leading international organizations like the International Council for Science and the International Social Science Council have even merged to bring about sustainable development at the global scale. According to McBean and Martinelli (2017), presidents of the two organizations in question that merged in October 2017 into the International Science Council, the merger:

"will provide a new institutional context for the long-called-for convergence to become a reality". It "should help foster meaningful interdisciplinarity that begins with the joint framing of problems; ensure that all disciplines are exploiting opportunities of the digital revolution, including for data integration; and unify scientific communities. It will be guided in its actions by the shared vision of advancing all sciences as a global public good. (McBean & Martinelli 2017, p. 975)

37.4 Conclusion

A major implication of the discussion above is that education needs to comply, in a systemic perspective, with neuroscience-based cognitive premises in order to empower students for lifelong learning and success in life. The works of Mario Bunge and the emerging field of Mind, Brain, and Education (MBE) are seminal to this end. Systemic Cognition and Education (SCE), a generic pedagogical framework for student and teacher education at all levels, is being developed in this direction. The framework comes with cognitive tenets, pedagogical principles, and operational rules that govern devising and implementing every component of any curriculum, from programs of study to means and methods of learning, instruction, assessment, and evaluation.[5]

References

Alberts, B. (2012). Failure of skin-deep learning. *Science, 338*, 1263.
Archambault, R. D. (Ed.). (1964). *John Dewey on education. Selected writings.* Chicago: The University of Chicago Press.
Assaraf, O. B.-Z., & Orion, N. (2005). Development of system thinking skills in the context of earth system education. *Journal of Research in Science Teaching, 42*(5), 518–560.
Bachelard, G. (1949). *Le Rationnalisme Appliqué.* Paris: Quadrige (6th ed., 1986)/Presses Universitaires de France.
Baddeley, A. D. (2012). Working memory: Theories, models and controversies. *Annual Review of Psychology, 63*, 1–29.
Baddeley, A. D., & Hitch, G. (1974). Working memory. In G. H. Bower (Ed.), *The psychology of learning and motivation: Advances in research and theory* (Vol. 8, pp. 47–89). New York: Academic.
Bunge, M. (1967). *Scientific research I. The search for system.* New York: Springer.
Bunge, M. (1973). *Method, model and matter.* Boston: D. Reidel.
Bunge, M. (1979). *Treatise on basic philosophy. Vol 4. Ontology II: A world of systems.* Dordrecht: Reidel.
Bunge, M. (1983). *Treatise on basic philosophy. Vol 6. Understanding the world.* Dordrecht: Reidel.
Bunge, M. (2000). Systemism: The alternative to individualism and holism. *The Journal of Socio-Economics, 29*, 147–157.
Cowan, N. (2014). Working memory underpins cognitive development, learning, and education. *Educational Psychology Review, 26*, 197–223.
Ekman, P. (1992). Are there basic emotions? *Psychological Review, 99*(3), 550–553.

---

[5]Related information can be found at http://www.halloun.net/sce/

Fischer, K. W., Daniel, D. B., Immordino-Yang, M. H., Stern, E., Battro, A., & Koizumi, H. (2007). Why mind, brain, and education? Why now? *Mind, Brain, and Education, 1*(1), 1–2.

Garcia, P., Armstrong, R., & Zaman, M. H. (2014). Models of education in medicine, public health, and engineering. *Science, 345*(6202), 1281–1283.

Gee, B. (1978). Models as a pedagogical tool: Can we learn from Maxwell? *Physics Education, 13*, 287–291.

Gentner, D., & Stevens, A. L. (Eds.). (1983). *Mental models*. Hillsdale: Lawrence Erlbaum.

Giere, R. N. (Ed.). (1992). *Cognitive models of science* (Minnesota studies in the philosophy of science, Vol. XV). Minneapolis: University of Minnesota Press.

Glas, E. (2002). Klein's model of mathematical creativity. *Science & Education, 11*(1), 95–104.

Goleman, D. & Senge, P. (2014, August 15). Educating for the bigger picture. *Education Week Commentary, 34* (1).

Gregory, G., & Kaufeldt, M. (2015). *The motivated brain: Improving student attention, engagement, and perseverance*. Alexandria: ASCD.

Halloun, I. (2001). *Apprentissage par Modélisation : La Physique Intelligible*. Beyrouth: Phoenix Series/Librairie du Liban Publishers.

Halloun, I. (2004/2006). *Modeling theory in science education*. Dordrecht/Boston: Kluwer.

Halloun, I. (2007). Mediated modeling in science education. *Science & Education, 16*(7), 653–697.

Halloun, I. (2011). From modeling schemata to the profiling schema: Modeling across the curricula for Profile Shaping Education. In M. S. Khine & I. M. Saleh (Eds.), *Models and modeling: Cognitive tools for scientific inquiry* (Models & modeling in science education, Vol. 6, pp. 77–96). Boston: Springer.

Halloun, I. (2017a). *Mind, brain, and education: A systemic perspective* (Working Paper). Jounieh: H Institute.

Halloun, I. (2017b). *SCE taxonomy of learning outcomes* (Working Paper). Jounieh: H Institute.

Halloun, I. (In preparation). Systemic cognition and education.

Harré, R. (1970). *The principles of scientific thinking*. Chicago: The University of Chicago Press.

Hernández, A., Calva, J. C., Matus, M., Gutiérrez, R., Acevedo, R., Leff, P., & Torner, C. (2002). Understanding the neurobiological mechanisms of learning and memory: Memory systems of the brain, long term potentiation and synaptic plasticity. *Salud Mental, 25*(4), 78–94.

Hesse, M. B. (1970). *Models and analogies in science*. South Bend: University of Notre Dame Press.

Hmelo-Silver, C. E., Marathe, S., & Liu, L. (2007). Fish swim, rocks sit, and lungs breathe: Expert-novice understanding of complex systems. *Journal of the Learning Sciences, 16*(3), 307–331.

Jensen, E. (2005). *Teaching with the brain in mind* (2nd ed.). Alexandria: The Association for Supervision and Curriculum Development (ASCD).

Johanessen, J. A., Olaisen, J., & Olsen, B. (1999). Systemic thinking as the philosophical foundation for knowledge management and organizational learning. *Kybernetes, 28*(1), 24–46.

Johnson-Laird, P. N. (1983). *Mental models. Towards a cognitive science of language, inference, and consciousness*. Cambridge: Cambridge University Press.

Johnson-Laird, P. N. (2006). How we reason. Oxford: Oxford University Press.

Kandel, E. R., Schwartz, J. H., Jessel, T. M., Siegelbaum, S. A., & Hudspeth, A. J. (Eds.). (2013). *Principles of neural science* (5th ed.). New York: McGraw-Hill.

Kiefer, M., & Trumpp, N. M. (2013). Embodiment theory and education: The foundations of cognition in perception and action. *Trends in Neuroscience and Education, 1*, 15–20.

Knox, R. (2016). Mind, brain, and education: A transdisciplinary field. *Mind, Brain, and Education, 10*(1), 4–9.

Kubesch, S., Walk, L., Spitzer, M., Kammer, T., Lainburg, A., Heim, R., & Hille, K. (2009). A 30-minute physical education program improves students' executive attention. *Mind, Brain, and Education, 3*(4), 235–242.

Lakoff, G. (1987). *Women, fire, and dangerous things. What categories reveal about the mind*. Chicago: The University of Chicago Press.

Lakoff, G., & Johnson, M. (1980). *Metaphors we live by*. Chicago: The University of Chicago Press.

Laszlo, A. (2015). Living systems, seeing systems, being systems: Learning to be the system that we wish to see in the world. *Spanda Journal Systemic Change, 6*(1), 165–173.

Liu, J., Mooney, H., Hull, V., Davis, S. J., Gaskell, J., Hertel, T., et al. (2015). Systems integration for global sustainability. *Science, 347*(6225), 963. Full article available at https://doi.org/10.1126/science.1258832.

Matthews, M. R. (2012). Mario Bunge, systematic philosophy and science education: An introduction. *Science & Education, 21*, 1393–1403.

McBean, G., & Martinelli, A. (2017). Blurring disciplinary boundaries. *Science, 358*(6366), 975.

Mehren, R., Rempfler, A., Buchholz, J., Hartig, J., & Ulrich-Riedhammer, E. M. (2018). System competence modeling: Theoretical foundation and empirical validation of a model involving natural, social, and human-environment systems. *Journal of Research in Science Teaching, 55*, 685–711.

National Academies of Sciences, Engineering, and Medicine (NASEM). (2017). *Seeing students learn science: Integrating assessment and instruction in the classroom.* Washington, DC: The National Academies Press.

Osgood-Campbell, E. (2015). Investigating the educational implications of embodied cognition: A model interdisciplinary inquiry in mind, brain, and education curricula. *Mind, Brain, and Education, 9*(1), 3–9.

Panksepp, J. (1998). *Affective neuroscience: The foundations of human and animal emotions.* New York: Oxford University Press.

Panksepp, J. (2006). The core emotional systems of the mammalian brain: The fundamental substrates of human emotions. In J. Corrigall, H. Payne, & H. Wilkinson (Eds.), *About a body: Working with the embodied mind in psychotherapy.* New York: Routledge.

Panksepp, J., & Biven, L. (2012). *The archaeology of mind: Neuroevolutionary origins of human emotions.* New York: Norton.

Pickering, S. (Ed.). (2006). *Working memory and education* (Educational psychology book series). Amsterdam: Elsevier.

Poincaré, H. (1902). *La Science et l'Hypothèse.* Paris: Flammarion.

Rodriguez, V. (2013). The potential of systems thinking in teacher reform as theorized for the teaching brain framework. *Mind, Brain, and Education, 7*(2), 77–85.

Rose, S. N., LaRocque, J. J., Riggall, A. C., Gosseries, O., Starrett, M. J., Meyering, E. E., & Postle, B. R. (2016). Reactivation of latent working memories with transcranial magnetic stimulation. *Science, 354*(6316), 1136–1139.

Saper, C. B., Iversen, S., & Frackowiak, R. (2000). Integration of sensory and motor function: The association areas of the cerebral cortex and the cognitive capabilities of the brain. In E. R. Kandel, J. H. Schwartz, & T. M. Jessel (Eds.), *Principles of neural science* (5th ed., pp. 349–380). New York: McGraw-Hill.

Schwartz, M. (2015). Mind, brain and education: A decade of evolution. *Mind, Brain, and Education, 9*(2), 64–71.

Sousa, D. A. (Ed.). (2010). *Mind, brain, & education. Neuroscience implications for the classroom.* Bloomington: Solution Tree Press.

Wartofsky, M. W. (1968). *Conceptual foundations of scientific thought.* New York: Macmillan.

Waters Foundation. (2010). *The impact of the systems thinking in schools project: 20 years of research, development and dissemination* (White Paper). Pittsburgh: Waters Foundation.

Part X
Varia

Chapter 38
Bunge's Requirement of Neurological Plausibility for a Linguistic Theory

José María Gil

Linguistic theories are tested against theories of data rather than against actual data.
Bunge (1984, p. 154)

Abstract Mario Bunge claimed in many places that core hypotheses of mainstream linguistics have been contrasted with theoretical assumptions, but not with empirical evidence. Bunge's criticism is not only correct for what was known in the middle-1980s, but it also remains valid today. First of all, the very distinction between "faculty of language in the broad sense" (FLB) and "faculty of language in the narrow sense" (FLN) is inconsistent, because it is sometimes presented as an empirical hypothesis and sometimes as a mere terminological or expository aid. Secondly, the Universal Grammar (UG) hypothesis, in any of its forms, is incompatible with biological evidence. Third, the hypothesis that language is a system capable of performing operations on some kind of objects is incompatible with basic neurological evidence, because it assumes (explicitly or implicitly) that the mind/brain is able to store and manipulate objects.

38.1 Introduction: Bunge's Conception of Mainstream Linguistic Theory

According to Noam Chomsky and the majority of the linguists all over the world, being a "biolinguist" presupposes being a generative linguist. The main reason for maintaining such a belief is the assumption that generative linguistics is the branch

J. M. Gil (✉)
CONICET & Universidad Nacional de Mar del Plata, Mar del Plata, Argentina
e-mail: josemaria@gilmdq.com

© Springer Nature Switzerland AG 2019
M. R. Matthews (ed.), *Mario Bunge: A Centenary Festschrift*,
https://doi.org/10.1007/978-3-030-16673-1_38

717

of linguistics that, with a substantial interdisciplinary cooperation, will help us to understand the faculty of language and the Universal Grammar.[1]

Generative linguistics assumes that human beings have an innate capacity for language (the language faculty), which contains a Universal Grammar: a mental organ which provides a grammatical blueprint for the development of the grammars of particular languages. The hypothesis of the existence of Universal Grammar has always been a cornerstone of generative linguistics (See Chomsky 1965, 1981, 1986, 1995, 2005).

However, by the middle of the 1980s, Mario Bunge stated that the hypothesis of Universal Grammar was not only rather fuzzy: He also considered that it had not been confirmed by empirical evidence (Bunge 1983, p. 79). This criticism is really valuable and impressive because such hypothesis has been accepted in mainstream linguistics on the basis of the overwhelming academic success of generative theory.

According to Bunge, the Chomskyan approach involves a curious reheating of centuries-old conceptions of learning.

> One might think that reheating the ideas of Socrates or Leibniz on learning, towards the end of of the twentieth century, requires not only boldness/, but also powerful reasons and amazing experimental findings. None of this: Chomsky offers only two reasons, none of which is enough. (Bunge 1983, pp. 82–83)

In fact, the hypothesis of the existence of Universal Grammar should be supported by powerful reasons or empirical evidence. Nevertheless, Chomsky only provides two weak arguments: His own attack on behaviorism and is assumption about the functional rigidity of the brain. Bunge (1983, pp. 83–87) considers that the two arguments are wrong. First of all, behaviorism and innatism must not be considered incompatible. Secondly, neurophysiological research has shown that the brain cortex has an astonishingly plasticity.

Contrary to Chomsky, Bunge (1983, p. 92) suggests that linguistics needs to get in touch with biological reality by testing hypotheses against what is known about the brain from neuroscience. Bunge also considers that linguistic knowledge must be represented in specific neuronal systems. In this sense, neurological research sheds light on the highly complex processes of linguistic production, linguistic understanding, language learning, and even disorders of speech. Bunge emphasizes that language has to be materially represented in the brain because it has been widely confirmed by aphasiology that brain damages caused by stroke, injury or other events produce linguistic deficits (Bunge 1983, p. 93).

In the next section a Bunge-style analysis will be developed in order to show serious inconsistencies with the popular and dominant idea of Universal Grammar (UG).

---

[1] See Hauser et al. (2002), Fitch et al. (2005), Chomsky (2005, 2006, 2016), and Berwick and Chomsky (2016).

38.2   A Bungean Analysis of the Last Version of UG

In their famous article published in *Science*, Hauser et al. (2002) maintain that advances in linguistics will be beneficial when framed in the joint work of evolutionary biology, anthropology, psychology and neuroscience. This interdisciplinary project should distinguish the two aspects of the faculty of language: The faculty of language in the broad sense (FLB) and the faculty of language in the narrow sense (FLN).

- FLB includes an internal computational system combined with at least two other organism-internal systems, which are called "sensory-motor" and "conceptual-intentional". It also includes the computational mechanisms of recursion.
- FLN only includes recursion and is the only uniquely human component of the faculty of language.

In fact, the recursive mechanisms of FLN are the only exclusive mechanisms of humans, and provide the ability to generate an infinite range of expressions from a finite set of elements: This hypothesis about the creative potential of the linguistic system goes back to the phase of the "systems of the rules" of the generativist program (Chomsky 1957, p. 27, 1965, pp. 3–4).

In order to contribute to the interdisciplinary enterprise, Hauser et al. (2002) present three hypotheses about the two dimensions of the faculty of language. The first two hypotheses involve FLB and the third is about the FLN. All three are related to comparative studies of "non-human animals":

Hypothesis 1: FLB is strictly homologous to animal communication.
Hypothesis 2: FLB is a derivative and exclusively human adaptation for language.
Hypothesis 3: Only FLN is exclusively human.

As a response to Pinker and Jackendoff (2005) and Fitch et al. (2005) claim that the contents of FLN will have to be determined empirically. Thus, such contents could be empty, if the empirical findings showed that none of the mechanisms involved is exclusively human or exclusive of language. In this case, only the form of integration would be specific to human language. However, Fitch et al. (2005, p. 181) admit that the distinction between FLB and FLN is "a terminological aid to the discussion and interdisciplinary exchange and obviously does not constitute a testable hypothesis".

The following quote reveals that Fitch, Hauser and Chomsky first say that their hypothesis about the faculty of language is empirical (and therefore testable) and *immediately afterwards* they say that the hypothesis about the faculty of language is not empirical (and therefore not testable!):

> [W]e denoted "language" in a broad sense, including all of the many mechanisms involved in speech and language, regardless of their/overlap with other cognitive domains or with other species, as the "faculty of language in the broad sense" or FLB. This term is meant to be inclusive, describing all of the capacities that support language independently of whether they are specific to language and uniquely human. Second, given that language as a whole is unique to our species, it seems likely that some subset of the mechanisms of FLB is both

unique to humans, and to language itself. We dubbed this subset of mechanisms the faculty of language in the narrow sense (FLN). Although these mechanisms have traditionally been the focus of considerable discussion and debate, they are neither the only, nor necessarily the most, interesting problems for biolinguistic research. *The contents of FLN are to be empirically determined*, and could possibly be empty, if empirical findings showed that none of the mechanisms involved are uniquely human or unique to language, and that only the way they are integrated is specific to human language. *The distinction itself is intended as a terminological aid to interdisciplinary discussion and rapprochement, and obviously does not constitute a testable hypothesis.* (Fitch et al. 2005, pp. 180–181; emphasis added JMG)

According to Chomsky and his colleagues, the FLB-FLN distinction is fundamental to keep hypothesis (3) undistorted, because it refers only to the faculty of language in a narrow sense (and not to "language as a whole"). The distinctive feature of FLN is recursion, and the hypothesis that recursion is the substantial part of FLN should be supported by the following reasons:

- Most linguists agree that recursion is the indispensable computational skill that underlies syntax and, with it, language.
- No animal communication system has shown evidence of recursion.
- Apes cannot process structured phrases, much less recursion.
- There are no manifestations of recursion in other human cognitive domains, with exceptions that clearly depend on language, such as mathematical formulas or computer programs.

In their view, most of the data that have been presented as refutations or disconfirmations of *Hypothesis 1*, *2*, and *3*, for example those maintained by Pinker and Jackendoff (2005), only deal with mechanisms that are part of FLB by definition, because related mechanisms exist in other species or other cognitive domains. Fitch et al. (2005, p. 204) conclude that such data are thus irrelevant to *Hypothesis 3*, which concerns FLN. According to them, the conclusion supported by Pinker and Jackendoff is based on misreading of their hypothesis postulating that FLN consists of the core computational capacities of recursion as they appear in narrow syntax and the mappings to the interfaces. In fact, when considering criticisms, Fitch, Chomsky, and Hauser point out that they refer to FLB and that, consequently, they are irrelevant concerning FLN.

In the context of this debate, the prominent generativist linguists Steven Pinker and Ray Jackendoff have argued that "the empirical evidence for the hypothesis of only-recursion is too weak" because for example "the hypothesis of only-recursion" implies that there is no natural selection for the production of speech in the human species (Pinker and Jackendoff 2005, p. 217). For example, supra-laryngeal vocal tract control is much more complex in human language than in primate vocalizations. In their reply, Fitch, Hauser and Chomsky say that this data cannot be considered as a disconfirmation that recursion is characteristic of FLN, because it is a fact about FLB. It is worth noting that Fitch, Hauser and Chomsky blame Pinker and Jackendoff for not having understood well the fundamental distinction between the two types of language faculty. The most important area of controversy revolves around the computational apparatus that underlies language, especially syntax.

Anyway, Fitch, Hauser and Chomsky admit a crucial suggestion by Pinker and Jackendoff: FLB may be an evolutionary adaptation for communication, although they reject that same hypothesis applied to FLN.

In relation to all this, Chomsky's individual position is that biolinguistics will be able to explain the design of language. In fact, if it is assumed that the faculty of language has the same general properties as other biological systems, we must analyze "the three factors" that participate in the development of language in the individual (Chomsky 2005, p. 6):

(i) Genetic endowment, apparently nearly uniform for the species, which inter-prets part of the environment as linguistic experience, a nontrivial task that the infant carries out reflexively, and which determines the general course of the development of the language faculty.

(ii) Experience, which leads to variation, within a fairly narrow range, as in the case of other subsystems of the human capacity and the organism generally.

(iii) Principles not specific to the faculty of language.

In "the third factor" it is possible to identify principles of data analysis that can be used in language acquisition, and other domains. Lorenzo González (2006, p. 91) points out that, in the third factor, there are biological forces which are capable of creating natural designs outside the only force contemplated by natural selection in the strict sense: adaptation. But beyond the debate there are very important points around FLB in which the agreement prevails:

• FLB evolved and works as a specific adaptation of humans for many areas that are now useful, one of which is communication.
• FLB must be divided into separate subcomponents, each of which may have experienced different evolutionary processes.

It is worth noting that the criticisms of Pinker and Jackendoff (2005) do not imply, under any point of view, that these two authors want to abandon generative theory. On the contrary, they have both written passionate expositions of the overall program, defending core assumptions such as that language is a combinatorial, productive, and partly innate mental system.[2] Pinker and Jackendoff are convinced that their adaptationist hypothesis, according to which language evolved for commu-nication, does not cause a crisis in biology, but rather the opposite: It contributes to the harmony between biology and linguistics. "Rather than being useless but perfect, language is useful but imperfect, just like other biological systems" (Pinker and Jackendoff 2005, p. 229).

Now, from all the above, it seems that there is a striking inconsistency in the distinction FLN-FLB: The very hypothesis that there are the language faculty in a narrow sense (FLN) and the faculty of language in the broad sense (FLB) is first pre-sented as an empirical hypothesis and immediately afterwards as a terminological aid. This inconsistency allows us to present the following considerations:

---

[2] See Jackendoff (1994, 2002), Pinker (1994, 1999), and Pinker and Jackendoff (2005).

1. The very thesis that there must be two aspects of the faculty of language seems rather vague or confused. Not even linguists of the stature of Steven Pinker and Ray Jackendoff understand the distinction well. In fact, according to Fitch et al. (2005) the criticisms that the two of them make to hypotheses about FLN would be about FLB.

2. As we have seen, Fitch et al. (2005) say that the contents of the language faculty should be determined empirically. There are two alternative roads here:

   • The empirical evidence will allow to expand the contents of FLN, with which its existence is taken for granted.
   • The empirical evidence may show that FLN "is empty". Consequently, it would not exist.... (In Russell's terms, everything that was said about 'the language faculty in the narrow sense' would be false, because there would be no such thing).

3. However, a few lines after suggesting that the contents of the language faculty in narrow sense should be determined empirically, Fitch, Hauser and Chomsky admit that the distinction between the two dimensions of the faculty of language is a "terminological aid", and that it does not constitute ("obviously") a testable hypothesis. If things were like this, the desired reconciliation between linguistics and biology would have to be based on non-empirical definitions of "language".

Thus, generative linguistics assume there is a faculty of language, although this hypothesis does not admit or need empirical testing. Within the faculty of language (whose existence, as we said, is taken for granted) the distinction between "broad sense" and "narrow sense" is inconsistent because sometimes it is considered empirical and sometimes not. At this point we can ask whether a genuinely biolinguistic theory can be developed from a hypothesis whose formulation manifests such inconsistencies. Bunge's criticism seems to apply very well here: Mainstream linguistic hypotheses are tested against other hypotheses rather than against actual data.

38.3 A Bungean Analysis of Biological Evidence Against UG

Bunge has suggested that there is plenty of evidence that argues against (any form of) Universal Grammar. It is the case that children instead appear to learn most aspects of language using the general processes employed in learning to walk or to play the violin. In addition, it seems that there is a manifest continuity of evolution in the basic or "primitive" features of language that we human beings share with other creatures, and the "derived" features that would differentiate human language faculty. In spite of Chomskyan claims, other species possess some elements of human linguistic abilities. Comparative studies have shown that apes have some lexical and syntactic abilities (Gardner et al. 1989). It has also been shown, for example, that primates can communicate, to a limited degree, by means of vocal signals (Cheney and Seyfarth 1980; Slocombe and Zuberbuhler 2005), and that dogs can comprehend words without formal training (Kaminski et al. 2004).

The fact that animals have some language is not incompatible with the fact that no other living species can form a virtually unbounded set of texts on the basis of a finite set of speech sounds and words, or learn a vocabulary exceeding 10,000 words. But, as Philip Lieberman says (2008, p. 218), it is also the case that no other species has the creative cognitive capacities of humans, nor can any other species write or play the violin.

In order to account for the existence of UG, Chomskyan linguistics accepts the following assumptions. In other words, it considers that the following auxiliary hypotheses are true:

(i)   There are modules in the human brain that are exclusive to language.
(ii) Syntax is autonomous since it confers the human linguistic faculty.
(iii) Universal Grammar (UG) is a genetically determined neural organ that controls the range of possible syntactic operations for all particular languages.

Within the generative framework, children do not learn syntax: Instead children activate certain constraints governing syntactic instructions that are already present in their brains. In Lieberman's words (2008, p. 219), "humans are born with preloaded syntax". The shape and the global structure of particular aspects of the genetically determined neural organ that controls the range of possible syntactic operations for all particular languages has been modified along the time: Chomsky has talked about a "language acquisition device" (Chomsky 1965), the Universal Grammar (Chomsky 1981, 1986, 1995), and the "narrow faculty of language" (NFL) (Chomsky 2005, 2006, 2013, 2016). Nevertheless, the fundamental hypothesis according to which children do not learn their native language but activate innately acquired syntactic operations of some kind, remains immutable.

It is well known that Chomsky has always stated that only human beings possess the faculty of language:

> There is no structure similar to UG in non-human organisms ... the capacity for free, appropriate, and creative use of language as an expression of thought, with the means provided by the language faculty, is also a distinctive feature of the human species, having no significant analogue elsewhere. (Chomsky 1975, p. 40)

However, the experiments carried out by Beatrix and Alan Gardner have shown that experts in American Sign Language (ASL) could interact with chimpanzees, and they also filmed and photographed unrehearsed human-chimpanzee interactions (Gardner and Gardner 1973, 1984; Gardner et al. 1989). There were subsequent replications of their work using computer keyboards (Savage-Rumbaugh et al. 1985, 1986; Savage-Rumbaugh and Rumbaugh 1993). It has also been stated that chimpanzees are not able to talk and that they do not progress beyond the linguistic ability of a 2–3 year-old child. Concretely, they can learn and produce approximately 150 words, can form new words and extend the semantic referents of words. Their comprehension is much larger than their active vocabulary and they can also understand spoken syntactically simple English sentences (Lieberman 2008, p. 222).

According to the neurolinguist Sydney M. Lamb, in the case for the apes and dolphins, instead of asking whether they can be taught human-like languages, we should be asking questions like, "How many lexemes can a dolphin learn?", "How many complex lexemes can a chimpanzee learn?", "Can some dolphins or apes learn new lexical and semantic structures on the basis of lexemes and meanings they have already learnt?" (Lamb 1999, p. 285). Actually, investigators at the Marine Mammal Laboratory of the University of Hawaii have shown that dolphins *can* learn new meanings.[3]

In fact, many features of human speech are shared with other animals. However, human being have an exclusive upper airway which allows them to produce speech sounds such as the vowels [i] (as in *see*), [u] (as in *do*), and [a] (as in *ma*), thanks to which it is possible to yield optimal calibration signals for the recognition of speech. The vocal tract of a newborn infant is similar to those retained throughout life in nonhuman primates and most mammals. For example, the newborn tongue and larynx are in the oral cavity, the larynx can lock into the nose and forms a closed pathway for air into the lungs. Newborn infants are able to suckle and breathe at the same time. The exclusively human supra-laringeal vocal tract, i.e., the airway between larynx and lips, gradually takes shape over the first 6–8 years of life. The human vocal tract enables human beings to produce the vowels that the most frequents in the languages of the world, [i], [u] and [a], which have formant frequency patterns that make them resistant perceptual confusion, and also are easier to produce because they yield stable formant frequency patterns in spite of errors in articulation (Stevens 1972).

On the other hand, according to Chomskyan linguistics, UG is domain specific and it determines the features of grammar that can take place in every human language. During the decade of 1980 and the beginnings of the 1990s, the UG determined the complex and countless principles and parameters (Chomsky 1981, 1986). Within the context of the Minimalist Program, through the decade of the 1990s, UG turned out to be a system of negative constraints. As we have seen, in the dawn of the twentieth century, Chomsky and his colleagues proposed that FLN together with lexicon guide the rewriting of sentences that have been recursively inserted into the frame of a model sentence into the words that a person actually hears or reads.

Independently from the theoretical differences of generative models, all versions of UG strongly assume humans beings have a genetically determined knowledge of the range of syntactic operations of all human languages. In other words, if the knowledge of language was not genetically determined, children would not be able to acquire the highly complex syntactic operations of (any) language. But it seems that such genetic or innate determination is implausible. For example, when discussing variation and natural selection, Philip Lieberman accounts for the implausibility of UG when being contrasted with genetic evidence (Lieberman 2006, pp. 347–354).

---

[3]See Herman (1980), Hermann (2008), Herman and Morrel-Samuels (1990), and Herman and Uyeyama (1999).

In Darwinian terms, it is well known that that the feedstock for natural selection is variation. For example, natural selection for life at extreme altitudes has developed an enhanced capacity to deliver oxygen to the bloodstream in the people of Tibet. Since natural selection has been acting during 40,000 years on this pool of variation, it can be said that virtually all Tibetans have this capacity enhanced (Moore et al. 1998). It is not surprising that Tibetan children whose parents emigrated from Tibet and live at low altitudes, retain this enhanced capability, simply because it is part of their genetic endowment. During 40,000 years of life at extreme altitude, natural selection acted on the existing pool of genetic variation.

A comparison with language learning could be established here. As it has been said, there are natural selectional constraints regarding the respiratory capacity, i.e., respiratory capacity is genetically determined. Thus, for example, Tibetan children who are born in England "inherit" the respiratory capacity of their parents. In similar terms, if language were genetically determined by an innate UG, then there should be natural selectional constraints determining that Tibetan children who are born in England should inherit the ability to learn Tibetan more easily than other languages. But it is apparent that it is not the case: Independently from ancestry, any normal child learns the language/s to which s/he is exposed in the first years of life. For instance, Tibetan children who are born in England are perfectly able to learn English if they grow in a sociocultural context where such language is predominantly used. In other words, if there were a genetic determination for language learning, the opposite of what Chomsky suggests with his idea of GU should occur: Natural selectional constraints should favor the acquisition of a specific language.

Considering a concrete case, Lieberman (2008, p. 223) points out that Pinker's (1994) claim about the existence of (a) language gene is false because it is based on an erroneous interpretation of the affliction of KE family, whose members had a rare impairment of language caused by a dominant allele of a single gene, FOXP2 gene (Lai et al. 2001). The members of KE family lacked regular plurals and past tenses but had normal syntax (Gopnik and Crago 1991). On that basis, Pinker and other generative linguists inferred that the FOXP2 counted as genetic evidence for the gene of language. Nevertheless, such hypothesis is false because the KE family members affected by such defective gene experienced also a suite of cognitive, motor and general syntactic deficits (Vargha-Khadem et al. 2005).

In conclusion, it seems that the ubiquitous nature of genetic variation rules out the hypothesis of Universal Grammar.

38.4   A Bungean Analysis of Neurological Evidence Against UG

It has been said that that "there are three things that aren't worth chasing after: a woman, a bus and a theory of transformational grammar; another one is bound to come along soon" (Aitchison 1993, p. 189). But all the ephemeral generativist models have something in common: They are based on the hypothesis according to which the language system stores and manipulates symbols. Such hypothesis involves a transparent analogy between the brain and the computer: The linguistic system "contains" objects, like lexical items, syntactic objects, and operational devices that manipulate such objects, for example the operation Merge (refs). In this sense, we may consider the following passage, written by Steven Pinker:

> The representations that one posits in the mind have to be arrangements of symbols, and the processor has to be a device with a fixed set of reflexes ... [A representation] just has to use symbols to represent concepts, and arrangements of symbols to represent the logical relations among them, according to some consistent scheme. (Pinker 1994, p. 78)

However, following Bunge's criticisms, it could be suggested that there has never been found any kind of empirical evidence, not even indirect evidence, which exhibits the existence of symbols, syntactic objects, or operations in the brain. Rather, the belief that syntactic objects (like sentences), words, phonemes, and so on are materially located in the "mind/brain", which is a duality or equivalence suggested by Fodor (1983), is supported by a not yet confirmed and also implausible hypothesis: The words that some person utters or writes must have been previously in the brain/mind of such person.

The hypothesis according to which *there are* objects or symbols manipulated by specific operations has its origins in the cognitive metaphor that compares the "mind/brain" with the computer. But in reality, the functioning of the brain cannot be seriously compared to that of a computer. Among other things, in the brain there is no workspace, no deposits, no transducers, no entry systems, no central processing system, no storage places (Lamb 2005, pp. 156–157). There is also no need for complete connectivity or computational efficiency (Anderson 1995, p. 304).[4]

The neurological evidence at the microscopic level seems very strong and allows discarding the hypothesis that in the linguistic system there are syntactic structures and operations that act on the supposed structures. As it has been suggested, the hypothesis of storage and manipulation of linguistic items requires complementary equipment, for example, a buffer in which the input item is stored as the recognition process takes place, a mechanism to make the comparison, and, the most important, another mechanism of some kind (such as a "homunculus") that executes the process. Of course, the hypothesis of storage and processing of linguistic items is not saved if it is presented as a simple "functional metaphor", and not as a structural one. Even as a metaphor, it would be unnecessary, because it is incompatible with the most obvious neurological evidence: The brain has neither the structures nor the devices of a computer.

The hypothesis of storage and processing of linguistic items has not only been used by generative theory, but also has been accepted by other important linguistic theories, for example the theory of relevance, which aims at accounting

---

[4]The foundational work of Pylyshyn (1984) can be consulted for an alternative formulation of the computational account of mind.

for communication and cognition (Sperber and Wilson 1995, 2005). In fact, leading neuroscientists like Churchland and Sejnowski (1992) have promoted the idea that the brain stores and manipulates symbols, and such an idea is still relevant and valid in computational neuroscience.[5]

According to such hypothesis, every kind of information (including linguistic information) is stored in the brain as binary combinations (or perhaps as other types of symbols). This proposal may seem reasonable and may also fit with our habitual assumption that information is stored or represented in some kind of physical support like blackboards, paper, or pen drives. However, although the linguistic information is represented by means of symbols in some physical support, it is not confirmed by this that the hypothesis that these symbols are stored in the brain. In order to have some neurological support, the hypothesis of the storage and processing of linguistic information should show how neurons or groups of neurons are able to store binary digits (or other types of symbols) and how such symbols are manipulated during observable processes like speech production and verbal comprehension.

It is well known how recognition works on a computer: It depends on a comparison process. If an entry item appears, then a strategy is used to find candidates among items stored in memory. Then each of these candidates is compared with the item that entered. When a candidate and the entry item coincide, a successful recognition is given. Obviously, the same does not happen in the brain.

In short, generative linguistics claims to be biolinguistic.[6] It aims at accounting for how language (which is conceived as a combinatorial, productive, and partly innate mental system) is represented in the mind/brain. In one of his fundamental books, *Knowledge of Language* Chomsky had said that linguistics is an underdeveloped branch of psychology (Chomsky 1986, p. 42, note 15). In this book, Chomsky also claims to be both realist and empiricist, and he says that neurosciences will be the judges in the context of justification of rival linguistic theories.

Faced with two (or more) theories of UG, the study of the mechanisms of the brain should decide which is correct. Now, if the considerations in this section are adequate, then it may be suggested that generative theory does not offer plausible hypotheses about a person's actual linguistic system. Thanks to Chomsky, we can ask if it is possible to reconcile linguistics and biology in a genuinely empiricist way. But, although Chomsky believed that conciliation was going to occur within the generative theory, it seems that alternative paths should be explored.

Here, Bunge's account helps us to understand that a realistic theory of language should meet, among others, the requirement of neurological plausibility: A realistic linguistic theory has to be compatible with what is known about the brain from neurosciences (Bunge 1983, 1984, 1986a, b, 1999). It has been argued that there is a good amount of relevant indirect evidence for the neurological plausibility of relational network theory, a linguistic theory that is incompatible with Chomskyan

---

[5]See Adaszewski et al. (2013), Bower (2013), and Friston et al. (2014).
[6]See Chomsky (1986, 1995, 2000, 2005, 2006, 2013, 2016).

hypotheses because it states that the linguistic system is nothing but a network of relationships represented in the brain (Lamb 1999, 2000, 2005, 2006, 2013, 2016).

In this sense, Hubel and Wiesel (1962, 1968, 1977) discovered that visual perception in cats and monkeys works in the ways that would be predicted by the relational network model, and the nodes of visual network are implemented as cortical columns. "The nodes are organized in a hierarchical network in which each successive layer integrates features from the next lower layer and sends activation to higher layers" (Lamb 2005, p. 168).

The eminent neuroscientist Vernon Mountcastle and other neuroscientists have described the columnar organization of the cerebral cortex.[7] Mountcastle explains that the basic unit of the mature neocortex is the cortical minicolumn, a narrow chain of neurons that extends vertically across cellular layers II–VI. Each minicolumn contains about 80–100 neurons and all the major phenotypes of cortical neural cells. His general hypothesis is that the minicolumn is the smallest processing unit of the neocortex, and he also claims that "every cellular study of the auditory cortex in cat and monkey has provided direct evidence for its columnar organization" (Mountcastle 1998, p. 181).

Within the framework of relational network theory, the neurolinguist Lamb (2005, p. 170) claims that, since speech perception is a higher-level perception process, it is permissible to suggest the following extrapolation: Each node in the neurocognitive system of an individual can be implemented as a cortical column. Regarding the linguistic system, every node/cortical column has a highly specific function. For example, there may be a node/cortical column corresponding to a single word like *cat*.

Lamb also suggests that the examination of neurological evidence shows that minicolumns and their interconnections have every one of the properties of the nodes and connections of relational networks. For example, connections can have varying strengths, and are strengthened through successful use: the learning process.

On the basis of previous remarks, we can provide an argument for the neurological plausibility of relational networks:

- Linguistic nodes represented in relational networks are implemented as cortical minicolumns.
- Linguistic connections represented in relational networks are implemented as neural fibers.
- Minicolumns and fibres integrate real cortical connections.
- Therefore, relational networks represent linguistic information in the brain.

In summary, Bunge noticed that Chomsky was mistaken because the generative theory did not satisfy the requirement of neurological plausibility. On the other hand, relational network theory can be considered as a realistic theory because its hypotheses are confirmed by the evidence that is known about the brain thanks to neurosciences.

---

[7]See Mountcastle (1957, 1997, 1998, 2005), Mountcastle et al. (1975), and Martin (2015).

38.5 Conclusions

The following conclusions, inspired by Bunge's general conception science, can be drawn about the current state of mainstream linguistics:

1. Generative linguistics accepts traditional semantic theories despite their complete failure to characterize the meaning of even simple words such as *table, dog,* or *piano*. The predicate logic employed in linguistic semantic theories fails to capture the meaning of even a simple word which varies with context and circumstance. Bronowski (1978) pointed out such deficiencies some years ago: He noted that it is impossible, for example, to define the word *table* in terms of a set of discrete primitive elements. Aside from the question of determining the primitives, once the word *table* has been defined, someone can easily use the table as a chair. In brief, linguistic semantic theory is just a word game in which words are "defined" by sets of other words: *bachelor* can be defined as +animate, +human, +male, +unmarried, etc. But then unmarried, male, human, etc. should also be defined, using another set of words that should be defined..., and so on. "When you finish the exercise, you still will not have the foggiest notion of how a word is represented in the human brain, or the brain of a chimpanzee or a dog" (Lieberman 2008, p. 224). We do not know whether these processes can be accounted for in neurological terms. But we do know the characterization of the meaning of a word like *bachelor* or table in terms of "primitives" or "compounds" is a mere reconstruction which could hardly account for the actual linguistic system of a single person.

2. The hermetic nature of formal linguistics does not even catch the data at the phonological level. As we have observed, the claims about UG are incompatible with empirical evidence. Its deficiency is not limited to quasi-biological assumptions around UG, or "semantic theories that reflect the state of human knowledge in ancient Greece" (Lieberman 2008, p. 224). Chomskyan linguistics evidences a hermetic disavowal of the physiology of speech. For example, it does not recognize that there are two phonological systems, one for recognition and other for production. Phonological recognition is located in Wernicke's area, whereas phonological production is located in Broca's area (Lamb 1999, p. 356).

3. The concepts of "language" and "language faculty" are too vague. Chomsky, Fitch and Hauser admit that the FLN-FLB distinction is terminological, not empirical in nature. However, empirical linguistics works with testable hypotheses. It does not need to analyze concepts created within the scope of a particular theory and that avoid empirical testing. Now, it is said that the hypotheses about what content there is in FLN should be determined empirically, but it is also said (simultaneously) that the very distinction between FLN and FLB is not a testable hypothesis. And if it is not testable, it is not an empirical hypothesis, it is not a scientific hypothesis.

4. Color deficits and color blindness are common examples of behavioral deficits corresponding to the pool of genetic variation that is present in all human populations. Regarding genetic variation, if UG existed, there should be individuals who

were unable to acquire some concrete aspects of syntax, but who were perfectly able to acquire other aspects. Since we all do not inherit exactly the same genes for anything, and if UG existed, it should be expected to discover children who were able to acquire Spanish but unable to acquire Guarani, simply because some genes were missing. In this scenario, there would be individuals who lacked the gene/s that regulated the Spanish plural for Adjectives, but who could handle other aspects of Spanish syntax.

5. Generative theory postulates that language is a productive computational system. In its various models it holds that in the mind/brain there are symbols and operations that manipulate those symbols, for example syntactic objects. But the neurological evidence allows to notice that in the brain there is no structure of symbol storage nor manipulation operations of these symbols. In addition, the hypothesis that there is a store of symbols and various operations requires that the linguistic system work in successive stages, taking certain syntactic objects in a given state to which certain operations are applied to finally produce syntactic objects "in conditions to be pronounced". In this sense, the hypothesis of storage and manipulation of objects lacks operational plausibility, because it does not offer a plausible characterization of how the proposed linguistic system can be operated in real time to produce and understand speech.

6. At least since *Aspects of the Theory of Syntax* (Chomsky 1965), Chomsky has emphasized the importance of two great requirements of linguistic theory. First, according to the requirement of descriptive adequacy, the linguistic theory has to be sufficiently broad so that all the syntactic constructions can be described. On the other hand, the explanatory adequacy requirement establishes that the theory must be restricted enough to allow only a small number of grammars (consistent with the data) so that the rapid process of language acquisition can be characterized. Generative theory is an area characterized by the tension between the two requirements (Chomsky 1986, p. 56). Thus, the succession of models within the theory responds to a large extent to the need to solve this tension.

   For example, a system of rules seems to be very thorough but, at the same time, something very difficult to acquire by the child in real time. But this difficulty applies not only to the systems of rules but to any theory that proposes a linguistic system where there are symbols of some kind and operations that allow manipulating these symbols. Therefore, generative theory does not meet the requirement of development plausibility, because it does not go hand in hand with a plausible description of how the child can learn the proposed linguistic system.

7. If a linguistic theory has no operative plausibility or plausibility of development, then neither can it have neurological plausibility. The brain is a very complex network of connections between neurons and groups of neurons. The hypothesis that there are symbols and operations applicable to symbols (for example, sentences) is incompatible with this elementary observation.

8. In this work, Bunge's criticisms of mainstream linguistics have been analyzed and developed. Consequently the following can be concluded:

- The core hypothesis about the very existence of the faculty of language and UG is incompatible with genetic evidence and neurological evidence.
- The core hypothesis about the very distinction FLN-FLB is not an empirical hypothesis

9. Perhaps, "language" is not what generative theory has characterized. Ultimately, "language" is perhaps just a term that can be used to refer to a particular configuration of brain subsystems interconnected with each other and with other cognitive systems (such as vision for example) and which we find comfortable to conceive as a unity. We cannot be sure that this hypothesis is true, but at least it presents the great merit that it can be evaluated in terms of real linguistic and neurological data. In this way, the empirical commitment claimed by Bunge in the epigraph of this paper would be preserved, because linguistic hypotheses would not be tested against dogmatic theoretical assumptions like those of Chomskyan linguistics, but finally against actual data, like those provided by what it is known about the brain thanks to the development of neurosciences.

References

Adaszewski, S., Dukart, J., Kherif, F., Frackowiak, R., & Draganski, B. (2013). How early can we predict Alzheimer's disease using computational anatomy? *Neurobiology of Aging, 34*(12), 2815–2826.

Aitchison, J. (1993). *Linguistics*. London: Moughton.

Anderson, J. A. (1995). *An introduction to neural networks*. Cambridge, MA: MIT Press.

Berwick, R., & Chomsky, N. (2016). *Why only us: Language and evolution*. Cambridge, MA: MIT Press.

Bower, J. M. (Ed.). (2013). *20 years of computational neuroscience*. New York: Springer.

Bronowski, J. (1978). *The origins of knowledge and imagination*. New Haven: Yale University Press.

Bunge, M. (1983). *Lingüística y filosofía*. Madrid: Ariel.

Bunge, M. (1984). Philosophical problems in linguistics. *Erkenntnis, 21*, 107–173.

Bunge, M. (1986a). A philosopher looks at the current debate on language acquisition. In I. Gopnik & M. Gopnik (Eds.), *From models to modules* (pp. 229–239). Norwood: AblexPubls. Co.

Bunge, M. (1986b). *Philosophical problems in linguistics*. Tokyo: Seishin-Shobo.

Bunge, M. (1999). Linguistics and philosophy. In H. E. Wiegand (Ed.), *Sprache und Sprachen in der Wissenschaften* (pp. 269–293). Berlin/New York: Walter de Gruyter.

Cheney, D. L., & Seyfarth, R. M. (1980). Vocal recognition in free ranging vervet monkeys. *Animal Behavior, 28*, 362–367.

Chomsky, N. (1957). *Syntactic structures*. Amsterdam: Mouton-De Gruyter.

Chomsky, N. (1965). *Aspects of the theory of syntax*. Cambridge, MA: MIT Press.

Chomsky, N. (1975). *Reflections on language*. New York: Pantheon.

Chomsky, N. (1981). *Lectures on Government and Binding*. Dordrecht: Clarendon Press.

Chomsky, N. (1986). *Knowledge of language. Its nature, its origin, and its use*. Cambridge, MA: MIT Press.

Chomsky, N. (1995). *The minimalist program*. Cambridge, MA/London: MIT Press.

Chomsky, N. (2000). *On nature and language*. New York: Cambridge University Press.

Chomsky, N. (2005). Three factors in language design. *Linguistic Inquiry, 36*(1), 1–22.

Chomsky, N. (2006). *Language and mind*. Cambridge: Cambridge University Press.

Chomsky, N. (2013). Problems of projection. *Lingua, 130,* 33–49.

Chomsky, N. (2016). *What kinds of creatures are we?* New York: Columbia University Press.

Churchland, P., & Sejnowski, T. (1992). *The computational brain.* Cambridge: MIT Press.

Fitch, W. T., Hauser, M. D., & Chomsky, N. (2005). The evolution of the language faculty: Clarifications and implications. *Cognition, 97,* 179–210.

Friston, K. J., Stephan, K. E., Montague, R., & Dolan, R. J. (2014). Computational psychiatry: The brain as a phantastic organ. *Lancet Psychiatry, 1*(2), 148–158.

Fodor, J. A. (1983). *The Modularity of Mind.* Cambridge: MIT Press.

Gardner, R. A., & Gardner, B. T. (1973). *Teaching sign language to the chimpanzee, Washoe (16 mm sound film and transcript).* State College: Psychological Film Register.

Gardner, R. A., & Gardner, B. T. (1984). A vocabulary test for chimpanzees (*Pan troglodytes*). *Journal of Comparative Psychology, 4,* 381–404.

Gardner, R. A., Gardner, B. T., & Van Cantfort, T. E. (1989). *Teaching sign language to chimpanzees.* Albany: State University of New York Press.

Gopnik, M., & Crago, M. (1991). Familial segregation of a developmental language disorder. *Cognition, 39,* 1–50.

Hauser, M. D., Chomsky, N., & Fitch, T. (2002). The faculty of language: What is it, who has it, and how did it evolve? *Science, 298*(5598), 1569–1579.

Herman, L. M. (1980). Cognitive characteristics of dolphins. In L. M. Herman (Ed.), *Cetacean behavior: Mechanisms and functions* (pp. 363–429). New York: Wiley Interscience.

Herman, L. M., & Morrel-Samuels, P. (1990). Knowledge acquisition and asymmetries between language comprehension and production: Dolphins and apes as a general model for animals. In M. Bekoff & D. Jamieson (Eds.), *Interpretation and explanation in the study of behavior. Vol. 1: Interpretation, intentionality, and communication* (pp. 283–312). Boulder: Westview Press.

Herman, L. M., & Uyeyama, R. K. (1999). The dolphin's grammatical competency: Comments on Kako (1998). *Animal Learning and Behavior, 27,* 18–23.

Hermann, L. M. (2008). Can dolphins understand language? In P. Sutcliffe, L. Stanford, & A. Lommel (Eds.), *LACUS forum 34: Speech and beyond.* Houston: LACUS.

Hubel, D., & Wiesel, T. N. (1962). Receptive fields, binocular interaction and functional architecture in the cat's visual cortex. *Journal of Physiology, 160,* 106–154.

Hubel, D., & Wiesel, T. N. (1968). Receptive fields and functional architecture of monkey striate cortex. *Journal of Physiology, 195,* 215–243.

Hubel, D., & Wiesel, T. N. (1977). Functional architecture of macaque monkey cortex. *Proceedings of the Royal Society of London, 198,* 1–559.

Jackendoff, R. (1994). *Patterns in the mind: Language and human nature.* New York: Basic Books.

Jackendoff, R. (2002). *Foundations of language: Brain, meaning, grammar, evolution.* New York: Oxford University Press.

Kaminski, J., Call, J., & Fisher, J. (2004). Word learning in a domestic dog: Evidence for "fast mapping". *Science, 304,* 1682–1683.

Lai, C. S. L., Fisher, S. E., Hurst, J. A., Vargha-Khadem, F., & Monaco, A. P. (2001). A novel forkhead-domain gene is mutated in a severe speech and language disorder. *Nature, 413,* 519–523.

Lamb, S. M. (1999). *Pathways of the brain: The neurocognitive basis of language.* Amsterdam: John Benjamins.

Lamb, S. M. (2000). Neuro-cognitive structure in the interplay of language and thought. In M. Pütz & M. Verspoor (Eds.), *Explorations in linguistic relativity* (pp. 173–196). Amsterdam: John Benjamins.

Lamb, S. M. (2005). Language and the brain: When experiments are unfeasible, you have to think harder. *Linguistics and the Human Sciences, 1,* 151–178.

Lamb, S. M. (2006). Being realistic, being scientific. *LACUS Forum, 33,* 201–209.

Lamb, S. M. (2013). Systemic networks, relational networks, and choice. In L. Fontaine, T. Bartlett, & G. O'Grady (Eds.), *Systemic functional linguistics. Exploring choice* (pp. 137–160). Cambridge: Cambridge University Press.

Lamb, S. M. (2016). Linguistic structure: A plausible theory. *Language Under Discussion, 4*(1), 1–37.

Lieberman, P. (2006). *Toward an evolutionary biology of language*. Cambridge, MA: Harvard University Press.

Lieberman, P. (2008). Old-time linguistic theories. *Cortex, 44*, 218–226.

Lorenzo González, G. (2006). El tercer factor: reflexiones marginales sobre la evolución de la sintaxis. *Teorema, 25*(3), 77–92.

Martin, K. (2015). Vernon B. Mountcastle (1918–2015) discoverer of the repeating organization of neurons in the mammalian cortex. *Nature, 518*(7539), 304.

Moore, L. G., Niermeyer, S., & Zamudio, S. (1998). Human adaptation to high altitude: Regional and life-cycle perspectives. *Yearbook of Physical Anthropology, 41*, 25–61.

Mountcastle, V. B. (1957). Modality and topographic properties of single neurons of cat's somatic sensory cortex. *Journal of Neurophysiology, 20*(4), 408–434.

Mountcastle, V. B. (1997). The columnar organization of the neocortex. *Brain, 120*, 701–722.

Mountcastle, V. B. (1998). *Perceptual neuroscience: The cerebral cortex*. Cambridge, MA: Harvard University Press.

Mountcastle, V. B. (2005). *The sensory hand: Neural mechanisms of somatic sensation*. Cambridge, MA: Harvard University Press.

Mountcastle, V. B., Lynch, J. C., Georgopoulos, A., Sakata, H., & Acuna, C. (1975). Posterior parietal association cortex of the monkey: Command functions for operations within extrapersonal space. *Journal of Neurophysiology, 38*(4), 871–908.

Pinker, S. (1994). *The language instinct*. New York: Harper Collins.

Pinker, S. (1999). *Words and rules*. New York: Morrow Press.

Pinker, S., & Jackendoff, R. (2005). The faculty of language: What's special about it? *Cognition, 95*, 201–236.

Pylyshyn, Z. (1984). *Computation and cognition*. Cambridge, MA: MIT Press.

Savage-Rumbaugh, S., & Rumbaugh, D. (1993). The emergence of language. In K. R. Gibson & T. Ingold (Eds.), *Tools, language and cognition in human evolution* (pp. 86–100). Cambridge: Cambridge University Press.

Savage-Rumbaugh, S., Rumbaugh, D., & McDonald, K. (1985). Language learning in two species of apes. *Neuroscience and Biobehavioral Reviews, 9*, 653–665.

Savage-Rumbaugh, S., McDonald, K., Sevcik, R. A., Hopkins, W. D., & Rubert, E. (1986). Spontaneous symbol acquisition and communicative use by pygmy chimpanzees (*Pan paniscus*). *Journal of Experimental Psychology, 115*, 211–235.

Slocombe, K., & Zuberbuhler, K. (2005). Functionally referential communication in a chimpanzee. *Current Biology, 15*, 1–6.

Sperber, D., & Wilson, D. (1995). *Relevance: Communication and cognition*. Oxford: Blackwell.

Sperber, D., & Wilson, D. (2005). Pragmatics. *UCL Working Papers in Linguistics, 17*, 353–388.

Stevens, K. N. (1972). Quantal nature of speech. In E. David & P. B. Denes (Eds.), *Human communication: A unified view* (pp. 51–66). New York: McGraw Hill.

Vargha-Khadem, F., Gadian, D., Copp, A., & Mishkin, M. (2005). FOXP2 and the neuroanatomy of speech and language. *Nature Reviews. Neuroscience, 6*, 131–138.

Lubbe, S. J. & Zhou, T. Jump in state tree: A plausible theory. Lecture Notes. Dissertation, MIT, p. 47.

Hofstra, H. (2000). Vision and estimation: A biology of structure. Cambridge, MA: Harvard University Press.

Kleinsmith, L. (1998). Physiological theories as review. 48, 235–248.

Lettvin, J. McCulloch, S. (2006). Brain figures reductions as predicates seen as volatile Ag. In Eckmiller (Eds.), 54 (1), 79–97.

Lorento, J. & N. (2013). Studies in Mechanism – 2, 2013 – In structure of the repeating organization of neurons in the cerebral cortex, 88. J. (18), 1580, 304.

Margrie, F. T., Brikgrove, S. & Zamudio, S. (1998). Role for adaptation to high altitude. S. global acclimatization response. Journal of Physiology, 41, 25–51.

Martinelson, V. H. (1957). Modeling and report on properties of single neurons of cell complex cerebral cortex. Annual of Neurophysiology, 22 (1), 408–434.

Maturana, H. R. (1970). The biological organization of the biosystem. Brain, 120, 801–822.

Meunissen, V. et al. (2002). Perception and neuroscience. The cognitive estate. Cambridge, MA: Harvard University Press.

Monnier, V. et al. (1975). The nervous brain. A self-reconstructing sound occupation. Cambridge, MA: Harvard University Press.

Mountcastle, V. B., Lynch, J. C., Georgopoulos, A., Sakata, H., & Acuna, C. (1975). Posterior parietal association cortex of the monkey: Command functions for operations within extrapersonal space. Journal of Neurophysiology, 38(4), 871–908.

Piaget, J. (1952). The origins of intelligence. New York: Harper Collins.

Piaget, J. (1971). The biology of intellect. New York: Harper Books.

Pinker, S. & Jackendoff, R. (2005). The faculty of language: What's special about it? Cognition, 95, 201–236.

Rybak, I. (1998). Computations in perception and cognition. Cambridge, MA: MIT Press.

Sacks, O. (1985). The man who mistook his wife for a hat. New York: Harper.

Sejnowski, T. J. (1994). The computational brain. In R. K. Gibson & T. Ingold (Eds.), Tools, language and cognition in human evolution (pp. 86–99). Cambridge: Cambridge University Press.

Snyder, L. H., Grieve, K. L., Brotchie, P., & Andersen, R. (1998). Examples of mixture in two spaces of some features in the cortex of reference. Nature, 394, 887–891.

Sterling, P. & Laughlin, S. (2015). Principles of neural design. Cambridge, MA: MIT Press.

Stevens, C. F. (1980). Neuronal systems in neural communication. New Directions in Child Development (pp. 311–336).

Sun, R. & Bookman, L. (Eds.) (1995). Computational architectures integrating neural and symbolic processes. Norwell, MA: Kluwer.

Syder, D. S. & Ingber, L. (1976). Some differences in brain structure and function in relation to a field theory, 43, 256.

Tanaka, K. (1993). Neuronal mechanisms of object recognition. Science, 262, 685–688.

Thelen, E. & Smith, L. (1994). A dynamic systems approach to the development of cognition and action. Cambridge, MA: MIT Press.

Ullman, S. (2000). High-level vision: Object recognition and visual cognition. Cambridge, MA: MIT Press.

Wright, Phillips, J., Coulton, D., Popp, A. & Johnston, M. (2001). EEG P3 and the recruitment of cortical field organization. Cognitive Neuroscience, 14 (3), 10–14.

Chapter 39
Mechanisms in Clinical Research and Medical Practice

Omar Ahmad

Abstract Mario Bunge's medical philosophy emphasizes the importance of mechanismic models in guiding the design, analysis, and practical application of clinical research. By contrast, the Evidence-Based Medicine (EBM) movement regards mechanismic hypotheses as "evidence" dissociable from, and of secondary importance to, the findings of experimental research. In agreement with Bunge, it is argued here that mechanismic models and mechanismic thinking play essential roles in both clinical research and practice. Mechanismic models in medicine view health and disease as emergent processes occurring in complex biological systems and draw upon established scientific knowledge from multiple disciplines to help identify and control parameters that have decisive effects on clinical outcomes. Models play an essential role in designing efficient and reliable population-based studies, and in detecting and correcting for random error and systematic bias in clinical research. They are important both for extrapolating the results of clinical research to novel contexts and for tailoring interventions to the specific circumstances of an individual case. Contrary to the subordinate status they are accorded by EBM, empirically-validated mechanismic models should constitute the foundation of a scientific approach to medicine.

39.1 Introduction

Mario Bunge's philosophy of medicine has as its central theme the importance of grounding clinical practice on a scientific understanding of the mechanisms of diseases and their treatments. In his *Medical Philosophy* (Bunge 2013), Bunge draws attention to the results this philosophical approach has yielded through centuries of effort by its proponents in biology and medicine. These achievements include:

O. Ahmad (✉)
Department of Internal Medicine, Stanton Territorial Hospital, Yellowknife, NT, Canada

© Springer Nature Switzerland AG 2019
M. R. Matthews (ed.), *Mario Bunge: A Centenary Festschrift*,
https://doi.org/10.1007/978-3-030-16673-1_39

735

(i) the renunciation of superstition and vitalism;

(ii) the adoption of the scientific method, at the core of which is the assumption that mechanisms are material processes governed by natural laws that can be comprehended through the combined application of reason and experiment; and

(iii) the practical demonstration that a scientific understanding of the mechanisms of disease and treatment is the engine of radical innovation in medicine (Bunge 2013).

The cell theory, the germ theory of infectious diseases, the models of molecular and population genetics, and the physico-chemical models of proteins and small molecules, are just a few examples of ideas that directly inspired the creation of not just isolated remedies, but systemic frameworks for treatment that explicitly harness knowledge of underlying mechanisms to influence the course of disease.

Despite the successes of the scientific approach to medicine, much of medicine in the twentieth century remained based on entrenched dogma and custom, rather than on scientific knowledge. In many branches of medicine, there were large gaps between firm scientific knowledge of underlying mechanisms and the practices in widespread use. Although biomedical scientists worked to produce useful clinical applications of basic discoveries, many clinical practices were not based on a rigorous understanding of mechanisms and were of uncertain effectiveness and risk. Some, such as those of Freudian psychiatry, were manifestly anti-scientific. Both practitioners and patients witnessed first-hand the discrepancies between the promises of well-accepted treatments and their often disappointing results. Much of what physicians offered was either based on thin evidence or, worse, had been proven to be ineffective (Eddy 2005).

It was in reaction to these inadequacies that a movement that came to be known as *Evidence-Based Medicine* (EBM) arose (Eddy 2005): instead of taking treatments at face value, clinicians would insist upon direct and scientifically compelling empirical demonstrations of efficacy and risk through population-based investigations of human subjects. Physicians and their patients had urgent decisions to make on whether to pursue or abandon a clinical practice. These decisions could not wait on the arduous translation of an understanding of fundamental biological mechanisms into clinical recommendations, and the effort to carry out this translation could not be accomplished at the bedside in the mind of the practitioner. Physicians did not reason through the pathophysiology and available data to derive from first principles a clinical plan; instead they used treatment algorithms that specified how to act in given circumstances; that is, they relied on guidelines from textbooks, teachers, and professional societies. These recommendations, rather than flowing directly and explicitly from the analysis of clinical research, often simply reflected the opinions and experiences of their authors. Research methods were needed that could provide scientifically credible knowledge upon which to base practices even in the absence of a deep and comprehensive knowledge of underlying mechanisms.

It was to address these needs that the randomized controlled trial (RCT), the so-called "gold-standard" of EBM, was introduced into clinical research. In an RCT, patients are randomly allocated to treatment and control groups, and outcomes in each group are then measured. By comparing the average outcome in the treatment group to that of the control group, the average effect of treatment can then be estimated. Importantly, such an estimate can be obtained independent of whether researchers understand (or can even enumerate) all of the causal factors at play in the evolution of a given disease. The reason for this is that randomization, if properly executed, on average balances out between groups the effects of all factors other than the intervention, so that the specific contribution of the intervention can be estimated. Moreover, the results of individual RCTs that are suitably similar in design can be pooled (or *meta-analysed*) using statistical methods to provide ever more precise estimates of the causal effects of an intervention. Random fluctuations in the balance of confounding causal factors that can distort the estimates from individual RCTs are averaged out through the processes of replication and meta-analysis.

The culmination of the EBM project has been the development and widespread use of clinical practice guidelines (CPGs). CPGs are codified bodies of evidence-based recommendations created through a structured process of appraising the scientific literature and weighing the risks and benefits of treatment in light of patients' circumstances and values. The purpose of a CPG is to enable physicians to bring the results of clinical research to bear on the important decision problems they face in day-to-day clinical practice.

Though the process of guideline development has been refined over the past three decades, it continues to have shortcomings that have been the subject of criticism and reform (Lenzer et al. 2013). These problems include the often uneven quality of the clinical research upon which they are based, the lack of transparency in the methods used to create them, the presence of conflicting recommendations from guidelines with overlapping domains, and the financial conflicts of interest that threaten the objectivity of guideline authors, to name a few (Ioannidis et al. 2017).

Though these problems are important, our purpose in this chapter is to examine a more fundamental and under-appreciated limitation with the EBM approach, one that is the primary focus of Mario Bunge's criticism of the EBM movement: namely, that it devalues the role of mechanistic thinking and mechanistic modeling in producing reliable recommendations. There is no standardized, much less mathematically precise, approach in clinical guideline development for explicitly incorporating mechanistic reasoning (Howick et al. 2010). Indeed, guidelines regard mechanistic hypotheses as a kind of low-ranking "evidence", on par with mere opinion, rather than as the core ideas around which the results of empirical research are organized. Although evidence-based guidelines generally require that the hypotheses underlying their recommendations be biologically plausible, they often leave unstated both the basis for judging plausibility, and the mechanistic assumptions underlying the interpretation of clinical data.

Even where mechanistic hypotheses do appear, they are typically articulated without the help of any mathematics besides statistics, perpetuating the misconception of the latter as an all-purpose mathematical tool that has a role wherever uncertainty of any kind arises. Contrary to engineering disciplines, where there is generally an explicit connection between models of underlying mechanisms and technological models that guide action, guideline development in medicine has instead a strongly empiricist character that relies predominantly on statistical inference. EBM has encouraged the view that clinical research and clinical guidelines can be designed and judged by largely formulaic procedures that treat knowledge of underlying mechanisms as merely a point of departure, rather than as a core component in need of precise articulation. In so doing, EBM leaves out of its framework the major philosophical innovations in medicine of the past two centuries: namely that health and disease are emergent processes occurring in complex biological systems, and that diagnostic methods and treatments are processes for interrogating and influencing the mechanisms of growth, maintenance, and dissolution of organisms (Bunge 2004, 2013).

In this chapter, it is argued, in agreement with Bunge, that mechanismic models play an essential role in the design, analysis, and application of clinical research. This chapter first discusses how mechanismic models help guide research programs by integrating knowledge across disciplines and levels of organization to identify within complex systems the key parameters that should be measured or controlled in pursuit of a particular outcome (Sect. 39.1). Next, it examines how mechanismic hypotheses enter as essential components into the design and analysis of population-based studies, and argues that mechanismic hypotheses should be articulated explicitly and with mathematical precision (Sect. 39.2). Finally, it considers how mechanismic hypotheses underlie the extrapolation of population-based research to novel situations, and the application of population-based research to decision-making in individual cases (Sect. 39.3).

39.2 Mechanismic Models, Not Informal Reasoning, Should Direct Research and Practice

In well-developed sciences, mechanismic models play a major role in helping researchers choose what to study in the first place. In medicine, mechanismic modelling serves to connect the results of clinical research with knowledge from other disciplines, such as biophysics, molecular biology, and physiology: by locating hypotheses within a network of scientific ideas from diverse disciplines at multiple levels of organization, researchers can identify promising directions for further study and innovation. Equally important, mechanismic models can protect both patients and resources from ill-conceived hypotheses that do not fit with the bulk of established scientific knowledge and therefore may not merit costly further testing (Bunge 2017).

Hypotheses in clinical research are often treated as isolated ideas; repetitive tests of the same hypothesis yield estimates of effect size that in turn are aggregated using statistical methods. Though replication of research is important to confirm preliminary findings and improve precision, the process of replication is not creative; where, then, should scientists look for inspiration about what to study in the first place? This is a complex topic, but part of the answer is that they should adopt an intellectually fertile philosophy: that health and disease are emergent processes that occur in complex biological systems, and that these processes can be explained, measured, and controlled with the help of testable mechanismic models whose predictions span multiple levels of organization (Bunge 2013).

For instance, models of pathophysiology can be used to predict both the dynamics of surrogate biological markers and clinically important outcomes. Consider a drug developed to reduce mortality due to myocardial infarction ('heart attack'). This treatment might work by improving myocardial perfusion, and consequently reducing the size of a myocardial infarction. A mechanismic model of this process might predict: (i) a smaller rise in the bloodstream of biomarkers related to heart muscle damage; (ii) a more orderly appearance to the cardiac contractions observed on ultrasound imaging; (iii) a more rapid resolution of the electrocardiographic abnormalities characteristic of heart injury; and (iv) fewer clinical findings of heart failure. The model might predict the interactions between the treatment and other medications for myocardial infarction. For example, the reduction in infarct size might reduce complications due to thrombosis but might increase the risk of major bleeding.

In short, rather than predicting lone findings in isolation, mechanismic hypotheses, by virtue of regarding organisms as complex systems, can generate entire families of predictions for further testing. Thus, they reduce the scope for data dredging in search of fortuitous associations, and put the focus not on measures of statistical significance (such as p-values) but on the underlying mechanisms and their implications for clinical outcomes.

Numerous methods have been developed to systematically determine the importance of the various parameters of a system (White et al. 2016; Snowden et al. 2017). For many systems, including biological systems, a small number of governing parameters have a decisive influence on the outcome of interest, while many other parameters have little discernible impact on the system's dynamics even if varied by orders of magnitude (Machta et al. 2013). Using methods such as perturbation theory and sensitivity analysis (Snowden et al. 2017), these governing parameters can be identified and selected as targets for deeper study and their influence on a system's output can be characterized. Using knowledge of how outcomes can feasibly be influenced, treatment approaches can be designed that use mechanisms such as feedback to ensure robustness to variation in parameters that are difficult to measure or control (Aström and Murray 2008). Models can be used to design experimental strategies for probing independent degrees of freedom, an important step in efficient parameter estimation. Furthermore, sequences of experiments can be designed so as to be complementary in the sense that distinct experiments reduce uncertainty along independent directions in the parameter space (Transtrum and Qiu 2016). The results of these efforts are simplified models that incorporate empirically measurable effective degrees of freedom and that give insight into the key features of the system that are relevant to its control.

In short, systems thinking in medicine has the potential to lead to organized sets of predictions across multiple disciplines. Rather than pursuing endless variations of the same population-level study in search of greater precision in aggregate estimates of effectiveness, researchers inspired to understand underlying mechanisms can devise compelling tests of hypotheses that lead to deeper and more useful insights applicable to individual patients.

39.3 Mechanismic Models Are Essential for Designing and Interpreting Population-Based Studies

Mechanismic hypotheses underlie causal inferences made using the results of population-based studies. This fact is often hidden from view in clinical research because mechanismic assumptions in that field are generally schematic, and rarely formulated mathematically. Indeed, the use of mathematics is typically limited to statistical analysis, though statistical concepts alone are insufficient to express causal hypotheses (Pearl et al. 2016). The reliance on statistics alone for analyzing population-based research has left the explication of causal notions to informal intuition. This is despite the fact that several approaches have been developed to clarify mechanismic notions in population-based research.

For example, graphical modeling affords a general and flexible approach for describing the causal structure of systems investigated with population-based methods. This approach, pioneered by Wright (1921) and later elaborated by Pearl (2009) and others, uses a graph to represent the putative causal relationships between variables affecting the outcomes of interest: the nodes of the graph represent causal factors, and the edges represent relationships of influence between them. Graphs provide a non-parametric representation of the causal relationships among the variables that influence an outcome. Mathematical methods for analyzing the graph structure then allow for studying the effects of perturbations to the network of connections between the variables. This modeling approach provides a precise and flexible, though minimalist, framework for characterizing the causal relationships investigated in clinical research, where the microscopic details of many mechanisms or the quantitative aspects of macroscopic relationships may be poorly understood (Pearl et al. 2016).

Evidence-based-medicine (EBM) promotes the idea that research can be evaluated more or less mechanically by an assessment of a study's design and execution without much need for assumptions. The quality of evidence of a particular study is specified by a hierarchy whose rankings are determined by study design. Near the top of this hierarchy are randomized controlled trials, a study design superseded only by pooled analyses (*meta-analyses*) of multiple RCTs. RCTs are population-based studies in which participants recruited from a target population are randomly allocated either to a treatment group or to a control group. Both patients and clinicians are ideally oblivious (*blinded*) to their group assignment in order to militate against the possibility that differences in care or evaluation will accumulate that do not flow from the intervention itself.

Because of the random assignment of subjects to exposure groups, the effects of both known and unknown factors on the outcome will on average be balanced across exposure groups, so that any difference noted in the average outcomes between groups can be attributed to the treatment. Meta-analyses pool together the results of individual RCTs to reduce through replication the random error due to chance fluctuations in the balance across exposure groups of causal factors other than the treatment of interest. Next in the hierarchy comes individual randomized controlled trials, whose status is lower than that of meta-analyses because their lone findings are in need of confirmation through replication.

Observational studies of various designs occupy a still lower status, because the procedure for accounting for confounders is not automatically achieved through deliberate randomization as in an RCT, but instead depends on specific assumptions about which factors besides the treatment can influence the outcome. The virtue of randomization is that it allows interventions to be assessed even in the presence of poorly understood or poorly controllable factors that also affect the outcome. Even in observational studies where deliberate randomization of treatment assignment is not feasible, researchers often seek circumstances that approximate those of a randomized experiment.

For example, in clinical genetics, an observational study design called 'Mendelian randomization' makes use of the random allocation of genetic variants during meiosis to study the effects of genetic endowment, using a logic similar to that of analysing RCTs (Davey Smith and Hemani 2014). In the absence of a 'natural experiment' in which randomization occurs without intervention, researchers resort to models that of necessity depend on hypotheses about the role of confounding and mediating factors that affect the outcome. Such hypotheses themselves require independent testing, and complicate the pooling of observational results. For this reason, observational studies are regarded as less compelling than RCTs.

Finally, the lowest position in the hierarchy is occupied by expert opinion and mechanistic justifications. Thus, EBM's process of grading evidence according to study design directly mirrors the perceived dependence of various study designs on mechanistic hypotheses, with greater perceived dependence constituting a liability. Since randomization is felt to obviate the need for detailed mechanistic modeling, RCTs and their meta-analyses are placed at the pinnacle of the hierarchy.

Mechanistic models also have an essential role in correcting for random and systematic errors arising in population-based studies. There are several ways an RCT can fail both prior to and following randomization. Prior to randomization, an RCT can fail to be representative of the target population of interest, because the process of selecting individuals for enrolment to the trial is non-random. For example, patients may choose to enter an RCT because they believe they stand to benefit from enrolment. Because the enrolment process is selective rather than random, the subjects enrolled in an RCT may be relevantly different from the patients to whom the results of the study are meant to apply. Post-randomization factors that can compromise an RCT include failure of blinding to treatment allocation, crossover between treatment groups, and missing data due to dropout from the study. Models of the data generation process are a necessary part of analyzing the impact of pre-randomization sampling bias and post-randomization distortions in treatment or measurement.

The randomized allocation of treatment is a powerful technique for population-based studies of systems in which unknown, unmeasurable, or uncontrollable parameters can influence the outcome of interest. Randomized controlled trials (RCTs) have the virtue that, even in the absence of knowledge of all the factors that determine an outcome, or the ability to measure or control known factors in individual subjects, estimates of the average effects of treatment can nevertheless be obtained from a population-level comparison of the responses of subjects randomly assigned either the treatment of interest or a control condition (Cartwright and Deaton 2017).

Although randomization enables the study of systems for which detailed mechanismic models are not available or for which parameters are difficult to measure and control, randomization is no substitute for mechanismic modelling (Bunge 2013). Indeed, the choice of basic design parameters for an RCT, such as treatment dosing and sample size depends on hypotheses about the individual response to treatment, as well as of the data generation process. Moreover, modelling can greatly improve the efficiency of a randomized study. In a block randomized experiment, for example, sampling is stratified according to the values of variables known or hypothesized to be important. This improves efficiency and statistical power when compared to fully randomized experiments, and reduces heterogeneity in the response to treatment, albeit at the expense of administrative complexity (Imai et al. 2008).

Randomization is successful only on average at removing the effects of unmeasured or uncontrollable parameters. In any particular experiment, randomization can fail (Cartwright and Deaton 2017). Moreover, systematic bias can affect the reliability of replication of RCTs. Positive findings are preferentially reported, and negative findings underreported (Ioannidis 2005; Dwan et al. 2008). This publication bias in turn affects the quality of meta-analyses of the primary research, though the magnitude of this bias is often difficult to discern, even with the help of specialized theoretical methods (Murad et al. 2014).

Checking the success of randomization by assessing the balance of baseline characteristics between treatment and control group can be carried out only for those parameters that have been measured, which may give only a partial picture of the degree of balance in an RCT. In the absence of a well-validated mechanismic model, there may be important parameters that remain unmeasured, and for which imbalance between treatment and control groups will pass unnoticed in a given experiment, despite the fact that on average over multiple trials these parameters will be balanced. The more complete an understanding of the underlying mechanisms, the more confident one can be ex post about the degree to which important covariates have been balanced by randomization.

Models of the data generation process are equally important in observational studies. Although in observational studies, treatments are not typically allocated at random, sampling is more representative of the target population than for RCTs. Despite this, observational data may well be collected or represented differently in the various treatment groups, a fact that can distort causal inferences. Mechanismic models are helpful in accounting for missing data, and in clarifying the additional assumptions that need to be checked in order for effect estimates from observational studies to be reliable.

In observational studies, mechanismic assumptions are needed to address the problem of confounding in order to estimate causal effects of the exposure of interest. For example, in an instrumental-variable study, variables (called instruments) are sought that are correlated with the exposure of interest but are otherwise without connection to the outcome, either directly or through confounders. The correlations between the exposure and instruments, and the outcome and instruments, are then measured and compared to one another to obtain an estimate of the effect of exposure on outcome. The dissociation between the instrument and possible confounders needs independent confirmation, a task that is possible only for known confounders. Methods for obtaining quantitative effect estimates require additional assumptions, such as linearity in the relationship between exposure and outcome (Angrist and Pischke 2009). The various assumptions necessary for causal inference are mechanismic in nature; and need to be independently tested. Observational studies, including post-hoc subgroup analyses of RCTs, are vulnerable to being "hacked": data-sets can be analysed in multiple ways, with various rules for treating the data, until a statistically significant, though possibly coincidental, association between variables is found and reported.

The design of population-based research, whether experimental or observational, presupposes mechanismic hypotheses that require independent empirical verification. The stronger the hypotheses, the more specific the predictions. Thus, a hypothesis of linearity in the relationship between variables may be necessary for quantitative estimates of effect size. Identifying these hypotheses, spelling them out explicitly with the help of mathematical modelling, analyzing their implications, and then testing these hypotheses against data, are all essential steps in clinical research.

The results of both RCTs and observational studies require explanation by appeal to mechanisms, especially when these results appear to disagree. It should not be assumed, as most guidelines do, that the results of RCTs somehow supersede those of the observational studies with which they are inconsistent. Instead, the same mechanismic model should explain the findings of both kinds of study. A mechanismic perspective on the data generation process reveals that experimental and observational studies have complementary strengths and limitations. Although RCTs allow for unbiased assessments of the efficacy of treatments in a sample, they cannot generally involve representative sampling of subjects (Imai et al. 2008; King et al. 2011), and can be inefficient at studying the long-term adverse of effects of treatment. By contrast, observational studies generally include a more representative sampling of the target population, and are useful for tracking adverse effects, but may be more vulnerable to distortion by confounding in their assessment of treatment efficacy.

Mechanismic models are essential for extrapolating results generated in clinical trials to the varied circumstances encountered in practice; and are frequently advanced in the context of making predictions regarding individual cases. Extrapolation from RCTs is generally required since study conditions are often very different from the conditions that prevail in the clinician's office. Comorbidities, drug-drug interactions, and physiological variation such as could be due to difference in gender and body size, are all examples of factors that can alter the expected risks and benefits of a treatment. Predicting how these factors will alter the balance of risk and benefit is not primarily a problem of statistical extrapolation, but of understanding the relevant mechanisms at work: if a comorbid disease affects the kidneys, and the treatment in question is eliminated by this organ, then the effective dose of the treatment may be increased in patients with renal disease. If a concomitantly administered drug induces hepatic metabolism of other substances, including the treatment, then exposure to that drug may reduce the effective treatment dose.

In short, there is no substitute for understanding the underlying mechanisms in order to address the effect of novel circumstances. And once the relevant mechanisms are understood, appropriate corrective steps, such as those prompted by negative feedback, can be designed in order to ensure that the treatment remains effective even under novel circumstances.

The EBM approach may be a helpful heuristic for busy clinicians interested in knowing what has been proven from the grab-bag of approaches accreted over time. Indeed, if most practices are ineffective, and the mechanisms of most diseases and remedies poorly understood, EBM and its hierarchy will decimate them quickly without expending much effort on chasing observational trends or convenient speculative hypotheses. Though this function can be helpful, as a general framework for clinical research, it is primarily critical rather than constructive, and as such can lead to derivative, repetitive research. It is better to model the data generation process explicitly, and in so doing address the problems selection bias, confounding, mediation by intermediate variables, post-treatment bias, and extrapolation to novel circumstances (Bareinboim and Pearl 2016). It is best to actually understand the mechanisms at work in the system under study, and then find ways of controlling the system that are provably effective under the range of circumstances encountered in practice.

39.4 Mechanismic Models Guide the Application of Research Findings to Individual Cases

A central challenge in using population-based clinical research is to understand how this research relates to individual cases. One of the reasons population-based research is appealing is that, even if the underlying mechanisms that generate individual responses are not well-understood, likelihoods for probabilistic outcomes can be inferred from measuring the relative frequencies of various events in data collections from similar individuals. Although mechanismic insights can sometimes connect probability estimates to observable quantities other than relative frequencies (such as, for example, to power spectra), this situation is uncommon in medicine. Estimating the probability of an outcome is clearly not possible just by observing the actual outcome in a given individual. A patient may have benefited from a risk-modifying treatment and yet may still be unlucky and develop the disease. Another patient may have upon treatment experienced no alteration in their risk of disease, and yet by chance remain healthy. Prior knowledge might point to surrogate biological markers that are correlated with probabilities of outcomes, but unless we know the underlying mechanisms of the disease and of the treatment's effect, there is no way of knowing to what extent treatments that influence these markers actually influence risks.

For example, some treatments may have clearly beneficial effects, but with little effect on known markers. This can happen if the treatment influences risk through multiple mechanisms, only one of which affects the marker. Other treatments can have major effects on markers, without also producing obvious benefit. This can happen if, for example, there is a window of opportunity for a treatment to be effective for a condition; after this window closes, the marker may be influenced by the treatment, without a corresponding effect on disease risk. In both cases, understanding the mechanisms connecting the treatment to both the marker and the outcome is important for using the marker as a measure of treatment effects.

Ignorance about whether a process is probabilistic, however, is not a sufficient basis for treating individual events as random. If the dynamics of the individual treatment response are not random but instead deterministic, there are no grounds for assigning probabilities to individual events (Bunge and Mahner 2001). If the process is in fact random, then an estimate of the single-event probability can be inferred from multiple replicates of the process. Moreover, the indeterminacy of the process gives rise to uncertainty in the outcome, so that probabilities can in fact quantify uncertainty in this case. This issue is commonly misunderstood: since an unexplained variability in clinical trials results, whatever the source, is often treated as grounds for regarding the individual response as random.

In population-based research, probability theory naturally enters into the design and interpretation of studies through several routes: (i) random variation in the characteristics of patients selected for inclusion in the study; (ii) random assignment of the treatment to individuals; (iii) measurement error; and (iv) randomness in the characteristics of the treatment delivered or of effects of treatment on an individual patient. Only the last source of variation, inherent randomness in the treatment delivered or the response dynamics, warrants regarding the individual response to treatment as random.

Many clinicians assume that summary results of population-based studies apply to individual cases, so that if a trial overall shows net benefit, and an individual meets the enrolment criteria for the trial, then that individual is also likely to benefit from the treatment. In reality, inferences about the relationship between individual responses to treatment and the summary results of a population-based study depend strongly on hypotheses about a treatment's mechanism of action.

The net benefit of treatment can vary tremendously within a study population, such that the average treatment effect for the study can show net benefit, whereas most patients, and indeed the typical patient, can suffer net harm (Kent and Hayward 2007; Varadhan et al. 2013). This is because the distribution of risks for the outcome can be skewed such that a small proportion of subjects are high risk and accrue benefits from treatment, which yields a measured overall benefit for the sample. If the relative risk reduction is constant at varying absolute risk levels, then as the risk for the outcome increases, so does the benefit of treatment. If the risk of harm from treatment stays roughly constant across risk levels, the average treatment response can show net benefit, even if many individuals will be harmed from the treatment.

The balance of risks and benefits in a given individual can depend strongly on individual factors, so that even if on average there is benefit to an intervention, a given treatment can be non-beneficial or even harmful for a given individual because of circumstances, such as comorbidities, that apply in that particular instance. The degree of heterogeneity in treatment effects among enrolled individuals is often either unknown or unreported. Consequently, as discussed later, the average treatment effects used to summarize trial results may be very different from or even opposite in direction to the net effects on typical subjects, even after accounting for random error (Kent and Hayward 2007; Cartwright and Deaton 2017).

The unsettling truth is that there are clear limits to what can be inferred from population-based studies about individual responses to interventions. We may not know whether the underlying process at the level of the individual is random. Even with this knowledge, the overall observed balance of benefit at the population level may not apply to all or even most individuals. And the many factors that distinguish patients in a study from those in clinic may mean the results of the study cannot be extrapolated to the case at hand. To overcome this impasse, scientists and clinicians must look deeper, and with the right tools to understand the dynamics of complex systems: those of modern applied mathematics, the factual sciences, and engineering, to unravel and control the mechanisms of health and disease.

39.5   Conclusion

The core idea of Evidence-Based Medicine (EBM) – that of subjecting received medical dogma to rigorous empirical testing – has led to major improvements in clinical practice over the past three decades. The EBM movement has advocated for having clear and exacting standards in the conduct, analysis, and reporting of clinical research. It has promoted statistical literacy among medical trainees and practicing physicians, and it has motivated many clinicians to adopt a spirit of healthy scepticism, and to think critically about the scientific basis for their practices. Above all, it has made clear the growing need for practice guidelines that harness clinical research to advance patient care. As a result of widespread adoption of the EBM paradigm, many outdated approaches to treatment, supported by little more than anecdote and tradition, have been supplanted by systematically developed, evidence-based recommendations validated through explicit confrontation with data.

The EBM paradigm may well be a useful heuristic in circumstances where most clinical practices are useless, most observational research findings are spurious, and most pathophysiological processes are poorly understood. EBM's hierarchy of evidence can protect patients by using RCTs to root out ineffective or dangerous practices, and by setting a high bar for potentially speculative research to be taken seriously.

However, a philosophy of clinical research for the purpose of quality control is not the same as one that encourages creative research and radical innovation. Much of medicine remains semi-scientific (Bunge 2013), in the sense that there is limited knowledge of the mechanisms underlying important clinical outcomes. It is hard to see how EBM's philosophy, with its fixation on RCTs, will help to overcome this.

Mechanistic modeling is a difficult but necessary step in making medicine scientifically rigorous. It requires expertise in other scientific and technological domains besides clinical medicine, and mathematical disciplines other than statistics. Indeed, to the extent that statistical reasoning is the center-piece of medical decision-making for individuals, and to extent that the justifications for clinical decisions are based on the summary results of population-based research rather than an understanding of underlying mechanisms, clinical practice amounts to a form of gambling, and an impersonal one at that. This is not a disparagement but a statement of fact about the semi-scientific status of modern medicine. In medicine, as in other spheres of life, even if we do not always have a scientific understanding of mechanisms upon which to base our decisions, we still are often forced to decide.

Even though the form such gambling takes appears to follow some system and occasionally uses mathematics, it falls short of being scientific. This would be obvious to most people in other contexts: pilots do not fly us from Los Angeles to Tokyo by averaging across successful flight plans for previous journeys in other seasons (however meticulously documented), closing their eyes, and hoping for the best. The need for mechanistic understanding in order to enable responsiveness to unexpected circumstances should be equally obvious in medicine. Yet the seeming precision of evidence-based evaluation in clinical research contributes to the reflexive, automatic application of guidelines in clinical practice (Greenhalgh et al. 2014). Despite their limitations, guidelines often make specific and categorical recommendations where room for disagreement or for alternative approaches would be reasonable and should be spelled out.

Ironically, the clinical practice guidelines that constitute the culmination of a purportedly evidence-based approach, amount to expert recommendations vulnerable to influence by financial conflicts of interest, and to differences in expert opinion on the ranking of evidence or weighing of risks and benefits. Unmoored as they are from mechanistic knowledge, the presence of tacit elements of subjectivity and expert opinion in clinical guidelines should not be surprising; and statistics can cloak them only so long as we do not insist upon guidelines based on a deeper understanding of mechanisms.

There are limits to what can be achieved even with well-validated mechanismic models of disease. Many decisions in medicine involve weighing quality of life against some chance of increased longevity. Personal decisions of this kind are difficult to make for a person who has little experience with the hardships of the illness he or she is about to face. How should a person decide whether to suffer through the ordeal of chemotherapy in the hope of living a few months longer with metastatic lung cancer? How should a patient with end-stage heart failure or cirrhosis decide whether palliation is more appropriate for her than aggressive interventions? The source of uncertainty here is neither ignorance of underlying causes nor the indeterminacy in the underlying mechanism, but in a person's own values. Although a model might be able to accurately predict the risks and benefits of various courses of action, simple calculations based on their outputs may be of little use for decision-making without first deciding upon what should be achieved.

Physicians have a responsibility to help patients understand the limits of what is known by modern medicine, and what is knowable in individual cases. The physician's role involves not just offering technical expertise but also compassion and advocacy. Rather than reflexively following clinical guidelines, they should listen to their patients with empathy and help them to discover what is most important for them in order to make personalized choices informed not by guidelines based on actuarial tables, but by a scientific understanding of the mechanisms of disease.

References

Angrist, J. D., & Pischke, J. S. (2009). *Mostly harmless econometrics: An Empiricist's companion.* Princeton: Princeton University Press.

Aström, K. J., & Murray, R. M. (2008). *Feedback systems: An introduction for scientists and engineers.* Princeton: Princeton University Press.

Bareinboim, E., & Pearl, J. (2016). Causal inference and the data-fusion problem. *Proceedings of the National Academy of Sciences USA, 113*(27), 7345–7352.

Bunge, M. (2004). How does it work?: The search for explanatory mechanisms. *Philosophy of the Social Sciences, 34*(2), 182–210.

Bunge, M. (2013). *Medical philosophy: Conceptual issues in medicine.* Hackensack: World Scientific Publishing Company.

Bunge, M. (2017). *Philosophy of science. Volume 2: From explanation to justification.* New York: Routledge.

Bunge, M., & Mahner, M. (2001). *Scientific realism: Selected essays of Mario Bunge.* Amherst: Prometheus Books.

Cartwright, N., & Deaton, A. (2017). Understanding and misunderstanding randomized controlled trials. *Social Science and Medicine.* https://doi.org/10.1016/j.socscimed.2017.12.005.

Davey Smith, G., & Hemani, G. (2014). Mendelian randomization: Genetic anchors for causal inference in epidemiological studies. *Human Molecular Genetics, 23*(R1), R89–R98. https://doi.org/10.1093/hmg/ddu328.

Dwan, K., Altman, D. G., Arnaiz, J. A., Bloom, J., Chan, A. W., Cronin, E., Decullier, E., Easterbrook, P. J., Von Elm, E., Gamble, C., Ghersi, D., Ioannidis, J. P., Simes, J., & Williamson, P. R. (2008). Systematic review of the empirical evidence of study publication bias and outcome reporting bias. *Public Library of Science One, 3*(8), e3081. https://doi.org/10.1371/journal.pone.0003081.

Eddy, D. M. (2005). Evidence-based medicine: A unified approach. *Health Affairs (Millwood), 24*(1), 9–17.

Greenhalgh, T., Howick, J., Maskrey, N., & Evidence-Based Medicine Renaissance Group. (2014). Evidence based medicine: A movement in crisis. *British Medical Journal, 348,* g3725. https://doi.org/10.1136/bmj.g3725.

Howick, J., Glasziou, P., & Aronson, J. K. (2010). Evidence-based mechanistic reasoning. *Journal of the Royal Society of Medicine, 103*(11), 433–441.

Imai, K., King, G., & Stuart, E. (2008). Misunderstandings among experimentalists and observationalists about causal inference. *Journal of the Royal Statistical Society, Series A, 171*(Part 2), 481–502.

Ioannidis, J. P. (2005). Why most published research findings are false. *Public Library of Science Medicine, 2*(8), e124. https://doi.org/10.1371/journal.pmed.0020124.

Ioannidis, J. P., Stuart, M. E., Brownlee, S., & Strite, S. A. (2017). How to survive the medical misinformation mess. *European Journal of Clinical Investigation, 47*(11), 795–802.

Kent, D. M., & Hayward, R. A. (2007). Limitations of applying summary results of clinical trials to individual patients: The need for risk stratification. *Journal of the American Medical Association, 298*(10), 1209–1212.

King, G., Nielsen, R., Coberley, C., Pope, J. E., & Wells, A. (2011). Avoiding randomization failure in program evaluation, with application to the Medicare Health Support program. *Population Health Management, 14*(Suppl 1), S11–S22. https://doi.org/10.1089/pop.2010.0074.

Lenzer, J., Hoffman, J. R., Furberg, C. D., Ioannidis, J. P., & Guideline Panel Review Working Group. (2013). Ensuring the integrity of clinical practice guidelines: A tool for protecting patients. *British Medical Journal, 347,* f5535. https://doi.org/10.1136/bmj.f5535.

Machta, B. B., Ricky Chachra, R., Mark, K., Transtrum, M. K., & Sethna, J. P. (2013). Parameter space compression underlies emergent theories and predictive models. *Science, 342*(6158), 604–607.

Murad, M. H., Montori, V. M., Ioannidis, J. P., Jaeschke, R., Devereaux, P. J., Prasad, K., Neumann, I., Carrasco-Labra, A., Agoritsas, T., Hatala, R., Meade, M. O., Wyer, P., Cook, D. J., & Guyatt, G. (2014). How to read a systematic review and meta-analysis and apply the results to patient care: Users' guides to the medical literature. *Journal of the American Medical Association, 312*(2), 171–179.

Pearl, J. (2009). *Causality: Models, reasoning and inference.* Cambridge: Cambridge University Press.

Pearl, J., Glymour, M., & Jewell, N. P. (2016). *Causal inference in statistics: A primer.* New York: Wiley.

Snowden, T. J., van der Graaf, P. H., & Tindall, M. J. (2017). Methods of model reduction for large-scale biological systems: A survey of current methods and trends. *Bulletin of Mathematical Biology, 79*(7), 1449–1486.

Transtrum, M. K., & Qiu, P. (2016). Bridging mechanistic and phenomenological models of complex biological systems. *Public Library of Science Computational Biology, 12*(5), e1004915. https://doi.org/10.1371/journal.pcbi.1004915.

Varadhan, R., Segal, J. B., Boyd, C. M., Wu, A. W., & Weiss, C. O. (2013). A framework for the analysis of heterogeneity of treatment effect in patient-centered outcomes research. *Journal of Clinical Epidemiology, 66*(8), 818–825.

White, A., Tolman, M., Thames, H. D., Withers, H. R., Mason, K. A., & Transtrum, M. K. (2016). The limitations of model-based experimental design and parameter estimation in sloppy systems. *Public Library of Science Computational Biology, 12,* e1005227. https://doi.org/10.1371/journal.pcbi.1005227.

Wright, S. (1921). Correlation and causation. *Journal of Agricultural Research, 20,* 557–585.

Chapter 40
Emergence, Systems and Technophilosophy

Byron Kaldis

> In general, systemic issues call for systemic and long-term
> solutions, not sectoral and near-sighted measures. This is the
> practical message of systemism
> Bunge (2004a, p. 190)

Abstract The chapter discusses the three intertwined notions central to Mario Bunge's thought, emergence, systems and mechanism. It draws lines that contact or diverge from his counterparts and wider uses in recent philosophical work. Bunge's status as an early pioneer of the systemic approach, emergence and mechanismic explanation means that his work may be fruitfully supplemented or complemented by these recent advances. The chapter then moves on to discuss Bunge's philosophy of technology as the additional central theme in his overall opus. His insistence on technology being inherently philosophical, as technophilosophy, and his related thesis, technoethics, are further analyzed and expanded upon. It is proposed that along with neurons, at the physical level, and what Bunge has dubbed "psychons" at the level of the mental, there may be a notion of "technon" at the emerging level of modern technological convergence.

40.1 Introduction

Mario Bunge, perhaps the last Renaissance mind of our times, has an impressive *palmarès* of achievements on display in a wide-ranging spectrum of humanistic and scientific fields. He can safely be regarded the philosophers' philosopher – or more precisely their scientific philosopher. Bunge has singlehandedly offered a wealth of philosophical, social-scientific as well as scientific work that is as thought-

B. Kaldis (✉)
Department of Humanities, The National Technical University of Athens, Athens, Greece

Moral Culture Research Institute, Hunan Normal University, Changsha, China
e-mail: bkaldis@central.ntua.gr

© Springer Nature Switzerland AG 2019
M. R. Matthews (ed.), *Mario Bunge: A Centenary Festschrift*,
https://doi.org/10.1007/978-3-030-16673-1_40

provoking as it is original. Spanning perhaps almost every subject in a really long and on-going career, his immense output has neither flagged nor failed to be of significance. His style, at once profound and stingingly witty, leaves no doubt that he has little time for foolish intellectual fads, university charlatans or pseudoscientific superficialities. One can almost hear the intelligent mechanism of his mind at work and the ironic undertones of his sustained assault against all forms of irrationality. Bunge's mind is in the right place and his head on his shoulders, as Joseph Agassi has quipped, and I would add that his heart is on the correct side of social and political issues and values.

Emergence, systems approach, mechanismic explanations,[1] and the inherent philosophical richness of technology have been central to Mario Bunge's thought and constant staples of his prolific output throughout his life. These notions, and what they are meant to serve, to illuminate or to explain (ontologically, epistemologically, and normatively), have been elaborated further by Bunge over the years. However, they retain a central and identifiable core in his own thought while at the same time they have played a central role in recent developments in several branches of philosophy. Emergence, along with complexity, is nowadays being discussed as central to a variety of problems in the philosophy of mind, in various applications of non-reductive physicalism, as well as in delimiting a separate kind of non-physically reducible special sciences – as well in life sciences as such. Mechanistic explanations of sorts have resurfaced in recent discussions primarily of social explanation and secondarily of social ontology, especially in what is called analytic sociology. It has also been developed on its own as a kind of philosophical analysis of mechanism with wide applicability to different scientific fields (e.g. Illari and Williamson 2012), suggesting talk of a "new mechanical philosophy" (Glennan 2017). Furthermore, moving on to our third ingredient, technology, some schools of contemporary philosophy of technology underline both the systemic character of the technological phenomenon as well as its inherent 'technoethics' (to use Bunge's term). Technology, unavoidably raises ethical concerns.

Despite these similarities in their respective agendas, and Bunge's status as one of the pioneers in all three fields, there are clear divergences between Bunge's treatment of these notions and that of his more recent counterparts. Bunge's world-view permeated by a pervasive, ever-present, web of systems leads him towards convergence and multidisciplinarity while some classic (British) and current views on emergence sometimes aim, among other things, at justifying special sciences (with epistemic distinctness and ontological autonomy).

This chapter discusses some facets of the notions of emergence and system underpinned by the third notion of mechanism, and then sees how they can help us enrich Bunge's view of technology. They are all central to Bunge's lifelong opus. Some similarities or divergences between his views and some contemporary ones expressed in philosophy, social sciences or agent modelling and computer-simulated approaches to cognition and system building (and its subfield network theory) will

---

[1] 'Mechnanismic' is the adjective preferred by Bunge distinguishing his theory from classical mechanistic or mechanical ones.

be developed. The unifying thread is the idea of mechanism: emergence and systems approach are underpinned by the idea of mechanism[2]: This is not to invoke systems theory – a phrase that, for Bunge, has the whiff of hidden holism. Emergence is the result of mechanisms operating in a non-linear fashion on complex composites while each system exhibits the mechanism that runs it.

Philosophy and especially philosophy of mind has begun using emergence to explain non-reductive physicalism. Analytical sociology has prioritized the use of mechanism as explaining causal links between micro-parts and macro-processes in social action yielding totally surprising and novel results or non-identical to their micro-base ones. This latter goes back to the classic case-study of residential segregation studied by Thomas Schelling whereby the aggregate phenomenon resulting from the individual actions and beliefs of each actor surprisingly did not bear any resemblance to them. Computer modeling and artificial intelligence, especially contemporary brain–like or neuromorphic computing, also aims at modeling mechanisms of many kinds: social, mental, or of a network in urban planning or ecological population. Emergent complexity, systems and mechanisms thus intertwine.

Moving on to technology – the other major theme of Bunge's thought – the philosophical discussion of it begins by addressing two principal questions: (i) what is technology – how is it to be understood? how is the artifact to be analysed ontologically? And (ii) what are the ethical concerns raised by technology? Bunge has famously – and for me rightly – argued in favour of a technophilosphy (philosophy being intrinsic to technology) and – because of this – he has espoused what he has dubbed 'technoethics'. The chapter therefore will look also into Bunge's view of technology from the standpoint of the prior analysis of emergence, systems and mechanism and then go on to expand on his view by proposing what is to be called "*technon*" alongside what he had coined as "psychon".

40.2 Emergence, Complexity, Systems

As an adequate brush-stroke for present introductory purposes, emergence denotes the appearance of qualitatively novel features, or properties, at the macro level, after a certain process of composition exceeds a certain level of complexity. The crucial point for Bunge is that the novelty in question cannot be reduced to either features or qualities of the parts or compositional phases out of which it arose. This is an important feature of emergence also shared by how systems are defined. A system bears certain relations and exhibits certain attributes, or alteratively, certain propositions are true of it, that are not duplicated or re-iterated by its component

---

[2] A system is said to have a mechanism *essential* to it denoting its specific function; or the essential mechanism of a given system is the totality of processes that occur exclusively in that system and its conspecifics (Bunge 2004a, p. 193). In general, no concrete system lacks a mechanism, unlike conceptual or semiotic systems that may do.

parts or subsystems. So clearly, emergence and systems are analogues in this respect: emergence may be said to be the process culminating in a system as long as criteria such as permanence and reiteration are met.

Emergence has roots in the so-called British 'emergentists' of the past, such as J. S Mill, C. D Broad, Samuel Alexander among others.[3] Emergence comes in diverse forms, sometimes dubbed epistemological as opposed to ontological; or as weak versus strong with the two pairs coupling in various parallel combinations. Distinctions are also drawn between synchronic or a-temporal views of emergence, where there is no causal relationship between prior phases or lower levels of a certain process of evolving compositional complexity. Supervenience relations, the competitor notion to emergence, may be seen as similar to the synchronic version: the higher levels are simply realized by the lower one(s) so that the two levels are jointly metaphysically necessary. Alternatively, diachronic or strongly causal emergence may involve links between prior or lower phases/levels to the higher or later or emergent ones. As it will be shown below, a modern version of the latter, causal or dynamic emergence, must be espoused by Bunge and a specific contemporary version of ontological emergence may be suitable for Bunge's world view.

Consider first his standard definition:

To say that P is an emergent property of systems of kind K is short for 'P is a global [or collective or non-distributive] property of a system of kind K, none of whose components or precursors possesses P. (Bunge 2003a, p. 25 cf. 1977)

Any attempt to locate Bungean systemism and emergence with respect to its classical precursors' view of reality as hierarchical has to deal with the (inter)relationship between higher and lower levels of reality, specifically whether both mental and social cases admit downward-causation. Depending on how one reads some of Bunge's pronouncements it appears that he might side with J. S. Mill or C. D. Broad insofar as the mind is concerned. Emergent (systemic) qualities require separate higher-level causation, i.e. from higher-level to higher-level intra-linking. However, Bunge's views appear closer to Samuel Alexander's position when it comes to systemism as transcending ontological and methodological individualism and holism.

The minds and actions of individual persons remain the basic and only building blocks of social phenomena, for organizations do not have minds, Bunge repeatedly reminds us. However, there is a sense in which the individual receives pressure from the social (via the action of individuals), as Bunge admits at various places. This is mostly construed as the "environment" in his CESM[4] model, wherein the E of CESM constitutes an entity outside the internal ingredients of any given

---

[3]Though J. Kim mentions a reference that identifies Galen as perhaps the first to embrace the general idea of emergence (Kim 2006, p. 189). See also McLoughlin (2008).

[4]Standing for Composition, Environment, Structure Mechanism

system, ontologically speaking.[5] Of course, Alexander was not dealing with social emergence, and there is a sense in which the social or macro-properties that Bunge allows, when he applies systemism against individualism and holism in social science, are not really totally extraneous or exogenous vis à vis the individual micro-ones, as if those macro-properties run a separate course of their own relative to atomic ones. This is also borne by the logical notation in Bunge's more precise formulations of these theses in his earlier writings (the *Treatises*).

C. D. Broad's so-called "trans-ordinal laws" of higher-order, emergent, properties arising in a sequence of levels of substance in increasing complexity of orders or levels, connect adjacent orders or levels. Bunge insists on a similar principle whereby posterior (but not higher – for Bunge does not recognize superior levels) phases contain the previous ones as is summarized in a logical formulation (below). Bunge's view is more complex and nuanced, denying some of Broad's assumptions:

> All talk of interlevel action is elliptical or metaphorical, not literal. Second, the relation between levels is neither the part-whole relation nor the set inclusion relation but a *sui generis* relation definable in terms of the former. Third, there is nothing obscure about the notion of level precedence as long as one sticks to the above definition instead of construing '$L_i < L_j$' as "the L_i's are inferior to the L_j's" or in similar guise. Fourth, it is mistaken to call a level structure $\mathcal{L} = \langle L, < \rangle$ a *hierarchy*, because the level order $<$ is not a dominance relation. Fifth, our concept is so far static: we are not assuming anything about the origin or mode of composition of systems in terms of evolution. (Bunge 1979a, p. 14)

Such ambivalence is sometimes present in his writings but it can be dispelled if we strictly follow Bunge's analysis of systems. The CESM ingredients are sufficient to position Bunge's emergentism as of a *third* kind, due mainly to his attendant robust systems approach, missing from the early British emergentists. This can be seen in the following.

When Bunge talks about properties emerging, he refers to the appearance of novel entities, composite objects of a systemic kind. It can be said, in contemporary terminology, that Bunge's emergence is both epistemological and ontological. The more appropriate term to describe his view of actual scientific theories and disciplines is, of course, convergence but epistemological emergence can be retained for describing explanatory tasks, rather that whole scientific disciplines and how they relate to each other. Then, it is clear that a major part of explaining through Bunge's systems approach unavoidably involves a kind of epistemological emergence. Compared to recent standard treatments of emergence, Bunge's contribution is thus a robust notion of system and the protagonist role that it plays together with its environment as a co-starring partner in the causal unfolding of a system's life. While recent discussions debate whether emergent properties are involved in downward causation or whether they are mere epiphenomena, Bunge's version allows interaction with environment, including other systems, as a crucial

---

[5]The same point will concern us in the next section when we introduce the methodological credentials of analytical sociology's mechanisms: it is basically the same venerable problem of the whole impacting on the part.

component of what it means to be an emergent system. This is mostly ignored in contemporary discussions.

Michael Silberstein's version of so-called systemic (downward) causation is an ontological type of emergence can be proposed as one complementing Bunge's view with a systemic embeddedness. This version integrates all kinds of systems, physical, chemical, biological, mental, social so that, in an anti-individualist vein, the social and physical worlds of the individuals combined constitutes a single cognitive system (Silberstein 2006, p. 211). Silberstein points out that the mental can have causal powers as long as we do not call this downward causation but instead systemic causation (ibid., p 201). How this may be useful to Bunge's systemism will emerge in the next section.

Paul Humphrey provides a sui-generis yet very interesting variety of emergence as fusion (Humphreys 2008, pp. 111ff). He aims to describe a real physical process (as Bunge understands concrete systems) on the basis of a hierarchy of levels (unlike Bunge: see above[6]). However, his version can be used to fine-tune Bunge's view: a fused event need not be considered 'higher' but it results from having survived the destruction of the properties of its parts. The latter no longer exist since they have been fused, so all the causal burden is carried by the unified whole. On this understanding physical emergence as a dynamical process is unlike supervenience since the subvenient base properties do not exist as realizers any more.

However, this view only works if we do not worry about special sciences that are at once non-reductive while still retaining the physics of material life or neuronal activity. Nonetheless, if we accept that the base from which the system arose ceases to exist, then the real emergent novel entity may be allowed to impact on another such higher-level entity *without the latter emergent entity being expected to depend on a lower base of its own from which it could have arisen*. On this view, there must thus be a higher-level to higher-level causation. Now given my point about the special sciences just made, and notwithstanding Bunge's penchant for not losing individual human beings or building blocks (i.e. not considering them as annihilated or totally ceasing to exist in any compositional process towards higher complexity, as Humphrey does) emergence by fusion as systemic activity of a higher kind may be of use to the other major Bungean idea, i.e. disciplinary convergence.

Furthermore, emergence and how it is understood within a Bungean systemic approach affects how technology and mechanism are understood too. The advanced emerging technologies of our times can be regarded as a system of sorts. The different kinds of emerging technologies namely, bio-nano-info-cogno technologies converge into a complex system that exhibits qualitatively different emergent properties, compared to what it would be the case if those technologies remained unconnected. Some of these properties are clearly social and ethical and so

---

[6]In his (1979a, p. 29), Bunge makes clear that the precursors of a system giving rise to an emergent entity do not blend but "keep their individuality to some extent". The latter qualification is obviously unclear.

Bunge's insight of philosophy at the heart of modern technology becomes plausibly justifiable.

This is one of the main points made in this paper. Modern (converging) technology, seen as a system in congruence with the overall Bungean systemic thinking, must exhibit qualitatively novel features. Analysis of these qualitatively new social and ethical features, will follow different paths, depending on the kind of emergence adopted. Ethical and social concerns may be features of parts of the system, individually in the separate technologies, and therefore only epistemically distinct. Alternatively, they may be seen as features of a complete system of totally converged technology, and therefore these ethical or social concerns are ontologically robust and entirely novel.[7] In either case the crucial point is that the philosophical or ethical kernel of technology asserted by Bunge – that technology is not nuts and bolts – is now safeguarded by the systemic nature of currently converging technology. So, emergence is pivotal: how it is to be construed affects what we do with it and that affects mechanism, the other ingredient in our analysis.

Mechanismic explanations depend on what kind of emergence is espoused. An Alexander-like construal suggests that emergence is a brute fact that cannot be explained: totally novel and unpredictable. On this construal emergent properties may be discoverable through mechanismic explanations, or by analytical sociological attempts to explain social facts and processes or by means of agent-modeling that reveal a hidden mechanism in various fields. This modeling may appear to depend on luck, but it requires special attention to the explanatory value of the particular mechanism and the models employed.[8] If modeling of bottom-up computation of increasing complexity leads our analysis from the micro to the macro and reveals qualitatively new kinds of properties, then it is worth pursuing it within the context of systemic approaches. However if no really *qualitatively* novel macro- or system-properties result after increasing degrees of computational complexity are run over by the model, then running such computational models will not help us gauge the explanatory power of the underlying mechanism they are targeted to reveal.[9] They will not help us judge the difference between supervenience and emergence in such cases, either. This problem can plague analytical sociology's modelling of social facts or contemporary brain-like computing simulating mental processes arising from physical substrates or complex processing by myriads of individual micro-parts, like neurons.

The following discussion of mechanism contains both parallels with and divergences from contemporary accounts of mechanism found in the relatively new

---

[7]This is for instance something more obviously appreciated, i.e. the totally non-predicable novelty of ethical issues arising in the case of human enhancement and transhumanism as I have argued in several places, more recently Kaldis (2018)

[8]Work is being done on this and guarded satisfaction with modeling and mechanisms leads to fine-tuning that is a welcoming balance to the opposite hype.

[9]This will be the case if they are simply higher level properties of the same qualitative kind, that is aggregative sums of the same kind extensionally understood, or even if intensionally different, but not radically new or entirely unpredictable properties.

social-scientific approach of analytical sociology.[10] It also introduces computer modeling as an agent-based modelling approach to explication of aggregate facts that are systemic or emergent.

40.3 Systemism and Mechanismic Explanation

Bunge advertised his systemism as a third way beyond the venerable opposition between social holism and methodological individualism in the philosophy of the social sciences.[11] In addition, his mechanismic explanation is an alternative to recent analytical sociology which has elevated mechanisms (in particular micro-mechanisms and micro-foundations) to the status of primary explanatory tools while remaining firmly individualistic in its philosophical orientation. Mechanisms and in particular micro-mechanisms can now be realized by computer modelling as in multi-agent or individual agent modelling. A so-called "mechanical philosophy" is supposedly emerging. If modelling is crucial to deciphering micro-mechanisms then we must understand the former before we resort to the latter. Especially so when what we are in fact doing is explaining. The upshot of this discussion is that Bunge's systemism and mechanismic explanation can be supplemented by models and how they have been used especially in recent discussions in the philosophy of mind.

There are a variety of different but kindred definitions of mechanism in a variety of disciplines. Bunge's (2004a) preferred one is "a mechanism is a process in a concrete system" However they are defined, mechanisms are seen as involving models. They are also understood as offering explanations that bring concealed entities to light, constituting a connected series of steps forming a process explaining *how* an outcome has come about. All proponents of mechanismic explanation wish to distance themselves from covering-law models of scientific explanation or statistical correlations and so this aligns with Bunge's view. I would suggest that the earliest classic account of this may be seen in mid-twentieth century analytic philosophy of history. The latter repudiated Hempelian covering-law models in favour of step-wise explanations of how a concatenation of historical events are connected in order to bring about a certain outcome, thus offering an explanation-how. A similar path appears explicitly and systematically for the first time for social science in analytical sociology. The recent proliferation of mechanism-based explanations has produced a spate of empirical case-studies and specific explanatory research outcomes.

---

[10] A dissenting voice is that of Pierre Demeleunaere who in his Introduction to his edited volume rejects the idea that mechanism-based explanations necessarily involve macro-entities which are to be considered higher or emergent or robustly systemic so that they obey the tenets of emergence: i.e. robust novelty that cannot be explained by constituent parts (Demeleunaere 2011, pp. 23–24).

[11] On early social ontology in the wake of systems theory, see Bunge (1974); systemism is a later version building on this.

Mechanisms are put forward as explanatory vehicles in system-building or in explaining emerging social phenomena arising from individual or micro-action. Then models are put forward as realizations of mechanisms.[12] How useful can adequate modelling be for explaining mechanisms that do the further explaining? Most of the illuminating work done on such modelling has been carried out in the philosophy of mind and it applies to analytical sociology too. Problems with modelling that we detect where it has most been used will therefore expose the strength and weaknesses of mechanisms before Bungean mechanismic explanation can be supplemented by such approaches as is proposed here.

The anti-reductionist idea of special sciences and the notion of mechanism links with the discussion of emergence in the previous sections. First, the individualist-holist or collectivist debate impacts the status of the social sciences. Bunge's *systemism* is an attempt to go beyond this debate as well as beyond reductionism: it is a *synergetics*. Social sciences study social systems and the latter are concrete things (Bunge 2000, p. 155). Bunge (1979b, 2004a, b) provides a full-blown exposition and defence of systemism against holism and individualism. Second, the notion of *mechanism* and its contemporary realizations in the form of modelling may provide the unifying thread that links up all sciences. Fodor introduced his well-known schema regarding special laws for the special sciences, such as psychology or biology. Social sciences might still count as special because of their separate subject matter.[13] Such status might not survive the introduction of a unified account of mechanism or in case models like multi-agent modelling or neural computations are applied equally to the social or the mental or the biological.[14]

Such use of models added to Bunge's view of mechanismic explanations is pivotal in redefining what is special about the special sciences and what, at the same time, unifies them after all. What could be said about the models that shoulder the burden of actual design and implementation of mechanisms? Can they be general or specific? Bunge says that they can sometimes be both. Can they be abstract or concrete? Is something vital lost if abstractions or generalities are introduced? Or are such things inevitable since when modelling certain variables are silenced as unimportant?

[12]For a proposed taxonomy of mechanisms into kinds that significantly alters how the New Mechanism or New Mechanical Philosophy is to be understood see Levy (2013). For an overview of mechanisms used in recent social-scientific explanation see Ylikoski (2018). For Bunge's canonical view on explanation and mechanism see Bunge (1997).

[13]Strictly speaking their special subject matter involves in the classic Fodorian sense the existence of special kinds (e.g. social, like markets and their functioning) connected by means of lawful regularities which can be bona fide special laws of that separate science due to the multiple physical realizability of these special (higher) kinds by their basal physical causal links – i.e. same special effect (economic or psychological) but different physical ways to achieve it. Hence the non-reducibility of special properties or entities to the physical ones underpinning them, i.e. there exists a type non-identity.

[14]The homogeneous use of mechanisms (along with the attendant systemic paradigm) ensures what Bunge preaches as convergence. It is these two major instruments that build converging patterns of scientific disciplines.

However, it is crucial to realize that talking about models and mechanisms introduces us surreptitiously to technology, for these are nothing but 'technemes', opening space for an integrated philosophy of technology within Bunge's systemism.[15]

Models in general, and computer-simulated models in particular, have been used widely in various disciplines to assist discovery of explanatory mechanisms for certain composite phenomena. Agent-based models are sometimes called "generative models" because they are meant to provide a precise, revelatory account, by duplication, of the process that generates a certain outcome. This is not far from Bunge's view.

Yet Bunge seems to shrink from all sorts of modelling by misconstruing or explicitly misdescribing it. Here is a classic example, less about computer than about mathematical modelling: "mathematical 'catastrophes' are singularities in manifolds, not social disasters; mathematical 'chaos' is the complexity involved in certain nonlinear differential equations; and the 'systems' that dynamical systems theory deals with are not concrete systems but systems of ordinary differential equations" Bunge (2004b, p. 378). However, this is not what the proponents of these modelling would claim.

Mark Bedau has argued in favour of weak emergence that unavoidably requires a specific kind of modelling as an explanatory route of phenomena. The basic idea is that only by tracing step by step all the nodes of a complex net by means of computer simulation models, can we follow the path that generated the explanandum outcome and, furthermore, such weakly emergent states can only be predicted if modelled in detail (Bedau 2008, p. 161ff). Such weak emergence requires computer simulations that both explain natural emerging phenomena but also reveal surprising mechanisms or what he calls incompressible generative explanations (Bedau 2010, pp. 51ff) Shorter explanations won't do; they betray any allegiance to realistic representations of natural dynamic processes by taking short-cuts. "Crawling the micro-causal web" is what guarantees generative explanations and this can be done only by computer simulation models. Such computer simulation models are therefore significant because they offer a clearer view of what Bunge is after. The behaviour we capture in computer modelling results from massively parallel populations at the micro-level, where autonomous agents interact independently with other such agents and with their environment (Bedau, ibid., p. 54) – a picture

---

[15]Bunge's criticism of computationalism in the past is out of date I think. And so is its anti-neural networks position. Both can safely be ignored in our attempt at a synthesis. Contemporary attempts at brain-like computing would meet with his approval. See for some fruitful cases illustrating such novel type of work e.g. on biologically motivated computer vision Bülthoff et al. (2002), or on brain informatics Yao et al. (2010), on computer vision Cipolla et al. (2013). These are all cases where computer programs carry the burden of revealing the underlying mechanisms. In the past Bunge has attacked artificial life (in its strong version) and has claimed that in the case of computer simulations we must distinguish between program simulation and process simulated, an imitation from what is imitated, and borrows the formulation that however life-like a simulation becomes after a certain degree of perfection it does not become a realization of life itself (Mahner and Bunge 1997, p. 152) Of course brain-like computing does not claim to realize life if that is meant to hold as a type identity.

that Bunge wishes to capture with his notion of "systemism". It is impossible to predict any future state of such a complex intertwining web of actions unless you use a model to go over each and every actual step. This final state is impossible to derive, either mathematically or deductively or game-theoretically by looking at prior steps.

Whether this is consistent with reductionism, or Bunge's own systemism, is important for the philosophy of mechanism and the technological means afforded by computer simulation. This is important for one of this chapter's subthemes: assertion of full-blown reductionism puts the autonomy of the special sciences in jeopardy. A classic dichotomy once dominant in the philosophy of mind claimed that mechanismic explanations, inviting reductionism, must be distinguished from functional explanations suitable for mental phenomena. This is also crucial for Bunge's systemism. Recent work has cast doubts on this dichotomy. The classic gambit against mechanismic explanations was that a higher property, entity or higher-level composite thing can be multiply realizable and hence pinpointing its underlying mechanism could only be showing its token-identity to a nomologically necessary but metaphysically contingent subvenient base of lower level relata. However, this brings us back to emergence. Recent work has shown that the distinctiveness of functional explanations is not sustainable (Kaplan 2017 and his Introduction to the same volume pp. 17ff). A weak sense of reductionism is compatible with autonomy as long as the organization of the component parts is seen as imposed on them (see Bechtel 2007). I contend that this requires an unacknowledged system's view and Bunge said as much since the 1970s.

Even earlier, Bunge indicated the importance of the way we need to characterize mechanisms if they are to do the required explanatory work. A mathematical formula describing a phenomenon, being applicable to data or even predicting future behavior of a system may be an ad hoc device tracing a curve on which we have plotted a system's series of phases. It does not really capture the explanatory task we desire. Mechanisms on the other hand do this to the extent that they are not simply descriptive or phenomenological, or "black boxes" (Bunge 1964). Such approaches describe the units of a phenomenon/system but they are devoid of structure. They simply summarize data, or input-output system processes, without showing how a phenomenon works. Agent modeling avoids this: it is not simply a mathematical formula capturing dots on a curve in space (see also Kaplan 2011).

Most social life consists of complex adaptive systems that have been shown to be explicable through computer agent-based modelling. If Bunge's vision of ever present systems is to gain traction beyond a general manifesto it must be supplemented by such computational methods that are appropriate to systems dynamics. Dynamic equilibria in certain systems (aggregates or macro-entities) or in complex (chaotic) structures arising from the interaction of individual agents/parts can better be explained following agent-based computer modelling.[16]

---

[16]A case in point is this: "In most networks, however, there are multiple routes between two nodes, and thus the probability increases with each additional path. This increase is roughly additive if the

Assessing and replying to certain criticisms against such agent-based modelling helps set misunderstandings straight while also refining such modelling. One of the criticisms that such computer modelling is limited by programming that embeds assumptions regarding the population aggregate of a certain social interaction can be answered. Such modelling often produces dynamic and surprising aggregate results that are not what is expected by mere intuition and consequently they are rich in explanatory relevance. Further, programming exposes hidden assumptions about the system that might otherwise mislead.

Another criticism is that computational models are overly complex compared to alternative methods and owing to their being sensitive to wired-in numerical values in a given parameters' set, they may gain in immense computational power (between individual actors and outcomes, say) at the expense of concealing the actual causation: i.e. the mechanism that was meant to explain is instead lost sight of. That is, a mesh of correlations obscures causal patterns. The solution offered is simplifying these models so as not to extinguish their illuminating power of causal connections but then the problem is how to calibrate such simplicity (see Macy et al. in Demeleunaere 2011, pp. 250ff).

Bunge cannot emphasize enough the use of mechanismic explanations:

> The days of phenomenalism and descriptivism are over in science ... the reason [we need mechanisms] is that most of reality is unobservable", i.e. mere description or denying the need to endorse unobservables (i.e. what makes mechanisms tick) bars one from truly scientific knowledge of the real world (physical or social) as opposed to merely descriptive or intuitive guesses. "No law, no possible mechanism; and no mechanism, no explanation." (Bunge 2004a, pp. 206–207, cf. 1964)

For Bunge, all kinds of systems share three features: they are composite; have structure and they exhibit a mechanism[17] (Bunge 2004a). In various places Bunge underlines that systems are concrete, and their structure is a set of inward and outward relations. Mechanisms are processes, and therefore the two must be kept apart: you can preserve the structure of a system while altering its mechanism. According to a major exponent of "mechanical philosophy" systems are to be distinguished from mechanisms but most of the time systemic entities require a certain mechanism operating; but systematic processes involving mechanisms can be distinguished from 'ephemerals' that are not repeatable, so there may be entities that are not systems under this criterion (Glennan 2017, p. 27). This is contrary to Bunge but it seems worthy of adoption.

Bunge is critical of analytical sociology (Bunge 2007) but some discussion of the engagement could be useful. Analytical sociology has gone through various metamorphoses and constitutes a large and growing terrain of theoretical pronouncements but mostly of empirical studies that have proved to be rather

---

paths are independent, and is also a key reason that multiple connectivity increases social cohesion" (Moody 2009, p. 456).

[17]Their structure can be endostructure whereby their internal arrangement is given or exostructure that describes a system's relation(s) to its surrounding context or environment.

fruitful.[18] At its core, the notion of mechanism and how far it may enjoy explanatory relevance are all disputed and shifting marshes, compared to the clarity and stability of Bunge's (albeit sparse) systemism. Analytical sociology also espouses so-called "structural individualism", distinguished from classical methodological individualism in the social sciences by placing emphasis on relational structures between the individuals' beliefs and actions (Hedström and Bearman 2009, p. 8). This is little different from Bunge's version of systemism and echoes his superior view recognizing the environment as impacting on the parts of a CESM-modelled system.

This point brings us to the related issue of downward causation: it refers to causal direction from the upper-level or wholes to lower levels or parts; or from a system to its components: for example, causation from the mental to the physical base or from social groups or collective entities to their members. This constitutes the hardest modern problem and only vague or outright doctrinal or even sometimes fantastical views have been put forward.[19] Bunge's own view, systemism, is also vague on this issue and blurs over important details (Bunge 1979b). The literature usually acknowledges either the impact of social facts explicitly being taken into account by an individual before acting or unrecognised influences of social context on individuals. Neither helps explain the actual relation.

40.4 *Ethica More Tecnico:* Technophilosophy and Technoethics

Bunge observes that: "There are no independent sciences or technologies. If a field of knowledge is disjoint from all the sciences, then it is nonscientific" (Bunge 2000, p. 156), urging that we should promote what he dubs *intersciences* and *intertechnologies* (ibid., p. 155). Quite early on, too, but via an indirect way, while talking about modern ontology having gone mathematical (and rightly so, as he says) Bunge has been celebrating the scientific turn in metaphysics since the early 1970s. This mathematical turn in ontology was increasingly cultivated by engineers and computer scientists. He calls the development of novel theories, such as machine

---

[18] See Hedström and Svedberg (1998), Hedström and Bearman (2009), and Hedström and Ylikoski (2010).

[19] Elder-Vass (2010, pp. 60–62) – in what is effectively the venerable conundrum about structure and agency – offers a vague claim marred by unhelpful ambivalence about the whole impacting on the parts and the latter on the whole or both the higher or composite entity – a star – and its lower level particles are simultaneously causally efficacious a claim that explains really nothing. Philip Pettit offers a rather fanciful explanation of how to account for whole-part non-causal relations by means of a model of levels of programming properties whereby the higher level holistic properties program for the individual ones that cause the actual event – i.e. emergent properties non-causally ensuring (or programming) the instantiating ones but without remaining epiphenomenal – see Macdonald and Macdonald (2010, pp 160–165)

theory, information theory, automata theory, switching theory, etc., as the "youngest metaphysical offspring of contemporary technology" (Bunge 1977, p. 7). The fusion of technology and philosophy is thus present in his thought quite early on: it is an ontological fusion at the start, turning into an ethical one in his later work.

> What are the metaphysical (ontological) presuppositions of technology? In particular, do technologists have to assume, if only tacitly, that the external world is real and lawful, and that assembly processes result sometimes in systems possessing emergent properties? (Bunge 1979c, p. 69)

40.4.1 Technoethics

Bunge has only recently widened the validity of his systemic approach to explicitly encompass technology in earnest (Bunge 2003a, p. 81). However, all the right ingredients were always there for his preaching of the gospel of systemics as applicable to technology. His basic ontological postulate that everything is a system or part thereof directly implies that technology itself is a system. His own justification for a systemic approach to technology starts with the idea of the environment, E, one of the four ingredients of his CESM model of systems, as he asserts that technological artifacts are surrounded by a social or a natural environment.

Bunge also points to the centrality of cybernetics and operational research as foundational stones of modern technology in the mid 1950s, and asserts that they are essentially cross-disciplinary. This squares with his complementary belief in convergence but contemporary emerging technologies are more than what Bunge suggests. An extension of his views, coherent with the basics of his thought, is needed to justify technology's systemic character.

In his 2000 paper on Systemism, Bunge explicitly identifies technology, like sciences and the market, as forming a system (Bunge 2000 p. 151). "[A]rtefacts are only tools, and hence they cannot be understood in purely technological terms" (Bunge 2003, p. 87). This prioritizes the systemic or environmental nature of technological inventions, but it distorts the gem of wisdom buried in it. Artefacts are never really tools that are later integrated into wider wholes or systems. Artefacts have no prior independent identity, for this would violate application of Bunge's own basic systemic postulate to emergent properties. No ingredient in a system can retain its original identity after the system has emerged (what Bunge 1979a, p. 30 calls total qualitative novelty – n – in a thing x as a symmetric difference over a time interval τ):

$$n_x\left(t, t'\right) = p_x(t)\Delta \bigcup_{t < \tau \leq t'} p_x\left(\tau\right)$$

Technology clearly has a social function for Bunge:

> systemism can be subversive merely by insisting that, because every thing is a system or part of one, some social and interdisciplinary borders are artificial or even harmful, and hence they should be trespassed by the technologists and consumer advocates. (Bunge 2003, p. 88)

Technoethics in tandem with philosophy as already embedded in technology is a highly useful and inspiring insight but in need of justification. A systemic reading of technology provides such further justification and analysis of the nature of technology by bringing it into contact with modern views of artefacts in their socio-ethical dimension.[20]

Bunge's early work on technoethics began developing what can be called a "seamless view of ethics-in-technology", rather than division of responsibility into kinds, jobs, parts or walks of life. (Bunge 1975, cf. 1980) Yet this early work is still marred by the naïve view of technological neutrality.[21] The artifacts are morally neutral and it is the people who are culpable: "instruments are morally inert and socially not responsible" (ibid., p 70). Not all technology is good but technology is nevertheless portrayed as *a source of inspiration for ethics*. Fact and value are intertwined because at the heart of every artefact there is a normative kernel built-in, an "ought" regarding its optimal operation. The rich bud of the Bungean philosophy of technology is already beginning to blossom: there is an *ethica more tecnico*. Moral codes *necessarily* combine science with *explicit* valuation (ibid., p. 75).

Philosophy itself contains technology in the form of ethics and action, which is equally important as engineering or any other technology. These branches of philosophy must therefore be evaluated in the same way ordinary technology is evaluated (Bunge 1998). This is a revolutionary claim of the "seamless view of ethics-in-technology". Despite differences that certainly exist, these philosophical branches and ordinary technology have a common aim: designing to get things done or avoided. Both technological and human action must be both instrumentally and morally rational.

---

[20]Nanotechnology, a technology or rather technoscience that bears explicitly on the need for mechanism as essential to understanding reality as is championed by Bunge is curiously not mentioned by him as a central case, as far as I know.

[21]This simplified view of technology as applied science is now dropped (see Bunge 2017) though retaining the initial idea about artifacts as things being morally neutral. (but see below this section and Bunge (2003b) (originally published in 1979) where it is acknowledged that methodologically there is no difference between scientific and technological research ibid., p. 174) A criticism of the always lurking naïve view separating applied from pure science is that as I have explained above modern scientific work is done in a novel way, employing models that are meant to reveal hidden mechanisms and unobservables, So – contra Bunge – I would claim that (a) engaging in such modelling as pure science is already fraught with moral decisions and normative judgments – even for deciding to do it to begin with and (b) the line between pure science and applied is now heavily blurred as a result of such computer modelling. So Bunge's view is untenable even on the basis of the use of explanatory mechanisms and their application(s) that he himself values so much. And (c) the view of technology as applied science is not true to his systemism and he must thus abandon it in the interest of self-consistency.

"Technophilosophy" is clearly a fusion of philosophical and technological knowledge dealing with the challenges of modern technology through its branches such as technometaphysics, technoaxiology, or technoepistemology and so forth. "A technophilosopher could be instrumental in clarifying a number of ideas, in digging up dubious presuppositions, in evaluating means and goals, and in alerting to possible undesirable side effects" (Bunge 1979c, p. 72). Technology is permeated by philosophy inherited from science and intimately connected to culture. It has a philosophical input partly controlled by a philosophical output (Bunge 2003b). Bunge emphasizes the inherent intellectual richness of technology and this leads him to assert that isolation of the technologist from all cultural and social decisions as a 'skillful barbarian' is a fundamental mistake. Technology's highest praise is that it should be allowed to "act as methodological model for the normative sciences, in particular ethics" (ibid., p. 179).

Bunge's systems approach could therefore enhance his views about the ethics of technology. If technology is viewed as a phenomenon exhibiting ever-present systemic characteristics, then it is plausible to re-visit technoethics as resting on such a systems-view of technology. This will add sophistication to the whole approach. From merely claiming or describing the inevitable philosophical presence at the heart of technology, we move to establishing its inevitability and offering a more sophisticated explanation of it:

> Just as Monsieur Jordain was unaware that he spoke prose, so the technologist may not realize that he is a part time technophilosopher, for he devotes some of his time to thinking of the difference between the artificial and the natural, or between technological knowledge and scientific knowledge, or between technical feasibility and moral desirability, or between subjective value and objective value, and so on and so forth. (Bunge 1979c, p. 73)

Artefacts are not neutral and contemporary philosophical schools also prioritize the inescapable ethical dimension of technology. Although these readings belong to different traditions in philosophy, they come close to what I have called the "seamless view of ethics within technology". There is a "morality of things" that we cannot avoid as we design and manipulate (See e.g. Verbeek 2011). These recent alternative views can be recast in systemic language.

What is crucial in these post-phenomenology readings of technology is that they wish to extend the notion of moral agency to non-cognitively endowed things. Technologies are moral agents themselves. Our life is constantly mediated by technology, which thus shapes our decisions. Bunge would accept the latter but on different grounds. Although he would reject any non-cognitive entities as bearing moral agency when they are not even minimally minded, there is a sense in which these views come close to Bunge's systemic vision. They both put forward a unified view. Convergence or a seamless system of ethics-within-technology or vice versa (technology inside ethics) can be harmonized with technology permeating or mediating our life as a carrier of moral value and pressing ethical issues in every aspect of life.

40.4.2 Technology as System

System is the middle term linking emergence first to mechanism and now to technology. Bunge's three foundational tenets of his all-pervasive systematicity amount to: Everything is interconnected forming a world (itself the supra system) of systems; there is a variety of systems; and no system except the world itself can escape change (Bunge 1979a). "Systemics" is distinct from systems analysis, down-playing or de-emphasizing the peculiarities or specificities of subject matter or component parts and keeping the discussion of *system* at the highest or most general level possible. "Systemics or general system theory, is a field of scientific and technological research and one of considerable interest to philosophy" (Bunge 1979a, p 3).

Bunge espouses a disjunctive taxonomy wherein a system is either conceptual (e.g. a theory) or concrete/material (e.g. a hospital). His refusal to accept what he terms mixed or hybrids by claiming that the clear links between components that are necessary to define a system are lacking with hybrids seems challenged by both computer science and metaphysics. Progress in the former has indicated that the technological realization of machine learning, or AI computation, couples both concept and hardware in its modelling of mental activities in brain-like computers.

The latter, metaphysics has been able to talk meaningfully about realms of our world or its possible instantiations in ways that combine physical facts with conceptual ingredients. For example, a law court in session is not merely furniture or human bodies or neural waves, it is determinate meaning patterns instantiated in brains or voices or on paper as legal decisions. Systemism need not shrink from mixture. Bunge's own demand of convergence and the realities of modern converging technologies strongly suggest that hybrids emerging from modern technology, or a seamless technoethics, should not be considered. After all, 'technoscience' is a fusion of systems, scientific theory and technology. Moreover, it is precisely the case of modern technoscience where we can observe the links between the component parts that Bunge rightly singles out as the prerequisites for understanding a system and its kind. Computation and modelling (as discussed in the previous section) are precisely such a mixture of abstract reasoning and material conditioning.

Emergence is naturally linked to its alter-ego, complexity. Its links with systems theory tie it into a unified theory that puts forward specially defined mechanismic explanations as crucial ingredients of scientific advance.[22] Emergence has to do

---

[22]Though Bunge champions mechanimsic explanation along with his systems view that both together countenance anti-reductionism and anti-individualism, he asserts, (e.g. Bunge 1977, p. 97) that unlike mechanism which is reductionist, his emergentist view requires properties of wholes to be novel (not had by their constituent parts – see logical notation above) and not simply hereditary or resultant – as he calls those properties that result by aggregation of the same tokens thereof had by their components. This is no inconsistency on his part. He has a novel, anti-reductionist, view of mechanism not to be confused with the classical one he puts aside. This divergence of Bunge's thought from classical mechanism and its attendant reductionism is important for understanding him.

with the appearance of novel items or properties. Systems theory prioritizes the idea of higher systems' properties or operations that are non-reducible to lower-level operations characterizing their components. Novelty often comes from combination. Emergence, systems and mechanism are therefore all forming a whole. They are applied to various scientific programs but only in Bunge's thought does one find all three underlined as present in technology alongside science. Bunge has always coupled science and technology and it was prescient of him to implicitly recognize that the emerging technologies of our times form a unified whole in the sense that they are intermingled in the form of a converging technology. This fact is now viewed as inescapable.

The Bungean philosophical thesis of convergence is thus a clear prediction of contemporary developments. Although he defines technology as ultimately applied science[23] his work contains a rich notion of technology that is congruent to recent views illuminating the epistemological drive behind emerging technology. This is something that distances technology from the rather simplistic idea of merely applied science. In fact, Bunge's insistence on the philosophical core of technology elevates technology to something above merely applied science.[24] Technology has never been devoid of philosophy, and thus never ethics-free. The commonality between technology and ethics is summarized in this:

> "[T] o face a technological problem in any depth necessitates invoking general praxiological concepts and principles. And to tackle a problem with social responsibility requires some ethical concepts and principles." (Bunge 1999)

As the title of this work indicates there is a two-way interaction between ethics and technology: *ethics and praxeology as technologies*. The task of philosophers is to "bridge ethics and praxeology to technology. I want to close the section by bringing in – in true systemic fashion and linking up with those recent views of technology as a moral agent mediating our lives – some Bungean considerations from neuroscientific explanation to bear upon technology and thus suggest an analogue to his idea.

Bunge in Voume 4 of his *Basic* Philosophy (Bunge 1979a, p. 128) introduces a curious term: *psychon*: "We shall call a *psychon* any neural unit capable of discharging mental functions of some kind". Contra outright holism in neuroscience, Bunge goes on to claim "Although there is plenty of evidence for the strong coupling among a number of neural systems, there is also evidence for localization of, e.g., pleasure and speech" (ibid). There is such localization but now we know that it

---

[23]This less than thought-through position leads him to embracing less interesting positions in the moral assessment of science and technology: scientists are innocent knowledge-seekers working for the sake of knowledge, technology as applied science can be good or evil (Bunge 1988). It has recently become clear that this view is not espoused any more in its earlier simplified form (Bunge 2017).

[24]However in Bunge (2003a, p. 86), he tacitly brings to light the hybrid nature of the outputs of designing artifacts whereby scientific theory merged with engineering or design, psychiatry with pharmacology and so forth – thereby admitting that technology is no mere applied science or denying that there is a schism between theory and practice in technological designing (if he insists in calling it applied science he at least adds the qualification 'multidisciplinary' now).

is complex, with the same circuit carrying out multiple functions. Bunge favours instead what he calls "psychosystem". The brain is neither a conglomerate of separate units nor a homogeneous jelly. It is a *system of specialized subsystems* or organs (ibid., p. 129), In the same work he provides more helpful, rigid definitions of "psychons" (ibid., p. 132).

DEFINITION 4.7. A neuronal system is *plastic* (or *uncommitted*, or *modifiable*, or *self-organizable)* iff its connectivity is variable throughout the animal's life. Otherwise (i.e. if it is constant from birth or from a certain stage in the development of the animal), the system is *committed* (or *wired-in*, or *pre wired*, or *preprogrammed)*.

DEFINITION 4.8. Every plastic neural system is called a *psychon.* (Bunge 1979a, p. 132)

And later:

We call a psychon of kind K the smallest plastic neuronal system capable of discharging a mental function of kind K. Every state or stage in a mental process – or, equivalently, every state of a psychon or of a system of psychons – is called a mental state of the animal. For example, the formation of purposes and plans appears to be a specific activity of psychons in the prefrontal cortex. (Bunge 2003a, p. 62)

These definitions allow construction of a Bungean vision of technology, technophilosophy and technoethics. We may call it a "system of technoscience" and expect it to meet stringent demands of an underlying moral philosophy. For this to work we shall have to construe technology – already accepted as a system – as more precisely a system of subsystems called *technons.* Concrete technoscience can be seen as a plastic supra-system if only because of its inherent converging nature, let alone because of its exhibiting a steady-state quasi-cosmological expansion ontologically. If it has also to exhibit a concurrent ethical feature tracing such plasticity we shall have to begin constructing such a system as consisting of subsystems of *technons.* These should be understood as subsystems or different but converging bio-, cogno-, info- or nano-technologies the identifying feature of each of which is its plasticity.

This plasticity should be understood first as emergent technological innovation that exhibits qualitative novelty with respect to its earlier sub-stages or subparts. This allows for ever-changing inter-connectivity of the ethical concerns, moral principles or even meta-ethical philosophical analyses of technology as a system consisting of parts in ever-shifting connecting patterns. In a way that is analogous to the psychon-definition above, such emergent ethical features of each level of technological advance follows upon the underlying technical plasticity. A case of ethical plasticity or total ethical novelty arises as emergent from the technical plasticity true of the technical level. This technical level has emerging ethical features once contemporary complex technologies are seen for what they are. Ethical plasticity or ethical novelty characterizes technologies to the extent that they are by nature plastic systems, what are here called 'technons'. So, assuming fusion of science and technology, we can build a Bungean emergent systemic hierarchy:

from neurons → to psychons → to technons

If praxeology and ethics must be fused with technology by philosophers as Bunge insists, the end-result should be a view of technology as not simply applied science or merely nuts and bolts. Its ethics should be an integral part of it in the systemic sense. The ethical dimension of technology – what I have called the "seamless view" – is therefore to be seen as an emerging property of its systemic plasticity, not an idle, impossible to ground moral epiphenomenon.

40.5 Conclusion

This chapter has tried to carve out of Bunge's rich and intricate thinking what I regard as the bare essentials. It discloses the basic thread that links emergence and its kindred themes, systems and mechanismic approaches, in order finally to establish a link with technophilosophy and technoethics. In all these areas I tried to enhance Bungean thought by bringing to bear some ideas from current discussions in these areas. My basic conviction is that all these Bungean themes together intertwine to form the fundamental core of his thought. That core remains essential in spite some of the supplementary modern extensions introduced in this chapter. It seems possible to coherently augment his thought by annexing a systems view of technology, although that has not been considered so far by Bunge himself.

References

Bechtel, W. (2007). Reducing psychology while maintaining its autonomy via mechanistic explanations. In M. Schouten & H. Looren de Jong (Eds.), *The matter of the mind: Philosophical essays of psychology, neuroscience and reduction* (pp. 172–198). Malden: Blackwell.

Bedau, M. (2008). Downward causation and autonomy in weak emergence. In M. A. Bedau & P. Humphreys (Eds.), *Emergence: Contemporary readings in philosophy and science* (pp. 155–188). Cambridge, MA: MIT Press.

Bedau, M. (2010). Weak emergence and context-sensitive reduction. In A. Corradini & T. O'Conor (Eds.), *Emergence in science and philosophy* (pp. 46–63). New York: Routledge.

Bülthof, H. H., et al. (Eds.). (2002). *Biologically motivated computer vision*. Berlin: Springer.

Bunge, M. (1964). Phenomenological theories. In M. Bunge (Ed.), *The critical approach* (pp. 234–254). Glencoe: Free Press.

Bunge, M. (1974). The concept of social structure. In W. Leinfellner & E. Kohelr (Eds.), *Developments in the methodology of social sciences* (pp. 175–216). Dordrecht: D. Reidel.

Bunge, M. (1975). Towards a technoethics. *Philosophic Exchange, 6*(1), 69–79.

Bunge, M. (1976). The philosophical richness of technology. *PSA: Proceedings of the Biennial Meeting of the Philosophy of Science Association, 2*, 153–172.

Bunge, M. (1977). *Treatise on basic philosophy, Vol. 3 – Ontology I: The furniture of the world*. Dordrecht: D. Reidel.

Bunge, M. (1979a). *Treatise on basic philosophy, Vol. 4 – Ontology II: A world of systems*. Dordrecht: D. Reidel.

Bunge, M. (1979b). A systems concept of society: Beyond individualism and holism. *Theory and Decision, 10*, 13–30.

Bunge, M. (1979c). The five buds of technophilosophy. *Technology in Society, 1*, 67–74.

Bunge, M. (1980). Technoethics. In M. Kranzberg (Ed.), *Ethics in an age of pervasive technology* (pp. 139–142). Boulder: Westview Press.

Bunge, M. (1988). Basic science is innocent, applied science and technology can be guilty. In G. E. Lemarchand & A. R. Pedace (Eds.), *Scientists, peace and disarmament* (pp. 245–261). Singapore: World Scientific.

Bunge, M. (1997). Mechanism and explanation. *Philosophy of the Social Sciences, 27*(4), 410–465.

Bunge, M. (1998). The philosophical technologies. *Technology in Society, 20,* 377–383.

Bunge, M. (1999). Ethics and praxiology as technologies. *SPT, 4*(4). https://scholar.lib.vt.edu/ ejournals/SPT/v4n4/bunge.html. Accessed 2 May 2017.

Bunge, M. (2000). Systemism: The alternative to individualism and holism. *Journal of Socio-Economics, 29,* 147–157.

Bunge, M. (2003a). *Emergence and convergence: Qualitative novelty and the unity of knowledge.* Toronto: Toronto University Press.

Bunge, M. (2003b). Philosophical inputs and outputs of technology. In R. Scharff (Ed.), *Philosophy of technology: The technological condition* (pp. 170–181). Malden: Blackwell.

Bunge, M. (2004a). How does it work? *Philosophy of the Social Sciences, 34*(2), 182–210.

Bunge, M. (2004b). Clarifying some misunderstandings about social systems and their mechanisms. *Philosophy of the Social Sciences, 34*(3), 371–381.

Bunge, M. (2007). Review of: Hedström P., Dissecting the social: On the principles of analytical sociology. *American Journal of Sociology, 113*(1), 258–260.

Bunge, M. (2017). Technology, science and politics. In M. Bunge (Ed.), *Doing science in the light of philosophy* (pp. 150–160). Singapore: World Scientific.

Cipolla, R., et al. (Eds.). (2013). *Machine learning for computer vision.* Heidelberg: Springer.

Demeleunaere, P. (2011). Introduction. In P. Demeleunaere (Ed.), *Analytical sociology and social mechanisms* (pp. 1–30). Cambridge: Cambridge University Press.

Elder-Vass, D. (2010). *The causal power of social structures: Emergence, structure and agency.* Cambridge: Cambridge University Press.

Glennan, S. (2017). *The new mechanical philosophy.* Oxford: Oxford University Press.

Glennan, S., & Illari, P. (2018). Varieties of mechanisms. In S. Glennan & P. Illari (Eds.), *The Routledge handbook of mechanisms and mechanical philosophy* (pp. 91–103). London/New York: Routledge.

Hedström, P. (2005). *Dissecting the social: On the principles of analytical sociology.* Cambridge: Cambridge University Press.

Hedström, P., & Bearman, P. (2009). What is analytical sociology all about? An introductory essay. In P. Hedström & P. Bearman (Eds.), *The Oxford handbook of analytical sociology* (pp. 3–24). Oxford: Oxford University Press.

Hedström, P., & Svedberg, R. (Eds.). (1998). *Social mechanisms: An analytical approach to social theory.* Cambridge: Cambridge University Press.

Hedström, P., & Ylikoski, P. (2010). Causal mechanisms in the social sciences. *Annual Review of Sociology, 36,* 49–67.

Humphrey, P. (2008). How properties emerge. In M. A. Bedau & P. Humphreys (Eds.), *Emergence: Contemporary readings in philosophy and science* (pp. 111–126). Cambridge, MA: MIT Press.

Illari, M. P., & Williamson, J. (2012). What is a mechanism? Thinking about mechanism *across* the sciences. *European Journal of Philosophy of Science, 2,* 119–135.

Kaldis, B. (2018). Concept nativism and transhumanism: Educating future minds. *Humana Mente: Journal of Philosophical Studies, 33,* 145–153.

Kaplan, D. M. (2011). Explanation and description in computational neuroscience. *Synthese, 183*(3), 339–373.

Kaplan, D. M. (2017). Neural computation, multiple realizability and the prospects of mechanistic explanation. In D. Kaplan (Ed.), *Explanation and integration in mind and brain science* (pp. 164–189). Oxford: Oxford University Press.

Kim, J. (2006). Being realistic about emergence. In P. Clayton & P. Davies (Eds.), *The re-emergence of emergence: The emergentist hypothesis from science to religion* (pp. 189–202). Oxford: Oxford University Press.

Levy, A. (2013). Three kinds of new mechanism. *Biology and Philosophy, 28*(1), 99–14.

Macdonald, C., & Macdonald, D. (2010). Emergence and downward causation. In C. Macdonald & D. Macdonald (Eds.), *Emergence in mind* (pp. 139–168). Oxford: Oxford University Press.

Macy, M. W., et al. (2011). Social mechanism and generative explanations: Computational models with double agents. In P. Demeleunaere (Ed.), *Analytical sociology and social mechanisms* (pp. 250–265). Cambridge: Cambridge University Press.

Mahner, M., & Bunge, M. (1997). *Foundations of biophilosophy*. Berlin/Heidelberg: Springer.

McLoughlin, B. P. (2008). The rise and fall of British emergentism. In A. Bedau & P. Humphreys (Eds.), *Emergence: Contemporary readings in philosophy and science* (pp. 19–59). Cambridge, MA: MIT Press.

Moody, J. (2009). Network dynamics. In P. Hedström & P. Bearman (Eds.), *The Oxford handbook of analytical sociology* (pp. 447–474). Oxford: Oxford University Press.

Silberstein, M. (2006). In defence of ontological emergence and mental causation. In P. Clayton & P. Davies (Eds.), *The re-emergence of emergence: The emergentist hypothesis from science to religion* (pp. 203–226). Oxford: Oxford University Press.

Verbeek, P.-P. (2011). *Moralizing technology: Understanding and designing the morality of things*. Chicago: The University of Chicago Press.

Yao, Y., et al. (Eds.). (2010). *Brain informatics*. Berlin: Springer.

Ylikoski, P. (2018). Social mechanisms. In S. Glennan & P. Illari (Eds.), *The Routledge handbook of mechanisms and mechanical philosophy* (pp. 401–412). London/New York: Routledge.

Part XI
Bibliography

Chapter 41
Mario Bunge Publications
(All Languages)

Marc Silberstein

Abstract This bibliography list all of Mario Bunge's known publications in all languages, including English, Spanish, French, Italian, German, Russian, Fasi, Chinese, Japanese, Hungarian and Polish. Sometimes translations are made and published without an author's knowledge or permission, so any of these may not be on the list. There are 150 books and 540 book chapters and articles listed.

This bibliography lists all of Mario Bunge's known publications in all languages, including English, Spanish, French, Italian, German, Russian, Fasi, Chinese, Japanese, Hungarian and Polish. Sometimes translations are made and published without an author's or publisher's knowledge or permission; so some of these unauthorised translations may not be on the list. There are 152 books and 543 book chapters and articles listed.

41.1 Books

1. *Temas de educación popular*. Buenos Aires: El Ateneo, 1943.
2. *La edad del universo*. La Paz: Laboratorio de Física Cósmica, 1955.
3. *Causality: The Place of the Causal Principle in Modern Science*. Cambridge, Mass.: Harvard University Press, 1959.
4. *Metascientific Queries*. Springfield Ill.: Charles C. Thomas, Publisher, 1959.
5. *Etica y ciencia*. Buenos Aires: Siglo Veinte, 1960.
6. *La ciencia, su método y su filosofía*. Translation of three chapters of #4. Buenos Aires: Siglo Veinte, 1960.
7. *Antología semántica* (Editor). Buenos Aires: Nueva Visión, 1960.
8. *Cinemática del electrón relativista*. [1952 Ph.D. dissertation]. Tucumán: Universidad Nacional de Tucumán, 1960.

M. Silberstein (✉)
Editions Matériologiques, Paris, France

© Springer Nature Switzerland AG 2019
M. R. Matthews (ed.), *Mario Bunge: A Centenary Festschrift*,
https://doi.org/10.1007/978-3-030-16673-1_41

9. *Causalidad.* Translation of #3. Buenos Aires: Editorial Universitaria de Buenos Aires, 1960.
10. *Intuition and Science.* Englewood Cliffs, N.J.: Prentice-Hall, 1962.
11. *Pritchinost.* Translation of #3. Moscow: Publishing House for Foreign Literature, 1962.
12. *The Myth of Simplicity.* Englewood Cliffs, N.J.: Prentice-Hall, 1963.
13. *Causality.* Paperback edition of #3, with new Foreword and Appendix. Cleveland and New York: The World Publishing Co., 1963.
14. *La ciencia, su método y su filosofía.* Second, enlarged edition of #6. Buenos Aires: Siglo Veinte, 1963.
15. *The Critical Approach: Essays in Honor of Karl Popper* (Editor). Includes a Preface. Glencoe: Free Press, 1964.
16. *Intuición y ciencia.* Translation of #10. Buenos Aires: Editorial Universitaria de Buenos Aires, 1964.
17. *Delaware Seminar in the Foundations of Physics* (Editor). Includes an Introduction. Berlin-Heidelberg-New York: Springer-Verlag, 1967.
18. *Scientific Research I: The Search for System.* Berlin-Heidelberg-New York: Springer-Verlag, 1967.
19. *Scientific Research II: The Search for Truth.* Berlin-Heidelberg-New York: Springer-Verlag, 1967.
20. *Foundations of Physics.* Berlin-Heidelberg-New York: Springer-Verlag, 1967.
21. *Quantum Theory and Reality* (Editor). Includes an Introduction. Berlin-Heidelberg-New York: Springer-Verlag, 1967.
22. *Az oksag.* Translation of #3. Budapest: Gondolat Kiado, 1967.
23. *Intuitsia i nauka.* Translation of #10 with a study by V. G. Vinogradov. Moscow: Progress, 1967.
24. *O Przycznowosci.* Translation of #3. Warsaw: Panstwowe Wydawnictwo Naukowe, 1968.
25. *La investigación científica.* Translation of #18 and #19. Barcelona: Ediciones Ariel, 1969.
26. *La causalità.* Italian translation of #3, with a new Preface and a study by E. Panaitescu: "La causalità e il determinismo secondo Mario Bunge". Torino: Boringhieri, 1970.
27. *Problems in the Foundations of Physics* (Editor). Berlin-Heidelberg-New York: Springer-Verlag, 1971.
28. Japanese translation of #3. Tokyo: Iwanami, 1972.
29. *Ética y ciencia.* 2nd revised ed. of #5. Buenos Aires: Siglo Veinte, 1972.
30. *Teoría y realidad.* Barcelona: Ariel, 1972.
31. Reprint of #25 with Preface by Eramis Bueno. La Habana: Instituto Cubano del Libro, 1972.
32. *Philosophy of Physics.* Dordrecht: Reidel, 1973.
33. *Method, Model and Matter.* Dordrecht: Reidel, 1973.
34. *Exact Philosophy* (Editor) Dordrecht: Reidel, 1973.
35. *The Methodological Unity of Science* (Editor). Dordrecht: Reidel, 1974.
36. *Filosofia fiziki.* Russian translation of #32. Moscow: Progress, 1974.

37. *Sense and Reference.* 1st vol. of *Treatise on Basic Philosophy.* Dordrecht: Reidel, 1974.

38. *Interpretation and Truth.* 2nd vol. of *Treatise on Basic Philosophy.* Dordrecht: Reidel, 1974.

39. *Philosophie de la physique.* Transl. of #32. Paris: Ed. du Seuil, 1974.

40. *Teoria e realidade.* Transl. of #30. São Paulo: Editora Perspectiva, 1974.

41. Reprint of #10. Westport, Conn.: Greenwood Press, 1975.

42. *Tratado de Filosofía Basica*, Vol. 1. Portugese trans. of #37. São Paulo: Ed. da Universidade de São Paulo & Ed. Pedagógica e Universitária. 1976.

43. *Tratado de Filosofía Básica*, Vol. 2. Portugese trans. of #37. São Paulo: Ed. da Universidade de São Paulo & Ed. Pedagógica e Universitária. 1976.

44. *Tecnología y filosofía.* Monterrey, México: Universidad Autónoma de Nuevo Leon, 1976.

45. *Ética y ciencia,* 3rd ed. New appendix: "Por una tecnoética". Buenos Aires: Siglo Veinte, 1976.

46. *The Furniture of the World.* 3rd vol. of *Treatise on Basic Philosophy.* Dordrecht: Reidel, 1974.

47. *Filosofía de la física.* Spanish transl. of #32. Barcelona: Ariel, 1978.

48. *A World of Systems.* 4th vol. of *Treatise on Basic Philosophy.* Dordrecht: Reidel, 1979.

49. *Causality in Modern Science.* 3rd edition of #3. Corrections and new Preface. New York: Dover Publications, 1979.

50. *Epistemología. Curso de actualización.* Barcelona: Ariel, 1980.

51. *The Mind-Body Problem.* Oxford and New York: Pergamon Press, 1980.

52. *Ciencia y desarrollo.* Buenos Aires: Siglo Veinte, 1980.

53. *Epistemologia.* Portugese transl. of #50. São Paulo: Queiroz and Editora da Universidade de S. Paulo, 1980.

54. *Ciência e desenvolvimiento.* Portugese transl. of #52. Bello Horizonte: Itaitia; S. Paulo: Editora da Universidade de S. Paulo, 1980.

55. *Materialismo y ciencia.* Barcelona: Ariel, 1981.

56. *Scientific Materialism.* Dordrecht-Boston: D. Reidel Publ. Co. 1981.

57. *Economía y filosofía.* Madrid: Tecnos, 1982.

58. *The Mind-Body Problem.* Japanese transl. of #51. Post-scriptum by Prof Hiroshi Kurosaki. Tokyo: Sangyo Tosho, 1982.

59. *Lingüística y filosofía.* Barcelona: Ariel, 1983.

60. *Epistémologie.* French transl. of #50. Paris: Maloine, 1983.

61. *Controversias en física.* Madrid: Tecnos, 1983.

62. *La investigación científica,* rev. ed. of #25. Barcelona: Ariel, 1983.

63. *Epistemologie.* German transl. of #50. Mannheim: Bibliographisches Institut, 1983.

64. *Exploring the World.* 5th vol. of *Treatise on Basic Philosophy.* Dordrecht: Reidel, 1983.

65. *Understanding the World.* 6th vol. of *Treatise on Basic Philosophy.* Dordrecht: Reidel, 1983.

66. *Das Leib-Seele-Problem.* German transl. of #51. Introduction by Prof. Bernulf Kanitscheider. Tübingen: J. C. B. Mohr (Paul Siebeck), 1984.
67. *Stiinta si filosofie.* Anthology of articles translated into Romanian. Preliminary study by Prof. Calina Mare. Bucharest: Editura Poltica, 1984.
68. *Economía y filosofía.* Second enlarged ed. of #57. Madrid: Tecnos, 1985.
69. *El problema mente-cerebro.* Spanish transl. of #51. Madrid: Tecnos, 1985.
70. *Philosophy of Science and Technology,* part I: *Formal and Physical Sciences.* Part of *Treatise on Basic Philosophy,* Vol. 7. Dordrecht: Reidel, 1985.
71. *Philosophy of Science and Technology,* part II: *Life Science, Social Science and Technology.* Part of *Treatise on Basic Philosophy.,* Vol. 7. Dordrecht: Reidel, 1985.
72. *Seudociencia e ideología.* Madrid: Alianza Editorial, 1985.
73. *Racionalidad y realismo.* Madrid: Alianza Editorial, 1985.
74. *Philosophical Problems in Linguistics.* Japanese transl. of #59. Tokyo: Seishin-Shobo, 1986.
75. *Intuición y razón.* Updated and enlarged version of No. 16. Madrid: Tecnos, 1986. Revised version: Buenos Aires, Sudamericana, 1996.
76. *Kausalität, Geschichte und Probleme.* German transl. of #3, revised and with a new Preface. Tübingen: J. C. B. Mohr, 1987.
77. *Philosophy of Psychology* (with Rubén Ardila). New York: Springer-Verlag, 1987.
78. *Vistas y entrevistas.* Buenos Aires: Ediciones Siglo Veinte, 1987.
79. *Filosofia de la psicología.* Spanish transl. of #77. Prologue by L. García-Sevilla. Barcelona: Ariel, 1988.
80. *Ethics: The Good and the Right,* Vol. 8 of *Treatise on Basic Philosophy.* Dordrecht-Boston: Reidel, 1989.
81. *Mente y sociedad.* Madrid: Alianza Editorial, 1989.
82. *Filosofia della fisica.* Italian transl. of #32. Abano Terme: Piovan Editore, 1989.
83. *Philosophie der Psychologie.* German transl. of #77, with author's new preface. Tübingen: J. C. B. Mohr (Paul Siebeck), 1990.
84. *Sociología de la ciencia.* Buenos Aires: Siglo Veinte, 1993.
85. *La science, sa méthode et sa philosophie.* Annotated and revised translation of #6. Paris: Vigdor, 1994.
86. Revised and augmented version of #6. Buenos Aires: Sudamericana, 1995.
87. *Sistemas sociales y filosofía.* Buenos Aires: Sudamericana, 1995.
88. *Finding Philosophy in Social Science.* New Haven CT: Yale University Press, 1996.
89. *Etica, ciencia y técnica.* Revised and augmented version of #5. Buenos Aires: Sudamericana, 1996.
90. *Intuition et raison.* Paris: Vigdor, 1996. French version of #75.
91. *Intuición y razón,* revised ed. of #75. Buenos Aires: Editorial Sudamericana, 1996.
92. *Foundations of Biophilosophy* (with Martin Mahner). Berlin-Heidelberg- New York: Springer-Verlag, 1997.

93. *Epistemología de las ciencias y técnicas naturales y sociales*: *Selección de textos*. Arequipa: Universidad Nacional de San Agustín de Arequipa, 1997. [Anthology].

94. *Ciencia, técnica y desarrollo*. Rev. ed. of #52. Buenos Aires: Editorial Sudamericana, 1997.

95. *Epistemología*. Rev. ed., with new Preface, of #50. México City-Madrid: Siglo Veintiuno, 1997.

96. *La causalidad*. Repr. of #9. Buenos Aires: Editorial Sudamericana, 1997.

97. Revised and augmented version of #78. Buenos Aires: Editorial Sudamericana, 1998.

98. *Social Science Under Debate*. Toronto: University of Toronto Press, 1998.

99. Revised and augmented version of #84. Buenos Aires: Editorial Sudamericana, 1998.

100. *Philosophy of Science*, Vol. 1: *From Problem to Theory* Updated version of #18. New Brunswick NJ: Transaction Publishers, 1998.

101. *Philosophy of Science*, Vol. 2: *From Explanation to Justification*. Updated version of #19. New Brunswick NJ: Transaction Publishers, 1998.

102. *Vigencia de la filosofía*. Lima: Universidad Inca Garcilaso de la Vega, 1998.

103. *Elogio de la curiosidad*. Buenos Aires: Editorial Sudamericana, 1998.

104. *Critical Approaches to Science and Philosophy*. Reissue of #15 with a new Introduction. New Brunswick NJ: Transaction Publishers, 1999.

105. *Dictionary of Philosophy*. Amherst NY: Prometheus Books, 1999.

106. *The Sociology-Philosophy Connection*. Foreword by Raymond Boudon. Brunswick NJ: Transaction Publishers, 1999.

107. *Buscar la filosofía en las ciencias sociales*. México, D.F.: Siglo xxi, 1999.

108. *Las ciencias sociales en discusión*. Buenos Aires: Editorial Sudamericana, 1999. Spanish translation of #96.

109. *La relación entre la sociología y la filosofía*. Madrid: Edaf, 2000. Spanish translation of #105.

110. *Philosophische Grundlagen der Biologie*. Translation of #91 plus preface by Gerhard Vollmer. Preface by Gerhard Vollmer. Berlin-Heidelberg-New York: Springer-Verlag, 2000.

111. *Fundamentos de la biofilosofía*. Translation of #91. México-Buenos Aires: Siglo Veintiuno Editores, 2000.

112. *Philosophy in Crisis*: *The Need for Reconstruction*. Amherst NY: Prometheus Books, 2001.

113. *Diccionario de filosofía*. Spanish transl. of #104. México: Siglo XXI Editores, 2001.

114. *Scientific Realism*: *Selected Essays by Mario Bunge*. Ed. Martin Mahner. Amherst, NY: Prometheus Books, 2001.

115. *Ser, saber, hacer*. México City: Paidós-Universidad Nacional Autónoma de México, 2001.

116. *Dicionário de filosofía*, Portuguese transl. of #104. Sao Paulo: Editora Perspectiva, 2002.

117. *Crisis y reconstrucción de la filosofía,* Barcelona: Gedisa, 2002. Translation of #119.
118. *Philosophical Dictionary,* enlarged ed. Amherst NY: Prometheus Books, 2003.
119. *Cápsulas.* Barcelona: Editorial Gedisa, 2003.
120. *Emergence and Convergence.* Toronto: University of Toronto Press, 2003.
121. *Matérialisme et humanisme: Pour surmonter la crise de la pensée.* French transl. of #110. Preface by Laurent-Michel Vacher. Montréal: Liber, 2004.
122. *Mito, realidad y razón.* Santa Fe: Universidad Nacional del Litoral; BuenosAires; Sudamericana, 2004.
123. *Ueber die Natur der Dinge,* with Martin Mahner. Stuttgart: S. Hirzel, 2004.
124. *Emergencia y convergencia.* Spanish transl. of #117. Barcelona: Gedisa, 2004.
125. *Chasing Reality: The Strife over Realism.* Toronto: University of Toronto Press, 2006.
126. *100 ideas.* Buenos Aires: Sudamericana, 2006.
127. *Una filosofía realista para el nuevo siglo.* Lima: Universidad Inca Garcilaso de la Vega, 2007.
128. *Deu assaigs filososòfics i una diatriba exasperada.* Girona: Documenta Universitaria, 2007.
129. *Filosofía y sociedad.* México, D.F.: Siglo XXI, 2008.
130. *Tratado de filosofía,* tomo 1: *Semántica I.* Spanish translation of #37.Barcelona: Gedisa, 2008.
131. *Foundations of Biophilosophy,* with Martin Mahner. Japanese translation of #91. Tokyo: Springer, 2008.
132. *Le matérialisme scientifique.* Paris: Syllepse, 2008. French Translation of #56.
133. *Teoría e realidade,* repr. of #40. Sao Paulo: Perspectiva, 2008.
134. *Political Philosophy: Fact, Fiction, and Vision.* New Brunswick, NJ: Transaction Publishers, 2009.
135. *Causality in Modern Science,* 4th ed. New Brunswick, NJ: Transaction Publishers, 2009.
136. *Tratado de filosofía,* tomo 2: *Semántica II.* Spanish translation of #38. Barcelona: Gedisa, 2009.
137. *Filosofía política.* Spanish translation of #131. Barcelona-Buenos Aires: Gedisa.
138. *¿Qué es filosofar científicamente?* Lima: Universidad Inca Garcilaso de la Vega, 2009.
139. *Vigencia de la filosofía,* 2nd ed. Lima: Universidad Inca Garcilaso de la Vega, 2009.
140. *Matter and Mind.* London: Springer, 2010.
141. *Ontología,* 4th vol. of *Tratado de Filosofía.* Barcelona: Gedisa, 2012.
142. *Provocaciones.* Buenos Aires: EDHASA, 2012.
143. *Evaluating Philosophies.* Dordrecht: Springer, 2012.
144. *Diálogos urticantes.* Lima: Universidad Garcilaso de la Vega, 2012.
145. *Filosofia de la tecnologia.* Lima: Universidad Garcilaso de la Vega, 2012.
146. *Filosofía para médicos.* Barcelona-Buenos Aires: Gedisa, 2012.
147. *Medical Philosophy.* Singapore: World Scientific Publishing, 2013.

148. *Entre dos mundos*: *Memorias de un filósofo científico*. Buenos Aires, Barcelona: Gedisa, 2015.
149. *Between Two Worlds*. *Memoirs of a Scientific Philosopher*. Springer International, 2016.
150. *Entre deux mondes*. *Mémoires d'un philosophe-physicien*. Paris: Editions Matériologiques, 2016. French Translation of #147.
151. *Doing Science in the Light of Philosophy*. Singapore: World Scientific Publishing, 2017.
152. *From a Scientific Point of View*, Cambridge Scholars Publications, Newcastle, UK., 2018.
153. *Philosophie de la médecine*. *Concepts et méthodes*, Paris: Editions Matériologiques, 2019. French Translation of #145.
154. *Dictionnaire philosophique*, Paris: Editions Matériologiques, 2019. French Translation of #118.

41.2 Translations

W. O. Quine, *El sentido de la nueva lógica*. Buenos Aires: Nueva Visión, 1958.
Max Born, *El inquieto universo*. Buenos Aires: Editorial Universitaria de Buenos Aires, 1960.

41.3 Papers

1. Introducción al estudio de los grandes pensadores. *Conferencias* (Buenos Aires) III: 105–109, 124–126, 1939.
2. El tricentenario de Newton. Buenos Aires: Universidad Obrera Argentina, tract, 8 pages, 1943.
3. *Significado físico e histórico de la teoría de Maxwell*. Texto de una conferencia dictada el 21 de junio de 1943 en la Facultad de Química Industrial y Agrícola de la Universidad Nacional del Litoral, en Santa Fe, 16 páginas, Buenos Aires, 1943.
3a. *El tricentenario de Newton*. Buenos Aires: Universidad Obrera Argentina, Instituto Científico. Seminario de Filosofía, 8 páginas, 1943.
4. La epistemología positivista. *Nosotros* (Buenos Aires) VIII, No. 93, 283–290, 1943.
5. A new representation of types of nuclear forces. *Physical Review* 65: 249, 1944.
6. Una nueva representación de los tipos de fuerzas nucleares. *Revista de la Facultad de Ciencias Físicomatemáticas* (La Plata), III: 221–239, 1944.
7. ¿Qué es la epistemología? *Minerva* (Buenos Aires), 1, 27–43, 1944.

8. Precursores, predecesores y predictores. *Minerva* (Buenos Aires), 1: 61–62, 1944.

9. Una de las posibles metafísicas. *Minerva* (Buenos Aires) 1, 167–168, 1944.

10. Auge y fracaso de la filosofía de la naturaleza. *Minerva* (Buenos Aires) 1, 213–235, 1944.

11. Una nueva interpretación de Rousseau. *Minerva* (Buenos Aires) 1, 274–278, 1944.

12. Nietzsche y la ciencia. *Minerva* (Buenos Aires) 2, 44–50, 1944.

13. Ludwig Boltzmann. *Minerva* (Buenos Aires) 2, 70–72, 1944.

13a. 'El spin total de un sistema de más de 2 partículas', *Revista de la Unión Matemática Argentina*, 10(1), 13–14, (1944).

14. Cómo veía el mundo Florentino Ameghino. *Minerva* (Buenos Aires) 2, 184–185, 1945.

15. El spin total de un sistema de más de dos partículas. *Revista de la Unión Matemática Argentina* X, 13–14, 1945.

16. Neutron-proton scattering at 8.8 and 13 MeV. *Nature* 156, 301, 1945.

17. Fenómenos de resonancia en la difusión de neutrones por protones. *Revista de la Unión Matemática Argentina* XI, 35, 1945.

17a. 'Difusión neutrón-protón a 8.8 y 13 MeV', (5th Meeting *AFA*, March-April 1945), Comunicaciones, *Revista de la Unión Matemática Argentina*, 1946, **11**(3), 103.

18. La fenomenología y la ciencia. *Cuadernos Americanos* (México) No. 4, 108–122, 1951.

19. Bemerkung über den Massendefekt des Wasserstoffatoms. *Acta Physica Austriaca* 5, 77–79, 1951.

20. Mach y la teoría atómica. *Boletin del Químico Peruano* 3, No. 16: 12–17, 1951.

21. What is chance? *Science and Society* 15: 209–231, 1951.

22. New dialogues between Hylas and Philonous. *Philosophy and Phenomenological Research* 15: 192–199, 1954.

23. Exposición y crítica del principio de complementaridad. *Notas del Curso Interamericano de Física Moderna*, pp. 27–36. La Paz: Laboratorio de Física Cósmica, 1955.

24. A picture of the electron. *Nuovo Cimento*, ser. X, 1, 977–985, 1955.

25. Strife about complementarity. *British Journal for the Philosophy of Science* 6, 1–12: 6, 141–154, 1955.

26. The philosophy of the space-time approach to the quantum theory. *Methodos* 7, 295–308, 1955.

27. A critique of the frequentist theory of probability. *Congresso Internacional de Filosofia*, Sao Paulo (Brasil), III, 787–792, 1956.

28. La interpretación causal de la mecánica ondulatoria. *Ciencia e Investigación* (Buenos Aires) 12: 448–457, 1956.

29. Nuevas constantes del movimiento del electrón. *Revista de la Unión Matemática Argentina y de la Asociación Física Argentina* XVIII, 25, 1956.

30. La antimetafísica del empirismo lógico. *Anales de la Universidad de Chile* CLLXIV, No. 102, 43, 1956.
31. Do Computers Think? *British Journal for the Philosophy of Science* 7, 139–148; 7, 212–219, 1956.
32. Beitrag zur Diskussion über philosophische Fragen der modernen Physik. *Deutsche Zeitrschrift für Philosophie* 4, 467–496, 1956.
33. A survey of the interpretations of quantum mechanics. *American Journal of Physics* 24, 272–286, 1956.
34. ¿Ha progresado la filosofía en el siglo XX? *Revista do Livro* (Rio de Janeiro) I, No. 3/4, 15–21, 1956.
35. El método científico. *Revista del Mar Dulce* (Buenos Aires), I, No. 3, pp. 1–7, 1956.
36. Ubicación de la física teórica. *Revista de la Universidad de Buenos Aires* 1, 405–409, 1956.
37. Las ideas fundamentales de la mecánica ondulatoria. *Ciencia y técnica* 123, No. 616: 3–21, 1957.
38. Lagrangian formulation and mechanical interpretation. *American Journal of Physics* 25, 211–218, 1957.
39. Filosofar científicamente y encarar la ciencia filosóficamente. *Ciencia e Investigación* 13, 244, 1957.
40. ¿Qué es la ciencia? Buenos Aires: Facultad de Ingeniería. 1958.
41. Sobre la imagen física de la partículas de spin entero. *Ciencia e Investigación* 14: 311–315, 1958.
42. On multi-dimensional time. *British Journal for the Philosophy of Science* 9: 39, 1958.
43. ¿Qué significa 'ley científica'? México: Universidad Nacional Autónoma de México, 1958, 13 pages.
44. A filosofia tem progredido durante o século XX? Translation of #34. *Revista Filosófica* (Coimbra) 8, No. 22, 1959.
45. Análisis epistemológico del principio de Arquímedes. Buenos Aires: Facultad de Filosofía y Letras, 12 pages, 1959.
46. Comentario crítico de algunas ideas de Poincaré sobre la hipótesis. Buenos Aires: Facultad de Filosofía y Letras, 13 pages, 1959.
47. ¿Qué es un problema científico? *Holmbergia* (Buenos Aires) VI, No. 15, 47–63, 1959.
48. ¿Cómo sabemos que existe la atmósfera? *Revista de la Universidad de Buenos Aires* IV, No. 2, 246–260, 1959.
49. La axiomática de Peano. Buenos Aires: Centro de Estudiantes de Filosofía y Letras, 1959.
50. On the connections among Levels. *Proceedings of the XIIth International Congress of Philosophy* VI, 63–70. Firenze: Sansoni, 1960.
51. Levels: a semantical preliminary. *Review of Metaphysics* 13, 396–406, 1960.
52. The place of induction in science. *Philosophy of Science* 27, 262–270, 1960.
53. Probabilidad e inducción. *Ciencia y Técnica* (Buenos Aires) 129, 240, 1960.

54. Are there timeless entitites? *Miscelanea de Estudos a Joaquim de Carvalho,* Figueira da Foz (Portugal) No. 3, 290–292, 1960.
55. Analyticity redefined. *Mind* LXX, 239, 1961.
56. The weight of simplicity in the construction and assaying of scientific theories. *Philosophy of Science* 28, 129–149, 1961.
57. Kinds and criteria of scientific law. *Philosophy of Science* 28, 260–281, 1961.
58. Laws of physical laws. *American Journal of Physics* 29, 518–29, 1961.
59. Causality, chance and law. *American Scientist* 49: 432–448, 1961.
60. Ley y determinación. *Scientia* 55,1, 1961.
61. Ethics as a science. *Philosophy and Phenomenological Research* XX: 139–152, 1961.
62. Significación del humanismo en el mundo contemporáneo. *Revista de la Universidad de Buenos Aires* VI, 563, 1961.
63. The complexity of simplicity. *Journal of Philosophy* LIX, 113–135, 1962.
64. Causality: A rejoinder. *Philosophy of Science* 29, 306–317, 1962.
65. La teoría del conocimiento en nuestro tiempo. *Ciencia e Investigación* 18, 60–65, 1962.
66. Cosmology and magic. *The Monist* 44, 116–141, 1962.
67. An analysis of value. *Mathematicae Notae* XVIII, 95–108, 1962.
68. Bertrand Russell y la teoría del conocimiento. In *La filosofía del siglo XX y otros ensayos*, pp. 89–100. Montevideo: Alfa, 1962.
69. Tecnología, ciencia y filosofía. *Revista de la Universidad de Chile* CXXI, No. 126, 64–92, 1963.
70. A general black box theory. *Philosophy of Science* 30, 346–358, 1963.
71. Phenomenological Theories. In Book #15, pp. 234–254.
72. Physics and Reality. *Dialectica* 19, 195, 1965.
73. Technology as applied science. *Technology and Culture* 7: 329–347, 1966.
74. Mach's critique of Newtonian mechanics. *American Journal of Physics* 34, 585–596, 1966.
75. Reprint of #72. *Dialectica* 20: 174–195, 1966.
76. Are there operational definitions of physical concepts? (In Russian). *Voprosi filosofii* No. 11, 66, 1966.
77. On null individuals. *Journal of Philosophy* 63, 776, 1966.
78. Reprint of #56. In M. H. Foster and M. Martin, eds. *Probability, Confirmation and Simplicity*. New York: Odyssey Press, 1966.
79. The structure and content of a physical theory. In Book #17, pp. 15–27.
80. A ghost free axiomatization of quantum mechanics. In Book #21, pp. 105–117.
81. Quanta and philosophy. *Proceedings of the 7th Inter-American Congress of Philosophy* I: 281–296. Québec: Presses de l'Université de Laval, 1967.
82. Physical Axiomatics. *Reviews of Modern Physics* 39, 463–474, 1967.
83. Quanta y filosofía. Translation of #81. *Crítica* (México) 1, No. 3, 41–64, 1967.
84. Analogy in quantum mechanics: from insight to nonsense. *British Journal for the Philosophy of Science* 18, 265–286, 1967.

85. Machs Kritik an der Newtonschen Mechanik. In *Symposium aus Anlass des 50. Todestags von Ernst Mach*, pp. 227–246. Freiburg i. Br.: Ernst-Mach-Institut, 1967.

86. The maturation of science. In I. Lakatos and A. Musgrave, Eds., *Problems in the Philosophy of Science*, pp. 120–137. Discussions by L. L. Whyte, K. R. Popper, and E. H. Hutton, and author's reply: pp. 138–147, Amsterdam: North Holland, 1968.

87. The nature of science. In R. Klibansky, Ed., *Contemporary Philosophy* II, pp. 3–15. Florence: La Nuova Italia Editrice, 1968.

88. Scientific laws and rules. In R. Klibansky, Ed., *Contemporary Philosophy* II, pp. 128–140. Florence: La Nuova Italia Editrice, 1968.

89. Philosophy and physics. In R. Klibanksy, Ed. *Contemporary Philosophy* II, pp. 167–199. Florence: La Nuova Italia Editrice, 1968.

90. On Mach's nonconcept of mass. *American Journal of Physics* 36: 167, 1968.

91. Problems and games in the current philosophy of science. *Proceedings of the XIVth International Congress of Philosophy* I: 566–574. Wien: Herder, 1968.

92. Physique et métaphysique du temps. *Proceedings of the XIVth International Congress of Philosophy* II: 623–629. Wien: Herder, 1968.

93. Towards a philosophy of technology. Reprint of #73. In S. Dockx, ed., *Civilisation technique et humanisme*, pp. 189–210. Bruxelles: Office internationale de librairie, 1968.

94. Conjunction, succession, determination, and causation. *International Journal of Theoretical Physics* 1, 299–315, 1968.

95. Physical time: the objective and relational theory. *Philosophy of Science* 35, 355–388, 1968.

96. La vérification des théories scientifiques. In *Démonstration, vérification, justification: Entretiens de l'Institut International de Philosophie*, pp. 145–159. Comments by N. Rescher, N. Rothenstreich, G. Hirsch, G. Granger, A. J. Ayer, J. Hersch, J. Hyppolite, A. G. M. van Melsen, and H. L. van Breda, and author's replies: pp. 160–179, Louvain-Paris: Nauwelaerts, 1968.

97. Les concepts de modèle. *L'âge de la science* I: 165–180, 1968.

98. Theory of partial truth: not proved inconsistent (L). *Philosophy and Phenomenological Research* 29, 297, 1968.

99. Arten und Kriterien wissenschaftlicher Gesetze. Translation of #57. In G. Kröber, Ed., *Der Gesetzbegriff in der Philosophie und den Einzelwissenschaften*, pp. 117–146. Berlin: Akademie-Verlag, 1968.

100. Filosofía de la investigación científica en los países en desarrollo. *Acta Científica Venezolana* 19, No. 3, 118, 1968.

101. Corrections to Foundations of Physics: Correct and incorrect. *Synthese* 19: 443–452, 1969.

102. Reprint of #100. *Mensurae* (Buenos Aires) 2, No. 11, 1969.

103. Machs Beitrag zur Grundlegung der Mechanik. *Philosophia Naturalis* 11: 189–213, 1969.

104. The metaphysics, epistemology and methodology of levels. In L. L. Whyte, A. G. Wilson, and D. Wilson, Eds., *Hierarchical Levels*, pp. 17–28, New York: American Elsevier, 1969.

105. Alexander von Humboldt und die Philosophie. In H. Pfeiffer, Ed., *Alexander von Humboldt: Werk und Weltgeltung*, pp. 17–30. München: Piper & Co. Verlag, 1969.

106. Analogy, simulation, representation. *Revue internationale de philosophie* 23: 16–33, 1969.

107. What are physical theories about? In N. Rescher, Ed., *Studies in the Philosophy of Science*: American Philosophical Quarterly Monograph No. 3: 61–99, 1969.

108. Azar, probabilidad y ley. *Diánoia* (México) 15, 141–160, 1969.

109. Models in theoretical science. *Proceedings of the XIVth International Congress of Philosophy* III: 208–217. Wien: Herder, 1969.

110. A covariant position operator for the relativistic electron (with A. J. Kálnay). *Progress of Theoretical Physics* 42, 1445–1459, 1969.

111. Four models of human migration: An exercise in mathematical sociology. *Archiv für Rechts- und Sozialphilosophie* 55: 451–462, 1969.

112. The arrow of time. *International Journal of Theoretical Physics* 3: 77–78, 1970.

113. Time asymmetry, time reversal and irreversibility. *Studium Generale* 23: 562–570, 1970.

114. Reprint of #100. *Folia humanística* (Barcelona) 8: 141–154, 1970.

115. La ciencia ¿es éticamente neutral? *Folia humanística* (Barcelona) 8: 241, 1970.

116. Alexander von Humboldt y la filosofía. Spanish translation of #104. *Folia humanística* (Barcelona) 8: 535–546, 1970.

117. The so-called fourth indeterminancy relation. *Canadian Journal of Physics* 48: 1410–1411, 1970.

118. Problems concerning intertheory relations. In P. Weingartner and G. Zecha, Eds., *Induction, Physics and Ethics*, pp. 285–315. Discussion with J. Bar-Hillel, G. Ludwig, W. Leinfellner, H. Margeneau, A. Grünbaum et al.: pp. 316–325, Dordrecht: Reidel, 1970.

119. Comments on Groenewold's paper. In volume mentioned in #118, pp. 202–207, 213–214.

120. Theory meets experience. In H. Kiefer and M. K. Munitz, Eds., *Contemporary Philosophic Thought*, Vol. 2, pp. 138–165, 1970.

121. The physicist and philosophy. *Zeitschrift für allgemeine Wissenschaftstheorie* 1: 196–208, 1970.

122. Reprint of #84. *Archives de l'Institut International des Sciences Théoriques* No. 16: *La symétrie*, pp. 35–58. Bruxelles: Office international de librairie, 1970.

123. Physik und Wirklichkeit. German translation of #72. In L. Krüger, Ed., *Erkenntnisprobleme der Naturwissenschaften*, pp.435–457. Köln-Berlin: Kiepenheur & Witsch, 1970.

124. Reprint of #106. *General Systems* 25: 27–34, 1970.
125. Space and time in contemporary science (in Russian). *Voprosi filosofii* No. 7, pp. 81–92, 1970.
126. Virtual processes and virtual particles: real or fictitious? *International Journal of Theoretical Physics* 3: 507–508, 1970.
127. Conjonction, succéssion, détermination, causalité. In J. Piaget, Ed., *Les théories de la causalité*, pp. 112–132. Paris: Presses universitaires de France, 1971.
128. A philosophical obstacle to the rise of new theories in microphysics. In E. W. Bastin, Ed., *Quantum Theory and Beyond*, pp. 263–273. Cambridge: Cambridge University Press, 1971.
129. The paradox of addition and its dissolution. *Crítica* (México) 3: 27–31, 1971.
130. Is scientific metaphysics possible? *Journal of Philosophy* 68: 507–520, 1971.
131. A mathematical theory of the dimensions and units of physical quantities. In Book #27, pp. 1–16, 1971.
132. "Scientific metaphysics": addenda et corrigenda. *Journal of Philosophy* 68: 876, 1971.
133. A new look at definite descriptions. *Philosophy of Science* (Japan) 4: 131–146, 1971.
134. Reprint of #111. *General Systems* 16: 87–92, 1971.
135. On method in the philosophy of science. *Archives de philosophie* 34: 551–574, 1971.
136. Space and time in modern science. *Il Bienal de Ciência e Humanismo* (São Paulo), pp. 21–34, 1971.
137. Seudociencia y seudofilosofía: dos monólogos paralelos. *Ciencia nueva* (Buenos Aires) No. 15, pp. 41–43, 1971.
138. Reprint of #113. In J. T. Fraser et al., Eds., *The Study of Time*, pp. 122–130. Berlin – Heidelberg – New York: Springer-Verlag, 1971.
139. A program for the semantics of science. *Journal of Philosophical Logic* 1: 317–328, 1972.
140. Metatheory. *In Scientific Thought*, a UNESCO project, pp. 227–252. Paris-The Hague: Mouton/UNESCO, 1971.
141. Modelo del dilema electoral argentino. *Ciencia nueva* (Buenos Aires) No. 21, pp. 52–54, 1972.
142. Reprint of one chapter of Book #19, Ch. 12. In Carl A. Mitcham and Robert Mackey, Eds., *Philosophy and Technology, pp. 62–76*. Riverside, N. J.: The Free Press 1972.
143. Reprint of #72. In Edward A. Mackinnon, Ed., *The Problem of Scientific Realism*. New York: Appleton-Century-Crofts, 1972.
144. Adevar. In *Mario Bunge* (Logicieni si filosofi contemporani, No. 2), pp. 65–115. Bucuresti: Centrul de Informare si Documentare in Scintele sociale si politice, 1973.
145. Meaning in science. *Proc. XVth World Congress of Philosophy* 2: 281–286, 1973.

146. Normative Wissenschaft ohne Normen – aber mit Werten. *Conceptus* VII: 57–64, 1973.
147. Bertrand Russell's regulae philosophandi. In Book #35, pp. 3–12.
148. On confusing 'measurement' with 'measure' in the methodology of the behavioral sciences. In Book #35, pp. 105–122, 1973.
149. The role of forecast in planning. *Theory and Decision* 3: 207–221, 1973.
150. A decision theoretic model of the American War in Vietnam. *Theory and Decision* 3: 328–338, 1973.
151. ¿Es possible una metafísica científica? Spanish translation of #130. *Teorema* III: 435–454, 1973.
152. Conceptul de structura social. *Informatica si model matematice in stiinstele sociale* (Bucharest) II, No. 2, pp. 5–57, 1973.
153. Reprints of #56 and #73 in Alex C. Michalos, Ed., *Philosophical Problems of Science and Technology*, pp. 28–47. Boston: Allyn & Bacon, Inc., 1974.
154. The relations of logic and semantics to ontology. *Journal of Philosophical Logic* 3: 195–210, 1974.
155. Reprint of #73. In Friederich Rapp, Ed., *Contributions to a Philosophy of Technology*, pp. 19–39. Dordrecht-Boston: Reidel, 1974.
156. The concept of social structure. In W. Leinfellner and E. Köhler, Eds., *Developments in the Methodology of Social Science*, pp. 175–215. Dordrecht & Boston: Reidel, 1974.
157. Les présupposés et les produits métaphysiques de la science et de la technique contemporaines. *Dialogue* 13: 443–453, 1974.
158. Metaphysics and Science. *General Systems* 19: 15–18, 1974.
159. Things. *International Journal of General Systems* 1: 229–236, 1974.
160. Teoria stiintifica. Romanian translation of Ch. 7 of Book No. 18. *Epistemologie: Orientari contemporane*, pp. 214–267. Ed. Ilie Parvu. Bucharest: Editura politica, 1974.
161. The methodology of development indicators. UNESCO, Methods and Analysis Division, Dept. of Social Sciences, 1974.
162. Crítica de la noción fregeana de predicado. *Revista Latinoamerican de Filosofiá* 1: 5–8, 1975.
163. Enscheidungstheoretische Modelle in der Politik: Vietnam. German transl. of #150. In R. Simon-Schaefer and W. Ch. Zimmerli, Eds., *Wissenschaftstheorie der Geisteswissenschaften*, pp. 309–323. Hamburg: Hoffman & Campe, 1975.
164. El significado en ciencia. Spanish transl. of #145. *Teoría* (México) 1, No. 1, 1975.
165. What is a quality of life indicator? *Social Indicators Research* 2: 65–80, 1975.
166. Towards a technoethics. *Philosophic Exchange* 2, No. 1, p. 69–79, 1975.
167. Welches sind die Besonderheiten der Quantenphysik gegenüber der klassischen Physik? (With Andrés J. Kálnay). In R. Haller & J. Götschl, Eds., *Philosophie und Physik*, pp. 25–38. Braunschweig: Vieweg, 1975.
168. La paradoja de la adición: respuesta al Maestro Margáin. *Crítica* 7: 105–107, 1975.
169. Ontología y ciencia. *Diánoia* 1975: 50–59.

170. La representación conceptual de los hechos. *Teorema* 5: 31; 7–360, 1975.
171. Reprint of #145. *Poznán Studies in the Philosophy of the Sciences and the Humanities* 1, No. 4, 55–64, 1975.
172. A critical examination of dialectics. In Ch. Perelman, Ed., *Dialectics/Dialectique*, pp. 63–77. The Hague: Martinus Nijhoff. Commentary by I. Narsky, "Bemerkungen über den Vortrag von Prof. Bunge", pp. 78–86, 1975.
173. Russian translation of #170. *Voprosi filosofii* No. 4, 1975.
174. ¿Hay proposiciones? *Aspectos de la Filosofía de W. V. Quine,* pp.53–68. Valencia: Teorema, 1975.
175. Reprint of #100. In Jorge A. Sábato, Ed., *El pensamiento latinoamericano en la problemática ciencia-tecnología-desarrollo-independencia*, pp. 44–51. Buenos Aires: Paidos, 1975.
176. French transl. of #167. *Fundamenta scientiae* No. 11, 1976.
177. Possibility and probability. In W. Harper and C. Hooker, Eds., *Foundations of Probability Theory, Statistical Inference, and Statistical Theories of Science*, Vol. III, pp. 17–33. Dordrecht-Boston: Reidel, 1976.
178. Reprint of #165 with slight changes. In J. King-Farlow and W. Shea, Eds., *Values and the Quality of Life*, pp. 142–156. New York: Neale Watson Academic Publications, 1976.
179. The relevance of philosophy to social science. In W. Shea, Ed., *Basic Issues in the Philosophy of Science*, pp. 136–155. New York: Neale Watson Academic Publications, 1976.
180. El método en la biología. *Naturaleza* (México) 7: 70–81, 1976.
181. El ser no tiene sentido y el sentido no tiene ser. *Teorema* VI: 201–212, 1976.
182. Differentiation, participation and cohesion. (With Máximo García Sucre). *Quality and Quantity* 10: 171–178, 1976.
183. Reprint of #169. *La filosofía y la ciencia en nuestros días*, pp. 27–40. México: Ed. Grijalbo, 1976.
184. Spanish transl. of #178. *La filosofía y las ciencia sociales*, pp. 43–69, México: Ed. Grijalbo, 1976.
185. A model for processes combining competition with cooperation. *Applied Mathematical Modelling* 1: 21–23, 1976.
186. Polish transl. of #145. *Poznanskie Studia z Filosofii Nauki*, Vol. 1, pp. 13–23. Warsaw-Poznan: Panstwowe Nydawnictwo Naudowe, 1976.
187. Reprint of #158. *Science et métaphysique*, pp. 193–206. Bruxelles: Office international de librairie, 1976.
188. Is science value-free and morally neutral? *Philosophy and Social Action* II, No. 4, pp. 5–18, 1976.
189. Examen filosófico del vocabulario sociológico. *Diánoia* XXII, pp. 56–75, 1976.
190. Greek transl. of #87. *Deukalion* 3: 351–363, 1976.
191. ¿Qué es y para qué sirve la epistemología? *Revista de la Universidad de México* XXXI, No. 2, 1–7, 1976.
192. Reprint of #166. *The Monist* 60: 96–107, 1977.

193. The interpretation of Heisenberg's inequalities. In H. Pfeiffer, Ed., *Denken und Umdenken: zu Werk und Wirkung von Werner Heisenberg*, pp. 146–156. München-Zürich: Piper & Co., 1977.

194. Reply to van Rootselar's criticisms of my theory of things. *Intern. J. General Systems* 3: 181–182, 1977.

195. A theory of properties and kinds. (With Arturo Sangalli). *Intern. J. General Systems* 3: 183–190, 1977.

196. Tres políticas de desarrollo científico y una sola eficaz. *Interciencia* 2: 76–80, 1977.

197. Reprint of #196. In Enrique Leff, Ed., *Primer Simposio sobre Ecodesarrollo*, pp. 88–96. México, D. F.: Asociación Mexicana de Epistemología, 1977.

198. Spanish transl. of #127. *In La teoría de la causalidad*. Salamanca: Ed. Sígueme, 1977.

199. A relational theory of physical space. (With A. García Máynez). *Intern. J. Theoretical Physics* 15: 961–972, 1977.

200. Emergence and the mind. *Neuroscience* 2: 501–509, 1977.

201. Levels and reduction. *Am. J. Physiology: Regulatory, Integrative and Compar. Physiol.* 2: 75–82, 1977.

202. The philosophical richness of technology. In F. Suppe and P. D. Asquith, Eds., *PSA* 2, pp. 153–172, 1977.

203. The GST challenge to the classical philosophies of science. *Intern. J. General Systems* 4: 29–37, 1977.

204. General systems and holism. *General Systems* XXII, 87–90, 1977.

205. States and events. In William Hartnett, Ed., *Systems: Approaches, Theories and Applications*, pp. 71–95. Boston and Dordrecht: Reidel, 1977.

206. Reprint of #154. In Edgar Morscher, Johannes Czermak, and Paul Weingartner, Eds., *Problems in Logic and Ontology*, pp. 29–43. Graz: Akademishces Druck und Verlagsanstalt, 1977.

207. ¿Qué es y a qué puede aplicarse el método científico? *Diánoia* 88–101, 1977.

208. A systems concept of the international system. In Mario Bunge, Johan Galtung and Mircea Malitza, Eds., *Mathematical Approaches to International Relations*, pp. 291–305. Bucharest: Romanian Academy of Social and Political Sciences.

209. Quantum mechanics and measurement. *International Journal of Quantum Chemistry* 12, Suppl. 1: 1–13, 1977.

210. The mind-body problem in the light of contemporary biology. (With Rodolfo Llinás). *16th World Congress of Philosophy: Section Papers*, pp. 131–133, 1978.

211. Restricted applicability of the concept of command in the neurosciences: dangers of metaphors. (With Rodolfo Llinás). *The Behavioral and Brain Sciences* 1: 3–31, 1978.

212. Physical space. In M. Svilar and A. Mercier, Eds., *Space*, pp. 133–148. Comments by H. Törenbohm, ibid., pp. 149–167. Other comments and author's replies on pp. 167–171. Bern-Frankfurt-Las Vegas: Peter Lang, 1978.

213. A model of evolution. *Applied Mathematical Modelling* 2: 201–204, 1978.

214. The limits of science. *Epistemología* 1: 11–32, 1978.

215. Iatrofilosofía. In F. Alonso de Florida, Ed., *Ensayos de Yatrofilosofía*, pp. 3–5. México: Academia Nacional de Medicina, 1978.

216. La enfermedad como estado o proceso. In #215, pp. 65–69.

217. Conocimiento objetivo y mundos popperianos. *Semestre de Filosofía* I, No. 21, pp. 7–25, 1978.

218. La cultura como sistema concreto. *Ciência e Filosofía* 1, No. 1, pp. 7–30, 1978.

219. A systems concept of society: Beyond individualism and holism. *Theory and Decision* 10: 13–30, 1979.

220. The five buds of technophilosophy. *Technology in Society* 1: 67–74, 1979.

221. The mind-body problem, information theory, and Christian dogma. *Neuroscience* 4: 453–454, 1979.

222. Philosophical inputs and outputs of technology. In G. Bugliarello and D. B. Donner, Eds., *The History and Philosophy of Technology* pp. 262–281. Urbana Ill.: University of Illinois Press, 1979.

223. Preface to Augusto Fernández Guardiola, Ed., *La conciencia*, pp. 5–8. México: Trillas, 1979.

224. La bancarrota del dualismo psiconeural. In #223, pp. 71–84. Comment by Carlos Pereda, ibid., pp. 85–87, 1979.

225. Russian translation of #224. *Filosofskie Nauki* No. 2, pp. 77–87, 1979. Comment by D. I. Dubrosvskii, ibid., p. 88–97, 1979.

226. El finalismo en biología, psicología y sociología. *Revista Latinoamericana de Filosofía* 5: 33–40, 1979.

227. ¿Ideologizar la ciencia o cientificizar la ideología? In Mario H. Otero, Ed., *Ideología y ciencias sociales*, pp. 41–51. México: Universidad Nacional Autónoma de México, 1979.

228. A model of secrecy. *Journal of Irreproducible Results* 25: 25–26, 1979.

229. The mind-body problem in an evolutionary perspective. In *Brain and Mind*, Ciba Foundation Series 69, pp. 53–63. Amsterdam: Excerpta Medica 1979.

230. Relativity and philosophy. In J. Bärmark, Ed., *Perspectives in Metascience* pp. 75–88. Göteborg, Regiae Societatis Scientiarum et Litterarum Gothoburgensis, Interdisciplinaria 2, 1979.

231. Reprint of #230. *Physics in Canada* 35: 105–111, 1979.

232. The Einstein-Bohr debate over quantum mechanics: Who was right about what? *Lecture Notes in Physics* 100: 204–219, 1979.

233. Some topical problems in biophilosophy. *Journal of Social and Biological Structures* 2: 155–172, 1979.

234. Reply to Craig Dilworth's review of the *Treatise on Basic Philosophy*, vols. 1–4. *Epistemologia* II: 425–428, 1980.

235. Reprint of #214. *The Physiologist* 23: 7–13, 1980.

236. From neuron to behavior and mentation: an exercise in levelmanship. In H. M. Pinsker and W. D. Williams, Eds., *Information Processing in the Nervous System* pp. 1–16. New York: Raven Press, 1980.

237. Valor biológico y valor psicolólogico. In J. J. E. Garcia, Ed., *El hombre y su conducta/Man and his Conduct*, pp. 102–11. Rio Piedras, P. R: Editorial Universitaria, 1980.

238. Technoethics. In M. Kranzberg, Ed., *Ethics in an Age of Pervasive Technology* pp. 139–142. Boulder, Colo.: Westview Press, 1980.

239. Reprint of #200 in A. D. Smith, R. Lllinas and P. K. Kostyuk, Eds., *Commentaries in the Neurosciences* pp. 633–641. Oxford: Pergamon Press, 1980.

240. Die Standardphilosophie der Physik. German translation of #121. *Physik und Didaktik* 8: 261–267 (1980).

241. La función de la ciencia básica en el desarrollo nacional. *Tecnología y Desarrollo* 4: 153–170, 1980.

242. Introduction to Dalbir Bindra, Ed., *The Brain's Mind*, pp. 1–5. New York: Gardner Press, 1980.

243. The psychoneural identity theory. In #242, pp. 89–108.

244. Una teoría materialista de la mente. Spanish translation of #243. *Episteme* II No. 4 pp. 43–50, 1980.

245. The geometry of a quantal system (with M. García-Sucre), *International J. of Quantum Chemistry* 19: 83–93, 1980.

246. Reprint of #216. *General Systems* XXIV: 27–44, 1980.

247. El problema mente-cuerpo. *Ciencia, Technología y Desarrollo* 4: 239–310, 1980.

248. Materialismo sin dialéctica. *Ciencia, Technología y Desarrollo* 4: 511–512, 1980.

249. Systems all the way. *Nature and System* 3: 37–47, 1981.

250. Development indicators. *Social Indicators Research* 9: 369–385, 1981.

251. Biopopulations, not biospecies, are individuals and evolve. *The Behavioral and Brain Sciences* 4: 284–285, 1981.

252. Four concepts of probability. *Applied Mathematical Modelling* 5: 306–312, 1981.

253. From mindless neuroscience and brainless psychology to neuropsychology. *Annals of Theoretical Psychology* 3: 115–133, 1981.

254. Polish translation of #220. In W. Gasparski and D. Miller, Eds., *Projectowanie i Systemy* pp. 131–139l. Warsaw: Wydawnictwo Polskiej Akademia Nauk, 1981.

255. Conceptual existence. In P. Cohn, Ed., *Transparencies: Philosophical Essays in Honor of J. Ferrater Mora* pp. 5–14. Atlantic Heights, N. J.: Humanities Press, 1981.

256. Half truths. In E. Morscher and G. Zecha, Eds., *Philosophie als Wissenschaft/Essays in Scientific Philosophy* pp. 87–91. Bad Reichenhall: Comes Verlag, 1981.

257. Relatividad y filosofía. Spanish translation, with some changes, of #230. In J. Chela-Flores, Ed., *Einstein*, pp. 43–61. Caracas: Equinoccio, 1981.

258. Systemism: a new cognitive paradigm. *Philosophon Agora* (Lublin) 1: 1–4. Polish translation, idem 1: 5–9, 1981.

259. Analogy between systems. *International Journal of General Systems* 7: 221–223, 1981.

260. Las funciones de la ciencia y de la técnica en el desarrollo nacional. In *Memoria del Primer Seminario Nacional sobre Política de Desarrollo Científico y Tecnológico* Vol. 1 pp. 115–138. Quito: Editorial Voluntad, 1981.

261. Ciencia básicas, ciencia aplicada, técnica y producción. Differencias y relaciones. In work #260, Vol. II pp. 51–68.

262. Reprint of #261. *El País* (Madrid), 18 and 20 June 1982.

263. Los determinantes de la moral humana. *El País* (Madrid). 19 April 1982.

264. The revival of causality. In G. Floistad, Ed., *Contemporary Philosophy* Vol. 2 pp. 133–155. The Hague: Martinus Nijhoff, 1982.

265. La comunicación. *Papeles de comunicación* (Madrid) 1: 11–39, 1982.

266. Is chemistry a branch of physics? *Zeitschrift für allgemeine Wissenschaftstheorie* 13: 209–233, 1982.

267. Desde una neurociencia sin mente y una psicología sin cerebro a una neuropsicología. Translation of #253. *Revista de Filosofía* (Chile) XX: 5–22, 1982.

268. Cómo desenmascarar falsos científicos. *Los Cuadernos del Norte* Vol. III, No. 15: 52–69, 1982.

269. A pszichoneuralist azonossag elmélete. Transl. of #243. *Magyar filozofiai szemle* 1982: 540–553. Comment by Szentagothai Janos, ibid., pp. 554–557.

270. La psicología como ciencia natural. *Actas del I Congreso de Teoría y Metodología de las Ciencias*, pp. 25–32. Discussion: pp. 33–49. Oviedo: Pentalfa, 1982.

271. El significado de la física cuántica. Ibid., pp. 363–366. Discussion: pp. 366–371.

272. Teoría económica y realidad económica. Ibid., pp. 11–151. Discussion: pp. 455–471.

273. Speculation: wild and sound. *New Ideas in Psychology*. 1: 3–6, 1983. Comment by Thomas Nickles, ibid., pp. 7–10.

274. Comment on a paper by Fedanzo. *J. Social and Biological Structures* 6: 159–160, 1983.

275. Demarcating science from pseudoscience. *Fundamenta Scientiae* 3: 369–388, 1983.

276. Prologue to F. Parra Luna, *Elementos para una teoria formal del sistema social*, pp. 13–15. Madrid: Editorial de la Universidad Complutense, 1983.

277. Solution to two paradoxes in the quantum theory of unstable systems (with A. J. Kálnay) *Nuovo Cimento* B 77: 1–9, 1983.

278. Real successive measurements on unstable quantum systems take nonvanishing time intervals and do not prevent them from decaying (with A. J. Kálnay) *Nuovo Cimento* B 77: 10–18, 1983.

279. Paradigmas y revoluciones en ciencia y técnica. *El Basilisco* No. 15: 2–9, 1983.

280. Sobre materialismo y dialéctica. *El Basilisco* 15: 94–95, 1983.

281. El arca hispánica en el nuevo diluvio internacional. In Instituto de Cooperación Iberoamericana. *Iberoamérica: Encuentro en la Democracia,* pp. 268–276. Madrid, 1983. Debate following the paper: pp. 276–280.

282. Lo que el físico espera de la filosofía. *Episteme* NS 1: 17–32, 1983.

283. El estilo de Russell. Preface to *Bertrand Russell, Significado y verdad,* pp. 7–14. Barcelona: Ariel, 1983.

284. Reprint of #281 in *Gaceta de Canarias* II, No. 5, pp. 76–80, 1983.

285. La crisis actual no data de ayer ni es exclusivamente económica. *Boletin de Estudios Económicos* (Spain) XXXVIII: 43–45, 1983.

286. De la neurologie sans âme et de la psychologie sans tête à la neuropsychologie. *Petite revue de philosophie* 5: 1–45, 1983. Translation of #253.

287. Reprint of #275. In M. Wahba, Ed., *The Unity of Knowledge,* pp. 63–88. Cairo: Ain Shams University Press, 1983.

288. Closing remarks: Philosophy and the fellah. In volume cited in #287, pp. 219–221.

289. Japanese translation of #253, *New Medical World Weekly* (Tokyo) 1983, 10, 31, pp. 6–7; 1983, 11, 7 p. 2, 1983, 11, 14, p. 9.

290. Metateoría. Spanish transl. of #140. In Y. Bar-Hillel et al. *El pensamiento científico,* pp. 225–265. Madrid: Tecnos/UNESCO, 1983.

291. La necesidad de mantener la dicotomía entre verdades de razón y verdades de hecho. *Revista Latinoamericana de Filosofía* 10: 63–69, 1983.

292. Philosophical problems in linguistics. *Erkenntnis* 21: 107–173, 1984.

293. Philosophical conditions of scientific development. *Philosophy and Social Action* 10: 9–25, 1984.

294. Hidden variables, separability, and realism. *Rev. Brasil. Fisica,* Volume especial os 70 anos de Mario Schönberg 150–168, 1984.

295. Die Wiederkehr der Kausalität. Transl. of #264. In B. Kanitscheider, Ed., *Moderne Naturphilosophie* pp. 141–160. Würzberg: Königshausen & Neumann, 1984.

296. What is pseudoscience? *The Skeptical Inquirer* IX, No. 1: 36–46, 1984.

297. Hacia la cooperación auténtica. Prologue to José Luis Pardos, *Crecimiento y desarrollo en la década de los 80,* pp. 11–14. Madrid: Tecnos, 1984.

298. Fundamentos y filosofía de la matemática. *Arbor* CXVIII Nos. 463–464: 7–39, 1984.

299. Albert Einstein, el célebre desconocido. Prologue to Banesh Hoffmann, *Einstein,* pp.11–17. Barcelona: Salvat, 1984.

300. Can science and technology be held responsible for our current social ills? In P. T. Durbin and C. Mitcham, Eds., *Research in Philosophy and Technology* 7: 19–22. Greenwich, Conn.: Jai Press, 1984.

301. Hungarian translation of #172. *Magyar Filozof Szemle* 4: 566–578, 1984.

302. Comment on Apostel's paper. *Studies in Soviet Thought* 29: 137–138, 1985.

303. Comment on Mark Klein Taylor's paper. *Current Anthropology* 26: 174, 1985.

304. Cajas negras y translúcidas y acción a distancia: Sánchez Ron. *Teorema* 15: 271–274, 1985.

305. Realismo y antirrealismo en la filosofía contemporánea. *Arbor* CXXI (473): 13–40, 1985.

306. From mindless neuroscience and brainless psychology to neuropsychology. *Annals of Theoretical Psychology* 3: 115–133. Comments "On being brainy" by M. C. Corballis, ibid., pp. 135–142, and "Is neuropsychology something new?", by P. C. Dodwell, ibid., pp. 143–150, 1985.

307. On research strategies in psychology. Reply to commentators. *Annals of Theoretical Psychology* 3: 151–156, 1985.

308. ¿Qué es un individuo concreto? *Theoria* 1: 121–128, 1985.

309. Correspondencia, analogía, complementaridad, superposición y realismo: homenaje a Niels Bohr (1885–1962). *Arbor* CXXII: 51–64, 1985.

310. Reprint of #174, with some additions. In J. E. Gracia et al., Eds., *El análisis filosófico en América Latina* (México: Fondo de Cultura Económica, 1985), pp. 580–592.

311. Types of psychological explanation. In J. McGough, Ed., *Contemporary Psychology: Biological Processes and Theoretical Issues*, pp. 489–501. Amsterdam: North Holland, 1985.

312. Spanish Translation of #273, *Revista de Filosofía* (Chile) XXV–XXVI: 7–11, 1985.

313. Preface to Lluís Garcia i Sevilla's *Anàlisi de la psicoanàlisi*, pp. 5–7. Barcelona: Institut d'Estudis Catalans, 1985.

314. Reprint of #309. *Física* (Buenos Aires) No. 1: 29–37, 1985.

315. A philosopher looks at the current debate on language acquisition. In I. Gopnik and M. Gopnik Eds., *From Models to Modules* pp. 229–239. Norwood NJ: Ablex Publs. Co., 1986.

316. Reportaje a Mario Bunge. *Física*, año 2, No. 2, pp. 6–13, 1986.

317. ¿Grados de existencia o de abstracción? *Theoria* 1: 547–549, 1986.

318. Individuos, conjuntos y sistemas. *Theoria* 1: 555–560, 1986.

319. Science, technology and ideology in the Hispanic World. *Free Inquiry* 6(3): 36–49, 1986.

320. Ciencia e ideología en el mundo hispánico. Translation of #319. *Interciencia* 11: 120–125, 1986.

321. Ideology and science. In M. Wahba, Ed., *Philosophy and Physics*, pp. 105–114. Cairo: Ain Shams University, 1986.

322. Reprint of #202. In M. Wahba, Ed., *Philosophy and Physics*, pp. 81–103. Cairo: Ain Shams University, 1986.

323. Considérations d'un philosophe sur l'économie du néo-conservatisme (néo-libéralisme). In J. Jalbert & L. Lepage, Eds., *Néo-conservatisme et restructuration de l'état*, pp. 49–70. Sillery, Qué.: Presses de l'Univeristé du Québec, 1986.

324. Two controversies in evolutionary biology: saltationism and cladism. In N. Rescher, Ed., *Scientific Inquiry in Philosophical Perspective*, pp. 129–145. Lanham, MD: University Press of American, 1987.

325. La psicología: ¿disciplina humanística, autónoma, natural o social? *Arbor* 126, No. 496: 9–30, 1987.

326. Chinese translation of #275. *Philosophical Research* (Beijing) 1987, No. 4: 46–51.
327. Borges y Einstein, o la fantasía en arte y en ciencia. *Revista de Occidente* No. 73: 45–62, 1987.
328. Le problème corps-esprit. *Médecine psychosomatique* 15: 85–94, 1987.
329. Seven desiderata for rationality. In J. Agassi & I. Jarvie, Eds., *Rationality: The Critical View*, pp. 5–15. Dordrecht: Nijhoff, 1987.
330. In defence of realism and scientism. *Annals of Theoretical Psychology* 4: 23–26, 1987.
331. El marxismo hoy. In R. Reyes, Ed., *Cien años después de Marx*, pp. 27–41. Madrid: Akal, 1987.
332. Reprint of #286. *Verhaltens Therapie* 1/1986 pp. 3–39.
333. Philosophie, science, politique (Entretien). *Concordia* 10: 51–67, 1987.
334. Existe el tiempo? *Revista de Occidente*, No. 76: 35–40, 1987.
335. Ten philosophies of mind in search of a scientific sponsor. *Proceedings of the 11th International Wittgenstein Symposium* pp. 285–293. Wien: Hölder-Pichler-Tempsky, 1987.
336. Supervivencia o extinción. *Papeles de Campanar* I, No. 2, pp. 18–24, 1987.
337. Eine Kritik des Mentalismus. Transl. of Ch. 5 of Book #77. *Zeits. f. klinische Psychologie, Psychopathologie und Psychotherapie* 35: 244–269, 1987.
338. Causality. *Systems and Control Encyclopaedia*, pp. 552–556, 1988.
339. Ideology and science. In G. L. Eberlein & H. Berghel, Eds., *Theory and Decision: Essays in Honor of Werner Leinfellner* pp. 79–89. Dordrecht-Boston: Reidel, 1988.
340. Filosofía de la ciencia y de la técnica. *Fisica* III, No. 8/9: 74–87, 1988.
341. Desarrollo y medio ambiente. *Ciencia energética*, No. 62: 20–28, 1988.
342. Two faces and three masks of probability. In E. Agazzi, Ed., *Probability in the Sciences*, pp. 27–50. Dordrecht-Boston: Reidel, 1988.
343. Why parapsychology cannot become a science. *Behavioral and Brain Sciences* 10: 576–577, 1988.
344. Analytic philosophy of society and social science: The systemic approach as an alternative to holism and individualism. *Revue internationale de Systémique* 2: 1–13, 1988.
345. Analytische Sozialphilosophie und Philosophie der Sozialwissenshchaften: der systemische Zugang als eine Alternative zum Holismus und zum Individualismus. *Proceedings of the 12th International Wittgenstein Symposim*, 166–175.1988.
346. The thirteen riders of the Apocalypse. *Free Inquiry* 8, no. 2, 9.1988 347La psicologia ¿ ciencia del siglo XXI? *Boletín Argentino de Psicología* 1: 8–9, 1988.
347. La psicologia ¿ ciencia del siglo XXI? *.Boletín Argentino de Psicología* 1: 8–9, 1988.
348. ¿Qué es la mente? *Il cannochiale* #3: 99–109 (1987).

349. Reprint of #277. In L. E. Ballentine, Ed., *Foundations of Quantum Mechanics Since the Bell Inequalities. Selected Reprints,* pp. 53–61.College Park, MD: American Associacion of Physics Teachers, 1988.

350. Modelos para procesos que combinan competencia y cooperación. Transl. of #185. *Revista Iberoamericana de Autogestión y Acción Comunal,* VI, No. 13–14, pp. 27–32, 1988.

351. The scientific status of history. In U. Hinke-Dörnemann, Ed., *Die Philosophie in der modernen Welt,* Vol. I, pp. 593–602. Frankfurt: Peter Lang, 1988.

352. Superposition of quantum states: fact or fiction? In N. Fleury et al., Eds., *Leite Lopes Festschrift: A Pioneer Physicist in the Third World,* pp. 135–142. Singapore: World Scientific, 1988.

353. The nature of applied science and technology. *Proceedings of the XVII World Congress of Philosophy,* Vol. II, pp. 599–604, 1988.

354. Scientific change: gradual or catastrophic? *Proceedings of the XVII World Congress of Philosophy,* Vol. IV, pp. 792–796, 1988.

355. The ambivalent legacy of operationism. *Philosophia naturalis* 25: 337–345, 1988.

356. Niels Bohr's philosophy. *Philosophia naturalis* 25: 399–415, 1988.

357. Relaciones de la lógica y la semántica con la ontología. Transl. of #154. In *Antologia de la lógica en América Latina,* pp. 577–591. Madrid: Fundación del Banco Exterior, 1988.

358. Basic science is innocent, applied science and technology can be guilty. In G. E. Lemarchand & A. R. Pedace Eds. *Scientists, Peace and Disarmament* pp. 245–261. Singapore: World Scientific, 1988.

359. El país puede salir del pozo. In *Argentina ¿tiene salida?,* pp. 221–225. Buenos Aires: Clarín-Aguilar, 1989.

360. Development and the environment. In E. F. Byrne & J. C. Pitt, Eds., *Technological Transformation: Contextual and Conceptual Implications,* pp. 285–304. Dordrecht-Boston: Kluwer, 1989.

361. From neuron to mind. *News in Physiological Sciences* 4: 206–209, 1989.

362. Gradualism vs. saltationism in evolutionary biology: From Darwin to Gould (with David Blitz). *Proceedings of the 13th International Wittgenstein Symposium* pp. 297–301. Wien: Hölder-Pichler-Tempsky, 1989.

363. Aprender ciencia y técnica o decaer. *Educa* (Spain) VI, No. 21: 10–16, 1989.

364. Toward a survival morality. In P. Kurtz, Ed., *Building a World Community,* pp. 36–41. Buffalo, N.Y.: Prometheus, 1989.

365. Reduktion und Integration, Systeme und Niveaus, Monismus und Dualismus. In E. Pöppel, Ed., *Gehirn und Bewusstsein,* pp. 87–104. Weinheim: VCH, 1989.

366. Entretien avec Mario Bunge, par Cécile Landry. *Philosopher* No. 8 pp. 25–33, 1989.

367. Il problema mente/cervello nella interpretazione della teoria emergentista. Conversazione con Mario Bunge. Interview by Silvano Chiari. *Psicologia italiana,* Vol. X, No. 1: 32–37, 1989.

368. Game theory is not a useful tool for the political scientist. *Epistemologia* 12: 195–212, 1989.

369a. The popular perception of science in North America. *Transactions of the Royal Society of Canada* Ser. V, Vol. V: 269–289, 1989.

369b. *Philosophia Naturalis*, 26(1), 121, 1989.

370. Des bons et des mauvais usages de la philosophie. Chapter 1 of Book #39. *L'Enseignement philosophique* 40(2) 97–110, 1990.

371. De la neurona a la mente. Translation of #360. *Boletín Argentino de Psicología* III, No. 3:1–7, 1990.

372. Mario Bunge: un filósofo que defiende la idea de progreso científico. Interview with Julio Abramczyk, *Arbor* CXXXV (No. 530): 9–18, 1990.

373. What kind of discipline is psychology: Autonomous or dependent, humanistic or scientific, biological or sociological? *New Ideas in Psychology* 8: 121–137, 1990. Commentaries by J. Panksepp (pp. 139–149), R. E. Mayer (pp. 151–154), J. R. Royce (pp. 155–157), and G. Cellerier & J.-J. Ducret (pp. 159–175).

374. The nature and place of psychology: A reply to Panksepp, Mayer, Royce, and Cellerier and Ducret. *New Ideas in Psychology* 8: 176–188, 1990.

375. La opinión pública y el desarrollo científico y técnico en una sociedad democrática. *Arbor* CXXXVI No. 534–535: 13–42, 1990.

376. Computerism – A brainless approach to cognition: A reply to Sloman. *New Ideas in Psychology* 8: 377–379, 1990.

377. El sistema técnica-ciencia-filosofía: un triángulo fértil. *Telos* No. 24: 13–22, 1991.

378. Reprint of #358. In D. O. Dahlstrom, Ed., *Nature and Scientific Method* pp. 95–105. Washington DC: The Catholic University of America Press, 1991.

379. La percepción popular de la ciencia en Norteamérica. Transl. of #369. *El ojo escéptico* Vol. 1 #2: 1–4, 1991.

380. The power and limits of reduction. In E. Agazzi, Ed., *The Problem of Reductionism in Science*, pp. 31–49. Dordrecht-Boston: Kluwer, 1991.

381. Una caricatura de la ciencia: la novísina sociología de la ciencia. *Interciencia* 16: 69–77, 1991. Replies by J. Requena, E. Quevedo V., L. C. Arboleda and M. Hernández, S. Schwartzman, J. M. Carvalho, and L. Velho in *Interciencia* 16: 229, 266–271.

382. A skeptic's beliefs and disbeliefs. *New Ideas in Psychology* 9: 131–149, 1991. Replies by J. E. Alcock, E. Bauer and W. v. Lucadom, D. Blitz, R. Boudon, P. Feyerabend, W. Harman, G. Kreweras, W. Laucken, S. Moscovici, M. Perrez, R. Thom, and J. Van Rillaer 9: 151–244.

383. What is science? Does it matter to distinguish it from pseudoscience? A reply to my commentators. *New Ideas in Psychology* 9: 245–283, 1991.

384. Charges against applied game theory sustained: Reply to Schmidt. *Epistemologia* 13: 151–154, 1991.

385. Le système technique-science-philosophie: un ménage à trois fécond. *Revue internationale de systémique* 5: 171–180, 1991.

386. A critical examination of the new sociology of science, Part 1. *Philosophy of the social sciences* 21: 524–560, 1991.

387. La producción y el consumo de leyes científicas. *Interciencia* 16: 173, 1991.

388. Dos caras y tres máscaras de la probabilidad. Spanish translation of #342. *Física* VI (2): 16–29, 1991.

389. Chinese translation of #345. *Journal of Philosophy in Translation* 2: 9–15, 1991.

390. Five bridges between scientific disciplines. In F. Geyer, Ed., *The Cybernetics of Complex Systems: Self-Organization, Evolution, and Social Change*, pp. 1–10. Salinas CA: Intersystems Publications, 1991.

391. Rights imply duties. In E. Groffier and M. Paradis, Eds., *The Notion of Tolerance and Human Rights: Essays in Honour of Raymond Klibansky*, pp. 47–54. Ottawa: Carleton University Press, 1991.

392. Why we cherish exactness. In *Advances in Scientific Philosophy: Esays in Honour of Paul Weingartner*, pp. 591–598. Amsterdam: Rodopi, 1991.

393. La investigación científica como empresa. *Interciencia* 16: 297, 1991.

394. La botanique est. venue après les fleurs. *Philosopher* #ll: 41–51, 1991.

395. A philosophical perspective on the mind-body problem. *Proceedings of the American Philosophical Society* 135: 513–523, 1991.

396. La vérité. In M. A. Sinaceur, Ed., *Penser avec Aristote* pp. 453–457. Paris: Erès, 1991.

397. Le lieu et l'espace. In M. A. Sinaceur, Ed., *Penser avec Aristote* pp. 483–488. Paris: Erès, 1991. Comments by P. Aubenque, p. 495–496, 519–520.

398. José Ferrater Mora (1912–1991) *Revista Latinoamericana de Filosofía* 17: 373–375, 1992.

399. A critical examination of the new sociology of science, Part 2. *Philosophy of the Social Sciences* 22: 46–76, 1992.

400. Systems everywhere. In C. Negoita, Ed., *Cybernetics and Applied Systems*, pp.23–41. New York: Marcel Dekker, 1992.

401. System boundary. *International Journal of General Systems* 20: 215–219, 1992.

402. La philosophie de Niels Bohr [French translation of #356] *Horizons philosophiques* 2, No. 2: 27–50, 1992.

403. A neurophysiological explanation of creativity. In J. Brzezinski, F. Coniglione, and T. Marek, Eds., *Science: Between Algorithm and Creativity* pp. 161–164. Delft: Eburon, 1992.

404. The scientist's skepticism. [Repr. of two fragments of #382]. *Skeptical Inquirer 16*: 377–380, 1992.

405. La percepción popular de la ciencia en Norteamérica. Spanish transl. of #369. *La alternativa racional* No. 24: 20–27, 1992.

406. Reprint of #74. In J. Blackmore, Ed., *Ernst Mach – A Deeper Look* pp. 243–261. Dordrecht-Boston: Kluwer Academic Publishers, 1992.

407. Subjetivismo y relativismo: Síntomas y causas de decadencia cultural. *Balcón* No. 8–9: 121–124, 1992.

408. Los pecados filosóficos de la nueva sociología de la ciencia. In C. A. di Prisco & E. Wagner, Eds., *Visiones de la ciencia. Homenaje a Marcel Roche*, pp. 33–42. Caracas: Monte Avila Editores Latinoamericana & Instituto Venezolano de Investigaciones Científicas, 1992.

409. Morality is the basis of legal and political legitimacy. In W. Krawietz & G. H. von Wright, Eds., *Oeffentliche oder private Moral?* pp. 379–386. Berlin: Duncker & Humblot, 1992.

410. Lógica y verdad (with J.-P. Marquis). In D. Sobrevilla and D. García Balaúnde, Eds., *Lógica, razón y humanismo: La obra filosófica de Francisco Miró Quesada C.* pp. 359–369. Lima, 1992.

411. Sette paradigmi cosmologici: L'animale, la scala, il fiume, la nuvola, la macchina, il libro e il sistema dei sistemi. *Aquinas* 35: 219–235, 1992.

412. Eine Kritik der Grundlagen der Theorie der rationalen Wahl. *Zeitschrift für Wissenschaftliche Forschung* 7/9: 19–33, 1992–93.

413. Vetenskapsmannens skepticism. [Swedish transl. of #404.] *Sökaren* No. 1: 29–31, 1993. Comment by Sven Magnusson (p. 31).

414. La anticiencia no tiene cabida en *Interciencia. Interciencia* 18 No. 2 p. 58, 1993.

415. Reprint of #305 in *Cuadernos de documentación filosófica* (Rosario) 1: 7–31, 1993.

416. Wissenschaft hautnah. Prologue to G. Vollmer, *Wissenschafsttheorie im Einsatz*, pp. xi–xix. Stuttgart: S. Hirzel, 1993.

417. Survival, rights and duties. In P. Morales, Ed., *Medio ambiente: El desarrollo y los derechos del hombre. Environment: Development and Human Rights*, pp. 20–24. Buenos Aires: Sagier & Urruty, 1993.

418. Chinese translation of #358. Zhexue Yicong No. 3 pp. 35–41, 1993.

419. Reprint of #402. In J. Brzezinski, S. Di Nuovo, T. Marek & T. Maruszewski, Eds., *Creativity and Consciousness* pp. 299–304. Amsterdam: Rodopi, 1993.

420. Die Bedeutung der Philosophie für die Psychologie. In L. Montada, Ed., *Bericht über den 38. Kongress der Deutschen Gesellschaft für Psychologie in Trier 1992*, Vol. 2 pp. 51–63. Göttingen: Hogrefe, 1993.

421. Inverosímil pero cierto. In *Por 100 años de democracia: 10° aniversario*, pp. 65–69. Buenos Aires: Eudeba-Prode, 1993.

422. Realism and antirealism in social science. *Theory and Decision* 35: 207–235, 1993.

423. Seven cosmological paradigms: Animal, Ladder, River, Cloud, Machine, Book, and System of Systems. [Original of Italian version #411]. In M. Sánchez Sorondo, Ed., *Physica, Cosmologia, Naturphilosophie: Nuovi Approcci*. Roma: Herder-Università Lateranense, pp. 115–131, 1993.

424. Repr. of #52, in A. P. Iannone, Ed., *Through Time and Culture: Introductory Readings in Philosophy* pp. 239–246. Englewood Cliffs NJ: Prentice-Hall, 1994.

425. Technoholodemocracy: An alternative to capitalism and socialism. *Concordia* No. 25: 93–99, 1994.

426. Counter-Enlightenment in Contemporary Social Studies. In P. Kurtz & T. J. Madigan, Eds., *Challenges to the Enlightenment. In Defense of Reason and Science*, pp. 25–42. Buffalo NY: Prometheus Books, 1994.

427. La filosofía es pertinente a la investigación científica del problema mente-cerebro. Translation of #420. *Arbor* CXLVV No. 580: 51–70, 1994.

428. The concept of a social system. In R. Rodríguez Delgado & B. H. Banathy, Eds., *International Systems Science Handbook* pp. 210–221. Madrid: Systemic Publications, 1994.

429. Survival, rights, and duties. In M. C. P. Morales, Ed., *Indigenous Peoples, Human Rights and Global Interdependence* pp. 109–114. Tilburg, Netherlands: International Centre for Human and public Affairs, 1994.

430. L'écart entre les mathématiques et le réel. In M. Porte, Ed., *Passion des formes* [Festschrift for René Thom] Vol. 1 pp. 165–173. Fontenay-St Cloud, E.N.S Editions, 1994.

431. Quality, quantity, pseudoquantity and measurement in social science. *Journal of Quantitative Linguistics* 2: 1–10, 1994.

432. Causality and probability in linguistics: A comment on "Informational measures of causality" by Juhan Tuldava. *Journal of Quantitative Linguistics* 2: 15–16, 1994.

433. The poverty of rational choice theory. In I. C. Jarvie & N. Laor Eds., *Critical Rationalism, Metaphysics and Science* Vol. I pp. 149–168. Dordrecht-Boston: Kluwer Academic, 1995.

434. Realismo y antirrealismo en las ciencias sociales. Translation of #422. *Mientras tanto* No. 61: 21–48, 1995.

435. Rational choice theory: A critical look at its foundations. In J. Götschl, Ed., *Revolutionary Changes in Understanding Man and Society* pp. 211–228. Dordrecht-Boston: Kluwer Academic, 1995.

436. Holotechnodemocrácia: A kapitalizmus és szocializmus egy alternatívjája. *Magyar Filozófiai Szemle Nos.* 1994/5 4–5:869–876. Hungarian translation of #425.

437. Tecnoholodemoracia: Una alternativa al capitalismo y al socialismo. In M. A. Paz y Miño, ed., *Filosofía social* pp. 9–23. Lima: Revista Peruana de Filosofía Aplicada, 1995.

438. Repr. of #294. In M. Marion & R. S. Cohen, Eds., *Quebec Studies in the Philosophy of Science* Vol. I, pp. 217–227. Dordrecht-Boston: Kluwer Academic, 1995.

439. Repr. of #427 in Francisco Mora, ed., *El problema cerebro-mente* pp.55–72. Madrid: Alianza Universidad, 1995.

440. Pobreza de la teoría de la elección racional. Spanish transl. of #433. *Revista de filosofía [Universidad de Chile]* XLV–XLVI: 7–25, 1995.

441. Reprint of #73 in D. F. Channell, ed., *The Relationship between Science & Technology*. Chicago: University of Chicago Press, 1995.

442. Is religious education compatible with science education? (with Martin Mahner) *Science & Education* 5: 101–123, 1996.

443. The incompatibility of science and religion sustained: A reply to our critics (with Martin Mahner) *Science & Education* 5: 189–199, 1996.
444. In praise of intolerance to charlatanism in Academia. *Annals of the New York Academy of Sciences* 775: 96–116, 1996.
445. Sociologías del conocimiento: científicas y anticientíficas. *Redes: Revista de Estudios Sociales de la Ciencia* 3: 125–128, 1996.
446. Los límites de la competencia y de la cooperación. Preface to Jorge Etkin, *La empresa competitiva*, pp. 11–15.Buenos Aires: McGraw Hill, 1996.
447. Mind-body problem. J. Graham Beaumont, M. Rogers, P. M. Kenealy & M. J. C. Rogers Eds., *The Blackwell Dictionary of Neuropsychology* pp. 488–492. Oxford: Blackwell, 1996.
448. The seven pillars of Popper's social philosophy. *Philosophy of the Social Sciences* 26: 528–556, 1996.
449. Hechos y verdades morales. In L. Olivé and L. Villoro, eds., *Filosofía moral, educación e historia: Homenaje a Fernando Salmerón.*, pp. 27–36. México: Universidad Nacional Autónoma de México, 1996.
450. Ciencias básicas y aplicadas, técnicas y servicios: similitudes y diferencias. El caso particular de las ciencias biomédics. *Actas de fisiología* [Uruguay] 4: 11–28, 1996.
451. El derecho [booklet]. Lima: Fondo Editorial de la Facultad de Derecho de la Universidad de San Martín de Porres, 1996.
452. Moderate mathematical fictionism. In E. Agazzi and G. Darwas, Eds., *Philosophy of Mathematics Today* pp. 51–71. Dordrecht & Boston: Kluwer Academic, 1997.
453. El derecho [booklet]. Lima: Fondo Editorial de la Facultad de Derecho de la Universidad de San Martín de Porres, 1997.
454. A humanist's doubts about the information revolution. *Free Inquiry*, 17(2), p. 24–28, 1997.
455. A new look at moral realism. In E. Garzón Valdés, W. Krawietz, G. H. von Wright and R. Zimmerling, eds., *Normative Systems in Legal and Moral Theory* pp. 17–26. Berlin: Duncker & Humblot, 1997.
456. Mechanism and explanation. *Philosophy of the Social Sciences* 27: 410–465, 1997.
457. Une caricature de la science: la nouvelle sociologie de la science. Transl. of #381. Internet, www.vigdor.com, 1998.
458. The end of science? *Philosophy of Social Action* 24: 19–26, 1998.
459. Semiotic systems. In G. Altmann and W. A. Koch, Eds., *Systems: New Paradigms for the Human Sciences* pp. 337–349. Berlin-New York: Walter de Gruyter, 1998.
460. Humanismo e informática. Translation of #253. *Razonamientos* (México), No. 7: 9–17, 1998.
461. The philosophical technologies. *Technology in Society* 20: 377–384, 1998.
462. La explicación en ecología (with Luis Marone). *Boletín de la Asociación Argentina de Ecología.*, No. 7(2): 35–37, 1998.

463. La energía entre la física y la metafísica. [Translation of #462]. *Revista de enseñanza de la física* 12(3), 1: 53–56, 1999.

464. ¿Qué es filosofar científicamente? *Revista Latinoamericana de Filosofía* XXV: 159–169, 1999.

465. The end of science? In D. J. Stlottje, ed., *Advances in Econometrics, Income Distribution and Scientific Methodology* pp.293–300. Heidelberg-New York: Physica-Verlag, 1999.

466. Riot, revolution and national breakdown. [Fragments of *Social Science under Debate]. Philosophy and Social Action* 25, No. 2: 17–28, 1999.

467. Ethics and praxiology as technologies. In E. Agazzi and H. Lenk, eds., *Advances in the Philosophy of Technology*, pp.375–379. Newark, DL: Society for Philosophy and Technology, 1999.

468. Status epistemológico de la administración. En E. R. Scarano, ed., *Metodología de las ciencias sociales*, pp. 349–356. Buenos Aires: Ediciones Maxcchi, 1999.

469. Linguistics and philosophy. In H. E. Wiegand, ed., *Sprache und Sprachen in der Wissenschaften*, pp. 269–293.Berlin-New York: Walter de Gruyter, 1999.

470. ¿Qué es filosofar científicamente? Repr. Of #39. *Revista Latinoamericana de Filosofía* 25: 159–169, 1999.

471. Energy between physics and metaphysics. *Science & Education* 9: 457–461, 2000.

472. Physicians ignore philosophy at their risk – and ours. *Facta philosophica* 2: 149–160, 2000.

473. Philosophy from the outside. *Philosophy of the Social Sciences* 30: 227–245, 2000.

474. Skeptisches zum Skeptizismus. *Skeptiker* 13 (1): 35–39, 2000.

475. Systemism: The alternative to individualism and holism. *Journal of Socio-Economics* 29: 147–157, 2000.

476. Absolute skepticism equals dogmatism. *Skeptical Inquirer* 24, No. 4: 34–36, 2000.

477. Ten modes of individualism—none of which works—and their alternatives. *Philosophy of the Social Sciences* 30: 384–406, 2000.

478. Euclides dos milenios después. Prologue to Beppo Levi, *Leyendo a Euclides*, 2nd ed., pp. 9–14. Buenos Aires: Libros del Zorzal, 2000.

479. Veinticinco siglos de teoría cuántica: De Pitágoras a nosotros y del subjetivismo al realismo. *Saber y tiempo* 10: 5–23, 2000.

480. Análisis del concepto de magnitud física. *Revista de Enseñanza de la Física* 13, No. 2: 21–24, 2000.

481. Function and functionalism: A synthetic perspective (with Martin Mahner). *Philosophy of Science* 68: 75–94, 2001.

482. Construyendo puentes entre las ciencias sociales. In *Desigualdad y globalización: Cinco conferencias*, pp. 47–74. Buenos Aires: Facultad de Ciencias Sociales (UBA), 2001.

483. Recuerdo de Francisco Romero. In José L. Speroni, ed., *El pensamiento de Francisco Romero*, pp. 175–182. Buenos Aires: E. Divern, 2001.

484. Recuerdo de Enrique Gaviola. Preface to Omar Bernaola, *Enrique Gaviola y el Observatorio Astronómico de Córdoba*, pp. xiii–xxi. Buenos Aires: Ediciones Saber y Tiempo, 2001.

485. Introducción to Robert K. Merton, *Teoría y estructura sociales*, pp. 1–8. México, D.F.: Fondo de Cultura Económica, 2002.

486. Velocity operators and time-energy relations in relativistic quantum mechanics. *International Journal of Teoretical Physics* 42: 135–142, 2003.

487. Philosophy of science and technology: A personal report. In Guttorm Fløistad, ed., *Philosophy of Latin America,* pp. 245–272. Dordrecht: Kluwer, 2003.

488. Twenty-five centuries of quantum physics: From Pythagoras to us, and from subjectivism to realism. *Science & Education* 12: 445–466, 2003. Preceded by Michael R. Matthews' "Mario Bunge: Physicist and philosopher", pp. 431–444. Followed by 6 papers, commenting on the target article, by Massimo Pauri, John Forge, Jean-Marc Lévy-Leblond, Alberto Cordero, Adrian Heathcote, and Marcello Cini.

489. Quantons are quaint but basic and real: Reply to my critics. *Science & Education* 12: 587–597, 2002.

490. How to handle the goose that lays golden eggs. *Graduate Researcher* 1: 41–42. 2003.

491. Los médicos ignoran la filosofía a su riesgo y al nuestro. [Transl. of #470]. *Desideratum* (Lima) 3, No. 3, 66–76, 2003.

492. Interpretation and hypothesis in social studies. In R. Boudon, M. Cherkaouki & P. Demeulenaere, eds., *The European Tradition in Qualitative Research*, Vol. IV, pp. 20–40 London: Sage Publications, 2003.

493. Systemic problems call for systemic studies and solutions: A philosophical reflection upon the Delphi Declaration. In Dennis V. Razis, ed., The Human Predicament II: 529–539. Athens GR: S&P Advertising, 2003.

494. Toward a systemic approach to disease (with G. Thurler et al.) *ComPlexUs* 1: 117–122, 2003.

495. The pseudoscience concept, dispensable in professional practice, is required to evaluate research projects. *Scientific Review of Mental Health Practice* 2: 111–114.2004.

496. The centrality of truth. In Evandro Agazzi, *Right, Wrong and Science*, pp. 233–241. Amsterdam: Rodopi, 2004.

497. How does it work? The search for explanatory mechanisms. *Philosophy of the Social Sciences* 34: 182–210, 2004.

498. Clarifying some misunderstandings about social systems and their mechanisms. *Philosophy of the Social Sciences* 34: 371–381, 2004.

499. Systemism: The alternative to individualism and holism. In A. van den Berg and H. Meadwell, eds., *The Social Sciences and Rationality*, pp. 109–123. New Brunswick NJ: Transaction Publishers, 2004.

500. Vers un nouveau matérialisme. In J. Dubessy, G. Lecointre & M. Silberstein, eds., *Les materialismes (et leurs détracteurs)*, pp.75–80. Paris: Syllepse, 2004.

501. Rol del generalista en un mundo de especialistas: Filosofía y ciencias empresariales. Reproduction of a chapter of book #119. In Roger Churnside, ed.,

Espacio y tiempo en gestión y análisis social, pp.1–14. San José: Universidad de Costa Rica, 2004.

502. Presentación. In M. A. Quintanilla, *Filosofía de la tecnología*, pp. xiii–xvi. Lima: Uivesidad Ica Garcilaso de la Vega, 2005.

503. A systemic perspective on crime. In P.-O. Wikström and R. J. Sampson, eds., *The Explanation of Crime*, pp. X–Y Cambridge: Cambridge University Press, 2006.

504. Enlightened solutions for global challenges. *Free Inquiry* Vol. 26, No. 2: 29–34, 2006.

505. Matérialismes et sciences. *Matière première* 1: 251–62, 2006.

506. Reprint of #329 in E. Suárez-Iñiguez, ed., *The Power of Argumentation*, pp. 131–142. Amsterdam/New York: Rodopi, 2006.

507. The philosophy behind pseudoscience. *The Skeptical Inquirer* vol. 30, No. 4: 29–37, 2006.

508. A systemic perspective on crime. In P.-O. Wikström and R. J. Sampson, eds., *The Explanation of Crime*: Context, Mechanisms, and Development, pp.8–30, Cambridge, UK: Cambridge University Press.

509. Max Weber did not practice the philosophy he preached. In Lawrence McFalls, ed., *Max Weber's "Objectivity" Revisited*, pp. 119–134. Toronto: University of Toronto Press. 2007.

510. The ethics of science and the science of ethics. In Paul Kurtz, ed., *Science and Ethics*, pp. 27–40. Amherst, N.Y.: Prometheus Books 2007.

511. Escepticismo politico. *El escéptico* (Spain), No. 24: 19–25, 2007.

512. Teoría y práctica del cooperativismo: De Louis Blanc a la Lega y Modragón. *Revista Iberoamericana de Autogestión y Acción Comunal* XXV, No. 50: 13–16, 2007.

513. Blushing and the philosophy of mind. *Journal of Physiology Paris* 101: 247–256, 2007.

514. Reprint of #508 in Fabio Minazzi, ed., Filosofia, scienza e bioetica nel dibattito contemporaneo, pp. 427–438. Roma: Presidenza del Consiglio dei Ministri.

515. Preface to Pierre Moesssinger, *Voir la société*: Le micro et le macro, pp. 11–14. Paris: Hermann, 2008.

516. Contribution to D. Ríos and C. Schmidt-Petri, eds., *Philosophy of the Social Sciences: 5 Questions*, pp. 31–42. London: Automatic Press, 2008.

517. Bayesianism: Science or pseudoscience? *International Review of Victimology* 15: 169–182, 2008.

518. ¿Personas, sociedades o ambas? El enfoque sistémico de los problemas sociales. Foreword by Ignacio Morgado Bernal. Barcelona: Fundació Ernest Lluch, 2009.

519. The failed theory behind the economic crisis. In M. Cherkaoui and P. Hamilton, eds., *Raymond Boudon: A Life in Sociology*, vol. 1. Oxford: Bardwell Press, 2009.

520. Advantages and limits of naturalism. In John R. Shook & Paul Kurtz, eds., *The Future of Naturalism*. Amherst, N.Y.: Humanity Books, 2009.

521. From philosophy to physics and back. In S. Nuccetelli, O. Schutte, and P. Bueno, eds., *A Companion to Latin American Philosophy*, pp. 525–539. Malden, MA: Wiley-Blackwell, 2010.

522. ¿Tiene porvenir el socialismo? SinPermiso 2010.

523. Filosofía del progreso científico. CAI en el Siglo XXI, Septiembre-Mayo 2008–2009: 153–168; 357–366, 2010.

524. The troubled relationship between physics and philosophy. In Juan Ferret and John Symons, eds., *Philosophy of Physics : 5 + 1 Questions*, pp. 19–35. Milton Keynes: Automatic Press, 2010.

525. Two unification strategies: Analysis or reduction, and synthesis or integration. In J. Symons, O. Pombo, and J.M. Torres, eds., *Otto Neurath and the Unity of Science*, pp. 145–157. Dordrecht, Heidelberg, London, New York: Springer, 2011.

526. Knowledge: genuine and bogus. *Science & Education* 20: 411–438, 2011.

527. ¿Es una filosofía el existencialismo? In M.A. Rodríguez Rea & N. Osorio Tejeda, ed., *La filosofía como repensar y replantear la tradición: Libro de Homenaje a David Sobrevilla*. Lima: Universidad Ricardo Palma, 2011, pp. 41–48.

528. Wealth and well-being, economic growth, and integral development. *International Journal of Health Services* 42: 65–76, 2012.

529. The correspondence theory of truth. *Semiotica* 188: 65–76, 2012.

530. Does quantum physics refute realism, materialism and determinism? *Science & Education* 21: 1601–1610, 2012.

531. Bruce Trigger and the philosophical matrix of scientific research. In S. Chrisomalis and A, Costopoulos, eds., *Human Expeditions Inspired by Bruce Trigger*, pp. 143–159. Toronto: University of Toronto Press, 2013.

532. Mechanism and mechanical explanation. In B. Kaldis, ed., *Encyclopedia of Philosophy and the Social Sciences*: Sage, 2013.

533. La physique quantique réfute-t-elle le réalisme, le matérialisme et le déterminisme? In Marc Silberstein (ed.), Matériaux philosophiques et scientifiques pour un matérialisme contemporain.Vol 1. Paris: Editions Matériologiques, 417–434, 2013.

534. In defense of scientism. *Free Inquiry* Vol.35, No. 1, 24–28, 2014.

535. Does the Aharonov-Bohm effect occur? *Foundations* of Science. *20:* 129–133, 2014.

536. Big questions come in bundles, hence they should be tackled systematically, *International Journal of Health Services* 44(4), 835–844, 2014.

537. Foreword to Dominique Raynaud. *Qu'est-ce que la technologie?*, pp. 5–12. Paris: Editions Matériologiques, 2016.

538. Sciences et philosophie, un dialogue. In Marc Silberstein (ed.), *Qu'est-ce que la science... pour vous? 50 scientifiques et philosophes répondent*. Paris: Editions Matériologiques, 39–42, 2017.

539. Why axiomatize? *Foundations of Science* 22(4), 695–707., 2017.

540. 'Evaluating Scientific Research Projects: The Units of Science in the Making', *Foundations of Science* 22(3), 455–469, 2017.

541. 'Reconceptualizing Mental Disorders: From Symptoms to Organs', *PsyCh Journal* (Institute of Psychology, Chinese Academy of Sciences), 6, 161–165, 2017.
542. 'Why Don't Scientists Respect Philosophers?'. In Nimrod Bar-Am & Stefano Gattei (eds.) *Encouraging Openness: Essays for Joseph Agassi on the Occasion of His 90th Birthday*, Springer, (*Boston Studies in Philosophy of Science* vol.325), Dordrecht, pp.3–12, 2017.
543. Foreword to Dominique Raynaud. *Sociologie des controverses scientifiques*, pp. 3–10. Paris: Editions Matériologiques, 2018 (new ed.)
544. Gravitational Waves and Space-Time, *Foundations of Science*, 23(2), 399–403, 2018.
545. The Dark Side of Technological Progress'. In R. Sassower & N. Laor (eds.) *The Impact of Critical Rationalism*, Springer, Dordrecht, pp.109–113, 2019.
546. Inverse Problems, *Foundations of Science* 2019, DOI: https://doi.org/10.1007/s10699-018-09577-1

Name Index

© Springer Nature Switzerland AG 2019
M. R. Matthews (ed.), *Mario Bunge: A Centenary Festschrift*,
https://doi.org/10.1007/978-3-030-16673-1

Subject Index

© Springer Nature Switzerland AG 2019
M. R. Matthews (ed.), *Mario Bunge: A Centenary Festschrift*,
https://doi.org/10.1007/978-3-030-16673-1

Printed in the United States
By Bookmasters